Lecture Notes in Earth Sciences

Edited by Somdev Bhattacharji, Gerald M. Friedman, Horst J. Neugebauer and Adolf Seilacher

18

N. M. S. Rock

Numerical Geology

A Source Guide, Glossary and Selective Bibliography to Geological Uses of Computers and Statistics

Springer-Verlag

Berlin Heidelberg New York London Paris Tokyo

Author

Dr. Nicholas M. S. Rock
Department of Geology, University of Western Australia
Nedlands, 6009, Western Australia

ISBN 3-540-50070-7 Springer-Verlag Berlin Heidelberg New York
ISBN 0-387-50070-7 Springer-Verlag New York Berlin Heidelberg

Printed in Germany

Printing and binding: Druckhaus Beltz, Hemsbach/Bergstr.
2132/3140-543210

ACKNOWLEDGEMENTS

This book was completed despite the best efforts of Malcolm (aged 6) and Duncan (aged 4). I would like to thank the University of Western Australia (especially my colleagues in the Geology Department) for giving me the opportunity to write the work, and for all those on and off campus who contributed comments on its various stages. Permission to use Fig.3.2 (ERCC), Figs.4.1–4.2 (WARCC), Fig.4.4 (ESRI) and Table 3.3 (Gordon & Breach publishers) was much appreciated. Owners of various trade marks and trade names which are used frequently in this book are also acknowledged below. Other names annotated ™ or ® in the text are trade names, trade marks or registered trade marks of various software and hardware suppliers, which are too numerous to be listed individually.

Since desktop publishing technology is very new and may be unfamiliar to some readers, brief comments on how this book was produced may be of interest. The text was 'desktop published' entirely by the author on Macintosh microcomputers, using MacWrite for text, Superpaint for figures, MacΣqn for mathematical formulae and ANUGraph for curves of statistical functions. It was output on an Apple Laserwriter Plus laserprinter. The laserwriter output was then used as camera-ready copy for reduction to 80% in final printing by the publishers. The main text is in Times Roman font, and the literature citations in Avant Garde, while Helvetica is used for annotation on figures and Symbol font (Σψμβολ φοντ) for Greek letters.

ANUGraph is a tradename of N.Smythe, M.Ward and the ANU.

Apollo, Apple, CLIRS, Cray, Cyber, DIALOG, ICL, NAG, Oracle, PLATO, Prime, SAS, SPSS, Sun, & Systat are trademarks, tradenames or registered trademarks of the eponymous companies or organizations.

ARC/INFO is a registered trademark of ESRI Ltd.

Avant Garde is a trademark of International Typeface Corporation.

CEEFAX is a trademark of Independent Television (UK).

DEC, UNIX, VAX & VMS are registered trademarks of Digital Equipment Corp.

Helvetica and **Times** are trademarks of Allied (Linotype) Corporation.

IBM, IBM-PC and PC-DOS are registered trademarks of International Business Machines Corp.

Macintosh, MacWrite, MacDraw, MacPaint, MacTerminal, Imagewriter, Laserwriter, Postscript, are registered or licensed trademarks of Apple Corp. or CLARIS Corp.

MacΣqn is a trade mark of Software for Recognition Technologies.

Microsoft, MS-DOS, PC-DOS and **XENIX** are registered trademarks of Microsoft Corp.

ORACLE is a trademark of the British Broadcasting Corporation and also of Oracle Corporation.

PRESTEL is a trademark of British Telecom.

Statview 512+ and **Statview II** are trade names of Brainpower Inc. and Abacus Concepts Inc.

Superpaint is a registered trademark of Silicon Beach Software Inc.

VIATEL is a trademark of Australian Telecom.

Perth, Australia Nicholas Rock

CONTENTS

SECTION I: INTRODUCTION TO GEOLOGICAL COMPUTER USE

SECTION II. THE BEHAVIOUR OF NUMBERS: ELEMENTARY STATISTICS

TOPIC 6. SCALES OF MEASUREMENT AND USES OF NUMBERS IN GEOLOGY

TOPIC 7. SOME CRUCIAL DEFINITIONS AND DISTINCTIONS

TOPIC 8. DESCRIBING GEOLOGICAL DATA DISTRIBUTIONS

TOPIC 9. ASSESSING VARIABILITY, ERRORS AND EXTREMES IN GEOLOGICAL DATA: SAMPLING, PRECISION AND ACCURACY

SECTION III: INTERPRETING DATA OF ONE VARIABLE: UNIVARIATE STATISTICS

SECTION V: SOME SPECIAL TYPES OF GEOLOGICAL DATA

TOPIC 15. SOME PROBLEMATICAL DATA-TYPES IN GEOLOGY

TOPIC 16. ANALYSING ONE-DIMENSIONAL SEQUENCES IN SPACE OR TIME

TOPIC 17. ASSESSING GEOLOGICAL ORIENTATION DATA: AZIMUTHS, DIPS AND STRIKES

SECTION VI: ADVANCED TECHNIQUES

TOPIC 21. CLASSIFYING OBJECTS FROM FIRST PRINCIPLES

TOPIC 22. COMPARING 'KNOWNS' AND ASSIGNING 'UNKNOWNS': Discriminant Analysis and related methods

TOPIC 23. EXAMINING STRUCTURE, RECOGNIZING PATTERNS & REDUCING THE DIMENSIONALITY OF MULTIVARIATE DATA

List of Symbols and Abbreviations Used

It is impossible to be wholly consistent with symbols, because many such as λ, κ, γ, α, W have been used in the statistical literature for several different (and conflicting) purposes. Conversely, different symbols are used in different texts for the same purpose (e.g. k is sometimes used for the same purpose as q in this book, but then conflicts with k as used for concentration). The following list adopts some symbols which appear to have been used consistently in the statistical literature (e.g. F for Fisher's F; ρ for Spearman's ρ), and will therefore be consistent with most of the sources listed under the various Topics; others will be consistent with some sources but not with others.

§	refers the reader to a particular Topic; e.g. §8b1 means section 8b1 in Topic 8.
[...]	numbers in square brackets refer to **equations**, e.g. [17.8] is equation 8 in Topic 17.
bold	words in bold type refer the Reader to entries in the Glossary (first appearance in text only).

A	Normal score (in Normal scores tests).
a,b,c,d	cell contents in 2 x 2 contingency tables (Table 13.6) and in defining matching coefficients.
b	number of blocks (in ANOVA-type tests).
C	mean cosine of angular data-values, i.e. $[\sum \cos(\theta_i)]/n_i$ (§17).
c	number of classification criteria/treatments/related groups (e.g. in ANOVA-type tests).
d	various differences; e.g. in paired t-test; also d_{max} = Kolmogorov–Smirnov test statistic.
D	bulk distribution coefficient (= element concentration in solid/concentration in liquid; §18).
df	degrees of freedom.
e	base of natural logarithms (2.7182818285....)
$E(x)$	an expected frequency of x.
$f(X)$	any function of X.
$\mathrm{F}(X)$	frequency distribution of X.
f	melt fraction in a system (§18).
$F_{\alpha,(\upsilon,\nu)}$	Fisher's F variance-ratio (at the α% significance level, with υ degrees of freedom in the numerator, ν in the denominator); or of a test statistic to be compared with Fisher's F.
i,j	suffices denoting individual objects (X_i applies to ith object of n_i), mainly in summations.
k	various constants; also concentration parameter for orientation data (§17 only).
l	lag in comparing sequences of data (§ 16).
M	population median, or other robust estimate of location.
n_i	number of objects in the ith of a number of groups of data; goes from n_1 to n_q.
N	total number of data-values in *all* groups of data being considered ($=\sum n_i$).
p	probability (thus $p = 0.05$ refers to a 5% probability or 95% significance level).
P	proportions, percentages and ratios of the form X_1/X_2, $X_1/[X_1+X_2]$, etc. (§15).
q	number of *unrelated* groups of data being considered in a multi-group test ($q > 2$).
®	rank of a value in an ordered set (takes integer or, for tied data, $\frac{1}{2}$ values); also sum of ranks.
R	value of *mean* resultant length for several groups of data combined (§17).
R_i	value of *mean* resultant length for the ith of q groups of data (§17).

r	value of Pearson's product-moment correlation coefficient.
r_o	null correlation against which r is tested for significance (generally, $r_o = 0$).
s	sample standard deviation.
S	mean sine of angular data-values, i.e. $[\sum \sin(\theta_i)]/n_i$ (§ 17).
SP_{XY}	sum of cross-products for X and Y, i.e. $\sum (X_i - \bar{X})(Y_i - \bar{Y})$; N times the covariance.
SS_x	sum of squares about X, i.e. $\sum (X_i - \bar{X})^2$; $N - 1$ times the variance of X; similarly SS_Y.
$t_{\alpha,\nu}$	Student's t statistic (at the α% significance level and with ν degrees of freedom, if given).
U	value of the Mann-Whitney statistic.
$U^{(2)}$	value of Watson's U^2 statistic.
U_i	value of Watson's U^2 statistic for the ith of q groups of data.
v	number of variables; e.g. $v = 2$ means bivariate, $v > 2$ means multivariate data.
w	weighting factors, e.g. in weighted means, kriging equations.
X, Y	individual data-values (X refers to dependent, Y to independent variables as appropriate).
$\bar{X}$	mean of a group of data (sample mean, in standard statistical terminology).
X_{med}	median of a group of data (sample median, in standard statistical terminology).
Z	(1) value of a standard Normal deviate; (2) a discriminant function score (§22 only); (3) value of Fisher's $\tanh^{-1}$ transformation of Pearson's r correlation coefficient (§13 only).

Greek symbols (in *Greek* alphabetical order)

α (alpha)	significance level (probability of Type I error); thus $\alpha = 95$ means that a test is is being carried out at the 95% significance level (probability 0.05 of error); α takes a % value between 90 and 99.99.., corresponding to p (see above) between 0.1 and 0.00..1.
ß (beta)	power of a test/probability of Type II error.
$ß_o$, $ß_1$ etc.	coefficients in regression equations (general linear model) $Y = ß_o + ß_1X_1 + ß_2X_2 + ... \varepsilon$
γ (gamma)	population slope for regression of the form $Y = \delta + \gamma X + \varepsilon$.
δ (delta)	population Y-intercept for regression of the form $Y = \delta + \gamma X + \varepsilon$.
$\partial (X)$	cumulative distribution function of X, especially as used in EDF tests.
ε (epsilon)	error term for regression and other statistical models of the form $Y = \delta + \gamma X + \varepsilon$.
η (eta)	value of the correlation ratio (§14).
θ (theta)	an individual angular measurement in degrees; for 3-D data applies to azimuth only (§17).
κ (kappa)	population concentration (§17).
μ (mu)	population mean.
ν (nu)	number of degrees of freedom (mostly applied to test statistics).
ρ (rho)	value of Spearman's rank correlation coefficient.
$\sum$ (sigma)	summation symbol.
σ (sigma)	population standard deviation.
τ (tau)	value of Kendall's rank correlation coefficient.
υ (upsilon)	uniform score in Mardia's test (§17); equals 2π®/N, where ® is the rank of an angle.
ϕ (phi)	dip direction for 3-D orientation data (§17).
$\chi^2_{\alpha,\nu}$	value of chi-squared (at α% significance level and with ν degrees of freedom, if given).
ω (omega)	range of a set of data-values.

INTRODUCTION — WHY THIS BOOK?

Why study Numerical Geology?

Although geologists have dabbled in numbers since the time of Hutton and Playfair, 200 years ago (Merriam 1981e), geology until recently lagged behind other sciences in both the teaching and geological application of mathematics, statistics and computers. Geology Departments incorporating these disciplines in their undergraduate courses are still few (particularly outside the USA). Only two international geomathematical/computing journals are published (*Computers & Geosciences; Mathematical Geology)*, compared with dozens covering, say, petrology or mineralogy. It also remains common practice for years (and $1000s) to be spent setting up computerized machines to produce large volumes of data in machine-readable form, and then for geologists to plot these by hand on a sheet of graph paper!

Despite this, the use of numerical methods in geology has now begun to increase at a rate which implies a revolution of no less importance than the plate tectonic revolution of the 1960's — one whose impact is beginning to be felt throughout the academic, commercial, governmental and private consultative geological communities (Merriam 1969, 1981c). Although a few pioneers have been publishing benchmark papers for some years, the routine usage of machine-based analytical techniques, and the advent of low-priced desk-top microcomputers, have successively enabled and now at last persuaded many more geologists to become both numerate and computerate. Merriam (1980) estimated that two decades of increasing awareness had seen the percentage of geomathematical papers (*sensu lato*) rise to some 15% of all geological literature; meanwhile, mineralogy-petrology and geochemistry had both fallen to a mere 5% each!

In these Notes, *geomathematics* and *numerical geology* are used interchangeably, to cover applications of mathematics, statistics and computing to processing real geological data. However, as applications which primarily store or retrieve numbers (e.g. databases) are included, as well as those involving actual mathematical calculations, 'Numerical Geology' is preferred in the title. 'Geomathematics' in this sense should not be confused with 'geostatistics', now usually restricted to a specialised branch of geomathematics dealing with ore body estimation (§20).

Reasons for studying Numerical Geology can be summarised as follows:

(1) **Volumes of new and existing numerical data:** The British Geological Survey, the world's oldest, recently celebrated its 150th anniversary by establishing a National Geoscience data-centre, in which it is hoped to store all accumulated records on a computer (Lumsden & Howarth 1986). Information already existing in the Survey's archives is believed to amount to tens or hundreds of Gb (i.e. $\approx 10^{10-11}$ characters) and to be increasing by a few percent annually. The volumes of valuable data existing in the *world's* geological archives, over perhaps 250 years of geological endeavour, must therefore be almost immeasurably greater. It is now routine even for students to produce hundreds or thousands of multi-element analyses for a single thesis, while national programs of geochemical sampling easily produce a million individual element values. Such volumes of data simply cannot be processed realistically by manual means; they require mathematical and statistical manipulation on computers — in some cases *large* computers.

(2) **Better use of coded/digitised data:** In addition to intrinsically numerical (e.g. chemical) data, geology produces much information which can be more effectively used if numerically coded. For example, relatively little can be done with records of, say, 'limestone' and 'sandstone' in a borehole log, but very much more can be done if these records are numerically coded as 'limestone = 1' and 'sandstone = 2'. Via encoding, enormous volumes of data are opened to computer processing which would otherwise have lain dormant. More importantly, geological maps – perhaps the most important tool of the entire science – can themselves be digitised (turned into large sets of numbers), opening up vast new possibilities for manipulation, revision, scale-change and other improvements.

(3) **Intelligent data use:** It is absurd to acquire large volumes of data and then not to interpret them fully. Field geologists observing an outcrop commonly split into two (or more) groups, arguing perhaps over the presence or absence of a preferred orientation in kyanite crystals on a schist foliation surface. The possibility of actually measuring these orientations and analyzing them statistically (§17) is rarely aired — at last in this author's experience! Petrologists are equally culpable when they rely on *X–Y* or, at maximum 'sophistication', *X–Y-Z* (triangular) variation diagrams, in representing the evolution of igneous rocks which have commonly been analyzed for up to 50 elements! Whereas some geological controversies (especially those based on interpretation of essentially subjective field observations) cannot be resolved numerically, many others can and should be. This is not to say (as Lord Kelvin did) that quantitative science is the *only* good science, but qualitative treatment of quantitative data is rarely anything but *bad* science.

(4) **Literature search and data retrieval:** Most research projects must begin with reviews of the literature and, frequently, with exhaustive compilations of existing data. These are essential if informed views on the topic are to be reached, existing work is not merely to be duplicated, and optimum use is to be made of available funding. The ever-expanding geological literature, however, makes such reviews and compilations increasingly time-consuming and expensive via traditional manual means. Use of the increasing number of both bibliographical and analytical databases (§3) is therefore becoming a prequisite for well-informed, high-quality research.

(5) **Unification of interests:** In these days of inexorably increasing specialisation in ever narrower topics, brought about by the need to keep abreast of the exploding literature, numerical geology forms a rare bridge between different branches not only of geology but of diverse other sciences. The techniques covered in this book are equally applicable (and in many cases have been in routine use for far longer) in biology, botany, geography, medicine, psychology, sociology, zoology, etc. Within geology itself, most topics covered here are as valuable to the stratigrapher as to the petrologist. 'Numerical geologists' are thus in the unique (and paradoxical) position of being both specialists and non-specialists; they may have their own interests, but their numerical and computing knowledge can often help *all* of their colleagues.

(6) **Employment prospects:** There is a clear and increasing demand for computerate/numerate geologists in nearly all employment fields. In Australia, whose economy is dominated by geology-related activities (principally mining), a comprehensive national survey (AMIRA 1985) estimated that A$40M per annum could be saved by more effective use of computers in geology. Professional computer scientists are also of course in demand, but the inability of some of their

number to communicate with 'laymen' is legendary! Consequently, many firms have perpetual need for those rare animals who combine knowledge of computing and mathematics with practical geological experience. Their unique bridging role also means that numerical geologists are less likely to be affected by the vaguaries of the employment market than are more specialised experts.

Rationale and aims of this book

This is a *highly experimental* book, constituting the interim text for new (1988) courses in 'Numerical Geology' at the University of Western Australia. It is published in the Springer *Lecture Notes in Earth Sciences* series precisely because, as the rubric for this series has it, "the timeliness of a manuscript is more important than its form, which may be unfinished or tentative." Readers are more than welcome to send constructive comments to the author, such that a more seasoned, comprehensive version can be created in due course. Readers' indulgence is meanwhile craved for the number of mistakes which must inevitably remain in a work involving so many citations and cross-references.

Emphasis is particularly placed on the word *Notes* in the series title: this book is *not* a statistical or mathematical treatise. It is *not* intended to stand on its own, but rather to complement and target the existing literature. It is most emphatically *not* a substitute for sound statistical knowledge, and indeed, descriptions of each technique are deliberately minimized such that readers *should never* be tempted to rely on this book alone, but should rather read around the subject in the wealth of more authoritative statistical and geomathematical texts cited. In other words, this is a synoptic work, principally about 'how to do', 'when not to do', 'what are the alternatives' and 'where to find out more'. It aims specifically: (1) to *introduce* geologists to the widest possible range of numerical methods which have already appeared in the literature; and thus (2) to *infuse* geologists with just sufficient background knowledge that they can: (a) *locate* more detailed sources of information; (b) *understand* the broad principles behind interpreting most common geological problems quantitatively; (c) *appreciate* how to take best advantage of computers; and thereby (d) *cope* with the "information overload" (Griffiths 1974) which they increasingly face. Even these aims require the reader to become to some extent geologist, computer scientist, mathematician and statistician rolled into one, and a practical balance has therefore been attempted, in which just enough information is hopefully given to expedite correct interpretation and avoidance of pitfalls, but not too much to confuse or deter the reader.

Despite the vast literature in mathematics, statistics and computing, and that growing in geomathematics, no previous book was found to fulfill these aims on its own. The *range* of methods covered here is deliberately much wider than in previous geomathematical textbooks, to provide at least an introduction to *most* methods geologists may encounter, but other books are consequently relied on for the detail which space here precludes.

These Notes adopt a practical approach similar to that in language guidebooks — at the risk of emulating the 'recipe book' abhorred in some quarters. Every Topic provides a minimum of *highly condensed* sketch-notes (fuller descriptions are included only where topics are not well covered in existing textbooks), complemented by worked examples using real data from as many fields of geology as space permits. Specialists should thereby be able to locate at least one example close to their problems of the moment. In the earlier (easier) topics, simple worked examples are calculated in full, and equations are given wherever practicable (despite their sometimes forbidding appearance), to enable readers not only to familiarise themselves with the calculations but also to experiment with their own data. In the later (multivariate) topics (where few but the sado-masochistic would wish to try the calculations by hand!), the worked examples comprise simplified output from actual software, to familiarise readers with the types of computer output they may have to interpret in practice.

Topics were arranged in previous geomathematical textbooks by *statistical* subject: 'analysis of variance', 'correlation', 'regression', etc., while nonparametric (rank) methods were usually dealt with separately from classical methods (if at all). Here, topics are arranged by *operation* (what is to be done), and both classical and rank techniques are covered together, with similar emphasis. When readers know what they want to do, therefore, they need only look in one Topic for all appropriate techniques.

The main difficulty of this work is the near impossibility of its goal— though other books with similarly ambitious goals have been well enough received (e.g. *J.Math.Geol.* **18(5),** 511-512). Some constraints have necessarily been imposed to keep the Notes of manageable size. *Geophysics,* for example, is sketchily covered, because (i) numerical methods are already far more integrated into most geophysics courses than geology courses; (ii) several recent textbooks (e.g. Camina & Janecek 1984) cover the corresponding ground for geophysicists. *Structural geology* is less comprehensively covered or cited than, say, stratigraphy, because (a) it commands many applications of statistics and computing unto itself alone (e.g. 3-D modelling, 'unravelling' of folds), whereas these Notes aim at techniques equally applicable to most branches of geology; (b) excellent comprehensive reviews of structural applications are already available (e.g. Whitten 1969,1981). *Remote sensing* is also barely covered, since comprehensive source guides similar in purpose to the present one already exist (Carter 1986).

For the sake of brevity, phrases throughout this book which refer to males are, with apologies to any whose sensitivities are thereby offended, taken to include females!

HOW TO USE THIS BOOK

The course is arranged into 6 sections and 23 topics, in order of increasing difficulty. Topics 6–7 should probably be regarded as essential reading. Under each topic, citations are listed to help users to find further information on a particular technique. Because citing the already vast literature comprehensively is impossible, selectivity had to be imposed. Citation therefore concentrates on books or full-length papers, in the widely available journals. Papers in local journals, abstracts and open file reports are cited only where they fill major gaps in the international literature. Theses are excluded. One or more of the information categories below is given, as appropriate. Geologically biased texts are preferentially cited under all categories; entries under the first two headings also appear in the *Glossary*:

Dedicated monographs & reviews: Cites entire books or reviews devoted to a given method.

Further discussion: Cites sources which give a useful, if less comprehensive, description.

Worked examples: Cites examples which work step-by-step through relevant calculations. These complement the many new worked examples provided in these Notes, so as to cover the widest possible geological spectrum. The disciplines concerned are specified (*palaeontology, petrology, etc.)* so that, with the aid of the Bibliography, readers should be able to follow up those papers closest to their own area of interest. In Sections II – V, a good method for achieving proficiency is thus as follows: (1) to work through one or more of these worked examples using the steps shown in the text; (2) to try a further example manually, using one's own, real geological data; and (3) to test one's calculations against a computer calculation on the same, real data. Steps may need to be repeated if the results from steps (2) and (3) do not agree! In Section VI, manual calculation is not generally feasible, due to the difficulty of the matrix manipulations, but readers should try to check the calculations using one of the suitable computer packages cited as shown below.

Selected geological applications: Cites descriptions of practical applications of the technique to as many (named) fields of geology as possible. These differ from the 'worked examples' above in that calculations are no longer set out in such detail (if at all); hence the beginner would be best advised to consult the 'worked examples' first. Disciplines are specified as for the worked examples.

Statistical tables: Cites sources of *extensive* tables of test statistics or distributions. Statistical tables are very scattered in the literature, and whereas versions for the commoner test statistics (t, F, $\chi 2$ etc.) are legion, more obscure statistics may be covered by one or two sources only. Since reproduction of all such tables in this book is impossible, citations here are as comprehensive as space permits. Tables consulted are: Beyer (1968); Harter & Owen (1970); Kellaway (1968); Mardia & Zemroch (1978); Murdoch & Barnes (1974); Neave (1978,1981); Owen (1962); Pearson & Hartley (1966); Pillai (1960); Powell (1982); Rohlf & Sokal (1969). This section also cites methods for calculating the *exact distribution* of test statistics (e.g. Kim 1969).

Stand-alone computer programs: 'Stand-alone' is used primarily to indicate that the *source code* is actually published (i.e. non-proprietary). No distinction is made between programs which are completely self-contained, and those which actually form parts of packages from which they may call subroutines: the point is that both can be copied, amended, and *made* to stand alone. Programs whose existence is merely mentioned (without actual code) are not in general cited, except where no other program serving the same purpose is known. Citations are not exhaustive, nor valid beyond early 1988; users are advised to check up-to-date sources of further information as listed in §2. Nearly all cited programs have been tried, and many results in this book were derived using them.

Mainframe packages & libraries: Cites the existence of at least one implementation among 6 of the best-known and most widely-distributed proprietary packages (Table 2.1). Here, actual computer code is not usually available, so users cannot change the software. Since there are over 120 mainframe statistical packages on the market (§2), with new ones or versions released regularly, citation is necessarily even less comprehensive than for the stand-alone programs. Where no citation is given, users should refer to Francis (1981) or to reviews in recent computer magazines (§2).

Microcomputer packages: Cites proprietary packages on the *Macintosh* preferentially. This is because: (1) 'Mac' packages have been used in preparing this book; (2) they represent a completely different (and therefore complementary) environment to large mainframe packages ('dinosaurs'!), and can be mastered even by novices in a much shorter time; (3) there are relatively few currently available, so that citations can be reasonably comprehensive; (4) the Macintosh is being increasingly widely adopted in geology teaching. Reviews of Mac statistical packages are available in Berk (1987), Lehmann (1986,1987), Rock et al.(1988), Shervish (1985) and Waltz (1986). To inveterate users of IBM PCs and 'clones', the author pleads that there are too many (at least 30) packages to cite comprehensively, and that many of these packages are merely scaled-down versions of the mainframe packages; they can, in any case, refer to Carpenter et al.(1984) for details! It is still worthwhile distinguishing in this way between packages specifically designed for microcomputers, and those which began their life before microcomputers even existed, not least because mainframe and micro environments are still rigidly separated in some institutions. Users should nevertheless realise that the distinction is already disappearing, as the power of micros rises: thus most of the long-lived mainframe packages (Table 2.1) are already available in micro versions.

Proprietary packages: This heading is mostly used where no computer implementations are known.

The following complementary list of specialised reviews will facilitate geologists' access to the literature on applications of computers and statistics *in general* to their own discipline (many of those listed below contain large bibliographies, which are not reproduced here for reasons of space):

Coal mining:	Armstrong (1983); Novak et al.(1983).
Geochemistry:	Howarth (1983a); Thornton & Howarth (1986).
Geomorphology:	Doornkamp & King (1971).
Geophysics:	Camina & Janecek (1984); Lee & Habermann (1986).
Hydrology:	Haan(1977).
Mathematics & modelling:	Fenner (1969); Ferguson (1987); Harbaugh & Bonham-Carter (1970).
Microcomputers:	Hanley & Merriam (1986); Krajewski(1986).
Mineral exploration/mining/ ore geology:	Agterberg(1967b); Hazen(1967); IMM(1984); Koch(1981); Krajewski(1986); Longe(1976); Lepeltier(1977); Li et al. (1987); Wignall & de Geoffroy (1987).
Palaeoecology:	Reyment (1971,1981).
Palaeontology:	Bayer (1985); Raup(1981); Reyment et al. (1984).
Petroleum geology:	Davis (1981); McCray (1975); Robinson (1982).
Petrology:	Le Maitre (1982).
Random processes:	Merriam(1976a).
Remote sensing:	Burns (1981); Carter (1986); Condit & Chavez(1979); Fabbri (1986); Taranik (1978); Yatabe & Fabbri (1986).
Sedimentology:	Gill & Merriam(1979); Griffiths(1967); Merriam (1972,1976c).
Stratigraphy/biostratigraphy:	Brower (1981); Harbaugh & Merriam (1968); Mann(1981); Gradstein et al. (1985).
Structural geology:	Crain (1976); Whitten (1969, 1981).

SECTION I. INTRODUCTION TO GEOLOGICAL COMPUTER USE

This section is merely an overview of aspects important for what the author has come to know as the 'typical' geologist or student in several countries. Experienced computer users can omit or 'skim' it.

TOPIC 1. UNDERSTANDING THE BASICS ABOUT COMPUTERS

1a. Background history of computer use in the Earth Sciences

Further geological discussion: Botbol (1986); McIntyre(1981); Merriam (1969, 1978b,c; 1981a,b), Thornton & Howarth (1986), Hanley & Merriam (1986).

Not long ago, Earth Scientists using computers were a small fraternity. They were also unevenly spread, computerate petrologists being fewer than, say, geophysicists. Now, however, the caucus of experts from the 1950s has been joined by an expanding army of geologists who expect computers to solve their problems. The computer revolution promises great improvements in geologists' wealth-producing and popularizing powers, and may persuade other scientists and even (dare we hope?!) politicians that geology is as quantitative and rigorous as her sister sciences, Physics and Chemistry.

Many organisations now flourish which did not exist a few years ago — computer-oriented geological societies (e.g. COGS in Denver, USA), international bodies (e.g. the International Association of Mathematical Geology, IAMG) and specialist groups (e.g. the Geological Information Group of the London Geol. Soc.) International databases are now being set up under the auspices of the International Geological Correlation Program, and both national and international conferences (e.g. the esoterically named Geochautaqua organized by IAMG) are now run in various countries. From the most elementary uses for computers (e.g. word-processing), many geologists have climbed the tree of computer sophistication into spread-sheets, databases, simple and multivariate statistics, to communications and even expert systems (this last by definition the 'highest' use of computers, speaking anthropocentrically, in that they are supposed to emulate human intelligence!)

Useful charts detailing the progress of computer usage in geology are given in the references listed above. It is, however, worth briefly stating some major reasons *why* this revolution has taken place:
(1) *The need*. The Introduction cited the inevitability and unavoidability of computer processing, if modern geologists wish to cope efficiently with the huge volumes of data now at their disposal.
(2) *The means*. Affordable, desktop computers now rival the power of mainframes a decade ago, and even of modern minicomputers. Gone are the days when computing consisted of handing a moth-eaten pack of punch-cards through the security screen of some "inner sanctum" in a monolithic computing centre, hoping to receive printout the next day (if at all). Nowadays, many geologists can do their data interpretation and plotting in their own offices, homes, or even field areas; they can produce professional-quality reports, complete with tables, figures and maps, often without any expert help. Given that geologists tend to be independent-minded, microcomputers have afforded them a means to control their own computing, and hence a significant psychological stimulus to become computerate.

1b. Hardware: computer machinery

Further sources of information: far too many books on computers are available to list meaningfully here but, for the complete novice, useful introductions are given in Bergerud & Keller (1987); Chandor et al.(1985); Connors (1987); Green (1985); Webster (1983); Webster et al. (1983).

1b1. Types of computers: accessibility, accuracy, speed and storage capacity

Many dictionaries define computers simply as devices which can accept and manipulate data, then present results in some intelligible form. This includes ordinary desktop calculators which, to this author at least, are not computers. A computer should be capable of understanding some kind of computing language, and preferably also, be capable of both reading data from such devices as cassette/ reel magnetic tapes or floppy disks, and writing it to other devices such as screens or printers. Many "programmable" calculators on the market, such as the slimline, fist-sized Casio FX702-P™ are indeed in this sense tiny computers, for they can do all these things.

Although the boundary between what is and what is not a computer may remain contentious at the 'low' end of the spectrum, most people are agreed that 'true' computers can be divided into several types as below. However, the boundaries between each type (Fig.1.1) are both poorly defined and extremely fluid, as the capabilities of each type are changing rapidly.

Micro- (desktop, laptop, personal) computers (PCs). 'Micros' are machines which can be run on their owner's desk or lap (in his office, home or field laboratory, or even in a car or plane), and are under his complete control. A user actually sits at the computer itself, and it interacts with him alone. Unlike larger computers, micros do not require any specialized staff or environment to operate them They are tolerant of dust, humidity and erratic power sources, and will work in climates ranging from alpine to subtropical and arid. Although, only a few years ago, they could cope only with simple computing operations like *word-processing,* and not with the complexities of geological data, that situation has now completely changed. Machines such as the Apple Macintosh II rival many minicomputers in power; they have 256 Kb ROM (i.e. 256,000 **alphanumeric** characters of instructions can be stored in the machine's inaccessible Read-Only Memory), with expandable RAM (Random Access Memory, the machine's accessible memory) of up to 8 Mb (8,000,000 characters), and can be linked to *hard disks* (§1b3) capable of storing (at present) up to 230 Mb — that is, the entire contents of a moderate-sized library. Complex mathematical and statistical manipulations (e.g. **matrix inversion, factor analysis**), formerly executable only on large computers, can also be performed, and the advent of desktop publishing (§4c), with devices such as laserwriters (§1b5), allows users to produce high quality documents from personal computers.

Work-stations. These are a recent development, dominated by firms such as SUN and APOLLO. They take over at the upper end of the microcomputer range and extend well into the minicomputer range (Fig.1.1). Smaller work-stations are just overgrown microcomputers, still small enough to sit on the user's desktop and operated in much the same way. More powerful (and physically larger) versions have to be placed 'deskside' (under/beside the user's desk) and, depending on which operating system is being used, may require specialized computing personnel (operators, etc.) to assist in running them.

Minicomputers. These and all higher categories definitely require specialized support staff, and have to be housed in air-conditioned, humidity-controlled, dust-free rooms with electrically stabilized power sources. The staff are employed to ensure the problem-free operation of both *hardware* (machinery) and *software* (programs), perform *backups* (safety copies) of users' files, etc. Users do not actually sit at a minicomputer, but interact with it via a remote, cable-linked *terminal*, which can be either a simple screen device *(visual display unit,* or *VDU)*— a 'dumb' terminal — or a micro fulfilling the role of a terminal — an 'intelligent' terminal. A minicomputer operates far more quickly than any individual user can respond to it, taking milliseconds or nanoseconds to answer a query; if the machine interacted with just one user, it would effectively spend much of its time (on its own time-scale) waiting for the user to respond. Most machines are therefore set up for *multi-user* (often called *time-sharing*) access. That is, many people (perhaps 5–10) are linked to the machine at once via separate terminals; as soon as the computer answers user *A*'s command, it searches for other users whose commands are still unanswered, while it awaits a new command from user *A*. In this way, the number of commands the machine actually answers approaches its actual capability more closely, though only 50% total capability — Central Processing Unit (CPU) utilization — might be regarded as satisfactory. Today's minicomputers (e.g. VAX) are also capable of being linked together as *networks* (§3a), whose combined capacity approaches that of a much larger computer. Such machines can also talk to one another across nations or even internationally, via satellite and other links which may or may not link with the normal telephonic networks (Figs.3.2–3). In geology, even those rare, highly sophisticated data-handling techniques which outgrow microcomputers can mostly be performed on minicomputers.

Mainframe computers (e.g. IBM, CYBER). These are one step larger still, but behave from the user's viewpoint like minicomputers. Most geologists will rarely require their full power, although they may find themselves using mainframes solely because some widely-used packages (e.g. SPSS: Table 2.1) are too large to fit on minicomputers and are thus usually implemented on mainframes.

Supercomputers (e.g. Cray 2). These are the most powerful machines of all, and many countries only have one or two. Their power is required mainly for the most complicated numerical techniques (e.g. modelling the weather or universe), and little use is at present made of them in Earth Sciences, except for the most sophisticated modelling techniques in geophysics. They are supposed to complement or even rival the human brain in some respects.

The following criteria are useful for comparing the power of different computers:

Accuracy is simplistically measured by the number of *bits* (binary digits, i.e. 0s or 1s) a machine uses to store a number or other piece of information (termed a computer *word).* A 16-bit machine reserves 16 bits per word, so that the highest *integer* number which can be represented, allowing for the sign, ±, which takes one bit, is ±111111111111111 in binary (§1b3) notation (= 32,767 in decimal notation). With real numbers *X.Y*, 11 bits are conventionally reserved for the mantissa *Y*, 4 for the exponent *X* and one for the sign. Translating the 11 bits for the mantissa into decimal notation — i.e. dividing by $\log_2(10)$ — this means that such a machine will store real data to ≈3–4 decimal places accuracy (the 4th decimal place and thereafter will be dubious). The table below shows the progression.

No.of bits	Highest integer	Bits for mantissa in reals	Bits for exponent	Approx.decimal accuracy
8-bit machine	$2^7-1 = 127$	5	2	≈1-2 decimal places
16-bit machine	$2^{15}-1 = 32{,}767$	11	4	≈3-4 decimal places
32-bit machine	$2^{31}-1 \approx 2 \times 10^9$	23	8	≈7-8 decimal places
64-bit machine	$2^{63}-1 \approx 9 \times 10^{18}$	52	11	≈15-16 decimal places

All this means that 16-bit or even 8-bit machines (i.e. small microcomputers) may be accurate enough for some geological applications, but for calculations involving matrix operations (as in Section VI), 32- or 64-bit machines (mainframes, or a few of the more powerful minis or micros) are likely to be required. Isotope geologists need to be particularly aware of computer accuracy, now that isotopic ratios are commonly quoted to 6 decimal places (e.g. 0.703158 ± 0.000008 for $^{87}Sr/^{86}Sr$). The results of complex statistical calculations involving such ratios will be heavily dependent not only on machine accuracy but also on correct statistical algorithms. (The 'correct' way to handle such data would in fact be to subtract, say, 0.703 from each measurement and then multiply by 10^4, so that 0.703158 becomes 158; neither computer accuracy nor algorithms would then be anything like so critical).

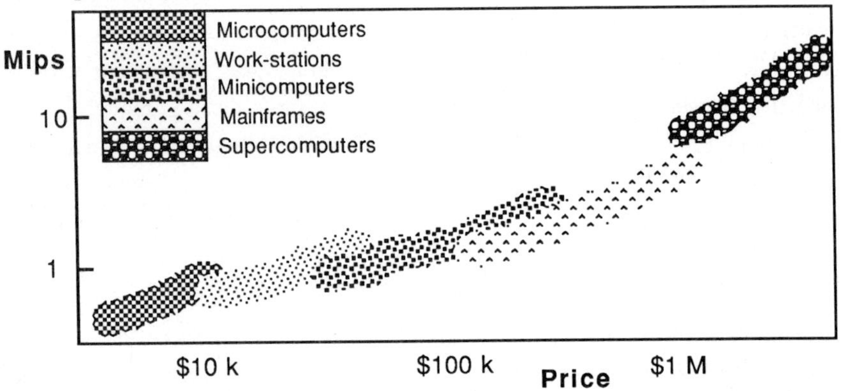

Fig.1.1.Highly generalized guide to types of computers as measured by their performance and price
(The scales are logarithmic, but not too much attention should be paid to them! — see text for explanation)

Speed is of course a requirement which only the geologist himself can judge. Furthermore, for many purposes, the extra time micros take to complete actual calculations is more than compensated by the greater accessibility of the results (e.g. one does not have to wait for operators to tear printout off a mainframe line-printer, and then go and fetch it). Nowadays, computer speed is commonly measured in *Mips* (Millions of instructions per second), or, for real numbers, *Mflops* (Millions of floating-point operations per second). Unfortunately, again, there are various ways of measuring Mips performance, but one commonly accepted way is to rate machine *relative* to a popular minicomputer, the VAX 11/780, rated approximately at 1 Mips, whereupon typical ranges of speed are shown on Fig.1.1. The differences are important: for example, the Macintosh II is rated by this method at ≈0.8 Mips, but is publicized by its manufacturers at 2-3 Mips; this is well into the mainframe range on Fig.1.1.

Storage capacity is again dependent on what the geologist wants to use the computer for, but today's microcomputers can comfortably handle documents or datasets as large as many geologists may feel capable of dealing with at once. Some minicomputers can expand their effective storage capacity via *virtual memory:* a hard disk (§1b3) is used to extend the machine's central memory, swapping data to and from the central memory and hard disk, to make room for currently active processing.

In general, the speed, accuracy and storage capacity of computers increases with price (Fig.1.1), but there is such considerable overlap that, for example, powerful microcomputers may nowadays outperform some minicomputers. Sub-categories such as 'supermicros' and 'superminis' further blur the divisions. Fig.1.1 therefore gives only a very generalized comparison.

1b2. Hardware for entering new data into a computer (converting it to machine-readable form)

Until the late 1970s, the main method of data-entry to computers was 80-column cardboard *punch-cards*, which were punched with holes using a special typewriter-like machine (a card punch). These are now little more than objects of derision to some younger computer 'buffs' (although many geology departments seem to have remarkably large stores of them accumulated from previous researchers!) Another obsolete medium formerly used in many geochronological laboratories was *paper-tape*, effectively a continuous role of punch-card. Nowadays, there are 5 main methods for entering new data.

(1) *Manually*. The user merely types the data in at a computer terminal.

(2) *Scanning*. Scanners are modern devices similar to photocopiers, except that they generate a machine-readable document from a paper copy. Cheaper scanners merely take a pixel image of the page — that is, graphics and text alike are converted into hundreds of boxes (pixels) which are either *on* (black) or *off* (white): a letter 'A' is stored as a picture, not a letter 'A' as such (see Fig.4.3), and so cannot be edited in the usual manner of word-processing. However, more expensive devices now becoming available are able to distinguish true graphics (lines, ornaments etc.) from alphanumeric text, and read the latter in as actual numbers and letters (*ASCII text:* §3a) so that they can be word-processed. At present, most such scanners can only understand simple, monospaced type-styles (fonts), such as Courier (a standard typewriter style), in which every letter occupies the same width (§3). Proportional fonts such as Times used in this book (and similar fonts in most geological journals) might be read with a much higher error rate, or not at all, because different letters occupy different widths ('i' is narrower than 'w'). Eventually, however, it should be possible to convert a published paper (text and maps alike) into a computer file as easily as the geologist at present takes a photocopy. The potential effort which would thereby be saved in, for example, compiling bibliographies and abstracts, is extremely enticing!

Progress toward realizing this dream has been made on other fronts. The contents list for the journal *Computers & Geosciences* is now issued in Softstrip® form (Fig.1.2), which can be read by a special reader (it is analogous to bar codes used at supermarket checkouts for reading prices).

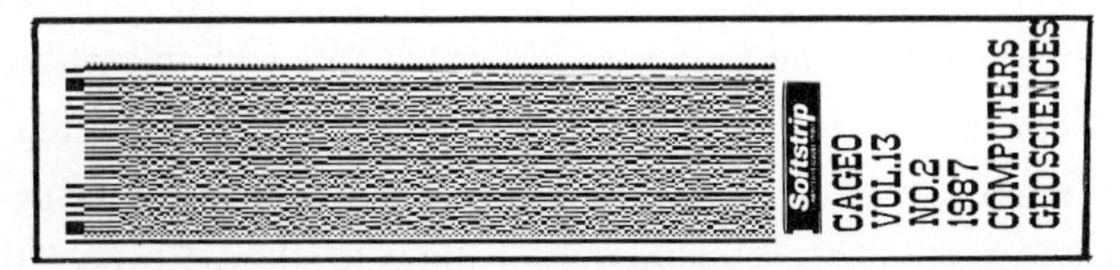

Fig.1.2.A piece of Softstrip® carrying the contents for one issue of the journal *Computers & Geosciences*

(3) *Digitizing*. Digitizers are devices which allow lines and curves (and hence entire maps) to be converted into machine-readable files. They are, not surprisingly, finding increasing use in geology. The user merely runs a small device with a cross-hair cursor along the lines/curves to be digitized, and the computer converts these movements into a continuous series of *X-Y* coordinates. Specialized systems for digitizing geological data have been developed (e.g. Unwin 1986).

(4) *Pointing devices*. Positional information, or responses to multiple-choice menus, can be made on certain systems using *cross-hair cursors* (as used on digitizers), *light pens* (pointed at the screen),

touch-sensitive screens, *touch pads*, *joysticks* (like car gear levers), or *mouses* (small devices with a clicking button, fundamental to the Apple Macintosh and now increasingly used with other micros).

(5) *Voice recognition.* Computers are already available which can, within limits, recognize speech and translate it into machine-readable files. Significant developments in this area are widely anticipated.

1b3. Storage media for entering, retrieving, copying and transferring pre-existing data

Nowadays, the main media are three, each of which has its associated reader (tape drive, disk drive):

(1) *Magnetic tapes* include reels similar to those on old-fashioned tape recorders, and cassettes as used in modern cassette recorders. The amount of data storable on a tape varies enormously, depending on the length of the tape (usually hundreds of feet), and on the packing density of the data (typically 1600 bits per inch (bpi) but up to 6250 bpi). The main disadvantages of magnetic tapes are: (a) *vulnerability:* they are easily torn or otherwise damaged, and if they become unreeled.......!; (b) *instability:* they tend to decay with time, and therefore have to be 'exercised' every so often to safeguard the data stored on them; (c) *inaccessibility:* access to stored data can be very slow; to read data at the end of a tape, for example, the whole of the tape has to be put through the tape drive; even mounting and dismounting a tape on a drive can be frustratingly slow, since it is usually performed by trained operators, and the user may have little control over when it is done. Despite all this, tapes are widely used for geological data gathered in the field (e.g. ocean drilling ships, land seismic surveys), and for backing up files (i.e. producing multiple copies in case originals are damaged or destroyed). Furthermore, notwithstanding considerable improvements in inter-computer communications, nationally and internationally, tapes often still constitute a failsafe way of transferring data from one mainframe to another, since various parameters of the tape (block length, parity, bpi, etc.) can usually be adjusted to suite any mainframe anywhere in the world.

(2) *Magnetic disks* can be divided into 3 sub-types: (a) *5.25" floppy disks*, as used by many brands of micros, now increasingly serve the same purpose as tapes, although most can as yet only store 360Kb of data; (b) *3.5" diskettes*, as used by Macintosh and now IBM-PC micros, are rigid disks, capable of storing up to 400 Kb (single-sided) or 800 Kb (double-sided); (c) *hard disks* are 'monster' diskettes, capable of storing 50-1,000 times as much data (up to 10 Gb = 10^{10}b in some cases). Floppy disks, diskettes and desktop hard disks available with micros are physically just as portable, robust and convenient as the micros themselves. By contrast, the largest hard discs (*Winchesters*), which store data on large mainframes such as government tax or health records, are highly sensitive, non-portable media which require careful handling in specialized environments. Disks suffer none of the above-mentioned disadvantages of tapes. For example, accessing data from a disk (especially hard disk) is much faster, because the computer can go direct to the relevant 'track' or 'sector' on which the required information is stored (just like a hi-fi needle can be placed straight on the track corresponding to the 4th movement of a symphony on a 12" hi-fi record). Disks are also far less vulnerable, although all micro users will have encountered 'duds'. The main *disadvantage* of disks is that, although all share a common basic pattern of concentric tracks divided into sectors, there are hundreds of incompatible disk formats defined by numbers of tracks or sectors on single- or double-sided disks at single, double or quadruple packing density. Hence one disk format may only

be useful for transferring data between micros of the same type because, unlike tapes, disks cannot be formatted freely by the user.

(3) *Laser disks* (CD-ROM = Compact Disk-Read Only Memory) are basically similar to the *audio* compact disks now familiar to the public for playing music; they use a laser-beam to read the recorded information. Disk drives which read CD-ROMs in the same way that traditional disk drives read floppy disks are available, and simply slot into a PC like a normal disk drive. Larger (12") laser disks are also now produced (e.g. Philips laservision) which can store *both* audio *and* visual information — and can hence be used to show films in the same way that compact disks can be used to play music. They have been used to archive information from some national studies (e.g. the UK's *Doomsday project)*, and have obvious potential for storing graphical information (geological maps, slide collections etc). Laser disks have all the main advantages of magnetic disks but to a greater degree; for example: (a) *huge storage capacity* : a standard 12 cm-diameter CD-ROM can store 550 Mb (or in some cases, 550 Mb *per side*), i.e. about 1500 floppy disks or 250,000 pages of A4 text — far more than currently available on microcomputer hard disks, and comparable to Winchesters on mainframe systems; (b) *invulnerability:* like domestic compact disks, they can (theoretically) be spread with jam (Vegemite in Australia!) and survive the experience. Their main disadvantage is that they only offer *permanent* data storage (information cannot be rubbed out and rewritten). This does, however, mean that they offer a superb potential medium for long-term storage in overstocked libraries (taking over from the much-maligned microfiche) or for retaining permanent copies of data from terminated geological projects. They are beginning to be used for storing large databases (§3).

1b4. Hardware for interacting with a computer: Terminals

Terminals now come with a wide range of shapes, sizes and capabilities, but all offer a typewriter-like keyboard and VDU (Visual Display Unit). The keyboard may offer from a few to dozens of special *function keys* in addition to the standard QWERTY and 123... layout. *Text terminals* can only handle text (including numbers, of course). Earlier models have the keyboard and VDU rigidly connected together, but later ones separate them for ergonomic reasons. In the simplest (and cheapest) types ('dumb terminals') lines of text roll up the screen as they are sent by the computer, until they are irretrievably lost off the top. This can be extremely irritating, especially if the lines are generated too fast for the user to interrupt the output at a chosen place. More sophisticated ('intelligent') examples have an internal memory (microprocessor) which enables *scrolling* (lines which scroll off the top are stored and can be recalled up to the memory limit), and/or *editing* (lines earlier than the current one can be selected, edited and then sent back to the computer). Programmable keys are also commonly provided. *Graphics terminals,* which can handle pictures as well as text, come in 2 main types: (1) *vector terminals* draw lines on the back of the VDU via an electronic gun; (2) *raster terminals* work on the same principle as TV sets, by dividing the screen into many (typically ≈400 x 200) tiny boxes (called *pixels*, or *rasters*), and then directing a cathode-ray tube repeatedly in row fashion over the screen, so as to light up or leave dark each individual pixel; the resulting image is thus a matrix of dots in row/column form.

With the staggering drop in their price, microcomputers are increasingly taking the place of both dumb and intelligent terminals as a means of interfacing with larger computers. Given a suitable *terminal emulation package*, most microcomputers can be made to imitate many of the popular types of terminals (text or graphics) and, of course, can still be used as computers in their own right (§1d). A single micro thus effectively takes the place of 3 earlier devices.

1b5. Hardware for generating (outputting) hard-copies

Output devices fall into many categories, including the following:

(a) *Traditional printers* come in many forms, but can be divided into two main types. (1) *Impact printers* are glorified typewriters bringing paper, ribbon and key hammer into brief, violent contact. In general, the faster printout is generated, the lower its quality. Among these devices, *line-printers*, one of the earliest computer print technologies, are still linked to most mainframes. They can produce fast (250–1800 lines per minute) but generally low quality output with a maximum of 136-characters per line (typically 124 or 132), at a fixed 10 characters per inch, on continuous, sprocket-fed, plain or lined paper. Some line-printers are still only capable of printing UPPER CASE LETTERS. Line-printers are fine for inspecting data and text but not for final copy. *Letter-quality (LQ) printers* achieve higher quality output (similar to a golfball typewriter) via carbon ribbons and at slower speed (12–80 characters per second). *Dot-matrix* printers form all the individual letters from closely spaced or overlapping patterns of tiny dots imprinted on the paper by fine needles. (2) *Non-impact printers* are far less noisy, and will probably dominate the technology in the future. They build up an image similarly to a dot-matrix printer, but use an *ink-jet* (ink is sprayed at the paper), an *electrostatic method* (the image is represented by charges on a line of stationary nibs), or a *thermal method* (heat-sensitive paper changes colour when moved across a heated pin). Both these types of traditional printer can only handle text effectively, although crude 'line-printer' graphics (e.g. *X-Y* plots) can be simulated by judicious use of characters such as = — I I \ / * o and x.

(b) *Modern printers*, by contrast, can handle text and graphics side-by-side. *Imagewriters* are a type of combined non-impact printer and plotter, generating text and pictures alike at a resolution of 72 dots-per-inch (better than a typical impact printer). Colour can be supported. *Laserwriters* work similarly, but use a sweeping laser-beam to produce far higher quality output still (≈300 dots-per-inch resolution) at about 8 A4 sheets/minute. They are actually powerful micros in their own right, and can understand specialised programming languages (e.g. Apple's Postscript™). *Linotronic typesetters* generate still higher quality output than laserwriters (typically ≈1200 dots-per-inch), and can generate bromide prints, but work on the same principles. Fig.1.3 shows the difference in output quality between these 3 types of printer.

(c) *Plotters* produce high-quality graphs, maps and pictures, often in several colours. Again, there are numerous types. (1) *Pen-plotters*, which use pens and inks, include *flat-bed plotters*, which use a flat piece of paper over which the pen moves, and are capable of generating very large maps, stratigraphic sections, etc. of the highest possible quality; *drum-plotters*, where the paper is on a rotating cylinder across which the pen moves; *belt-bed plotters*, which combine some advantages of flat-bed and drum types through use of a drum and flexible belt; and small *desktop plotters*, usually restricted to diagrams no larger than A3 or A4 paper. The first two are usually kept within the 'nerve-centre' of a computer centre, for remote generation of plots from mainframes, whereas, as the name suggests, the latter can sit on the geologist's bench, next to his microcomputer or graphics terminal. (2) *Electrostatic plotters* generate a plot quite differently, representing it as pixels rather than as lines (vectors). As with electrostatic printers, the configuration of the plot is converted to charges on a line of densely packed stationary nibs; the paper picks up the charges as it passes over the nibs, and then proceeds into a trough (several troughs for coloured plots) of liquid toner which deposits on it according to the pattern of charges acquired. Electrostatic plotters are quicker, and achieve higher resolution with greater colour density than pen plotters, but are more expensive, require special paper, and cannot be overplotted.

LINOTRONIC 300
FOR QUALITY

LINOTRONIC 300
FOR QUALITY

LINOTRONIC 300
FOR QUALITY

Fig.1.3. Difference in quality between an *imagewriter* (72 dots per inch), *laserwriter* (300 dpi) and *linotronic typesetter* (≈1200 dpi) when magnified 3 times

1b6. Modes of interacting with a computer from a terminal

There are two ways of interacting with a computer. (1) In *interactive time-sharing* (multi-user) mode, the user issues commands, and the computer obeys them more-or-less at the time they are typed in; the results are available almost immediately. (2) In *batch* mode, the user sends a job to the machine which joins a 'batch queue' with other jobs to be run later (e.g. at night, when the machine is less busy), and the user may not collect his output until the next day. Most minis and mainframes support both modes, but micros are pre-eminently interactive machines. Nevertheless, some micros can be networked to share output devices, so that several users can ask to print their files, whereupon the machine queues ('spools') and prints them in a predetermined order — a nascent form of batch mode.

The advantages of interactive use are obvious: the user can review his results, correct errors, and write his report as he goes. Probably most geologists aim for this type of usage. Batch use, however, is often predetermined by cost: interactive use during peak periods (i.e. daytime) is usually far more expensive than batch use overnight, so that for large jobs, batch use may become most cost-efficient.

Some popular packages mentioned throughout these Notes (e.g. SPSS: Table 2.2) are only designed for batch-use, whereas others (e.g. MINITAB) are primarily interactive.

1b7. The terminology of data-files as stored on computers

A single piece of data (e.g. the value for Rb in a rock) is here called a *data-value,* while many data-values form a *data-set.* Related data-values are usually stored in tabular form as rows and columns; values for several elements in the same rock would typically be listed on a VDU as horizontal rows, and are called *records* (or *card-images*, reflecting their historical origin in punch-cards). Some hardware and software prefers records not more than 80 characters long, to keep them within the width of a VDU screen, but far longer records can be handled by other machines. Mainframes typically allow up to 136 characters at least, to allow printing on a single line of standard computer paper on a line-printer.

Any collection of records stored as a single magnetic entity on a disk or tape is called a *file* or, less commonly, a *document.* Many computers enable files to be organised in the same way that one organizes paper files in an office, into *folders, catalogues* or *directories* arranged in a hierarchical structure. Thus in the 'Mac' hierarchical file system (HFS), files can occur within folders within other folders (Fig. 1.4). A more organized form of file is a *spreadsheet*, essentially a table which allows preset mathematical operations (totals of rows or columns, percentages, etc.) to be performed automatically as the data are updated. A stage beyond the spreadsheet is the *database*, discussed in §3.

1c. Software: programs and programming languages

1c1. Types of software

Software (the sets of instructions which make computers actually do things), can be divided for our purposes into five main types: (1) *Systems software* (§1c2) includes the operating system itself, without which the computer cannot operate, and associated utilities supplied by the computer manufacturer. (2) *Programming software* (§1c3) includes the quasi-English programming languages themselves, together with compilers/interpreters which translate them into binary code which computers can understand and obey. (3) *Communications software* enables computers to talk to other computers, or to other hardware devices (plotters, etc.) (4) *Applications software (packages and libraries*) are collections of programs or subroutines, which users can invoke; they are supported by the professional staff of a computing centre, who either write them themselves, or obtain them from outside vendors. (5) *Personal software* is what the user writes, provides or is otherwise responsible for.

Here, only brief comments on (1) and (2) are given. Types (3)–(5), with which most geologists would be primarily concerned, are covered in detail in §2–3.

1c2. Systems software (operating systems and systems utilities)

Operating systems are regularly updated and managed on-site (at least on mainframes) by central computing staff. They are essential at all times to all machines; they are immediately invoked when a machine is switched on, and continue running until it is switched off; they interpret all the user's commands when he is not running any other kind of software, and in the background even then. Nevertheless, the geologist only needs to learn as many commands as he requires, and keep up-to-date with any changes; he need rarely concern himself with details of the system. In fact, the hallmark of a good operating system is that the user should be virtually unaware of its existence. Operating systems may be *single-* or *multi-user* (i.e. capable of servicing one or many *users* signed onto the same machine at once), *single* or *multi-tasking* (i.e. capable of running one or several *programs* effectively at once). Most mainframe systems are multi-user, multi-tasking, but micro systems are generally single-user, single-tasking. An exception is the latest system for the Macintosh II microcomputer (System 4.2, Multifinder 6.0), which emulates multi-tasking.

Although most operating systems have commands to cover an essential range of operations (listing, printing, deleting, appending, copying of files, etc.), they perform them in very different ways. Some systems (e.g. VMS on VAX minicomputers) are relatively user-friendly, and offer simple quasi-English commands which actually string a series of smaller operations together; others (e.g. NOS on CYBER mainframes) are far more difficult to master, and only provide single-step commands which the user must string together to perform apparently simple operations (though this also gives him more precise control over what he can do). Ubiquitous systems utilities include *text editors*, *sorting* routines and *help* systems (invoked at the terminal and, in theory, explaining all the other available utilities).

Every make of computer has its own operating system, but 4 now approach the status of standards:
(1) *UNIX* was developed at Bell laboratories in the early 1970s, and enhanced by various US universities. It is an exceptionally portable system, which runs either in its original or a 'lookalike' form on the whole range of computers from micros (Apple Macintosh, Hewlett Packard) to minis (VAX), mainframes (ICL, Prime) and even supercomputers (Cray 2).
(2) *XENIX* is Microsoft's multi-user, multi-tasking implementation of UNIX, and is likely to be widely adopted on smaller machines, since it can be far more compact.
(3) *MS-DOS* (MicroSoft Disk Operating System), another single-user, single-tasking system, was launched in 1981 to coincide with the release of the IBM Personal computer (PC), which uses the machine-specific version *PC-DOS*. These two have now by far the widest implementation of any system on microcomputers, being available on all the numerous IBM-PC 'clones'. MS-DOS will probably in time move closer towards XENIX and effectively become a single-user version of UNIX.
(4) *Macintosh*. This system is totally different from any other and, to its proponents, incomparably easier to learn and use. The computer screen is a clever, electronic, graphical representation of an ordinary desktop (Fig.1.4). To perform systems operations, the user moves pictures (*icons*) around using a *mouse*, or chooses options from pull-down menu, instead of typing in strings of word-based commands. For example, to delete a document in most operating systems one types "DEL(ETE) <DOCUMENT NAME>", but on the Mac the document icon is 'thrown' into the Trash — which bulges accordingly as on Fig.1.4 — by dragging it over the Trash icon with the mouse. The Mac system is a godsend to geologists who cannot type, because many basic operations can be executed with one finger and no typing. More and more computers (e.g. Amiga) are emulating this type of system, and Mac-like mouses and menus are now available on other popular micros such as the IBM-PC. In essence, the Macintosh system transfers much of the effort of computing onto the

programmer: things are made far easier for the user, but there is a corresponding penalty in writing programs, because all the windows, icons and menus have to be programmed on top of what goes into a normal mainframe or IBM-PC program. For those interested in the machine's possibilities, the following are among the growing library of books by enthusiasts:

General guides: Cary(1982), Miller & Myers(1984), Morgan(1985), Poole(1984), Venit(1985).
Programming guides: Huxham(1986), Lewis(1986), Moll (1985), Twitty(1986),Willen(1985), Williams(1986).
Software guides: Adamis(1985), Cobb(1985), Flast & Flast(1986), Hoffman(1986), Stiff(1985), Van Nouhuys(1985).
Graphics & CAD: Hastings(1986), Schmieman(1985), Schnapp(1986).

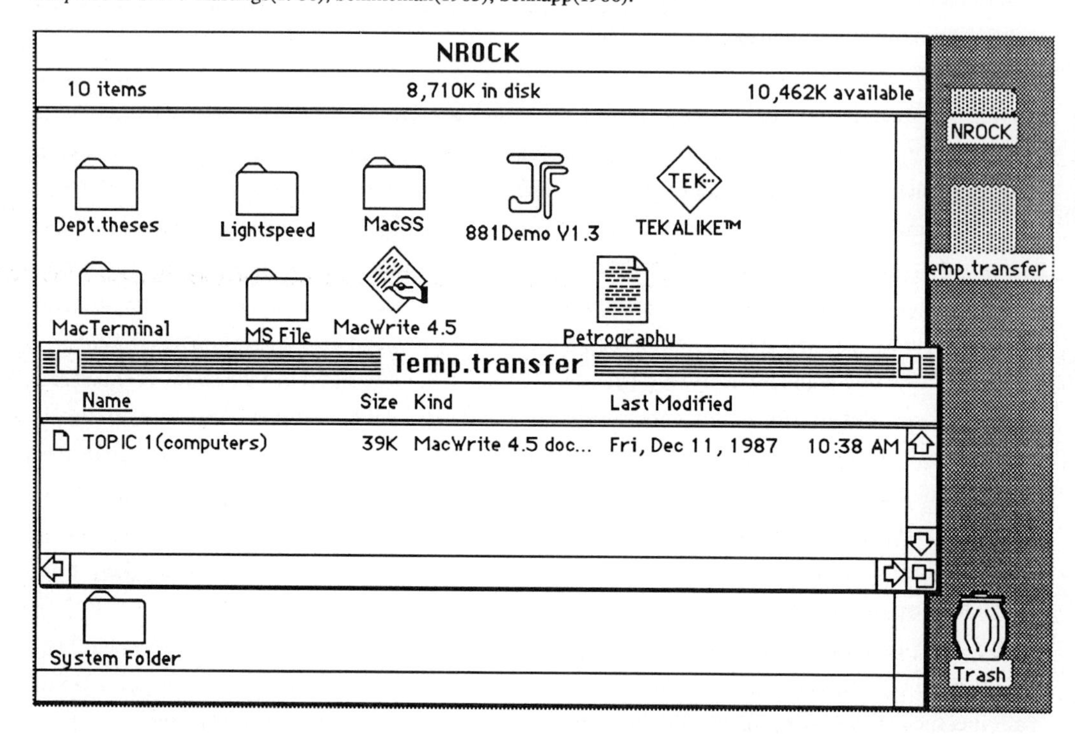

Fig.1.4. A typical Macintosh desktop, generated by its unique graphics-based operating system
Two windows are superimposed, showing the contents of the 20 Mb hard disk and 3.5" diskette (icons on right). In the upper window, different pictures (*icons*) distinguish different types of document: the suitcases (e.g. 'Dept.theses') are *folders* (directories), which can be opened to reveal their hierarchical structure; 'Petrography' indicates a word-processor document), and MacWrite and 881 Demo are actual executable programs. Within any window, files can be listed either pictorially (upper window) or in a more mainframe/PC-like fashion (centre window).

1c3. Programming languages

General guides to programming: Bergerud & Keller (1987); Helms (1980); Unwin & Dawson (1985).

Computers can actually only perform instructions written in *binary code* — i.e. on/off, 0/1. In the binary system, numbers are written to the $base_2$ instead of the $base_{10}$ as in the everyday decimal system, so that just as 111_{10} means $[1x10^2 + 1x10 + 1x1]$, so 111_2 $(1x2^2 + 1x2 + 1x1)$ is equivalent to 7_{10}, and so on. For the ease of scientists, however, computer instructions can also be written in quasi-English using *languages* such as BASIC, FORTRAN; these are then translated into binary code which the computer can understand, via programming software called a *compiler* (or *interpreter*).

FORTRAN is a *high-level language* because its instructions are at a far 'higher' (nearer English) level than binary code; low-level languages also exist which are nearer to binary code and hence quicker for computers to interpret, but they are not widely used (at least in geology).

Many different computing languages now exist. Their history and comparative advantages are outside the present scope, but a simplistic, highly generalized guide is given in Table 1.1. Readers should be warned that views may differ considerably from Table 1.1 not only between geologists, but also among computer professionals of different age-groups.

FORTRAN was the first high-level language, developed by IBM in the 1950s, and has a vast reserve of expertise worldwide, so that it is unlikely ever to disappear. Most widely-used mainframe packages, and nearly all programs in geological journals, are still in FORTRAN. The language keeps undergoing major revisions (currently used versions include FORTRAN 66, FORTRAN IV and FORTRAN 77, while a FORTRAN 88 is mooted); many of the changes are moving it towards greater compatibility with the next language, Pascal.

Pascal is considered a far more elegant language by computer scientists, partly because it was written by a single individual (Prof.N.Wirth), with the benefit of hindsight, in the late 1960s. At present, rather little specifically geological software written in Pascal is available, but many microcomputers (e.g. the Macintosh) work best in this language, so the situation is rapidly changing.

BASIC (another popular language, at least for schoolboy computer freaks) is less favoured by professionals than both the above, because of the larger number of dialects which make BASIC programs very machine-specific, and hence extremely tedious to move between computers.

For geologists tackling their first programming language, FORTRAN or Pascal are recommended. However, geologists intending to program *extensively* are best advised to learn *both,* as there are still many operations which can be performed in FORTRAN but not in Pascal, and rather fewer vice-versa. It is probably easier to start with FORTRAN, in that books which teach Pascal to FORTRAN programmers exist (e.g. Pericott & Allison 1984). From Pascal, it is only a fairly small step to learn 'C', which many professionals see as a language of the future. Other languages which have their main application in fields of limited interest to geologists (e.g. COBOL, in business studies), have become obsolete (e.g. ALGOL, effectively replaced by Pascal), or are specifically aimed at particular computer applications (e.g. Prolog in expert system work: Clocksin & Mellish 1981) are not listed in Table 1.1.

Among very few studies on the comparative performance of these various languages with geological data is Carr's (1987) comparison of FORTRAN, Pascal and C in calculating a simple semivariogram (§20e). Carr here concludes that there is little to choose between the three languages: in particular, those conversant in FORTRAN (probably the majority of computerate geologists) probably have little need to master Pascal, *provided* they are interested in language performance alone. Other considerations, however, may also be crucial. Those interested in the Macintosh, for example, have little choice but to master Pascal if they wish to employ the most user-friendly features of the machine; programs using FORTRAN (e.g. Systat) are restricted to a rather cumbersome user interface.

Running a program in FORTRAN or Pascal most commonly involves turning it into an *executable module* in binary code, which a computer can actually understand, as follows: (1) the source code is *compiled* using a *language compiler*, which isolates programming errors and generates a binary file (*object module*); (2) the object module is then *linked* ('link-edited', 'loaded', 'task-built'), a process which builds in all the necessary function calls and subroutines invoked; (for example, if the programmer has used the SQRT function in FORTRAN, which calculates a square-root (√), he does not actually provide FORTRAN code to do this, but the capability is built into the program at this stage using a systems FORTRAN library). A compiled program is self-contained: it runs independently of the compiler — though generally only on the original machine where it was compiled. If a compiled version will not work on another machine, it cannot be made to work unless source code is available.

An alternative method, used most commonly with old-style BASICs (and occasionally Pascal) on micros, is to *interpret* the program. Here, an *interpreter* reads the original BASIC code step by step, interprets it in machine-understandable form, and executes it. Interpreted programs generally run much more slowly than compiled programs, and are no longer stand-alone, requiring the interpreter at all times. However, they are more portable in the sense that if they fail to work on another machine, the original code is available and can be modified to make the program work.

Fourth-generation languages are a development designed to help programming, which are intimately interfaced to database management systems (§3). *Fifth-generation languages* (e.g. Prolog) are those concerned with **expert systems** (§5), which mimic human decision-making.

Table 1.1. Overview of selected computer languages of practical interest to geologists

FEATURE	FORTRAN[3]	BASIC[4]	Pascal[5]	APL[6]	C
Guides to programming in the particular language	Carlile & Gillett (1973); Dorn & McCracken(1972); Grout(1983); Law(1987); McCalla(1967); Metcalf(1985a,b); Ramden(1977)	Sinclair (1973)	Crandall & Colgrove(1986); Lecarme & Nebut(1984) Moll(1985); Pardee(1984); Pericott & Allison (1984)		Waite et al. (1984)
Availability of stand-alone geological software and mathematics algorithms	Excellent	Moderate	Improving	Poor	Poor
Availability of subroutine libraries	Excellent	Nil	Poor	Nil	Poor
Range of computer implementations	Excellent	Poor	Improving	Poor	Moderate
Availability on microcomputers	Excellent	Excellent	Good	V.restricted	Good
Availability of computer experts	Vast	Large	Improving[2]	Restricted	Improving
Availability of geoscience experts	Vast	Large	?Improving	Minimal	Poor
Protection against programming errors	Fair	Poor	Good	Slight	Poor
Precision on small computers[1]	Excellent	Poor	Poor	Good	Good
Brevity of computer programs	Moderate	Poor	Moderate	Excellent[7]	Moderate
Support of structured programming	Good	Poor	Excellent	Nil	Excellent
Portability between computers	Good	Poor	Good	Poor	Moderate
String(text)-handling capability	Poor	Poor	Moderate	Poor	Moderate

[1] Good precision implies the availability of equivalents to the DOUBLE PRECISION data-type of FORTRAN.
[2] Most university Computing Science departments nowadays teach Pascal (or C), with FORTRAN given only a quick introduction; however, expertise in FORTRAN remains very widespread not only among older practicing scientists (including geologists) but also in many computer centres.
[3] Stands for FORmula TRANSlation.
[4] Stands for Beginners' All-purpose Symbolic Instruction Code.
[5] Named after the French philosopher; other languages exist which are close to Pascal but have different names.
[6] Stands for A Programming Language.
[7] At the sacrifice of legibility (i.e. the language is far from English-like).

1c4. Graphics software standards

Further discussion: AMIRA (1985, vol.2); Brodlie (1985); ISO (1983).

There are 4 elements involved in generating a computer plot (graphic):

1. a *graphics program*, which actually codes the required operations in FORTRAN, etc;
2. a *graphics subroutine library*, a collection of commands which the graphics program invokes to draw lines, circles, letters of the alphabet, etc;
3. a *graphics device*, hardware (terminal or plotter) which actually generates the plot (§1b5);
4. a *device driver*, which tells the graphics device how to draw the lines, circles etc. as commanded.

Unfortunately, 2–4 vary enormously, even between different institutions operating the same computer system. All this makes graphics software on mainframes notoriously non-portable — to run a program on a different machine, a great deal more rewriting ('translation') is usually necessary than to translate a program which, say, calculates a CIPW norm (numerical data but no graphics). Attempts have been made to establish international graphics software standards, but the only one currently approved by the International Standards Organisation (ISO 1983) is GKS (Graphics Kernel System), widely used in the USA, but still relatively limited as regards 3-dimensional graphics. Other systems include GPGS (NORSIGD 1980), favoured in Europe, and PHIGS (Programmers' Hierarchical Interactive Graphics Standard), which goes further than GKS in 3-D graphics but is less widely available. The further that a given program departs from a standard such as GKS, the more effort will be involved in keeping it running should a new computer be installed, should GKS itself change, or should it be necessary to change to say, GPGS.

Hardware standards concerning device drivers are also formulated (e.g. CGI — Computer Graphics Interface), but these should be generally transparent to the geologist, who is rarely concerned with communications software.

Two different ways of generating graphics may be appropriate to different geological needs:

(a) *Spooling*. Here, the graphics subroutines invoked in the calling program do not actually generate the graphic itself, but a *spool file* (sometimes called a *metafile*). This file is then read by the *device driver*, which allows the actual plot to be produced on any one of the graphics devices supported by that driver. Spooling has 4 advantages, in that a plot can be (1) inspected (for errors), etc. on a screen before being generated as a hard-copy; (2) easily regenerated on any number of different devices; and (3) retained for future replotting without rerunning the program; furthermore, (4) the whole process can be performed from a non-graphics terminal, which can be useful where expensive graphics terminals are scarce.

(b) *Direct plotting*. Here the plot is produced as the graphics program is run; it cannot generally be viewed on a screen prior to plotting, and the program has to be rerun to plot on a different device.

1d. Mainframes versus micros — which to use?

Further discussion: Rock(1987f); Rock & Wheatley (1988).

It is impossible to generalize as to which scale of system will best suit the needs of a particular geologist (always assuming, of course, that the choice is not dictated by senior line management or other administrative considerations!) Choice will depend not only on what he wants to do, but also on existing computer set ups in his company or institution, local arrangements (availability of support and expertise, discounts etc.) At the most general level, however, the following advantages can be claimed for the mainframe versus microcomputer routes (the latter including work-stations):

Advantages of mainframes: expert personnel are employed to keep the system running, so the user does not have to worry about repairs, backing up files, new versions of systems, bugs, incompatabilities, etc.; greater overall computer power is available.

Advantages of microcomputers: the user retains more control of his data; the machine 'talks' to him alone, so the system does not slow down at times of heavy usage; adequate computing power and software for many geological purposes is available at a fraction of the cost on mainframes; on-going financial charges from many computer centres (e.g. CPU and disk storage charges) are avoided.

Fortunately, it is possible nowadays to combine the capabilities of *both* mainframes *and* microcomputers, since a microcomputer can be used as an intelligent terminal to log onto a mainframe. This requires a *terminal-emulation package*, which makes the microcomputer imitate a particular make of terminal. For example, using a standard emulation package such as MacTerminal™, a Macintosh emulates a VT100 terminal, so that any non-graphics program implemented on a mainframe can be run on that mainframe but *from* the Macintosh, so that the results appear on the Macintosh screen. They can then be captured very conveniently and processed immediately using any appropriate Macintosh software. At a higher level, Tekalike™ turns a Macintosh into a quasi-Tektronix *graphics* terminal, sc that a graph can be produced on the Macintosh screen by running a suitable mainframe program. This, too can then be 'captured' from the Macintosh screen, edited in a Macintosh graphics application, and then printed at high quality on a laserwriter, either as, say, a graphics document, or after incorporation at the appropriate position within a word-processor document. The time- and cost-saving opportunities of combining microcomputer and mainframe capabilities in this way are prodigious.

One problem is that some mainframes require different data-formats from some microcomputers. Two 'FORTRAN-style' data-formats are most commonly used on mainframes: (1) *Free format*, that is, individual data-values are separated by spaces or commas, and records are separated by a <CR><LF> (carriage-return, line-feed) character pair. (2) *Fixed format*, that is, the values for a particular variable are determined by the *range of columns* they occupy (columns 1-10 might denote the sample number, columns 11-15 SiO_2, etc.); there are no extraneous spaces and no commas, but records may spill over several actual lines of computer file (i.e. <CR><LF> may appear *within* as well as *between* records). Whereas some microcomputers can read both these formats, others cannot. For example, the Macintosh does not use <LF> characters (whereas the IBM-PC does), and individual Macintosh data-values are separated by <TAB> characters, not spaces or commas (whereas the PC and most mainframes understand <TAB> characters only with difficulty, if at all!) File conversion utilities may then become necessary for translating even straightforward ASCII data-files (let alone actual computer programs): the mere process of moving a file from one machine to another electronically is no longer sufficient to allow the second machine to read that file.

TOPIC 2. RUNNING PROGRAMS: MAKING BEST USE OF EXISTING ONES, OR PROGRAMMING YOURSELF

Programming software is here classified in order of increasing complexity as follows: (i) *subroutine libraries* (e.g. NAG: Table 2.2), which provide individual subroutines to perform particular mathematical operations, but cannot be used on their own; (ii) *stand-alone programs*, which are self-contained, and perform a restricted range of mathematical operations; (iii) *'everyman' packages*, consisting of suites of integrated programs which perform whole series of operations, and ranging from those with a sophisticated *command language* (e.g. Systat), which expect the user to set up his own operations but in exchange allow him considerable flexibility, to those which do everything for the user but allow him proportionately less flexibility; (iv) *data-systems* (e.g. SAS) which combine the role of both **databases** (§3) and packages. As a result, geologists wishing computers to solve specific problems can adopt 5 main approaches, listed below in order of decreasing input from the user, and dealt with in more detail in the succeeding sections:

(a) Write their own stand-alone programs completely from scratch, covering the entire process.
(b) Adapt stand-alone programs whose actual source code is published or otherwise available.
(c) Use subroutine libraries for fundamental operations, writing extensive new code around them.
(d) Write tailor-made macros in a command-rich 'everyman' package.
(e) Invoke commands in 'everyman' packages which perform the entire process on their own.

Table 2.1 compares the FORTRAN code needed to perform one statistical operation using these approaches. It is particularly important to distinguish between cases (a) to (c), where the user has the actual source code (which he can adapt as necessary, etc.), and cases (d) to (e), where he does not.

2a. Writing stand-alone programs from scratch

General guides to 'good' programming (and getting the correct answers!): Arthur (1987), Berk (1987a), Greenfield & Siday (1980); Knuth (1969); Law (1987); Longley (1967); Minoux (1986); Moshier (1986); Phipps (1976); Unwin & Dawson (1985); Wilkinson (1965).

No-one embarking on a new program should underestimate the amount of time required to debug or thoroughly test it on benchmark data. Nevertheless, this approach has the following advantages:

(i) The product is presumably as fully tailored as it can be to the user's current requirements.
(ii)The product can be adapted at a future date to some other, related requirement.
(iii)The product is entirely the user's, so there are no copyright restrictions, etc.

This approach is unavoidable for geological applications not previously programmed (e.g. Rock 1987b,c,f,g). As time goes on, however, the likelihood of no program being available for a particular application must decrease, and it is worth spending time (before embarking on new software), to determine whether a product already exists which could be adapted more cost-efficiently for the purpose (§2b). Some authorities are already stating that scientists should no longer consider writing their own programs until they are absolutely sure no program already exists to do what they want!

2b. Sources of software for specialised geological applications

Crystallography and X-ray diffraction: Frizado (1985); Taylor & Henderson (1978).
Electron microprobe data-processing Frost (1977b).
Geochemistry (anomaly recognition, etc.): Koch(1987).
Geothermometry : Bowers & Helgeson (1985); Ghiorso & Carmichael (1981).
Mineral formula recalculations: Ikeda (1979); Knowles(1987); Lapides & Vladykin (1975); Rieder (1977); Rock (1987g).
Petrological: *Norm/classification programs (CIPW, Niggli, Rittman, etc.):* (Andar 1976; Bickel 1979; Carraro & Gamermann 1978; Cerven et al. 1968; Currie 1980; Fears 1985; Glazner 1984; Iregui 1972; Kosinowski 1982; Miranda 1977; Piispanen 1977; Smith & Stupak 1978; Smyth 1981; Stormer et al. 1986; Timcak & Horoncova 1978; Usdansky 1986; Verma et al. 1986); *Projections* (Henkes & Roettger 1980; Lenthall et al. 1974)
Graphics/plotting /structural geology software: see separate list under §4c.

The quantity of *specifically geological* software available is already prodigious, and rising rapidly. Although various attempts have been made, and another is currently mooted by the UK Geological Information Group, it is essentially impossible to build or maintain a comprehensive, up-to-date list, because mountains of unpublicized programs exist in many institutions. The necessarily brief lists above merely illustrate two points: (1) There is already a plethora of published programs in some areas, underlining a lack of awareness of what is already available; the list of norm programs is a classic case of overkill! (2) Most of the routine, numerical operations in geological work have been computerized by someone, somewhere. Furthermore, many programs written originally for other disciplines can be used in geology with minimal adaptation; this certainly applies to most of the statistical programs listed in this book, and increases the quantity of available software by several further orders of magnitude.

Users interested in keeping abreast of developments are well advised to join the *International Association for Mathematical Geology* (Merriam 1978c), the *Computer-oriented Geological Society*, (COGS, Box 1317, Denver, CO 80201, USA), and/or the appropriate geomathematical group of their national geological society. Meanwhile, there are several general sources for both geologically-oriented and general software of which geologists should be aware; readers in some countries (especially the USA) will have others specific to them. The following provide users with actual program code:

(a) Computing/geomathematical journals: *Computers & Geosciences, Computers & Mining, Mathematical Geology* and the *Journal of Geological Education* . The *American Mineralogist* has also recently started a 'Software Notices' section which advertises appropriate programs.

(b) Textbooks like Sokal & Rohlf(1969), Davis (1973) and Mather(1976) which give large, integrated suites of programs. The effort of getting these operational however may be less and less worthwhile as time goes on. [Note that the new edition of Davis (1986) no longer gives actual code, but was the first geological textbook to include a floppy disk in its packaging— a trend now established with books such as Koch (1987)].

(c) Transmissions over networks (§3) and so-called 'bulletin boards'. Much public domain software and **shareware** is now transmitted daily in this form. However, potential users should beware that such software can be untested and, at worst, hugely destructive: there are, regrettably, some astonishingly maladjusted individuals around who have transmitted 'virus' or 'trojan-horse' programs, the sole purpose of which is to foul up other people's work. Other programmers have had to resort to 'antibiotic' programs to get rid of the 'viruses' spread around by such 'germware'.

The following additional sources sometimes yield actual code, but otherwise only executable versions:

(d) Public domain disks of software obtainable from organizations like COGS.

(e) Advertisements in geological and mining journals, geological society newsletters, commercial newspapers, and (more rarely) the computing magazines.

(f) Specially compiled geological catalogues and program collections, e.g. AMIRA(1985, vol.4); Burger (1983-1984, 1986); COGS (1986); Koch et al.(1972); Mayo & Long(1976); Moore et al.(1986), Roloff & Browder(1985); Smith & Waltho (1988), Sowerbutts(1985), Till et al.(1972) and Watterson (1985). Care should again be exercised in using these catalogues, however, as they rapidly become out of date; regularly updated catalogues are most trustworthy. For example, a regularly updated catalogue of specialized mapping software has been available in Australia (Burns 1979). A catalog of computer program catalogs is also published (Chattic 1977).

(g) National and international serial catalogues, computerized listings and journals, which include:

(1) Elsevier's *Software Catalog* which comprises 3 subcatalogs: (i) *Microcomputers,* (ii) *Minicomputers* and (iii) *Science and Engineering*, the latter covering >4,000 programs and including an Earth Science section;

(2) the biennial *International Directory of Software* (Computer Publications, London);

(3) the annual *DP Index and Software register* (Peter Isaacson Publs.Pty);

(4) the *Directory of Mining programs* and *EZ Search-Petroleum* (a DOS diskette of over 700 current petroleum application programs which is updated quarterly); both of these are available in machine-readable form from Gibbs Associates (PO Box 706, Boulder Co 80306, USA);

(5) Macmillan's *Omni catalog* (Davies 1984).

(h) On-line databases, such as the *Microcomputer Hardware & Software Guide* and *MENU™— the International Software Database* (both available on the DIALOG system — §3).

The user who obtains existing source code enjoys all the advantages of the "write your own program" approach, but *usually* involves himself in far less work. The code can normally be tailored to special requirements and adapted to run on his own computer, especially if written in portable languages like FORTRAN. Moreover, if bugs are found, they can (in theory!) be corrected.

In summary, a book like this one cannot begin to answer every geologist's software availability questions, and there is little alternative but for individuals to keep their ear to the ground!

2c. Using proprietary or published subroutine libraries

Subroutine libraries are collections of fully tested code which enable a vast array of mathematical and statistical operations to be performed. Perhaps the best-known are NAG and IMSL, implemented on mainframe computers, but FORTRAN subroutine libraries for micros have also been available since 1984 (Table 2.2). Subroutines are not self-contained; they must be invoked from specially written programs which call the subroutine at particular stages. Nevertheless, they can save a great deal of work in carrying out standard or repeated operations within a specialized stand-alone program: in effect, they prevent the user from having to reinvent the wheel, and ensure that critical stages of his

Table 2.1. Calculating a median using four different computer approaches
(All code assumes that the data have already been read into a one-dimensional real array (*X*) of size *N*)

(a) The "write your own code from scratch" approach (all in FORTRAN)

```
      I = 1                                 Algorithm to sort array (X) into ascending order
1     I = I + I
      IF (I.LE.N) GOTO 1
      M = I - 1
2     M = M/2
      IF (M.EQ.0) GOTO 10
      K = N - M
      DO 4 J = 1,K
      L = J
5     IF (L.LT.1) GOTO 4
      IF (X.(L+M).GE.X(L)) GOTO 4
      Z = X(L+M)
      X(L+M) = X(L)
      X(L) = Z
      L = L - M
      GOTO 5
4     CONTINUE
      GOTO 2                                End of algorithm
10    M = (N+1)/2                           Determines position of middle value for N odd or even
      L = (N+2)/2
      MEDIAN=0.5*(X(M)+X(L))                Calculates median value
```

(b) Using NAG subroutines (all in FORTRAN); 3 lines identical to last 3 in option 1 follow those shown below

```
      CALL M01ANF(X,1,M,IFAIL)              Calls sorting routine on array X
      M = (N+1)/2                           Determines position of middle value for N odd or even
      L = (N+2)/2
      MEDIAN=0.5*(X(M)+X(L))                Calculates median value
```

(c) Using SYSTAT command language (NB this is NOT actual FORTRAN, though FORTRAN-like)

```
      SORT X                                Calls sorting routine on array X
      LET N=CASE                            Determines position of middle value for N odd or even
      LET N1=INT(N/2)+1
      LET N2=N-N1+1
      RUN
      IF CASE=N1 THEN LET MEDIAN=MEDIAN+X/2 Calculates median
      IF CASE=N2 THEN LET MEDIAN=MEDIAN+X/2
      RUN
```

(d) Using SPSS commands

```
      STATISTICS = MEDIAN                   Uses MEDIAN keyword within STATISTICS
                                            subcommand to perform entire operation
```

calculations are performed by fully tested code and correct algorithms. Rock & Duffy (1986), for example, used NAG routines to perform sorting and calculate rank correlation coefficients, in a program implementing a nonparametric regression technique .

The disadvantage of this approach is its limited portability. If the library ceases to be available (the users changes job, or his institution no longer support it), his program will no longer work. Even if the library *is* available, it may be implemented in a different way (e.g. NAG is sometimes implemented in *single precision,* sometimes in *double precision* FORTRAN, the two differing in the number of accurate decimal places), requiring extra effort to reimplement in a different computer environment.

For this reason, collections of subroutine algorithms *published* as source code are often very useful. These can actually be incorporated fully into programs, rather than simply being called, so making that program wholly self-contained, and more readily translatable onto other computers. Press et al. (1986), Press (1986) and Griffiths & Hill (1985) have provided large sets of such algorithms — the latter based on studies published over the years in *Applied Statistics (J.R.Stat.Soc.London Ser.C)* — which cover many of the commonplace (but critical) mathematical and statistical operations (e.g. calculating summary statistics, inverting matrices, percentiles of the Normal Distribution, etc.)

2d. Using 'Everyman' packages

Reviews and listings of 'everyman' statistical packages: Berk(1987b), Cable & Rowe (1983), Carpenter et al. (1984), Francis (1981), Fridlund (1986), Lehmann (1986,1987), Marascuilo & Levin (1983), Neffendorf (1983), Rock et al.(1988); Shervish(1985), Siegel (1985); Webster & Champion (1986).

Geologically-oriented packages HARDROCK (Till 1977), ROK-DOC (Loudon 1974), GEOIC (Parker 1981), KEYBAM (Barr et al. 1977), PETCAL (Binsler et al. 1976); PETPAK (Fitzgerald & Mackinnon 1977), NOROCK (Miranda 1977), STATPAC (Van Trump & Miesch 1977), STATCAT (David 1982).

Geological data- and mapping-systems: *generalized* (G-EXEC: Cubitt 1976; Jeffrey & Gill 1976a,b); *geochemical exploration* (Koch 1987); *petrology* (CLAIR: Le Maitre 1973, 1982; Le Maitre & Ferguson 1978); *well-logging* (KOALA: Doveton et al.1979); *mineral exploration/mining* — see text.

To complement the above reviews, Table 2.2 gives brief details only of the most widely-distributed and best-known statistical packages (commonly known as the 'mainframe dinosaurs'!), together with all packages actually used in the preparation of this book. Such packages take most of the work out of the user's hands, but some still allow significant flexibility, via use of their own command language. This enables fairly complex operations to be invoked by one-line commands, or sets of commands to be concatenated into larger 'macros', to perform still more complex processes. Although many sophisticated statistical techniques useful in geology (e.g. those covered in Section VI) are not specifically programmed in most smaller packages, it may be possible to program them via these macros: MINITAB macros can for example be used to perform discriminant analysis (§22), plot triangular diagrams, etc. The amount of code required to be written by the user is consequently much less than using subroutine libraries. On the other hand, discriminant analysis is fully executable via a few relatively simple commands in SPSS and SAS!

Table 2.2. Examples of major computer packages

Mainframe	Micro	Source	Language	Sites	Machines	Comments (subjective!)	References
BMDP (BioMeDical Programs)	BMD/ PC	Health Science Faculty, UCLA, California,USA	FORTRAN	>1500	>25 mainframes; now available for some microcomputers	A package of >40 programs, generally regarded as one of the most accurate and versatile available. Recent reviews of PC version rather unenthusiastic, however.	Manuals: Dixon(1975, 1981); Dixon & Brown(1979)
GENSTAT (GENeral STATistics)	—	Rothampstead Station, Harpenden, UK (distributed by NAG: see below)	FORTRAN + some assembler	>220	Most mainframes and minicomputers	Generally regarded by professional statisticians as the 'soundest' set of algorithms. Elegant but arcane, 'gothic' complexity for the occasional user. Manuals generally regarded as 'highly difficult'!	Manuals +Weekes(1985); Alvey et al. (1982)
GLIM-3 (Generalized Linear Modelling)	—	Royal Statistical Society (distributed by NAG: see below)	FORTRAN + some assemblage	>600 (>24 countries)	>20 (most mainframes and minicomputers)	Specialized high-precision library for regression, ANOVA, contingency tables, etc.	Manuals
IMSL (International Mathematics & Statistics Library)	—	IMSL Inc., NBC Building, 7500 Bellaire Blvd., Houston, TX 77036, USA	FORTRAN	>2000 (>36 countries)	>20 (most mainframes and minicomputers)	Library of >540 mathematical and statistical subroutines	Manuals
MASS (Microprocessor Applied Statistics System)	MASS	Westat Associates, PO Box 247, Nedlands, Perth, W. Australia	Pascal		Cyber,DEC,VAX mainframes; IBM PC, Z-80, Macintosh micros	Specially designed for maximum mathematical precision; research and teaching aid.	Manuals
MINITAB	MICRO-TAB	Prof. T.Ryan, Pennsylvania State University Stats Dept.	FORTRAN	>1000	IBM and other micros as well as most mainframes	Very user-friendly, and particularly suitable for students; highly interactive. Some PC versions only implement a subset of the mainframe version.	Manuals: Ryan et al. (1976,1982)
NAG (Numerical Algorithms Group)	—	NAG, 7 Banbury Rd., Oxford, OX2 6NN, UK	FORTRAN, ALGOL60	>400	>30 (CDC, DEC, GEC, Honeywell, HP, ICL, PRIME...)	Mathematically precise library of ≈500 subroutines, available in either FORTRAN or ALGOL60	Manuals: NAG (1987)

Table 2.2. Examples of major computer packages (contd.)

Mainframe	Micro	Source	Language	Sites	Machines	Comments (subjective!)	References
SAS (Statistical Analysis System)	PC SAS	SAS Institute Inc., Box 8000, Cary, N.Carolina, 27511, USA	PL/1 + assembler	>3500	IBM, DEC, VAX, Prime mainframes; IBM micros	Colour graphics; full-screen editing; wide range of techniques. Supplemented by SAS/GRAPH, SAS/FSP, SAS/AF, SAS/IML, SAS/OR modules. Current PC version supposedly sluggish and not well-reviewed.	Manuals: SAS(1980; 1985a, b,c; 1986); Lewis & Ford (1983)
SPSS (Statistical Analysis for the Social Sciences)	SPSS/PC +	SPSS Inc., 444 N.Michigan Ave., Chicago, IL 60611, USA	FORTRAN	>4000 (60 countries)	> 30 (IBM, CDC,VAX,HP, ICL, Honeywell; IBM PC)	Sometimes considered less accurate than SAS; covers more options than most; only usable batch-mode; supplement SPSS/GRAPH. Manuals generally regarded as among best available. PC version rather a dinosaur but well reviewed.	SPSS (1986), Hull & Nie (1981); Nie et al.(1975); Norusis (1983)
—	Statview 512+	Brainpower Inc., 24009 Ventura Boulevard, Calabasa 91302, CA, USA	Pascal	?	Apple Macintosh	An exceptionally user-friendly, powerful package, used for many examples in this book; has one or two minor bugs in the version originally released (v1.0)	Reviews: Berk (1987); Lehmann (1986,1987); Rock et al.(1988)
—	Statworks	Cricket Software, 3508 Market St., Philadelphia PA 19104, USA	Pascal	?	Apple Macintosh	One of the most user-friendly statistics packages available, unfortunately with numerous bugs in the versions so far released (v1.1 & 1.2)	Reviews: Lehmann (1986,1987); Rock et al.(1988)
Systat (the SYStem for STATistics)	Systat	Systat Inc., 1127 Asbury Ave., Evanston, IL, 60202, USA	FORTRAN	?	IBM PC, Apple Macintosh micros	One of the most sophisticated packages currently available; PC version avoids many of the pitfalls of mainframe to micro conversions and has been well reviewed; particularly comprehensive manual explaining statistical aspects.	Manuals: Wilkinson (1986); Reviews: Berk (1987), Rock et al. (1988)
(Micro Mathematical and Statistical Library)	µ2SL	Micro State Software, PO Box 125, Queanbeyan, NSW 2620, Australia	FORTRAN-80	?		The first library of FORTRAN 80 subroutines to be released for microcomputers	Manuals

Packages in Table 2.2 are evolving organisms, several now supported by large commercial organisations, and generalizations (e.g. as given in Francis 1981) rapidly become out of date as new releases appear. At present, however, the packages form a spectrum ranging from those in which the command language is highly sophisticated but relatively few operations pre-programmed (e.g. GENSTAT), to those with the converse (e.g. SPSS). GENSTAT thus allows the geologist more detailed control over his statistical operations, but SPSS minimizes the programming he has to do. Again, sophisticated file-handling, graphics, database and other facilities are available on top of the statistical operations in SPSS, while SAS has add-on modules (SAS/GRAPH etc.) to the same purpose. Both are nowadays better described as data-systems than 'statistical packages'.

Although the more geologically-oriented 'everyman' packages and data-systems listed above do not begin to cover the range of operations available in the larger packages of Table 2.2, actual source code may be available for subsequent tailoring to an individual's needs. However, the most comprehensive of the specifically geological data-systems, G-EXEC (Geologists'-EXECutive) is unfortunately no longer being developed or supported by its parent organisation, the British Environment Research Coucil.

'Everyman' packages specifically aimed at the minerals/petroleum exploration industry have also proliferated. Some, developed in-house by large oil companies, are not available to outside users, but others are marketed by software houses. In Australia, where mining now accounts for almost half the country's GNP, a particular flowering of such software has occurred. Some of the more popular packages include (in alphabetical order): DATAMINE™ (James Askew Associates), GEOS™ (Siromath), MEDSYSTEM™ (MeTech), MINEX™ (Exploration Computer Services), MOSS™ (Trit Systems), and SURPAC™ (Surpac mining systems). All these are implemented at one or more of Australia's hundreds of mining operations, and some are available in microcomputer as well as mainframe versions. Capabilities include advanced graphics, database, (geo)statistical manipulation, 3-D modelling, cartographic and photogrammetry, mapping and networking; these allow such applications as mine planning, orebody modelling, recoverable reserve calculations, site surveying, haul road design, earthworks calculations, production scheduling, and pit design. The AMIRA (1985, vol.4) catalogue contains outlines of over 150 such packages distributed by some 50 companies. Further compilations are provided by Salomon & Lancaster (1983) and by Sandford et al. (1978,1979). Some specific geological uses for these packages are considered in §4c and §5b.

2e. A comparison of options

Unfortunately, there is no simple way of identifying the best option for a particular purpose, among such a multitude of available software. Those users who are forced by cost considerations into using whatever is implemented on their own institution's computers are in some ways lucky, since the difficult decision is taken out of their hands! Some published help is available: for example, Berk(1987) and Brownstein & Kerner(1982) provide a thoroughly general methodology for evaluating and selecting software, guiding the user through the entire process by providing a system of checks and balances. Meanwhile, 'in-depth' reviews of individual packages (including not only factual information but also attempts to assess their capabilities and performance) appear all the time in computer and other magazines, to complement the larger, comparative surveys listed earlier. Just as when buying a car, the user should check all the available consumer sources. However, there is no substitute in the case of software for direct experience with a package, or discussion with someone having such experience.

The regrettable truth is that 'bug-free' software probably never did exist, but that market pressures nowadays greatly exacerbate the problem. Many new packages are underfunded, and are released for sale before they have been properly tested (they are so-called *ß-releases*). One expensive word-processing package (Microsoft Word 3.0) was released for sale in 1987 with over 400 bugs in it (see *1st 1988 update* to Coleman & Naiman 1987), and some software houses are now even releasing what they *admit* to be ß versions, at reduced prices, in the hope that the general public will do their ß testing for them! Again, Rock et al.(1988) documented statistical packages for the Apple Macintosh, costing several hundred dollars, which failed to perform calculations listed in their manual (or gave different answers from the manual), produced statistically impossible answers for even simple calculations, gave inconsistent results depending on how the data are entered, etc. 'In-depth' reviews in journals sometimes fail to point out such 'bugs', and it is wise to examine several reviews in different magazines if possible — one magazine may have a commercial connection with the software house concerned, or other factors which result in a less than disinterested assessment. Even the large mainframe packages (notably SPSS) are known to have had bugs in years gone by, and at least one recently published package of immense geological potential (CODA in Aitchison 1986) is similarly infected (Woronow 1988). As already mentioned, the problem with all proprietary packages is that the user cannot correct bugs himself, but must rely on the goodwill and commercial sense of the software house to do so. In the author's own experience, this can be time-consuming and frustrating!

On balance, a mix of the five approaches outlined at the start of this Topic may be required for different projects and people. For competent programmers, perhaps the greatest flexibility would be achieved by modernising such suites of programs as those in Davis (1973) and Mather (1976), since these are both wide-ranging and earth-science oriented, but can be readily adapted to individual users' requirements. However, writing one's own programs is certainly no longer an efficient option for any of the commonplace geological calculations (e.g. norms), and should *only* be entertained even for esoteric geological applications after careful search through software already available. For non-programmers, 'everyman' packages offer the most versatile option, but they are a double-edged sword, also representing by far the *most dangerous* option for the computer and/or statistical novice, because they tend to produce overwhelming reams of output replete not only with the actual results of interest, but also with a plethora of (mostly useless) system information from the computer on which the program was run. The mammoth range of options presented by everyman packages can also be a death-trap for the unwary, since novices tend to rely on the 'default' options, which may not be at all suitable for a particular set of data.

TOPIC 3. COMPUTERS AS SOURCES OF GEOSCIENCE INFORMATION: NETWORKS & DATABASES

3a. Communicating between computer users: Mail and Network Systems

Tremendous recent advances in global communications offer hitherto undreamt-of facilities for geologists. Unfortunately, actual practice changes so spectacularly and rapidly, and is so installation-dependent, that no discussion here could be either exhaustive or up-to-date. Only a few facilities (varyingly implemented in different countries) which the author has found particularly useful, are therefore mentioned here. Geologists interested in communications should seek *expert local* help.

(1) *Networked mainframe computers*. The possibility of networking computers (especially those of the VAX family) was mentioned in §1. Networks are now in place in many countries (e.g. JANET in the UK, CSIRONET and AUSINET in Australia: Fig.3.3) which allow a user to access remote machines from his office, to leave messages (next section), transfer files, or use programs implemented on those remote machines to analyze his own data. Some networks are set up such that if a user's nearest machine is busy or out of action, his job can be routed to another machine capable of executing the job, much improving the turnround time. For example, the British Environment Research Council's network serves most of the university and governmental geological departments, so that any networked user can use machines in Edinburgh, Liverpool, etc. from one terminal in, say, London. Local networks, connecting different departments or campuses, are more common still, but perform the equivalent electronic job to national networks over shorter physical distances (Fig.3.2).

(2) *Mail networks*. This is one extremely useful facility available via many of the above networks. Indeed, some networks can *only* be used for mail, and not for more complex executable jobs. Usually, two facilities are available: 'conversation' (usually termed CHAT or the like), which allows two users actually signed onto the network to send and reply to messages to each other; and 'post' (usually called POST, MAIL or the like), which allows one user to leave a message for another user not actively signed onto the network, who will receive the message as soon as he next signs on. Diagrams as well as written text can usually be sent via both methods, together with much longer documents, as appropriate. Indeed, this is a very good way to send machine-readable documents to one's colleagues and collaborators. It is often better than telephone messages, since it maximises the chance of the recipient actually receiving the message!

(3) *Communications over telephone networks*. It is also possible nowadays to link most computers into standard telephone networks, by using **modems**, which convert computer signals into those transmitted via telephone lines. After one user has phoned a collaborator on the other side of the world, and after both have linked their computers to their own telephones, their computers can 'talk' to one another. For example, with Macintosh microcomputers, software such as MacChat™ allows one user actually to work his colleague's machine over the telephone; diagrams can be drawn while the other user watches, corrections made to documents, etc. Endless possibilities are opened up.

Binary transmission codes must be considered when using computer communications. The two most common are *ASCII* (American Standard Code for Information Interchange), an 8-bit code, in which one figure is reserved for parity, leaving the other 7 to represent $2^7 = 128$ characters, and *EBCDIC*, an IBM standard which lacks a parity (±) bit and can thus represent $2^8 = 256$ characters. Generally speaking, files which consist only of ASCII or EBCDIC characters can be freely transmitted between different computers worldwide. (This will usually apply to files consisting of alphanumeric data only). However, files containing embedded characters idiosyncratic to the original host machine will either fail to transmit properly or cause errors. This is a particular problem with word-processed documents, whether from mainframe text formatters, microcomputer packages or dedicated word-processing machines (§4a), which all define the fancy and varied formats of text (upright, *italic* and **bold** type, superscripts and subscripts, √§¶ symbols and different typefaces etc.), by mutually incompatible control characters which are invisible on screen except in their effect. For example, the documents of this text could not easily be read by a mainframe computer (if at all) in the form in which they appear here; they must first be saved in *text-only* form, which means all the embedded control characters are removed to leave ASCII-only text. Unfortunately, this means that once the document is transmitted to another computer, all the 'fancy stuff' has to be put back manually!

Although networks are by definition multi-user, and therefore necessarily implemented on mainframes, geologists devoted to microcomputers can easily 'hook' themselves into networks by using *terminal-emulation packages,* which turn micros into intelligent terminals for interfacing with larger computers (§1b4). One of the most versatile and widely distributed of these, fast approaching the status of an international standard, is KERMIT, from Columbia University center for Computing activities (New York, 10027). This can already link most popular mainframes and micros together.

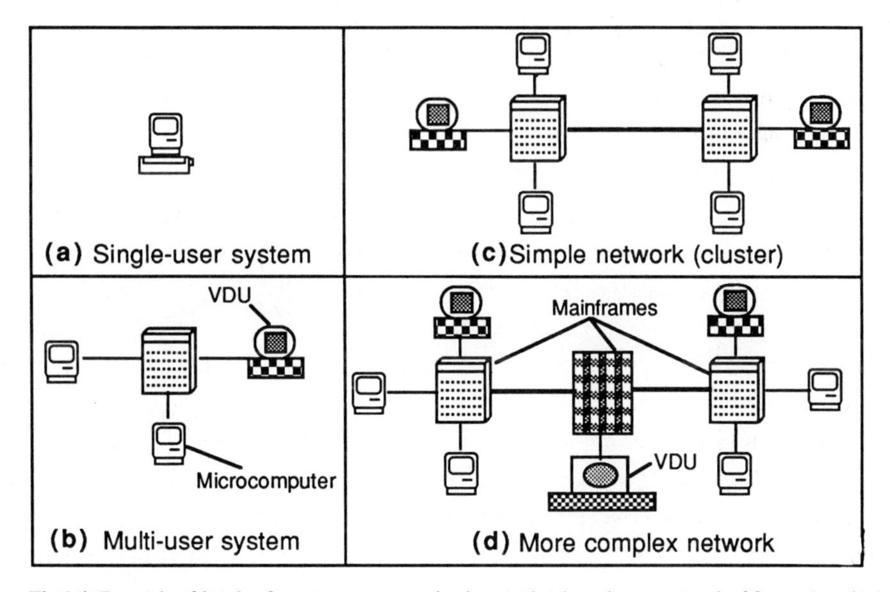

Fig.3.1. Example of levels of computer communications technology (see text overleaf for explanation)

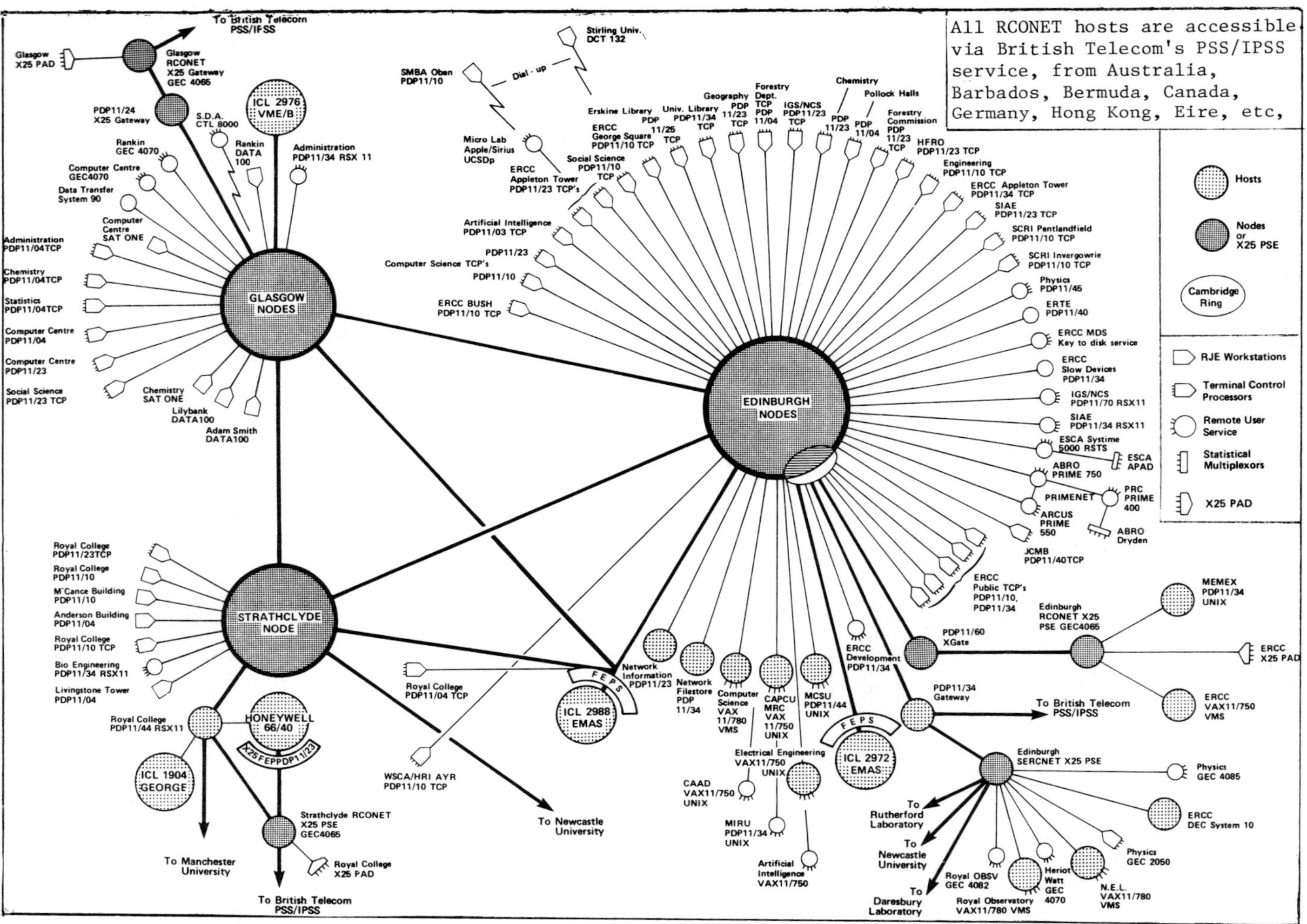

Fig.3.2 Communications network between various Scottish universities. Used by permission.

Fig.3.1 illustrates hypothetical computer communications of increasing complexity. In **(a)**, a single user interacts with his microcomputer, perhaps with a hard disk to increase storage capacity. In **(b)**, many users interact with one mainframe; individuals may log into the mainframe via microcomputers or via 'dumb' terminals (VDUs). In **(c)**, several mainframes of the same type, each capable of interacting with many users, are linked in a simple network or 'cluster'; this architecture typifies VAX networks now run by some companies and universities. In **(d)**, *different* mainframes are linked together. In both **(c)** and **(d)**, users at any terminal or microcomputer can send information to and from any other user. We can envisage far more complex networks than this. Fig.3.2 for example shows a real, local example from the UK, while Fig.3.3 shows a national communications system.

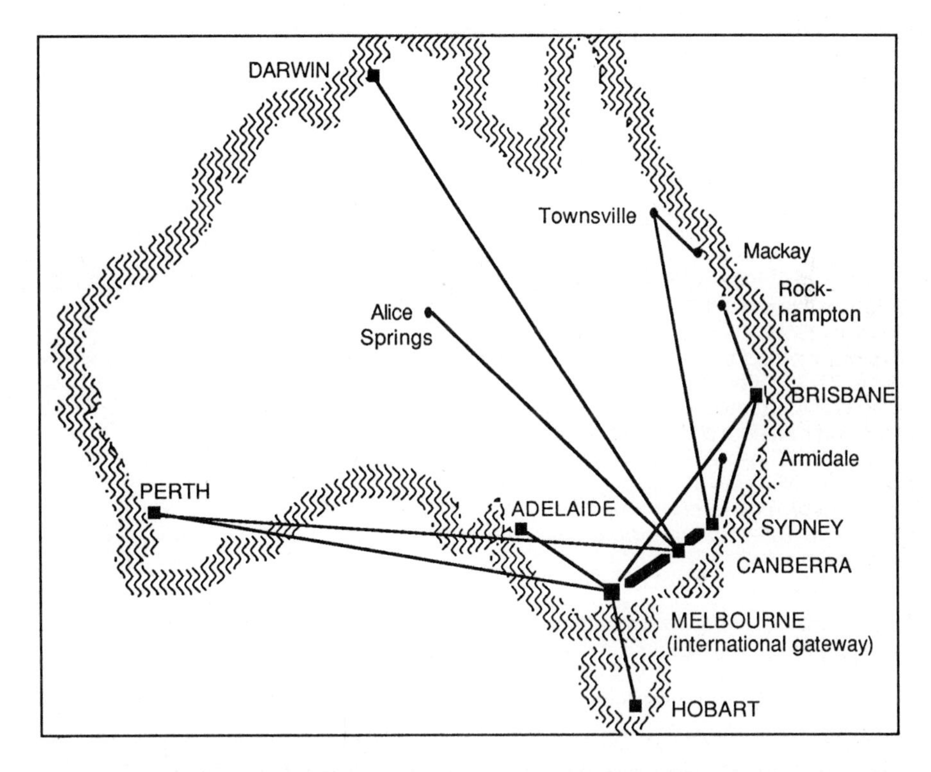

Fig.3.3. Connections involved in the CSIRONET and AUSINET computer networks in Australia.
Mainframe computers of various makes (IBM, PDP, VAX etc.) are present in the regional centres. Via the international 'node' or 'gateway' in Melbourne, which accesses global satellite communications, users in any other centre can communicate with the 'global village'. Many large mining companies have their own networks in addition to the above, which, for example, link computers at remote mine sites into their head office systems

3b. Archiving and Compiling Large Bodies of Information: DATABASES and INFORMATION SYSTEMS

Definitions of *data-value, data-set, data-file, database,* and *record* (§1b7; Glossary) are assumed here.

3b1. Progress with Databases and Information Dissemination in the Geoscience Community

Reviews of progress and predictions for the future: Jones (1986), Le Maitre(1982,ch.12), Pruett (1986), Robinson(1974), Sutterlin (1981), Ward & Walker (1986), Zwillenberg (1986).

Further geological discussion:

Database maintenance: Hawkins & Kew (1974);

Encoding and integrating descriptive with numerical data: Lessells & Webster(1984), Morgan & McNellis (1971), Rassam & Gravesteijn (1986);

Experiences with specific databases and networks in particular organizations or countries: Bliss (1986a,b), De Backer & Kuypers (1986), Grandclaude et al. (1976), Horder(1981), Laycock & Staines (1979), Loudon (1969), Tellis & Gerdes (1986), Wadatsumi et al. (1976), Walker (1986);

Policy implications: Price (1986);

Professional implications: Wolfenden (1978);

Storage, handling and retrieval of geological information: Bergeron et al.(1972), Burk (1973, 1976); Chayes (1978), Gill (1975), Gilliland & Grove (1973), Goubin (1978), Grandclaude(1976), Henderson(1986); North (1986), Sutterlin et al. (1974);

Teaching implications: Kohut (1978).

Catalogue of geological databases: Wigley (1984).

Descriptions and uses of existing geological databases (see also Table 3.1):

Applied geochemistry (Bary et al. 1977; Plant et al. 1984,1986; Van Trump & Miesch 1977).

Bibliographical (Graffenreid & Cable 1978; Herbst 1977; Hruska & Burk 1971; Lea 1978; Leigh 1981; Martin 1978; Moore 1985; Rassam 1983; Walker 1977).

Field data-recording systems (Burns et al.1978; David & Lebuis 1976; Garrett et al.(1980); Hudson et al. 1975; Hutchison 1975; Odell 1976,1977a; Page et al. 1978; Sharma et al. 1972; Sutterlin & Sondergard 1986).

Geochronology and isotope geology (Doe & Rohrbough 1979; Siegenthaler 1986).

Geothermal: (Bachu et al. 1966; Bliss 1986b; Bliss & Rapport 1983).

Igneous petrology & Volcanology (Angelis et al. 1977; Carr & Rose 1987; Chayes 1979-1980,1982, 1983b; Chayes et al.1982,1983; Griffin et al. 1985; LeBas et al. 1983; Mutschler et al.1976; Rock 1987a,e; Unan 1983; Zi & Chayes 1983).

Map collections: Cobb (1986); Lea et al. (1978).

Mineral exploration and mining (Anon 1978a,b; Chatterjee et al. (1977); Fabbri et al.1975; Mock et al. 1987; Placet 1977; Porter et al. 1986; Romaniuk & Macdonald 1979).

Mineral identification (Smith & Leibowitz 1986).

National and international databases: (Anon 1978a; Burk 1978; Chayes 1979-1980,1982, 1983b; Chayes et al.1982,1983; Ghosh & Rajan 1986; Gill et al. 1977; Lumsden & Howarth 1986; Neilson 1980; Ningsheng 1986; Nishiwaki 1986; Robinson 1966; Shelley 1985).

Specimen (mineral, rock & fossil) collections (Nishiwaki 1986; Smith 1978).

Stratigraphy & sedimentology —boreholes, basins etc. (Bachu et al. 1966; Reinhardt & Van Driel 1978; Selby & Day 1977).

Many people consider a machine-readable collection of data (a data-file) to deserve the appellation 'database' merely when it becomes sufficiently large: any computerized telephone- directory, for example. Mere size can certainly confer greater 'relevance' and 'significance' on any database, in the sense of making it of interest to a greater number of individuals. However, this is only the crudest form of database, more sophisticated forms of which can integrate huge volumes of many different types of information in a way offering maximum retrievability. A true database (syn. 'databank') is best defined as a formalized collection of inter-related data with the following characteristics:

— it is managed and accessed via software termed a **database management system** (DMBS);
— it holds data in the form of **records,** which consist of one or more **fields;**
— it protects data in some way from unauthorized access, either by record or by field;
— its overall logical structure can be changed without changing the DBMS;
— its physical storage (tape, disk etc.) may be changed without effecting its logical structure;
— it stores data in such a way as to allow the fastest possible random access to selected fields;
— it holds data with minimum redundancy (duplication of information);
— it may contain several different data structures stored together and inter-related;
— it can be accessed by several users concurrently;

— it stores not only the data themselves, but also the relationships between data-items.

Databases are already very widely used by governments to keep tabs on their populations (health department records, tax records). The larger and more competitive mining and exploration companies have also been setting up large databases for some years, containing drillcore information, chemical data, etc. Many of these have never been publicized, and some have been restricted in their operation (especially in the multi-nationals) to particular divisions, national groups, or local projects, rather than implemented consistently across the entire organization. Furthermore, many have failed to outlive the lifetime of the particular commercial project for which they were designed. Probably the same can be said of many of the computerized systems erected for large-scale reconnaissance field-mapping projects by governmental surveys (e.g. Page et al. 1978), although these have undoubtedly had considerable success and advantages during the duration of the project concerned.

On the other hand, many geological Surveys, museums and academic departments have been attempting to erect *permanent* databases for their geological specimen collections of rocks, minerals and fossils — with varying degrees of success. On a *national* scale, a further step has been taken with the setting up of national geoscience information centres (e.g. the Keyworth centre of the British Geological Survey, designed to computerize outcrop, borehole, hydrogeological and museum collection records accumulated over the past 150 years). On an *international* scale, geoscientists from various disciplines have been getting together to pool their individual database efforts. For example, after petrologists in various countries had erected their own personal databases of igneous rock analyses (see references above), a prototype international system IGBA (IGneous dataBAse) was implemented under the auspices of the International Geological Correlation Program (IGCP Project 163), by arrangement between groups from Australia, Bulgaria, France, Hungary, India, Portugal, Spain, Turkey, the UK and USA. An international system was long overdue, since the number of *published* analyses of igneous rocks stood at 100,000+ over 15 years ago (Chayes & Le Maitre 1972a,b) and must have multiplied severalfold since then, with the increasing use of rapid, modern analytical techniques.

National versions of IGBA also exist (e.g. Le Bas et al. 1983), and the first international training course for petrologists in the use of databases was held during September, 1987, in Kuwait, under the aegis of the continuation IGCP project 239. The idea is that contributors to most petrological and mineralogical journals will eventually be asked, at the time of acceptance of their papers, to submit their raw data (in machine-readable or hand-written form) for incorporation into IGBA. Future researchers in the same field thus benefit. Several major journals (e.g. *J.Petrol., Mineral. Mag.*) have already adopted this scheme. Equivalent databases for metamorphic and sedimentary rocks are mooted.

Two types of commercial database are useful to many or all geoscientists. (1) *Bibliographic databases*, of which there are many national examples (e.g. AESIS in Australia), as well as three giants, GEOARCHIVES, GEOBASE and GEOREFS (Table 3.1). Most organisations charge for retrievals according to the number of items printed from the database. (2) *General purpose public information databases,* set up by national telecommunications bodies in several countries (e.g. PRESTEL in the UK, VIATEL in Australia). These are accessible via domestic telephone networks using either ordinary domestic TV sets, or microcomputers. Thousands or even millions of 'screens' of information are available, and users pay a nominal charge for each screen they choose to call up, the charge varying

Table 3.1. A selection of large geological databases

Name	References	Information contained	Records	Last updated	Back to	Comments	Availability
AESIS (Aust.Earth Sci.Inf.Sys)	Tellis & Gerdes (1986)	Bibliographic database for literature concerning (or generated in) Australia and Papua New Guinea	50,000	Current; about 4,000 records added annually	1975 (some 1965); 1979 on covers foreign journals	Unpublished (open-file reports etc.) as well as published information	CLIRS, Level 39, MLC Centre, Martin Place, Sydney 2000, Australia
ALKY	Griffin et al. (1985)	Descriptive + chemical N. American alkaline rocks	4,208	1984			ASCII files on 9-track 1600 bpi tape from Eastern Washington University, USA
BIB. ALKY	Mutschler et al. (1986)	Bibliography on N. American Cordillera alkaline rocks	1,123	1987			Printout; ASCII files on 9- track 1600 bpi tape or disk from Eastern Washington University
BIB.MOLY	Steigerwald et al. (1983a)	Bibliography on stockwork Mo deposits	1,187	1986			Printout; ASCII files on 9- track 1600 bpi tape or disk from Eastern Washington University
CLAIR	Le Maitre (1973, 1976a, 1982)	Published analyses of igneous rocks worldwide	26,373	Current	c.1900	Since merged with PETROS	Dr.R.W.Le Maitre, Dept. Geology, Univ. Melbourne, Parkville, Australia 3052
FERROS	Carter (1987)	Descriptive + chemical data for BIFs and related rocks associated with gold deposits	1,320	1984			As ASCII files on disks from Eastern Washington University
GEO-ARCHIVE	Published as "Geotitles"; see also Lea (1978)	On-line bibliographic database on general geology (including geophysics, hydrology & oceanography)	542,000	Current; about 60,000 records added annually	1974	Indexes > 5,000 journals, books, theses and >100,000 maps	Geosystems, London, UK
GEOBASE	—	On-line bibliographic database to geography,ecology,climatology,remote sensing,mineralogy,palaeontology etc.	230,000	Current	1980	More geographical bias than GEOREF or GEOARCHIVES	Available on DIALOG (§3#); contact Geo Abstracts, 34 Duke St., Norwich, NR3 3AP, UK
GEOREF	Published as "Bibliography & Index of Geology"	On-line bibliographic database on general geology (including geophysics)	1.2 million	Current; about 85,000 records added annually	1785 (N.American material); 1967 (worldwide)	Indexes > 4,500 journals, books, theses etc.	American Geological Inst., 4220 King St., Alexandria, Virginia 22302, USA
GOLDY	Mihalasky et al. (1987)	Summary information on N. American gold camps (> 106 oz Au)	112	1985		Requires dBASE III+ program	On disks from Eastern Washington University
GRANNY	Steigerwald et al. (1983b)	Descriptive + chemical data for Laramide and younger acid igneous rocks from Colorado/New Mexico	507	1983			Hardcop from USGS; 9-track 1600 bpi tape or disks from Eastern Washington University

Note: many of the above are trade marks

Table 3.1. A selection of large geological databases (contd.)

Name	References	Information contained	Records	Last updated	Back to	Comments	Availability
IGBA and UK-IGBA	Chayes (1979-80); Le Bas et al. (1983)	Descriptive geological, mineralogical and chemical information for igneous rocks worldwide	>10,000	Current	Mid-1970s for new data; older data sporadic at present	Supported by UNESCO/IUGS projects 163 & 239	World Data Center-A for Solid Earth Geophysicccc Code E/GCI, Boulder, Colorado 80303, USA
IMMAGE	—	Commercial bibliographic database for Economic Geology	>25,000	Current			IMM, Portland Place, London
LAMPS	Rock (1987e)	Whole-rock and mineral data for lamprophyres	>5,000	Current	c.1900	Requires relational DBMS	Contact the author
MARLA/ PETROS	Mutschler et al. (1976)	Descriptive + chemical data for igneous rocks worldwide	37,297	1983		MARLA is enhanced version of PETROS	ASCII files on 9-track tape from Eastern Washington Univ. or World Data Center-A (see IGBA)
MINDEP	Mock et al. (1987)	Published data on Western Australian gold deposits	80	1986		Implemented in ORACLE DBMS	Bureau of Mineral Resources, Box 378, Canberra, Australia 2601
MinIdent	Smith & Leibowitz (1986)	Physical, optical and chemical properties of minerals		Current		Integrated expert-type system for identifying unknown minerals	Dr.Dorian G.W.Smith, Dept.Geology, Univ.Alberta, Edmonton, Canada T6G 2E3
MINPRI	—	On-line access to daily prices for metals, ores, etc. Monthly prices for industrial minerals and other statistics	Hundreds of price series	Daily	Generally 1971	Maintained by BMR (Australia) Information and Statistics Section	Worldwide access via I.P. Sharp Associates, Carlton Centre, 55 Elizabeth St, Sydney NSW 2000
NESY	Rock (1987a)	Geological data for alkaline rock occurrences worldwide linked to chemical data for syenitoids	>4,000	1986	c.1900	Requires relational DBMS	Contact the author
NGDB/ RGRP	Plant et al. (1984, 1986); Plant & Moore (1979)	Regional stream sediment analyses over northern UK for up to 35 elements	>50,000	Current	Project started early 1970s	Is being used to produce 1:250,000 geochemical atlases	Purchasable on tape from British Geological Survey, Keyworth, Notts, UK

Note: many of the above are trade marks

according to the type and commercial value of the information. Geologically significant information available via such systems may include up-to-date prices for primary products (gold, tin, etc.), and the latest developments in computer software and hardware. Geologists with wanderlust can also price and purchase airline tickets on some national systems, or manipulate their personal bank accounts from their lounge armchair! An alternative, more passive, type of database is available in the UK and one or two other countries, based on TV transmissions rather than telephone networks, and called TELEDATA (two UK versions are put out: the BBC's CEEFAX and ITV's ORACLE). Here, a special type of TV set is needed to unscramble the signals, which are broadcast repeatedly in a pre-set sequence; the user must await the 'next time around' to view a particular screen of interest. The amount of information on such systems is usually far less than via telephonic systems, and the opportunity to respond to the system (e.g. order tickets) is not available; nevertheless, the same types of information of interest to geologists are usually broadcast. Some TV companies are also able to broadcast home computer software using the same system: the user plugs his machine into his TV set, and what comes over the air as apparently jumbled noises is actually perfectly understandable to a computer.

It is conventional in all the above to distinguish **reference** from **source databases**: the former merely indicating where information can be found, the latter actually giving the information. However, the distinction very commonly breaks down: bibliographical databases and IGBA clearly belong to the reference and source categories respectively, but the index to a museum collection is equally clearly a hybrid: it contains *some* of the vital information (locality details, names of collectors, etc.), but the rest is contained in the specimens (rocks and minerals) themselves.

It is extremely difficult for the individual geologist to keep track of what databases are available, since the field changes so rapidly. Fortunately, librarians are usually trained intensively in database work, and should always be consulted for the latest information. Source guides such as the *Directory of Online databases* (published quarterly by CUADRA/Elsevier) may also be of help. Two fairly recent developments are also worth mentioning, not least for their future implications:

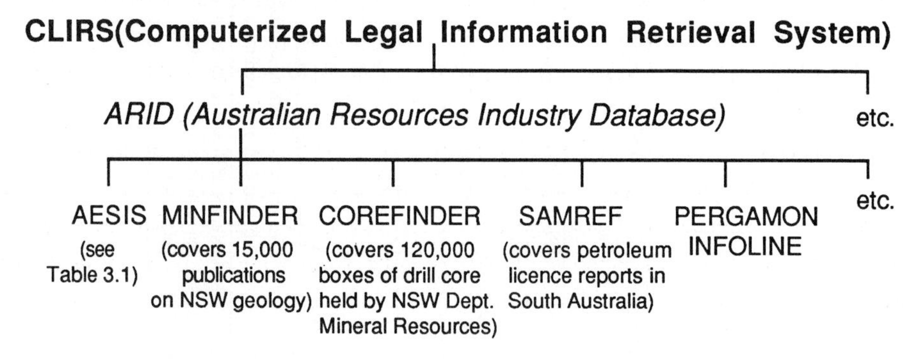

Fig.3.4. Hierarchical structure of the Australian CLIRS system allowing access to a large number of databases

(1) *Integrated systems.* Hierarchical 'data-systems' ('superdatabases') have now become available, which allow users to access individual databases within them. Part of the structure of an Australian example is depicted in Fig.3.4. To access AESIS, for example, one first dials up CLIRS, then chooses

ARID within CLIRS, and finally AESIS within ARID. From CLIRS, one can also access many other databases (both Australian and international) covering State and Commonwealth legislation, political statements from Hansard, and other items of general relevance to the resources industry.

The top level of the CLIRS system in Fig.3.4 also gives access to probably the largest information system of international interest: DIALOG®, operated by DIALOG Information Services Inc. from the USA. This gives 24-hour weekday (and more restricted weekend) access to some 300 different databases totalling over 100 million records. As well as specifically geological databases such as GEOARCHIVES, GEOBASE and GEOREF (Table 3.1), and sources of computer information such as MENU™ (an international software database) and the *Microcomputer Software and Hardware Guide*, DIALOG covers catalogues of major publishing houses (e.g. Wiley), corporate directories, current affairs, encyclopoedias, economic data, international business and financial news, product listings and announcements, telephone directories, various Abstracts (Analytical, Dissertation, Meteorological and Geoastrophysical, Nonferrous metals, Oceanic), Who's Who, etc. — and even the King James Bible! For those faintly overwhelmed by this surfeit of information, a database of databases is provided. Such all-embracing systems are good places to start many generalized searches.

(2) *CD-ROM databases.* Large databases have traditionally been stored on central mainframes and accessed remotely via satellites and computer networks. Such databases may be prohibitively large and costly for single institutions to purchase or, more especially, to store on their microcomputers. A number of major databases in other disciplines, however, are now becoming available on CD-ROMs (§1b3), whose storage capacity far exceeds that so far available on microcomputer hard disks. One system of interest to geology teachers is ERIC, which covers the abstracting publications CIJE (Current Index to Journals in Education) and RIE (Resources in Education), and covers millions of journal articles and research reports from 1966 onwards. ERIC is available (at a cost measured in only thousands of $) on 3 CD-ROMs, and access to it only requires an IBM-PC equipped with a CD-ROM drive. Instead of paying for each search of a database, therefore, the user now has the possibility of purchasing his own copy outright. This greatly facilitates 'browsing' through a database without incurring vast on-line fees. Of course, the CD-ROMs must be constantly updated if they are not to become obsolete, and commercial systems such as ERIC typically issue new CD-ROMs at regular intervals (e.g. quarterly) — the cost of the actual disk being only a few dollars.

Although no specifically *geological* databases appear yet to be available on CD-ROM, databases on agriculture ('Agricola') and other sciences are already available, while DIALOG is placing a subset of its own databases onto the medium. Those interested in geoscience information should therefore keep their ears to the ground, perhaps by joining national organizations with an information bias (e.g. the US or Australian Geoscience Information Associations).

3b2. Implementing and running databases: DataBase Management Systems (DBMS)

Dedicated monographs and reviews: Date (1981), Fidel (1987), Laurie (1985), Martin (1975,1981), Ullman (1980), Van Duyn (1982)
Further geological discussion: Burk(1973), Hruska (1976), Martin & Gordon(1977).
Geological file-handling, encoding and database software: Ayora (1984); Barr et al. (1977); Botbol & Botbol (1975); Bowen(1986); Burwell & Topley (1982a); Butler(1986); Hage(1983); Jeffrey & Gill(1976a,b); Kremer et al. (1976); Lessells & Webster (1984); Parker (1981); Yarka & Cubitt(1977).

A database is usually taken to comprise raw data only; a DBMS is required to actually manage and retrieve information from these data on a computer. Most large databases require the full-time attention of a specialist *database manager* to oversee the DBMS. Like other software discussed in §2, DBMS fall into two main groups, as regards their development histories: (i) proprietary, commercially available software; (ii) specific, 'home-grown' software designed from scratch for the task in hand. Some arguments in favour of both options are summarised in Table 3.2:

Table 3.2. Advantages and disadvantages of using proprietary versus 'home-grown' database software

	Advantages	Disadvantages
Home-grown	—Fully adaptable to the specific purpose in hand. —Experts in the software are automatically at hand	—Development cost may greatly exceed cost of buying a commercial package.
Proprietary	—In theory, professional backup is available at all times, independent of geological staff changes; the database should not 'die' if any geologist goes away	—Some products are very hard to learn —Local support and expertise may not be readily available

The relative *portability* of the two alternatives depends on exactly how they are implemented. In the *best* case, a home-grown package (for which the source code is available) can in theory be implemented on any machine, though the effort of conversion may be great unless it is written in, say, standard FORTRAN or Pascal. In the *worst* case, a database depending on commercial software may have to be revamped from scratch if moved to another machine on which the particular software is not implemented. On the other hand, some home-grown packages themselves cease to be viable for various reasons, in which case the effort of reimplementing with other software may be far greater than of transferring between different proprietary packages. Proprietary packages conforming to one of the various international standards — e.g. COGEODATA for geological DBMS software (Burk 1973, Hutchison 1975) — are particularly easy to transfer between.

Although the market is fluid, it is probably fair to say that the availability of cheap, sophisticated database software has now reached a point where the development of a 'home-grown' DBMS is less and less cost-effective. A Macintosh II microcomputer with 80Mb hard-disc (i.e. capable of storing 10^6 80-character records), and DBMS software such as *Omnis 3+™* or *4th Dimension™* can now be purchased for only a few thousand dollars; such a system rivals the speed and storage capacity of smaller minicomputers, and may be entirely satisfactory for local databases (e.g. of departmental rock collections). For larger databases, DBMS designed to cope with specific data-types are readily available. For example, STATUS, a mainframe database originally designed to cope with millions of words spoken at a public enquiry into the siting of an English nuclear power station, has been used to implement one prototype database for the British Geological Survey's rock collections; most of the data stored are words and descriptions which require no mathematical manipulation. Widely distributed DBMS more biased towards numerical data include ORACLE™, available on both minis and micros.

The history of the UK version of IGBA (Le Bas et al. 1983) is informative here. This was originally implemented at the British Geological Survey (BGS) under the *home-grown* data-system G-EXEC (§2d) on a Honeywell mainframe computer, but when the computer itself reached the end of its working life, the effort of reimplementing G-EXEC on the replacement computer was too great, and the system 'died', presenting the UK-IGBA database manager with an enormous problem. Meanwhile, elsewhere in BGS, a far larger database for borehole, rock specimen, fossil, photograph, hydrological and other information had been set up under a *proprietary* DBMS developed by the Swedish Geological Survey and called MIMER. This DBMS, written in FORTRAN, accepts ordinary FORTRAN format data-files, as used for normal processing, so that the whole database could be readily converted to run under another DBMS. Thus when the host PDP computer was withdrawn, the effort of converting to a different machine was far less than for IGBA.

A typical DBMS will offer the following features:

— its own high-level, quasi-English, command language, used both to define data relationships and to selectively retrieve and report on the data (cf. SYSTAT command language in Table 2.1c); this is a *non-procedural* language, which defines the result required and leaves the system to determine how to produce this result, whereas FORTRAN and others in Table 1.1 are *procedural* languages, which give the system exact instructions as to how to achieve a given result.

— an interface, enabling programs written in FORTRAN etc. to access the data;

— automatic back-up and checking for duplicates and for internal logical consistency of the data;

— a **data dictionary,** maintaining information on formal names, codes, pseudonyms and abbreviations, compositions of records within a file, and relationships between files; this, for example, can be used to translate free-text rock-names into short, more consistent codes, when performing a retrieval, and then to translate back again when presenting the results.

— independent maintenance of three views of each set of data (the user's view, the logical structure used to design the database, and the electronic layout of the data on the storage medium) in such a way that a change in the definition of one does not affect the others; thus the user need only know that data with a certain name exist *somewhere* in the base, in order to be able to retrieve them.

3b3. Database architecture — types of database structure

Databases can be divided into four main types according to the variety of DBMS they use:

(1) **Hierarchical databases** were the earliest types developed. They have their data arranged in a fixed kind of tree structure, in which the only allowed relationship is one-to-many (1:*N*), between an "owner" set and a hierarchically lower "member" set (cf. stratigraphical Groups, Formations, Members). They have two disadvantages which have limited their popularity of late: (a) in order to reach information on a different branch, the system must move to the 'trunk' of the tree and back along the other branch, which is cumbersome; (b) if the data structure has to be changed, the entire database has to be rebuilt. Nevertheless, the fixed architecture means that retrieval is particularly fast, so if a database is likely to be stable over a long period, this type is worth considering.

(2) **Network databases** are a superset of hierarchical, which overcome rigidity by establishing links ('pointers') between data-items, thus allowing many-to-many relationships (i.e. multiple "owners" of one "member" set). Many successful geological databases have been implemented on this model.

(3) **Relational databases** are by far the easiest to conceptualise for non-experts, which accounts for their current pre-eminence. Data are arranged in the form of **relations**, which can be thought of merely as computerized two-dimensional tables like those in any geological publication. A relational

database can comprise one relation (table) — as might be the case for a database cataloging rock specimens in a collection according to their specimen number, locality, rock-type, etc. — or many relations (e.g. Table 3.3, Fig.3.5). In the latter case, a 1: N relationship between relations is general. In Table 3.3, a relational database for the chemistry of alkaline rocks, one record in the *occurrence* relation codes a description for a particular alkaline body, which corresponds to N (> 1) records in the *rock specimen* relations, which contain chemical data for the many samples analyzed from that body; this in turn corresponds to many records in the *mineral* relation, which contain chemical data for the many minerals analyzed from each rock specimen. Critical in relational database architecture are the concepts of **primary keys** (unique identifiers to each record in each relation, which link information in different relations), and **third normal form** (which determines whether the database is set up in the best structure for ready updating, changing, etc.) Fig.3.5 illustrates how primary keys link different relations together.

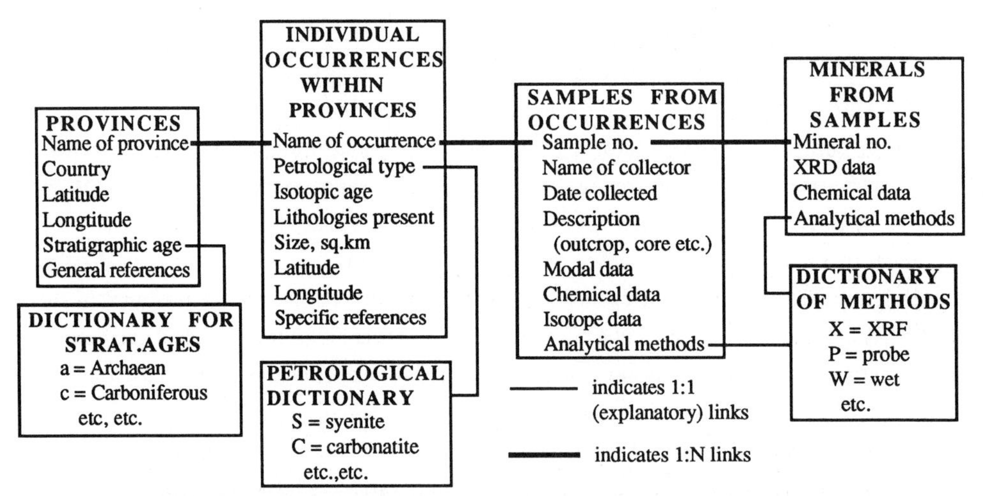

Fig.3.5. Structural layout of a database similar to that whose extract appears in Table 3.3
The four main tables (**PROVINCES...MINERALS**) are linked via *primary keys;* each 'province' in the **PROVINCES** table corresponds to many 'occurrences' (intrusions, volcanoes, lava-fields etc.) in the **OCCURRENCES** table. The 3 data-dictionaries shown merely provide explanations for the coded entries within the main tables. Other dictionaries/links would exist in the complete structure of such a database, but are omitted for clarity

Relational databases do not require data relationships to be rigidly defined. This flexibility has its associated negatives: namely, some data redundancy, and slower access times. However, common types of queries can still be optimised. Hence relational structures provide the best means for storing geological data where:

— data relationships are not well understood or fixed at the time data are entered into the database;

— use of the data (and hence data structures) may change over time;

— certain well-defined queries are likely to be made frequently, with only occasional 'ad hoc' queries;

— access will generally be by geologists rather than by computer professionals;

— speed of access to large volumes of data is not critical;

Table 3.3. A sample relational database structure

The database contains whole-rock and mineral chemistry for alkaline rocks (relations d,e,f,g) together with bibliographical sources (c), and a global listing of alkaline provinces (a) and individual complexes (b). The primary 'key' field links equivalent records in the various relations, which are mostly on a 1: N **basis; it builds up a coded description of each record on a hierarchical basis. For full details see Rock(1987a).**

(a) Alkaline province/geographical area dictionary relation

Code	Name of alkaline province/geographical area	Age
EU	Miscellaneous occurrences in Western and Central Europe	—
FE	Miscellaneous occurrences of the Far East (China, etc.)	—
FR	Franspoort line alkaline province, S. Africa	Proterozoic
GD	Gardar alkaline Province, SW Greenland	Proterozoic
OL	Ontario Labrador nepheline syenite belt, Canada	Grenvillian
IB	Iberian alkaline Province, Spain/Portugal	Cretaceous
IN	Miscellaneous occurrences in Indian subcontinent	—
KA	Kapuskasing High alkaline Province, Canada	Proterozoic
KP	Kola Peninsula alkaline Province, USSR	Caledonian

1:N

(b) Individual alkaline occurrence details relation

Code	Name of occurrence	Lithological details	Age	Area
IBG	Gorringe Bank, Atlantic	M	60	
IBM	Monchique, Portugal	M• BBBCCC DDDDDDD	72	65
IBN	Sines, Portugal	M	72	
IBO	Ortegal Penin, N. Spain	M• B		
IBP	Pyrenees (Fitou, Pouzac)	M• BB D D CR	82	0.5
IBS	Sintra, Portugal	M B CCCD D D	82	

1:N

(c) Bibliographic source relation

Key	Source
IBM01	Rock (1983)
IBM02	Rock (1983)
IBM03	Rock (1983)
IBM04	Rock (1983)
IBM05	Washington (1917)

(d) Whole-rock major element relation

Key	SiO_2	Al_2O_3	Fe_2O_3	FeO	MnO	CaO	Na_2O	K_2O	H_2O-	TiO_2	P_2O_5	MnO	CO_2	LOI
IBM01MNI	55.46	22.05	1.23	0.67	0.01	0.36	8.48	6.36–1.00		0.15	0.01	0.17–1.00		2.53
IBM02MNI	56.27	22.80	1.71	1.31	0.33	1.80	8.74	6.67–1.00		0.82	0.04	0.14–1.00		1.06
IBM03MPM	56.42	22.64	1.63	1.35	0.01	1.46	7.40	7.13–1.00		0.56	0.05	0.17–1.00		1.99
IBM04MPM	56.49	22.46	5.25	0.67	0.05	0.37	8.51	6.15–1.00		0.15	0.01	0.00–1.00		2.53
IBM05MNI	53.71	21.82	0.78	2.47	0.56	1.90	8.52	7.07–1.00		1.03	0.01	0.19–1.00		0.00

1:<1

(e) Whole-rock trace element relation

Key	La	Ce	LiBeSc	V	Cr	Co	Ni	Cu	ZnGa	Rb	Sr	Y	Zr	Nb	Sn	Cs	Ba	Hf	Pb	Th	U
IBM01	31	45	-1-1-1	49	171	-1	33	-1	6251	1911	922	32	392	226	-1	3	154	7	-1	10	8
IBM02	35	55	-1-1-1	31	30	-1	80	-1	-171	2531	356	23	876	214	-1	-1	647	15	-1	31	24
IBM04	42	84	-1-1-1	53	14	-1	6	-1	5454	175	641	11	483	111	-1	1	117	-1	-1	22	4

1:N

(f) Whole-rock isotope data relation

Key	Sr87/86	18
IBM01	0.7033	
IBM04	0.7016	5.70

(g) Mineral analysis relation

Key	SiO_2	Al_2O	Fe_2O_3	FeO	MnO	CaO	Na_2O	K_2O	H_2O+	TiO_2	P_2O_5	MnO
IBM01P1PR	51.62	2.09–1.00		9.20	11.63	22.78	0.67	0.00–1.00		0.51–1.00		0.80
IBM02P1G	51.53	1.86–1.00		7.65	12.84	22.88	0.73	0.00–1.00		0.82–1.00		0.57
IBM02P2G	51.39	1.93–1.00		14.00	7.68	20.50	1.94	0.00–1.00		0.40–1.00		1.89
IBM03P1PC	51.01	2.13–1.00		13.80	8.65	21.52	1.30	0.00–1.00		0.47–1.00		1.26
IBM03A1PR	44.41	8.08–1.00		16.34	10.16	9.42	4.31	1.57–1.00		1.24–1.00		2.33
IBM01B1PC	40.07	11.38–1.00		21.64	7.38	0.15	0.00	9.89–1.00		2.47–1.00		2.02
IBM02B1PR	37.33	14.33–1.00		20.93	11.13	0.00	0.00	9.82–1.00		2.92–1.00		1.03
IBM03B1PC	37.94	10.87–1.00		25.67	6.25	0.00	0.00	9.35–1.00		3.95–1.00		2.01

(4) **Geographic information systems** (GIS): combine a traditional (usually relational) database with a graphics system; they are dealt with fully in §4 since they are more concerned with production of geological graphics than databases *per se*.

3b4. Facilitating exchange of data: standard formats and procedures

Further geological discussion: *Geochemistry* (de la Roche et al. 1986; Grandclaude 1978; Grandclaude & Stussy 1978); *Geophysics* (Dampney et al. 1986); *Exploration and mining* (Anon 1978b, 1979; Cargill & Clark 1977; Hexagon Software 1987; Longe et al. 1984; Porter et al. 1986; Smith 1986); *Computer aspects* (Henley 1977).

A major obstacle towards greater national and particularly international exchange of geological data of all types is the diversity of ways in which individual geologists, institutions or firms format their data. There are four main components to such variations:

(1) *Inherent inter-computer differences.* Some have already been mentioned (§1d): for example, standard Apple Macintosh files delineate data-values using <TAB> characters, which many mainframes do not understand; to convert a Macintosh to a mainframe file requires either addition or stripping out of these <TABS>. These differences are irrevocable and have simply to be lived with.

(2) *Reals versus integers, numbers of decimal places, etc.* These kinds of differences can also be important if one institution uses *fixed format* and another *free format* to define data-items (§1d; i.e. variables are defined by ranges of columns in fixed format, but merely separated by spaces or commas in free format). For example, if one organization defines the element Ni as an three-digit integer, format (I3), data from another firm with 4 or more digits for Ni, or with Ni cast as a real, may be mis-read or not read at all, depending on the flexibility of the software involved.

(3) *Code versus free text.* This is a very major problem, requiring international assessment. For example, the rock-type 'granite' would be listed in some large petrological databases simply as the free-text word GRANITE, whereas in IGBA (§3b1) it would be specified numerically as 1420, in the coded system of Harrison & Sabine (1970) as 1AA0, etc.etc. Unfortunately such codes do not even have a 1:1 correspondence between the different systems, so that the tedious job of constructing translation software (data-dictionaries) to convert one format to the other is made the more difficult (and probably quite impossible). Full discussion of these aspects would require a complete Topic!

(4) *Order of variables.* There is no standard element order for published analyses. Two common major element orders (the first supposedly used by the British Geological Survey in all its publications, the second by IGBA), are as follows; other papers use many other orders for the minor elements:

(a) SiO_2, Al_2O_3, Fe_2O_3, FeO, MgO, CaO, Na_2O, K_2O, $H_2O\pm$, TiO_2, P_2O_5, MnO, CO_2

(b) SiO_2, TiO_2, Al_2O_3, Fe_2O_3, FeO, MnO,MgO, CaO, Na_2O, K_2O, P_2O_5, CO_2, $H_2O\pm$

The situation is even worse with trace elements, even though *only* two orders are logical: (1) *alphabetical* (As to Zr); or (2)*atomic number* (Li to U). However, many authors use the most amazing orders, which seem only to reflect the way in which the data came out of their XRF machine!

Although software does exist to inter-convert data-formats (e.g. Rock & Wheatley 1988), conversion can add to confusion and error. Given the plethora of ways geological journals format references, international standards for actual data would seem to be a distant hope! Nevertheless, the efforts listed above represent valuable attempts to bring greater order out of the present chaos.

TOPIC 4. WRITING, DRAWING & PUBLISHING BY COMPUTER

Nearly all geologists must write reports, papers or books to get their message across. Recent improvements in computer capabilities make all this vastly less time-consuming than in the past.

4a. Computer-assisted writing (word-processing)

There are four main ways of producing a computer-readable document (geological or otherwise):

1.*Dedicated word-processors.* 'Stand-alone' word-processors are glorified electronic typewriters. Like many microcomputers, they store text on floppy discs, but unlike microcomputers they cannot draw graphics or perform mathematical calculations. Although most modern offices have such machines, their days are probably numbered since the advent of advanced word-processors on microcomputers, for 4 reasons: (a) many geologists find it quicker to type and edit a document themselves than to manually edit and then proof-read a typist's work, but few are allowed access to word-processors, which remain the secretary's exclusive domain; (b) even cheap microcomputer word-processing packages (see 3 below) have far more advanced capabilities than many word-processors (e.g. in handling many levels of superscripts and subscripts, global edits, italics and bold fonts, Greek mathematics and complex formulae); (c) the quality of output from many word-processor printers is inferior to that of microcomputer printers (especially laserwriters); (d) dedicated word-processors may cost 3–4 times as much as microcomputers which can perform not only this one job, but also many others. The present trend in geological institutions, in the writer's experience, is for existing word-processors to become white elephants. Most can now be linked to mainframes or microcomputers, so that documents prepared on one machine can be edited or printed on the other (although some firms do not encourage such traffic, for obvious commercial reasons!)

2.*Mainframe text editors.* These fall into two main groups: *line-editors*, in which the text is displayed and edited one line at a time, and *full-screen editors*, in which a screenful of text is displayed at once. Some editors can be used in either mode. Simple editors are really only useful for files in which each typed line is a discrete entity — e.g. FORTRAN programs and data-files where the <CR><LF> (carriage-return, line-feed) at the end of each line is an essential part of the text, separating individual lines of code, or data for different specimens.

3.*Mainframe text-formatters.* Ordinary written documents, reports, letters, papers, etc. cannot conveniently be edited or printed using a simple editor, because the <CR><LF> at the end of each line cannot easily be removed. Text-formatters allow this to be done, and constitute the next most advanced word-processing step. Again, however, many constitute a somewhat elephantine method.

To use a text-formatter, a text file is first created using an editor. This can be typed in 'any old how', with lines of any length, any number of space between words, etc. This file is then peppered with special command lines which give instructions as to how the eventual text should be formatted. When input to the text-formatter, these commands disappear, along with the <CR><LF>, and 'final' text appears (hopefully in the expected format)! Table 4.1 shows a typical input and output file, using perhaps the most sophisticated of the text-formatters, T$_{E}$X™ — a huge mainframe system of great sophistication, which is probably capable of producing the closest approach, on a computer, to traditionally typeset material from a traditional publishing house.

Unfortunately, text-formatters can be extremely frustrating (one wrong command can produce dire results, like a page-throw after every word!) The ideal text-editing program is WYSIWYG (What You See Is What You Get), but text-formatters are very far from this ideal (Table 4.1), and commensurately harder to use. T$_{E}$X has a dedicated following, but few geologists may feel inclined to master its 5-volume set of manuals, merely to generate the extra sophistication of its output over the incomparably more user-friendly word-processor on many microcomputers.

Table 4.1. An example input and output document to the text-formatter mainframe program, T$_{E}$X

Input document (fragments beginning '%' are comments)	*Output text*
/document style (article) % specifies the document style	
/title {Output text from} % specifies the document title	**Output text from**
/author {TEX program}	TEX program
/date {May 11th,1988}	May 11th, 1988
/begin {document}	
/maketitle % produces the title centred	
This is to show that TEX is not WYSIWYG compared with Macintosh's MacWrite, used to produce this book.	This is to show that TEX is not WYSIWYG as compared with Macintosh's MacWrite, used to produce this book.
/section {This is how to produce a heading}.	**This is how to produce a heading**
You can type any number of spaces between words, one and one or more blank lines denote the end of a paragraph.	You can type any number of spaces between words, and one or more blank lines denote the end of a paragraph.

4. *Microcomputer word-processing packages.* Many of these are nearly or completely WYSIWYG. Unlike all three previous methods, word-processors treat *paragraphs* rather than lines of text as blocks; that is, <CR><LF> characters are only inserted between paragraphs, rather than between lines. This can at first cause confusion for experienced typists, used to pressing the "return" key on a manual or electric typewriter at the end of every line. Once mastered, however, word-processors are the most 'user-friendly', quick and efficient means of producing documents.

4b. Computer-Assisted (Desktop) Publishing (CAP/DTP)

Dedicated monographs and reviews: Hartnell(1986); Baxter (1986a,b).
Dedicated journal: Desktop Publishing (Sky Business Publications Pty.).

Desktop publishing refers to an individual publishing himself, via a micro and laserwriter or linotronic typesetter. Many steps involved in traditional publishing (Fig. 4.1) are thereby obviated. A simple DTP setup is now within the reach of many professionals, costing perhaps US$5,000 for both hardware and software; an ever-increasing volume of material is now being produced in this way (e.g.

Ho & Groves 1987). More sophisticated setups such as that depicted in Fig.4.2 facilitate productions which rival those of traditional publishing. There is increasing divergence between the terminology (font, typeface, etc.) used by DTP devotees and that of the ancient and honourable company of typographers. The following, however, follows DTP publications such as those listed above.

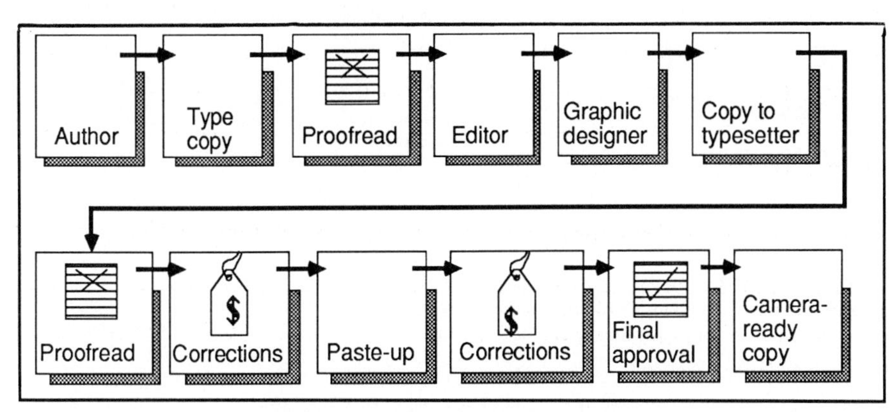

Fig.4.1. The traditional publishing process

Fonts. DTP addicts are now faced with literally hundreds of fonts, varying *from the elegant (Zapf Chancery)* to the gothic (London) and purely frivolous (San FranSisco). Two main font families are: (1) *serif fonts*, like Times, used here (following the the revered Times newspaper of London), in which each letter has small delimiting lines at top and bottom (the 'serifs') and (2) *sans-serif fonts*, lacking these lines, like Avant Garde, used here in lists of sources. Surveys (e.g. Wheldon 1986) have found serif fonts easier to read in blocks of text, although sans serif have a more 'modern' look, and are perhaps easier to read in figures, like Helvetica used here. A further division is into *monospaced (typewriter)* fonts like Courier and Monaco, in which each letter occupies an equal width of text, and *proportional* fonts like Times, where wide letters like M, W occupy more space than narrow ones like I, J. Proportional fonts give a more professional appearance, but can be extremely annoying in tables of figures, since they may prevent numbers from being vertically aligned. Other useful fonts include *pictorial* (merely typing FGHJ in Cairo font on a Macintosh generates [pictorial symbols]), *foreign language* (abcdefg becomes αβχδεφγ...in Symbol font) and *symbol/mathematical* (abcdefg becomes [symbols] in Systat font).

Styles. Most fonts exist in numerous different forms, specified by the *point- (pt) size* and *style*. A point is 0.00138" (≈ 1/72"). The full description of the font used in this sentence is Times 12 pt Roma (but now reduced to 80% size). This is Times 14 pt and this is Times 10 pt. For most geological purposes, lettering smaller than about 6 pt is illegible, while lettering larger than 18 pt is rarely required other than on posters and notice-boards. 'Roman' is equivalent to 'upright'; other *italic*, **bold**, shadow, underline and outline styles are also usually available in each of the point-sizes.

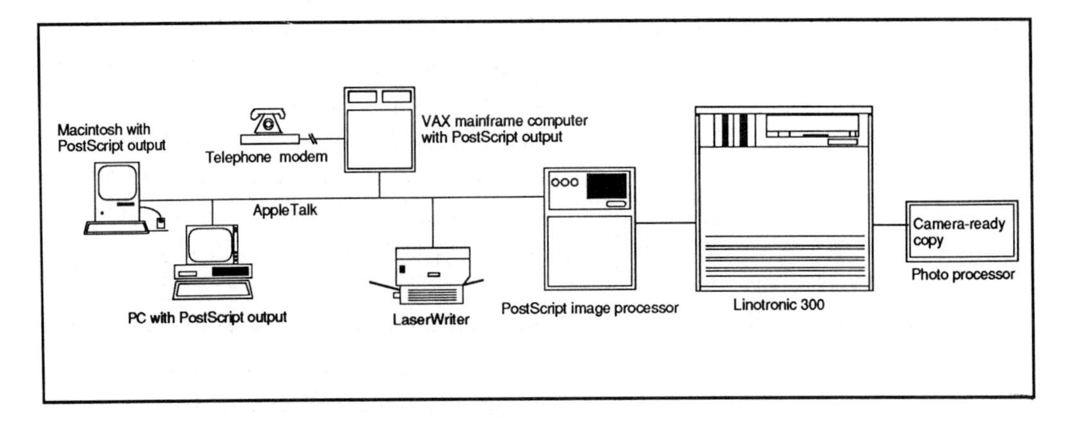

Fig.4.2. A sophisticated desktop publishing system operated by the Western Australian Regional Computing Centre. Various brands of micros and minis are linked to laserwriters and linotronic typesetters; the final output is camera-ready copy as in Fig.4.1, but the correction, proofreading, graphic design, and paste-up stages have been eliminated

4c. Producing Maps & Plots: Computer-Assisted Drafting (CAD) and Mapping

Dedicated monographs and reviews: Foley & Van Dam (1982); Myers(1986); Rogers & Adams(1976); Schnapp(1986).
Further geological discussion: Boyer (1986); Burns et al. (1969); Hodell & Estep(1985); Miranda & Roquette(1977); Olson(1986); Pavlidis(1982); Sowerbutts(1982); Tipper(1979b); Howarth & Turner (1987).
Geological plotting software: *Generalized* (Andrew & Linde1981; Butler1972);*Triangular diagrams* (Blencoe 1976; Burwell & Topley 1982b; McHone 1977; Rock & Carroll 1988; Topley & Burwell 1984); *Structural geology—stereonets, fabric diagrams,* (Bonyum & Stevens 1971; Cubitt & Celenk 1976; Lafontein 1970; Spencer & Clabhugh 1967;Tocher 1978a,b, 1979); *maps and cross-sections* (Duncan 1985b; Langenberg et al. 1977; Tipper 1976, 1977b); *folds, faults & fractures* (Charlesworth et al.1975; Jeran & Mashey 1970; Kimberley 1986; Serra 1973); *Stratigraphic column/borelog construction:* (Cairncross & McCarthy 1986; McCarthy 1981; Odell 1977b); *Contour plots* —see §20b3; *petrological/geochemical plots* (Daost & Gelinas 1981; Lenthall et al. 1974; Ludwig & Stuckless 1979; Nichol et al. 1966; Palacios & Faria 1982; Rock 1987f; Till 1977; Wheatley & Rock 1988).
Mainframe graphics subroutine libraries: *GKS* (Graphics Kernel System), *GPGS* (General Purpose Graphics System), *PHIGS* (Programmers' Hierarchical Interactive Graphics Standard); see §1c4.
'Everyman' mapping/CAD systems: SURFACE II (Sampson 1975); *evaluation of 22 mapping systems:* (CEED 1986); some packages in Table 2.1 have strong graphics capabilities.
Specialized geological mapping systems: see catalogues provided by AMIRA (1985, vol.4), Salomon & Lancaster (1973), Sanford et al. (1978,1979), and further discussion in §2d and §5b.

The proportion of graphics in geological publications is higher than in many other sciences (Howarth & Turner 1987). After many decades in which traditional pen-and-ink methods have been the only means of production, there is now an overwhelming profusion of both hardware and software for generating graphics of publishable quality. Adverts appear regularly, and national exhibitions are mounted regularly for companies to demonstrate their new products. As in §2, software ranges from that for which code is available for the user to tailor to his own needs, through proprietary subroutine libraries to 'everyman' packages (usually called *CAD* or *mapping systems*). On the whole, there is even less call for the geologist to program his own graphics software from scratch than with numerical software. Only a few very general comments are proferred below. Systems and hardware aspects of generating graphics (graphics standards, plotting devices, etc.) were dealt with in §1c4.

(1) *Using modern printers with suitable graphics software* This combines the "do-it-yourself" and "find existing software" approaches in §2a & 2b. A complete document (text + diagrams) can

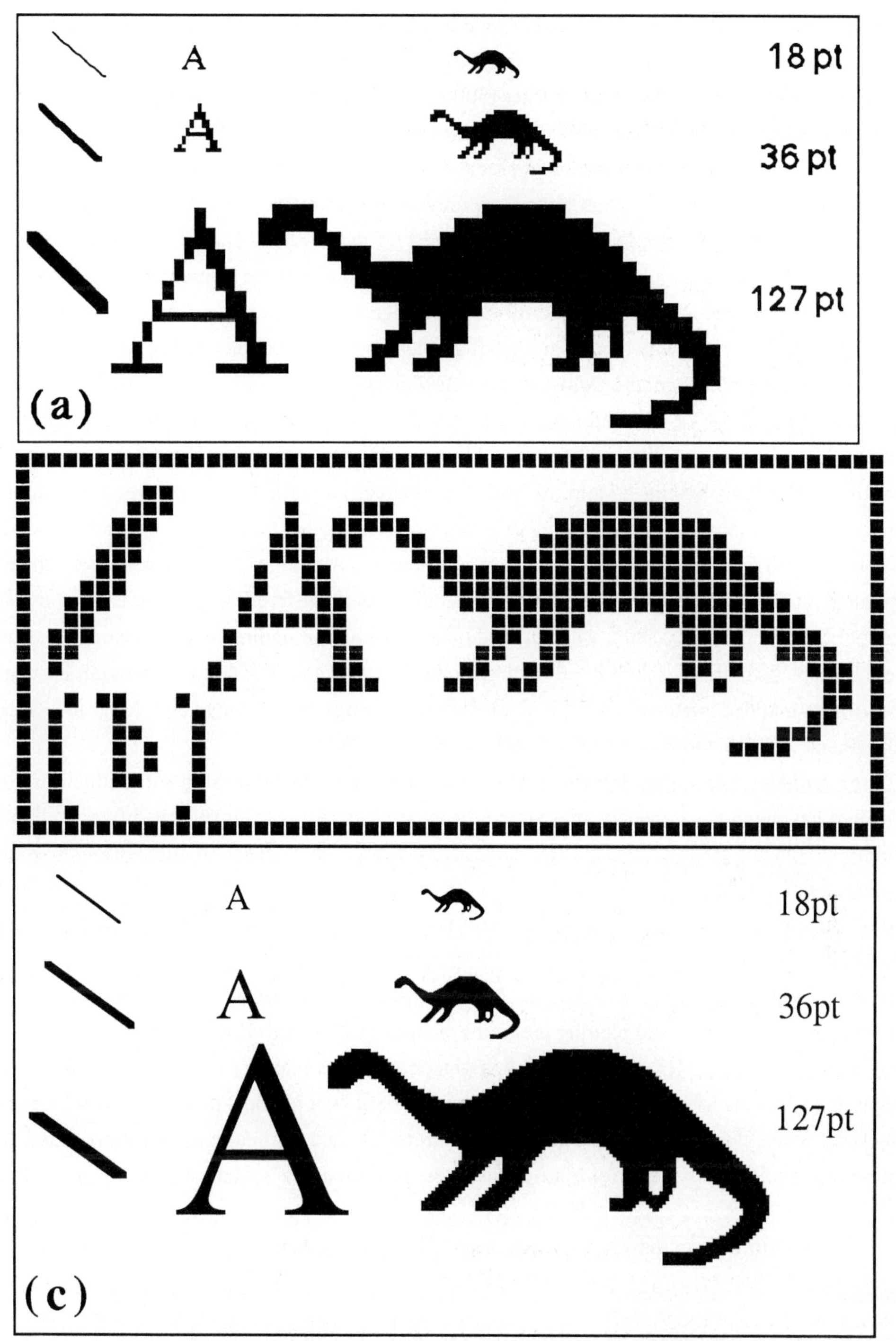

Fig.4.3. Bit-mapped (a,b) versus object-mapped (c) characters, printed on a laserwriter at 300 dpi resolution; (b) magnifies 18pt characters in (a) to show individual pixels; quality in (a) deteriorates as point-size increases, but in (c) remains virtually unchanged

nowadays be held on computer file, and printed as a single entity on a laserwriter, as was this book. The final quality of diagrams depends on the resolution of the printer, and increases consequently in the order imagewriter (72 dpi) < laserwriter (300 dpi) < linotronic (up to 2450 dpi). When used with the elegant software designed specifically to maximise the graphics capabilities of machines like the Macintosh, these printers can generate geological maps and diagrams formerly costing hundreds or thousands of dollars of draftmens' time, can now be drawn even by an unartistic geologist in a fraction of the time and cost. Diagrams as large as 10 A4 pages can be printed at one time. The main limitation at present is on colour printing, which is cumbersome. Hence figures for most professional papers can be generated in this way, but full-scale geological maps are beyond its present capabilities.

Fig.4.3 illustrates how two main groups of graphics software produce rather different results on these machines. (a) *Bit-mapping software* (e.g. MacPaint) stores each character, number or picture as a pattern of pixels, at 72 dpi, and even when printed on a laserwriter or linotronic, the resolution remains 72 dpi, so the quality (particularly of lettering) is only moderate (Fig.4.3a). Diagonal lines are also a problem, having a staircase like appearance. Nevertheless bit-mapped diagrams are relatively simple to edit. (b) *Object-mapping software* (e.g. MacDraw) stores characters as ASCII characters, with lines and curves as mathematical equations. Although on the micro's screen the result looks identical to a bit-mapped diagram, it is edited in a quite different way (which, being dynamic, cannot be illustrated easily here); when printed on a laserwriter or linotronic, the machine recalculates the images to its higher resolution, producing a much better result (Fig.4.3c). In general, bit-mapped and object-mapped material are suited to different things in geology, and programs such as SuperPaint™ which combine them are particularly powerful.

(2) *Using graphics subroutine libraries:* This approach is equivalent to using subroutine libraries in §2c, and has much the same advantages and disadvantages. Any reasonable mainframe computing facility will have a graphics library, subroutines from which can be linked to tailor-made programs in conventional languages. For example, the user might call a routine LINE(X,Y) which draws a straight line between two points *X* and *Y*, and so on. Although lack of portability (§1c4) is a severe drawback, coloured graphics are now no problem, and can be generated on the range of mainframe-linked plotters described in §1.Examples of software using this approach are the geochemical and geophysical plotting programs of Rock (1987f) and Busby (1987).

(3) *Using 'everyman' CAD systems.* Many sophisticated packages are now available for drawing maps, and allow the user tremendous flexibility and control over his final product. Three-dimensional block diagrams, borehole logs, stratigraphic columns, etc. are also commonly programmed. More specialized geological systems for mine planning etc. are discussed in §2d and §5b.

4d. Combining graphics with databases: Geographic Information Systems (GIS)

Dedicated journal: *International Journal of Geographic Information Systems* (published by Taylor & Francis, 1st issue February 1987); see editorial review in first issue (Coppock & Anderson 1987).
Further geological discussion: Kelly & Phillips (1986); Nystrom(1987); Robinove(1986).
Further non-geological discussion: Calkins & Tomlinson(1977); Green et al.(1985).
Selected geological applications: *Environmental geology* (Nystrom 1987); *earthquakes* (Rich 1987); *hydrogeology* (Augenstein & Wolf 1987); *economic geology* (Aubrey 1986; Cary 1987).
Proprietary software: ARC/INFO (ESRI 1985a,b,c; 1986); GeoVision; Tydac.

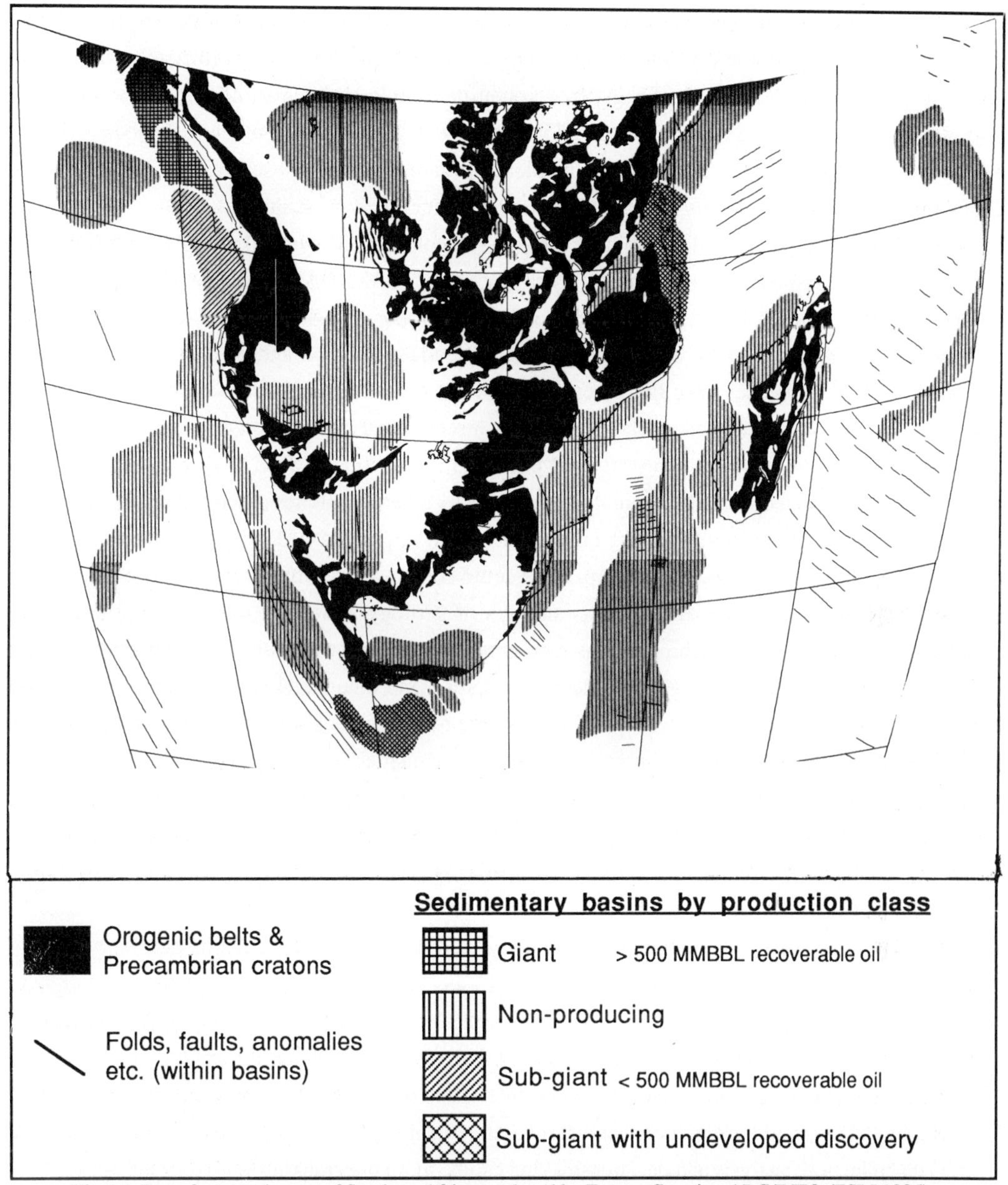

Fig.4.4. Part of a tectonic map of Southern Africa produced by Exxon Co.using ARC/INFO (ESRI 1986). The map overlays selected tectonic features with the producing status of known basins from two large existing databases

Geological research often requires the representation of large volumes of data on maps: (e.g. environmental impact surveys, mine documentation, mineral resource estimation). Thus far we have dealt separately with methods of storing, manipulating and retrieving the large volumes of data (in databases), and with computer-assisted map production. GIS are a relatively recent approach, going a stage further and combining a conventional database with a graphics package: that is, they can analyze numerical data which have associated spatial coordinates. Both the database and graphical parts of a

GIS can be used separately, but the user can also extract information from his database in *graphical* as well as numerical form. For example, using a simple DBMS for a museum collection, a user could obtain a straightforward list, say, of registered komatiite samples from Australia, together with such information as latitudes and longtitudes (or grid coordinates). Using a GIS, however, the locations (and other details, as appropriate) could actually be plotted on a map at any chosen scale, with any desired amount of topographical information (roads, geological boundaries, mine sites, etc.)

ARC/INFO is a typical GIS, combining ARC (a full colour CAD system) with INFO (a relational DBMS). It is mentioned because one of its major users is the US Geological Survey (USGS). Nystrom (1987) enthuses as follows: "The use of GIS technology has the potential for revolutionizing the way the USGS, State and other Federal agencies conduct research and present research results. Advanced GIS technologies have the potential to greatly enhance the USGS's ability to perform its traditional missions of earth science data collection, research and information delivery". ARC/INFO can also interface with well-known statistical packages (SAS, SPSS, MINITAB, etc.: Table 2.2) so that data can be extracted from INFO and then analyzed by other techniques described in this book.

In addition to simple location maps of sampling sites, mines, etc., ARC/INFO can analyze maps divided into polygons: on a geological map, these could correspond to different outcrops, soil types, or even political boundaries. As an example, surveys of geology as related to population health are becoming fashionable. Such information as rock-type, and the soil or rock contents of health-affecting elements (e.g. I, Cd, Hg, Pb) might be held within INFO, and then ARC used to query the database and plot a coloured map displaying only the areas where rock polygons with Pb > 500 ppm, overlapped soil polygons with Cd > 100ppm, further distinguishing the latter as *green* < 200 ppm in green, *pink* < 500 ppm in pink, and *red* > 500 ppm. The red areas might then be identified as health hazards!

The advantages of a GIS over a traditional CAD graphics system can be summarised as follows:

(1) A GIS can analyze, manipulate and manage spatial data as well as merely display them.

(2) Data can be maintained, updated and extracted more efficiently.

(3) Spatial coordinates are stored mathematically, allowing techniques such as nearest neighbour analysis (§20), terrain analysis, slope angles, and watershed analysis to be carried out.

(4) Interactions between many types of information can be effectively displayed and quickly examined; cartographical, hydrological, geological, and geographical information have all been integrated into existing GIS implementations cited at the beginning of this section.

(5) Rapid and repeated testing of conceptual models is facilitated.

(6) Data collection, analysis and decision-making can be more integrated.

(7) Drafting becomes automated, error-free, interactive and hence more cost-effective.

TOPIC 5. USING COMPUTERS TO BACK UP HUMAN EFFORT: COMPUTER-ASSISTED TEACHING, EXPERT SYSTEMS & ARTIFICIAL INTELLIGENCE

Topic 2 briefly covered situations where computers reduce the drudgery of mechanical tasks: norm calculations and the like. In this Topic, more advanced uses are considered, in which computers assist or begin to imitate the inventive, perceptual or rational processes formerly carried out only by humans.

5a. Computers as teachers: computer-aided instruction (CAI)

Further geological discussion: Bartholomew & Jonas(1974); Davis(1975); Kennedy(1982); Laudon(1986); McKenzie(1984); Merriam(1976b); Ousey(1984); Pirie(1982); Reeves & Jucas(1982); Rustagi & Wolfe(1982).

CAI applied to specific teaching courses: *PLATO* (Mann 1976); *Exploration* (Hayes & Allard 1981; Searight 1985; Shackleton 1977; Vannier & Woodtli 1976); *Structural geology* (McEachran & Marshak 1986); *Geophysics* (McCann & Till 1976); *Igneous petrology* (Frost 1977a); *Geobiology* (Dodd & Immega 1974); *Environmental geochemistry* (Brown 1984); *Photogeology* (Jones 1978).

Many geology teachers have devised microcomputer CAI programs to complement traditional teaching: programs which, for example, test students' knowledge of rock classification or terminology. Many examples are described in the *Journal of Geological Education*, and some in the bulletins and newsletters of national geomathematical groups. Some such programs are public domain, others are distributed as small commercial enterprises; it is not feasible to provide a comprehensive list here.

National computeracy campaigns in some countries have had significant geological spinoff. For example, the British Broadcasting Corporation's campaign of the early 1980s led to widespread adoption in schools and colleges of the graphics-rich BBC "B" micro, and the dissemination of large volumes of mostly BASIC software. Computer programs were broadcast after radio and TV transmissions, and could be picked up by ordinary cassette recorders for loading into the computer. Among resulting geological programs are those by Knight et al.(1985), which model volcanic eruptions and lava-flows. Unfortunately, like all BASIC software, but more so because of their strong graphics content, these programs would be difficult to implement on any other microcomputer.

A UK *Computers in Teaching Initiative* has also been launched, whose support service (at the South-West Universities Regional Computer Centre, Bath University) publishes a quarterly magazine (*CTISS File*). Among 139 projects listed in the Sept.'87 issue, relevant examples include the following:

	University	Department(s) participating	Nature of CTI project
G	Keele	Geology/Geography/Physics	Computer graphics for geology teaching
E	Leicester	Geology/Physics/Geography	Remote sensing and image analysis
O	Edinburgh	Geology	Simulation, modelling and 3-D graphics
L	Reading	Geology	Microcomputer facilities for geology teaching
S	Heriot-Watt	Civil Engineering	Geological map interpretation
T	Lancaster	Applied Statistics/Mathematics	Statistics teaching using graphics workstations
A	Glasgow	Statistics	Computer-illustrates texts for teaching statistics
T	Cardiff	Statistics	Computer-assisted statistics teaching
S	London	Statistics/Government	Computer teaching of social statistics

A more international form of CAI is represented by large mainframe systems such as PLATO (Programmed Learning for Automated Teaching Operations), originally developed at the University of Illinois over 25 years ago. PLATO now contains over 50,000 lessons in many arts and science disciplines, and is in use on campuses worldwide. Mann (1976) gives a useful introduction to the system for earth scientists, although the geological lessons his paper mentions were never fully implemented (Mann, pers.com. 1987). Access to PLATO can be either via special terminals (with touch-sensitive screens), or via microcomputers ported to the mainframe on which PLATO normally resides (e.g. Macintosh owners can use PLATO via a communications package called MacPad™).

PLATO users can choose either individual lessons, or work through structured courses (e.g. in FORTRAN programming); the system may refuse to allow a student to proceed to more advanced lessons until questions from simpler lessons have been correctly answered! When entering PLATO, the user can find what lessons are available in a particular subject via subject and author catalogs, and in some implementations via local catalogs which take the user from the general to the specific (e.g. from 'mathematics' to 'statistics' to 'correlation'). More sophisticated management facilities for student classes are also available: a 'group' can be set up, for example, to which each individual students and one or more instructors have signons. A menu of selected lessons available to this group is set up by the instructor, and then automatically presented when the student signs on, the machine keeping a record of how long each student has spent on the system, which lessons he has completed, etc. Students and instructors can also then send messages to one another.

PLATO has proved invaluable in backing up formal instruction in geology at the University of Western Australia. Even students with little or no computing experience have been able to master the system sufficiently within an hour or so of first use, and this in turn has great value in overcoming their innate fear of computers. Aspects of formal lectures and tutorials not understood at the time can be followed up in the student's own time and at his own speed. The student can also be sure how well he understands a topic, by the proportion of answers PLATO tells him are correct! Psychologically, PLATO also has advantages to in that no-one else need know if a student gives a wrong answer.

At present, *geological* lessons in PLATO are limited to a few first-year undergraduate topics (e.g. composition of the earth; also a handy mineral identification quiz). However, a wider range of mathematical-statistical and computing lessons are available, covering programming in BASIC, FORTRAN, and Pascal; various lessons in understanding what computers do; lessons in matrix algebra; and statistical topics ranging from Normal distribution theory to analysis of variance and factor analysis. One difficulty in determining what PLATO material already exists is that there is no central clearing house, even in the USA; one has to rely on personal contact and other sources.

The alternative, for those working in institutions supporting PLATO, is to write their own lessons in PLATO's own language, TUTOR. Policy decisions are then necessary as to whether PLATO should be used 'front-end', i.e. to replace more traditional lectures and practical classes, or 'back end', i.e. merely to supplement these methods. Although PLATO obviously cannot substitute for practical classes teaching rock, mineral or fossil identification, in traditional geological disciplines, its potential in numerical geology courses such as the present one is far greater. Indeed, a well-written PLATO course, improved and updated over the years, could in some respects be superior to a traditional teaching course, especially if the alternative is a lecturer whose quality of teaching is undermined by his intrinsic ability, poor health or excessive workload. Students would be able to work at their own pace, and ensure their understanding of each topic before proceeding to the next one. Lecturers would be freed from some of the drudgery of marking (done by the computer), and would be freer to concentrate on specific topics of difficulty, or to tutor smaller student groups. This approach has already been taken by one faculty (Business Studies) at UWA — partly out of sheer necessity, because no existing PLATO material in this subject was available; it has, however, proved highly successful in practice.

5b. 'Humanoid computers' in geology: Artificial Intelligence (AI) and Expert Systems

Dedicated monographs and reviews: Buchanan & Shortliffe(1984); Forsyth (1984); Hayes-Roth et al.(1983); Li et al. (1987); Nanninga & Davis(1984); Royer (1986); Shafer(1986).
Further geological discussion: Bichteler (1986); Henderson(1986); Kasvand(1983).
Selected geological applications: *mineralogy* (Bernhardt 1979; Smith & Leibovitz 1986; West 1985); *mineral exploration & mining* (Campbell et al. 1982; Eyre 1983; Maslyn 1986; Skelton & Franklin 1984; Smith et al. 1978); *petrology* (Bkouche-Palen 1986; Hawkes 1985; Pearce 1987; Wright & Hamilton 1978a,b); *point-counting* (Dunn et al. 1985); *X-ray diffraction* (Fabregat 1977; Glazner & MacIntyre 1979).
Applications in data-analysis: (Chambers et al. (1981).

In some respects, this topic overlaps with CAI. For example, the computer identification of igneous rocks (from modal and/or chemical analyses) can be used equally as a teaching or applied tool, depending on exactly how the system is implemented. Potentially, however, the capabilities of artificial intelligence (AI) go far beyond teaching requirements. As anyone who has seen the capabilities of modern robots or computerized chess games must know, it is now possible to computerize very complex thought or manual processes — traditionally the reserve of humans — provided they are logical; such operations can then be repeated flawlessly *ad infinitum*.

An expert system is a computer system which not only stores information but attempts to make decisions from it. It can be expressed as a simple equation:

Expert System = Knowledge Database + Inference Engine

The inference engine is a series of decision-making rules based on the knowledge base. Special computer languages (e.g. Prolog) have been developed to facilitate this process; they contrast with traditional languages like FORTRAN in that the programmer can effectively ask questions rather than simply telling the computer to execute a series of defined steps. For example, old maxims assert that gold deposits tend to occur with lamprophyres (Boyle 1979, p.250) on the one hand, or at the ends of rainbows (pots of gold) on the other. To build an expert system for gold exploration using these (and presumably other!) maxims, the knowledge base would accumulate statistics on the number of associations between gold, lamprophyres, rainbows and other relevant factors Decision rules would then be based on these statistics, leading (in highly simplified terms) to an equation of the form:

$$\textit{Probability of gold deposit} = \alpha \times \textit{rainbow factor} + \beta \times \textit{lamprophyre factor} + ..(\textit{other factors})$$

where α, β are weights and the explanatory factors may be yes/no variables (as with rainbows) or proper numbers. Presumably α would be one of the smallest weights, in this example!

Such an expert system is an evolving organism. The knowledge base would initially comprise a 'training set' of known gold deposits, with information on all factors geologists felt to the important (a 'first shot'). The inference engine would be set up in such a way that its predictive efficiency was maximised for the training set (by iterative trial and error). Gradually, the training set would be expanded, and new factors introduced as knowledge demanded (or others possibly eliminated if found unimportant), until the expert system correctly accounted for most known gold deposits. It could then be used to predict new gold deposits when provided with the appropriate information.

Geological use of AI is at present very limited. Senior management and bureaucrats often share the fears of society at large concerning computer imitations of human skills. These fears are exacerbated by the lack of experience many senior managers and academics in geological institutions still have of computers, which can lead to misinformation and misunderstanding. Hopefully, these fears will disappear with time as the next generations of geologists become more computerate. The present writer still considers these fears to have less foundation in geology than in practically any other area of human endeavour: most geological disciplines are heavily dependent on tasks which cannot be (and almost certainly never will be) executable by computers (e.g. field mapping, borehole logging, map interpretation, microscope work, hand specimen examination, report writing). Again, although *parts* of some tasks (e.g. igneous rock identification) can be performed by computers, others cannot (e.g. obtaining the original modal or chemical analysis on which the identification is based). Even the progressive automation of machine-based analytical techniques has its obvious limits.

The arguments for and against AI have attracted media attention for years, and cannot be aired in depth here. Advantages of *supplementing* the work of geologists with AI include the following:

(1) AI systems are not prone to 'off days', personal problems, sickness, divorce, etc.

(2) AI systems can engender more consistency than is usually possible within a team of geological personalities (geologists being noted for their idiosyncracies), let alone within the global tribe of the all too often mutually antagonistic specialists in many geological disciplines!

(3) Many geologists are notorious for having an encyclopædic knowledge of some subject which they never manage to put in written form; such knowledge is consequently lost on their death or retirement. AI systems can accumulate such knowledge so that it is *never* lost (in theory).

Specific geological case histories should help illustrate these advantages.

Map contouring. This is a good example of a process equally readily carried out by man or machine. Dahlberg(1975) provides a useful comparative study of the results, concluding that computers can be more efficient when information is sparse, though geologists catch up as more data are presented.

Mineral identification. Although most minerals are well-defined species which can be objectively identified, given the right kind of evidence, most forms of routine geological examination (e.g. petrographical) only furnish *part* of this evidence. Expert systems have therefore been devised which incorporate a large body of data on diagnostic features of many hundreds of minerals (refractive indices, chemical compositions, X-ray spectra, etc.) The user supplies what he knows about the mineral, and the system responds with an identification, usually in the form of a 'shortlist' of the most likely candidates. Obviously in some cases, the identification will be certain (a mineral formed of 56% CaO and 44% CO_2 can only be calcite), but some systems can also deal with more subjective factors. *MinIdent* (Smith & Leibovitz 1986), developed at the University of Alberta (Canada) is a particularly sophisticated example (Table 2.2). The user inputs as many properties of the mineral as he knows, and MinIdent then generates a list of possible minerals which most closely approximate these properties, assigning each possibility a quantitative 'similarity index' (from 100% for identical properties, downwards). The final decision, therefore, still remains with the geologist.

Rock identification. Though superficially related, this is a more subjective field than mineral identification, since rocks are not natural 'species' with fixed structure and range of composition, but arbitrary, subjective subdivisions of a natural spectrum. (This is why the history of petrology is such a miasma of contradictions!) Now that an internationally agreed igneous rock nomenclature has at last been formulated — on both modal and chemical criteria (Streckeisen 1976,1979; Le Bas et al. 1986) — expert rock identification systems have more *raison d'être*, and may have a greater chance of reducing the continual irritations of nomenclature in the petrological literature.

Tectonics: Ways of inferring the tectonic setting of ancient igneous rocks from their trace element chemistry (e.g. Pearce & Cann 1971), have attracted a growing literature. Recognising the many associated problems and uncertainties, Pearce(1987) attempted to integrate chemistry, mineralogy, regional setting and other factors into an expert system which would predict tectonic settings more reliably than chemistry alone. Each step necessitated weighting the contribution of each factor to the overall pattern: actual numbers had to be assigned, for example, to the various indicators of a subduction regime (associated trench sediments, geophysical evidence, negative Nb anomalies on 'spidergrams', etc.) Whilst this may seem curiously arbitrary, it is really no different from the mental processes a geologist normally goes through in weighing up different forms of evidence: he, too, assigns weights intuitively, save only that he does not write the numbers down. Moreover, once set up, this expert system system should allow the efficient prediction of tectonic setting to be gradually approximated in a systematic fashion, as knowledge accumulates.

Mineral exploration and mining. The location of mineral deposits still depends on a rich blend of relatively objective data interpretation, coupled with subjective 'hunches'. Increasingly, sophisticated multivariate techniques (e.g. **discriminant analysis** — §22) are being introduced to quantify as much of this interpretation as possible and make it all more objective. Although 'hunches' will probably continue to play a role well into the future, these are still based on logical thought processes which can be imitated by a computer. The more successful the 'hunch', the more logical and quantifiable is likely to have been the basis on which it was formed. Mining companies can therefore only stand to gain if they attempt to quantify these thought processes in computer programs. It is a moot point whether many of the economic applications briefly mentioned below should be regarded merely as rather complex applications of computer statistics or graphics, or as 'true' artificial intelligence — certainly, they are both assisting and mirroring human endeavour in a very real way!

At the exploration stage, computers have been used to estimate resources and identify undiscovered deposits, via quantifying variables on geological maps (Agterberg 1981; Chung & Agterberg 1980). They can also help to home in on target areas (Agterberg & David 1979; Bates 1959; Botbol et al.1978), and contribute to the more objective parts of feasibility studies (Erskine & Smith 1978). One particularly notable success was scored by the expert system PROSPECTOR (Kasvand 1983), in locating a molybdenite deposit at Mt.Tolman (New York) which had been missed by traditional methods (Campbell et al. 1982). Again, regional geochemical (e.g. stream-sediment) surveys inevitably produces enormous volumes of data which can *only* be handled by computer; an extensive literature has consequently built up around prospecting via computer analysis of such data (Howarth 1983a, 1984; Howarth & Martin 1979; Rose et al.1979; Thornton & Howarth 1986; Webb et al. 1978). Interactive computer modelling of large geochemical databases can now be highly sophisticated, and some regional geochemical maps have been produced which reproduce conventional geological maps of the area to a very remarkable degree (e.g. Plant et al. 1984, 1986). Excellent examples of the way computers can help assess geochemical data are given by Saager et al.(1972) and Saha et al.(1986): many of the later topics in this book are encompassed therein.

Once mines have actually been opened, computers can become major tools in surveying and planning. For example, computers were recently used to assist with a major problem in the world's largest iron ore mine, at Mt.Whaleback in the Pilbara region of Western Australia (*Australian Mining*, March '88, p.20-21). The mine is structurally very complex and a major fault zone was known to exist behind the 2.6 km-long north wall. It was therefore feared that continuing mining to 350 m below the natural surface might undermine this wall. Unfortunately, traditional methods of face mapping or drilling were inapplicable for various reasons, so a mechanical engineering package called MEDS was used to produce a 3-D structural model of the critical parts of the mine from existing geological and structural data (borehole logs, face maps, surface outcrops). This was then used to assess various pit wall design operations and the resultant stability of the wall, and so to find the most cost-effective means of safely recovering the maximum possible quantity of iron ore.

Another Western Australian mine, Boddington, now expected to produce 300,000 ounces of gold per year from auriferous laterites and hence already one of the three largest mines in Australia, is also using computers in a big way to overcome the problems provided by its very low ore-grade (1.9 g/t). The system erected illustrates many of the topics so far discussed in this book. A comprehensive combined database–interactive 3-D graphics system called Vulcan™ is being used to plan drilling activities, collect and evaluate geological data. Vulcan runs on a VAX 8500 minicomputer at the mine site, which has associated work-stations, plotter and digitizer, but is also linked via a small network to further hardware at a nearby bauxite mine 25 km away, and is expected to run both gold and bauxite operations in due course. Both mines have large numbers of drill-holes because of the complexity of the geology and grade estimation, and 500 samples are gathered daily from the gold operation. It is claimed that engineers can now design operational pits in < 2 days which formerly would have taken weeks or months (*Australian Mining*, November '87, p.46). Furthermore, the system integrates geological, engineering and analytical efforts, with laboratory assays in particular passed in standard AMIRA (1985) format from chemist to computer database.

Future geologists will have to look harder to find less, so combining human and computer skills will become increasingly vital (Koch 1981). It is not necessary to delve into the most arcane methods to achieve some return: Davis (1981) shows how a combination of simple computer techniques can be quite successful. Harbaugh (1981) emphasises both the challenge and need for computerate geologists to influence future policy decisions, release 'frozen' data and provide sustained information throughput, if the resource demands of future generations are to be met.

SECTION II. THE BEHAVIOUR OF NUMBERS: ELEMENTARY STATISTICS

TOPIC 6. SCALES OF MEASUREMENT AND USES OF NUMBERS IN GEOLOGY

Further geological discussion: Cheenoy(1983, p.8); Davis(1986, p.7); Till(1974, p.3).
Further non-geological discussion: Blalock(1972); Conover(1980); Siegel(1956).

Numbers in one form or another pervade everyone's daily life. Even those who have no contact with formal mathematics have to cope with simple arithmetic operations in shopping, and with a profusion of numbers imposed on them by various organisations (bank account numbers, credit card numbers, insurance policy numbers, telephone numbers, etc.) The layman develops an intuitive feeling that one can add $2.45 to $2.45, or multiply $2.45 x 2, to get $4.90, and also that one can perform similar additive and multiplicative operations on weights (kg, lbs) and distances (km, miles), but that equivalent operations on other types of numbers are meaningless (e.g. bus route no. 103 + bus route no. 10 does *not* 'equal' bus route no. 113; telephone number 123-4321 x 2 ≠ 246-8642).

In geology and other sciences, numbers are also used in several ways, mirroring everyday usage. No geologist can use numerical or computing techniques correctly or effectively until he understands these different ways, for each type of number calls for quite distinct techniques. Take the mineral headings for probably the most widely-used mineralogical textbook — 'DHZ' (Deer *et al.* 1962). Each mineral description includes at least 6 classes of numerical data, some of which (e.g. α, β, γ = refractive indices; D = specific gravities) require entirely different mathematical manipulation from others (e.g. H = Moh hardness, or $2V_\gamma$ angles). These 6 classes of data are more formally defined below; the differences between them will be referred to again and again during this book. Some classes are referred to by several different names in the literature; these are given in brackets below, after the name adopted in this course. Table 6.1 compiles common examples.

Table 6.1. Summary and geological examples of different scales of measurement

Scale	Geological examples	Permissible operands	Appropriate statistics†
Dichotomous	Presence or absence of a fossil in a rock	= ≠	Frequencies
Nominal	Rock-types coded as '1 = limestone' etc.	= ≠	Modes
Ordinal	Moh's scale of hardness	= ≠ < > ≤ ≥ « »	Median, quartiles, rank correlations
Interval	Centigrade temperatures	= ≠ < > ≤ ≥ « » + – [* +]‡	Means and standard deviations
Ratio	Chemical composition, mass, volume	= ≠ < > ≤ ≥ « » + – * +	Geometric mean, coefficient of var^n
Angular	Strikes, extinction angles	= ≠, sometimes < >	Circular or spherical statistics

† These are cumulative, i.e. appropriate statistics for each entry include all those listed higher in the table.
‡ * + under certain circumstances only.

6a. Dichotomous (binary, presence/absence, boolean, logical, yes/no) data

These constitute the lowest form of measurement, and indicate merely whether something is 'present' or 'absent' (e.g. whether or not a particular mineral occurs in a thin section). Dichotomous data can therefore take only two values, which can be represented by numbers, words or symbols. They are easily recognised by satisfying oneself that "I can represent these data fully and equally well with 0/1, true/false, pass/fail, yes/no, +/–, male/female, etc". In high-level computer languages (§1), they are called LOGICAL (FORTRAN) or *boolean* (in Pascal), and are the most economical to store.

All one can do mathematically is to compare two dichotomous values and say whether they are 'equal' or 'not equal'. Consequently, the only permissible statistics calculable from them are based on frequencies — the raw counts of numbers of objects which are 'present' or 'absent'. However, even the **modal** (most frequent) class is fairly meaningless where there are only two possible categories!

6b. Nominal (multistate, identification, categorical, grouping, coded) data

Some textbooks make no distinction between dichotomous and nominal data, but the differences not only in practical geology but in computer treatment are important. Nominal data are, in effect, dichotomous data which can take more than 2 values; they can be represented by numbers, words, symbols *or letters*, which merely differentiate one object from another. Examples include: numbers on football players' jerseys, field/museum numbers on rock specimens, and many identification numbers we are lumbered with in everyday life: driving licence numbers, telephone numbers, bank account numbers, etc. As with dichotomous data, we can only state whether 2 nominal data-values are 'equal' or 'not equal'. Two rock specimens either have the same number, or a different number. In no sense can we add 'specimen number 45' to 'specimen number 50', to get 'specimen number 95'. In addition to frequencies of objects in each category, however, the modal class now becomes meaningful.

Nominal data have an important role in information storage and retrieval. As well as being used for actual numbers (e.g. telephone numbers in a national database), they can also be used to store intrinsically non-numerical information, via *coding*. For example, in describing the borehole log of a rock succession, limestone might be represented by '1' or 'A', sandstone by '2' or 'B', and shale by '3' or 'C', so that 123321123321 or ABCCBAABCCBA represent quick and economical codings for a rhythmic sequence, which can be much more readily handled by computer than the full geological names (§16). Such codes are entirely arbitrary; shale (='3') is in no sense 3 times limestone (='1'), and the use of letters may therefore be preferable. The codes merely represent a series of mutually exclusive *states* (hence the alternative term *multistate data)*. Of course, bank account and other numbers are themselves just codes, representing the name, address etc. of the person concerned. Nominal data are easily recognised by satisfying oneself that "I can represent these data fully and equally well with 1,2,3....; I,II,III....; A,B,C.....,a,b,c.....; α,β,γ.....; or even '#', '+', '*'...".

6c. Ordinal (ranking) data

These contain more information than nominal or dichotomous data, and attempt to represent the *order* of objects or properties in a rough and ready way. One can thus speak of the 'ordinal (or ranking) *scale*' whereas 'nominal scale' (though encountered in textbooks) is somewhat self-contradictory. The best-known ordinal scale in geology is Moh's scale of hardness, H1–H10. In this, the hardness of each successive mineral exceeds the preceding one: for example, gypsum (H2) is harder than talc (H1), but is *not* twice as hard. In absolute terms, the hardness difference between H9 (corundum) and

H10 (diamond) is actually far greater than that between gypsum and talc, though both are adjacent on the Moh scale. Thus we can write H1 < H2, H10 > H9 and also, more subjectively, H10 » H1, as well as H10 ≠ H9, but we *cannot* write [H3 – H2 = H2 – H1]; mathematically, we have added '<', '>'and very probably '»', '«' to the list of permissible operands, but *not* yet *, ÷, +, or –. The *distances* (intervals) between the points on the scale have no numerical significance as yet.

Another ordinal scale is that of meteorite classes 1–6, which are based on *degrees of recrystallisation* (class '1' being least, and class '6' most recrystallised, i.e. 6 > 5 > ...1).

Ordinal scales can only be represented by numbers — not by words, letters or symbols. They can thus be recognised by satisfying oneself that "I can represent these data fully by 1,2,3...*N*, in which '*N*' >...> '3' > '2' > '1', but '3' ≠ 3 x '1' ; I cannot represent them as 'A,B,C...' or '#', '+', '*', because it is then unclear whether 'A' > 'B' or 'B' > 'A' etc." In fact, the choice of 1,2,3... *N* is made purely for convenience; one could equally well use 0,1,2... *N*, 10,20,30... *N* or even - *N*...-3,-2,-1.

Sometimes, intermediate values are assigned to ordinal scales: Deer *et al.* (1962) quote many fractional Moh values, e.g. **H** $5\frac{1}{2}$. This is merely a convenient way of representing a value *between* **H5** and **H6**, but *not* necessarily halfway. Its correct representation is certainly **H** $5\frac{1}{2}$, and *not* **H 5.5**.

Ordinal data are closely linked with the concept of *ranks,* which play an important role in this course. Ranks are simply ordinal representations of some measurement. For example, university examination marks from 1-100% are generally used to give students Class 1, 2 or 3 degrees, in which the '1', '2' and '3' are 'reversed' ranks: Class 1 is better than Class 3, but *not* 300% (or 33%!) better.

Some authors (e.g. Coxon 1982) divide ordinal data into further sub-types. On Moh's scale, for example, the *differences* between the scale points are random, measured in terms of absolute hardness; it is a **simple ordinal scale**. But if the *difference* between H10 and H9 also exceeded the *difference* between H9 and H8, and so on, this would represent a slightly higher degree of ordering (**ordered metric scale**). Many of the semi-quantitative trace element data in the geological literature of the 1950's and 1960's (some would perhaps add a good deal of more recent data also!) probably constitute ordered metric data. Thus values quoted as, for example, 'Ba 200 ppm' really only mean 'Ba is somewhere between 150 & 250ppm', i.e. that a value of 200ppm > 100ppm, but is not known to be *exactly* twice as great. More obvious characteristics of ordinal data are actually implied in, for example, published values of '> 500 ppm' or '< 10 ppm'.

A still higher ordering (**higher-ordered metric scale**) requires rank ordering between all possible pairs of points on the scale (not just adjacent ones). Richter's scale of earthquake intensities appears to be an example of this. Here, the points on the scale are defined *logarithmically*, i.e. R_2 is 10 times more intense than R_1, R_3 is 10 times more than R_2, etc., so that the *differences* $(R_n - R_{n-1})$ between successive points change, but the *ratios* (R_n/R_{n-1}) remain constant.

6d. Interval data

Here, the relative *distances* (intervals) between the scale points are now expressed exactly, i.e. their numerical values are significant; however, their *ratios* are still not meaningful. This comes about because the numbers are expressed in units with an *arbitrary zero point*, i.e. the notation '0' does not actually symbolise a zero quantity of the measured variable. The most commonly encountered interval scales in geology are the Fahrenheit and Celsius temperature scales, in which '0°C' and '0°F' are wholly arbitrary. Time and position (relative to some fixed point), but *not* length, are also interval variables, since only *intervals* are meaningful (there is no absolute zero in either time or space – theological questions excepted)! Other common interval measurements include: (1) 'net sand' thicknesses from self-potential curves of electric logs; (2) all the common descriptive isotopic parameters expressed relative to some standard, such as ∂O^{18}, ∂C^{13},ε_{Sr}, ε_{Nd}. The "shale-line" and various standards (SMOW, PDB, bulk earth, CHUR, etc.) employed here are all arbitrary zero-points.

Interval data can be recognised by satisfying oneself that both +ve and –ve values are meaningful, but that ratios are not, and that '0' does *not* mean zero. Negative °C and °F temperatures are of course regularly experienced in some parts of the world! As regards ratios, we can write '100°C > 50°C', *and* '100°C – 50°C = 50°C – 0°C = 50 *degrees Celsius'*, but *not* '100°C – 50°C = 50°C', or '50°C x 2 = 100°C'. Nevertheless, the concept of 'average temperatures' is clearly meaningful, so interval data *must* be capable of being added and divided. We have thus added '+' , '–' , 'x' and '÷' as permissible mathematical operands, but x and ÷ only under certain restrictions.

6e. Ratio data

The Kelvin temperature scale, by contrast, does have an *absolute* zero, so that 600°K *does* equal 2 x 300°K (it represents exactly twice as much energy). It is an example of the highest, **ratio** scale of measurement, in which the numerical values of the ratios between all data-values, *as well as t*heir differences, are now significant. Mass, volume, and many other everyday geological measurements, such as oxide weight percentages in rock analyses, are all measured on ratio scales. All the mathematical operands are unrestrictedly permissible on such data. Ratio data *cannot* take –ve values: one cannot have negative mass, volume, %SiO_2, or a temperature below 0°K (absolute zero).

6f. Angular (orientation) data

Measurements of dips and strikes constitute the final major geological data-type, which must be handled differently from all others (§17). Angular data are in fact *vectorial,* representing directions, whereas all 5 previous data-types are linear, representing pure numbers. Actually, angular data mimic different properties of the other scales of measurement. For example, it is clearly meaningful to speak of the 'average azimuth of kyanite porphyroblasts' on a foliation surface, so that the '+' operand must be permissible (in order to calculate an average); this is a property of **interval** and **ratio** scales alone. Yet one can only say '090° ≠ 045°', and *not* '090° > 045°' or '090° = 2 x 045°' (consider an alternative 'azimuth E–W ≠ NE–SW' is meaningful, whereas 'E-W > NE-SW' and 'E-W = 2 x NE-SW' are not); these are properties of **nominal** scales. Dip data must also be additive, since 'average dip' is meaningful. Here, however, it *is* equally meaningful to say '80° > 40° dip', but not that 80° is 'twice as steep' as 40°, because, on the same basis, 1° dip would be infinitely steeper than 0°! Dips thus mimic properties of both the **ratio** and **ordinal** scales. Note that several geomathematical textbooks (e.g. Krumbein & Graybill 1965, p.37) incorrectly describe angular data as *ratio*; in fact, the ratio of two strikes (e.g. 90°/45° or, alternatively written, E-W/NE-SW) is, as stated, clearly meaningless.

6g. Alternative ways of classifying of scales of measurement

Three other ways of classifying measurements in the literature must also be briefly considered.

(1) **Numeric versus alphanumeric:** *Numeric* data consist of numbers (1,2,3...) only, and can therefore be measured on any of the above scales. *Alphanumeric* data cover both numbers and letters (A,B,..1,2...); if used interchangeably, then these can only represent *nominal* measurements.

(2) **Qualitative, semi-quantitative** and **quantitative**: These broadly correspond to nominal, ordinal and ratio. However, the analogy is not exact; some 'semi-quantitative' data, as already mentioned, are higher forms of ordinal measurement than others. Meanwhile, qualitative data quoted as 'absent', 'trace', 'present', and 'abundant' would constitute an ordinal (rather than nominal) scale, equivalent to 0 (absent) < 1 (trace) < 2 (present) < 3 (abundant).

(3) **Continuous** versus **discontinuous (integer,counted** or **discrete):** Ratio/interval data are usefully subdivided into **continuous** measurements, which can take any numerical value (whole-number, fraction or decimal), and **discontinuous** measurements, which are restricted to whole-numbers (1,2,3..) On this basis, for example, Koch & Link (1971, p.9) distinguished 4 other scales in geology: **measurement** (corresponding to 'continuous ratio' here), **counting** ('integer ratio'), **ranking** (ordinal) and **identification** (nominal). Continuous values are stored on computers as REAL data-types, discontinuous as INTEGER, and are handled quite differently. It is very important to distinguish *intrinsically* discontinuous numbers (e.g. anything involving counting — numbers of mineral grains in point-counting), which can nevertheless be added, subtracted, etc., and thus constitute a **discrete ratio** scale, from numbers which are *arbitrarily* chosen as integers. Moh's original hardness scale, for example, comprises the integers 1-10, but is *ordinal*, not ratio, and could equally well have been represented as decimals (see above). Similarly we must distinguish fractions or decimals which imply underlying continuity of measurement on a ratio scale, from arbitrary fractions (such as the intermediate points on Moh's scale like 'H 5½') which do not. Overall, the continuous/discontinuous distinction has *no meaning* for angular data (which are always continuous), or for scales lower than interval (which are arbitrary).

Another scale occasionally encountered in geology is termed **continuous data forced into a dichotomy**: mineral optic signs '+' and '–', for example, are dichotomies derived from continuous angular (2V) measurements, '+' corresponding to $0 < 2V_\gamma < 90°$. Similarly derived, but more immediately relevant to students of this course, are examination marks: these are continuous (ratio) measurements, from 0% to 100%, forced into the fateful dichotomy of 'pass' and 'fail'!

Finally, in addition to the hierarchy of measurement scales themselves, certain geological data form hierarchies as to what can be measured and done with them. In stratigraphy, for example (Krumbein 1974), the only variables measurable on a single sediment grain are its size and shape, but as the sedimentary unit being studied expands to a hand specimen or a single bed, its composition, porosity, permeability, packing, fabric, geophysical properties can be measured, and the frequency distribution and summary statistics of grain properties can be calculated. For a Member, Formation, or a Group, the thickness and sequence of individual beds are additionally determinable, with their inter-correlations and areal trends. Finally, for the largest manageable sedimentary unit (a complete depositional basin), the area, thickness, volume, shape of the basin itself can also be measured.

TOPIC 7. SOME CRUCIAL DEFINITIONS AND DISTINCTIONS

Although most of the terms covered in this Topic are defined in the Glossary, for users' convenience, the distinctions between them are sufficiently crucial as to require further comment.

7a. Some Distinctions between Important but Vague Terms

Students are often confused by the plethora of everyday words in geomathematical and statistical textbooks which are used either with special meanings, or synonymously. Some important antitheses are explained below and in later sections. Synonyms common in the literature are given in brackets:

Variable (measurement, property, quantity) versus **Attribute**

A **variable** is anything (tangible or intangible) that varies and can be measured: length, mass, time, velocity, *pH*, %SiO_2, $\delta^{18}O$, specific gravity, etc. It is useful to distinguish variables, measured on the ordinal, interval, ratio or angular scales, from **attributes**, measured on dichotomous or nominal scales; otherwise 'variable' may refer to properties which are *not* freely variable (i.e. they can only take the values '0' or '1')!

Object (case, event, individual, specimen) versus **Entity**:

An **object** is the single basic unit of study (usually tangible) on which variables can be measured: a fossil, a lump of rock, a mineral grain, a stratigraphical horizon, the Earth, etc. Obviously, it has no inherent size connotation. Hence when we analyze an igneous rock, the rock is the *object*, and the elements SiO_2 etc. are the *variables*. When we perform morphological measurements on fossils, the fossils are the objects and the lengths, etc. are the variables.

The distinct concept of a geological **entity** is most important in the discussion of errors (§9), but is briefly introduced here for completeness. An **entity** (following Rock 1988) here covers geological objects which show *no natural variation*, that is, on which we would expect measurements of a particular variable to be *constant* (within experimental error). That is, a rock which represents a *geochronological* entity must have formed instantaneously (geologically speaking), so that measurements of its age should give the same answer. A mineral grain which represents a *compositional* entity has a constant composition throughout the grain. A given object may thus be an entity in one sense but not in another: a dyke, for example, might be a geochronological but not a compositional entity (it would be expected to vary in composition but not in age across the dyke), whereas a polyphase granite batholith would not usually be an entity in *either* sense.

Data versus **Information**

Data are raw (unprocessed) sets of measurements relating to objects, either numeric (e.g. the counts or chemical analyses from an XRF machine), or alphanumeric (the rock-names in a borehole log). **Information** refers to *processed* data (e.g. a variation diagram plotted from the chemical analyses, or a correlation chart of several borehole logs).

Sample and **Group** (class, file, set, type) versus **Population**

There are two quite distinct uses of the word **sample** in the geological literature. The normal geological 'sample' is synonymous with **object** above (e.g. *one* pebble from a beach). Statisticians, however, use 'sample' for a *set* of objects (e.g. a *collection* of pebbles from the beach). The statistician's 'sample' is, in turn, taken from a **population** (Krumbein 1960), i.e. the total number of all possible specimens in the study (e.g. *all* pebbles on the beach). Confusion arises because most geology textbooks and papers use 'sample' in the geological sense, whereas some geomathematical textbooks (e.g. Le Maitre 1982, p.2; Cheeney 1983, p.7) use it in the statistical sense. Moreover, the recommendation by Cheeney and Le Maitre of using **specimen** for the geologist's **sample**, and sample in the statistical sense, is not so good for some geological objects: a bag of stream sediment collected for a regional geochemical survey, for example, has to be called a 'specimen' on their definition, whereas geochemists would universally term it a 'stream sediment *sample*'. For these reasons, 'sample' is avoided where possible in these notes, but where not otherwise annotated it refers to a *geological* sample (*single* object): **group of data** is used for a statistical 'sample' (*many* objects).

Statistic versus **Parameter**

A *statistic* refers to something calculated from a given set of data (a mean, standard deviation, t value, etc.) A *parameter* usually refers to a property of a parent *population*, which cannot in practice be measured directly, but only estimated by a statistic. Statisticians conventionally use different symbols for these: $\overline{X}$ and s are usually used (as here) for the mean and standard deviation statistics for a group of data, whereas μ and σ refer to the corresponding population parameters.

7b. Parametric versus robust, nonparametric and distribution-free methods

Monographs and reviews — Robust Statistics: Andrews et al.(1972); Beran(1977); Hampel et al.(1986); Hoaglin et al.(1983); Hogg(1974,1977); Huber(1972,1973,1977,1981); Launer & Wilkinson (1979).

Monographs and reviews — Nonparametric Statistics: Bradley(1968); Conover (1980); Conover & Iman(1981); Gibbons (1971); Hollander & Wolfe(1973); Lehmann(1975); Meddis (1984); Savage (1962); Siegel(1956); Singer(1979); Sprent(1981).

Most geomathematical textbooks concentrate on **parametric** tests, such as the t-test and F-test (§10a). These assume that the data are distributed in a certain way (usually following a **Normal Distribution**: §8a1); that is, they assume certain *parameters* of the population are known. As shown later, real geological data are rarely as well-behaved as these tests assume, and geomathematicians are increasingly turning to the following alternative types of tests, which make fewer assumptions about the underlying distributions, or are less affected by departures from these assumptions.

Robust statistics and tests are still parametric, and usually based on the Normal Distribution, but are designed to be 'robust' to limited departures from the assumed distribution — that is, they give stable (or consistent) answers both for data which follow the distribution and for data which do not.

Parametric and robust tests are *only* applicable to data measured on an interval, ratio or angular scale.

Nonparametric *(distribution-free)* tests make no assumptions whatever about the form of the underlying distribution, and are therefore unaffected by it. They are relatively poorly covered in the geomathematical literature, but are increasingly recognised as important in geology (Brown 1985; Cain & Cain 1968; Cheeney 1983; Hall 1983). Many of them are based on ranking, and are therefore applicable to ordinal, as well as to interval, ratio or angular data (§6). A few are applicable even to nominal data. Some writers make a fine distinction between *nonparametric* and *distribution-free*, because there are one or two rather obscure tests (e.g. **Chebyshev's inequality**) which can be described as *both* parametric *and* distribution-free (see further discussion in Conover 1980, p.91-3). For our purposes, Conover's interchangeable use of these two terms is entirely appropriate.

Overall, therefore, the sensitivity of these 3 main types of tests to departures from a specified (e.g. Normal) distribution decreases in the order *parametric > robust >> nonparametric* (Fig.7.1).

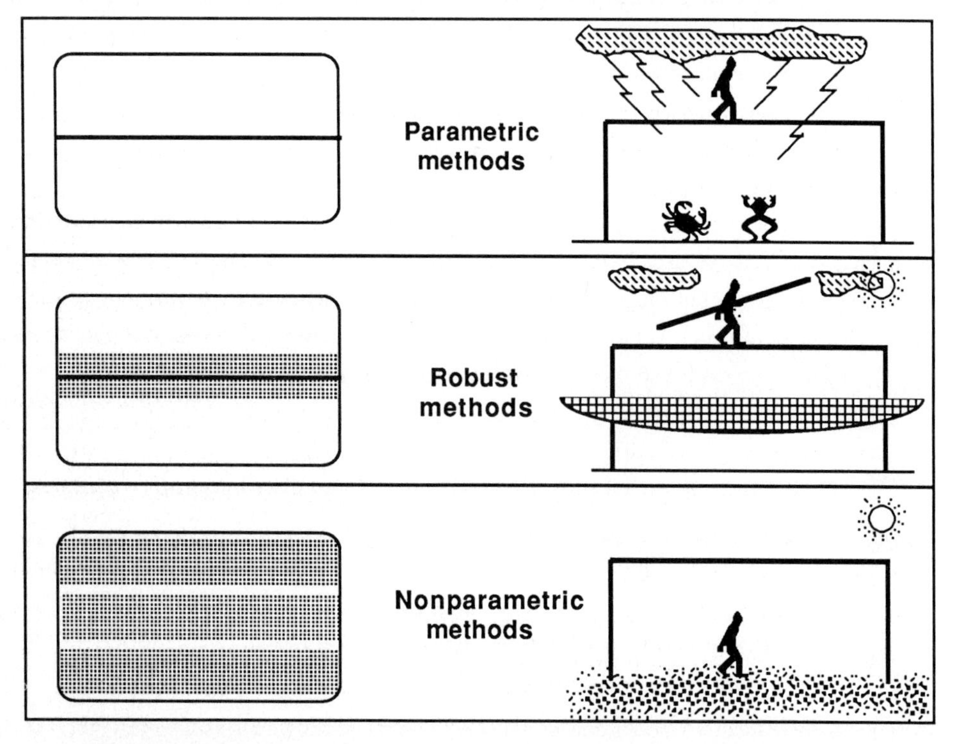

Fig.7.1. Parametric, robust and distribution-free methods — a diagrammatic illustration
Parametric statistics tolerate little or no departure from the tightrope of the Normal Distribution; although the forces pressing the geologist to violate this assumption (e.g. badly behaved data) are fairly unrelenting and inexorable, the penalties for falling off the tightrope can be severe! *Robust statistics* allow a greater margin for error, and provide a safety net; the forces arrayed against the geologist are also emasculated. *Nonparametric statistics* allow 'all' possible distributions, so the geologist no longer needs to tread the tightrope; he is walking on soft sand in the sunshine!

The structure and methodology of these Notes gives considerably more prominence to nonparametric methods than most previous geomathematical textbooks (many of which — e.g. Le Maitre 1982 — omit them completely). Reasons for this include the following:

—As shown by the increasing number of benchmark statistical texts (listed at the beginning of this section), nonparametric statistics are no less inherently 'respectable' than parametric statistics.

—Nonparametric methods are based on simple mathematical ideas, and their calculation often involves only elementary manipulation of ranked numbers, which most geologists can comprehend; parametric methods, by contrast, may involve integrals, matrices, exponents, Bessel functions, complex summations or other features which (in the author's experience) block the comprehension of the 'typical' geology student (and maybe even professional?!)

—Nonparametric methods are intrinsically more suitable for a vast range of non-Normally distributed geological data in which **outliers** are present (§9).

—Nonparametric methods may be far more cost-effective for geological data interpretation, because ranks can be far more quickly and cheaply measured than actual data-values. Take a geomorphologist interested in areas occupied by lakes, for example. *Relative* (i.e. ranked) areas could be correctly estimated from a map in a few minutes, whereas *actual* areas (in km^2) would be far more difficult. Again, take a palaeontologist wishing to compare the typical size of two fossil populations. For a parametric test, he must measure several parameters (length, breadth etc.) accurately; for a nonparametric test, he need only arrange the fossils in order of size, then rank them by eye. Finally, take a petrologist wishing to determine the volcaniclastic input into an accretionary greywacke sequence in the form of, say, detrital pyroxene grains. A parametric test would require difficult and extremely laborious point-counting of numerous thin sections; for a nonparametric test, the relative (i.e. ranked) modal quantities of pyroxene could probably be no less accurately estimated by rapid thin-section inspection. Furthermore, the most powerful nonparametric tests are often *more* efficient (see below) than the classical parametric tests, for distributions departing even slightly from Normality, and even for truly Normal data, as good a result can be obtained from a nonparametric test on 20 measurements as a parametric test on 19. Since many times more ranks could be measured in a given time, nonparametric tests may in fact be vastly more efficient, per unit cost.

Possibly the sole former disadvantage of nonparametric tests was that, in requiring ordering of data, they could be laborious to calculate by hand for large data-sets, while in computer calculations they still required storage and manipulation of large arrays. To calculate an arithmetic mean, for example, each data-value only has to be taken note of and then discarded, whereas to calculate a median, all data-values have to be stored simultaneously. Both of these disadvantages have nowadays effectively disappeared, with the prodigious increases in both storage and execution powers not only of mainframes but also of microcomputers. Furthermore, iterative procedures for calculating medians have been developed (e.g. Press et al. 1986). Microcomputer software covering nonparametric statistics is consequently now available even on 'budget'-priced machines.

Some students also complain that parametric tests are covered by very few names (*t, F),* whereas names for nonparametric tests are confusing and legion (Mann-Whitney, Kolmogorov, Klotz, Quade, etc.) For those with a phobia towards eponymous tests, however, Meddis (1984) has provided an integrated and name-free approach to the whole field of nonparametric tests. It is also important to note that nonparametric tests can be imitated in most respects by calculating parametric tests on the ranks of the data rather than on the raw data themselves (Conover & Iman 1970); for example, Spearman's ρ

nonparametric correlation coefficient is *precisely* equivalent to Pearson's *r* parametric coefficient in this way (§13).

7c. Univariate, Bivariate and Multivariate methods

These are simply methods applicable to data consisting respectively of *one*, *two* or *many* variables. Perhaps the majority of geological data are **multivariate** — that is, it requires many variables (%SiO_2, Al_2O_3 etc. for chemical analyses; length, breadth etc. in quantitative fossil morphology) to describe a single object. **Bivariate** methods (e.g. regression: §14) are commonly employed, as in the ubiquitous Harker variation diagrams in igneous petrology, but these are often mere subsets of truly multivariate data. Bivariate data are perhaps most commonly encountered in structural geology, where the variables strike and dip fully describe orientations in 3 dimensions.

Multivariate methods are the most complex, requiring matrix manipulations, and are therefore dealt with last in this course. Multivariate *nonparametric* methods are virtually unknown in the geological literature, and very poorly covered by proprietary computer packages; they are therefore only briefly mentioned here, although they would be exceptionally useful in geological interpretation.

7d. Q-mode versus R-mode Techniques

Q-mode techniques examine relationships between *objects* (in conventional computer files: rows), whereas R-mode techniques analyse relationships between *variables* (columns). Q-mode analysis of rocks, for example, would examine relations between rock-types or specimens, whereas R-mode would examine relations between oxides. A few geological papers (e.g. Jones & Facer 1982), get these terms the wrong way around.

7e. One-group, Two-group and Many-(multi-)group tests

Further geological discussion: Cheeney(1983,p.41ff)

Referred to in the literature as 'one-sample', 'two-sample' etc. (see §7a re. *sample*), these merely refer to the number of groups or sets of measured objects being compared. 'One-group' tests are sometimes interested in the internal structure of a set of data (e.g. testing for the presence of a trend), but more commonly compare the measured data-set with some absolute value or *standard*, in answer to such questions as: 'Is this distribution Normal?', 'Does the mean value of this set of data equal 1.5?' 'Two-group' and 'multi-group' tests respectively compare 2 and > 2 measured data-sets.

7f. Related (paired) and independent (unpaired) data/groups

Further non-geological discussion: Siegel (1956).

Related data are data in which some parameter is varied, while another is common or is held constant.

They are very important in sociology (e.g. measurements from twins brought up in different homes, or from 'before' and 'after' tests on the same people). Their relative rarity in geology reflects the difficulty of conducting controlled laboratory experiments in this science. They might arise, however, in experimental petrology (e.g. comparisons of glass compositions from the same sets of starting materials, before and after some common treatment), in geochemical sampling (e.g. analyses of stream sediments from exactly the same locations before and after a storm, to determine the effect of weather on composition), or in petrology (e.g. comparisons of pairs of weathered and unweathered rock samples from each of several individual outcrops, to assess the mobility of elements).

Independent data are data in which no particular factor is in common or constant. This can cause confusion. For example, analyses of different sedimentary horizons from the same formation, or of intrusive units from a single granite batholith, are independent, *not* related, because the data are in no way paired. To avoid confusion, simply ask: "Can I obtain a different *number of measurements* from each group of data?" The answer 'no' means the data must be related; 'yes' that they are independent. For example, only two measurements of the same variable can be made on a set of twins (one on each individual), whereas any number of measurements could be made on two sedimentary horizons.

7g. Terminology related to hypothesis testing

Further geological discussion: Cheeney(1983,p.33ff); Davis(1986, p.54); Koch & Link(1970,p.105); Till(1974,p.61).
Non-geological discussion: Siegel(1956,p.6).

Statistical tests are formally made via a procedure known as *hypothesis testing*, which bases decisions about populations from tests carried out on *statistical* samples. Testing involves 7 steps (NB: some texts combine the first two below into a single step, leaving only 6 steps overall):

(1) Formulate the **null hypothesis**, H_0; this is usually some equality referring to the parent population such as 'Cr content in unit A equals Cr content in unit B', or, formally expressed, H_0: $Cr_A = Cr_B$, and is generally formulated with the expectation of being rejected.

(2) Formulate the **alternative hypothesis**, H_1; this is a corresponding inequality which may take two forms: a **two-sided (tailed) test** uses 'Cr content in unit A does not equal Cr content in unit B', or H_1: $Cr_A \neq Cr_B$; while a **one-sided test** uses 'Cr content in unit A exceeds Cr content in unit B', or H_1: $Cr_A > Cr_B$ (or the converse H_1: $Cr_A < Cr_B$).

(3) Decide which statistical test (and hence test statistic) will be used.

(4) Decide how large the **critical region** will be (equivalently, what the **significance level** is); the smaller the critical region, the more statistically significant the result.

(5) Obtain from tables or theoretical equations the critical value of the appropriate test statistic corresponding to this significance level (and the appropriate **degrees of freedom** (Walker 1940) or numbers of data-values involved, where applicable). A test statistic which is significant at the 95% level (equivalently, has an associated probability of 5% or 0.05), would be expected to arise by chance alone in one out of every 20 trials of a repeated experiment.

(6) Calculate the value of the test statistic from the group of data (statistical sample).

(7) According to whether the calculated value from (5) falls within or outside the critical region from (4),that is, departs from the critical value from (5), reject or accept the null hypothesis H_0.

There are two ways of expressing the result formally. Using all 7 steps above, one might say "the null hypothesis H_0: $Cr_A = Cr_B$ is rejected (or accepted) at the 95% level". Equivalently, step 4 may be effectively bypassed, and the test statistic quoted with its associated significance level: e.g. 'test statistic = 13.95, which is significant at the 95% level (has associated probability 0.05), indicating unequal Cr contents'. This is further discussed below.

There are 4 possible outcomes from a hypothesis test (Table 7.1), two correct and two incorrect:

Table 7.1. Types of results and errors associated with hypothesis testing

Real result	H_0 accepted	H_0 rejected
H_0 is actually true	*Correct result*	Type-I (α) error (significance level)
H_0 is actually false	Type-II (ß) error	*Correct result*

The ideal is, of course, to reduce the probability of *both* α and ß errors, but these are controlled quite differently. The user sets the α level *himself* in step (4) above, so that if he rejects H_0, he also accepts that his result has an associated probability α of being wrong (1 chance in 20 for a 95% significant result). The smaller α (the higher the significance level), the less likely his result is to be wrong. On the other hand, if he accepts H_0, the associated error is *not* controlled by him, but by the statistical test itself. The more **powerful** (efficient) a test, the lower its associated ß, and the more confident a user can be in accepting H_0. The power of a test depends partly on intrinsic properties of the test itself, and partly on the data it is used to assess (in other words, it is different in every case). Some tests are readily seen to be more powerful because they use more of the available information in data. The relative power of others is only testable empirically. However, the power of *all* tests increases as: (a) the data conform more closely to any assumptions underlying the test; (b) the number of data- values, *N*, rises. Classical parametric tests are based on the properties of the Normal Distribution, and are thus most powerful when used on Normally distributed data. Although nonparametric tests are intrinsically less powerful for such data, because they lose some of its information when converting the data to ranks, geological data are rarely Normally distributed, so that nonparametric tests prove to be as powerful as parametric tests (or more so) for a wide range of geological data-sets.

As an example hypothesis test, suppose we use a one-sided Mann-Whitney test (§10d3) to assess H_0: $Cr_A = Cr_B$ against H_1: $Cr_A > Cr_B$, where data-set A has 10 measurements and B has 15. The critical region corresponds to a U test statistic $\leq$ critical values of 34 (99% level), 45 (95%) or 52 (90%). Thus if we calculate a U statistic from our data of 25 (which is < 34), we can reject H_0 at the 99% level (associated α error 0.01), whereas if U is 35, we would accept H_0 at the 99% level but reject it at the 95% level, and so on. Although there is no objective datum for the size of critical regions, most statistics textbooks adopt a convention of 95%. However, it is often better to compare the calculated test statistic with the critical value and state the result as fully as possible. For example 'H_0 accepted at the 95% level' could refer to a calculated U of 46 or 406, and is particularly misleading in the former case because H_0 is actually very close to being rejected. Far better to state here that 'the associated U statistic of 46 is significant at ≈94%', which gives the reader a far clearer idea of how firm the conclusion actually was. Perhaps the only generally accepted convention is that significance levels <90% (corresponding to more than 1 incorrect decision in 10) are not significant at all! Hence in the above example, a *U* value above 52 would almost universally be regarded as non-significant (H_0 would be accepted). The probability that this was correct would then depend on the power of the test for the particular data, which could only be estimated theoretically.

A common error in hypothesis testing is contained in a statement such as: "application of the *XYZ* test shows with 90% confidence that the two distributions are the same". This implies that accepting H_0 at the 90% level yields a corresponding 1 in 10 chance of being wrong. But if a test fails to reject H_0 at 90% then it must also fail at 95%, 99% and all higher levels — so what is really the chance of being wrong? In fact, it is the ß, *not* the α error, which determines how confident we can be in the result, once we have failed to reject H_0; the α significance level we have set ourselves and it tells us *nothing*. The correct formulation is therefore: "application of the *XYZ* tests fails at the 90% level to reject the null hypothesis that the two distributions are the same"; this leaves the ß error open.

This distinction is often crucial in geology, with the small and/or 'badly-behaved' data-sets used to test hypotheses. Suppose, at the extreme, that we are foolish enough to test whether sets of 3 Ba determinations on 2 granites: [1, 800, 900] and [1, 8, 900] come from the same population (i.e. indicate equal average Ba contents). Even though median Ba in one granite is 100 times the other (800 as against 8), all the common tests (*t*-test, Mann-Whitney etc.) will fail to reject H_0 at 90% here. In fact, rejection fails because the data-sets are probably non-normal (for the *t*-test), and are certainly too small (given their great variability) to detect any difference in Ba content. We learn little about the two granites from these tests; *least of all* can we be 90% confident that they have equal Ba contents.

7h. Stochastic versus Deterministic Models

Monographs and further geological discussion: Krumbein & Graybill(1965); Krumbein(1974); Mann(1978); Parzen(1962); Schwarzacher (1978).

Deterministic models are those in which variation of a dependent variable *Y* is *completely* controlled by one or more independent variables *X:* e.g. the potential energy of a boulder is completely determined by its elevation above datum. **Stochastic models** are those which allow for an element of random error or fluctuation: the relationship of *Y* to one or more *X*'s ranges from direct dependence to complete independence, depending on the action and intensity of the random components. Most of the techniques discussed in this book are stochastic, for many of them ask the question: "what is the probability of these geological results being produced by chance?"

When combined with the different scales of measurement in §6, the distinctions outlined in this Topic define nearly all the methods in the remainder of this book. For example, the **Kruskal-Wallis test** (§11e2) is a parametric, multi-group test for univariate, nominal to ratio data; **discriminant analysis** (§22c) is the corresponding parametric, multi-group method for multivariate, ratio data. Wherever possible, each topic includes simple tables showing how the different techniques covered in it are applicable to the different scales of measurement and circumstances contrasted above.

TOPIC 8. DESCRIBING GEOLOGICAL DATA DISTRIBUTIONS

Two types of distribution must be distinguished. 'Statistical' distribution (§8b) refers here to theoretical sampling distributions of test statistics (t, F, etc.), which do not govern real geological data as such. Real data may approximate a number of hypothetical distributions, to which we turn first.

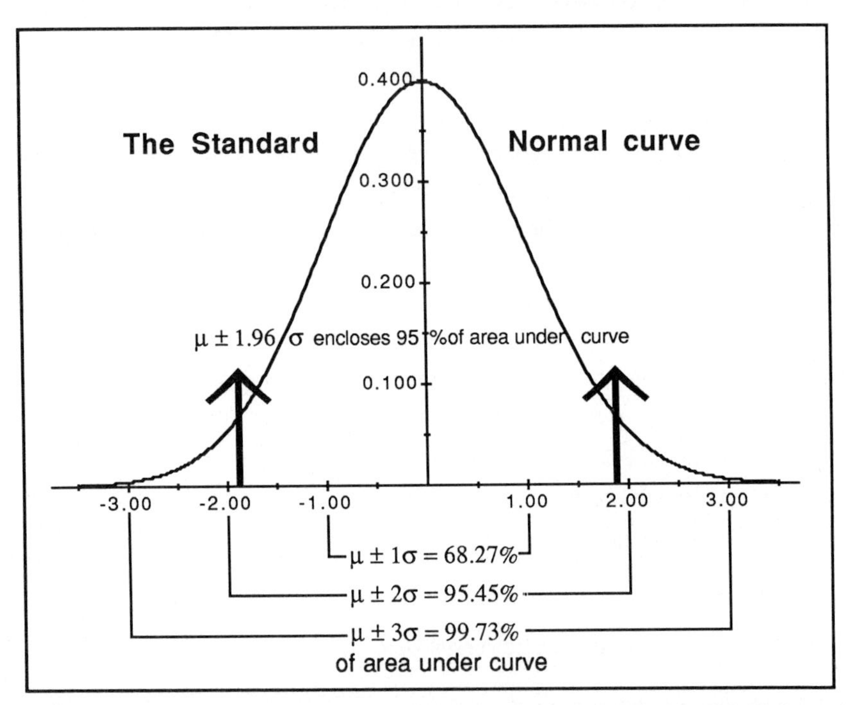

Fig.8.1. The Standard Normal distribution ($\mu = 0, \sigma = 1$). Note the symmetry and 'bell' shape.

8a. The Main Types of Hypothetical Data Distribution encountered in Geology

General monograph: Patil et al.(1975).

8a1. The Normal (Gaussian) distribution (Figs.8.1, 8.2 & 8.5a)

Dedicated monographs and reviews: Jorgensen (1982); Patel(1982).
Further geological discussion: Miller & Kahn (1962,p.33); Koch & Link (1970, p.34); Till (1974, p.30); Lewis (1977, p.239ff); Le Maitre (1982, p.8); Cheeney (1983, p.59); Davis (1986, p.44ff).
Statistical tables: see Fig.8.2 caption; also Jorgensen(1982).
Stand-alone computer programs: McCammon (1977); Press et al. (1986).
Mainframe software: distribution function and/or inverse programmed in most large statistics packages (e.g. SPSS) and libraries (e.g. NAG routines G01CEF, S15ABF, S15ACF).
Microcomputer software: in many statistics packages (e.g. Statview 512+™).

This is by far the most important theoretical distribution in statistics, and has a familiar symmetrical bell-shaped form (Fig.8.1-2). For reference only, its mathematical equation is as follows:

$$Y = \frac{1}{\sigma\sqrt{2\pi}} e^{-(X-\mu)^2/2\sigma^2} \quad \text{for } -\infty < X < +\infty \qquad [8.1]$$

This formula is characterised by two parameters — the *mean* (generally symbolised μ), measuring the position of the peak on the curve, and the *standard deviation* (σ), measuring its spread. Commonly, one summarises a quasi-Normally distributed set of data by quoting a mean and standard deviation (hereafter symbolised $\overline{X}$ and s), which are calculated by the well-known formulae:

$$\overline{X} = \frac{\sum_{i=1}^{i=N} X_i}{N} \quad [8.2a] \qquad \text{and} \qquad s = \sqrt{\frac{\sum_{i=1}^{i=N}(X_i - \overline{X})^2}{(N-1)}} \quad [8.2b]$$

where N is the number of data-values. These are only *estimates* of the actual underlying population parameters μ and s, but they become more reliable as N increases.

The *Standard Normal Distribution* has $\mu = 0$, $\sigma = 1$ (Fig.8.1). 'Standardization' refers to converting original X data-values to **standard Normal deviates (snd)**, or **Z scores**:

$$Z = \frac{X - \overline{X}}{s} \qquad [8.3]$$

The *Circular Normal distribution* is the Normal distribution as applied to angular data (strikes, azimuths). It is discussed in §17. Throughout this book, Normal, Normally, Non-Normal and other words with a capital 'N' all refer to the Normal (Gaussian) distribution, not their mundane meanings.

Normally distributed data are pervasive both in everyday life — (e.g. intelligence, heights and weights of people) — and in geology (Table 8.1). It is important to note, however, that the Normal is *not* the only bell-shaped distribution. Koch & Link(1970,p.35) show that the *Cauchy Distribution* has a very similar shape to the Normal. Many geological data-sets, moreover, are *mixtures* of several Normal Distributions with different means and/or variances (Koch & Link 1970,p.243ff).

The Normal Distribution is also crucial because of the **Central Limit Theorem** (Koch & Link 1970; Till 1974, p.53), which states that the sampling distribution of *any* statistic derived from *any* data-distribution becomes more Normal as N increases. Hence even if a geological population follows a Binomial, Poisson, LogNormal or other distribution, if we take repeated means (or medians, or standard deviations, or any other statistic) from large sets of measurements, the *means* etc. will be roughly Normally distributed. This, too, is the reason for the 'Normal approximations' given for many statistical tests in the sections which follow. Although the sampling distribution of most test statistics for small groups of data requires special tables, statistics are approximately Normal for larger groups, and hence can be converted by [8.3] into Z scores, and thence into probability levels.

8a2. The LogNormal Distribution (Figs.8.3 & 8.5b)

Dedicated monographs and reviews: Aitchison & Brown(1957);*Gold mining* (Koch & Link 1971,ch.16)
General geological discussion: Agterberg & Divi(1978), Ahrens(1954,1957,1963); Chapman (1976-7,1977); Chayes(1954c); Davis (1986, p.87); Durovic(1959), Govett et al.(1975); Jizba(1959); Le Maitre (1982, p.11); Miller & Goldberg(1955), Miller & Kahn (1962,p.33); Till (1974, p.43).
Worked geological examples: *Economic Geology* (Koch & Link 1970, p.213).
Selected geological applications: too numerous to list (especially in geochemistry and mining).
Statistical tables: use Normal Distribution tables (see Fig.8.2 caption).

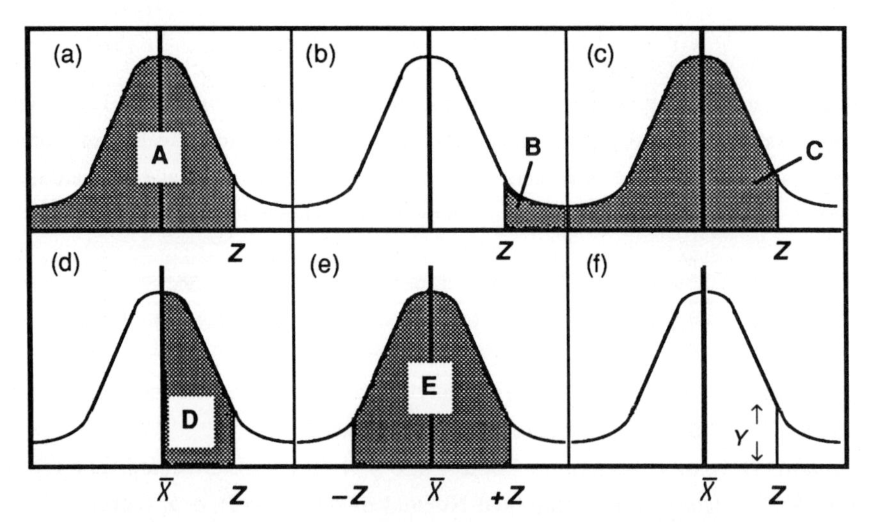

Fig.8.2.Information contained in different tables of the cumulative standardized Normal Distribution.
As this can be extremely confusing, the novice is advised to examine the table he intends to use very carefully.

(a) *Lower tail probabilities*, sometimes called *Standard Normal probabilities*, or just *the Normal Distribution Function* (e.g. Cambridge tables, p.37). These show the areas **A** under the Normal curve which lie to the *left* of a given Z value (snd). For example, 0.9772 or 97.72% of the area under the curve lies left of a value 2 standard deviations above the mean (Z = 2.0). Sometimes, only –ve Z values are tabulated (e.g. Bhattacharyya & Johnson 1977, p.597), sometimes +ve values also (e.g. Krumbein & Graybill 1965, p.415).

(b) *Upper tail probabilities.* As **(a)** but showing areas to the *right* of a given Z value, **B** (e.g. Till 1974, p.34; Le Maitre 1982, p.236; Lewis 1977,table 1; Powell 1982, p.31). For example, 0.0228 or 2.28% of the area under the curve lies above Z = 2. Most tables quote **B** for Z values between 0 and *c.* +3.

(c) *'Upper quantiles' or 'percentage points'* (e.g. Conover 1980, p.429; Koch & Link 1970, p.343; Powell 1982, p.30). The inverse of **(a)**, i.e. given **C** yields Z. Generally, the area given is that to the *left* of Z, so that for **C** = 50% (0.5), Z = 0; for **C** = 1% (0.01), Z = –2.3263; for **C** = 99% (0.99), Z = +2.3263; and in general $Z_C = -Z_{100-C}$.

(d) *Areas from* $\overline{X}$ *to* +Z (e.g. Downie & Heath 1974, p.298).

(e) *Areas between –Z and* +Z (Krumbein & Graybill 1965, p.417). These give the *central* area under the curve: thus 68.27, 95.45 and 99.73% of the curve lies within ±1, ±2 and ±3 standard deviations of the mean respectively.

(f) *Ordinates* or *probability densities* (Krumbein & Graybill 1965, p.417; Powell 1982, p.30). These give the *Y* value for a given Z, and are of relatively little use for statistical testing purposes.

Some tables (e.g. Downie & Heath 1974, p.298ff) include several of the above alternatives within the same set of tables. If a different value is required than that given in available tables, use the following relationships:

A = [1 – B]; C = A; D = [A – 0.5]; E = 2D.

The logNormal is a straightforward variant of the Normal distribution, described by the equation:

$$Y = \frac{1}{X\sigma\sqrt{2\pi}} e^{-(\ln X - \mu)^2/2\sigma^2} \quad \text{for } -\infty < X < +\infty \qquad [8.4]$$

It is distinguished graphically from the Normal distribution by its positive **skew** — that is, its right-hand side or 'tail' is more spread out than its left (Fig.8.3). It is distinguished numerically from the Normal, not least by the fact that its *geometric* (rather than arithmetic) *mean* equals its *median.*

LogNormally distributed data are also very common in geology (Table 8.1), and apply generally where most values are 'low' but there are also a relatively few 'extreme' values (as in earthquakes). There has, however, been a tendency to assume that *all* positively skewed distributions are logNormal when, in fact, they may represent overlapping mixtures of two or more Normal distributions (Govett et al.1975), or may be more closely approximated by one of the more esoteric theoretical distributions. For example, bed-thicknesses in stratigraphic successions have long been taken as logNormal, but

Table 8.1. Examples of geological data distributions

Normal	Lognormal	Poisson	Binomial/multinomial	Gamma	–ve binomial
Means of large data-sets from other distributions	Contents of trace elements in rocks	No.of ß particles emitted by a radioactive source in a given time	Probability of grain-to-grain contacts in thin sections	Thickness of some sedimentary beds	Areal distributions of mines in some regions
Pebble sphericity for fixed particle/pebble size	Production of certain types of mines	Numbers of major earthquakes in a given time interval	Geographical distribution of mines over certain areas	Sand/shale ratio of stratigraphical units	Certain mixtures of Poisson distributions
Water levels in a well through time	Sediment particle-size distributions	Numbers of meteorite FALLS (NOT finds) over a given area	Occurrence of cross-beds in sandstone	% of organic matter in some rocks	
Miles of streams per unit area of drainage basin	Heights of floodwaters in a river	Sizes of invertebrates in a "death" population (thanatocenose)	Numbers of abundant fossil specimens in rock samples of fixed	Thicknesses of some sedimentary beds	
Percentages of moisture in sediments	Gold assay values from many mines	Numbers of grains of a mineral per unit area in an isotropic rock	Numbers of grains of minor minerals in a bag of sediment, rock or section	Particle sphericity over large size-range	
Topographic relief	Permeability of some sediments	Numbers of grains of accessory minerals in sediment subsamples	Numbers of pebbles (out of say 100) of a particular rock-type from gravel		
Densities of specimens from an intrusion	Areas of river placer deposits	Numbers of particles emitted per unit time by radioactive materials			
Major elements (e.g. SiO2) in some rocks	Magnitudes of volcanic eruptions				NOTE: the relative number of examples
Lengths of fossils (and of geologists!)	Magnitudes of earthquakes				of each distribution is intended as a rough
Percentages of major minerals in rocks					indication of its geological importance

Nishiwaki(1979) indicates that Gamma or other distributions may not only give a closer fit, but also have a sounder theoretical justification. The consequences of incorrectly assuming logNormality are debated, but can be serious (cf. Link & Koch 1975; Chapman 1976–7,1977; Miesch 1976a).

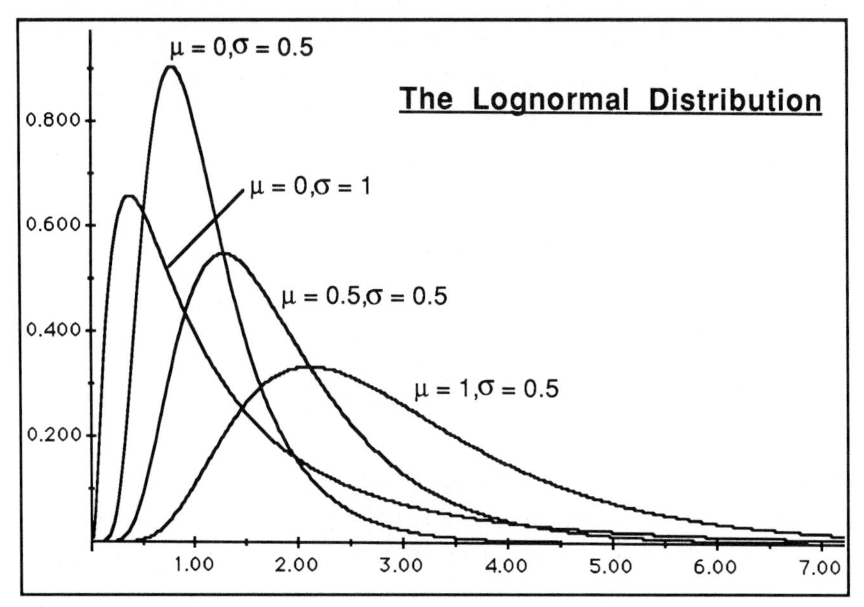

Fig.8.3. The logNormal distribution for various values of μ and σ. Note the positive skew (longer tail to the right)

8a3. The Gamma (Γ) Distribution

This is only occasionally encountered with real geological data (Table 8.1), but is involved in the definition of the main statistical distributions (§8b).We first define a **gamma function:**

$$\Gamma(y) = \int_0^\infty x^{y-1} e^{-x} dx \qquad y > 0 \qquad [8.5]$$

This has certain peculiar properties, for example $\Gamma(1) = 1$, $\Gamma(0.5) = \sqrt{\pi}$, $\Gamma(y + 1) = y\,\Gamma(y)$ and, where y is a non-negative integer, $\Gamma(y) = (y - 1)! = (y - 1)(y - 2)(y - 3)....3x2x1$. Thus it is the generalization of the factorial, $y!$, to a continuous distribution, applying for *all* real $y > 0$ rather than to integers only (1! = 1, 2! = 2, 3! = 3; 4! = 12......and so on).

The gamma *distribution* has parameters φ, λ corresponding to μ, σ for the Normal distribution:

$$y = \frac{\lambda^{\varphi}}{\Gamma(\varphi)} x^{\varphi-1} e^{-\lambda x} \qquad x > 0, \varphi > 0, \lambda > 0 \qquad [8.6]$$

8a4. The Binomial Distribution

Dedicated monographs and reviews: Molenaar(1970).

Further geological discussion: Miller & Kahn (1962,p.20); Koch & Link (1970, p.202); Till (1974, p.25); Le Maitre (1982, p.12); Cheeney (1983, p.29ff); Davis (1986, p.13ff).

Statistical tables: *Upper tail probabilities* (Powell 1982,p.14); *Lower quantiles* (Powell 1982, p.24); *Cumulative binomial probabilities* (Bhattacharyya & Johnson 1977,p.588; Lewis 1977,table 3).

Stand-alone computer programs: *Cumulative binomial probabilities* (Press et al.1986,p.169); *fitting discrete data to a binomial distribution* (Gates & Ethridge 1973).

This is best known as the distribution controlling the probability that a given number of heads will be obtained from tossing coins. Where there are only two possible outcomes to some event, A and B, it maps the probability of a given number of A's or B's. In geology it determines for example the abundance of fossils or essential minerals in some rocks (expressed as number of grains in fixed size subsamples: Table 8.1). If the probability of an event occurring is p, then the probability P of n such events occurring in r trials (assuming p remains constant all the time) is given by:

$$P = {}^{n}C_{r}(1-p)^{n-r}p^{r} \quad [8.7a]$$

$$\text{where } {}^{n}C_{r} = \frac{n!}{r!\,(n-r)!} = \frac{n(n-1)(n-2)(n-3)...(n-r+1)}{r(r-1)(r-2)(r-3).....1} \quad [8.7b]$$

Suppose the probability of discovery of a gold vein from a single drill-core is 0.4, in a certain area. The binomial probability that 3 out of 17 holes will intersect gold veins (n =17, r = 3) is then:

$$^{17}C_{3} \times (0.6)^{14} \times (0.4)^{3} = 0.034104$$

8a5. The Multinomial Distribution

Further discussion : Lewis(1977,p.64); Miller & Kahn(1962, p.26).

This extends the binomial distribution to cases where there are more than 2 possible outcomes. If an experiment can result in ν possible outcomes, with the probability of each being $p_1, p_2....p_\nu$, then the probability P that the first outcome occurs n_1 times, the second n_2 times, and so on, is:

$$P = \left[\sum_{i=1}^{i=\upsilon} n_i\right]! \prod_{i=1}^{i=\upsilon}\left\{\frac{(p_i)^{n_i}}{(n_i)!}\right\} \quad [8.8]$$

where Π represents the *product* (instead of Σ the sum) of the items in brackets. Suppose we are interested in basalt chips from a roadstone quarry, which should fall within a certain size-range. There are 3 possible outcomes of a sampling: chips too small (p_1 = 0.05), the right size (p_2 = 0.85), and too large (p_3 = 0.1). In a sample of 100 chips, the probability that 5 will be too small, 80 the correct size, and 15 too large, is therefore as follows:

$$P = 100!\,\frac{(0.05)^5}{5!} \times \frac{(0.85)^{80}}{80!} \times \frac{(0.1)^{15}}{15!}$$

8a6. The Hypergeometric Distribution

Further discussion : Davis(1986, p.19); Lewis(1977,p.67); Molenaar(1970).

This is another extension of the binomial distribution, applying where the sampling experiments are done *without replacement:* that is, an event is not allowed to recur once it has happened. It tends to the binomial distribution for infinitely large data-sets. In §8a4 we assumed that the probability of intersecting gold veins remained constant through a drilling programme, but intuitively we would expect further discoveries to be less and less likely (in a given patch of ground) as each new discovery is made (i.e. as the ground is 'bled dry'). If we have some ***independent*** estimate of the likely proportion of successful drill-holes, we can use the hypergeometric distribution to estimate the probability that specified numbers of veins will be located, if only some of the known prospects (due to budgetary restrictions) are drilled. The probability P of making n discoveries (events) in a drilling program of r holes (trials), when sampling from v prospects of which m are believed to be prospective, is in fact:

$$P = {}^{m}C_n \times {}^{v-m}C_{r-n} / {}^{v}C_r \qquad [8.9]$$

(see [8.7b] for definition of ${}^{n}C_r$). The probability of 2 out of 3 drill-holes successfully locating veins among 20 prospects, of which we independently expect 4 to be prospective, is thus:

$$P = {}^{4}C_2 \times {}^{16}C_1 / {}^{20}C_3 = 0.084$$

8a7. The Poisson Distribution

Dedicated monographs and reviews: Haight(1967); Molenaar(1970).
Further geological discussion: Miller & Kahn (1962,p.366ff); Koch & Link (1970, p.205); Till (1974, p.45); Lewis (1977, p.69ff); Le Maitre (1982, p.13); Davis (1986, p.167, 299 etc).
Statistical tables: *upper tail probabilities* (Powell 1982, p.26); *cumulative probabilities* (Lewis 1977); *general* (GEC 1962).
Stand-alone computer programs: *Cumulative Poisson probabilities* (Press et al.1986,p.165); *fitting discrete data to a Poisson distribution* (Gates & Ethridge 1973).

This is a limiting case of the binomial distribution, which again relates the probability P of n events:

$$P = \frac{m^n e^{-m}}{n!}, \text{ where m is the rate of occurrence of the event} \qquad [8.10]$$

It applies where the probability of individual events is very small, but the time (or space) for them to occur is large. (Though not intuitively obvious, it can in fact be shown (Lewis 1977, p.70-4) that [8.7a] evolves into [8.10] as $p \to 0$ and $r \to \infty$). The Poisson distribution consequently applies for example to the distribution of accessory minerals in rocks (expressed as the number of grains in fixed-size subsamples: Griffiths 1960), to the number of a particles emitted per unit time from a rock, to the sizes of invertebrate fossils in a thanatocenose ('death population'), and, possibly, to the number of meteorites falling over a given area per unit time (Table 8.1).

8a8. The Negative Binomial Distribution

Further geological discussion: Koch & Link (1970, p.208); Davis(1986, p.16).

This relates to the *inverse* (rather than negative) of the problem in §8a4, i.e. the number of trials required to obtain n successes. The probability P that exactly m failures (dry holes, barren core) will be suffered before n discoveries are made, where the individual success has probability p, is:

$$P = {}^{n+m-1}C_m (1-p)^m p^n \qquad [8.11]$$

Hence the probability that a 3-hole programme will make 2 discoveries ($m = 1; n = 2$) where p is 0.4 is:

$$P = {}^{2}C_1 \; (0.6)^1 \, (0.4)^2 = 0.0192$$

If it is of more interest to determine the likelihood of *more than m* holes having to be drilled before n discoveries are made, then the *cumulative* negative binomial distribution can be used.

Certain mixtures of Poisson distributions with different means can also generate a negative binomial, which can be an important control over areal distributions of certain types of spatial data in geology (e.g. numbers of mines per unit area), where these depart from true randomness.

8a9. How well are the hypothetical data distributions attained by real geological data?

Dedicated monographs: & reviews: Bagnold (1983); Hastings & Peacock (1975).
Further discussion: *Oil exploration* (McCray 1975);*Geochemistry* (Choubert & Zanone 1976; Tennant & White 1959; Tolstoy & Ostafiychuk 1979); *General* (Philip & Watson 1987a,b; Gubac 1986).

Many geomathematical studies are based *either* on the assumption that data follow the Normal distribution, *or* that a suitable transformation (§8f2) — commonly logarithmic — can make them Normal. The name 'Normal' of course assumes that this distribution is in some sense the 'norm'. In fact, real geological data rarely follow *any* of the above hypothetical distributions. Indeed, Bagnold (1983) summarizes a more extreme view many statisticians are moving towards, namely that "no indisputable evidence has so far been found for the existence of the...'Normal' distribution" in nature. In other words, the Normal distribution is a hypothetical, idealized concept (like a 'perfect gas' in physics, or an 'ideal solution' in chemistry), which no natural distributions actually attain. Philip & Watson (1987a) have speculated that this is because Normally distributed data can only be obtained from perfectly disordered (maximum entropy, random) systems, so that natural distributions become more Normal as they become more disordered. Meanwhile, *most* natural data distributions which might be expected to be Normal (as opposed to Poisson, binomial, etc.) are in fact **heavy-tailed** (i.e. have more numerous extreme relative to 'typical' values) than a Normal distribution. All this has very important implications for the way geologists should analyze their data statistically, as discussed later.

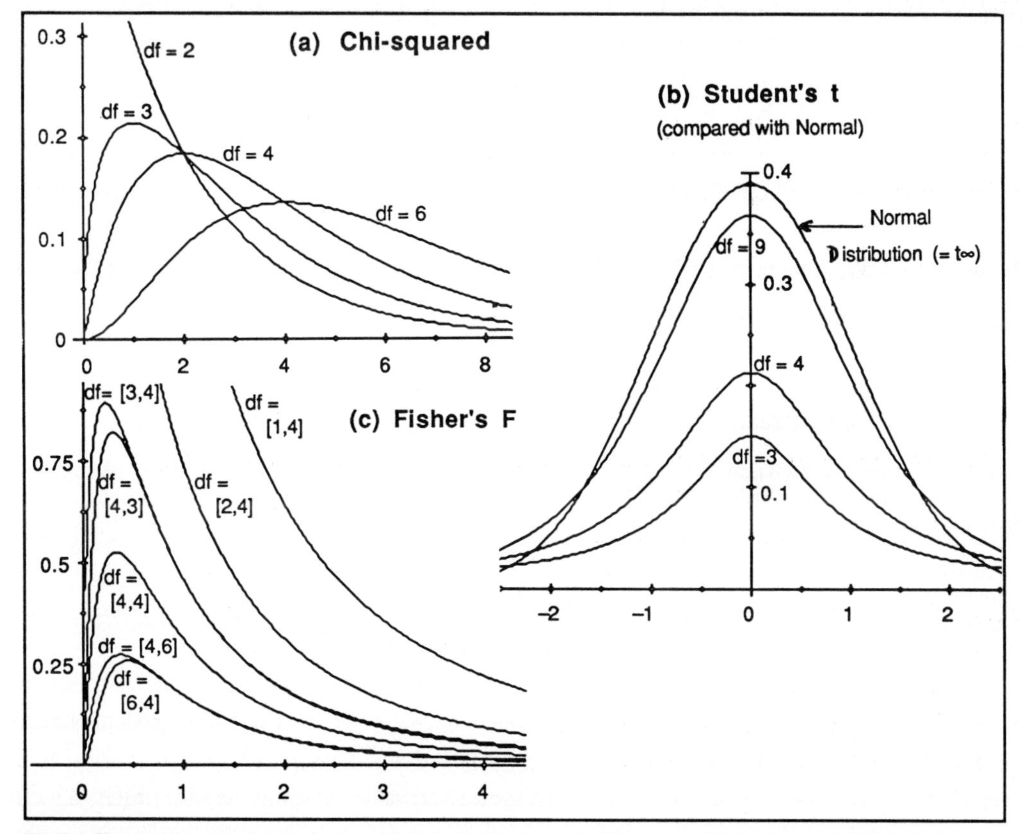

Fig.8.4.Shapes of curves for the main statistical distributions (a) Chi-square; (b) Students *t*; (c) Fisher's *F*

8b. The Main Theoretical Sampling Distributions encountered in Geology

The following statistical distributions are used widely for testing hypotheses, and will be referred to repeatedly in later sections. Their forms are shown in Fig.8.4. They are the distributions of sampling statistics (point estimators) derived from data, rather than of the actual data.

8b1. χ^2 (chi-squared, pronounced ky-squared) distribution (Fig.8.4a):

Mathematical basis of statistical distribution: Miller & Kahn (1962, pp. 60 & 458).
Geological discussion: Koch & Link(1970,p.73); Till(1974,p.68);
Statistical tables: Cheeney(1983, p.56); Krumbein & Graybill (1965, p.418); Koch & Link (1971, p.344); Le Maitre (1982, p.242); Powell(1982,p.38).
Stand-alone computer programs: Press et al.(1986,p.165).
Mainframe software: distribution function and/or inverse programmed in most large packages (eg.) and libraries (e.g. NAG routine G01BCF/G01CCF).

χ^2 is defined as a ratio of *statistical* sample to population variances:

$$\chi^2_{N-1} = \frac{(N-1)s^2}{\sigma^2} \qquad [8.12]$$

If the variances are Normally distributed, this ratio follows $\chi^2$with $(N-1)$ **degrees of freedom**, *df*. Alternatively expressed, if $Y_1, Y_2.....Y_N$ represent random samples from a standard Normal Distribution, then the sampling distribution of $[Y_1{}^2+ Y_2{}^2 +....Y_N{}^2]$ is $\chi^2{}_N$, with mean N and variance $2N$. The actual formula for the curves in Fig.8.4a (for information only) is as follows:

$$\chi^2(X)_n = \frac{1}{2^{\frac{n}{2}}\Gamma(n/2)} X^{\left[\frac{n}{2}-1\right]} e^{\frac{-X}{2}}, \quad X > 0 \qquad [8.13]$$

where $\Gamma(y)$ was defined in [8.5]. χ^2 is used to test the extent to which sets of 'observed' values differ from 'expected' values, and hence whether for example a data distribution fits a particular model (population) distribution (§8e1), whether two or more distributions differ from each other (§10d1), or whether occurrences are random or connected (§13d1). It has one associated *df*. It is the only one of the 3 sampling distributions in this section which can be used with dichotomous, ordinal or ratio data.

8b2. Student's t distribution (Fig.8.4b):

Further geological discussion: Miller & Kahn (1962,p.96); Koch & Link (1970, p.85); Till (1974, p.56); Cheeney (1983, p.66); Davis (1986, p.59ff).
Statistical tables: Beyer(1968,p.289); Bhattacharyya & Johnson(1977,p.608); Cheeney(1983,p.69); Krumbein & Graybill (1965, p.421); Koch & Link (1970, p.346); Le Maitre (1982, p.237); Powell(1982,p.39).
Stand-alone computer programs: Press et al.(1986,p.168).
Mainframe software: in large packages (e.g. SPSS) & libraries (e.g. NAG routine G01BAF/G01CAF).

This is the distribution of the mean/variance ratio, in data-sets from a Normal Distribution. Mathematically expressed, t_n is the distribution of the ratio $Z/\sqrt{\{\chi^2{}_n/n\}}$, where Z is a standard Normal variable. The integer n is the one associated *df*, and the distribution has mean 0, variance $n/(n-2)$. It is used primarily to test an observed mean against some fixed value, or against another observed mean

(§§8g,9e1,10a2). Its shape is similar to the Normal Distribution (Fig.8.1) and approaches it asymptotically as $n \rightarrow \infty$. The formula for the curves in Fig.8.4b (for information only), is as follows:

$$t(X)_n = \frac{\left[\frac{(n-1)}{2}\right]!}{\sqrt{n\pi}\left[\frac{(n-2)}{2}\right]!}\left\{1+\frac{X^2}{n}\right\}^{-(n+1)/2}, \quad -\infty < X \leq +\infty \qquad [8.14a]$$

which works where n is 2, 4, 6..... More generally, for all integer $n > 0$:

$$t(X)_n = \frac{\Gamma\left(\frac{n+1}{2}\right)}{\sqrt{n\pi}\,\Gamma\left(\frac{n}{2}\right)}\left\{1+\frac{t^2}{n}\right\}^{-(n+1)/2} \qquad [8.14b]$$

8b3. Fisher's (Snedecor's) F distribution (Fig.8.4c):

Mathematical basis of statistical distribution: Miller & Kahn (1962, p. 62).
Geological discussion: Koch & Link(1970,p.137); Till(1974,p.66)
Statistical tables: Cheeney(1983,p.105); Krumbein & Graybill (1965, p.422); Koch & Link (1970, p.348); Le Maitre (1982, p.238); Mardia & Zemroch (1978); Powell(1982,p.40); Steel & Torrie(1980,p.580).
Stand-alone computer programs: Press et al.(1986,p.169).
Mainframe software: distribution function and/or inverse programmed in most large packages (e.g. SPSS) and libraries (e.g. NAG routine G01BBF/G01CBF).

This represents the distribution of the ratio of variances of two independent data-sets from a Normal Distribution. Alternatively expressed, $F_{m,n}$ is the distribution of $(vm)/(wn)$ if v and w are independent random variables with distributions χ^2_m, χ^2_n. F has two associated *df*: one for the nominator, one for the denominator. Hence $F_{1,2}$, $F_{[1,2]}$ and $F(1,2)$ are alternative, commonly used notations for F with 1 *df* in the numerator and 2 in the denominator. F is used to compare two variances (§10a1), to test for the equality of several means simultaneously (§11b), and very widely in multivariate analysis (§18–23), to compare matrices. The actual formula for the curves in Fig.8.4c (for information only) is:

$$F(X)_{m,n} = \frac{\left[\frac{(m+n-2)}{2}\right]!\left\{\frac{m}{n}\right\}^{\frac{m}{2}} X^{\frac{(m-2)}{2}}}{\left[\frac{(m-2)}{2}\right]!\left[\frac{(n-2)}{2}\right]!\left\{1+\left[\frac{m}{n}\right]X\right\}^{\frac{(m+n)}{2}}} \qquad X > 0 \qquad [8.15]$$

As with [8.14] for Student's t, this has an equivalent form involving $\Gamma(m,n)$, which we will eschew!

8b4. Relationships between the Normal and statistical sampling distributions

Further discussion: Koch & Link (1970, p.198).

The Normal, t, F and χ^2 distributions are related in the following ways:
(1) $\chi^2/n = F_{(n-1,\infty)}$ — (i.e. bottom row of most F tables); (2) t_n is the same as $\sqrt{F_{(1,n)}}$;
(3) t_∞ is the Normal Distribution, and consequently, $F_{1,\infty}$ is the square of the Normal Distribution.

8c. Calculating summary statistics to describe real geological data distributions

8c1.Estimating averages (measures of location, centre, central tendency)

Dedicated monographs and reviews: Andrews et al.(1972); Hampel et al.(1986); Hoaglin et al.(1983).
Further geological discussion: Abbey(1983); Cheeney (1983, p.18); Davis (1986, p.28); Ellis et al.(1977); Ellis & Steele (1982); Hawkins(1981b); Lister(1982, 1984, 1985, 1986); Miller & Kahn (1962,p.68,70); Rock (1988); Rock et al.(1987); Shaw(1969); Yufa & Gurvich(1964); Zizba (1953).
Further non-geological discussion: Hoff(1973); Rocke et al.(1982); Sankar(1979); Stigler(1977).
Worked geological examples: Koch & Link (1970, p.48); Le Maitre (1982, p.20).
Selected geological applications of robust estimates: *Geochemical exploration* Garrett et al.(1980); too numerous to list for classical estimates!
Stand-alone computer programs: *Classical estimates* (Davis 1973, p.64,60,74,80; Press et al.1986,p.458); *Robust estimates* (Andrews et al.1972; Rock 1987c).
Proprietary software: ubiquitous for classical estimates (mean, median ,etc); limited selection of robust estimates programmed in BMDP and Data Desk only.

Geological data are conventionally summarised using the **arithmetic mean** [8.2a]. Sometimes the modified, **weighted mean** is calculated to take the importance of different values into account:

$$\overline{X_w} = \frac{\sum_{i=1}^{i=N} w_i X_i}{\sum_{i=1}^{i=N} w_i}, \text{ where } 0 \le w_i \le 1 \text{ is the weight for the } i\text{th value } X_i \quad [8.16]$$

In geochemistry, the **geometric mean** has also been widely used:

$$\overline{X_{geom}} = \sqrt[N]{\prod_{i=1}^{i=N} X_i} = \sqrt[N]{X_1 \times X_2 \ldots\ldots \times X_N} \quad [8.17]$$

Calculating the arithmetic mean of the *logarithms* of a data-set gives the antilog of the geometric mean. A third, **harmonic mean**, has received little attention in geology, but is useful for averaging rates:

$$\overline{X}_{harmonic} = \frac{N}{\sum_{i=1}^{i=N} \frac{1}{X_i}}, \text{ i.e. the reciprocal of the mean reciprocal} \quad [8.18]$$

Other simple estimates which can be used include the **median** (50% percentile), which divides the ordered (ranked) data-set exactly in half (i.e. it is the 3rd value in an ordered set of 5, or the average of the 3rd and 4th in a set of 6). The **mode** is simply the most frequently occurring value; there will be *more than one* mode (but only one mean or median) if the histogram or frequency distribution shows more than peak (Fig.8.7b): hence the terms **bimodal, polymodal**. The **midrange** is a third estimate, equal to $[X_{max} - X_{min}]/2$. These are compared in Fig.8.5. For *any* symmetrical distribution (including the Normal), the arithmetic mean, median and mode are identical. For a logNormal distribution, the geometric mean equals the median.

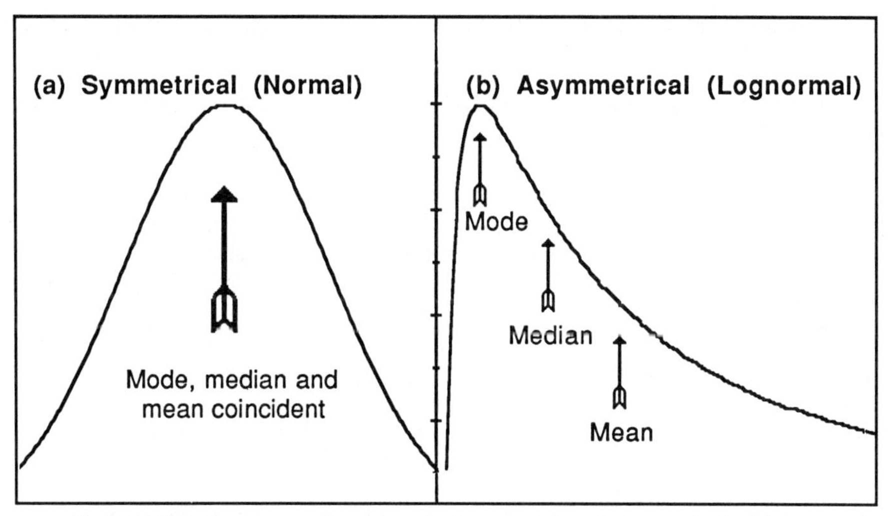

Fig.8.5. Arithmetic mean, median and mode for symmetric and asymmetric distributions

These various estimates illustrate how many different ways a data-set can be legitimately summarised. The midrange and median rely entirely on only one or two of the individual data-values: the median on the central one or two, the midrange on the outer two. Hence an infinite range of data-sets will have the same median (midrange) provided they have the same middle (extreme) values. The three means, by contrast, use all the data-values equally, so a change even in one individual value will change all three means. Given that the extreme values in most geological data-sets are the most questionable, the midrange is most likely to be 'wrong' (i.e. affected by 'bad' extreme values), the three means will be affected to some extent, but the median will be altogether unaffected. Put statistically, the median is a highly **robust** estimate, the midrange is highly **non-robust**, and the means are intermediate. As an extreme example, data-values of 2,4,5,8 and one totally erroneous value of 1000 ppm, yield a median of 5, an arithmetic mean of 204, and a midrange of 501ppm. The order of adulteration by the 'bad' value (i.e. of non-robustness) is median « arithmetic mean « midrange.

Statisticians have given much attention to the 'best' ways of estimating real data-sets, but few geologists have so far paid much attention to their work. If data are Normally or logNormally distributed, the arithmetic and geometric means are respectively the most statistically efficient ('maximum likelihood') averages. The geometric mean is superior to the arithmetic mean with near-logNormal geochemical data, but the universal taking of logarithms sometimes practiced in geochemistry by no means cures its disadvantages. The geometric mean also suffers from the disadvantage that it vanishes or becomes imaginary if zero or negative values occur in the data (and hence with all standardized data). This can be serious in the common geochemical situation where values below detection limits are present, as discussed below, and is catastrophic where real negative data-values are commonplace (e.g. δO^{18} or δC^{13} values in stable isotope geochemistry).

The problem with these conventional means is that they are *parametric,* that is, they assume the form of the underlying (population) distribution for the given data-set is known (§7) — mostly, they assume distributions are Normal. In geology, the population distribution can rarely be predicted or known on theoretical grounds, and all too commonly, none of the hypothetical distributions discussed in §8a (let alone the Normal) fit the data sufficiently well for statistical purposes. In truth, it does not *really* matter in most geological studies whether the distribution is in fact Normal, Poisson, binomial or thoroughly abnormal! In these circumstances, as emphasised by many writers (e.g. Philip & Watson 1987b; Rock et al. 1987; Rock 1988), parametric methods are highly inappropriate.

Robust averages (estimates) are a useful compromise, whose potential for geological data is only just beginning to be realised. They are less sensitive than conventional means to extreme values or to departures from Normality, but use more information in the data than the median. By objectively reducing the influence of noise in data, they provide averages which are statistically more 'efficient' — that is, both more accurate (i.e., nearer to the 'true' value) and more precise (i.e., smaller confidence intervals, §9e). These features are illustrated diagrammatically in Fig.7.1.

At least 70 *robust* location estimates have been proposed previously. Extensive tests by statisticians seem to have revealed the following as the most powerful:

Trimmed means: These are based on **censored** data, formed by trimming a pre-determined percentage of values from both ends of the data-set, and calculating the arithmetic mean of the remainder. The 0% trimmed mean is thus the arithmetic mean, and the 50% is the median; the 25% is sometimes called the **midmean**. The 10%, 20% and 25% trimmed means seem to be most favoured by statisticians —the 25% being most suitable as an omnibus measure (especially for heavy-tailed distributions), but the 20% is preferable for somewhat less heavy tails.

Combined, '*L*-estimates': These are simple linear combinations of selected trimmed/order statistics. They include the **Gastwirth median** (0.4 x median + 0.3 x each of the **tertiles**), and the **Trimean** (0.5 x median + 0.25 x each of the **hinges**, which approximate the **quartiles**).

Adaptive trimmed means: These are trimmed means whose trimming percentage is determined from the data themselves rather than arbitrarily by the user. The chosen percentage minimises the variability (variance) of the estimate itself.

One step Huber '*W*-estimates': These begin with preliminary estimates for the location (θ) and **scale** (s), and then refine them. The median, and interquartile range divided by 1.35 (its expected value for a unit Normal distribution), are used in this case for θ and s. The residuals of each X value from θ are then calculated, and all X values whose absolute residuals exceed ks ('bad' values), where k is a chosen constant, are replaced by ks x [sign of residual]; others are left alone. The estimate is then the sum of revised X values/number of unadjusted ('good') values, or the median if all values are 'bad'. Parameters for favoured examples are $k = 1.2$–2.0, θ = median and s = median deviation from the median (§8c2).

Hampel's and Andrew's *M*-estimates: These are defined as the solution M to the equation:

$$\Sigma\, \varphi\, (X_i - M)/s = 0 \qquad [8.19]$$

where, for Hampel's estimates, the function φ is a polygon defined by 3 parameters A,B,C, and for Andrew's it is the sine curve. These estimates fulfill all the basic intuitive requirements for robustness: the influence of 'wrong' data is restricted, outliers are eliminated smoothly, rounding and grouping errors are minimised, and efficiency is maximised (for Normally-distributed data) relative to conventional means. Lister (1984) gives a fully worked example of the calculations.

Huber's biweight (bisquare) estimates: These, similarly, are solutions to the equation:

$$\sum_{i=1}^{i=N}\Psi(u_i) = 0, \text{ where } \Psi(u_i) = u(1-u^2)^2 \text{ for } |u| \le 1, 0 \text{ for } |u| > 1, \text{ and } u_i = \frac{X_i - M}{c\theta} \quad [8.20]$$

θ is a scale estimate; usually the median absolute deviation from the median (§8c2) is used. These estimates reduce the influence of data-values smoothly as they diverge from the centre, to an extent determined by the 'tuning constant' c. With a Normal data-set, for example, values > $0.667c$ from the median are given zero weight. Typical values of c used in [8.20] are 6 or 9. Biweight estimates are popular among statisticians, and have good overall performance with a wide range of data-types.

Multiply skipped estimates: Here, outliers are rejected iteratively using a simple scale estimate based on the 'central portion' of the data. One 'multiply-skipped mean' takes the 'central portion' as that between two **fences** (Fig.8.9). If K is the number of outliers deleted after a single skipping at this level, a further L values are then deleted from each end of N original values, where $L = \min\{\max(1,2K); 0.6N-K\}$. The mean of the remaining values is then taken as the average.

Johns' (1974) adaptive nonparametric estimate: This is a complicated combination of two trimmed means, which attempts more than some above to estimate actual characteristics of the underlying distribution. It is only calculated for $N > 6$, otherwise degenerating to the simple median.

Shorth (shortest half) estimate: This is merely the mean of the shortest half of the data, i.e., the mean of $X_{(k)} \ldots X_{(k+N/2)}$, where k minimises the function $\{X_{(k+N/2)} - X_{(k)}\}$.

Dominant cluster mode: This involves eliminating all results > k standard deviations from the mean, recalculating the mean and standard deviation, and repeating the process iteratively with k decreasing asymptotically towards unity (from 4 at cycle 1 to 1.001 at cycle 20). Calculations proceed until remaining results are identical, or until about 5 results remain, the mean of which is taken as the average. Such estimates are widely used in software linked to mass spectrometers for the processing of peak height measurements in isotope geochemistry ($^{87}Sr/^{86}Sr$, etc.)

Faced with such an array of estimates recommended by statisticians, the following generalizations suggest how the geologist can provide the most meaningful 'averages' for any given data-set:

(1) For the small data-sets ($N \le 10$) encountered regularly in, for example, isotope geoscience, use the **median**. One of the median's few traditional disadvantages relative to the mean — that it can be expensive in terms of computer calculation (requiring the whole data-set to be stored and sorted, the number of necessary operations rising with N^2) — can now be overcome by iterative search methods (e.g. Press et al. 1986,p.459).

(2) For larger data-sets, use the **arithmetic mean** *only* if the data fit a Normal distribution with high confidence, or the **geometric mean** *only* if they fit a LogNormal distribution (for tests, see §8e3).

(3) If a larger data-set is asymmetrical, heavy- or light-tailed, or contains doubtful outlying values (for tests, see §9f3), quote *at least two* 'averages', preferably including one or more of the robust estimates, together with the arithmetic mean and median if desired.

(4) Notwithstanding (1)–(3), include some kind of **weighted mean** if data are in any way referable to *volumes of rock*, as for example in determining the average mineral content of a drill-core.

(5) For summarizing substantial quantities of data plotted as density or contoured diagrams (e.g. sandstone compositions on a ternary diagram), the **mode** has much to recommend it (Philip & Watson 1988), particularly because it identifies separate populations where other averages do not.

The importance of (4) can be seen from Fig.8.6. The correct average in fact requires the *integral* of the area under the curve in Fig.8.6a to be obtained. If the simple arithmetic mean is used on N (3) point samples along the core, the actual area obtained is as shaded in Fig.8.6b, since each sample is then given equal weight, and taken to constitute $1/_N$ (here $\frac{1}{3}$) of the core length. Assuming the mineralized zones are actually analyzed, they are greatly over-represented in the final average, giving a serious overestimate for the ore grade. On the other hand, if the trimmed mean, median or other robust estimates are used, the two mineralized zones and the zone of lowest mineral content on Fig.8.6c would be eliminated during the calculations as outliers; the accepted values would then be expanded along the whole core, to give an analogous plot to Fig.8.6b but with the average grade now severely *under*estimated. The best estimate is achieved by a weighted mean, in which the determined values at each point in the core are multiplied by the *lengths* of core over which those values are believed to apply (Fig.8.6d). Of course, the correctness of the final average does depend on taking a sufficiently large number of point samples so as to assess the fine detail of the curve in Fig.8.6a, and hence to calculate these estimated lengths correctly.

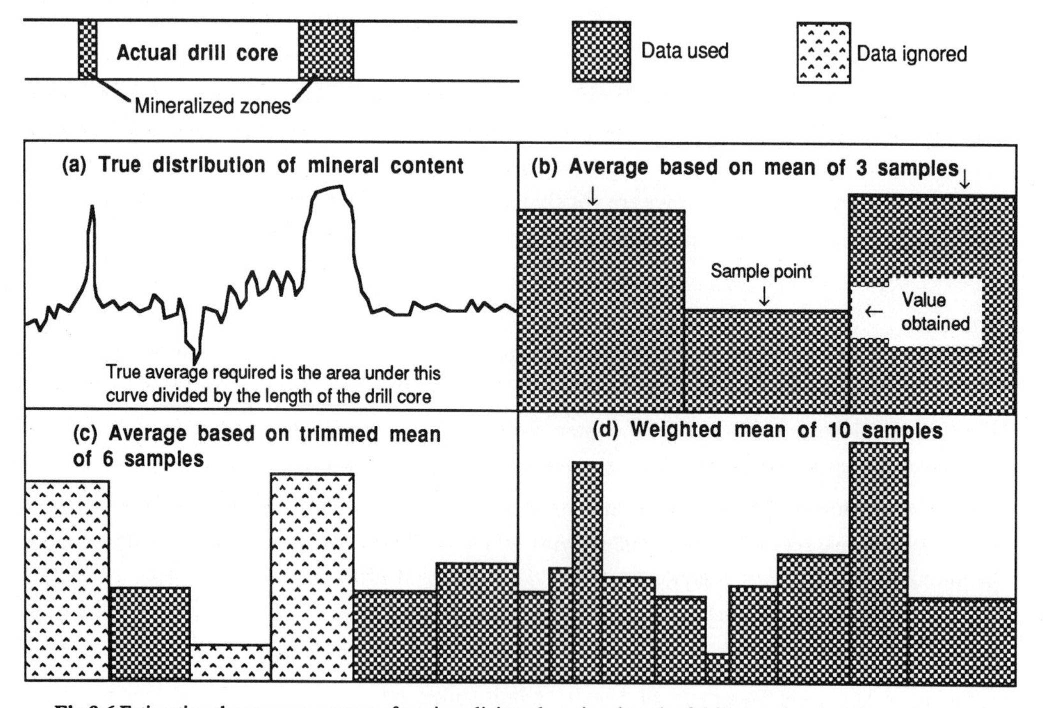

Fig.8.6.Estimating the average content of a mineralizing phase in a length of drill-core (see text for explanation)

8c2. Estimating spread (dispersion, scale, variability)

Further geological discussion: Butler(1979d); Rock (1988); Rock et al.(1987).
General non-geological descriptions: Hoaglin et al.(1983).
Worked geological examples: Koch & Link (1970, p.48ff); Le Maitre (1982, p.20); Cheeney (1983, p.20); Davis (1986, p.30ff).
Worked non-geological examples and selected applications: too numerous to list.
Software: *classical estimates* (Davis 1973, p.64,60,74,80); *exploratory and robust estimates*: (Rock 1987c; Velleman & Hoaglin 1981); *percentiles*: (Howell 1983); see also Press et al.(1986).

A recurrent problem in the literature concerns summary statistics in the form $A \pm B$, where A is an average and B a measure of variability. Howarth and Turner (1987) found that the exact nature of B is rarely specified — it may be a simple scale estimate (considered here) or a confidence interval (§9e).

The classical measure of spread, complimentary to the mean, is the **standard deviation:**

$$s = \sqrt{\frac{\sum_{i=1}^{i=N}(X_i - \overline{X})^2}{N-1}} = \sqrt{\frac{\sum_{i=1}^{i=N} X_i^2 - \frac{1}{N}\left\{\sum_{i=1}^{i=N} X_i\right\}^2}{N-1}} = \sqrt{\frac{M_2}{N-1}} \quad [8.21]$$

where $\overline{X}$ is the arithmetic mean from [8.2a] and M_p is defined as:

$$M_p = \sum_{i=1}^{i=N}(X_i - \overline{X})^p \quad [8.22]$$

The **variance** equals s^2, and the **coefficient of variation,** which allows for the absolute magnitude of the data, is $\overline{X}/s$. In geological $A \pm B$ summary statistics, B is sometimes quoted as s and sometimes as $2s$. Like the mean, s is non-robust to extreme values, so other scale estimates have increasingly been advocated by statisticians. In order of robustness, they include (1):

$$\text{The } \textbf{mean deviation from the mean} = \frac{\sum_{i=1}^{i=N}|X_i - \overline{X}|}{N} \quad [8.23]$$

(2) the **mean deviation from the median,** which replaces $\overline{X}$ in [8.23] with the median M, and (3) the **median deviation from the median** (MAD), the middle value of all the absolute deviations from the median. Note that these 3 estimates all use|*absolute*| deviations (without regard to sign).

The **interquartile range,** between the first and third **quartiles** (25th and 75% **percentiles**), and the almost equivalent **hinge-width** between the two **hinges,** are also excellent scale estimates used in boxplots (Fig.8.9). For real geological data-sets (especially small ones), MAD is often the best of the simple scale estimates. A more complex robust **biweight estimate** is also available, claimed to be more efficient over a wider range of distributions than the above. As with the corresponding location estimate (§8c1), c is typically chosen to be 9 or 6 in the formula below:

$$s_{bi} = \frac{\sqrt{N\sum_{|u_i|<1}(X_i - X_{med})^2(1-u_i^2)^4}}{\left|\sum_{|u_i|<1}(1-u_i^2)(1-5u_i^2)\right|}, \text{ where } u_i = \frac{X_i - X_{med}}{c\text{MAD}} \quad [8.24]$$

8c3. Estimating symmetry (skew) and 'peakedness' (kurtosis)

Further geological discussion: Butler (1979d); Cheeney (1983, p.21); Mardia (1970).
Stand-alone computer programs: Rock(1987c); Press et al.(1986).
Mainframe software: in many packages (e.g. SPSS) or libraries (e.g. NAG subroutines G01AAF etc.).
Microcomputer software: programmed in some statistics packages (e.g. Statview 512+™).

A number of estimates of **skew** are again available (see Glossary), but the commonest is:

$$\sqrt{b_1} = \sum_{i=1}^{i=N}(X_i - \overline{X})^3 \sqrt{\frac{N}{\left[\sum_{i=1}^{i=N}(X_i - \overline{X})^2\right]^3}} = \frac{M_3}{M_2^{1.5}} \qquad [8.25]$$

where M_2, M_3 are defined by [8.22]. Any symmetric distribution has a value of zero; $\sqrt{b_1} > 0$ indicates +ve skew (a long tail to the right — Fig.8.5b), and $\sqrt{b_1} < 0$ indicates –ve skew (long tail to the left). A LogNormal distribution (Fig.8.3) has $\sqrt{b_1} = +4$ (strong positive skew).

Kurtosis is usually estimated by the following value:

$$b_2 = \frac{N\sum_{i=1}^{i=N}(X_i - \overline{X})^4}{\left[\sum_{i=1}^{i=N}(X_i - \overline{X})^2\right]^2} = \frac{NM_4}{M_2^2} \qquad [8.26]$$

This coefficient has the value 3 for a Normal distribution, so many computer programs calculate the value $b' = [b_2 - 3]$, sometimes known as the **coefficient of excess**, instead of b_2 itself. *It is important to check this, since it is by no means obvious from some output, or even manuals,which of these two statistics is being calculated* (Rock et al. 1988)! Distributions with $b_2 > 3$ ($b' > 0$) are termed **leptokurtic**, have fewer values than the Normal distribution around the centre (and hence more in the tails), while **platykurtic** distributions have $b_2 < 3$ ($b' < 0$), and **mesokurtic** have $b_2 \approx 3$ ($b' \approx 0$). Kurtosis is always > 1 (i.e. coefficient of excess > – 2), and a logNormal distribution has $b_2 = 41$.

Skew and kurtosis are useful Normality tests (§8e3), and can also be used to detect outliers (§9f).

8d. Summarising data graphically: EXPLORATORY DATA ANALYSIS (EDA)

Dedicated monographs and reviews: Fisher(1983); Tukey (1977); Velleman & Hoaglin (1981).
General geological discussion: Howarth & Turner(1987)
Stand-alone computer programs: Velleman & Hoaglin (1981) and Press et al.(1986) both give libraries of programs and routines for various summary statistics and graphics.
Mainframe software: partially covered in some large packages (e.g. SPSS).
Microcomputer software: programmed in many packages (e.g. Statview 512+™, DataDesk, Systat).

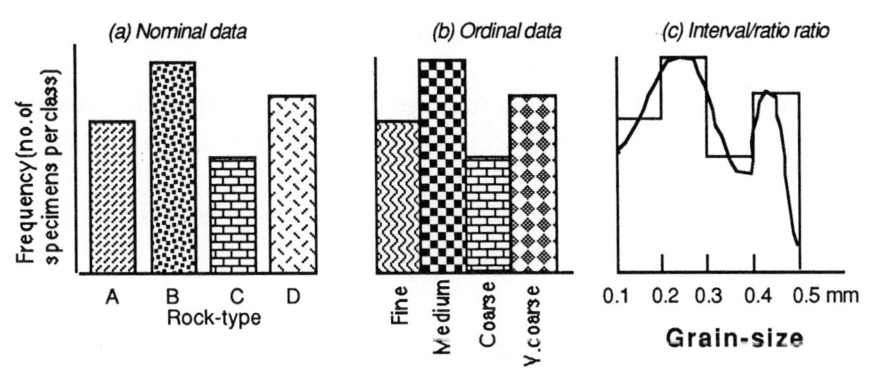

Fig.8.7. (a) Bar-chart, (b) histogram (bimodal) and (c) frequency distribution curve (step and continuous). Though rarely distinguished, these should be used for (a) nominal, (b) *at least* ordinal, and (c) interval/ratio data.

Most readers will probably be used to conventional histograms as a means of displaying data distributions. Three different kinds, dependent on the scale of measurement of the data being displayed are all too often confused in the geological literature (Fig.8.7). Histograms readily show the position of the **mode** (the longest bar), and indicate whether a distribution is bimodal or polymodal, but the positions of other summary statistics (mean, median, etc.) are by no means so obvious. Histogram shapes can also be extremely sensitive to the choice of bar width and cutoff values.

For these reasons, **stem-and-leaf** plots are preferred by some statisticians (Fig.8.8). These are essentially rotated histograms which display the actual data-values, thus having a higher, more objective, information content, and allowing the user to recast them as desired.

0	5	represents one value of 5
1	359	represents 3 values of 13, 15, 19
2	000	represents 3 values of 20
3 *H*	022455568	includes position of lower hinge plus 9 data-values
4	6688	represents 2 values of 46 and 2 of 48
5	3779	
6 *M*	2269	includes position of median
7	79	
8	488	
9 *H*	24456	includes position of upper hinge
10	17	
11	156778	
12	27	
13		
14	6	
15	3	represents a value of 153

Fig.8.8. A typical stem-and-leaf plot

A complementary form of display is the **box-and-whisker plot** (Fig.8.9). The central 'box' extends between the two **hinges**, with a dividing line at the median. The 'whiskers' extend from the hinges to the **fences**, which lie 1.5 times the **hinge-width** beyond the two hinges. Any outliers beyond the hinges are displayed individually. A boxplot can also be written numerically as a '5-number summary' (lower fence, lower hinge, median, upper hinge, upper fence) although the outliers are then ignored. Boxplots are particularly effective in revealing asymmetry and the presence of extreme values in one data-distribution, or in comparing two or more data-distributions. A more sophisticated modification is the **notched boxplot** (Fig.8.9a), in which the 'notches' indicate the width of the confidence intervals on the median (§9e). These allow an actual statistical test for the equality of two medians to be performed graphically (Fig.10.2).

Graphical methods can also be used to detect how closely a given data distribution conforms to a model distribution such as a Normal, a problem to which we now turn.

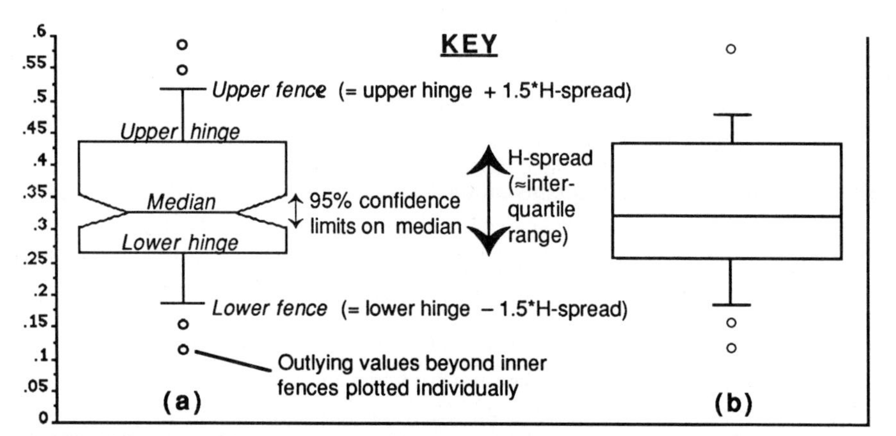

Fig.8.9.Box-and-whisker plots used in Exploratory Data Analysis to summarise distributions simply: (a) is a *notched* boxplot (McGill et al.1978), (b) is a standard boxplot (Tukey 1977)

8e. Comparing real with theoretical distributions: GOODNESS-OF-FIT TESTS

Further discussion: Miller (1978), Stephens (1974).

An oft-neglected, but vital part of statistical analysis is to determine the type of data distribution one is dealing with. Previous sections in this topic, and §7b, have emphasised that most of the classical summary statistics and tests assume Normally distributed data, and become at best unreliable if this assumption is not met. Perhaps the most common need, therefore, is to determine whether data follow a Normal or, indeed, some other defined distribution. These aspects are considered in this section. If not, one must determine whether the data are capable of being adjusted (transformed) in some way to become Normal, which is considered in the final section (§8f).

Goodness-of-fit tests are those which test to what extent a given data distribution approximates (fits) a particular model distribution — Normal, LogNormal, Poisson or whatever. The term **one-sample test** used in many statistical textbooks as an synonym often causes confusion to geologists — particularly if they thinking in terms of one *rock* sample. One-sample (**one-group** in this book) merely means that only one *measured* data distribution is being compared with a *theoretical* distribution; the latter has not, of course, been measured but calculated. Here, goodness-of-fit tests (dealt with in this section) are distinguished as those which test fit to *distributions*, and one-group tests (dealt with in §8g) as those which go deeper, testing fit to particular *parameters* of distributions (i.e.

determining whether a given data-set has a specified mean, standard deviation, etc).

The most important groups of goodness-of-fit test are dealt with in the next 3 sections. All of them are necessarily nonparametric, for a parametric test (which assumes a particular distribution from the start), to test whether a data-set fitted such a distribution, would be a circular statistical argument!

8e1. A rather crude omnibus test: χ^2 (chi-squared)

Further non-geological discussion: Miller(1978); Slakter (1965).
Worked geological examples: Miller & Kahn (1962,p.114); *Igneous Petrology*(Till 1974, p.70; Le Maitre 1982, p.27); *Metamorphic petrology:* (Cheeney 1983, p.51); *Remote sensing* (Ammar et al. 1982); *Sedimentology* (Davis 1986,p.85; Koch & Link 1970, p.212).
Selected geological applications: *Sedimentary petrology* (Griffiths 1960); *Geochemistry* (Vistelius 1960; Burrows et al. 1986); *Economic geology* (Bozdar & Kitchenham 1972).
Statistical tables: see under §8b3.
Stand-alone computer programs: Press et al.(1986,p.471); Romesburg et al. (1981).
Mainframe software: covered in most large packages (e.g. SPSS).
Microcomputer software: programmed in many statistics packages.

χ^2 has already been introduced (§8b1). *Observed* data-values O_k are broken into k classes, as in a histogram, and compared with *expected* values E_k according to the theoretical distribution, using:

$$\chi_1^2 = \sum_{i=1}^{i=k} \frac{(O_k - E_k)^2}{E_k} \qquad [8.27]$$

If the result exceeds the critical χ^2, the data are taken *not* to conform to the assumed distribution.

This is a perfectly general test, that is, it can be used to test goodness-of-fit to *any* theoretical distribution, by judicious choice of E; it is perhaps most advantageous where the theoretical distribution is uncertain. For example, in petrography, the distribution of minor and accessory minerals can follow a Normal, binomial or Poisson distribution under different circumstances. Three χ^2 tests could be readily performed using one set of O_k values and three sets of E_k values for these three theoretical distributions. Its disadvantage is its requirement to split the data up into arbitrary classes — the result is sometimes unacceptably sensitive to the class intervals chosen, especially where there are < 5 data-values in one or more classes.

8e2. A powerful omnibus test: the Kolmogorov ('One-sample Kolmogorov-Smirnov') Test

Further non-geological discussion: Abrahamson (1967); Harter (1980); Slakter (1965);
Worked geological examples: *Fieldwork* Cheeney(1983, p.42); *Sedimentology* (Davis 1986, p.101).
Statistical tables: Beyer(1968, p.428); Birnbaum (1952); Cheeney(1983,p.46); Conover(1980, table A14); Lewis (1977,table 16); Siegel(1956,table L); Steel & Torrie (1980,p.620); Smirnov (1948).
Stand-alone computer programs: Press et al.(1986,p.474).
Mainframe software: in most packages (e.g. SPSS) or libraries (e.g. NAG routine G08CAF).
Microcomputer software: programmed in many statistics packages (e.g. Statview 512+™).

This book follows Conover (1980) in adopting the terms **Kolmogorov Test** and **Smirnov Test** for the one-group and two-group versions of what some textbooks combine under the umbrella term 'Kolmogorov-Smirnov test': also affectionately known as the 'vodka test'! These tests also belong to a wider family of tests called **EDF (empirical distribution function) tests.**

The Kolmogorov Test is based on the *maximum separation, d_{max}*, between the cumulative step-functions of the data-set and compared distribution (Fig.8.10); obviously, the larger this separation, the less exact the fit, and the more likely it is that the null hypothesis (that the data-set follows the theoretical distribution) will be rejected. For small data-sets ($N \leq 40$), exact tables of critical values are consulted; note that in different tables, the critical region variously *includes* or *excludes* the tabulated value, requiring careful inspection of the 'small print' for the particular table used. For bigger N, the approximation to the critical value is $A/\sqrt{N}$ or, better still, $A/\sqrt{[N + \sqrt{(N/10)}]}$, where A increases with the significance level from 1.07 @ 90% to 1.63 @99.5% (one-sided).

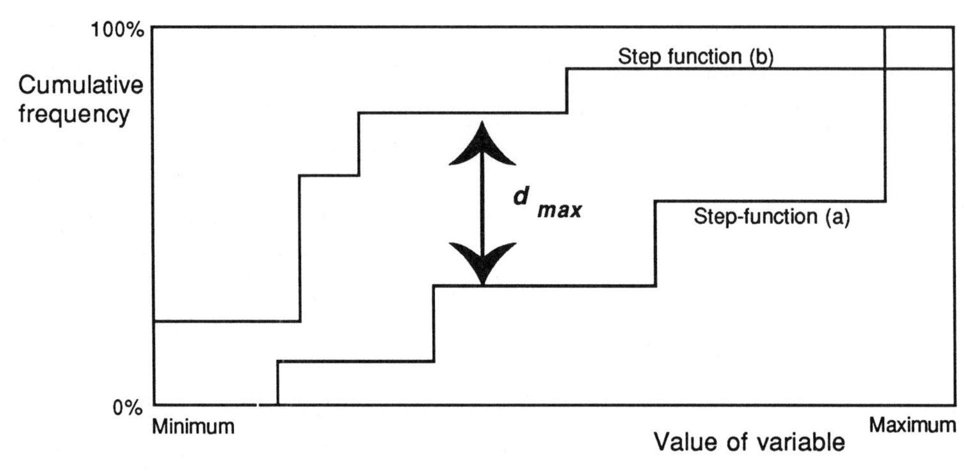

Fig.8.10. Test statistic for the Kolmogorov goodness-of-fit (one-group) and Smirnov two-group tests. Two cumulative step functions (a) and (b) are determined. For the Kolmorogov test these compare the data-set with the distribution it is being tested against; the latter may be drawn as a continuous curve instead of a step function, if preferred. For the Smirnov test (§10f2), the step functions of the two data-sets are compared. The maximum separation d_{max} between (a) and (b) is the test statistic.

8e3. Testing goodness-of-fit to a Normal Distribution: specialized NORMALITY TESTS

Further discussion: Lister(1982); Rock(1987c); Rock et al.(1987); Rock (1988); Gnanadesikan(1977, p.162ff) gives several additional tests to those listed below.
Selected geological applications: Burch(1972) and further examples for individual tests below.
Stand-alone computer programs program in Rock(1987c) implements tests b, d, e & f below; see also Dhanoa (1982); Wainwright & Gilbert (1981).
Mainframe software: wide range available in the larger statistics packages (e.g. SPSS) or libraries (e.g. NAG routine G01DDF, G01AHF); one or two sometimes covered in smaller packages.

As mentioned earlier, the χ^2 and Kolmogorov Tests can be used as Normality tests. Because of the importance of the Normality assumption in classical statistics, however, certain additional properties of the Normal Distribution have been exploited to provide a host of further tests, outlined below.

(a) **Graphical methods.** On standard *linear* graph paper, a plot of Normally distributed data in order of size yields a characteristic S-shaped curve. Departures from the S-shape can be judged with experience, and are recommended by Lister (1985). Specialized *Normal probability graph paper* is designed such that Normally distributed data yield a straight line when their relative cumulative frequencies are plotted (Koch & Link, 1970, p.237); this is even easier to judge! If this specialized paper is unavailable, a linear plot is also obtained by plotting data-values in order against their **rankits** (expected values from a Normal distribution: Powell 1982, p.34). This procedure is implemented in NAG(1987, routine G01AHF). Yet another type of paper is available for testing logNormality (Koch & Link 1970, p.238), and is used by Lepeltier(1969). Sinclair(1974,1977) provides overviews on graphical methods based on probability paper. The importance of looking at graphs of data cannot be over-emphasised; the human eye can often quickly and efficiently detect departures from Normality on a plot, which it would otherwise miss in the mass of statistics from a purely numerical calculation.

(b) **Filliben's probability plot correlation coefficient.** This is the mathematical equivalent of Normal probability paper, testing the fit of the data to the theoretical straight line by calculating a linear correlation coefficient (§13a), which should have a value of unity if the data are in fact Normal. Filliben (1975) provides full details, including tables of critical values.

(c) **The Shapiro-Wilk *W* test.** This test involves a rather cumbersome calculation procedure set out in Conover (1980, p.365). It is essentially testing the same property as Filliben's test. Computer implementations are provided by Mark (1978) and NAG(1987, routine G01DDF), and tables are available in Conover(1980,p.466) and Wetherill(1981,p.378). The formula for the test statistic W involves the *ordered* X observations:

$$W = \frac{\left[\sum_{i=1}^{i=N} w_i X_i\right]^2}{\sum_{i=1}^{i=N}(X_i - \overline{X})^2} \qquad [8.28]$$

where $X_1 < X_2 ... X_N$ and the w_i are weighting coefficients (obtained from large tables) calculated from the Normal distribution itself; (formally, the w_i are related to expected values of standard Normal order statistics, while the numerator of W is the squared best linear unbiased estimate of the standard deviation for a Normal data-set, and its denominator is merely $N-1$ times the variance: cf. [8.2b]). Small W corresponds to non-Normal data.

(d) **Skew ($\sqrt{b_1}$) and kurtosis (b_2).** For a Normal distribution these equal 0 and 3 respectively (§8c3). Both are convertible into standard Normal deviates (Pearson 1936) to provide Normality tests

$$Z = \sqrt{b_1}\sqrt{\frac{(N+1)(N+3)}{6(N-2)}} \approx \sqrt{b_1}\sqrt{\frac{N}{6}} \text{ for large } N; \quad Z \approx (b_2 - 3)\sqrt{\frac{N}{24}} \qquad [8.29]$$

The terms in $\sqrt{N}$ are the standard deviations of skew and kurtosis, used in the $Z = (X - \overline{X})/s$ formula [8.3], with $\overline{X}$ equalling 0 and 3 respectively. Equations [8.29] work well for skew, but the distribution of kurtosis is complex (D'Agostino & Tietjen 1971), making it work well only for $N >$ 1000. Rather incomplete tables for both parameters are alternatively available (e.g. Lister 1982). Although skew ≈ 0 or kurtosis ≈ 3 *individually* are insufficient proof that a distribution is Normal (many other symmetrical or mesokurtic distributions exist), the two together provide a fairly sensitive Normality test (Shapiro et al.1968), and can in fact be combined into a single χ^2 test

(Gnanadesikan 1977, p.164). Malmgren (1979) uses both values to test some palaeontological data.

(e) **Geary's A ratio.** Geary (1935, 1936) showed that the following ratio can be used:

$$a = \frac{\sum_{i=1}^{i=N} |X_i - \overline{X}|}{\sqrt{N \sum_{i=1}^{i=N} (X_i - \overline{X})^2}} = \frac{\text{mean deviation}}{\text{standard deviation}} \qquad [8.30]$$

The value for a Normal Distribution is $\sqrt{3/\pi} = 0.7979$. Critical values are tabulated in Lister(1982).

(f) **Test based on biweight scale estimator** (§8c2). The ratio (Hoaglin et al. 1983, p.425-6):

$$I = \frac{\sum_{i=1}^{i=N} (X_i - X_{med})^2}{(N-1)s_{bi}^2} \qquad [8.31]$$

is particularly powerful in detecting heavy-tailed symmetric distributions, but performs well for a wide variety of distributions ($I = 1$ indicates Normality). Critical 90% and 95% values are given by:

$$I_{crit\ 90\%} = 0.6376 - 1.1535\tilde{N} + 0.1266\tilde{N}^2 \qquad [8.32a]$$

$$I_{crit\ 95\%} = 1.9065 - 2.5465\tilde{N} + 0.5652\tilde{N}^2 \text{ for } N < 50 \qquad [8.32b]$$

$$I_{crit\ 95\%} = 0.7824 - 1.1021\tilde{N} + 0.1021\tilde{N}^2 \text{ for } N \geq 50 \qquad [8.32c]$$

where $\tilde{N} = \log_{10}(N-1)$. Direct calculation like this is advantageous for computer implementation.

(g) **Lilliefors' test.** This is a variant of the Kolmogorov Test (§8e2). It uses the same method for determining d_{max} (Fig.8.10), but the data are first **standardized** after calculating the mean $\overline{X}$ and standard deviation s according to equations [8.2a,b]; the maximum separation is then found between the cumulative frequency step-function for the *standardized* data, and the curve for a Normal distribution with the same $\overline{X}$ and s. Cheeney(1983, p.63-4) and Lewis(1977, p.220) give worked examples, although only Lewis names the test as such. Lilliefors' test has its own tables of critical values for $N \leq 30$ (e.g. Conover 1980, p.463), but a similar large-N approximation $A'/\sqrt{N}$ to the Kolmogorov test, where A varies from 0.9773 @ 90% to 1.2743 @ 99%.

(h) **Coefficient of variation.** Koch & Link(1970,p.296) show that, at least for compositional (e.g. major and trace element) data which total 100%, the coefficient of variation (mean/standard deviation) provides a rough Normality guide. According to them, values > 0.5, which typically occur for minor and trace elements, indicate that Normality is unlikely or impossible, whereas values < 0.5 indicate Normality is possible (though not necessary). Shaw(1961) prefers a cutoff of 0.2, suggesting that logNormality is more likely than Normality where coefficients are higher than this. Size(1987b) recommends this statistic in a number of other contexts, as well as Normality testing.

The bemused geologist is entitled to ask which of this array of Normality tests he should use. Many statisticians have asked the same question, and provided detailed studies on their relative power-efficiency. The consensus appears to be that the Shapiro-Wilk test is probably the most powerful under the widest range of circumstances, although without a computer program its execution is time-consuming and error-prone. The test does however have the valuable advantage for geological data that if various data-sets are too small to test for Normality individually, they can be combined into a single Shapiro-Wilk test (Conover 1980, p.365); none of the other tests offer this option. Meanwhile, the Kolmogorov test has much to recommend it for simplicity of calculation, availability of computer programs, and power. At the other end of the scale, χ^2 appears to be one of the *least* effective. The skew and kurtosis tests are simple and, when combined, quite effective (Shapiro et al.1968), but they can only show that a data-set has 'Normal' skew and kurtosis (rather than overall shape); a whole range of non-Normal, but symmetrical and mesokurtic distributions (e.g. those with longer or shorter tails than a Normal Distribution) could thus, misleadingly, 'pass' these tests.

8f. Dealing with non-Normal distributions

As indicated in §7b, many of the classical statistical methods mentioned throughout this book are *parametric*, assuming that data are indeed Normally distributed, whereas Normal distributions never actually occur with real geological data. What, therefore, are geologists to do where this assumption is not met? All too commonly the response of geologists has been to ignore this problem. (We will encounter the same nonchalance again in the problem of data-closure: §15). It is merely pleaded that parametric tests are **robust** — that is, not too seriously affected by departures from the Normality assumption. Koch & Link(1970, p.297) justify this with stance with an empirical rule that no action need be taken unless the coefficient of variation exceeds 1.2. However, strong statements *against* such masterly inactivity can be easily found in the statistical literature. Four main alternative courses of action are available for tackling the problem properly and more effectively.

8f1. Use nonparametric methods, which are independent of the Normality assumption

As elaborated in §7b, nonparametric statistical techniques make few or no assumptions about the form of the population distribution, and are thus ideally suited to data whose distribution form is either unknown, or known not to be Normal. This approach is illustrated *throughout* this book by detailing parametric and nonparametric tests side-by-side for each technique. Since most nonparametric methods involving transforming raw data into ranks, they are effectively a variant on the next approach (c).

8f2. Transform the data to approximate Normality more closely

Further geological discussion: Chayes (1972,1983a); Howarth (1982); Koch & Link(1970,p.231); Le Maitre(1982,p.14).
Selected geological applications: *Geochemistry* (Christie & Alfson 1977; Mancey & Howarth 1980).
Stand-alone computer programs: Chung et al. (1977); Howarth & Earle(1979); Lindquist(1976).
Proprietary software: covered in most packages (e.g. SPSS, Statview 512+, Systat).

As nonparametric versions of the more complex statistical techniques are either not yet available, or not yet widely implemented on computers, geochemists often assume than their data are *logNormally* distributed, and indeed, logNormal distributions (§8a2) are commonly encountered for trace elements in stream sediment samples, ores and rocks (Table 8.1). A logNormal distribution, which shows +ve skew, can be confirmed by carrying out two Normality tests: one on the *raw* data, which will lead to rejection of the null hypothesis [H_0] of Normality, and the other on the *log-transformed* data, i.e. the logarithms of the data, which will lead to acceptance of H_0. A LogNormal Distribution can also be quickly suspected by comparing location estimates: for the raw data, the *median* will approximate the *geometric* mean; after log-transformation, it will equal the *arithmetic* mean; this is because the geometric mean is equivalent to the arithmetic mean calculated on the logarithms of the data (§8c1). Data identified as logNormally distributed can thus be analyzed by all the classical methods provided they are transformed to logarithms first; logarithms to base *e* or base 10 are equally suitable.

However, assuming universal logNormality can be as dangerous as assuming universal Normality. Lindqvist (1976) suggests a <u>SEL</u>ective <u>LO</u>garithmic (SELLO) transformation of the form:

$$y = \log (a + x) \qquad [8.33]$$

SELLO allows for variables which show +ve skew both before *and* after simple log-transformation — they will more closely approximate Normality with $a < 0$; while those which show +ve skew before and –ve skew after simple log-transformation will be Normalized with $a > 0$. Data which show –ve skew before transformation will of course be made even worse by log-transformation; therefore, they are first *mirror-transformed* by the equation:

$$V = C - X \qquad [8.34]$$

where C is an arbitrary constant $> X_{max}$. Variable V can then be treated by [8.33].

Other geological data-sets may approximate Normal Distributions only after taking square-roots (√), exponents (e^x), or reciprocals ($1/X$). Because transformation possibilities are legion, a **generalized power transform** (the **Box-Cox transform**) is very popular; it takes the form:

$$Y = \frac{X^\lambda - 1}{\lambda} \qquad [8.35]$$

This already includes the square-root ($\lambda = 0.5$), reciprocal ($\lambda = -1$) and logarithmic ($\lambda = 0$) transformations, as well as squares, cubes, etc.....

<u>8f3.Separate the distribution into its component parts</u>

Further geological discussion: M.W. Clark(1976); Lepeltier(1969); Le Maitre (1982,p.14).
Worked geological example: *Mining:* (Koch & Link 1970,p.243).
Selected geological applications: *Geochemistry & mineral exploration:* (Christie & Alfson 1977; Govett et al. 1975; Mancey & Howarth 1980; Otsu et al.(1985).
Stand-alone computer programs: Bridges & McCammon(1980), I.Clark(1977b), M.W.Clark (1977,1984); Howarth & Earle(1979); Hsu et al.(1986); Mundry (1972).

Many geological distributions fail to respond to transformation because they are in fact *mixtures* of several (Normal and/or logNormal) distributions. As elaborated in §9, this arises because it is often impossible to sample from one parent geological population at a time. The presence of two or more **modes (bimodality, polymodality)** on the histogram of a distribution is a sure sign that several distributions are obfuscated within the one data-set. Numerical techniques for separating mixed distributions include the **method of mixtures** and **least-squares decomposition**. Graphical methods are also available (e.g. Clark & Garnett 1974; Sinclair 1974, 1977).

8g. Testing whether a data-set has particular parameters: ONE-SAMPLE TESTS

Often, researchers may be interested not in whether a data-set is Normally distributed, but whether it has a specified mean, standard deviation, etc.; i.e. they wish to test not merely *distributions* but *parameters* of distributions. One-group tests do this, but presuppose that the researcher has already tested goodness-of-fit to the appropriate distribution. For example, it is of little use to test whether a data-set has a mean of 4.5 and a standard deviation of 2.3, if the data-set is not Normally distributed in the first place: this should be examined *first* using the tests in §8e3.

8g1.Testing against a population mean μ (population standard deviation σ known): the *zM* test

Further non-geological discussion and worked examples: Langley(1970).
Statistical tables: uses Standard Normal tables (see Fig.8.2 caption).
Computer software: none known.

This test is only applicable where the population against which a given data-set is being compared, is *completely* defined by its mean μ and standard deviation, σ. The test statistic is then simply:

$$Z = \frac{\overline{X} - \mu}{\sigma/\sqrt{N}} \qquad [8.36]$$

This is merely the standard Z formula [8.3] with $\overline{X}$ replacing X, μ replacing $\overline{X}$ and $\sigma/\sqrt{N}$ replacing s.

Suppose 9 fossils yield a mean length of 4.5 cm. Could they have come from a Normal fossil population known to have a mean length 5.0, with standard deviation 0.5 cm? The resulting Z is [4.5 – 5.0]/[0.5/3] = –3, which is almost significant at the 99.9% level. H_0 is therefore rejected: the set does *not* come from the specified population. This is a slightly surprising result in the sense that $\overline{X}$ is only 1 standard deviation from μ, and illustrates that intuitive conclusions are likely to be incorrect!

8g2.Testing against a population mean μ (population standard deviation σ unknown): one-group *t*-test

Further non-geological discussion: Blalock(1972, p.192).
Worked examples: *Geophysics* (Miller & Kahn 1962, p.97); *Sedimentology* (Davis 1986, p.61).
Statistical tables: uses *t* tables (§8b2).
Computer software: usually programmed in both small and large statistical packages.

Where σ is unknown, we substitute t for Z and the unbiased scale estimate s for σ in [8.36]:

$$t_{N-1} = \frac{\overline{X} - \mu}{s/\sqrt{N-1}} \qquad [8.37]$$

With the same example as in §8g1, suppose the fossil population is now only known by a far larger collection (but still a statistical sample rather than the true universal population), but the measurements are the same. We now obtain $t_8 = 0.5/[0.5/\sqrt{8}] = -2.828$. If H_0 was that $\overline{X} < \mu$ (a one-tailed test), we would reject H_0 at the 99% level ($t_{8,99} = 2.896$), but if H_0 was that $\overline{X} \neq \mu$ we could only reject at 98%. The confidence of rejection is less than in the *zM* test example because we have less precise information about the population (i.e. we are estimating it rather than knowing it beforehand).

This test is intimately related to the concept of **confidence intervals** (discussed fully in §9e), and can alternatively be performed by finding whether $\overline{X}$ lies within the confidence interval for μ.

TOPIC 9. ASSESSING VARIABILITY, ERRORS AND EXTREMES IN GEOLOGICAL DATA: SAMPLING, PRECISION & ACCURACY

9a. Problems of Acquiring Geological Data: Experimental Design and other Dreams

Dedicated monographs & reviews: Duckworth(1968); Hicks(1973).
Further geological discussion: Flinn(1959), Krumbein & Miller(1953); Philip & Watson (1987b).

Statistical textbooks aimed at life and physical scientists (biologists, chemists, etc.), sociologists and psychologists usually include extensive discussion on **experimental design,** that is, the different ways of setting up laboratory or outdoor studies to achieve particular ends. Such scientists are generally able to plan their experiments, choose their subjects carefully, operate controls (blanks, placebos, twins and the like), sample as widely as they wish, and repeat their results at will if they at first prove unsatisfactory. Geologists, by contrast, have few of these advantages, and many of the principles of experimental design are quite simply impractical for them. Instead of, say, choosing a litter of rabbit twins to subject to some variable stimulus (where the zoologist has controlled the parentage, the number of subjects, and the stimuli) the geologist can only sample what Mother Nature has provided: namely, outcrops or (if available) underground exposures. Whereas the zoologist can obtain another litter of rabbits cheaply, should his first one proves troublesome, the geologist who wishes to study a formation which does not outcrop, can do very little to change the situation unless finance is unlimited. The geologist is thus a far more 'passive' experimenter than most other scientists; overall, the situation can be summed up simply by saying that geology is *not* an experimental science (as witness the non-existence of the phrase "experimental geology"). Even experimental petrology, one of the few branches of geology where subjects (rocks) are indeed chosen and experimented with (melted) under varying laboratory conditions (temperature, pressure and composition), is quite different from, say, experimental psychology, because the results do not usually require interpretation by statistical means, and the principles of experimental design are simply not relevant to them.

9b. Sources of Variability & Error in Geological Data, and the Concept of 'Entities'

Dedicated monographs and reviews: Clifford(1973); Jaech (1985).
Further geological discussion: Aitchison & Shen(1984); Clanton & Fletcher(1976); Dennison (1962, 1972); Engels & Ingamells (1970); Garrett (1969); Jaffrezic et al.(1978), C.J.James(1970); Kleeman (1967); Koch & Link (1970, ch.8); Kretz(1985); Le Maitre (1982, p.1 & ch.11); McGee & Johnson(1979); Miesch(1967); Rock et al.(1987); Rock(1988);Thompson & Howarth (197 8); Till (1974, p.75ff); Vasilenko et al.(1978).
Stand-alone computer programs: *error-checking of igneous rock analyses:* (Stewart et al.1983).

Relationships between three everyday terms as used here — *error, mistakes* and *variability* — must be clarified from the outset. Geologists undertake sampling and analysis programmes in order to assess the *natural variability* of rock units. However, the various procedures they have to go through between, say, granite batholith and granite whole-rock analyses, introduce multi-source *errors*, so that the variability of a final set of analyses is substantially more than that of the parent batholith. 'Error' does not simply mean *mistakes* (e.g. samples mixed up, numbers wrongly read or typed in, flux incorrectly mixed with rock powder) but also covers unavoidable errors (e.g. available outcrop not representative of batholith, random statistical 'spikes' in XRF machine). Statistically, errors (whether mistakes or not) merely introduce further *variability* into data, so the two terms error and variability are here more-or-less synonymous. Errors are thus usually expressed by the conventional estimate of variability, the standard deviation, s, or some function of it such as a confidence interval (§9e).

It is important to distinguish **accuracy** (systematic error), namely how close measured values are to 'true' values, from **precision** (random error), namely how much repeated measurements vary.

Although more specific types of errors can be (and have been) classified in multitudinous different ways, the simple framework in Fig.9.1 sets out most possible sources of error in relation to the everyday production of a whole-rock analysis from a geological specimen on a machine such as an XRF. It is firstly useful to distinguish three main, composite sources of variation as follow, which together generate **total error** (*TE*) — the overall variability in the final data-set.

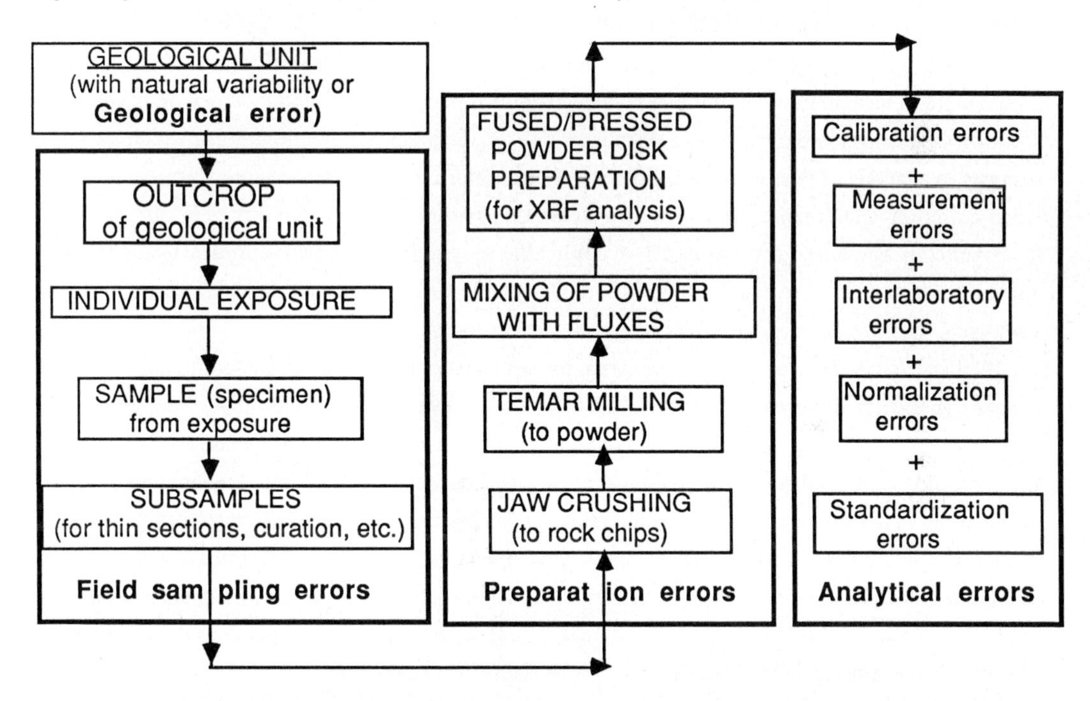

Fig.9.1. The four main, composite sources of error involved in machine-based analyses of geological rock unit. Total error (*TE*) is the sum of errors at every stage in the process (see text for explanation of individual errors)

Geological error *(GE)* covers all the inherent, natural variability in the unit being studied. In a batholith, it may include variation between different phases of intrusion, cryptic variations within individual intrusive phases, banding, etc. It is the real *signal* being measured, whereas all the other errors below are *noise* which obscure this signal. Although alteration and weathering effects would also be regarded as 'noise' by most igneous petrologists, to some individuals they may be the primary object of interest, and in any case they still constitute part of the *natural* variability. It is the job of the field sampling stage to either concentrate on them, or to attempt to eliminate their influence.

Field sampling errors *(FE)* and **Preparation errors** *(PE)* both concern the various stages of reduction in size of the material to be analyzed — from a thoroughly unmanageable granite batholith, to a manageable bottle of rock powder. At each stage of reduction, further errors are introduced because the smaller amount can never be exactly representative of the larger. Although these two sources are both aspects of *sampling errors* in general, it is useful to separate them, since entirely different methods are required to control them, and indeed, they are subject to quite different degrees of control. For example, it should be possible to devise a sophisticated laboratory programme which ensures that the final XRF discs are as close as humanly possible to representing the chunks of rock which come into the laboratory. But if available exposures of the granite batholith do not in fact represent the underlying, unexposed, rock-body, there is not a lot (bar a highly expensive drilling programme) which the geologist can do about it.

Analytical error (non-reproducibility), *AE,* varies considerably in make-up between geological disciplines (geochemistry, geochronology, geophysics, mineralogy, etc.), because of the varying techniques and machines used, but generally involves *calibration errors* (setting up the machine), which apply to all analyses, together with *measurement errors* (fluctuations in actual counting, peak-height determination etc., plus unwanted random 'spikes') and *machine errors* (e.g. for a mass spectrometer, errors in the extraction system, or due to filament mass fractionation), which vary from analysis to analysis. These errors arise from replicate determinations on the same specimen as well as from different specimens. It is usually assumed (without proof) that the *AE* obtained for a homogeneous specimen (or standard) is applicable to similar unknown samples using the same apparatus. Although some authors (e.g. Cox & Dalrymple 1967), have estimated *AE* for each step of a geological measurement, we generally only estimate *AE* for an entire analysis.

If an attempt is made to combine analyses, further errors are introduced, including *interlaboratory errors* (between different laboratories using the same analytical method — see §9c) and/or *normalization errors* (between different techniques, e.g. discordant K-Ar hornblende and biotite ages in geochronology). A final source of analytical error arises because machine-based techniques all depend on machine calibrations using 'accepted' values for international standard rocks and minerals. If the standard values are not in fact the 'true' values, then all analyses based on them must also all be wrong: this is *standardization error.* It might be regarded as separate from analytical error, since it is out of the control of the individual laboratory, but its magnitude is, of course, dependent in the first place on various laboratories' own analytical errors.

Normally, total error contains a contribution from *all* 4 major sources of error above (though the contributions of individual components in Fig.9.1 to each source vary considerably). There are, however, two special cases in which, *GE* and *FE/PE* can theoretically be reduced or eliminated.

It is useful to define an **entity** as a geological unit/body within which *GE* (for the particular parameter being measured) is in fact zero — that is, there is no real variation, measurements of the parameter are estimates of a *single number*, and *TE* is composed entirely of 'noise'. An *isotopic* or *geochemical* entity would be completely homogeneous with respect to the particular isotope or chemical element; a *geochronological* entity would be emplaced or formed, geologically speaking, instantaneously. Hence a granite batholith is in no sense an entity, because different batholithic phases often show subtle or even marked isotopic and/or age differences. Even far smaller units (e.g. fine-grained dykes), may still not constitute *isotopic* entities, because isotopic differences usually exist between constituent minerals. Even a single, zoned crystal is still not a chemical entity, precisely because it is zoned. But all these small units *would* constitute *geochronological* entities, if they formed rapidly. Note, therefore, that the same geological body may be an entity for one measurable parameter, but not for another. But for those parameters where it is an entity, *TE* consists only of *FE, PE* and *AE.*

Even where *GE* is not naturally zero, we can alternatively eliminate both it and *FE*/*PE* by obtaining **replicate data**, that is, multiple measurements of the *same* parameter on the same fused disk, spiked solution etc. using the same apparatus. Here, *TE* consists of *AE* only.

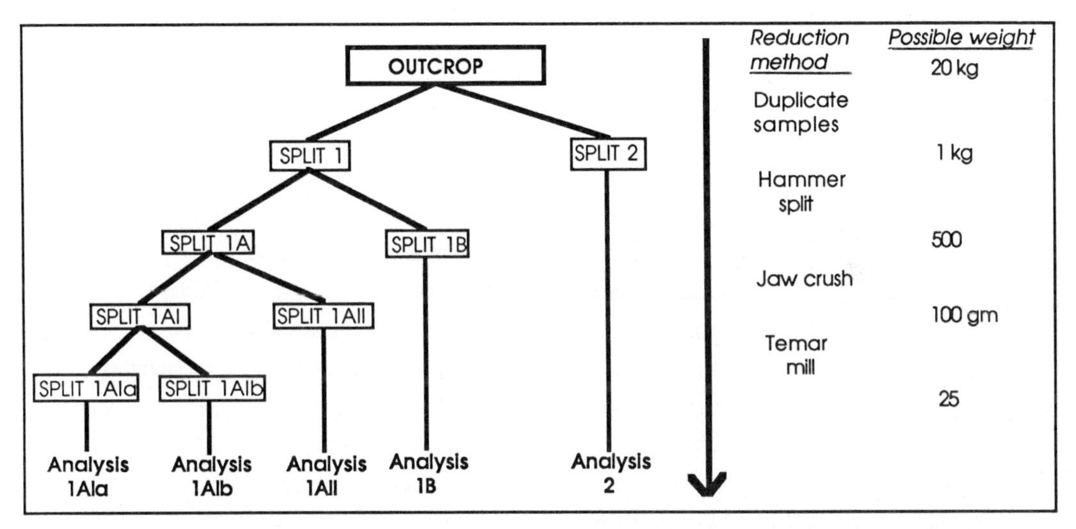

Fig.9.2. Experimental scheme for separating some of the different sources of preparation errors from Fig.9.1

9c. The Problems of Geological Sampling

Dedicated monographs & reviews: Cochran (1977).
Further discussion: Baird et al.(1964); Carras(1987); Cheeney (1983, p.133ff); Clanton & Fletcher(1976); Griffiths (1967); Kleeman (1967); Koch & Link (1970, ch.3 & 7; 1971, ch.15); Krumbein & Graybill (1965,p.146); Miesch (1972); Philip & Watson (1987b); Size(19878b); Sutherland & Dale (1972); Till (1974, p.48).

The discussion in this section is equally applicable to any geological sampling programme, whether from outcrops, drill-core, stream sediments, soils, etc.

Sampling lies at the basis of most geological studies, but the term itself is used rather variably and loosely. The major components of the sampling problem are usefully divided into the following:
(1)*Statistical component:* the numbers and sizes of samples (large or small, few or many).
(2)*Geological component:* the orientation and spacing of samples (e.g. along or across strike).
(3)*Physical component:* the processing of the sample into an analyzable state (e.g. crushing/splitting).
(4) *Laboratory component:* the actual analysis of the sample (e.g. to determine its Au content).

Different geological studies require data of different precision and accuracy, which in turn require different sampling plans, with different amounts of attention focussed on these 4 components. In regional geochemical stream sediment programmes, for example, accuracy is far less important than precision, whereas in determining the economic value of a mine prospect, accuracy becomes important, while in assessing interlaboratory standard materials, *both* are vital.

As regards (1) above, geological data are uniquely subject to the **volume-variance relationship**, which states that *the variance (or coefficient of variation) of measured values decreases as sample size increases* (large samples yield greater precision than small). This means that statistical studies should, as far as possible, be based on samples of the same size. Also, measurements (e.g. average ore grades) can be estimated with a given precision from a few large samples or from many small ones. The small samples will, however, give a better idea of the grade over the whole body, whereas the large ones will give information primarily on the areas of the body from which they were taken.(NB: 'large sample', as usual, is employed here in the *geological,* not statistical sense, to imply a *single,* but heavy, chunk of rock, not a set of N measurements where N is large).

As every geochronologist knows, the bulk required to constitute a reasonably representative sample of a rock-type also increases dramatically with its grain-size. For cherts and basalts a fist-size specimen may suffice, but for pegmatites a truck-load may be necessary. This consideration is particularly important in studying lunar rocks, meteorites and drill-core, where the amount of material an individual has to study may be strictly rationed (whether by Nature or Committee!) Whole-rock analyses of a coarse-grained meteorite based on < 1g specimens may unfortunately be statistically meaningless.

Component (2) above involves choice of both *orientation* and *spacing* of samples, relative to the local geology. This choice can be the most critical in some sampling programmes, and the source of the largest (order of magnitude) errors if chosen wrongly. For example, sampling veins and host-rock in an ore-body may require drillholes in very different orientations, if the veins are cross-cutting. Spacing must take account of repetition, banding, thickness of veins, intrusions or beds, and so on. The few detailed studies which have been carried out (notably Baird et al.1964; Size 1987b) have emphasised that small-scale variations even in the largest geological bodies (e.g. granite batholiths) can completely mask larger-scale variations, unless sampling is done carefully: in other words, that there is no point in even looking for *regional* trends unless and until *local* variations have been carefully delineated.

Conventional methods of physical processing (component 3) all have associated problems, which vary with the geological environment, so that generalizations are difficult and local practice is usually established by experience. In drilling, for example, length and (to a lesser extent) diameter of core affect resulting errors. If the rock being drilled contains minerals of very different hardness or brittleness (e.g. in shear zones or ultramafic sequences), what comes up may not even be an unbiased sample of what is down below, for the soft materials may be 'ground away'. In some mining areas, assay grades are found to vary with the grain-size of the sample, particularly in brittle materials. In the Golden Mile, Western Australia, the Au grade obtained from traditional grab samples of broken ore from a truck often proves higher than the true grade (i.e. the average Au content of the whole truck) because the grab samples are taken from the finer ore fractions. Expensive but vital quantitative studies have to be undertaken to establish regression relationships (§14) between 'grab grade' and 'true grade'.

Crushing and grinding also have associated sampling problems (see also next section, §9d). Different geological materials may require quite different crushing procedures, though the finer the sample before splitting, the better usually the result. Soft or 'sticky' materials (e.g. gold, clays) have problems of their own, as they can even agglomerate, rather than break up, during crushing.

Component (4) has already been discussed (§9a); its importance varies in different geological sampling scenarios, but it should *never* be ignored.

9d. Separating and Minimizing Sources of Error — Statistically and Graphically

Further geological discussion: Blasi (1979); Gy(1979); Ingamells (1974); James & Radford (1980); Radford (1987).

Interlaboratory tests on standard materials: Abbey(1983), Blattner & Hulston(1978); Brooks et al.(1978), Chayes(1969,1970a), Clanton & Fletcher(1976), Clemency & Borden(1978), Ellis(1976), Fairbairn (1953), Flanagan(1976a,b), International Study Group (1982); Lister(1985,1986), Schlecht(1951), Schlecht & Stevens (1951), Steele(1978), Steele et al. (1978), Stevens & Niles (1960), Vistelius(1970,1971). See also numerous papers in *Geostandards Newsletter* (Groupe International de Travail de l'Association Nationale de la Récherche Téchnique, Paris).

In the 1950s, an attempt was made to study a crucial geochemical problem: the accuracy and precision of rock analyses, then performed by wet methods. Large blocks of granite ('G1') and diabase ('W1') were crushed and analyzed by laboratories worldwide, with depressingly disparate results. Despite a *post mortem* in the literature over 20 years, the main source of the imprecision (or inaccuracy? — the two here blur because the 'true' values will never be known!) was never resolved. Some participants exasperatedly attributed it to preparation errors (specifically, incomplete homogenization of the powdered sample during crushing and grinding, such that different bottles of 'G1' and 'W1' in fact had different compositions), whereas others attributed it to interlaboratory errors. The G1–W1 experiment was quickly overtaken by events — namely, the replacement of wet methods by machine-based methods (XRF, INAA, AAS, ICP etc.). Unfortunately, its lessons were never been applied to modern methods, so that the depressing inconsistency of results from 'standard' samples continues unabated.

In fact, all the main sources of error can be separated by studies of the kind outlined in Fig.9.2. Given that independent errors (if expressed as standard deviations) are quadratically additive, i.e.:

$$TE^2 = GE^2 + FE^2 + PE^2 + AE^2 \qquad [9.1]$$

variances can be calculated from the duplicate data by [8.2b], which allow the % contributions of each E to the overall error for any given variable to be determined. A typical result might look as below:

Element error	Core sampling errors		Sample preparation errors				Analytical errors		Total
	Value	% of total	Stage 1	Stage 2	Subtotal	% of total	Value	% of total	
Au(ppb)^2	0.323	55.9	0.0567	0.0421	0.0988	17.1	0.156	27.0	0.5778

Overall error can thus be reduced by concentrating on the most error-prone stage (here, core sampling).

Rather than conventional variance, some experts prefer to use % differences:

$$\% \text{ difference} = \left[\frac{\text{routine value} - \text{duplicate value}}{\text{routine value} + \text{duplicate value}}\right] \text{x}200 \qquad [9.2]$$

A graphical method for separating errors in the sampling and preparation stages for an ore mineral is illustrated in Fig.9.3. The errors associated with the sub-sampling stages **B, D, F** are read off the diagonals — in this case ≈40%, ≈ 30% and ≈15% yielding total $\sqrt{[40^2 + 30^2 + 15^2]} = 52\%$.

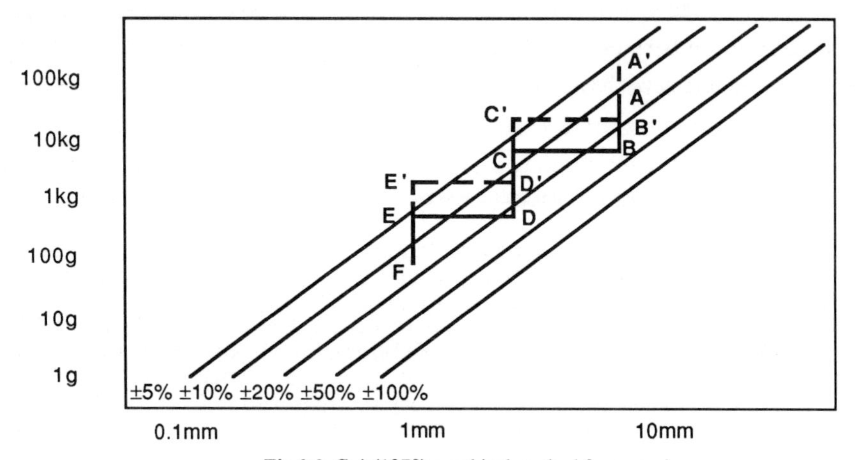

Fig.9.3. Gy's(1979) graphical method for assessing errors
(A field sample **A** is sub-sampled to **B**, jaw-crushed to rock-chips **C**; a further sub-sample **D** is then Temar milled to **E**, and an aliquot **F** is used for analysis)

The diagonals are positioned on such a diagram by means of the equation:

$$M = \frac{d^3 f.g.l.(1-a)[(1-a)r + at]}{s^2 a} \qquad [9.3], \text{ where}$$

M = mass of sample (gms); a = mineral content ($0 < a < 1$); s^2 = variance of mineral content; d = diameter of largest piece sampled (cms); f = shape factor (1.0 for spheres, 0.2 for flakes); g = particle size distribution factor (commonly set to 0.25 but to 0.5 in even-grained materials); l = liberation factor (conservatively set to 1.0 but often less); r = mineral sg; t = gangue sg.

This equation assumes not least an even distribution of mineral and gangue through the particle size spectrum, and in practice may overestimate total error. A diagram such as Fig.9.3 is therefore often 'calibrated', using a known error value determined via a flow-chart such as Fig.9.2. For example, if the total error for the route A-B-C-D-E-F in Fig.9.3 is known to be 20%, rather than the 52% calculated, the 20% error diagonal is positioned slightly to the right of point B, representing the largest of the three contributions (**B, D, F**) to that error.

Once set up, Fig.9.3 can obviate the complex experiments of Fig.9.2. For example the route **A'-B'-C'-D'-E'-F** substantially reduces the total error from **A-B-C-D-E-F**, merely by increasing the (sub-)sample sizes taken at each stage. No expensive equipment or other modifications are required. This also emphasises that the number of sub-sampling stages should be kept to a minimum: the field sample can be relatively small, if the final powder is produced directly from it. In practice, diagrams like these have shown that poor analytical reproducibility (often blamed on laboratory ineptitude or on inhomogeneous mineralization!) is more often attributable to field or sub-samples being too small.

9e. Expressing errors: CONFIDENCE LIMITS

Errors on scientific measurements are usually expressed in the form '$A \pm B$', where A is the measurement and B the error estimate. Unfortunately, in the geological literature, the exact nature of B is all too rarely stated (Howarth & Turner 1987), which severely reduces its value because there is no general convention. B may in fact simply be the standard deviation, s, or it may be $2s$, or even $s/\sqrt{N}$, the standard error (SE), which is an estimate of the *population* variance based on the measured data-set (sample variance). Hence a commonly quoted value such as 400 ± 10 Ma for a K-Ar age measuremen is rather meaningless unless it is stated which of these 3 conventions (or any other) the '± 10' refers to.

If '± 10' in fact represents one standard deviation, s, all we in fact imply is that 68.27% of repeated measurements *on the same specimen* should lie within $1s$ of 400, that is in the range 390–410 Ma, 95.45% should lie within $2s$ (380–420), etc., *always assuming the parent population from which we derive our measurements is in fact Normal.* What we really require is a '$\pm B$' which indicates with, say, 95% confidence that the *true population value* from which our measurements were made lies in the range 380–420 Ma. In conventional parametric estimation, this involves two adjustments to the simple standard deviation s as our $\pm B$ figure, now dealt with.

9e1. Parametric confidence limits for the arithmetic mean and standard deviation

Further geological discussion: Lister(1982); Rock et al.(1987).
Further non-geological discussion: Bhattacharyya & Johnson(1977,p.250 & 266).
Worked geological examples: *Economic Geology* (Koch & Link 1970, p.89 & 99); *Igneous petrology* (Le Maitre 1982, p.22); *Sedimentology* (Cheeney 1983, p.67); *Volcanology* (Till(1974, 56ff).
Applied geological examples: too numerous to list.
Stand-alone computer programs: Rock(1987c).
Mainframe software: covered in most statistics packages (e.g. SPSS) or libraries (e.g. NAG).
Microcomputer software: programmed in many statistics packages (e.g. Statview 512+™).

The Central Limit Theorem (§8a1) states that the mean $\overline{X}$ of data-sets drawn from *any* finite variance population (whether Normal or not) approaches a Normal distribution with mean $\mu = \overline{X}$ and scale SE = $s/\sqrt{N}$, as N rises. Hence the first adjustment is to replace s with SE to estimate population rather than sample (measured data-set) variance. The second adjustment is to introduce Student's t, which allows for the additional uncertainty of substituting the known standard deviation, s, for the unknown population standard deviation, σ (as in §8g2). Total error is then conventionally estimated by $\overline{X} \pm \text{SE} \times t_{\alpha/2,N-1}$, where α is the chosen significance level. The '±' factor is known as a **confidence interval**, and expresses the level of confidence we have that $\overline{X}$ actually lies within this ± interval. Confidence intervals are exactly equivalent to performing a one-sample t-test (§8g2) on the null hypothesis that μ has *any* value within the *limits* $[\overline{X} - s/\sqrt{N} \times t_{\alpha,N-1}]$ to $[\overline{X} + s/\sqrt{N} \times t_{\alpha,N-1}]$, which are consequently known as **confidence limits**. For example, if the ±10 in 400 ± 10 Ma represents 95% confidence intervals, the 95% confidence limits are 390 and 410, and we are 95% confident that the true value lies in the range 390–410; or, equivalently, a one-group t-test of H_0: $\mu = M$ would be accepted at the 95% level if M lies within the range 390–410.

Confidence intervals for the standard deviation are based on [8.12]. At the 95% level:

$$s\sqrt{\frac{N-1}{\chi^2_{.025}}} < \sigma < s\sqrt{\frac{N-1}{\chi^2_{.975}}} \qquad [9.4]$$

and so on. For example, to obtain 90% limits for $s = 0.4$ where $N = 10$, we use $\chi^2_{.05} = 3.325$ and $\chi^2_{0.025} = 16.919$, which indicates that σ lies in the range 0.29–0.66, or, equivalently, has 90% confidence *intervals* $\sigma = 0.4\ \{+0.26, -0.11\}$. These are markedly asymmetric about σ, unlike intervals for the mean. We can again use these in a one-group test: again, s can be equated with a given population value σ if it lies within the confidence interval for σ at the assumed significance level.

9e2. Robust Confidence Intervals for the Mean, based on the Jackknife

Further geological discussion and worked examples: Schiffelbein(1987).
Further non-geological discussion: Efron(1982); Gray & Schucany (1972); Miller (1974).
Computer software: none known.

The **Jackknife** is a fairly generalized statistical concept which is mentioned here only in outline and for one specific application, but can be applied equally well in regression (§14) and elsewhere. It has not so far been very popular in geology, perhaps because of the lack of widely available software, but has considerable potential for relatively large but non-Normal geological data-sets, especially those with outliers (§9f). Let Φ be our estimate of some unknown parameter ψ, based on N measurements $X_1 .....$ X_N (for example, Φ might be $\overline{X}$ and ψ might be μ, the sample and population means). The jackknife proceeds by dividing the N measurements into q groups of n measurements each (i.e. $n = N/q$), that is: $\{X_1 ..... X_n\}, \{X_{n+1} ..... X_{2n}\} ..... \{X_{(q-1)(n+1)} ..... X_N\}$; it then deletes each of these q groups in turn from the original data-set, and calculates q further estimates of ψ, which we shall call ϕ_ι (i = 1 to q), based on each of the q remaining subsets of $[N-n]$ measurements. We then define:

$$j_i(\phi) = q\Phi - (q-1)\phi_\iota, \qquad i = 1,2,.....q \qquad [9.5a]$$

$$J(\phi) = \frac{1}{p}\sum_{i=1}^{i=q} j_i(\phi) = q\Phi - \frac{q-1}{q}\sum_{i=1}^{i=q}\phi_i \qquad [9.5b]$$

The estimators $j_i(\phi)$ are termed the "pseudovalues" of the jackknife, and $J(\phi)$, their mean, is the jackknife itself. The jackknife can thus be regarded as a resampling scheme, which examines how a statistic behaves when subsets of the original data-set are taken. It is a robust estimate because values which would severely influence Φ, which is based on all N measurements, are only incorporated into *some* of the subsets on which the ϕ_ι are based, and hence affect $J(\phi)$ much less than Φ.

To estimate robust confidence intervals, we require the jackknife estimate of the mean, $J(\overline{X})$ which has its own associated standard deviation, $s_{J(\overline{X})}$, the two being calculated as follows:

$$\text{Confidence intervals} = \pm t_{N,\alpha} s_{J(\overline{X})} = \pm t_{N,\alpha}\sqrt{\frac{1}{q-1}\sum_{i=1}^{i=q}[j_i(\overline{X}) - J(\overline{X})]^2} \qquad [9.6]$$

and $J(\overline{X})$ itself is calculated as in [9.5], substituting $\overline{X}$ for Φ (and μ for ψ).

The jackknife is a particularly useful procedure for data where natural subsamples already exist (e.g. where measurements have been taken from, say, 4 outcrops, 4 dykes, 4 horizons within a stratigraphical unit, or from 4 drill-cores intersecting the same rock unit). The subsets can then be defined as these natural subsamples. Otherwise, however, an arbitrary definition of q is entirely valid.

9e3. Robust Confidence Intervals for location estimates, based on Monte Carlo Swindles

Further geological discussion: Rock et al.(1987).
Further non-geological discussion: Gross(1976); Hoaglin et al.(1983, p.424); Mosteller & Tukey (1977).
Computer software: none known.

These are considerably simpler to visualise and calculate than the Jackknife, but somewhat less well-established in theoretical terms. They replace the classical parametric formula above with $\overline{X} \pm st^*$ for confidence limits around the mean, or $X_{med} \pm \mathrm{MAD}t^*$ for the median, where MAD is the median absolute deviation from the median (§8c2), and t^* belongs to a family of *empirically* adjusted t multipliers — so-called "Monte Carlo swindles". Formulae for t^* variously suggested include (1) $t\{\sqrt{\pi/2N}\}$ — the large N standard deviation of the median being $\approx\sqrt{\pi/2}$ that of the mean; (2) $t_{0.7(N-1)}$; and (3) $t_{N-1}/1.075$. Intuitively acceptable errors are produced only for $N > 4$, where t^* falls below 2.

Specially designed confidence intervals for the various robust location estimates in (§8b) are also available (e.g. Ellis 1981). Overall, however, neither theory nor practice for robust *scale* estimation are as well-developed as for robust *location* estimation (cf. §8c1), and the applicabilities of the Central Limit Theorem, the Jackknife and "Monte Carlo swindles" to many geological data-sets are still dubious. For very small ($N < 10$) data-sets in particular, the wholly nonparametric error estimates in the next section are probably preferable.

Table 9.1. Nonparametric estimation of nominal asymmetric 95% confidence intervals for the median of small ordered data-sets, $X_1 < X_2 ... < X_N$ (based on Sign Test)

N	Median, X_{med}	Upper confidence interval, U	Lower confidence interval, L	Actual % confidence, A
3[a]	X_2	$X_{med} - X_1$	$X_3 - X_{med}$	75.0
4[a]	$[X_2+X_3]/2$	$X_{med} - X_1$	$X_4 - X_{med}$	87.5
5[a]	X_3	$X_{med} - X_1$	$X_5 - X_{med}$	93.8
6	$[X_3+X_4]/2$	$X_{med} - X_1$	$X_6 - X_{med}$	96.9
7	X_4	$X_{med} - X_1$	$X_7 - X_{med}$	98.4
8	$[X_4+X_5]/2$	$X_{med} - X_2$	$X_7 - X_{med}$	93.0
9	X_5	$X_{med} - X_2$	$X_8 - X_{med}$	96.1
10	$[X_5+X_6]/2$	$X_{med} - X_2$	$X_9 - X_{med}$	97.9
11	X_6	$X_{med} - X_3$	$X_9 - X_{med}$	93.5
12	$[X_6+X_7]/2$	$X_{med} - X_3$	$X_{10} - X_{med}$	96.1
13	X_7	$X_{med} - X_3$	$X_{11} - X_{med}$	97.8
14	$[X_7+X_8]/2$	$X_{med} - X_4$	$X_{11} - X_{med}$	94.3
15	X_8	$X_{med} - X_4$	$X_{12} - X_{med}$	96.5
16	$[X_8+X_9]/2$	$X_{med} - X_4$	$X_{13} - X_{med}$	97.9
17	X_9	$X_{med} - X_5$	$X_{13} - X_{med}$	95.1
18	$[X_9+X_{10}]/2$	$X_{med} - X_5$	$X_{14} - X_{med}$	96.9
19	X_{10}	$X_{med} - X_6$	$X_{14} - X_{med}$	93.6
20	$[X_{10}+X_{11}]/2$	$X_{med} - X_6$	$X_{15} - X_{med}$	95.9
21	X_{11}	$X_{med} - X_6$	$X_{16} - X_{med}$	97.3
22	$[X_{11}+X_{12}]/2$	$X_{med} - X_7$	$X_{16} - X_{med}$	94.8
23	X_{12}	$X_{med} - X_7$	$X_{17} - X_{med}$	96.5
24	$[X_{12}+X_{13}]/2$	$X_{med} - X_7$	$X_{18} - X_{med}$	97.7
25	X_{13}	$X_{med} - X_8$	$X_{18} - X_{med}$	95.7
>25	X_{med}	$X_{med} - X_d^*$	$X_{N+1-d} - X_{med}^{\dagger}$	≈95.0

The A% confidence interval is given by $X_{med}[+U, -L]$. X_d and X_{N+1-d} are the A% confidence *limits*.
† $d = 0.5[N + 1 - 1.96\sqrt{N}]$, or alternatively, choose the value nearest $0.98\sqrt{N}$ *ranks* from the median as the ≈95% confidence limit.

9e4. Nonparametric Confidence Limits for the Median based on the Binomial Model

Further geological discussion: Lister(1982); Rock et al.(1987).
Further non-geological discussion: Bhattacharyya & Johnson(1977); Noether(1971); Sprent(1981).
Stand-alone computer programs: Rock(1987c).
Proprietary software: none known.

A graphical method of displaying nonparametric confidence intervals for the median is the **notched boxplot** (Fig.8.9) which, once again, can be used to determine whether the median of a given data-set has a particular value *M*, according to whether or not *M* lies between the 'notches'.

The simplest numerical nonparametric confidence intervals, which make no assumptions whatever about the form of the distribution concerned, involve choosing actual values from the *ranked* data-set, based on a Sign Test (§10c1) — i.e. the distribution of binomial probabilities. Table 9.1 identifies the values nearest to yielding 95% intervals for all *N*. As an example, with the ranked data-set 25, 27, 31, 35, 39, 45 ($N = 6$), the median, upper and lower 96.9% confidence limits are respectively given by: the average of the 3rd/4th values (33), the 6th value (45) and the 1st (25) value; that is, 33 [+12, –8]. Using Binomial Distribution tables for $p = 0.5$ (§8a4), confidence limits nearer to 95% for $N \le 25$ can also be interpolated, if geologically meaningful. In this same $N = 6$ example, the 2nd and 5th values yield ≈78.2% limits, i.e. 33 [+6, – 6], so that ≈95% limits can be made by weighting 0.1 x 78.2% set with 0.9 x 96.9% set, i.e. 33 [+11.4, –7.8]. Asymmetric intervals of this kind are more plausible than the arbitrary symmetric $A \pm B$ usually quoted for geoscience results, especially where data are few.

Table 9.2. Nonparametric estimation of *symmetric* confidence limits for the median (based on Wilcoxon's signed-rank statistic)

(a) Tabulating the pair-medians

	25	27	31	35	39	45
25	**25**	**26**	28	30	32	35
27		27	29	31	33	36
31			31	33	35	38
35				35	37	*40*
39					39	**42**
45						**45**

(b) Critical values of the Wilcoxon statistic, *S*

N	*S*	*A*.%	*N*	*S*	*A*.%	*N*	*S*	*A*.%	*N*	*S*	*A*.%
3[a]	0	75.0	9	6	94.5	15	25	95.2	21	59	95.0
4[a]	0	87.5	10	8	95.1	16	30	94.9	22	66	95.0
5[a]	0	93.8	11	11	94.6	17	35	94.9	23	74	94.8
6	1	93.7	12	14	94.8	18	40	95.2	24	81	95.1
7	2	95.3	13	17	95.2	19	46	95.1	25	90	94.8
8	4	94.5	14	21	95.1	20	52	95.2	>25	*b*	95.0

Partly adapted from tables in Bhattacharyya and Johnson (1977), Noether(1971) and Sprent (1981).
The *A*% confidence *limits* are given by the outer remaining values, after *S* values have been deleted from both ends of the ordered set of $N(N+1)/2$ pair-medians, $(X_i + X_j)/2$, $1 \le i,j \le N$, derived from the original data-set. Note *S* values here are one less than *d* values in some published tables (e.g. Noether, 1971, table F), which give the confidence limits on the dth value in the set, rather than after S values have been eliminated.
[a] Symmetric and asymmetric methods identical for $N \le 5$ (cf. Table 9.1).
[b] For $N > 25$, $S \approx 0.5[0.5N(N+1) - 1 - 1.96\sqrt{\{N(N+1)(2N+1)/6\}}]$.

For $N > 20$, the Normal approximation to the binomial distribution $\sqrt{N}/2$ can also be used. When $N = 25$, for example, 95% confidence limits are given by ordered data-values $[t_{0.05,N-1}\sqrt{25}/2]$, i.e. approximately 5 ranks on either side of the median; ≈95% confidence intervals are therefore X_{13} $[+(X_{18}-X_{13}), -(X_{13}-X_{8})]$, wholly consistent with the more exact values given in Table 9.1.

If we are willing to assume merely that the population is symmetric, an alternative procedure is available as follows. First tabulate the $N(N+1)/2$ pair-medians between each pair of data-values, such as in Table 9.2a. Then find an appropriate critical value for Wilcoxon's signed-rank test statistic, S (§10c3) from Table 9.7b. For $N = 6$, $S = 2$ at the 90.6% level, so eliminate the 2 highest and 2 lowest entries (**emboldened**) from Table 9.2a. The 90.6% confidence limits are then given by the highest and lowest values remaining *(italicized)*, so the corresponding confidence intervals are now 33 [+7,–6] — far more symmetric than before. By careful arrangement of the data in an ordered table, the number of pair-medians actually calculated can of course be minimised.

9f. Dealing with outliers (extreme values): should they be included or rejected?

Dedicated monographs and reviews: ASTM(1969); Barnett & Lewis (1984); Beckman & Cook(1983); David (1981); Hawkins (1981a).
Further geological discussion: Bjorklund(1983); Harvey (1974); Lister (1982); Powell (1985); Rock et al. (1987); Rock(1987c,1988).
Further non-geological discussion: Anscombe (1960); Dixon (1953); Dybczynski(1980); Ferguson(1961); Gnanadesikan & Kettering (1972); Grubbs (1950,1969); Grubbs & Beck(1972); McMillan (1971); McMillan & David (1971); Neave(1979); Proschan (1953); Sheeley(1977); Tietjen & Moore(1972).
Worked geological examples: *sedimentology* (Miller & Kahn 1962,p.130); *mining* (Koch & Link 1970, p.238); *geochemistry* (Lister 1982); *isotope geochemistry* (Golding et al. 1987; Rock et al. 1987).
Statistical tables: Dixon (1953); Grubbs (1969); Harvey (1974); Lister (1982).
Stand-alone computer programs: Dhanoa (1981); Rock (1987c); Wainwright & Gilbert (1981).
Mainframe and microcomputer software: rarely programmed.

Outliers are aberrant (high or low) values which differ from most of a data-set, and may lie beyond the 'reasonable' limits of error. They are a complex and, to some extent, controversial subject which can only be touched on here. Before applying statistical tests, (§9f3) we need to consider what different types of outliers exist (§9f1), and how we recognize them in different geological contexts (§9f2).

9f1.Types of statistical outliers: true, false and bizarre, statistical and geological

Four types come under various different names in the literature, but are defined here as follows:

True (geological) outliers (discordant observations) are a real part of the data, have a sound geological explanation, and can be reproduced by different analytical techniques: for example, a locally sulphide-bearing horizon within an otherwise sulphide-poor succession could well yield 'true outliers' for such elements as Cu, Pb and Zn. Coherent sets of true outliers constitute **anomalies.**

Bizarre outliers (contaminants) are equally reproducible and geologically accountable, but do *not* belong to the data at all; for example, 2 misidentified carbonatites among an analyzed batch of 12 limestones may well yield 'bizarre outliers' for elements like Sr, Ba and Nb.

False outliers (mistakes) are non-reproducible, inaccurate, maverick data tarnished by machine failure or error, 'spikes', bad sampling or sample preparation, etc.; they have no geological value.

Statistical outliers are those identified by *purely statistical* means (e.g. the tests in §9f3); their identification as true, false or bizarre has not yet been made.

Generally speaking, true, false and bizarre outliers will all be statistical outliers as well. However, geological contaminants or mistakes will not always yield bizarre or false outliers (e.g. our carbonatites among limestones would be unlikely to yield outliers for most major elements). Nor does it follow that all statistical outliers should be eliminated from a data-set, for geology is an unusual discipline in which true outliers (and sometimes even bizarre outliers) can actually be be the object of prime interest. For example, true outliers in mineral exploration would usually outline the anomalies actually sought by mining companies, while bizarre outliers might outline something not anticipated in the exploration programme at all (e.g. our carbonatite among limestones). It is only false outliers which are to be eschewed under *all* geological circumstances.

9f2. Types of geological data: the concept of 'data homogeneity'

Whether summary statistics are meaningful for a given data-set, and how outliers are recognized, depends on a combination of both geological and statistical properties of that data-set. Calculation of the mean Sr content from all 14 specimens in our collection of 12 limestones and 2 carbonatites would, for example, be geologically meaningless, however easily calculable statistically, but we would intuitively feel that the carbonatites would generate outliers (in this case, bizarre outliers). On the other hand, the ore grade within a particular body might vary extremely irregularly; data-summaries would then be statistically difficult, but geologically would be crucial.

To reflect the interplay of both statistical and geological effects, we need two concepts. (1) A geological **unit** is a single, well-defined geological object on which the calculation of summary statistics is *geologically* meaningful (e.g. a single rock, mineral or fossil specimen; a recognised stratigraphical horizon; a particular rock-type such as syenite; a discrete igneous intrusion such as a dyke or pluton). (2) A **homogeneous** data-set can be meaningfully represented by a *single* location (or scale) estimate; this usually means it samples a single geological unit *with a unimodal population distribution*. An **inhomogeneous** data-set, by contrast, might sample either a mixture of several geological units, when summary statistics would be *geologically* meaningless, or one geological unit with a bimodal or polymodal population, when data-summaries would be *statistically* questionable.

Replicate data-sets are by definition homogeneous, with zero *GE*, and should in theory be Normally distributed with a mean value equal to the true value. Once an element of *GE* is introduced, however, assessment becomes more complicated: whether a data-set is homogeneous or not depends most importantly on what constitutes a single geological unit in the particular case. For example, a homogeneous sill or sandstone horizon is clearly a geological unit and might be expected to yield a homogeneous data-set. However, is a *differentiated* igneous sill or *graded* greywacke unit, showing a continuum of compositions, best regarded as a single geological unit or as a mixture of end-members? Do single, even intrinsically invariable, rock-species (e.g. dunite), or individual members of solid-solutions (e.g. labradorite), constitute single units, when such 'species' are themselves are based on artificial divisions in compositionally continuous space?

These questions can be partly resolved by recognising *degrees* of inhomogeneity: 'average crust', 'average igneous rock', 'average granitoid', 'average granite', 'average calc-alkaline granite' and 'average S-type granite' constitute a series of increasing homogeneity, based on successively finer sub-sets, presumably with fewer and fewer modes. All however are accepted as geologically reasonable concepts (given that they appear repeatedly in the literature). Homogeneity is thus seen in part to reflect *size* of data-sets: the wider the net is thrown, the less the likely homogeneity. There may be a size-limit beyond which inhomogeneity precludes geologically *or* statistically meaningful averages (?'average planet'!), but it has not yet been recognised in the literature. Such concepts as 'average chondrite', 'bulk earth' and 'average primordial mantle' are now fundamental in geochemistry, but whether they are statistically reasonable (or how they are best estimated) has barely been considered.

With these two concepts in mind, we can begin to see the complexities of the problem of recognizing outliers. The only simple case is with homogeneous (replicate) data-sets, where *all* outliers *must* be false. Hence the tests in §9f3 can be rigorously applied to eliminate outliers, using purely statistical reasoning, and so home in on a 'better' location or scale estimate. However, as more and more geological error is introduced, outlier recognition and treatment requires more and more geological input. Returning to our 12 limestone + 2 carbonatite collection, we could confirm that the carbonatite values did not represent false outliers by reanalyzing the samples. (Conversely, if extreme values had proved analytically non-reproducible, they would *by definition* constitute false outliers). However, we would then have no way of knowing whether extreme values were true or bizarre outliers, unless we had *independent geological evidence* that both limestones and carbonatites were present — i.e. that the data-set would be inhomogeneous.

Many geological scenarios are far more complicated. For example, 'best' estimates for the age of igneous intrusions are commonly derived from disparate sets of published K-Ar, Rb-Sr ± Nd-Sm or Pb-Pb ages, possibly on both whole-rocks and minerals, carried out by different laboratories over a period of years, using different techniques and apparatus. Let us say one K-Ar age is identified as a *statistical* outlier. If *FE, PE* or *AE* is the source of its extreme status, then we have good reason to identify this age as a *false* outlier, and so to eliminate it. However, if *GE* is the source of its status, eliminating it would *harm* our so-called 'best' estimate, because some of the real variability in the intrusion would have been deliberately (and erroneously) overlooked. All too commonly, published papers give us insufficient evidence on which to decide which case is true, and we can do no better than to base our geological decision at least in part on the statistical tests in the next section.

9f3. Tests for identifying statistical outliers manually (one-by-one)

In §8c1, various robust estimates were described which eliminate outliers *automatically* in calculating an 'average' value. This section considers *manual* outlier identification, in which the geologist inspects his data and chooses the individual values for rejection. Manual elimination is often vital in geology, since there may be good geological reasons for retaining statistical outliers, or, conversely, good geological reasons for rejecting values which statistics do not so identify.

Minimum and *maximum* values, the obvious outlier candidates, can be tested by the following criteria (NB: X here refers to *ordered* data, where $X_1 < X_2 ... < X_N$; X_1 = min., X_N = max.):

Mann-Whitney test. This nonparametric test can be used provided $N \geq 20$ (see §10d2 for details).

Dixon's (1953) test. This nonparametric test uses simple ratios, dependent on N (≤ 25):

$3 \leq N \leq 7$; for *minimum* value: $\dfrac{X_2 - X_1}{X_N - X_1}$, for *maximum* value: $\dfrac{X_N - X_{N-1}}{X_N - X_1}$, [9.7a]

$8 \leq N \leq 10$; $\dfrac{X_2 - X_1}{X_{N-1} - X_1}$; $\dfrac{X_N - X_{N-1}}{X_N - X_2}$ [9.7b]

$11 \leq N \leq 13$; $\dfrac{X_3 - X_1}{X_{N-1} - X_1}$ $\dfrac{X_N - X_{N-2}}{X_N - X_2}$ [9.7c]

$14 \leq N \leq 25$; $\dfrac{X_3 - X_1}{X_{N-2} - X_1}$; $\dfrac{X_N - X_{N-2}}{X_N - X_3}$ [9.7d]

Grubbs' maximum studentized residuals. This parametric test uses the statistics $[X_N - \overline{X}]/s$ and $[X_1 - \overline{X}]/s$. $\overline{X}$ is based on the *whole* data-set, including the value in question. It is available for all N.

Harvey's (1974) test. This uses the same ratios as Grubbs', but $\overline{X}$ *excludes* the value in question. It was designed originally for geochemical data, but critical values are available only for $N \leq 7$.

Studentized range test. This allows *both* the maximum *and* minimum values to be tested simultaneously, by comparing the value $[X_N - X_1]/s$ with critical values of the Studentized range.

Box-and-whisker plots (Fig.8.9). As described in §8d, individual values are considered as outliers when they plot beyond the end of the 'whiskers' on such plots. However, as the whiskers are defined as an arbitrary number (1.5) of interquartile spreads beyond the hinges, no statistical tests are applicable to provide confidence limits for such an assessment.

Full Normal Plots (FUNOPS). This is similar to making a test for Normality on Normal probability paper (§8e3), but a parameter known as a **Judd** is calculated for each value as follows:

$$\text{Judd} = \frac{X - X_{med}}{\text{standard normal deviate corresponding to \% relative cumulative frequency}} \quad [9.8]$$

For Normal data, the median Judd equals the standard deviation. Usually, Judds corresponding to the upper and lower one-third of the data are calculated, and any Judds exceeding 2 are taken as outliers.

Once again, it is emphasised that all the above tests only identify *statistical* outliers. Only in replicate data, where there is no geological error, can statistical outliers be automatically denounced as false, and so unequivocally eliminated from the data-set. As soon as geological errors are present, geological judgement is concomitantly required to decide whether the statistical outliers identified actually include true outliers, which should *not* be eliminated.

9f4. Avoiding Catastrophes: Extreme Value Statistics

Dedicated monographs and reviews: Gumbel(1958); Kinnison (1984).
Worked geological examples: *Hydrology* (Koch & Link 1970, p.239ff).
Selected geological applications: *Volcanology* (Mulargia et al. 1985).
Statistical tables: NBS (1953).
Software: none known.

Circumstances where the treatment of extreme values becomes vital to the general public are common in geology. Engineering geologists considering the best site for, say, a dam or a railway, have to be concerned with the probabilities of extremely rare events (e.g. failure of the rock holding the dam or railway), and with the behaviour of rock bodies under extreme circumstances such as earthquakes, volcanic eruptions, or floods. Seismologists and volcanologists predicting earthquakes and volcanic eruptions are primarily concerned with extreme value statistics. *Any* set of data selecting only the *minimum* or *maximum* measurements of a particular variable (e.g. highest rainfall over a period of years in a given locality), is a set of extreme-value statistics, and may be expected to follow an **extreme-value distribution**, which depends only upon the *tails* of the frequency distribution from which the measurements are drawn. The distribution most pertinent in geology is defined as:

$$f(x) = e^{-e^{-\alpha(X-\beta)}} \qquad \alpha > 0 \qquad [9.9]$$

where $f(\beta) = 0.3678$ and $f(\beta - \alpha^{-1}) = 0.066$. This distribution has been used to generate *extreme-value graph paper*, on which ordered observations are plotted against a plotting percentage, just as on Normal probability paper. This can be used to predict the reappearance of critical events (e.g. earthquakes of greater magnitude than a structure can withstand).

9f5. Identifying Anomalies: Geochemical Thresholds and Gap Statistics

Geological discussion, statistical tables, software & worked examples: Koch(1987); Miesch(1981b).

We mentioned above that in some circumstances, extreme values can symptomise not failure but triumph. In mineral exploration, detecting and following up geochemical outliers and anomalies (hopefully to discover exploitable mineral deposits) is indeed the whole object of the exercise. Whereas §9f1 dealt essentially with *one* outlier at a time, regional geochemical surveys aim to detect whole clusters of outliers (i.e. anomalies) over the site of a potential deposit. This had led to the concept of geochemical **thresholds,** namely values which separate normal background values from the whole range of higher values, representing regional and/or local anomalies. Thresholds may be obvious — as in the case of a bimodal distribution of Au values, with a higher mode over an orebody — or very far from obvious, just representing a small hiatus or 'gap' in a regional frequency distribution.

The method typically assumes logNormality, and proceeds as follows:

(1) Transform data using SELLO [8.33], i.e. $\ln(X + a)$ or $\ln(a - X)$.

(2) Standardize the transformed data using [8.3].

(3) Sort the standardized data into ascending order.

(4) Adjust each value for the expected frequency from a Normal population (the **rankit**).

(5) Find the gap statistic — the maximum gap between adjacent ordered observations.

(6) Take the midpoints of this gap as the threshold, and use it to separate the 'background' values (those below the threshold) from the 'anomalous' observations (those above the threshold).

SECTION III: INTERPRETING DATA OF ONE VARIABLE: UNIVARIATE STATISTICS

TOPIC 10. COMPARING TWO GROUPS OF UNIVARIATE DATA

This topic shows how, given two groups (data-sets) of geological data, it is possible to determine whether they are the 'same': namely, to test the null hypothesis that they are statistical samples from the same parent population. The applications of the various tests can be summarised as follows:

(1) *Type of data*. To carry out *any* of the statistical tests under this Topic, data must be measured on *at least* an ordinal scale, i.e. rankable (see §6). *All* the tests (parametric and nonparametric: §7b) can be carried out on ordinal data, but the parametric tests can only be carried out on interval or ratio data.

(2) *Data distribution*. Even if the groups are interval or ratio, but are (i) known not to be Normally distributed (§8e3); (ii) known to contain outliers (§9f); or (iii) of entirely unknown distribution, it may be preferable to use the nonparametric tests, for reasons already discussed in §7b.

(3) *Size of groups of data*. Parametric tests are, in general, less reliable on 'small' data-sets, whereas nonparametric tests become more relatively efficient; this is partly because it becomes more difficult to test Normality as the data-set gets smaller. Although the critical N varies between circumstances, $N <$ 50 would usually be taken as 'small'. There are, however, variants of classical parametric tests specifically designed for small groups (§10b), though their performance is rather undervalued in their name of 'inefficient statistics'.

(4) *Comparison criterion*. The two groups might differ in (a) **overall distribution;** (b) **location**; or (c) **scale** (Fig. 10.1). Each of these comparison criteria might be of chief interest in different geological circumstances: (a) is usually of interest in exploratory situations, where there is little basis for predicting in what respect two groups might differ. With a little more knowledge, it may be preferable to 'home in' on the predicted type of difference: for example, if one were interested in the relative *accuracy* of two analytical methods, (b) would be the appropriate criterion, whereas if *precision* were of concern, then (c) would be appropriate.

Using the nonparametric tests in this and succeeding topics involves 'degenerating' ratio data into ranks: for example, ratio measurements of gold values in an ore would be converted from, say, 567, 346, 965, 243 ppb into ordinal ranks of 3, 2, 4, and 1. Numerically inexperienced geologists often feel that this must result in a much less 'reliable' test. This is not in fact so. Nonparametric tests are conventionally compared with the equivalent classical parametric tests via the concept of **asymptotic relative efficiency (ARE)**. The Mann-Whitney test (§10d3) turns out to have a power-efficiency of $3/\pi$ (95.5%) relative to the most powerful parametric test available, the t-test (§10a2), even where the data are Normally distributed (i.e. where the t-test is at its *most* powerful). That is, a Mann-Whitney test should correctly accept the null hypothesis on groups of data of size 21 no less frequently than a t-test would on groups of size 20; or, put another way, one only needs 5% more data (in the *worst possible case*) to get as reliable a result from a Mann-Whitney test as from a t-test. For non-Normal distributions, the Mann-Whitney test is far *more* reliable than the t-test, and indeed, can be *infinitely* more reliable.

For those still not convinced by these arguments, however, a further class of tests is available which adroitly combine the advantages of *both* parametric *and* nonparametric techniques: **Normal scores tests** (see Glossary and §10c4, §10d4, §10e3).

Readers new to statistical tests will probably still be confused by the variety and nomenclature of nonparametric tests, which rival that of alkaline rocks in obtuseness! However, there is one *exact* nonparametric equivalent for each parametric test described below: for example, taking a group of data, converting to ranks and then performing a paired t-test (§10a2), is exactly equivalent to performing a Wilcoxon test (§10c3). Similarly, a Mann-Whitney test (§10d3) is equivalent to performing an unpaired t-test (§10a2) on the ranks of a group of interval or ratio data, rather than on the raw data themselves. For those who prefer even more order in their lives, and eschew eponymous statisticians like Messrs. Mann and Whitney, Meddis' (1984) valuable and systematic treatise barely mentions such gentleman but sets all the medley of nonparametric tests in relative, anonymous perspective.

The following decision rules should help decide which test from those in Table 10.1 to use under different circumstances; they also apply to the tests described in later Topics, as far as §17:

(a) If the data are numerous (say, $N > 25$), *and* Normally distributed (or transformable to Normality) *and* measured on an interval/ratio scale, use the appropriate **parametric** test or statistic (t, F, r, etc.)

(b) If the data are relatively few (say $10 < N < 25$), *or* non-Normally distributed, *or* measured only on a nominal scale, *or* include outliers, use the appropriate **nonparametric** test; use **Normal scores** tests in preference to less powerful ones based on ranks, or the least powerful based on signs.

(c) If the data are very few (say $N < 10$), use either the tests in (b) or the appropriate *substitute* test based on the range (Table 10.1).

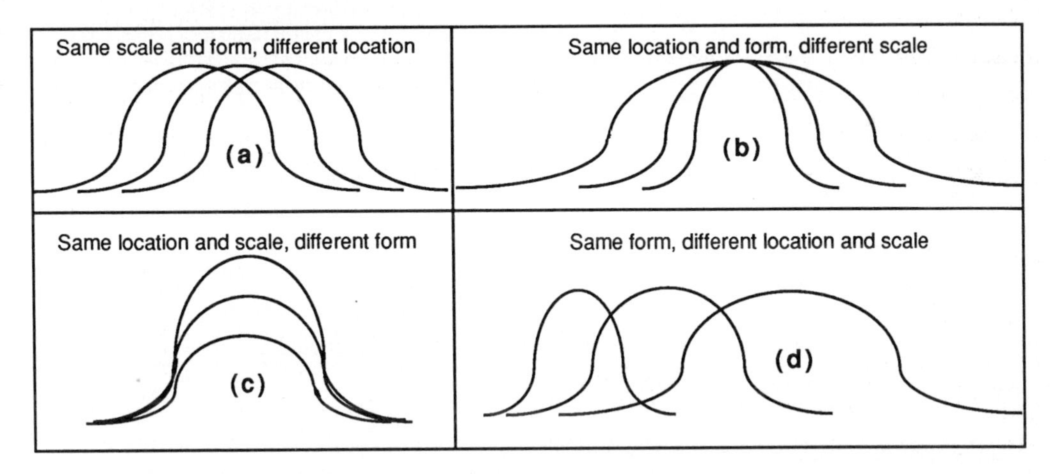

Fig.10.1.Differences of location, scale and form between three data-distributions relevant to tests in Topics 10–11
The tests of §10 can compare only 2 curves, those of §11 all three. For example, the Mann-Whitney and Kruskal-Wallis tests (§10d2,11e2) are sensitive to cases (a) and (d) but will not detect (b) or (c), whereas the Squared ranks test (§10e2,11f) will detect (a) and (d) but not (b) or (c). The Smirnov test (§10f2,11g) is sensitive to *any of* (a), (b), (c) or (d). It is advisable either to inspect histograms or frequency curves of the data-sets first, and choose tests accordingly from Tables 10.1/11.1, or to perform more than one test. Thus if data-sets differ as in (a), they will give significant Mann-Whitney/Kruskal-Wallis and Smirnov (but not Squared ranks) statistics; if they differ as in (b), they will give significant Squared ranks and Smirnov (but not Mann-Whitney/Kruskal-Wallis) statistics; and if they differ as in (c), they will give a significant Smirnov statistic *only*. Case (a) *alone* is in principle amenable to a parametric t-test (§10a) or one-way ANOVA (§11b), and even then only if *all* distributions are demonstrably Normal (which is not the case above!)

Readers will soon appreciate that the nonparametric tests in Table 10.1 are extremely easy to calculate — they involve only simple, 'back-of-envelope' manipulation of small numbers. The Normal scores and parametric tests, by contrast, require more sophisticated use of statistical tables and mathematics, but yield commensurately greater efficiency.

Table 10.2 gives the raw data which are used in all worked examples of tests covered in this Topic.

Table 10.1. A quick guide to tests for comparing two groups of data

	Parametric tests		*Nonparametric tests*	
Data-type / Comparison	Interval/ratio data, Normally distributed	Small groups of interval/ratio data	Ordinal data; interval/ratio data with outliers or non-Normal†	Nominal or dichotomous data*
LOCATION‡ (unpaired data)	t-test (§10a2)	Substitute t-test (§10b)	1.Median test (§10d1) 2.Tukey's quick test (§10d2) 3.Mann-Whitney test (§10d3) **4.Terry-Hoeffding test** (§10d4)	1.χ^2 test (§13a1)
LOCATION (paired data)	t-test (§10a2)		1.Wilcoxon test (§10c3) **2.Van Eeden test** (§10c4)	1.Sign, McNemar tests (§10c1-2)
SCALE	F-test(§10a1)	Substitute F-test (§10b)	1.Mood,David etc.tests (§10e1) 2.Squared ranks test (§10e2) **3.Klotz test** (§10e3)	—
OVERALL FORM			1.Wald-Wolfowitz Runs test(§10f1) 2.Smirnov test (§10f2)	—

† Power of nonparametric tests increases as 1 < 2 < 3 < 4. Normal scores tests (the most powerful) are **emboldened.**
‡ It should be noted that the comparison criterion varies according to which test is used: parametric tests of location are comparing *means*, whereas nonparametric tests are comparing *medians;* parametric tests of scale are comparing *variances /standard deviations)* whereas nonparametric tests are comparing *median deviations.*
* Comparisons of nominal/dichotomous data are tests for *association* (dealt with in §13) rather than for difference of location or scale (nominal data do not have locations or scales); the tests are listed here merely to show their relationship to the others; for example, the median test is equivalent to a χ^2 test on ordinal rather than nominal data.
NB. Each data-type can be analyzed by the tests in the column below it, *or in the columns to the right.* Thus interval/ratio data can be analyzed by *any* of the tests, but nominal data only by those in the *rightmost* column.

10a. Comparing Location (mean) and Scale (variance) Parametrically: t- and F-tests

These two tests are considered together because, unlike those considered later under this Topic, they are interdependent: although this is very often neglected in the literature, execution of a t-test does presuppose that an F-test has already been performed, and its result should be taken into account in determining which of the various available t-tests to use.

10a1. Comparing variances parametrically: Fisher's F-test

Worked non-geological examples: available in most statistics texts.
Worked geological examples: *Field geology* (Miller & Kahn 1962, p.109); *Igneous petrology* (Le Maitre 1982, p.25); *Sedimentology* (Davis 1986, p.68); *Volcanology* (Till 1974, p.68).
Applied geological examples: *Metamorphic petrology* (Ward & Werner 1963-4); *Palaeontology* (Krinsley 1960); *Petrography* (Chayes 1950).
Statistical tables see under §8b2.
Stand-alone computer programs: Press et al.(1986,p.468).
Mainframe software: available in most statistical packages (MINITAB, SPSS, BMDP, etc.).
Microcomputer software: available in most statistical packages (Statview, etc.).

This straightforward test considers H_o: $s_1^2 = s_2^2$. It uses Fisher's F ratio, because F is itself the distribution of the expected ratio of two variances from a Normal population. The larger variance (the numerator), say for group 1, is divided by the smaller (the denominator), for group 2, and the resultant ratio s_1^2/s_2^2 compared with $F_{\{n1-1,n2-1\}}$. Large values imply rejection of H_o.

For the data of Table 10.2, we obtain $F = (0.066/0.059)^2 = 1.251$, which is well below the critical $F_{9,8}$ (90%) value of 2.56, indicating equal variances for the specific gravities of the two beds.

Table 10.2. Data used in worked example tests in §10
(Specific gravities of two beds of limestone from a cement quarry)

	Bed 1	Bed 2
	2.78	2.86
	2.66	2.69
	2.64	2.85
	2.63	2.74
	2.75	2.75
	2.71	2.79
	2.80	2.72
	2.60	2.73
	2.69	2.81
	2.67	•
Mean	2.693	2.771
Standard deviation	0.066	0.059
Range	0.20	0.17
$\sqrt{b_1}$ skew	+0.33	+0.30
b_2 kurtosis −3	−1.095	−1.227
Lilliefors' statistic	0.136	0.194
n_i	10	9

Note: the skew, kurtosis and Lilliefors' statistic all lead to acceptance of the null hypothesis that the two data-sets are Normally distributed.

10a2. Comparing Two Means Parametrically: Student's *t*-test (paired and unpaired).

Worked non-geological examples: available in most statistics texts.
Worked geological examples: *Igneous petrology* (Le Maitre 1982, p.25); *Sedimentology* (Cheeney 1983, p.68; Davis 1986, p.64; Koch & Link 1970, p.95); *Volcanology* (Till 1974, p.61).
Selected geological applications: *Igneous petrology* (Rao & Sial 1972); *Geology teaching* (Kern & Carpenter 1986); *Geomorphology & Glaciology* (Doornkamp & King 1971); *Palaeontology* (Hayami 1971; Reyment 1966); *Remote sensing* (Ammar et al. 1982).
Stand-alone computer programs: Press et al.(1986,p.465).
Statistical tables: see under §8b1.
Mainframe software: available in most statistical packages (MINITAB, SPSS, BMDP, etc.).
Microcomputer software: available in most statistical packages (Statview 512+, etc.)

This test is so named because it was devised by W.S.Gossett in 1908, writing under the pseudonym of "Student". It has four main variants depending on: (a) the result of the prior *F*-test (i.e. equality or inequality of the two variances); (b) whether the data are **paired** (related) or **unpaired** (unrelated — see §7f). Each variant can also be conducted in either a **one-tailed** or **two-tailed** manner (§7g).

(1) *Unrelated data, equal variances* (**unpaired *t*-test**). If the *F*-test (§10a1) indicates *equal* variances, we *pool* them, to get a better estimate of the population variance than either group variance would provide alone. The appropriate formula (the pooled variance being under the √ in the denominator), is:

$$t = \frac{\overline{X_1} - \overline{X_2}}{\sqrt{\dfrac{\Sigma(X_1-\overline{X_1})^2 + \Sigma(X_2-\overline{X_2})^2}{n_1+n_2-2}\left[\dfrac{1}{n_1}+\dfrac{1}{n_2}\right]}} \qquad [10.1]$$

X_1 here refers to group 1, X_2 to group 2. The resulting t value is looked up in tables in the usual fashion, against Student's $t_{\{n1+n2-2\}}$. In a one-tailed test, the two means are of course subtracted in the appropriate direction. For the data in Table 10.2, we obtain $t = 2.69$, which exceeds the 98% two-tailed critical t_{17} of 2.567, indicating that the mean specific gravities of the two beds are different.

(2) *Unrelated data, unequal variances* **(unpaired *t*-test):** If the F test (§10a1) indicates *unequal* variances, [10.1] is still used, but t must then be compared with an *adjusted* critical t calculated from:

$$t = \frac{s_1^2 t_{n_1-1} + s_2^2 t_{n_2-1}}{s_1^2 + s_2^2} \qquad [10.2]$$

A specialized **Welch statistic** is also available for this situation.

(3) *Related data, equal variances* **(paired *t*-test).** Here, *individual differences,* $d_i = [X_{1i} - X_{2i}]$ between the N paired observations (in the appropriate direction, for a one-tailed test) are used to give t:

$$t_{N-1} = \frac{\overline{d}}{\sqrt{\dfrac{\Sigma d^2 - \dfrac{(\Sigma d)^2}{N}}{N(N-1)}}} = \frac{\Sigma d}{\sqrt{\dfrac{N\Sigma(d^2) - (\Sigma d)^2}{N-1}}} \qquad [10.3]$$

where all summations are from 1 to N. By comparing [10.3] with [8.2b] and [8.37], it can be seen that this is in fact a test of H_0: $\overline{d} = 0$, i.e. that the mean difference $\overline{d}$ equals zero.

Table 10.3. Raw data and calculations for tests comparing related (paired) means or medians (from Table 10.2, but rearranged in order of increasing |*d*|)

Raw data			Paired *t*-test		Wilcoxon and van Eeden tests				
Bed no.	s.g. (before)	s.g. (after)	d	d^2	$\lvert d\rvert$	Rank of $\lvert d\rvert$	Signed rank of $\lvert d\rvert$, R	$\{(1+R/(N+1))\}/2$ quantile	Normal score, A_i
1	2.80	2.80	0	0	0	•	Tied values ignored		•
1	2.64	2.64	0	0	0	•	Tied values ignored		•
2	2.85	2.86	-.01	.0001	.01	1	**−1**	.474	-.0652
1	2.67	2.65	.02	.0004	.02	2.5	2.5	.566	.1662
1	2.63	2.61	.02	.0004	.02	2.5	2.5	.566	.1662
2	2.86	2.83	.03	.0009	.03	4.5	4.5	.618	.3002
1	2.75	2.72	.03	.0009	.03	4.5	4.5	.618	.3002
2	2.72	2.68	.04	.0016	.04	6	6	.658	.4070
2	2.75	2.70	.05	.0025	.05	7.5	7.5	.697	.5158
2	2.79	2.74	.05	.0025	.05	7.5	7.5	.697	.5158
2	2.74	2.68	.06	.0036	.06	9	9	.737	.6341
2	2.69	2.62	.07	.0049	.07	10	10	.763	.7160
1	2.69	2.77	-.08	.0064	.08	11	**−11**	.211	-.8030
1	2.60	2.69	-.09	.0081	.09	12.5	**−12.5**	.171	-.9502
1	2.66	2.57	.09	.0081	.09	12.5	12.5	.829	.9502
2	2.73	2.62	.11	.0121	.11	14	14	.868	1.117
1	2.71	2.59	.12	.0144	.12	15	15	.895	1.2536
1	2.78	2.65	.13	.0169	.13	16	16	.921	1.4118
2	2.81	2.67	.14	.0196	.14	17	17	.947	1.6164
			Σd= .78	Σd^2=.1031			$\Sigma R-$ = 24.5	ΣA =	8.252

Suppose the limestone samples in Table 10.2 are subjected to some purification treatment, and the quarry manager is interested in whether this *decreases* their specific gravity. Such 'before' and 'after' represent an archetypal application of the paired t test. The difference between the mean s.g. of the beds is here irrelevant, and the test can be calculated as in Table 10.3 on all 19 samples. This yields $\bar{d}$ = 0.041, $t = 0.78/\sqrt{\{19 \times 0.1031 - 0.78^2)/18\}} = 2.842$, which almost reaches the 99.5% (one-tailed) value of $t_{18}(2.898)$, indicating a highly significant decrease in the s.g. after treatment.

There are many examples in the geological literature of inappropriate use of these t-tests. For example, Rao & Sial (1972) use t-tests on chemical analyses of phonolites, one oxide at a time, where a multivariate approach (e.g. discriminant analysis, §22) would have been more appropriate.

10b. Comparing two small samples: Substitute Tests based on the Range

Worked geological examples: Miller & Kahn (1962, p.103);
Selected geological applications: *Palaeontology:* Krinsley (1960).
Statistical tables: Beyer (1968); Dixon & Massey (1969); Langley (1970, p.367). Note that some tables give critical values for *twice* the test statistic of others, i.e. $2T$ as defined by (10.6).
Mainframe and microcomputer software: rarely programmed; easy to calculate manually.

For very small samples, the *range* [i.e. $\omega = X_{max} - X_{min}$] competes increasingly effectively with the standard deviation, s, as an estimate of population *variance*, σ, having a relative efficiency of about 91% for $N \leq 7$ and 85% for $7 < N \leq 10$ (David 1981). The range is even more relatively efficient for non-Normal data. A series of substitutes for the tests of §10a have consequently been developed, which replace s with ω. Though sometimes rather derogatorily referred to as 'inefficient' tests, these are actually perfectly valid and extremely easily calculated, if used in the correct statistical context. Range ratios ω_1/ω_2 have for example been used as substitutes for ratios of variances in a 'substitute F' test, but the most widely used of these substitute tests is that of the standard t-test, now considered.

The substitute test statistic for comparing the means of two very small data-sets, is:

$$T = \frac{\overline{X_1} - \overline{X_2}}{\omega_1 + \omega_2} \qquad [10.4]$$

For the data of Table 10.2, this yields $T = (2.771 - 2.693)/(0.20 + 0.17) = 0.21$. Critical values of T are tabulated only for $n_1 = n_2 \leq 20$; but as $n_1 \approx n_2$ here, the test statistic can reasonably be compared with the 99.5% critical values of ≈ 0.2 for $N = 9$ and 10. Since it exceeds these, we conclude (as with the unpaired t test of the previous section) that the s.g. of the two limestone beds differs strongly.

10c. Comparing Medians of Two Related (paired) Groups of Data Nonparametrically

It now becomes more appropriate to separate the various comparison criteria, bases of calculation, and data-types than in the previous two sections, because various nonparametric tests have quite different names and apparently quite different calculation procedures, unlike the variants of the t-test. To impose some semblance of order, tests are described under each section below in order of increasing power, from the 'quick-and-dirty' to the most powerful (those based on Normal scores).

10c1. A crude test for related medians: the Sign Test

Worked geological examples: *Environmental geology* (Lewis 1977, p.205).
Statistical tables: Lewis (1977,table 3b), Powell(1982), Siegel (1956), Sprent (1981): critical values are actually lower quantiles of the binomial distribution with p = 0.5.
Mainframe software: in most packages (BMDP, SPSS, etc.) and libraries (e.g. NAG routine G08AAF).
Microcomputer software: rarely programmed.

As mentioned in §9e4, the Sign Test is also used for generating confidence limits on medians, when its very simple rationale is not a drawback. Here, however, the differences between the various pairs of data-values are merely assigned '+' or '–', depending on which value is larger; the total number of pluses then becomes the test statistic. If the two medians are equal, the probability of both '+' and'–' is 0.5; the probability of r pluses is therefore ${}^{N}C_{r}(0.5)^{N}$, by [8.7a]. Because no account is taken of the *magnitude* of the differences, the test is a weak one, with an ARE of only $2/\pi$ = 63.7% relative to the equivalent parametric test (the paired t-test in §10a).

Note that because of its mathematical basis and divergent uses, the Sign Test may appear in statistical tables and textbooks under 'Binomial Distribution' (e.g. Powell 1982) or under 'confidence limits' (e.g. Sprent 1981) as well as under 'two-group comparisons'.

For the data in Table 10.3, we have 3 '+', 14 '–' differences and 2 ties (unchanged values) among N = 19 pairs. A binomial distribution table for p = 0.5 shows that 3 or fewer '+' has an associated probability of 0.0022, so we conclude with ≈98% confidence that the medians of the 'before' and 'after' data are different. This agrees with the paired t test, but the confidence level is less, because the Sign Test uses less of the information in the data. Nevertheless, the Sign Test could be used on far cruder data: for example, we could cut an original limestone sample into two halves, and subject only one half to the treatment. Assuming no volume change, we could then merely balance the two (untreated and treated) samples on scales and assign a '+' or '–' merely from which side of the balance went down, without actual calculation of s.g. values. This would be vastly quicker for a large batch of samples, and the cost saving would probably far outweigh the 1% confidence loss from the test.

10c2. A test for 'before-and-after' situations: the McNemar Test for the Significance of Changes

Worked and Selected geological applications: None known.
Worked non-geological examples: see glossary entry.
Statistical tables: Conover (1980), Lewis (1977), Siegel (1956). See also Powell(1982,p.25).
Mainframe software: SPSS (p.817-8).
Microcomputer software: None known.

This test is most commonly used for analyzing 'before and after' situations: changes in the same individual students' scores on a particular test before and after a teaching course, or the quality of analytical results before and after some technical modification is introduced. It can also be used for matched pairs (e.g. to compare male and female twins doing the same test).

This test uses a type of **contingency table** (§13d1), and is analyzed in much the same way.

Here, the following is tested for significance against χ^2 (for definitions of b, c see Table 13.5):

$$\chi^2_1 \approx \frac{(|b - c| - 1)^2}{b + c} \qquad [10.5]$$

The –1 is a **correction for continuity.** Formula [10.5] should only be used if $b,c > 5$; otherwise the probability is given exactly by [8.5a] with $r = b, N = [b + c], p = 0.5$.

10c3. A more powerful test for related medians: the Wilcoxon (matched-pairs, signed-ranks) Test

Worked geological example and statistical theory: *Environmental geology* (Lewis 1977, p.207ff).
Worked non-geological examples : Conover(1980,p.280); Bhattacharyya & Johnson(1977,p.519).
Statistical tables: Conover (1980), Dixon & Massey (1969), Lewis (1977), Siegel (1956), Sprent (1981).
Mainframe software: in most statistics packages (e.g. SPSS) or libraries (e.g. NAG routine G08ABF).
Microcomputer software: programmed in many statistics packages (e.g. Statview 512+™).

This test is equivalent to performing a *paired* t-test on the ranks of the data, rather than on the raw data themselves; it has an ARE of $3/\pi = 95.5\%$ relative to the paired t-test. As indicated under §9e4, it is also used to estimate nonparametric confidence intervals for medians from *symmetric* populations.

The *differences* between the raw data-values, d, are first calculated (as in the paired t-test, §10a2), carefully noting their signs, and then ranked. Then $R+$, the sums of ranks of the positive differences, and similarly $R-$ are obtained. For small samples, $\min\{R+,R-\}$ is the test statistic and is compared with exact critical values in tables; $R+$ should equal $R-$ if the null hypothesis of equal medians is true. For larger samples, the test statistic is converted into a standard Normal deviate, using its theoretical mean and standard deviation according to the **Central Limit Theorem** (§8a1), and the following:

$$Z = \frac{\min\{R+,R-\} - \dfrac{N(N+1)}{4}}{\sqrt{\dfrac{N(N+1)(2N+1)}{24} - \dfrac{\Sigma(t^3 - t)}{2}}} \qquad [10.6]$$

'No change' data-pairs are ignored. In this formula, 't' is the number of observations tied for a given rank: that is, if there had been two equal specific gravities 2.78 in Table 10.3, which both had corresponding ranks of $3\frac{1}{2}$, then 't' for this rank would have been 2. The reason for this adjustment (which applies to most other nonparametric tests discussed in later sections) involves almost the only assumption made by nonparametric tests: namely that the underlying distribution is *continuous*. If a variable can take *any* real value (7.437628184515... etc.), then ties are by definition impossible, but conversely, the more ties that occur, the *less* continuous can the underlying distribution be. The 't' adjustment allows for this minor violation of assumptions. Note that an *even* number of tied values ('t' even) must all have $\frac{1}{2}$ ranks, and an *odd* number of ties must all have integer (whole) ranks.

From the calculations in Table 10.3, we obtain $R+ = \mathbf{128\frac{1}{2}}$, $R- = \mathbf{24\frac{1}{2}}$, which is *less than* the critical 99.5% value of 33 for = 19, so once again we conclude that the treatment decreases the s.g. of the limestones. Note that this and the Mann-Whitney test (§10d2) are among very few where the test statistical has to be *less than* (rather than to exceed) the tabulated value, to be significant at that level.

10c4. The most powerful test for related medians, based on Normal scores: the Van Eeden test

Worked non-geological examples: Conover (1980, p.320-1).
Statistical tables: uses Normal Distribution tables (§8a1).
Worked geological examples , geological applications and software: none known.

This **Normal scores** analogue of the Wilcoxon signed-ranks test examines exactly the same hypothesis under the same assumptions, but replaces the signed-ranks R_i with *signed Normal scores,* A_i, which here are $[1 + R_i/(\nu + 1)]/2$th quantiles from a Normal distribution, where ν is the number of *non-zero* differences between the two sets of ranks. A_i has the same sign as R_i. The test statistic is:

$$Z = \frac{\sum_{i=1}^{i=\nu} A_i}{\sqrt{\sum_{i=1}^{i=\nu} A_i^2}} \qquad [10.7]$$

From Table 10.3, we obtain a test statistic of $8.252/\sqrt{11.728} = 2.41$ ($\nu = 17$), which compares with the 99% critical Z value of 2.3263, and hence once again leads us to infer a decrease in s.g.

10d. Comparing Locations (medians) of Two Unrelated Groups Nonparametrically

This is perhaps the most common situation in geology, so that the tests included here are of potentially great importance. As in the case of paired data, both crude and more powerful tests are available, the latter taking more information in the data into account.

Probably the quickest and visually most effective way of comparing medians is via **notched box-plots** (Fig. 10.2; cf. Fig.8.9):

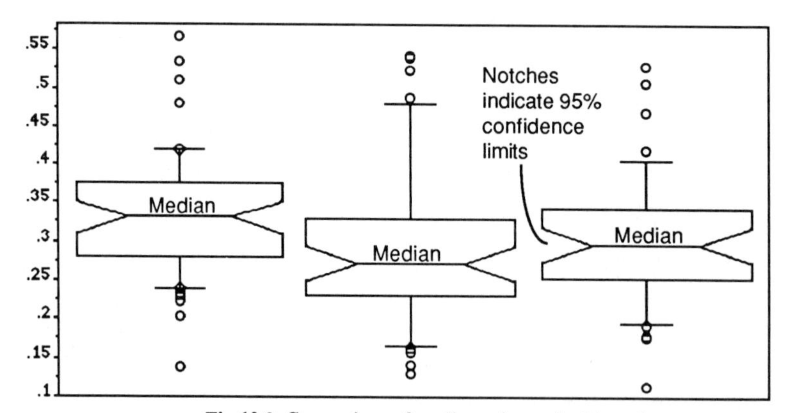

Fig.10.2. Comparison of medians via notched box-plots
The medians of **(a)** and **(b)** differ, because the notches do not overlap,
whereas the median of **(c)** equals both **(a)** and **(c)** statistically, because they *do* overlap.

However, the next 4 sections describe tests which perform the same function *numerically*.

10d1. A crude test for unrelated medians: the Median Test

Further discussion: Siegel(1956,p.111).
Worked geological examples: *Sedimentology* (Cheeney 1983, p.75).
Statistical tables: uses χ^2 tables (see §12a).
Stand-alone computer programs: Press et al.(1986,p.472).
Mainframe software: some packages (e.g. SPSS^X, p.824-5) and libraries (e.g. NAG routine G08ACF).
Microcomputer software: rarely specifically programmed owing to similarity with χ^2 test .

This simple test uses the *overall* median for the two groups of data combined to dichotomize each group, to generate a 2 x 2 contingency table (§13d1). For the data in Table 10.2, the overall median is 2.73, which yields the following table (one value equalling the overall median in group 2 is lost):

	Number of values > combined median	Number of values < combined median	Totals
Group1	*a* (3)	*b* (7)	N_1 (10)
Group 2	*c* (6)	*d* (2)	N_2 (8)
Totals	(9)	(9)	N (18)

The test statistic is then calculated by [13.2a] and compared with χ^2 for 1*df*. For the data in Table 10.2, χ^2 = 3.6, which is significant at ≈94.6%, leading to acceptance of H_1 (unequal medians). The basis of the test is that *ab* and *cd* should be equal under the null hypothesis (see §13d1 for details). Like the Sign Test, this relatively crude test has an ARE of only $2/\pi$ = 63.7% relative to the unpaired *t*-test (when used on Normally distributed data), because it virtually ignores the numerical values of the data, effectively reducing them to '+' or '–' values relative to the overall median. It can nevertheless be useful where data can only be measured to this level, and cannot even be individually ranked.

10d2. A quick and easy test for unrelated medians: Tukey's *T* test

Worked non-geological examples: Sprent(1981, p.130).
Statistical tables: Conover(1980,p.408); for $N \le 20$ and $N_1 \approx N_2$, critical values (two-tailed) are $T \ge 7$ (95%), ≥10 (99%) and ≥13 (99.9% levels).
Computer software: test is too straightforward to warrant computer implementation.

This is one of the quickest tests available, and probably the only one where the critical values can be remembered without consulting tables! Provided the two data-sets are of subequal size and < 20 values, the test statistic is simply the number of non-overlapping values, when the two data-sets are set out in overall rank order. For the data of Table 10.2, distinguishing the data-sets as 2.69/2.69, we obtain:

<u>2.60 2.63 2.64 2.66 2.67</u> 2.69 2.69 2.71 2.72 2.73 2.74 2.75 2.75 2.78 2.79 2.80 <u>2.81 2.85 2.86</u>

Hence T = 8 (the <u>underlined</u> values), which is significant at the 95% level, so H_0 (equal medians) is rejected. If ties occur, they score ½. Of course, if there is no overlap at all, H_1 is accepted at whatever critical level corresponds to N (95% for N = 7–9, 99% for N = 10–12, 99.9% for N ≥13).

If we instead make a one-tailed test of H_0: A < B, we can stop the test and accept H_0 before even determining T, if either (i) the smallest value in group A exceeds the smallest value in group B; or (ii) the largest value in A exceeds the largest value in B. In the above case, the test would proceed, and T = 7 would lead to rejection of H_0 at the 97.5% level. If H_0 had been B > A, however, the test would have been aborted (and H_0 accepted), before determining T.

10d3. A powerful test for unrelated medians: the Mann-Whitney test.

Further discussion: Basu (1967); Conover(1980); Iman (1976); Siegel(1956, p.116ff).
Worked geological examples: *Geochronology* (Miller & Kahn 1962, p.104); *Petroleum exploration* (Davis 1986, p.93-4); *Sedimentology* (Miller & Kahn 1962,p.131; Cheeney 1983, p.78); {*Sedimentary geochemistry:* (Till 1974, p.124ff: **NB.** Till's formula (7.4) and succeeding calculations incorrect}!
Selected geological applications: *Chemostratigraphy* (Rock 1986c,e); *Geomorphology* (Doornkamp & King 1971); *Igneous petrology* (Al-Turki & Stone 1978).
Statistical tables: Auble (1953); Beyer(1968, p.408); Cheeney(1983,p.84); Powell(1982,p.64); Siegel (1956,p.271).
Stand-alone computer programs : Rock (1986f).
Mainframe software: in most statistics packages (e.g. SPSS) or libraries (e.g. NAG routine G08ADF).
Microcomputer software: programmed in many statistics packages (e.g. Statview 512+™).

This is one of the most powerful nonparametric tests, with a power-efficiency of $3/\pi$ (≈95.5%) relative to the two-group unpaired *t*-test (to which it is exactly equivalent, if the calculation is made on the ranks rather than raw data). It is unfortunately referred to under several different names in the literature: e.g. **Wilcoxon rank-sum test, Wilcoxon-Mann-Whitney test**, and ***U*-test.** Some of these variants use slightly different calculations, but all are in the end equivalent.

The test involves ranking the two groups of data as one combined set, and then obtaining the sums of ranks for each group, $®_1$ and $®_2$. The *smaller* of two values, U_1 and U_2 is then taken, where:

$$U_1 = n_1 n_2 + n_1(n_1 + 1)/2 - ®_1 \qquad [10.8a]$$

$$U_2 = n_1 n_2 + n_2(n_2 + 1)/2 - ®_2 = n_1 n_2 - U_1 \qquad [10.8b]$$

If H_0 is true, U_1 and U_2 should be equal. For n_1 and $n_2 \leq 20$, tables of critical values are consulted; a significant value of U is *less than* critical values in these tables, since the *smaller* the value of U, the *wider* the divergence between the respective rankings. For larger groups of data, U is converted to a standard Normal deviate (§8a1), using the following formula based on [8.3]:

$$Z = \frac{U - \frac{n_1 n_2}{2}}{\sqrt{\frac{n_1 n_2}{12N(N-1)}\left[N(N^2-1) - \sum t(t^2-1)\right]}} \qquad [10.8c]$$

where, as before, 't' refers to the number of tied values for a given rank. In this case, significant values are *larger than* critical values of the standard Normal deviate (i.e. 2.326 at 99%, etc.) This difference between the small and large sample methods can cause confusion to the unwary. [Note that different textbooks give confusingly conflicting indications of the values of n_1 and n_2 at which this large sample approximation becomes appropriate; the above recommendation follows Rock (1987f)].

From the calculations in Table 10.4, we obtain $®_1 = 72$, $®_2 = 118$, $U_1 = 17$, $U_2 = 83$, so U_1 is taken, which lies between the 95% (20) and 99% (13) critical values of U. We therefore infer unequal median s.g. for the two limestone beds at somewhere around the 97% significance level.

The Mann-Whitney test is also useful as an additional, nonparametric test to those discussed in §9f3 for identifying **outliers** (see Miller & Kahn 1962, p.131 for worked example). The possible outlier is considered as a group of size 1 (i.e. $n_1 = 1$), so that $®_1$ is also the rank in the *combined* data-sets of this one value. The procedure will only work in practice where n_2 exceeds ≈19, so that either formula [10.8c] is applicable, or tabulated critical values for smaller data-sets (with $n_1 = 1$) are non-zero.

Table 10.4. Calculations for the Mann-Whitney and Terry-Hoeffding tests (data of Table 10.2)

Bed 1			Bed 2		
S.g.	Rank†	Rankit¶	S.g.	Rank†	Rankit¶
2.78	14	.5477	2.79	15	.7066
2.69	6.5	.5477	2.86	19	1.8445
2.71	8	.2637	2.85	18	1.3799
2.80	16	.8859	2.81	17	1.0995
2.75	12.5	.4016	2.72	9	.1307
2.63	2	1.3799	2.69	6.5	.4016
2.60	1	1.8445	2.73	10	0
2.64	3	1.0995	2.75	12.5	.2637
2.67	5	.7066	2.74	11	.1307
2.66	4	.8859	ΣRanks =	118	$\Sigma E(N,R)$ = 5.957 for smaller group
ΣRanks =	72				

† NB. These are the *overall* ranks of each s.g. in the *combined* data-set (Bed 1 + Bed 2).
¶ See explanation in §10d4 below.

10d4. The Normal scores tests for unrelated medians: the Terry-Hoeffding test

Further discussion: and statistical tables: $N \leq 20$: Powell(1982,p.73); $N > 20$ use (10.9b).
Geological examples and computer programs: none known.

This performs the same task in exactly the same way as the Mann-Whitney test, but the test statistic T is the sum of **rankits** corresponding to the *smaller* group of data, rather than the sum of raw ranks:

$$T = \sum_{i=1}^{i=n_1} E(N,R_i) \qquad [10.9a]$$

where $E(N,R)$ is the rankit for the Rth rank out of N values. This gives the test greater power. T as in [10.9a] is used for a one-sided test, testing that the *smaller* group has the *larger* median (for the converse we use $-T$). For a two-sided test, $|T|$ is used. For $N > 20$, the approximate critical value is:

$$T = r_{N-2}\sqrt{\frac{n_1 n_2 \sum_{i=1}^{i=N}\{E(N,i)\}^2}{N}} \qquad [10.9b]$$

where r is the critical value of Pearson's r (§13a), provided n_1/n_2 and n_2/N are not too small.

From Table 10.4 we obtain T for Bed 2 (the smaller group) = 5.957, which exceeds the 99.9% critical value of 5.929, confirming once again that there is a strong difference in the two s.g. medians.

10e. Comparing the Scale of Two Independent Groups of Data Nonparametrically

10e1. The Ansari-Bradley, David, Moses, Mood and Siegel-Tukey tests

Further non-geological discussion: Ansari & Bradley(1960); Kotz & Johnson(1982–8); Laubscher et al. (1968); Mood(1954); Moses(1952); Siegel & Tukey(1960).
Worked examples, selected geological applications and microcomputer software: none known.
Statistical tables: can be generated from computer program in Dinneen & Bradley(1976).
Mainframe software: implemented in NAG(routine G08BAF).

These tests use variants of similar systems for assigning ranks, but differ in that some assume equality of medians, whereas others makes allowance for differences of median. We illustrate with the Ansari-Bradley only. The two groups combined are ranked and then, working from the outer values, we assign new ranks of 1,2... successively inwards. Hence for N even we obtain a set of ranks 1,2...N...2,1 whereas for N odd we obtain 1,2...$(N-1)/2$, $(N+1)/2$, $(N-1)/2$...2,1. For the data in Table 10.2, differentiating the two groups as 2.63 and 2.75, we obtain the following:

Data:	2.60	2.63	2.64	2.66	2.67	2.69	2.69	2.72	2.72	2.73	2.74	2.75	2.75	2.78	2.79	2.80	2.81	2.85	2.86
Rank:	1	2	3	4	5	6	7	8	9	10	9	8	7	6	5	4	3	2	1

The test statistic W is the sum of new ranks for the *smaller* group (here, the underlined set of ranks for bed 2, which equals 49). For very large samples W is converted to a standard Normal deviate:

$$Z = \left[W - \frac{n_1(N+2)}{4}\right]\sqrt{\frac{48(N-1)}{n_1 n_2(N+2)(N-2)}} \quad \text{where } N \text{ is even and}$$

$$Z = \left[W - \frac{n_1(N+1)^2}{4N}\right]\sqrt{\frac{48N^2}{n_1 n_2(N+1)(N^2+3)}}, \quad \text{where } N \text{ is odd} \qquad [10.10]$$

Table 10.5. Calculations for the Squared Ranks test on the data of Table 10.2

Bed 1				Bed 2			
s.g.	$\lvert$s.g.–2.693$\rvert$‡	Rank†	Rank²	s.g.	$\lvert$s.g.–2.771$\rvert$‡	Rank†	Rank²
2.78	.087	16	256	2.86	.089	17	289
2.66	.033	7	49	2.69	.081	15	225
2.64	.053	11	121	2.85	.079	14	196
2.63	.063	13	169	2.74	.031	6	36
2.75	.057	12	144	2.75	.021	4	16
2.71	.017	2	4	2.79	.019	3	9
2.8	.107	19	361	2.72	.051	10	100
2.6	.093	18	324	2.73	.041	9	81
2.69	.003	1	1	2.81	.039	8	64
2.67	.023	5	25				
		Σ squared ranks =	1454				

‡ *Absolute* deviation from the mean of the *particular* group of data (i.e. 2.693 for Bed 1, 2.771 for Bed 2).
†*Overall* rank obtained from all 19 *absolute* deviations, i.e. ignoring sign (cf. Table 10.3).

10e2. The Squared Ranks Test

Worked geological examples and proprietary software: none known.
Worked non-geological examples: Conover (1980, p.239).
Selected geological applications: *Metamorphic geochemistry:* Rock (1986e).
Statistical tables: Conover(1980,p.454).
Stand-alone computer programs: Rock (1986f).

This test first requires the absolute deviations of all data-values from the mean of *each* group:

$$U_i = |X_i - \overline{X}_1| \text{ for group 1, and } V_i = |X_i - \overline{X}_2| \text{ for group 2;}$$

these are then combined and ranked, to yield $®(U_i)$ *and* $®(V_i)$. If there are no ties, the test statistic T is then simply $®(U_i)$; if there are ties, the following formula is used:

$$T = \frac{\sum_{i=1}^{i=n_1}[®(U_i)]^2 - n_1\overline{®^2}}{\sqrt{\frac{n_1 n_2}{N(N-1)}\left[\sum_{i=1}^{i=n_1}[®(U_i)]^4 + \sum_{i=1}^{i=n_2}[®(V_i)]^4\right] - \frac{n_1 n_2}{(N-1)}\left(\overline{®^2}\right)^2}} \quad \text{[10.11a], where}$$

$$\overline{®^2} = \frac{1}{N}\left\{\sum_{i=1}^{i=n_1}[®(U_i)]^2 + \sum_{i=1}^{i=n_2}[®(V_i)]^2\right\} \quad \text{[10.11b]}$$

For n_1 and $n_2 \leq 10$, T is compared with critical values in tables. A *two-tailed* test is significant at, say, the 95% level if T exceeds the 95% critical value *or is less than the 5% critical value*. A *one-tailed* test is significant if T exceeds the 95% critical value in the predicted direction, or is less than the 5% critical value in the other direction. For $N > 20$, the critical value T_c is given by:

$$T_c = \frac{n_1(N+1)(2N+1)}{6} + Z\sqrt{\frac{n_1 n_2(N+1)(2N+1)(8N+11)}{180}} \quad \text{[10.11c]}$$

where Z, as usual, is a standard Normal deviate at the appropriate level (i.e. 2.3263 @ 99%, etc.).

From the calculations in Table 10.5, we obtain $T = 1454$ for Bed 1. The 10% and 90% *two-tailed* critical values for $n_1 = 10$, $n_2 = 9$ are 883 and 1715, so T lies between these and is not significant — the same result as from the parametric F test in §10a1.

10e3. The Normal scores approach: the Klotz Test

Worked geological examples, geological applications and proprietary software: none known.
Worked non-geological examples: Conover (1980, p.321).
Statistical tables: uses Normal Distribution tables (see §8a1).

For Normal scores as usual notated A_i, the test statistic here is defined by the following formula:

$$Z = \frac{\sum_{i=1}^{i=n_1} A_i^2 - \frac{n_1\sum_{i=1}^{i=N} A_i^2}{N}}{\sqrt{\frac{n_1 n_2}{N(N-1)}\left[\sum_{i=1}^{i=N} A_i^4 - \frac{1}{N}\left\{\sum_{i=1}^{i=N} A_i^2\right\}^2\right]}} \quad \text{[10.12]}$$

From the Normal scores as calculated in Table 10.6, we obtain:

$$Z = \frac{8.967 - \frac{10\text{x}14.114}{19}}{\sqrt{\frac{10\text{x}9}{19\text{x}18}\left[23.962 - \frac{(14.114)^2}{20}\right]}} = 0.8015$$

which, being far below the 90% standard Normal deviate value of 1.2816, is again non-significant.

Table 10.6. Determination of Normal scores for the Klotz test on the data of Table 10.2

Bed 1					Bed 2				
Data	Mean dev‡	Rank, R†	$R/(N+1)$	Normal score, A_i	Data	Mean dev‡	Rank, R†	$R/(N+1)$	Normal score A_i
2.67	-.023	9	.45	-0.1257	•	•	•	•	•
2.66	-.033	7	.35	-0.3853	2.81	.039	14	.7	0.5244
2.75	.057	15	.75	0.6745	2.86	.089	18	.9	1.2816
2.71	.017	12	.60	0.2533	2.85	.079	16	.8	0.8416
2.69	-.003	11	.55	0.1257	2.73	-.041	6	.3	-0.5244
2.78	.087	17	.85	1.0364	2.72	-.051	5	.25	-0.6745
2.63	-.063	3	.15	-1.0364	2.69	-.081	2	.1	-1.2816
2.8	.107	19	.95	1.6449	2.79	.019	13	.65	0.3853
2.6	-.093	1	.05	-1.6449	2.75	-.021	10	.5	0
2.64	-.053	4	.20	-0.8416	2.74	-.031	8	.4	-0.2533

‡ *Absolute* deviation from mean of the *particular* group (i.e. 2.693 for Bed 1; 2.771 for Bed 2), as for squared ranks test.
†*Overall* rank obtained from all 19 deviations, i.e. *taking account of sign* (cf. squared ranks test in Table 10.5).

10f. Comparing the overall distribution of two unrelated groups nonparametrically

The preceding tests have been sensitive to differences *either* of location (mean, median) *or* of scale (standard or median deviation). The following two-group tests are sensitive to *any* differences between the two distributions — including skew, kurtosis, etc. (Fig.10.1). Once again, they have different efficiencies depending on how much of the original information in the data they use.

The Smirnov (§10f2) test complements the Kolmogorov Test (§8e2), and is another **empirical distribution function (EDF) test.** A third EDF test, the **Cramér-Von Mises,** appears in some textbooks. However, it is far less widely used, and provides no useful facilities above those inherent in the Smirnov Test. Though omitted here, a variant of it will be encountered in §17.

10f1. A crude test: the Wald-Wolfowitz (two-group) Runs Test

Worked geological examples: *Glaciology* (Miller & Kahn 1962, p. 341; Lewis 1977, p.204); *Sedimentary petrology* (Cheeney 1983, p.76);
Statistical tables: Dixon & Massey (1969,p.513), Downie & Heath (1974), Lewis (1977), Powell (1982,p.76), Siegel (1956).
Mainframe software: programmed in some large statistics packages (e.g. SPSS).
Microcomputer software: programmed in some statistics packages (e.g. Statview 512+™).

This depends on determining the number of *runs, R,* for two groups ranked together in one row. Thus

$$\overline{AA}\underline{BBB}\overline{A}\underline{BBBBB}\overline{AAAA}\underline{BBBBB}$$

contains 6 runs. The lowest possible R is clearly 2, when the two groups show no overlap whatever ($\overline{AA}...\overline{AA}\underline{BB}...\underline{BB}$), while its highest possible value is N, where two groups of equal size are fully interspersed ($\overline{A}\underline{B}\overline{A}\underline{B}\overline{A}\underline{B}...\overline{A}\underline{B}$). Hence the *smaller* R, the more significant it is. For small N, tables of critical values are consulted, whereas for larger N, R is converted to a standard Normal deviate thus:

$$Z = \frac{R - \dfrac{2n_1n_2}{N} - 1}{\sqrt{\dfrac{2n_1n_2(2n_1n_2 - N)}{N^2(N-1)}}} \qquad [10.13]$$

As for Tukey's quick test (§10d2), we rearrange the data in Table 10.2 in rank order, to obtain:

2.60 2.63 2.64 2.66 2.67 2.69 2.69 2.71 2.72 2.73 2.74 2.75 2.75 2.78 2.79 2.80 2.81 2.85 2.86

which gives a *minimum* of 8 runs (moving the tied values around actually increases R). For a 95% significant result, we require 5 runs *or fewer*, so this result is non-significant, indicating equal medians.

The Runs Test will reappear under our discussion of sequences and orientation data (§16-17).

Table 10.7. Exact method of calculating Smirnov test statistic (data of Table 10.2)

Bed 1		Bed 2		Cumulative difference
Data in order	Cumulative proportion	Data in order	Cumulative proportion	
2.60	0.1 (1 out of 10)			
2.63	0.2			
2.64	0.3			
2.66	0.4			
2.67	0.5			
2.69	0.6	2.69	0.1111 (1 out of 9)	0.4889
2.71	0.7		0.1111	**0.5889** = d_{max}
	0.7	2.72	0.2222	0.4778
	0.7	2.73	0.3333	0.3667
	0.7	2.74	0.4444	0.2556
2.75	0.8	2.75	0.5555	0.2445
2.78	0.9		0.5555	0.3445
	0.9	2.79	0.6666	0.2334
2.80	1.0		0.6666	0.3333
	1.0	2.81	0.7777	0.2223
	1.0	2.85	0.8888	0.1112
	1.0	2.86	1.0000	0.0000

10f2. A powerful test: the Smirnov (two-group Kolmogorov–Smirnov) Test

Further non-geological discussion: Abrahamson (1967); Lewis (1977, p.212ff); Steck (1969).
Worked geological examples: *Data segmentation* (Levine et al. 1981); *Igneous petrology* (Till 1974, p.126ff); *Palaeontology* (Miller & Kahn 1962,p.464); *Structural geology* (Cheeney 1983, p.45).
Selected geological applications: *Applied geochemistry* (Howarth & Lowenstein 1971); *Isotope geochemistry* (Burrows et al. 1986); *Metamorphic stratigraphy* (Rock 1986e).
Statistical tables: Beyer(1968); Borovkov et al. (1964); Kim (1969); Kim & Jennrich(1970); Massey (1951,1952); Powell(1982,p.68); Siegel(1956); Steel & Torrie(1980,p.621-3); approximate graph also available in Miller & Kahn(1962,p.468).
Stand-alone computer programs: Press et al.(1986,p.475); Rock (1986f).
Mainframe software: programmed in some large statistics packages (e.g. SPSS).
Microcomputer software: programmed in many statistics packages (e.g. Statview 512+™).

This is the two-group equivalent of the one-group Kolmorogov Test considered in §8e2. It is calculated in exactly the same way, but this time the maximum difference d_{max} between the cumulative frequency distributions of the *two* data-sets is of course determined. The literature is regrettably contradictory concerning the 'correct' decision rule for testing d_{max}, because of the complicated small-sample distribution of the text statistic. There is also a medley of confusing and mutually inconsistent tables. The rules here follow the recommendations briefly justified in Rock (1986f, p.762):

(1) $n_1 = n_2 < 41$; consult tables of exact critical values (listed above).

(2) $n_1 = n_2 > 40$; use the large sample approximation $A/\sqrt{n_1}$, where A varies with the significance level (from 1.52 @ 90% to 2.30 @ 99.5%; e.g. in Steel & Torrie 1980, p.621).

(3) $80 < n_1 < 100 < n_2$; use the large sample approximation $bN/\sqrt{n_1 n_2}$, where b again varies with the significance level (from 1.22 @ 90% to 1.95 @ 99.9%; see Siegel 1956, table M; Steel & Torrie 1980,p.622). *Highly approximate* graphs of critical values are alternatively available in Miller & Kahn (1962, p.468-9).

(4) For other cases, most statistical texts only provide exact critical values for *some* combinations of $n_1 \neq n_2$ (Powell 1982, p.68 gives the most complete); these are still regrettably inconsistent. However, Kim & Jennrich (1970) give *exact* values for *all* n_1 , n_2, together with a FORTRAN algorithm.

Care is needed with these tables, since in different cases the critical region may variously *include* or *exclude* the tabulated value (cf. respectively Siegel 1956, table I; Steel & Torrie 1980, table A.23A).

For the data in Table 10.2, we can calculate d_{max} exactly by tabulating the two sets of data as in Table 10.7 (shown in full for clarity, but only the top few lines are strictly necessary). Thus d_{max} = 0.5889, which lies between the two-tailed 95 and 99% critical values of 52/90 = 0.5777 and 62/90 = 0.6889 respectively; H_0 of identical distributions is thus rejected at about the 97% level.

For larger *N*, d_{max} would normally be obtained graphically, as in Fig.8.10, although this can make the precise value somewhat sensitive to the choice of class limits.

10g. A Brief Comparison of Results of the Two-group Tests in Topic 10

Table 10.8 compares the significance levels with which we have rejected the null hypothesis of equal location, scale or overall form using the various equivalent tests on the data of Table 10.2. With the single exception of the Wald-Wolfowitz Runs test result (which confirms the problems with this test already indicated by the tied data problem in §10f1), the results are congruent (i.e. the null hypotheses of equal locations are all firmly rejected, but those of equal scales accepted). The simpler tests generally give lower significance levels than the more powerful ones, as expected. The parametric tests give similar significance levels to the nonparametric, because this is a marginal situation in which the Normality suggested in Table 10.2 (which favours the parametric tests) is cancelled by the small numbers of values (which favour the nonparametric). In other words, the Normality tests have low power on these small data-sets, so that the Normality assumption is still insecure. Note particularly also that the Terry-Hoeffding test gives the highest significance level for the unrelated location comparisons because it is probably the most powerful test in this marginal situation, while the substitute *t*-test gives a higher level than the standard *t*-test precisely because it is designed for small numbers of data-values.

Table 10.8. Significance levels for rejection of various null hypotheses on the data of Table 10.2 & 10.3

Comparing unrelated location		Comparing related location		Comparing scale		Comparing overall form	
Test	Sig.lev.	Test	Sig.lev.	Test	Sig.lev	Test	Sig.lev.
Substitute *t*-test	≈99.5%	Sign Test	≈98%	Squared ranks test	<90%	Runs test	<95%
Median test	94.6%	Wilcoxon Test	≈99.5%	Klotz test	<90%	Smirnov test	≈97%
Tukey's quick test	≈97.5%	Van Eeden Test	≈99.5%	*F* test	<90%		
Mann-Whitney test	≈97%	Paired *t*-test	≈99.5%				
Terry-Hoeffding Test	>99.9%						
Unpaired *t*-test	≈98%						

TOPIC 11. COMPARING THREE OR MORE GROUPS OF UNIVARIATE DATA: One-way Analysis of Variance and Related Tests

In §10, we were able to test whether the means of 2 data-sets were statistically identical (given equal variances). To compare q (≥ 3) means parametrically, a *three-stage* procedure is required:

(1) test for **homogeneous** variances, i.e.

H_o: ($\sigma^2_1 = \sigma^2_2 = \sigma^2_3 = ... \sigma^2_q$) against H_1: (*at least one* σ^2_i differs from the others) [11.1a]

(2) if the H_o from (1) is *accepted*, test for homogeneous means, i.e.

H_o: ($\mu_1 = \mu_2 = ... \mu_q$), against H_1: (*at least one* μ_i differs from the others) [11.1b]

(3) if the H_o from (2) is *rejected*, test further to see *which* μ_i differ from one another (§10c).

Nonparametric methods are also available, though less well-developed for stage (3) (Table 11.1). Because there is no obvious translation for most of the archetypal, **related** data-types ('before' and 'after'; 'male' and 'female') into a q-group situation, the difference between **related** and **independent** data takes a somewhat different form from that in §10, and will be considered in §12.

As with two-group t-tests, parametric comparisons of *means* (§11b–c) assume that: (a) the q samples are random; (b) the q populations are Normally distributed; (c) the q populations have the *same variance*. They should thus be *preceded* by parametric tests for homogeneity of variance to which we turn first (§11a). The data used to illustrate all the tests in this Topic are given in Table 11.2.

Table 11.1. A quick guide to tests covered in Topics 11–13, for comparing three or more groups of data

	Parametric tests		*Nonparametric tests*	
Data-type / Comparison	Interval/ratio data, Normally distributed	Small sets ratio data	Ordinal data; interval/ratio data with outliers, or non-Normal†	Nominal or* dichotomous data
LOCATION‡ (unrelated data)	One-way ANOVA (§11a-b)	Lord's test (§11d)	1.q-group Median test(§11e1) 2.Kruskal-Wallis test (§11e2) 3.**Van der Waerden test** (§11e3)	1.χ^2 test (§13d1)
LOCATION‡ (related data)	Two-way ANOVA (§12c1) and subsequent sections	—	2.Friedman test (§12c2) 3.**Quade test** (§12c3)	1.Cochran's Q test (§13f1)
SCALE‡	1.Hartley's F test (§11a1) 2.Cochran's C (§11a2) 3.Bartlett's M test (§11a3)	—	Squared ranks test (§11f)	—
OVERALL FORM	—	—	Birnbaum-Hall test (§11g1) k-sample Smirnov tests (§11g2)	—

† Power of tests increases in the order 1 < 2 < 3. **Normal scores tests are emboldened.**

‡ It should be noted that the comparison criterion varies according to which test is used: parametric tests of location are comparing *means*, whereas nonparametric tests are comparing *medians;* parametric tests of scale are comparing *variances /standard deviations)* whereas nonparametric tests are comparing *median deviations.*

* Comparisons of nominal/dichotomous data are tests for *association* (dealt with in §13) rather than for differences of location or scale, which last such data do not possess; the tests are listed here merely to show their relationship to the others; for example, the median test is equivalent to a χ^2 test on ordinal rather than nominal data.

See Fig.10.1 for further information on tests appropriate to particular circumstances.

11a. Determining parametrically whether several groups have homogeneous variances

Exactly as the F test (§10a1) was a necessary preparation for the t test (§10a1), so the present group of tests are alternative forms of an essential preparation for one-way ANOVA in §11b, which compares q means on the assumption of equal variances.

11a1.Hartley's maximum-F test

Worked geological examples: *Hydrology/sedimentology*(Till 1974,p.105)
Statistical tables: Horton(1978,p.263); graph of critical values also given by Till(1974, p.106).
Mainframe software: covered in large statistics packages (e.g. SPSS^X).
Microcomputer software: none known.

This test is simply based on the statistics $F_{max} = [s^2_{max}/ s^2_{min}]$, i.e. the ratio of largest to smallest variances. Strictly, it requires equal size data-sets, i.e. $n_1 = n_2 = ...=n_q$, giving F_{max} $\{N-1\}$ *df,* but can also be applied where $n_1 \approx n_2 \approx ...\approx n_q$, giving $\nu = \{[n_1 + n_2 + ...+n_q]/q\}$ *df.* H_o is rejected if F_{max} *exceeds* the chosen critical value. For the data in Table 11.2, $F_{max} = (6.362/2.588)^2 = 6.04$ with $\nu \approx (22/4) = 5.5$ *df.* Till's(1974, p.106) chart indicates this is well below the 95% critical value, so H_o (equal variances) is accepted.

Table 11.2. Raw data for Topic 11: percentages of CO_2 in gases from 4 volcanic vents

	Vent 1	Vent 2	Vent 3	Vent 4	
Raw data	27	31	30	16	
	28	34	38	20	
	31	35	42	21	
	32	36	43	26	
	33	39		27	
		40		29	
				35	
n_i	5	6	4	7	$N = 22$
Mean	30.2	35.833	38.25	24.857	
Std.deviation	2.588	3.312	5.909	6.362	

11a2. Cochran's C Test

Further discussion & worked geological examples: Till (1974, p.107);*petrology*: Le Maitre (1982, p.35).
Statistical tables: uses χ^2 tables (see §8b3); Till(1974,p.107) provides a graph of critical values.
Mainframe software: covered in large statistics packages (e.g. SPSS^X).
Microcomputer software: none known.

This equally simple test, not to be confused with Cochran's Q test (§13f1), is based on the ratio:

$$C_{N-1} = \frac{s^2_{max}}{\sum_{i=1}^{i=q} s_i^2} \qquad [11.2]$$

Here, the data-sets *must* have $n_1 = n_2 = ...=n_q$, giving C $\{N-1\}$ *df.* H_o is again rejected if C *exceeds* the chosen critical value. The test cannot of course be applied to the data of Table 11.2, since $n_i \neq n_j$.

11a3.Bartlett's M Test

Further discussion: Bartlett(1937); Horton(1978,p.42); Shaw(1969,p.347).
Worked geological examples: *Igneous petrology* (Le Maitre 1982, p.35).
Selected geological applications: *Remote sensing* (Ammar et al. (1982).
Statistical tables: uses χ^2 tables (see §8b3), or F tables for $n_i < 10$.
Proprietary software: covered in large statistics packages (e.g. SPSS^X,Systat).

This is a more complex and sensitive version of Hartley's maximum-F test (§11a1).The statistic is:

$$\chi^2_{q-1} \approx \frac{(N-q)\ln\left[\sum_{i=1}^{i=q}\frac{(n_i-1)s_i^2}{(N-q)}\right] - \sum_{i=1}^{i=q}(n_i-1)\ln(s_i^2)}{\left[1+\frac{1}{3(q-1)}\left\{\sum_{i=1}^{i=q}\frac{1}{n_i-1}-\frac{1}{N-q}\right\}\right]} \qquad [11.3]$$

For the data in Table 11.2, this gives:

$$\chi^2_{q-1} \approx \frac{18\ln\left[\frac{4x6.7 + 5x10.97 + 3x34.92 + 6x40.48}{18}\right] - (4x1.902 + 5x2.39 + 3x3.553 + 3.701)}{\left[1+\frac{1}{3x3}\left\{\frac{1}{5}+\frac{1}{6}+\frac{1}{4}+\frac{1}{7}-\frac{1}{18}\right\}\right]}$$

= 4.644, which is not significant, agreeing with Hartley's F test above. A small data-set F approximation for $n_i < 10$ in the above case yields $F_{3,506} = 1.412$, which is again not significant.

Although very widely used in other disciplines, the complexity of calculations for Bartlett's test is not rewarded for most geological data, since the test is "so sensitive to departures from the underlying assumption of Normality as to be extremely misleading in practical applications" (Horton 1978, p.42). Furthermore, the test is not especially appropriate for small data-sets ($N/q < 5$). In general terms, the sensitivity of the F test itself in ANOVA (§11b) is not sufficiently high to warrant so sensitive a measure of departure from the assumption of homogeneous variances.

Table 11.3. ANOVA for Completely Randomized One-Way Design (using the data of Table 11.2)

(a) Comparison of separate and 'grand' (combined) means, variances and sums of squares

Vent	n	$\bar{X}$	SS	df = $n-1$	s^2
1	5	30.2	26.791	4	6.698
2	6	35.833	54.835	5	10.969
3	4	38.25	104.751	3	34.916
4	7	24.857	242.856	6	40.475
Weighted mean of 4	=	31.5	429.24 = SS_w	=	23.740
'Grand' (1+2+3+4 combined)	22	31.5	1041.5 = SS_t	21	49.595

(b) General formulation

Source:	DF:	Sum Squares:	Mean Square:	F-test:
Between groups	$q-1$	SS_b	MS_b	MS_b/MS_w
Within groups	$N-q$	SS_w	MS_w	
Total	$N-1$	SS_t		

(c) Results for data of Table 11.2

Source:	DF:	Sum Squares:	Mean Square:	F-test:
Between groups	3	612.26	204.087	8.558
Within groups	18	429.24	23.847	p = .001
Total	21	1041.5		

11b. Determining Parametrically whether Three or more Means are Homogeneous: One-Way ANOVA (the completely randomized design)

Further geological discussion: Miller & Kahn(1962, p.134ff); Shaw(1969); Size (1987b).
Further discussion: Bhattacharyya & Johnson(1977,p.456ff) and many other statistics textbooks.
Worked geological examples: *Igneous petrology* (Le Maitre 1982, p.31); *Marine geology* (Till 1974, p.108); *Sedimentology* (Davis (1986, p.75; Koch & Link 1970, p.133; Krumbein & Miller 1953, p.513).
Worked non-geological examples: Steel & Torrie (1980, chapter 7).
Selected geological applications: *Igneous petrology* (Flinn 1959; Chinadi et al.1972); *Isotope geochemistry* (Golding et al.1987); *Remote sensing* (Ammar et al. 1982).
Statistical tables: uses *F* tables (see §8b2).
Mainframe software: in most statistics packages (e.g. SPSS) or libraries (e.g. NAG routine G04AEF).
Microcomputer software: programmed in many statistics packages (e.g. Statview 512+™).

It is often far from obvious to students why a method analyzing *variance* should be used as a test for comparing *means*. The answer is based on our ability to add independent sources of variability together, as already discussed under §9b (see e.g. [9.1]): this means that the total variation (expressed as variance) in a set of data can be broken down as follows into its components:

Total variance of a system = Within-groups variance + between-groups variance + error term [11.4]

That is, the total variance is made up of the variation *within* the groups of data (in which we are not here especially interested), and that *between* the groups. Means can only be distinguished only if the groups are separated more than they are individually spread out: that is, if the *between*-groups variance greatly exceeds the *within* groups variance (Fig.11.1). This ratio of variances is tested using Fisher's F (the expected ratio of variances of Normal data-sets). Close-knit, well-separated groups will give a large, highly significant F, while fuzzy, overlapping groups will give a small, non-significant value.

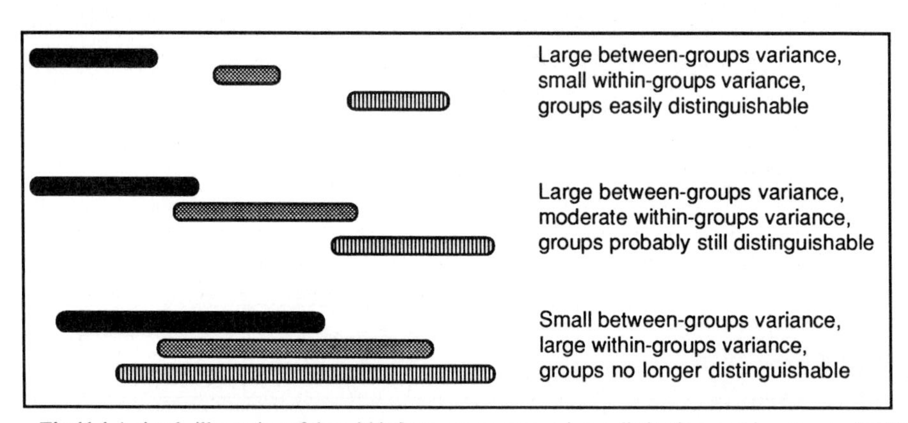

Fig.11.1. A simple illustration of the within/between-groups variance distinction used in one-way ANOVA

We now very briefly examine the data in Table 11.2, to see how these variances are derived and separated. A **sum of squares** is simply a sum of squared differences of a given set of values from its arithmetic mean (i.e. the variance equation [8.2b] without the denominator):

$$\text{Sum of Squares (SS)} = \Sigma(X_i - \overline{X})^2 \qquad [11.5]$$

For the data in Table 11.2, Table 11.3a compares the means, sums of squares and variances for the 4 vents individually and combined. The sum of squares of 26.791 for Vent 1, for example, is simply

$\Sigma\{(27 - 30.2)^2 + ... (33 - 30.2)^2\}$ or, alternatively, $s^2 \times df = 6.698 \times 4$. The crucial observation is that, whereas the weighted mean of the 4 individual means (i.e. $\{30.2 \times 5 + ... 24.857 \times 7\}/22$) is the same as the 'grand' mean of the 4 groups combined together (i.e. $\{27 + 28 + ... 29 + 35)/22$), the sums of squares and variances are *not* additive in the same way. Here, the 'grand' variance exceeds the mean of the 4 separate variances, while the SS for the 4 vents combined — the total sum of squares, SS_t — is substantially *larger* than the sum of the SS for the vents individually — the *within*-group sum of squares, SS_w. The difference between these sums of squares, namely [1041.5 – 429.24] = 612.26, is that attributable to the *differences* between the 4 vents — the *between*-group sum of squares, SS_b.

These values are conventionally laid out in the general form called an ANOVA table (Table 11.3b), and the results for the data of Table 11.2 are then as in Table 11.3c. The final F ratio here of 204.087/23.847 = 8.558 is compared with a critical $F_{between\ df,\ within\ df}$ — that is, $F_{q-1,N-q}$ or $F_{3,18}$ here, which has a value of 6.03 for the 99.5% significance level. Our value exceeds this critical value, so we would reject H_0 of equal mean CO_2 contents between the 4 vents.

Of course, a significant F like this is only a *statistical* result which still has to be interpreted *geologically*. Size (1987b, p.12) gives a good example of an ANOVA where a large F merely reflects a small SS_t and especially SS_w; it would be very sensitive to small changes in the data, and it would not be realistic geologically to infer a difference between the groups, even though F is highly significant statistically. In our vent example, however, the sums of squares are large and unequivocal: we can confidently assume that the significant F in fact means that the 4 vents have different CO_2 contents.

Sometimes students find it difficult to grasp why two completely different statistics appear to be used to compare different numbers of means: Student's t to compare *two* (§10a2), and Fisher's F for *three* or more (this section). In fact, one-way ANOVA can be used perfectly well for 2-group comparisons: Koch & Link(1970, p.145) give aneconomic geology worked example. We showed in §8b4 that $t_\nu = \sqrt{F_{1,\nu}}$, so that $t = 2.69$ in §10a2 equals the square-root of the calculated F statistic on the same data in Table 11.4: i.e. $t = \sqrt{F} = \sqrt{7.236} = 2.69$.

Table 11.4.ANOVA table for completely randomized one-way design on data of Table 10.2 (cf.§10a2)

Source:	DF:	Sum Squares:	Mean Square:	F-test:
Between groups	1	.029	.029	7.236
Within groups	17	.068	.004	p = .0155
Total	18	.097		

Our q-group comparison is called **one-way** ANOVA because only one 'factor' is involved: namely that defining the q groups in the first place (here, the four separate vents). The method is also called the **completely randomized design** because, in ideal experimental studies, subjects are allocated to one of the q groups completely randomly. In studying the effect of different fertilisers on the growth of a crop, for example, the experimenter would assign batches of crop to fields, later to be differently fertilized, according to a random number table. Such well-controlled situations are rare in geology, but perhaps the nearest approach might occur if a geologist were wishing to check whether four analytical methods produced the same gold assays for a series of grab samples from the conveyor belt at a mine;

he would then randomly assign samples to the four methods as they came off the belt. This, however, would add geological as well as analytical sources of error (§9b) due to inhomogeneity in the source ore-body, and might require a more complex form of ANOVA (§12). In the far more common geological situation where the groups are pre-defined on some classification variable (rock-type, mineral, vent, analytical method, etc.), rather than by randomization of subjects, randomness can still fortunately be assumed without invalidating the ANOVA (Steel & Torrie 1980, p.138).

Readers should note that the actual significance of our F ratio in Table 11.3c is dependent on three major assumptions; only if these are met is our comparison with the critical F value above correct:
(1) The sources of variation (within-groups and between-groups) are *independent*.
(2) The compared groups have equal variances (which we have already demonstrated in §11a).
(3) The errors are Normally distributed, which we have *not* here demonstrated.

Reams have been written about the effects of departures from these assumptions (e.g. Horton 1978, p.37ff). These cannot be detailed here, particularly as different conclusions are reached by different statisticians! Most texts do, however, seem to agree that assumption (1) is the most important — but it is also the one which is most difficult to visualise and obey geologically! In experimental designs in other sciences, it is generally obeyed by ensuring random assignation of subjects to groups, and by measuring in such a way that the response from one subject does not effect the response of another. In our vent problem, the groups are already determined by Nature (we sample the 4 vents, or we leave them!), but we can at least take our multiple samples from the vents in a random fashion. If sampling over several days, for example, we should *not* take all our samples from vent 1 on day 1, those from vent 2 on day 2, and so on, but should randomly change between vents at each sampling. If sampling from one vent in any way affected the CO_2 output of another vent, then assumption (1) *would* be violated and the significance of our result would be different from that asserted above. Such an effect is difficult to envisage for this particular problem, but were we draining a 'pool' of CO_2 of fixed size, then removal of the CO_2 from one vent would clearly affect the possible CO_2 output of the other.

If our F had been non-significant, the ANOVA would have terminated at this point, as the means would have been declared homogeneous. Here, however, F has told us that one *or more* means differ from the others, and the analysis needs to continue, using the multiple comparison techniques in the next section, to determine *which* means in fact differ from each other.

11c. Determining which of several means differ: MULTIPLE COMPARISON TESTS

Further discussion: Hochbert (1987); Tukey (1953).

Simple pairwise t-tests cannot be used to detect which means are different, because the wrong hypothesis with the wrong assumptions would then be tested. Six multiple comparison procedures are encountered sufficiently commonly in the literature and in statistical packages to require discussion. Five of these have similar aims, while Dunnett's (§11c6) is more specifically designed to compare a series of means with a *control*. These tests are often referred to as **post hoc** (Latin = 'after that') tests, since they assess an extension of the original null hypothesis *after* it has been rejected.

Given 6 competing procedures, which is the bemused geologist to choose? Einot & Gabriel (1985) suggest that Tukey's w (§11c3) outperforms the others, and it is certainly easy to calculate. However, multiple comparisons are a decidedly contentious subject among statisticians: Kotz & Johnson (1982-8, p.424, vol.1) state "it is not easy to assess relative merits — indeed it is difficult even to calculate the expected properties of even a single procedure". Since most computer packages implement more than one test, it is probably safest to perform all those available. Unfortunately, the results will not often be the same: for example, in Table 11.5, four of six tests give the same result, but two others a *slightly* different one, in which one fewer pair of means is identified as differing. This result is consistent with known properties of these two tests, which are more **conservative** (i.e. tend to find fewer differences). The only answer for the user is to decide whether being conservative suits his purpose!

Results from these tests are popularly summarised with the means ordered by size, and statistically identical ones jointly underscored by the same line; any two means not underscored by the same line are thereby denoted as statistically different at the chosen significance level. In the following example:

$$\overline{X}_1 \quad \underline{\overline{X}_2 \quad \overline{X}_3} \quad \underline{\overline{X}_4 \quad \overline{X}_5}$$

$\overline{X}_1$ differs from all the others, and the pair $\{\overline{X}_2—\overline{X}_3\}$ is statistically distinguishable from the pair $\{\overline{X}_4—\overline{X}_5\}$, whereas each of these pairs is statistically indivisible within itself.

Table 11.5. Multiple comparison tests for the data of Table 11.2

Comparison pair	Gap order, g	Difference between means	Fisher's PLSD*	Scheffé's test*	Dunnett's test*	Tukey's test†	SNK test‡	Duncan's test¶
Vent 1 vs. Vent 2	1	−5.633	6.213	1.210	1.905	NS	NS	NS
Vent 1 vs. Vent 3	2	−8.05	**6.883**	2.013	2.457	NS	Sig	Sig
Vent 1 vs. Vent 4	1	+5.343	6.008	1.164	1.869	NS	NS	NS
Vent 2 vs. Vent 3	1	−2.417	6.623	0.196	0.767	NS	NS	NS
Vent 2 vs. Vent 4	2	10.976	**5.708**	**5.441**	**4.040**	Sig	Sig	Sig
Vent 3 vs. Vent 4	3	13.393	**6.431**	**6.382**	**4.376**	Sig	Sig	Sig

* **Emboldened** values are significant at the 95% level.
Sig = significant at the 95% level; NS = not significant at the 95% level.
† Difference of means > 8.328 required for significance.
‡ Difference > 6.183 (gap order g = 1); > 7.519 (g = 2); > 8.325 (g = 3) for 95% significance.
¶ Difference > 6.183 (gap order g = 1); > 6.496 (g = 2); > 6.684 (g = 3) for 95% significance.

11c1. Fisher's PLSD (= Protected Least Significant Difference) test

Worked geological examples: *Marine Geology* (Till 1974, p.111).
Worked non-geological examples: Steel & Torrie (1980, p.173).
Selected geological applications: *Isotope geochemistry* (Golding et al. 1987).
Statistical tables: uses t tables (see §8b2).
Mainframe software: covered in some larger statistics packages (e.g. SPSS).
Microcomputer software: programmed in some statistics packages (e.g. Statview 512+™).

This test examines the means pairwise. The difference $[\overline{X}_i - \overline{X}_j]$ is regarded as significant if:

$$\overline{X}_i - \overline{X}_j > \mathrm{PSLD}_\alpha = t_{\alpha,N-q}\sqrt{\mathrm{MS_W}\left\{\frac{1}{n_i}+\frac{1}{n_j}\right\}} \qquad [11.6]$$

The α level must be the same as used in the precursor F test, and t_α refers to *two*-tailed t. Comparing vents 1 & 2 from Table 11.2, we obtain $\mathrm{PLSD}_{95} = 2.101 \times \sqrt{(23.847)}\{1/5 + 1/6\} = 6.213$, whereas $[\overline{X}_1 - \overline{X}_2]$ is only 5.633, so the means are indistinguishable. By contrast, for vents 2 & 3, we obtain

$PLSD_{95} = 2.101 \times \sqrt{(23.847)\{1/5 + 1/4\}} = 6.883$, whereas $[\overline{X}_1 - \overline{X}_3] = 8.05$, so this difference *is* significant. Notice that a slightly different calculation is involved for each pairwise comparison, though MS_w and t remain the same in each case. The overall result from this test (Table 11.5) is:

```
Vent 4          Vent 1          Vent 2               Vent 3
                ____________________
```

11c2.Scheffé's *F* Test

Worked geological examples: *Metamorphic petrology* (Koch & Link 1970, p.196; *Sedimentary petrology* (Miller & Kahn 1962, p.172).
Worked non-geological examples: Steel & Torrie (1980, p.183).
Selected geological applications: *Isotope geochemistry* (Golding et al. 1987).
Statistical tables: uses *F* tables (see §8b2).
Proprietary software: covered in some statistics packages (e.g. SPSS, Statview 512+) .

This test again considers pairwise mean differences. The difference $[\overline{X}_i - \overline{X}_j]$ is significant if:

$$\frac{[\overline{X}_i - \overline{X}_j]^2}{MS_w(q-1)\left\{\frac{1}{n_i} + \frac{1}{n_j}\right\}} > F_{[\alpha,q-1,n]} \qquad [11.7]$$

where $n = \{1/n_i + 1/n_j\}$. Strictly, it is only valid for $n_i = n_j$. Comparing vents 1 & 2, we obtain $5.633^2/\{23.847 \times 3 \times 0.3667\} = 1.210$, which compares with $F_{0.05,3,4}$ of 6.59 and $F_{0.05,3,5}$ of 5.41, so this pairs of means does not differ significantly. The other calculations again differ slightly. The final result is now slightly different from that of the PLSD test (Table 11.5), as follows:

```
Vent 4          Vent 1          Vent 2               Vent 3
                ____________________________________________
```

In other words, the difference between vents 1 & 3, assessed as significant in the PLSD test, is here no longer assessed as significant. Scheffé's test has proved to be more **conservative**, i.e. it is has rejected one fewer null hypothesis of the form H_o: $\overline{X}_i = \overline{X}_j$ than has the PLSD.

11c3.Tukey's *w* (HSD = Honestly Significant Difference) Test

Worked geological examples: *Metamorphic petrology* (Koch & Link 1970, p.196); *Sedimentary petrology* (Miller & Kahn 1962, p.173).
Worked non-geological examples: Steel & Torrie (1980, p.185).
Statistical tables: uses the *Studentized Range* (Koch & Link 1970,p.354; Powell 1982,p.36; Steel & Torrie 1980, p.588).
Proprietary software: covered in some larger statistics packages (e.g. SPSS, Systat).

In contrast, to the previous two tests, this only calculates a *single* statistic, *w*, against which *all* adjacent ordered means are compared. Differences are significant if they exceed:

$$w = SR_{\alpha,q,n(q-1)}, \sqrt{\{MS_w/n\}} \qquad [11.8]$$

where $SR_{\alpha,q,n(q-1)}$ is the **Studentized range** at the $\alpha\%$ significance level, for q treatment means and $n(q-1)$ is the *df*. Here, we obtain $SR_{0.05,4,18} = 4.00$, $w = 4.00\sqrt{\{23.847/5.5\}} = 8.328$. Hence *any* pairwise difference exceeding 8.328 is significant. The overall result in terms of mean comparisons is thus identical to that of Scheffé's test (Table 11.5).

11c4.The Student-Neuman-Keuls' (S-N-K) Test

Worked non-geological examples: Steel & Torrie(1980, p.186).
Statistical tables: uses the *Studentized Range* (see §11c3).
Mainframe software: covered in some larger statistics packages (e.g. SPSS).
Microcomputer software: covered in SYSTAT.

This works in a rather different way to the previous three tests, comparing the means in successively smaller blocks. In our usual example, we first arrange the 4 means in ascending order as follows:

```
24.857(vent 4)    30.2(vent 1)    35.833(vent 2)    38.25(vent 3)
```

We first compare the smallest and largest means, here referred to as a *block of gap-order* $g = 3$ $\{q{-}1\}$. If these are the same, the means are declared homogeneous and no further testing is done; however, an initial ANOVA should have told us that already! If they are different, the next stage is to compare the smallest with the second largest, and the second smallest with the largest mean — i.e. compare ordered mean {1} with mean $\{q{-}1\}$, and mean {2} with mean $\{q\}$, which is gap-order $g = 2$ $\{q-2\}$. If all of these differences are non-significant, the means within that block are declared homogeneous and testing stopes. If any are significant, we proceed to the next smallest gap-order, and so on, either until no further significant differences are obtained, or until $g = 1$ (when adjacent means are being compared).

The test statistic used for comparing each of these blocks of means, equal to:

$$T_g = SR_{\alpha,g+1,n(q-1),}\sqrt{\{MS_w/n\}} \qquad [11.9]$$

where, as above, SR is the Studentized Range. This test usually assumes all groups have equal n, but minor departures are not too serious, so we here take $n = 5.5$, $n(q{-}1) = 18$. In stage 1 ($g = 3$), we have $SR_{0.05,4,18} = 4.00$, so $T_3 \approx 4.00\sqrt{\{23.847/5.5\}} = 8.325$. The difference of gap-order 3 (means 4 vs. 3) is 13.393 (Table 11.5), greatly exceeding this value, so we proceed to the next stage, $g = 2$. Here, $SR_{0.05,3,18} = 3.612$, whereupon $T_2 \approx 7.519$. The gap-order 2 differences are now 10.976 (2 vs.4) and 8.05 (1 vs. 3), which again both exceed T_2, so we proceed to the final stage, $g = 1$. This yields $SR_{0.05,2,18} = 2.97$, and thus $T_1 \approx 6.183$; none of the adjacent mean differences now exceeds this critical value, so we can underline each *adjacent* pair of means. The overall result is thus the same as in Fisher's PLSD test (§11c1), but differs slightly from the Scheffé and Tukey w results (Table 11.5).

11c5.Duncan's New Multiple Range Test

Worked non-geological examples: Kotz & Johnson(1982-8); Powell(1982,p.45); Steel & Torrie(1980,p.187).
Selected geological applications: *Isotope geochemistry* (Torquato & Frischkorn 1983).
Statistical tables: Beyer(1968,p.368); Powell(1982,p.44); Steel & Torrie(1980, p.588).
Stand-alone computer programs and microcomputer software: none known.
Proprietary software: programmed in SYSTAT.

This resembles the S-N-K test in using multiple ranges, but is less conservative. By modifying the Studentized Range statistic to a slightly different significance level for each stage, the test allows for the fact that the means are more likely to be equal as their number increases. We begin in the same way as with the S-N-K test. The initial statistic is then:

$$T_g = DU_{\alpha,g+1,q(n-1),}\sqrt{\{MS_w/n\}} \qquad [11.10]$$

where DU is the modified statistic (the upper $100[1-(1-\alpha)^{g-1}]$ percentage point of the Studentized

Range), and g is the gap-order as before. Equal n_i are again assumed, but we can use $n = 5.5$, $q(n-1) = 18$ as before. If any block of means at any stage has a *smaller* range than T_g, we conclude it is homogeneous, underline it, and stop there. However, any block whose range *exceeds* T_g we dissect further, considering smaller gap-orders of size $(g-1)$, and so on until we reach pairs of means.

For $g = 2$ here (omitting $g = 3$ for brevity), the value for $DU_{0.05,3,18} = 3.12$, so we obtain $T_2 \approx 3.12 \times \sqrt{\{23.847/5.5\}} = 6.496$. Both mean differences for $g = 2$ exceed T_2, so we proceed to the final stage of pairwise comparisons ($g = 1$), obtaining $DU_{0.05,2,18} = 2.97$, so $T_1 = 6.183$. Now, none of the ranges of *adjacent* means are significant, so we can underline each pair, giving a final result exactly as in the Fisher's PLSD and S-N-K tests (Table 11.5).

Note that when comparing adjacent means ($g = 1$), Duncan's test statistic is always the same as in the LSD and S-N-K tests, but for larger g it is always smaller than for the S-N-K test (see statistics for g = 2 & 3 in Table 11.5), and so will detect smaller differences between means (i.e. reject H_0 more often).

11c6.Dunnett's Test.

Worked geological examples: *Sedimentology* (Miller & Kahn 1962, p.173).
Statistical tables: Steel & Torrie(1980,p.590).
Microcomputer software: programmed in a few statistics packages (e.g. Statview 512+™).
Mainframe software: none known.

This is more commonly used to determine whether particular groups of data differ from some standard or control, rather than to compare them. It might, for example, be used to ascertain whether a series of new analytical methods gave different replicate determinations of some element in a set of standard rocks *from some well-established method.* In our example, it might be used to compare the individual vents with a composite derived from all four groups of data. However, some texts use the test in the same way as the other five above, so that the test statistic is variously and rather confusingly defined. The results in Table 11.5 are based on a statistic of:

$$\text{Dunnett } t = \frac{\text{Mean difference}}{\sqrt{\text{MS}_w\left\{\frac{1}{n_1}+\frac{1}{n_2}\right\}}} \qquad [11.11]$$

which is compared with special tables of Dunnett's t. Comparing vents 3 and 4 in Table 11.5, for example, we obtain Dunnett $t = 2.417/\sqrt{\{23.847(\frac{1}{4}+\frac{1}{6})\}} = 0.767$, which is not significant.

11d. A quick parametric test for several means: LORD'S RANGE TEST

Worked examples and statistical tables: Langley(1970,p.367). Critical values only available for $q = 2$ ($n_1 = 2, n_2 = 2\text{-}7$; or $n_1 = 3$; $n_2 = 3\text{-}4$; or $n_1 = n_2 = 4$); $q = 3$ or 4 ($n_1 = n_2 = 2\text{-}4$).
Computer software: too simple to warrant computerization.

This q-group extension of the substitute t-test (§10b), is a quick and very useful, if crude, parametric test for comparing 2–4 very small groups of data ($2 < n_i < 4$). It is based on the ratio of the 'range of the group means' to the 'sum of the group ranges', that is:

$$L = \frac{\overline{X_{max}} - \overline{X_{min}}}{\omega_1 + \omega_2 + \dots \omega_q} \qquad [11.12]$$

where $\overline{X}_{max}$ is the highest group mean, and ω_q is the range for group q. Since the data in Table 11.2 are beyond the limit of available critical values, we choose a subset for illustration:

	Vent 1	Vent 2	Vent 3	Vent 4
Raw data	27	31	30	16
	28	34	38	20
	31	35	42	21
	32	36	43	26
Mean CO_2 %	29.5	34	38.25	20.75
Range	5	5	13	10

This yields $L = (38.25 - 20.75)/(5 + 5 + 13 + 10) = 0.5303$, which exceeds the 99% critical value of 0.33, indicating a strong inhomogeneity between the 4 mean CO_2 contents of this subset.

11e. Determining nonparametrically whether several groups of data have homogeneous medians

11e1. The q-group extension of the Median Test

Worked non-geological examples: Siegel(1956,p.179); Conover(1980, p.171).
Statistical tables: uses χ^2 tables (see §8b3).
Mainframe software: covered in most large statistics packages (e.g. SPSS^X).
Microcomputer software: none known.

The 2-group Median Test was explained fully in §10d1. The q-group extension proceeds in exactly the same way, by dichotomizing the data about the grand median of all the values, to derive a q x 2 contingency table such as Table 11.6 (see §13d1 for further details of contingency tables).

Table 11.6. Contingency table for a Median Test on the data on Table 11.2.
(numbers of values < and > grand median of 31.5)

	Vent 1	Vent 2	Vent 3	Vent 4	Totals
<31.5	3	1	1	6	11
>31.5	2	5	5	1	11
Totals	5	6	4	7	22

The test statistic is calculated by [8.27], giving the following from Table 11.6:

$$\chi_3^2 = \frac{(22)^2}{11x11}\left\{\frac{\left[3 - \frac{5x11}{22}\right]^2}{5} + \dots + \frac{\left[6 - \frac{7x11}{22}\right]^2}{7}\right\} = 7.439$$

which has $p \approx 0.06$, leading to rejection of H_o at ≈94% level.

11e2.A more powerful test: The Kruskal–Wallis One-way ANOVA by Ranks

Further non-geological discussion: Conover(1980); Gabriel & Sen (1968); Iman & Davenport (1976); Iman et al. (1975); Siegel(1956,p.184).

Further geological discussion: Cheeney (1983, p.80); Davis (1986, p.97).
Worked geological examples: *palaeontology* (Miller & Kahn 1962,p.182).
Selected geological applications: *metamorphic petrology* (Rock et al.1986a; Rock 1986e).
Statistical tables: for $q = 3$, or all $n_i \leq 5$: Conover(1980); Downie & Heath(1974); Powell(1982,p.70); Siegel(1956,p.282); for $q > 3$, or any $n_i > 5$, compare H with χ^2_{q-1} (see §8b3).
Stand-alone computer programs: Rock (1986f).
Mainframe software: in most packages (e.g. SAS,SPSS) or libraries (e.g. NAG routine G08AFF).
Microcomputer software: programmed in many statistics packages (e.g. Statview 512+™).

Parametric ANOVA and multiple comparison tests are fairly robust to *small* departures from their assumptions of equal variances and data-Normality, while inhomogeneity in the variances indicated by the tests in §11a can be compensated to some extent by making $n_1 = n_2 = ...n_q$. However, if departures from these assumptions are large, or if the data are only measured on an ordinal scale, this test (the equivalent, nonparametric form of one-way ANOVA) should be used.

The Kruskal-Wallis test is a q-group extension of the Mann-Whitney Test (§10d2), and is executed in the same, back-of-envelope, way, by ranking all the data together and calculating the test statistic as:

$$H = \frac{12}{N(N+1)} \sum_{i=1}^{i=q} \frac{(\Sigma®_i)^2}{n_i} - 3(N+1) \qquad [11.13]$$

where $\Sigma®_i$ is the sum of ranks for the ith group among *all q groups combined.* This formula merely detects the extent to which the rank sums (and hence the q medians) are unequal. The reader can easily verify that if the q medians are in fact equal, so that $\Sigma®_i$ is merely a function of n_i/N, then $H = 0$. If ties are present, H is divided by the factor $[1 - \{\Sigma(t^3 - t)\}/(N^3 - N)]$, where 't' is as usual, the number of tied observations in a tied group of ranks and Σ directs to sum over all such groups of ties. This correction factor is usually > 0.99, making only a small difference to H.

The rank calculations for the vent CO_2 data above appear below; [11.13] yields:

$$H = \frac{12}{22x23}\left[\frac{48^2}{5} + \frac{96^2}{6} + \frac{70^2}{4} + \frac{39^2}{7}\right] - 3x23 = 12.559 \text{ (uncorrected for ties)}$$

which corrects for ties to $H = 12.581$. Both Hs exceed the 99% value of χ^2_3 (= 11.34) and thus again lead to rejection of H_o. Notice, however, that the Kruskal-Wallis rejects H_o at a much higher confidence level than the Median test, because it uses far more of the information present in the data.

Table 11.7. Calculations for Kruskal–Wallis and Van der Waerden tests (data of Table 11.2)

Vent 1			Vent 2			Vent 3			Vent 4		
Raw score	Rank	Normal score	Raw	Rank	Normal score	Raw	Rank	Normal score	Raw	Rank	Normal
27	5.5	−0.709118	31	10.5	−0.109200	30	9	−0.275921	16	1	−1.711675
28	7	−0.511939	34	14	+0.275921	38	18	+0.781059	20	2	−1.360370
31	10.5	−0.109200	35	15.5	+0.450746	42	21	+1.360370	21	3	−1.124542
32	12	+0.054519	36	17	+0.640676	43	22	+1.711675	26	4	−0.938886
33	13	+0.164211	39	19	+0.938886				27	5.5	−0.709118
			40	20	+1.124542				29	8	−0.391197
									35	15.5	+0.450746
Σ®	48			96			70			39	
ΣA_{ij}		−1.111527			+3.321571			+3.580513			−5.785042

11e3. The most powerful nonparametric test based on Normal scores: the Van der Waerden Test

Worked non-geological examples: Conover (1980, p.318).
Selected geological applications: *metamorphic petrology* (Rock 1986e).
Statistical tables: uses χ^2 tables (see §8b3).
Stand-alone computer programs: Rock (1986f).
Mainframe and microcomputer software: none known.

This is the Normal score equivalent of the Kruskal-Wallis test. Its formula is:

$$T = \frac{(N-1)}{\sum_{i=1}^{i=N} A_{ij}^2} \sum_{i=1}^{i=q} \frac{\left[\sum_{j=1}^{j=n_i} A_{ij}\right]^2}{n_i} \qquad [11.14]$$

where A_{ij} is the Normal score for the ranks calculated as in the Kruskal-Wallis test, j = [1 to n_i] directs to sum over group i, i = [1 to q] over all groups, and i = [1 to N] over all individual values. For the CO_2 vent data, the first value (27), with rank 5.5 out of N = 22, converts to the (5.5/23) = 0.2391 quantile of a Normal distribution, A_{11} = –0.709118, and so on as tabulated in Table 11.7.The calculated T is 12.50 — almost identical to the Kruskal–Wallis H of the previous section.

11f. Determining Nonparametrically whether Several Groups of Data have Homogeneous Scale: THE SQUARED RANKS TEST

Worked non-geological examples: Conover (1980, p.241).
Selected geological applications: *metamorphic petrology* (Rock 1986e).
Statistical tables: uses χ^2 tables (see §8b3).
Stand-alone computer programs: Rock (1986f).
Mainframe and microcomputer software: none known.

The 2-group Squared Ranks test was explained fully in §10e2. Each value is first converted into an absolute deviation from its own group mean, then *all* these deviations are ranked, to yield:

$$d_{ij} = \text{®}\ |X_{ij} - X_j|, \quad D_j = \sum_{i=1}^{i=n_j} (d_i)^2 \qquad [11.15a]$$

The test then becomes:

$$\chi^2_{q-1} \approx \frac{(N-1)\left[\sum_{i=1}^{i=q} \frac{D_i^2}{n_i} - \frac{1}{N}\left\{\sum_{j=1}^{j=q} D_j\right\}^2\right]}{\sum_{i=1}^{i=N} d_{ij}^4 - \frac{1}{N}\left[\sum_{j=1}^{j=q} D_j\right]^2} \qquad [11.15b]$$

For the data of Table 11.2, calculations are as in Table 11.8. This yields a test statistic of 4.96, which is less than χ^2_3 at the 90% level (= 6.25), and so indicates equal variances, agreeing with the results of Hartley's and Bartlett's parametric tests (§11a1 & 11a3).

Table 11.8. Calculations for the q-group squared ranks test (data of Table 11.2)

Vent 1			Vent 2			Vent 3			Vent 4		
Raw value	Deviation from mean	Rank of abs.dev d_{ij}	Raw value	Deviation from mean	Rank of abs.dev. d_{ij}	Raw value	Deviation from mean	Rank of abs.dev. d_{ij}	Raw value	Deviation from mean	Rank of abs.dev. d_{ij}
27	-3.2000	12.0	31	-4.8333	18.0	30	-8.2500	20.0	16	-8.8571	21.0
28	-2.2000	9.0	34	-1.8333	7.0	38	-0.2500	2.0	20	-4.8571	19.0
31	0.8000	3.0	35	-0.8333	4.0	42	3.7500	13.0	21	-3.8571	14.0
32	1.8000	6.0	36	0.1667	1.0	43	4.7500	17.0	26	1.1429	5.0
33	2.8000	10.0	39	3.1667	11.0				27	2.1429	8.0
			40	4.1667	16.0				29	4.1429	15.0
									35	10.1429	22.0
$\overline{X}$ =	30.2000		$\overline{X}$=	35.8333		$\overline{X}$=	38.2500		$\overline{X}$=	24.8571	

11g. Determining Nonparametrically whether Several Groups of Data have the same Distribution Shape

We turn finally in this Topic to the multi-group extensions of the two-group EDF tests of §10f. Although potentially of considerable use, both tests below can be carried out only if all groups of data are of equal size, i.e. $n_1 = n_2 = ...n_q$, and such a situation is rare in geology. Because of this restriction, we are obliged to use different data for our worked example. Like their 2-group equivalents, these tests are still applicable to *continuous*, ordinal, interval or ratio data.

11g1. The 3-group Smirnov Test (Birnbaum–Hall Test)

Worked non-geological examples and statistical tables: Conover (1980, p.377 & 474).
Computer software: none known.

This extends the Smirnov Test (§10f2) to 3 groups of data, and similarly compares the observed cumulative distribution functions $\partial(X_1)$, $\partial(X_2)$, ...$\partial(X_q)$. The test statistic is also exactly analogous: namely, the maximum separation between *all three* ∂ functions without regard to sign.

For example, do the thicknesses of coal, limestone and sandstone beds in a cyclical succession (Table 11.9) indicate any difference between the thickness distributions for the 3 sediment types?

Table 11.9. Raw data and calculations for the Birnbaum-Hall test

Raw thickness data (m)	Coal	Limestone	Sandstone
	1.2	1.4	1.8
	1.3	2.0	2.2
	1.5	2.1	2.5
	1.9	2.3	2.7

Cumulative differences	Coal-Limestone		Limestone-Sandstone		Coal-Sandstone	
	0.25–0	= 0.25	0.25–0	= 0.25	0.25–0	= 0.25
	0.5 – 0	= 0.5	0.25–0.25	= 0	0.5–0	= 0.5
	0.5–0.25	= 0.25	0.5–0.25	= 0.25	0.75–0	= 0.75
	0.75–0.25	= 0.5	0.75–0.25	= 0.5	0.75–0.25	= 0.5
	1.0–0.25	= 0.75	0.75–0.5	= 0.25	1.0–0.25	= 0.75
			1.0–0.5	= 0.5		
Max.cum.diff		0.75		0.25		0.75

As further illustrated in Fig.11.2, d_{max} is here 0.75, which equals the 90% critical value of 0.75 for $N = 4$. Hence we cannot actually reject H_0 (to do so d_{max} must *exceed* the critical value), but we are effectively only prevented from doing so because of the small size of the data-sets.

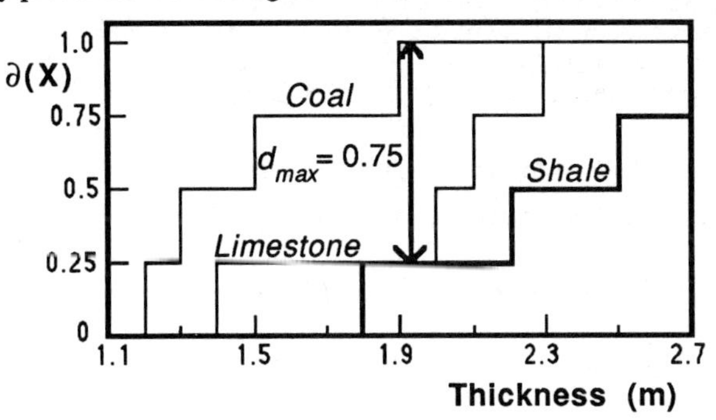

Fig.11.2. Obtaining the test statistic d_{max} in the Birnbaum–Hall test

11g2. The *q*-group Smirnov Test

Worked non-geological examples and statistical tables: Conover (1980, p.379 & 475).
Computer software: none known.

This is more useful that the previous test, not only because critical values are tabulated for a wider range of q ($2 \leq q \leq 10$), but also because it has a one-sided as well as two-sided form.

In the *one-sided* test, H_0 assumes that all groups come from identical populations, whereas H_1 is that the *i*th group has *smaller* values than the *j*th, for some $i < j$; that is, $H_0: \partial(X_1) \leq \partial(X_2) ... \leq \partial(X_q)$ against $H_1: \partial(X_i) > \partial(X_j)$ for some $i < j$. The test statistic is the largest vertical distance d_{max} between $\partial(X_1) \leq \partial(X_{i+1})$, i.e. the adjacent groups are compared as in the 2-group Smirnov test (Fig.8.10) as i is increased from 1 to q, and the largest difference is taken as d_{max}.

For example, let us add a 4th sediment type (shale), to the data of the previous section (Table 11.10): do these now support H_0 that the thicknesses are in the order coal > limestone > sandstone > shale: i.e. are the *cumulative functions* in the order ∂(coal) ≤ ∂(limestone) ≤ ∂(sandstone)≤ ∂(shale)?

Table 11.10. Raw data and calculations for *q*-group Smirnov test

Raw data	Coal	Limestone	Sandstone	Shale
	1.2	1.4	1.8	2.6
	1.3	2.0	2.2	3.0
	1.5	2.1	2.5	3.1
	1.9	**2.3**	**2.7**	**3.2**

Cumulative differences	Coal-Limestone		Limestone-Sandstone		Sandstone-Shale	
	0.25–0	= 0.25	0.25–0	= 0.25	0.25–0	= 0.25
	0.5 – 0	= 0.5	0.25–0.25	= 0	0.5–0	= 0.5
	0.5–0.25	= 0.25	0.5–0.25	= 0.25	0.75–0	= 0.75
	0.75–0.25	= 0.5	0.75–0.25	= 0.5	0.75–0.25	= 0.5
	1.0–0.25	= 0.75	0.75–0.5	= 0.25	1.0–0.25	= 0.75
Max.cum.diff		0.75		0.25		0.75

Hence $d_{max} = 0.75$, which is just significant at the 90% level, indicating that at least one of the thicknesses is in the opposite sense. (In this case, the relative thicknesses are *obviously* in the reverse sense, but because of the small sample size the test only just achieves the correct result).

To perform a *two-sided test,* that is, H_o: $\{\partial(X_1) = \partial(X_2) ...= \partial(X_q)\}$ against H_1: $\{\partial(X_i) \neq \partial(X_j)\}$ for some i, j, we identify the largest observation in each group, X_{max} (**emboldened** above), and then compare the ∂ functions of just two groups: (1) the group with the *largest observation of all*, that is, shale above; (2) the group with the *smallest* X_{max}, that is, coal. There is no overlap between these groups, so $d_{max} = 1$, which leads to rejection of H_o at >95% level.

11h. A brief comparison of the results of multi-group tests in Topic 11

Table 11.11 reviews the results from equivalent tests in this Topic. The tests of §11g are omitted since they were necessarily used on different data. Some of the variation in significance levels is purely mechanical, reflecting the lack of tabulated critical values at certain significance level. This notwithstanding, the results are closely comparable with those in Table 10.8, in that the parametric and nonparametric tests give congruent results, with the least powerful test (the q-group Median) giving the least significant result.

Table 11.11. Comparison of significance levels from equivalent tests for rejection of the null hypotheses of equal locations or scales on the data of Table 11.2

Comparing locations	Sig.lev.	Comparing scales	Sig.lev.
One-way ANOVA	>99.9%	Hartley's F test	< 95%
Lord's range test	> 99%	Bartlett's M test	< 95%
q-group Median test	≈ 94%	Squared ranks test	< 90%
Kruskal-Wallis test	> 99%		
Van der Waerden test	> 99%		

TOPIC 12. IDENTIFYING CONTROLS OVER DATA VARIATION: Analysis of Variance in more sophisticated forms

12a. A General Note on ANOVA and the General Linear Model (GLM)

Dedicated monographs and reviews: Guenther(1966); Hicks(1973); Horton(1978); Huitson(1966); Iversen & Norpoth (1976); John(1971); Krishnaiah (1980); Lewis (1971); Scheffé (1959).
Further geological discussion of ANOVA and the GLM: Flinn(1959); Griffiths(1967,ch.18-20); Krumbein & Miller(1953); Krumbein & Graybill(1965, ch.9 & 12); Koch & Link(1970, ch.5 & 7, p.184); Le Maitre(1982,ch.3); Miller (1986); Miller & Kahn(1962,ch.7 & 10); Shaw(1969); Tatsuoka (1975); Till(1974, ch.6).
Lists of geological applications of various ANOVA designs: Miller & Kahn (1962,p.178ff), Krumbein & Graybill (1965, p.214).
Stand-alone computer programs: See under individual designs below.
Proprietary software: Most statistics packages and the NAG library implement at least some basic (one- or two-way) designs; larger packages (BMDP, MINITAB,SAS, SPSS) generally cover a wider range. Some (e.g. ANOVA procedure in GENSTAT; GLIM; MASS; CLR-ANOVA; SYSTAT) set up the ANOVA in terms of the general linear model (see below), which can in theory handle all conceivable designs, including complex, unbalanced ones with missing data.

Topic 9 indicated how the difficulty of experimental control has tended to result in less extensive use of ANOVA in geology than in other sciences, at least in recent years. Early enthusiasm in the 1950s and 1960s for its geological potential seems to have waned, and publication of papers using ANOVA is nowadays rare compared to those using, say, regression (§14) or discriminant analysis (§22). Nevertheless, there are many recently published data-sets which could have been more rigorously interpreted using ANOVA, so an outline of the method is very much in order here.

We have already met the simplest (one-way) form of ANOVA in §11b as a method of comparing 3 or more means. The generalisation of its role is to identify which of any likely factors is important in controlling the variation of some **dependent variable**, by separating out sources of variation attributable to each factor. A 'factor' is any **independent**, dichotomous or nominal, variable (§6a–b: rock-type, fossil species, borehole, mine, tectonic province, and so on), which can subdivide data.

All forms of ANOVA work like the simple one-way form in §11b, by separating out the total data variability (variance, or what statisticians term 'error'), into components associated with each factor:

Total error = error associated with Factor A + error with Factor B +........unattributable error [12.1]

The similarity of [12.1] to [9.1] should be particularly noted.

ANOVA is also just one form of a fully generalized, umbrella technique which statisticians call the **General Linear Model** (or, equivalently, the **general linear hypothesis**). Other forms of the GLM include regression (§14), multiple regression, canonical correlation (§19) and discriminant analysis (§22). At its simplest, the GLM can be expressed in the following form:

$$Y_i = \beta_1 X_{i1} + \beta_2 X_{i2} + \beta_3 X_{i3} + \ldots\ldots \beta_m X_{im} + \varepsilon_i \qquad [12.2]$$

That is, the dependent variable Y_i is considered to be a linear function of m independent variables $X_{i1} \ldots X_{im}$, multiplied by m unknown constants $\beta_1 \ldots \beta_m$, plus a random error term ε_i. This type of equation will reappear again in §14, 19 and 22. Its application to ANOVA already dealt with in §11b or this Topic is at first sight less obvious, although the reader may sense a connection between [12.1] and [12.2], and may even be prepared to take on trust that **treatments** in ANOVA (§7) and **regression coefficients** in regression (§14) both correspond exactly to the β terms. The problem is that [12.2] is usually expressed in matrix form (where the ***bold italic type*** conventionally indicates matrices):

$$\boldsymbol{Y} = \boldsymbol{X\beta} + \varepsilon \qquad [12.3]$$

and anything beyond the most elementary treatment of the GLM in statistical textbooks goes on treat the problem in matrix notation, with ever more serious attendant problems for most geology students' comprehension! At the present time, there is no simple way of explaining the GLM without esoteric mathematics, which is probable why no geomathematical textbook has yet attempted to do so, and why virtually no geological papers on the GLM have yet appeared. Therefore, although the inter-relationships of statistical techniques covered in the remainder of this book would be 'simpler' (and certainly less apparently haphazard) if understood in the framework of the GLM, this approach is probably not practicable for geology students at the present time. Geologists intending to use some of the larger computer packages written by statisticians (notably Systat) should nevertheless be warned that they will have little alternative but to understand something of the GLM if they are to perform *any* ANOVA or other techniques subsumed within the model.

12b. What determines the range of designs in ANOVA?

This Topic merely presents variants of the volcanic vent CO_2 problem from §11, to illustrate *a limited range* of specific ANOVA designs. An infinity of other ANOVA designs are possible, but those covered below are felt most likely to have geological application. Readers who feel unsure of their ANOVA technique are strongly advised not to stray too far from the models outlined here, even though the dangers of following them parrot-fashion are no less real. All forms have as their end result an expansion of the standard ANOVA table already met (Table 11.3b), in which each source of variation is isolated and statistically tested via an F ratio (§8b3): just as one-way ANOVA led to one F ratio, so m-factor as in [12.2] ANOVA leads to at least m such ratios (more, if the data are designed to look at how these factors interact with one another). Since only three forms are covered by nonparametric tests, these are dealt with in the same sections as the equivalent parametric forms. The nonparametric tests have the usual advantages for many geological problems over classical, parametric methods (§7).

ANOVA terminology is horrendously complex, and inconsistent between texts. Few geologists would surmise intuitively that all the following yield an ANOVA table with exactly the same structure:

(1) single-factor ANOVA on related samples; (2) randomized complete block design;
(3) two-way classification with single observations; (4) two-factorial design with no replication.
(5) two-way single-entry model with fixed effects; (6) two-strata design.

To exacerbate matters, different software may require data to be set up in quite different formats to execute the *same* type of ANOVA; for example, data to be processed via the 6 equivalent designs above might require an input data file with c columns and b rows, or a totally different one with $c \times b$ rows and only 3 columns (Table 12.1). If the data are entered in the wrong format, the program may misinterpret the design and generate a result quite different from the one the user intended.

Table 12.1. Alternative data-formats required by some software packages to perform a b block x c treatment ANOVA

Block	Treatment 1	Treatment 2	.....	Treatment c
1	$X_{1,1}$	$X_{1,2}$	.....	$X_{1,3}$
2	$X_{2,1}$	$X_{2,2}$	.....	$X_{2,3}$
.....	.....	.....	.....	.....
b	$X_{b,1}$	$X_{b,2}$	.....	$X_{b,c}$

Block	Treatment	Data
1	1	$X_{1,1}$
1	2	$X_{1,2}$
.....	.....	.....
1	c	$X_{1,c}$
2	1	$X_{2,1}$
.....	.....	.....
b	c	$X_{b,c}$

The user must therefore take care to determine *what* he is analyzing and *how*. For our purposes, the most important ANOVA designs can be considered as representing permutations of *six* parameters (further definitions of which are given in the Glossary and §7). Their different roles are best seen via the examples in the sections which follow, although not all permutations can possibly be shown here.

(1) *Number of factors:* Only designs of two and three factors are considered here: data-sets in which there are up to 3 definable sources of variability (excluding random error). Each factor is usually also referred to as having so many *levels* — that is, different values it can take. Hence a dichotomous factor can have only two levels, whereas a nominal factor can have many. The statements "6 rock-types separate the data into 6 groups" and "the rock-type factor has 6 levels" are synonymous.

(2) *Factorial* versus *block-and-treatment* and *repeated-measures*. The difference between factorial and block-and-treatment designs is much the same as that between independent and related measurements in the previous 2 topics. Whereas factorial designs aim to isolate the role of *each and every* independent grouping factor, and expect to *add* to the overall variability with each added factor, block-and-treatment designs aim to home in on the treatment factor(s) via use of the block factor(s), the latter *not* being of primary concern. Kotz & Johnson (1982–8) in fact refer to blocks as "a convenient garbage can where an ingenious experimenter can dispose of unwanted effects such as spatial position, different observers, sources of material, and much else that would disturb the experiment if not controlled". From [12.1], it follows that if the variation associated with one factor (or group of factors) — the blocks — can be substantially reduced, variation associated with the other factor(s) — the treatments — will be commensurately magnified as a proportion of total variation. In other words, measurements are expected to be much more similar between blocks than between treatments, so that the blocks allow *smaller* variations to be detected between the treatments than would otherwise be possible for a given amount of overall variation in a data-set. See also §7, §12c and the Glossary for further comments on the relationship between blocks and treatments.

'Repeated measures' ANOVA (also known as 'within-subjects' ANOVA, and closely related to 'split-plot' designs), serves a fairly similar purpose to block-and-treatment designs. Suppose, for example, we were interested in how a group of 10 geology students is coping with 4 different computer operating systems, and we assess their relative performance by the time it takes them to successfully complete the same set of tasks on each system. In such a design, the students are the subjects (blocks) and the operating systems are the repeated-measures (treatments, also known as the 'within-factor'). Clearly, we would expect different students to have varying intrinsic abilities at such a task, and both a block-and-treatment and repeated-measures design would enable us to tell whether in fact the 10 students varied among themselves in their overall performance. The difference is that in a block-and-treatment design, we could only isolate the effect the different operating systems have on the students *as a whole*; whereas in the repeated-measures design, we could home in on its effect on the *individual* students, taking their differing abilities into account. In other words, the repeated-measures design should identify differences due to the computer systems which a block-treatment design might fail to detect if the students varied too much among themselves.

(3) *Replication* versus *single observations*. Each measurement on a particular object may be made once only (usually termed 'without replication') or it may be repeated (usually termed 'with replication' or 'duplicate observations' if there are two). The advantage of replications is that they allow us to look at the **interaction** between different factors — that is, to test whether the factors themselves are truly independent or interfere in some way. If there is no interaction, then no particular combination of different levels of these factors will generate a greater variance than any other combination.

Note that a factor with replications is *not* the same as a repeated-measures factor: for example, if we carry out our computer test on each individual student two or more times, or analyze exactly the same rock specimen two or more times for the same element, these are duplications or replications; but if we carry out the same test on two or more students from the same group, or analyze two or more different samples from the same outcrop, these are repeated-measures experiments. In geology, the difference once again involves the concept of 'entities' (§9b). Each measurement of a particular variable on a geological entity is an estimate of the *same number* — a replicate — and the error associated with the replication should ideally be zero. Measurements of the same variable on objects which are *not* geological entities, however, are not replicates but repeated-measures, and their associated error is never expected to be zero.

(4) *Complete* versus *incomplete*. Missing values can give rise to unbalanced and/or incomplete designs, depending on exactly how they are distributed. If there is a measurement for each and every factor combination, the design is complete, but if there are combinations for which no measurements are available, it is incomplete.

(5) *Balanced* versus *unbalanced*. A balanced design contains the *same* number of measurements for each and every factor combination; an unbalanced design contains different numbers of measurements. A balanced incomplete design has the missing data distributed symmetrically among the factors, so that there are exactly the same number of missing values associated with each possible factor combination. The more unbalanced and incomplete a design becomes, the more difficult it is to analyze objectively, and the more dependent the result may be on the computer or program used.

(6) *Crossed* versus *nested (hierarchical)* factors. Whereas crossed factors are independent, nested factors represent subdivisions of some broader category (e.g. in sampling, errors introduced at the outcrop, hand specimen, jaw crushing stages — §9b).

Generally, larger statistical packages can handle more varied designs: more factors, greater imbalance or incompleteness, greater complexity, etc. Simple packages can usually only handle 2-3 factors, with perhaps one within-factor, and are limited as regards imbalance or incompleteness.

Each of the designs below has a general formulation, in which the contents of each cell in the ANOVA table are defined mathematically (cf. Table 11.3b). As such calculations are comprehensively set out in several of the texts cited above, however, they are omitted below for brevity's sake.

It is important also in this general sketch of ANOVA to underline the role of *randomization* in the technique, which the names of designs (*completely randomized, randomized block)* already themselves stress. In controlled experiments in, say, agriculture, which are to be assessed using ANOVA, subjects (e.g. individual crop varieties) are assigned *randomly* to treatments (e.g. fertiliser brands) and blocks (e.g. fields), and much of the interpretation of the statistics depends on that randomization being successful. As we have emphasised several times already, randomization in geology is rarely so easy, and often quite impossible. While the geologist may be able to randomize if conducting an experiment involving other geologists (e.g. in assessing their mutual consistency in point-counting via different techniques), our example of sampling from volcanic vents is a very different matter: the vents are provided by Mother Nature as and when She wills! The ramifications of incomplete randomization in geology are legion, but are briefly touched on below under individual techniques, as space permits.

Table 12.2. Raw CO_2 data and rank calculations for Randomized Complete Block Design

	(a) Raw data and summary statistics	TREATMENTS				(b) Ranks for Friedman test			
B	Collectors	Vent 1	Vent 2	Vent 3	Vent 4	®(Vent 1)	®(Vent 2)	®(Vent 3)	®(Vent 4)
L	1	27	37	35	22	2	4	3	1
O	2	28	34	38	18	2	3	4	1
C	3	31	35	42	21	2	3	4	1
K	4	32	36	43	25	2	3	4	1
S	5	25	39	35	27	1	4	3	2
					Σ® =	9	17	18	6
					$[\Sigma®]^2$ =	81 +	289 +	324 +	36 = 730

NB. These data could represent *means* of determinations over, say, a period of time, as well as individual analyses.

12c. Two-way ANOVA on several groups of data: RANDOMIZED COMPLETE BLOCK DESIGNS and TWO-FACTORIAL DESIGNS WITHOUT REPLICATION

The *assumptions* involved in these and all the more complicated models covered here are as follows:

(a) Each combination of factors is a random sample drawn from different Normal populations;

(b) Each parent population has the same variance;

(c) There is no interaction between different treatments and different samples.

As already hinted, these assumptions are all too rarely met in full by geological data.

The randomized block design is the most widely-used in scientific experiments, because it is much more sensitive to small differences than one-way ANOVA (§11b), and yet only slightly more complex to execute. Suppose gas samples have been taken by 5 different 'collectors' (blocks) from each of our 4 vents (treatments: Table 12.2). 'Collectors' might represent 5 different geologists using the same apparatus; 5 different techniques or pieces of equipment used by one geologist; replicate analyses of the same 5 gas samples by 5 laboratories; or analyses by 5 techniques or 5 analysts within a single laboratory. We can thus replace 'collector' with 'laboratory', 'method', 'geologist', 'apparatus' etc. in Table 12.2 without altering the design. In all cases, however, we expect variation *within* blocks — 5 geologists sampling one vent, or 5 laboratories analyzing one gas sample — to be *less* than *between* blocks — one geologist measuring 5 different vents, or one laboratory analyzing 5 different gas samples. Because neither the blocks themselves, nor the interaction between blocks and treatments, are of primary concern, a *single* CO_2 measurement per cell is adequate for a randomized block design.

The data of Table 12.2 could also equally well represent designs where the blocks and treatments were interchanged, or where 'block' and 'treatment' became meaningless terms. For example, if 4 vents on one volcano were sampled simultaneously (blocks), at 5 stages during the course of a single eruption (treatments), the variability between the vents might be far less than that of the stages, assuming that the vents all tapped the same gas reservoir in the volcano at any particular moment. Alternatively, if we replaced 'collector' with 'year' or with 'volcano' itself, the block/treatment antithesis would disappear, because we would no longer have any *a priori* reason to suppose that the vents would vary less with time (or across volcanoes) than they would between themselves. The design would then be more accurately termed a *2-factor ANOVA without replication.* Which design we aim for does make a difference when carrying out the measurements, because in the 2-factor case we *should* be concerned with the interaction between the two factors (see Hicks 1973 and below); a single measurement would no longer be adequate, and *replicates* should be taken, as in §12e.

Irrespective of whether this is a randomized block or 2-factor-without-replication design, the hypotheses being tested and the resulting ANOVA have exactly the same form:

(1) H_0: {%CO_2 does not vary between the collectors}; against H_1: {%CO_2varies};

(2) H_0: {%CO_2 is the same for the various vents}; against H_1: {%CO_2 varies between vents}.

In the randomized block version, we are expecting to accept H_0 (1) — though we may of course be disappointed! — but in the 2-factor form we cannot anticipate the outcome from *either* hypothesis.

12c1. The parametric approach

Further geological discussion: Miller & Kahn (1962,p.144ff & 411ff).
Further non-geological discussion & worked examples: Bhattacharyya & Johnson(1977,p.471); Steel & Torrie (1980,p.196ff).
Worked geological examples: *Economic geology* (Koch & Link 1970, p.281 & 235); *Interlaboratory comparisons* (Le Maitre, 1936, p.33; Till 1974, p.114 — but note 'methods' mean square and *F* ratio in Till table 6.4 should read 0.001439 and 5.58 respectively); *Sedimentology* (Koch & Link 1970,p.158).

Parametric randomized block ANOVA is a multi-group extension of the paired *t*-test of §10a2. In Table 12.3–12.13, each *F* is calculated as the ratio of the mean-square for the particular factor to the residual (error) mean-square, and has *df* corresponding to the same rows. For example, the 'between vents' *F*

Table 12.3. Results of parametric ANOVA for Randomized Block Design of Table 12.2a
(results obtained using Statview 512+ package — see Table 2.2)

Source	Sum of Squares	Deg. of Freedom	Mean Squares	F-Ratio	Prob>F
Between Vents	800.60000	3	266.86667	30.52812	0.000
Between Collectors	49.50000	4	12.37500	1.41563	0.288
Error	104.90000	12	8.74167		
Total	955.00000	19			

in Table 12.3 equals 266.87/8.74 = 30.53 and has *df* [3,12], while the 'between collectors' *F* equals 12.38/8.74 = 1.42 and has *df* [4,12]. The associated probabilities indicate that the CO_2 contents vary between the vents but not between the collectors.

The 'error' row (sometimes termed the *residual variance*), represents the random, uncontrolled variance within the data, and may also include factors which were not recognized when setting up the ANOVA design. A large error or *df* almost certainly indicates an intrinsically inadequate design.

12c2. The simple nonparametric approach: the Friedman two-way ANOVA test

Worked geological examples: Lewis (1977, p.158).
Worked non-geological examples: Conover (1980, p.301).
Statistical tables: Siegel(1956,p.280); Powell(1982,p.74); uses χ^2 or *F* tables for large *N*.
Mainframe software: in large packages (e.g. SPSS) or libraries (e.g. NAG routine G08AEF).
Microcomputer software: programmed in some statistics packages (e.g. Statview 512+™).

As shown in Table 11.1, this is both the multi-group extension of the Sign Test (i.e. for $q = 2$ it is equivalent to the Sign Test), and the nonparametric equivalent of the two-way ANOVA in Table 12.3. It is extremely simple to perform, requiring only that the raw data be ranked by block (row), and the treatment (column) sums of squared-ranks combined (Table 12.2b). There are then two alternative forms of the test statistic, the first (comparison with χ^2) being given in most textbooks, but the second (comparison with *F*) being preferred by, for example, Conover(1980):

$$\chi^2_{c-1} \approx \frac{12}{bc(c+1)}\sum_{i=1}^{i=c} ®_i^2 - 3b(c+1) = \frac{12\sum_{i=1}^{i=c}(®_i - b\overline{®})^2}{bc(c+1)} - \frac{\Sigma(t^3 - t)}{c-1} \quad [12.4a]$$

The second part includes the usual correction for ties [cf. 10.8 et seq.]. Note the similarity of the first part to [11.12] for the Kruskal-Wallis test, the equivalent test for *one*-way rather than 2-way ANOVA.

$$F_{(c-1),(b-1)(c-1)} \approx \frac{(b-1)\left[\frac{1}{b}\sum_{i=1}^{i=c}®_i^2 - \frac{bc(c+1)^2}{4}\right]}{\frac{bc(c+1)(2c+1)}{6} - \frac{1}{b}\sum_{i=1}^{i=c}®_i^2} \quad [12.4b]$$

For Table 12.2b, with $b = 5$ and $c = 4$ in this case, we obtain from [12.4a]:

$$\chi^2_{c-1} \approx \frac{12}{4x5x5}x730 - 3x5x5 = 12.6$$

which compares with the critical 99% value of 11.35 for χ^2_3, confirming the difference indicated by the parametric method of the previous section for the *second* hypothesis about differences between the vents (treatments). Note that Friedman test does *not* test any hypothesis about the blocks, and so is better suited to randomized block than 2-factor designs. Multiple comparisons (§11c) can however be made if desired, assuming a significant treatment effect has been identified (Conover 1980, p.300).

12c3. A more complex nonparametric approach: the Quade Test

Worked non-geological examples: Conover(1980,p.297).
Statistical tables: uses *F* tables (§8b3).
Computer software: none known.

As shown in Table 11.1, this test is the multi-group extension of the Wilcoxon Test (§10c3). Therefore, just as the Wilcoxon Test is more powerful than the Sign Test, the Quade Test is theoretically more powerful than the Friedman Test just discussed. The Quade Test does, however, require that the values in the different blocks can be compared (ranked), and in practice it appears to be *less* powerful than the Friedman Test for $q > 4$.

Initially, it proceeds like the Friedman test, by ranking the data by block to give $®_j$. Then the *range* in each of the *b* blocks is obtained from the *raw* (not ranked) data, and the ranks P_i, from 1 to *b*, of these *b* ranges are obtained. The product $S_{ij} = P_i[®_i - (c+1)/2]$ is then obtained, the difference between the rank-within-block and the average rank-within-block, $(c + 1)$; this represents the relative size of each observation, adjusted according to the relative significance of the block in which it occurs.

Table 12.4. Calculations for the Quade test on the data on Table 12.2a

Block	% range in	Rank of	$S_{ij} = P_i[®_j - (c+1)/2] = P_i[®_j - 2.5]$†			
Vent/volcano	(raw data)	range P_i	S(vent 1)	S(vent 2)	S(vent 3)	S(vent 4)
1	15	2	−1	+3	+1	−3
2	20	4	−2	+2	+6	−6
3	21	5	−2.5	+2.5	+7.5	−7.5
4	18	3	−1.5	+1.5	+4.5	−4.5
5	14	1	−1.5	+1.5	+0.5	−0.5
		ΣS_{ij}	−8.5	+10.5	+19.5	−21.5
† $®_i$ from Table 12.2b		$(\Sigma S_{ij})^2$	72.25	110.25	380.25	462.25

The test statistic (assuming there are no ties) is then:

$$F_{(c-1),(c-1)(b-1)} \approx \frac{(b-1)B}{\dfrac{b(b+1)(2b+1)(c^3-c)}{72} - B}, \text{ where } B = \frac{1}{b}\sum_{i=1}^{i=c}\left[\sum_{i=1}^{i=b} P_i\left\{®_i - \frac{c+1}{2}\right\}\right]^2 \quad [12.5]$$

From Table 12.4, we obtain:

$$B = \frac{72.25 + 110.25 + 380.25 + 462.253}{5} = 205; \; F_{3,12} \approx \frac{4\text{x}205}{\dfrac{5\text{x}6\text{x}11\text{x}60}{72} - 205} = 11.71$$

which exceeds the critical $F_{3,12}$ value of 8.74 (95%), once again rejecting the null hypothesis, though the significance level is somewhat lower than in the parametric ANOVA or Friedman test of the previous two sections. Multiple comparisons can again now be made if desired (Conover 1980, p.297).

12d. Two-way ANOVA on several related but incomplete groups of data: BALANCED INCOMPLETE BLOCK DESIGNS (BIBD)

The situation here is as in §12c, except that we do not have a measurement for all possible combinations of blocks and treatments: the table of data is *incomplete*. This is an extremely common occurrence in geology, where some critical localities may be unexposed, or data otherwise unavailable. ANOVA is always possible when the incompleteness is **balanced,** generally possible in *partially* balanced incomplete designs, but increasingly difficult in thoroughly unbalanced designs (which are not covered here). A design of b blocks and c treatments is a balanced incomplete block design if: (1) we apply a ($< c$) treatments to *every* block (i.e. we compare only a treatments at once); (2) each treatment is applied d ($< b$) times; (3) every treatment appears the same number of times as every other treatment.

Returning to our volcanic vent data, a balanced incomplete block design might arise if a similar experiment is being conducted to that in Table 12.2a, but there is only sufficient time, manpower, collecting apparatus or protective equipment available to sample 4 out of 7 active vents at once. Therefore, 4 out of the 7 vents are chosen *randomly*, and rotated regularly during the experiment, so as to give a balanced design with CO_2 measurements such as those in Table 12.5.

Table 12.5. Raw data for Balanced Incomplete Block Design (BIBD)

	Vents (treatments)						
Collectors (blocks)	1	2	3	4	5	6	7
1			33		28	35	30
2	17			26		29	20
3	46	41			49		51
4	22	25	29			27	
5		34	42	40			38
6	19		28	17	25		
7		29		30	32	35	

Note that if the *vents* had imposed the restriction — if only some were active at any one time — this would be a subtly different problem, since it would *not* then be possible to sample the vents randomly. Even if 4 and only 4 vents happened to be active on every occasion sampling was undertaken, and the resulting data looked *exactly* as Table 12.5, one of the major (randomization) assumptions of ANOVA would have been violated, and the results would have to be treated with considerable caution.

12d1. The parametric approach

Worked geological examples: *Sedimentology* (Krumbein & Miller 1952,p. 518 – unbalanced design)
Statistical tables: uses *F* tables (§8b3).
Computer software: covered by more sophisticated ANOVA implementations.

Table 12.6 shows several ANOVA results from the BIBD in Table 12.5. In Table 12.6a, the correct result, both 'collectors' and 'vents' prove to be significant: in other words, the CO_2 content depends not only on which vent is sampled but also on who is sampling! In Table 12.6b–c, the *same* data are handled by a one-way ANOVA (§11b) — i.e. as 4 replicates from each of 7 'vents' *or* 'collectors'— whereupon only the 'collectors' factor emerges as significant. This emphasises how important it is to treat *all* discernible factors simultaneously. Notice also that the total sum-of-squares for the two-way and one-way treatments are not quite identical as they should be: this reflects a small (≈ 1%) error in the calculations consequent on the incompleteness of the design. Incomplete designs are particularly susceptible to inconsistent results using different statistical packages, since the ANOVA depends on exactly how the package treats 'missing' data and decomposes the total sum-of-squares into its component parts. Table 12.6d illustrates a set of results from one package which has attempted to *interpolate* data for all 21 missing values in Table 12.5. This would be the correct way to proceed for a design with only one or two missing values (or where the missing data were distributed unsystematically), but here it fails to recognize the balanced nature of the design and the results are incorrect. The total sum-of-squares has increased substantially because of the additional contribution from the interpolated values.

Table 12.6. ANOVA results on balanced incomplete block design of Table 12.5 (results (a) to (c) calculated using SYSTAT package — see Table 2.2)

Source	Sum of squares	df	Mean square	*F* ratio	*p*
(a) Two-way BIBD using 'vents' and 'collectors' as factors; *no* missing value interpolation					
Collectors	1879.857	6	313.310	41.083	0.000
Vents	222.857	6	37.143	4.870	0.006
Error	114.393	15	7.626		
Total	2217.107	27			
(b) One-way ANOVA using 'collectors' as only factor					
Collectors	1888.857	6	314.810	19.603	<0.001
Error	337.250	21	16.060		
Total	2226.107	27			
(c) One-way ANOVA using 'vents' as only factor					
Vents	231.857	6	38.643	0.407	0.866
Error	1994.250	21	94.964		
Total	2226.107	27			
(d) Two-way BIBD using 'vents' and 'collectors' as factors; missing values interpolated					
Collectors	3584.1599	6	597.360	76.59	0.000
Vents	401.6701	6	66.9450	8.58	0.001
Error	116.9940	15	7.7996		
Total	4102.8242	27			

12d2. The nonparametric approach: the Durbin Test

Worked non-geological examples: Conover(1980, p.311).
Statistical tables: uses χ^2 tables (§8b1).

This reduces to the Friedman Test (§12c2) if $a = c$ and $d = b$. The test statistic is:

$$\chi^2_{c-1} \approx \frac{12(c-1)}{cd(a-1)(a+1)} \sum_{i=1}^{i=c} \left[\sum_{j=1}^{j=b} ®_{ij} - \frac{d(a+1)}{2} \right]^2 \qquad [12.6]$$

where $®_{ij}$ ranks the *i*th treatment in the *j*th block. In Tables 12.5 & 7, $b = c = 7$; $a = d = 4$, yielding:

$$\chi^2_6 \approx \frac{12x6}{4x7x3x5}\left[(6-10)^2+(5-10)^2+(15-10)^2+(9-10)^2+(10-10)^2+(15-10)^2+(10-10)^2\right] = 15.77$$

Table 12.7. Calculations for Durbin test on balanced incomplete block design on Table 12.5

	Vents (treatments)						
Collectors (blocks)	1	2	3	4	5	6	7
1			3		1	4	2
2	1			3		4	2
3	2	1			3		4
4	1	2	4			3	
5		1	4	3			2
6	2		4	1	3		
7		1		2	3	4	
$\Sigma ®_{ij}$	6	5	15	9	10	15	10

This lies between the 97.5% and 99% critical values of χ^2_6 (14.45, 16.81), and thus confirms that the 4 vents yield significantly different CO_2. This test again tells us nothing about the block effect.

Table 12.8. Multi-factor ANOVA with replication: CO_2 by collector and vent level of activity

(a) BALANCED, COMPLETE DESIGN				(b) UNBALANCED DESIGN			
Collector	Vent no.	Activity	$\%CO_2$	Collector	Vent no.	Activity	$\%CO_2$
1	1	Quiet	27	1	1	Quiet	27
2	1	Quiet	28	2	1	Quiet	28
1	1	Quiet	31	1	1	Quiet	31
2	1	Quiet	30	•	•	•	•
1	1	Active	32	1	1	Active	32
2	1	Active	33	2	1	Active	33
1	1	Active	31	1	1	Active	31
2	1	Active	35	2	1	Active	35
1	1	Violent	34	1	1	Violent	34
2	1	Violent	35	2	1	Violent	35
1	1	Violent	36	1	1	Violent	36
2	1	Violent	37	2	1	Violent	37
1	2	Quiet	30	•	•	•	•
2	2	Quiet	32	2	2	Quiet	32
1	2	Quiet	29	1	2	Quiet	29
2	2	Quiet	31	2	2	Quiet	31
1	2	Active	35	•	•	•	•
2	2	Active	33	2	2	Active	33
1	2	Active	36	1	2	Active	36
2	2	Active	38	2	2	Active	38
1	2	Violent	39	•	•	•	•
2	2	Violent	37	2	2	Violent	37
1	2	Violent	41	1	2	Violent	41
2	2	Violent	38	2	2	Violent	38

12e. Some Simple Crossed Factorial Designs with Replication

As we have now exhausted all the designs covered by common nonparametric tests, remaining designs in this Topic are all dealt with by parametric ANOVA tables.

In Table 12.8 we extend our volcanic vent data to cover a new factor — the level (intensity) of activity in each vent. Notice that this in strictly an *ordinal* variable (§6c), but can be used just as well in ANOVA as a categorical variable (§6b). As before, the CO_2 data could equally well be mean values over a set period, or individual analyses of samples taken at one time. We now have three factors, 2 with two levels and one with three, so there are 12 possible combinations of the 3 factors (2 x 2 x 3). In Table 12.8a, we have exactly two measurements for each of these combinations, as shown by the incidence Table 12.11b, so the design is both balanced and complete. (Each factor combination in incidence tables like this is commonly referred to as a *cell*, so that we would speak of this design as having '2 replicates per cell'). In Table 12.8b, on the other hand, 4 observations have been lost (vents were inactive, analyses spoilt, etc.); the design is now unbalanced (Table 12.12b) and its solution requires considerably more calculation — the results may in fact depend on the exact algorithm used.

Table 12.9. Two-factor consideration of balanced design in Table 12.8a, using 'vents' and 'activity' as factors

(a) ANOVA table

Source:	df:	Sum of Squares:	Mean Square:	F-test:	P value:
Vent (A)	1	37.5	37.5	13.366	.0018
Activity (B)	2	220.083	110.042	39.223	.0001
AB	2	3.25	1.625	.579	.5704
Error	18	50.5	2.806		

(b) Means and incidence table (number of measurements per cell)

	Activity:	Quiet	Active	Violent	Totals:
Vent	level 1	4 29	4 32.75	4 35.5	12 32.417
	level 2	4 30.5	4 35.5	4 38.75	12 34.917
	Totals:	8 29.75	8 34.125	8 37.125	24 33.667

12e1. Two-factor Crossed Complete Design with Replication: Balanced and Unbalanced

Further geological discussion: Shaw (1970, p.350); Davis (1986, p.78).
Further non-geological discussion & worked examples: Steel & Torrie (1980,p.343ff).
Worked geological examples: *Rock analysis* (Le Maitre 1982, p.33).

We first ignore the 'collector' factor and consider the vents and their activity only. In both Table 12.9 and Table 12.10, we are testing *three* separate hypotheses:
(1) H_0: {%CO_2 does not vary with vent activity}; against H_1: {%CO_2 varies with the level of activity};
(2) H_0: {%CO_2 does not vary between the 2 vents}; against H_1: {%CO_2 varies between the 2 vents};
(3) H_0: {vent and activity do not interact}; against H_1: {vent and activity do interact}.

In Table 12.9a, both factors' F ratios are highly significant (very low associated P values). We thus conclude that the CO_2 in these vents *does* vary both between the vents (factor A) and with the activity of each vent (factor B). The interaction between AB, by contrast, is not significant (very high associated probability); this means that we can consider the vents and activities as quite separate factors. If the interaction *had* been significant, there would have been no point in examining the main effects of A and B separately, since the response to A would depend on the level of B. Turning to the unbalanced design (Table 12.10), the F ratios remain very close to those from the balanced design, and the same hypotheses are accepted. This will not always be the case, of course.

Table 12.10. Two-factor consideration of unbalanced design in Table 12.8b, using 'vents' and 'activity' as factors

(a) ANOVA Table

Source:	df:	Sum of Squares:	Mean Square:	F-test:	P value:
Vents (A)	1	35.64	35.64	10.306	.0063
Activities (B)	2	177.558	88.779	25.671	.0001
AB	2	1.186	.593	.171	.8442
Error	14	48.417	3.458		

(b) Means and incidence table (number of measurements per cell)

	Activities:	Quiet	Active	Violent	Totals:
Vents	level 1	3 28.667	4 32.75	4 35.5	11 32.636
	level 2	3 30.667	3 35.667	3 38.667	9 35
Totals:		6 29.667	7 34	7 36.857	20 33.7

12e2. Three-factor Crossed Complete Design with Replication: Balanced and Unbalanced

Further non-geological discussion and worked examples: Steel & Torrie (1980,p.348ff).

We now consider *all three* factors in Table 12.8a, adding in the 'collector' factor. With 24 cases and 12 possible factor combinations, we now have only 2 measurements for each combination of factors — the minimum necessary for assessing all possible interactions. Therefore the data in Table 12.8b pose computational problems, for we have no replicates for 4 combinations of factors (Table 12.12b), which makes assessment of the interactions more difficult. Despite this, the unbalanced design is still *complete*, because every factor combination still has at least one associated measurement. However, if

Table 12.11. Results of three-factor consideration of balanced design in Table 12.8a

(a) ANOVA Table

Source:	df:	Sum of Squares:	Mean Square:	F-test:	P value:
Vent (A)	1	37.5	37.5	13.636	.0031
Activity (B)	2	220.083	110.042	40.015	.0001
AB	2	3.25	1.625	.591	.5692
Collector (C)	1	1.5	1.5	.545	.4744
AC	1	2.667	2.667	.97	.3442
BC	2	4.75	2.375	.864	.4463
ABC	2	8.583	4.292	1.561	.2498
Error	12	33	2.75		

(b) Means and incidence table (number of measurements per cell)

	Activity:	Quiet		Active		Violent		Totals:
	Collector:	level 1	level 2	level 1	level 2	level 1	level 2	
Vent	level 1	2 29	2 29	2 31.5	2 34	2 35	2 36	12 32.417
	level 2	2 29.5	2 31.5	2 35.5	2 35.5	2 40	2 37.5	12 34.917
Totals:		4 29.25	4 30.25	4 33.5	4 34.75	4 37.5	4 36.75	24 33.667

we were additionally to remove analysis 2 from Table 12.8b, we would be left with *no* measurements for the factor combination 'collector 2–vent 1–activity quiet'; some statistical packages (including the one used here) would still attempt to compute the ANOVA, but garbage results would usually be obtained, because the design is then both unbalanced and incomplete: highly disorganized, in fact!

The results in Tables 12.11–12 show the vents and activity factors remain highly significant, but that the collecting method does not contribute to the CO_2 variation. Similarly, none of the four possible factor interactions are significant. Again, the missing values do not alter our conclusions.

Table 12.12. Results of three-factor consideration of unbalanced design in Table 12.8b

(a) ANOVA table

Source:	df:	Sum of Squares:	Mean Square:	F-test:	P value:
Vents (A)	1	36.125	36.125	10.321	.0124
Activities (B)	2	187.265	93.632	26.752	.0003
AB	2	2.935	1.468	.419	.6711
Collectors (C)	1	.125	.125	.036	.8548
AC	1	2	2	.571	.4714
BC	2	4.771	2.385	.682	.533
ABC	2	12.818	6.409	1.831	.2214
Error	8	28	3.5		

(b) Means and incidence table (number of measurements per cell)

	Activities:	Quiet		Active		Violent		Totals:
	Collectors:	level 1	level 2	level 1	level 2	level 1	level 2	
Vents	level 1	2	1	2	2	2	2	11
		29	28	31.5	34	35	36	32.636
	level 2	1	2	1	2	1	2	9
		29	31.5	36	35.5	41	37.5	35
Totals:		3	3	3	4	3	4	20
		29	30.333	33	34.75	37	36.75	33.7

12f. A Simple Repeated Measures Design

Further non-geological discussion: Steel & Torrie (1980,p.377ff).
Microcomputer software: programmed in some statistics packages (e.g. Statview 512+™).

We now consider different designs of *exactly the same data* as in Table 12.8, in which we are focussing on finer differences between the vents. Notice that the sums-of-squares and mean-squares for 'activity' and 'vent' are exactly the same in Table 12.13b as in Table 12.9a, while many of the rows in Table 12.13c are similar to Table 12.11a. However, error and interactions are broken down differently. The interpretation of Tables 12.13b-c is that whereas the 'collectors' do not affect the CO_2 results, the CO_2 output of the vents *at constant activity levels* does differ.

Table 12.13a is the sting in the tale to this section: a good example of a perfectly reasonable-looking ANOVA table which can be generated from many computer programs, which is actually geologically meaningless. We have here reduced the data to only 'vents' and 'samples, so geologically the only meaningful comparison is the difference between 'vents' — a *one-way* ANOVA which should have been analyzed as in §11b to give an ANOVA like Table 11.3b. What the computer has in fact done is to assume the same set of 'samples' have been subjected to the treatment 'vent'; in other words, the **dependent variables** (treatments) and **independent variables** (subjects) are the wrong way round. This kind of design would be perfectly valid for examining, say, the effect of 2 drugs (treatments) on a set of *people* (subjects, where the *same* person would undergo both treatments), or even for examining the CO_2 content of several vents (subjects) at different *times* (treatments), but in this particular case such a design has no meaning. ANOVA is unfortunately one of the easiest techniques to misuse on computers; Table 12.13a is just a more subtle case of which further, rather more blatant examples are given by Rock et al. (1988).

Table 12.13. Repeated-measures designs on the data of Table 12.8a

(a) Single within-factor ANOVA, with 'vents' as repeated-measures factor: *caveat computor!*

Source:	df:	Sum of Squares:	Mean Square:	F-test:	P value:
Between samples	11	247.333	22.485	4.216	.01
Within samples	12	64	5.333		
vents(treatments)	1	37.5	37.5	15.566	.0023
residual	11	26.5	2.409		
Total	23	311.333			

Reliability Estimates for- All treatments: .763 Single Treatment: .617

(b) Two-factor ANOVA, with 'activity' as grouping factor and 'vents' as repeated-measures factor (cf. Table 12.9a)

Source:	df:	Sum of Squares:	Mean Square:	F-test:	P value:
Activity (A)	2	220.083	110.042	36.344	.0001
subjects w. groups	9	27.25	3.028		
Repeated Measure (B)	1	37.5	37.5	14.516	.0042
AB	2	3.25	1.625	.629	.555
B x subjects w. groups	9	23.25	2.583		

(c) Three-factor ANOVA, 'activity' & 'collector' are grouping factors, 'vents' is repeated-measures factor (cf. Table 12.11a)

Source:	df:	Sum of Squares:	Mean Square:	F-test:	P value:
Collector (A)	1	1.5	1.5	.429	.537
Activity (B)	2	220.083	110.042	31.44	.0007
AB	2	4.75	2.375	.679	.5424
subjects w. groups	6	21	3.5		
Repeated Measure (C)	1	37.5	37.5	18.75	.0049
AC	1	2.667	2.667	1.333	.2921
BC	2	3.25	1.625	.812	.4872
ABC	2	8.583	4.292	2.146	.1982
C x subjects w. groups	6	12	2		

12g. Analyzing data-within-data: HIERARCHICAL (NESTED) ANOVA

Further geological discussion: Flinn(1958); Krumbein & Graybill(1965,p.205ff); Shaw(1969,p.352).
Worked geological examples: *Economic Geology* (Koch & Link 1971, p.148ff); *Geomorphology* (Krumbein & Graybill 1965, p. 210); *Sedimentology* (Miller & Kahn 1962,p.418).
Selected geological applications: *Igneous petrology* (Dawson 1958); *Metamorphic petrology* (Ward & Werner 1963,1964); *Sedimentology* (Krumbein & Tukey 1963).
Stand-alone computer programs: Garrett & Goss(1980).
Mainframe software: covered in specialised ANOVA packages (e.g. ANOVA procedure in GENSTAT; GLIM; MASS); also NAG routine G04AGF.

This design focuses on data which can be arranged in some form of logical hierarchy. In stratigraphy, for example, it could be used to compare the variation between lithostratigraphical Groups, Formations (within Groups) and Members (within Formations). In petrology, it could compare different intrusive phases within one batholith, or different batholiths within an igneous province. In our volcanic vent example, we examine variations among 6 vents sampled in duplicate from each of 4 volcanoes (all within the same oceanic island), to determine whether the major source of variation is between the volcanoes, or between the vents within one volcano. The data and ANOVA results are in Table 12.14.

Table 12.14. Hierarchical ANOVA: duplicate CO_2 values from 6 vents from each of 4 different volcanoes

(a) Raw data

	Vent 1	Vent 2	Vent 3	Vent 4	Vent 5	Vent 6
Volcano 1	15 14	11 13	16 17	19 20	21 22	13 16
Volcano 2	23 26	25 27	24 26	20 22	25 27	30 32
Volcano 3	9 9	16 17	10 13	12 14	15 19	20 23
Volcano 4	31 32	35 36	29 31	25 27	23 25	28 30

(b) ANOVA table‡

Strata & Effects	Sum of squares	DF	Mean square	*F*	prob.
Between volcanoes	1793.2	3	597.71	20.08	< 0.001
Between vents within volcanoes	595.3	20	29.76	13.87	< 0.001
Residual †	51.5	24	2.15		
Total	2440.0	47			

† Equivalent to "error between measurements within vents within volcanoes"
‡ Results calculated using MASS package — see Table 2.2

In hierarchical designs, *F* ratios are calculated slightly differently, though the ANOVA table is structured as before. In the present problem, *df* for the "between volcanoes" row is, as usual, $n - 1$, where n is the number of levels for the 'volcano' factor (here, $n - 1 = 3$). However, *df* for the "between vents within volcanoes" row is $n(m - 1)$, where m is the number of levels for the vents factor (here = $4 \times 5 = 20$). The *lower F* ratio in Table 12.14b is derived as for earlier ANOVA tables, by dividing the MS for that row by the residual MS, but the *upper F* ratio uses the ratio of MSs for the *top two rows*. In general, *F* is the MS ratio for *adjacent* rows in a nested design. This is because each successive row adds one more source of variation, so that rows two apart differ by *two* sources of variation, not just one as in earlier ANOVA tables; their ratio is *not* therefore distributed as *F*.

The results in Table 12.14b indicate that there are highly significant variations both between the volcanoes *and* between vents within one volcano. Clearly, this could be expanded to further levels, investigating differences between, say, different oceanic islands; or, if the vents were clustered into fumarole fields, an intermediate level of nesting between the volcanoes and vents could be studied.

SECTION IV: INTERPRETING DATA WITH TWO VARIABLES: Bivariate Statistics

So far we have dealt statistically with only one variable (of whatever type) at a time. This section now considers relationships between *two* variables: the relatively familiar topics of **correlation** and **regression.** We are concerned, in fact, with quantifying relationships between X and Y in scatter plots of the type in Fig. 13.1, with which the geological literature is saturated. *Correlation* (§13) is concerned only with measuring the *strength* of the X-Y relationship, and is just one aspect of *regression* (§14), which defines the actual *line* (or curve) on a scatter plot showing the 'best fit' between X and Y, and so allows Y to be predicted from any given value of X (or vice-versa).

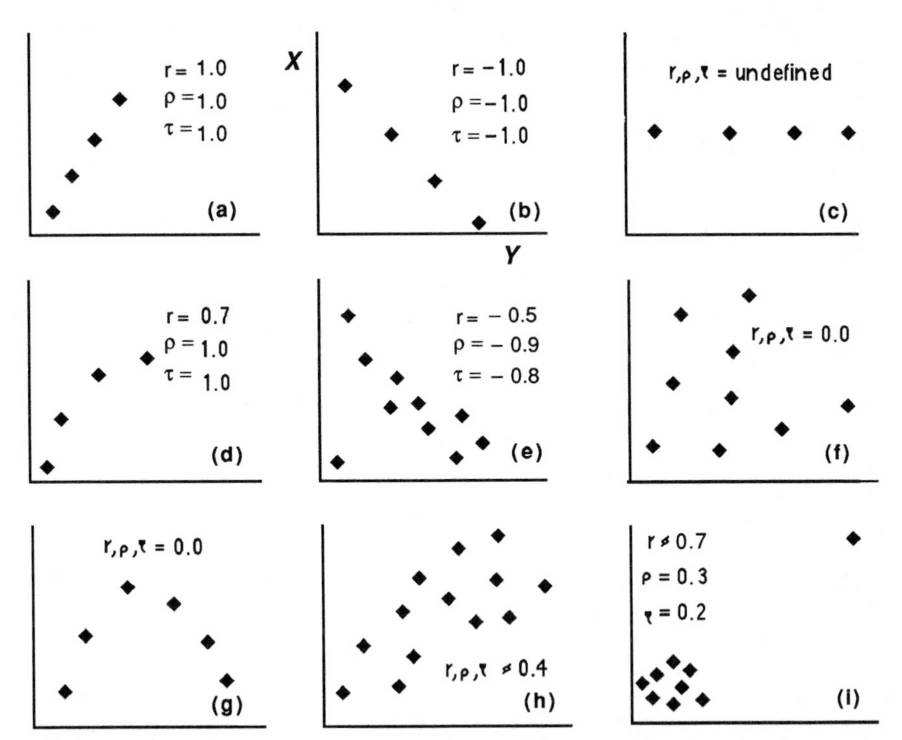

Fig.13.1. Values of the 3 most important correlation coefficients (Pearson's r, Spearman's ρ and Kendall's τ, §13a-b) for various X-Y relationships. **(a)** Perfect positive linear; **(b)** perfect negative linear; **(c)** horizontal (r = 0/0 i.e. undefined); the same applies to a vertical line; **(d)** perfect positive monotonic and **(e)** strong negative monotonic with one outlier; **(f)** random; **(g)** curvilinear (**polytonic**); **(h)** moderate positive; **(i)** spurious correlation induced by single outlier. Plots (d), (e) and (i) are best measured by ρ or τ (§13b), (g) by η (§13g). See also Fig.13.4.

TOPIC 13. TESTING ASSOCIATION BETWEEN TWO OR MORE VARIABLES: Correlation and concordance

Dedicated monographs and reviews: Carroll(1971); Open University (1974b).

We consider available measures below from the *highest* to *lowest* scales of measurement (reverse order to that listed in Table 13.1). It is logical to deal first with the best-known correlation coefficient, **Pearson's *r***, since this is the basis for many of the others. All of the measures to be described are dimensionless numbers; nearly all range between 0 (indicating no relationship) to ±1 (indicating a perfect, positive or negative correlation). With a perfect correlation, *Y* is completely dependent on *X* (or vice-versa). Fig.13.1 illustrates some extreme and intermediate stages. Readers are advised to work through the example given for the first three measures, even though, with the widespread programming of these measures even on hand-held pocket calculators, it is unlikely in practice that they will ever need to calculate any of the three by hand.

Table 13.1. Coefficients used to test the strength of association between various types of data

Variable types	*Dichotomous or Nominal*	*Dichot.*	*Ordinal*†	*Interval/Ratio*	*Concordance*§
Dichotomous or Nominal	**R-mode:*** χ^2, Cramer's *V*, Fourfold (φ)‡, Pearson's *C*, Tschuprow's *T*, Yules' *Q* ‡ **Q-mode:¶** Dice, Hamman‡, Jaccard, Kulczynski, Ochiai, Otsuka, Rogers-Tanamato, Russell-Rao, Simpson, Sokal-Sneath	Biserial	Kendall's *W*	Point biserial	Cochran's *Q*
Dichotomized	—	Tetrachoric	—	—	
Ordinal	—	—	Kendall's τ †, Spearman's ρ†, Somer's *d* †‡, Guttman's μ_2 ◊, Goodman & Kruskal's γ ◊	—	Goodman & Kruskal's τ
Interval/ratio	—	—	—	Covariance, Pearson's *r*‡	—

* **Contingency coefficients**, quantifying association between *V* (≥ 2) variables measured on *N* (≥ 2) objects; most range from –1 to +1; see §13d3.

¶ **Matching coefficients,** quantifying similarity between 2 objects measured on *V* (≥ 2) variables; most range from 0 to +1; see §21a1.

† **Rank correlation coefficients,** generally used R-mode to measure **strong monotonicity** between 2 variables measured on *N* (≥ 2) objects; all range from –1 to +1; see §13b and Figs.13.1, 13.4.

◊ Coefficients generally used R-mode to measure **weak monotonicity** between 2 variables measured on *N* (≥ 2) objects; all range from –1 to +1; see Fig.13.4.

§ **Tests of concordance**, used R-mode to correlate *V* (> 2) variables measured on *N* objects; range from 0 to 1; see §13f.

‡ Have been used both R-mode (to correlate 2 variables measured on *N* (≥ 2) objects) and Q-mode (to measure similarity between 2 objects measured on *V* (≥ 2) variables).

For individual definitions of all these coefficients, see Glossary.

WARNING: The literature is exceptionally confusing and inconsistent over the names and applicabilities of the above coefficients; the above is an attempt at a consensus, but some references in the Bibliography use some terms differently.

13a. Measuring Linear Relationships between Two Interval/ratio Variables: PEARSON'S PRODUCT-MOMENT CORRELATION COEFFICIENT, *r*

Further geological discussion: Harrell(1987); Kowalski (1972); Skala(1977).

Worked geological examples: *Geomorphology* (Till 1974, p.85); *Igneous petrology* (Cheeney 1983, p.89); *Mining* (Koch & Link 1971, p.16); *Palaeontology* (Davis 1986, p.40); *Tectonics* (Le Maitre 1982, p.43).

Selected geological applications: *Economic geology* (Silichev 1976); *Igneous petrology* (Aleksandrov 1985; Bailey & Macdonald 1975; Gerasimovsky et al.1981; Yoder & Chayes 1982,1986); *Metasomatism* (Ferguson & Currie 1972, Rock 1976); *Metamorphic petrology* (Rock 1987c,e; Rock & Waterhouse 1986; Rock & Macdonald 1986); *Sedimentology* (Till 1971).

Statistical tables: Downie & Heath(1974); Le Maitre(1982,p.244); Powell(1982,p.78) and many others.
Stand-alone computer programs: Jones & Facer (1982); Rock(1987d); Press et al.(1986,p.487).
Mainframe software: covered in most packages (e.g. SPSS) or libraries (e.g. NAG gives numerous routines G02BnF where n is from A to M, for different options).
Microcomputer software: programmed in many statistics packages (e.g. Statview 512+™).

This is simply the ratio of the **covariance** of X and Y to the product of their standard deviations, i.e:

$$r_{xy} = \frac{\mathrm{cov}_{xy}}{s_x s_y} = \frac{\Sigma(X-\overline{X})(Y-\overline{Y})}{\sqrt{\Sigma(X\quad\overline{X})^2\Sigma(Y-\overline{Y})^2}} = \frac{\Sigma XY - \dfrac{\Sigma X \Sigma Y}{N}}{\sqrt{\left\{\Sigma X^2 - \dfrac{(\Sigma X)^2}{N}\right\}\left\{\Sigma Y^2 - \dfrac{(\Sigma Y)^2}{N}\right\}}} \qquad [13.1]$$

where all summations are from 1 to N (there must, of course, be equal numbers of X and Y values).

Tables of critical r values are actually based on conversion into a value of Student's t:

$$t_{N-2} = r\sqrt{\frac{(N-2)}{(1-r^2)}} \qquad [13.2]$$

If t or r are not significant, we accept H_0: $r = 0$; if t or r are larger, we accept H_1: $r \neq 0$.

Suppose we are interested in assessing the dependence of Cr content with height in a differentiated sill (data in Table 13.2):

Table 13.2. Example data and calculations for Pearson's r, Spearman's ρ and Kendall's τ correlation coefficients

Raw data		Calculations for r			Calculations for ρ				Calc.for τ
Height (m) in sill,X	Cr (ppm),Y	X^2	Y^2	XY	®$_X$	®$_Y$	D (®$_X$ —®$_Y$)	D^2	Score
1.21	1300	1.464	1690000	1573.0	1	6	5	25	–5+0 = –5
2.42	1156	5.856	1336336	2797.5	2	4	2	4	–3+1 = –2
3.44	1124	11.83	1263376	3866.6	3	3	0	0	–2+1 = –1
5.87	1245	34.46	1550025	7308.2	4	5	1	1	–2+0 = –2
7.93	789	62.88	622521	6256.8	5	2	3	9	–1+0 = –1
10.35	546	107.1	298116	5651.1	6	1	5	25	0 0
ΣX=31.22	ΣY=6160	Σ=223.6	Σ=6760374	Σ=27453.2				ΣD^2= 64	S = –11

Thus in this example:

$$r = \frac{27453 - \dfrac{6160\text{x}31.22}{6}}{\sqrt{\left\{223.6 - \dfrac{(31.22)^2}{6}\right\}\left\{6760374 - \dfrac{(6160)^2}{6}\right\}}} = -0.8906$$

which is 99% significant. This means the Cr content *decreases* significantly with height ($r < 0$).

A value for r can also be calculated where individual X and Y values are not available, but data are presented in the form of a bivariate frequency distribution (i.e. grouped data), as in Table 13.3. Thus the X values are grouped into I cells with midpoints m_{X1}, ...m_{XI}, with $f_{i\cdot} = \Sigma f_{ij}$, the frequency of X values in the class with midpoint m_{Xi}, and so on. Means and variances are calculated via [8.2] as:

$$\overline{X} = \frac{1}{N}\sum_{i=1}^{i=I} m_{Xi} f_{i.},\quad \overline{Y} = \frac{1}{N}\sum_{j=1}^{j=J} m_{Yj} f_{.j},\quad s_X^2 = \frac{1}{N}\sum_{i=1}^{i=I} m_{Xi}^2 f_{i.} - \overline{X}^2,\quad s_Y^2 = \frac{1}{N}\sum_{j=1}^{j=J} m_{Yj}^2 f_{.j} - \overline{Y}^2 \qquad [13.3]$$

while the covariance is:

$$s_{XY} = \frac{1}{N}\sum_{\text{all cells}} m_{Xi} m_{Yj} f_{ij} - \overline{XY} \qquad [13.4]$$

Then r can be calculated from [13.1].

Table 13.3. Notation for calculating r from grouped data

Cell midpoints of X	Cell midpoints of Y m_{Y1}	m_{Y2}	.... m_{Yj}....	m_{YJ}	Total
m_{x1}	f_{11}	f_{12}	.... f_{1j}....	f_{1J}	$f_{1\cdot}$
m_{x2}	f_{21}	f_{22}	.... f_{2j}	f_{2J}	$f_{2\cdot}$
m_{xi}	f_{i1}	f_{i2}	.... f_{ij}	f_{iJ}	$f_{i\cdot}$
m_{xI}	f_{I1}	f_{I2}	.... f_{Ij}	f_{IJ}	$f_{I\cdot}$
Total	$f_{.1}$	$f_{.2}$	.... $f_{.j}$	$f_{.J}$	N

13b. Measuring Monotonic Relationships between Two Ordinal/Ratio Variables: RANK CORRELATION COEFFICIENTS

Dedicated monographs & reviews: Kruskal (1958).

These are correlation coefficients based on ranked data, and have several advantages over r (§13a):

—They are simpler to calculate (as witness Table 13.2).

—They can be used for semi-quantitative (ordinal) as well as quantitative (ratio) data.

—They are not so deleteriously affected by outliers (witness the large numbers accumulating in the calculations for r from the larger values in Table 13.2, and also compare r, ρ,τ in Figs.13.1e, i).

—They can detect strongly **monotonic** relationships (e.g. Fig.13.1d; Fig.13.4d) as well as strictly *linear* (Fig.13b,h) relationships.

— They are unaffected by any of the commonly used monotonic transformations (e.g. linear, logarithmic, square, square-root, reciprocal); r changes with *all* non-linear transformations.

As a result, they are finding increasing use in geology.

13b1. Spearman's Rank Correlation Coefficient, ρ

Dedicated monographs & reviews: Kendall(1955)
Further geological discussion: Brown (1985); Fieller et al. (1957).
Worked geological examples: *Geology teaching* (Till 1974, p.132); *Igneous petrology* (Miller & Kahn 1962,p.335); *Sedimentology* (Davis 1986, p.98).
Selected geological applications: *Economic geology* (Bates 1959; Ghosh 1965); *Igneous petrology* (Cawthorn & Fraser 1979); *Isotope geochemistry* (Golding et al. 1987); *Metamorphic stratigraphy* (Rock 1987c,e).
Statistical tables: Beyer(1968,p.447); Conover(1980,p.456); Lewis(1977,table 12); Powell(1982,p.80); Siegel (1956,p.284);
Stand-alone computer programs: Demirmen (1976); Press et al.(1986,p.490); Rock(1987d).
Mainframe software: covered in most statistics packages (e.g. SPSS) or libraries (e.g. NAG gives numerous routine G02BnF where n is from N to S, covering different options).
Microcomputer software: programmed in many statistics packages (e.g. Statview 512+™).

Provided there are no ties, as above, the formula for ρ is relatively simple, as follows:

$$\rho = 1 - \frac{6\sum_{i=1}^{i=N} D_i^2}{N(N^2 - 1)}, \text{ where } D_i = [\text{rank of } X_i - \text{rank of } Y_i] \qquad [13.5a]$$

It can be shown that [13.5a] is exactly equivalent to calculating r on the *ranks* of the data rather than the raw data themselves. Where there are ties, the formula becomes:

$$\rho = \frac{X_t - Y_t - \sum_{i=1}^{i=N} D_i^2}{2\sqrt{X_t Y_t}}, \text{ where } X_t = \frac{N^3 - N}{12 - \sum T_x}, T_x = \frac{t^3 - t}{12} \qquad [13.5b]$$

where t is the number of values tied for a given rank, and Y_t, T_y are defined analogously to X_t, T_x. For $N \leq 30$, tables are consulted; like those for r, they are based on conversion to a Student's t statistic:

$$t_{N-2} = \rho\sqrt{\frac{(N-2)}{(1-\rho^2)}} \qquad [13.6]$$

[cf. 13.2]. For higher N, the value $\rho\sqrt{(N-1)}$ is compared with a standard Normal deviate.

For the above sill example, [13.5a] gives $\rho = -0.829$, which is just significant at the 90% level. Note that ρ would have been the same if the height corresponding to the 1300 Cr content was *any* value below 2.42, and if that corresponding to 1156 was *anywhere* in the range 1.21–3.44, and so on. A similar latitude is allowed on the Cr values, illustrating the robustness of the ρ statistic to outliers in both variables. Suppose these X height measurements were made on a vertical (and perhaps rather inaccessible!) cliff formed by the sill, with a tape measure: though quoted to the nearest cm, the real accuracy would of course be rather less. In such situations, ρ is a more realistic statistic to use than r, because it does not require exact values . Even the smallest change in one X or Y value will affect r in the nth decimal place, but quite large changes can leave ρ unaffected.

13b2. Kendall's Rank Correlation Coefficient, τ

Dedicated monographs and reviews: Kendall(1955).
Worked geological examples: *Geomorphology* (Miller & Kahn 1962,p.288); *Sedimentology* (Cheeney 1983, p.81).
Selected geological applications: *Chemostratigraphy* (Rock 1987c,e); *Isotope geochemistry* (Golding et al. 1987);
Statistical tables: Conover(1980,p.458); Lewis(1977,table 13); Powell(1982,p.81); Siegel(1956,p.285).
Stand-alone computer programs: Demirmen (1976); Press et al.(1986,p.493); Rock (1987d).
Mainframe software: covered in most statistics packages (e.g. SPSS) or libraries (e.g. NAG gives numerous routine G02BnF where n is from N to S, covering different options).
Microcomputer software: programmed in many statistics packages (e.g. Statview 512+™).

This statistic complements Spearman's ρ, covering the same kinds of situations but being calculated by quite a different method. It measures how far the data are disordered from perfect concordance of their ranks (i.e. the value with the nth rank is the same in both sets for all n). If not already done, we merely reorganize the data such that one set, say X, is in rank-*order*; the actual ranks do not need to be determined. We then consider each Y_i value (or rank) in turn, consider all values lying *beyond* this value in the ordered sequence, and count up the number which are greater and less than Y_i, scoring +1 or –1 appropriately. In the sill example of Table 13.2, 5 values are below 1300, so we score –5 against it. For the second measurement, 1156, we score –3 + 1 = –2 against it. The total score (positive minus negative) is known as Kendall's S statistic. Kendall's τ is the ratio of S to the total number of paired comparisons we have made in this process, i.e. ${}^{N}C_2$ or $N(N-1)/2$. Thus, in the absence of ties, τ is defined by the relatively simple formula:

$$\tau = \frac{2S}{N(N-1)} \qquad [13.7a]$$

With ties present, the formula becomes:

$$\tau = \frac{S}{[0.5N(N-1) - \Sigma T_X][0.5N(N-1)\Sigma T_Y]}, \text{ where } T_X = \frac{t^2 - t}{2} \text{ and similarly } T_Y \qquad [13.7b]$$

An alternative formula for τ defines the following for paired values $\{X_i,Y_i\}$ and $\{X_j,Y_j\}$:

$$\alpha_{ij} = \begin{cases} 1 \text{ if } X_i > X_j \\ 0 \text{ if } X_i = X_j \\ -1 \text{ if } X_i < X_j \end{cases} \text{ and similarly } \beta_{ij} = \begin{cases} 1 \text{ if } Y_i > Y_j \\ 0 \text{ if } Y_i = Y_j \\ -1 \text{ if } Y_i < Y_j \end{cases} \qquad [13.8]$$

whereupon τ can be redefined as:

$$\tau = \frac{\sum_{i=1}^{i=N}\sum_{j=1}^{j=N}\alpha_{ij}\beta_{ij}}{N(N-1)} \qquad [13.9]$$

For $N \le 60$, exact tables (usually of S rather than τ) are available, but for N above *about* 30 (different texts differ as to the exact limit), the value:

$$\tau\sqrt{\frac{9N(N-1)}{2(2N+5)}} \qquad [13.10]$$

is compared with the standard Normal deviate.

In the sill example, S is –11, which is again 99% significant, while τ is – 0.733. Almost always, τ will give a *numerical* value somewhat nearer zero than ρ, but its significance level will usually be the same. τ does have some advantages over ρ, however, which warrant its complementary use: notably, it can be used as a *partial* correlation coefficient (§19a).

13c. Measuring Strengths of Relationships between Dichotomous and Higher-order Variables: POINT-BISERIAL & BISERIAL COEFFICIENTS

Worked non-geological examples: Downie & Heath (1974, p.101).
Statistical tables: graph for determining r_{pb} available in Downie & Heath(1974,p.106).
Computer software: none known.

These coefficients have to date have received virtually no attention in geology, but have considerable potential, especially for grouped data. The point-biserial is available where one variable is a true dichotomy, the biserial where it is a continuous variable forced into a dichotomy (e.g. pass/fail exam results obtained from continuous 0–100% marks). However, Downie & Heath (1974) greatly favour the point-biserial coefficient, so only this is treated here.

For example, the point-biserial could assess correlation between students' performance on two tests: one a simple 'pass/fail', 'yes/no' type, and the other marked continuously from 0-100. It could also assess, from the hypothetical data below, whether the occurrence of Mississippi-type Pb/Zn deposits (a 'yes/no' dichotomy) is dependent on the percentages of some target areas covered by limestone outcrop (a 0–100 continuous variable):

Table 13.4. Calculation of the point-biserial correlation coefficient

% of area covered by limestone, l	No.of areas with deposit, y	No.of areas without deposit, n	Total areas measured, t	lt	l^2t	ly
80-100 (i.e. 90)	11	2	13	1170	105300	990
60-79 (70)	9	5	14	980	68600	630
40-59 (50)	6	10	16	800	40000	300
20-39 (30)	2	7	9	270	8100	60
0-19 (10)	0	25	25	250	2500	0
Σ	28	49	77	3470	224500	1980

The point-biserial formula can be written in various ways, such as:

$$r_{pb} = \frac{\text{Mean}_{yes} - \text{Mean}_{to}}{\text{Standard deviation}}\sqrt{\frac{p}{1-p}} \qquad [13.11a]$$

where 'Mean$_{yes}$' is the mean value on the continuous variable for those items giving a 'yes' score on the dichotomous variable, 'Mean$_{to}$' and standard deviation refer to the continuous variable overall, and p is the proportion of all items having a 'yes' score on the dichotomous variable. For grouped results notated as in the above example, it can also be written as:

$$r_{pb} = \frac{\Sigma t\Sigma(ly) - \Sigma y\Sigma(lt)}{\sqrt{\Sigma y\Sigma n\{\Sigma t\Sigma(l^2t) - [\Sigma(lt)]^2\}}} \qquad [13.11b]$$

which gives:

$$r_{pb} = \frac{77.1980 - 28.3470}{\sqrt{28.49\{77 \times 224500 - (3470)^2\}}} = 0.6519$$

Yet another way of estimating r_{pb} is via a special graph (Downie & Heath 1974, p.106). To do so, we work out the overall median for the continuous variable (≈ 43 in the above example), and the proportion of cases in the 'yes' and 'no' columns with values above the median (≈ 23/28 and 10/49 respectively). The graph confirms $r_{pb} \approx 0.65$.

The point-biserial is in fact a special case of Pearson's product-moment correlation coefficient, and can be tested for significance using the same tables (see §13a). The value 0.651 is 99.9% significant: the presence of Pb-Zn deposits *does* correlate in the above data with limestone abundance.

13d. Testing whether Dichotomous or Nominal Variables are Associated

13d1. Contingency Tables (cross-tabulation), χ^2 (Chi-squared) tests, and Characteristic Analysis

Dedicated monographs and reviews: Agresti(1984).
Worked geological examples: *Petrography* (Cheeney 1983, p.48ff; Till 1974, p.118 & 121);

Sedimentology (Krumbein & Graybill 1965, p.184).
Selected geological applications: *Mineral exploration* (Botbol et al. 1978; Peach & Renault 1965); *Palaeontology* (Hayami 1971).
Statistical tables: uses χ^2 tables (see §8b3).
Stand-alone computer programs: Press et al.(1986,p.479ff); Romesberg & Marshall(1985).
Mainframe software: in most packages (e.g. SPSS) or libraries (e.g. NAG routines G01AFF, G11).
Microcomputer software: programmed in many statistics packages (e.g. Statview 512+™).

Contingency tables (Table 13.5) offer a simple way of determining whether or not a statistically significant association between two dichotomous or nominal variables exists.

Table 13.5. Makeup and adopted notation of contingency tables

(a) Dichotomous data (2x2 table)	Variable *X* 'present'	Variable *X* 'absent'	Totals
Variable *Y* 'present'	*a*	*b*	*a+b*
Variable *Y* 'absent'	*c*	*d*	*c+d*
Totals	*a+c*	*b+d*	*a+b+c+d=N*

a,b,c,d represent numbers of cases found corresponding to each cell; i.e. *a* is the number of cases with *both X* and *Y* are 'present'. 'Present' and 'absent' can be substituted with 'yes' and 'no', 'true' and 'false', 'male' and 'female', or any other appropriate dichotomous states (§6).

(b)Nominal data (ΩxM table)	Variable *X*,state 1	Variable *X*,state 2...	Variable *X*,state *M*	Totals
Variable *Y* in state α	$a = n_{1\alpha}$	$b = n_{2\alpha}$...............	$n_{M\alpha}$	$a+b+....n_{IX}$
Variable *Y* in state ß	$c = n_{1\beta}$	$d = n_{2\beta}$	$n_{M\beta}$	$c+d+....n_{M\beta}$
....	....	....	....	....
Variable *Y* in state Ω	$n_{1\Omega}$	$n_{2\Omega}$	$n_{M\Omega}$	$n_{1\Omega}+n_{2\Omega}+...n_{M\Omega}$
Totals	$a+c+...\ n_{1\Omega}$	$b+d+...\ n_{2\Omega}$.......	$n_{M\Omega}$	$a+b+c+d+....=N$

This table sets out a logical nomenclature, in which $n_{M\Omega}$ represents the number of cases with variable *X* in state *M* and variable *Y* in state Ω. However, *a,b,c,d* are conventionally used as above for the 2x2 case. This can be confusing, since *c* is also often used for the number of columns (i.e. a ΩxM table is referred to as an '*rxc* table'). α, ß...Ω and 1,2...M are differentiated, to indicate that the states of variables *X* and *Y* need not be the same in either type or number: for example, *X* might represent rock-type (nominal states: limestone, shale, etc.), while *Y* might represent mineralogy (nominal states: 'pyroxene-rich',' amphibole-rich' etc.) In the text, *N* always indicates the total number of all cases in all states.

As a simple example, Sabine (1953) found numbers of pyroxene-bearing and pyroxene-free dykes in a suite from NW Scotland, in relation to the limestone they intrude, to be as in Table 13.6. If the limestone host-rock has no effect on the presence or absence of pyroxene, then the cell-numbers *a,b,c,d* should be randomly distributed: the ratio of pyroxene-bearing dykes in limestone to the total number of pyroxene-bearing dykes, $a/(a+b)$, should be the same as the ratio of pyroxene-bearing dykes to all dykes, $(a+c)/N$. The expected value for *a* would thus be $(a+c)*(a+b)/N$, i.e. 93.5 x 51/141 = 33.8. Similarly, the expected value for *b* would be $(b+d)(a+b)/N$ = 17.1. However, the observed value of *a* above is larger and of *b* smaller than these expected values; in general, the larger *a* and/or *d* relative to *b* and/or *c*, the less the figures seem to be random, i.e. the more probable is an association.

χ^2 is the statistic used to compare 'observed' with 'expected' values (§8b3). In Table 13.5 the summation formula for χ^2 [8.27] reduces to:

$$\chi_1^2 = \frac{N(ad - bc)^2}{(a+b)(c+d)(a+c)(b+d)} \qquad [13.12a]$$

In Table13.6, χ^2 calculates to 3.69, which is significant at the 94% level. We would therefore reject H_0

that the emplacement of dykes into limestone is unrelated to the appearance of pyroxene in them.

One problem with this method is that χ^2 is a *continuous* distribution, here being used to analyze *discontinuous* (discrete) data. When any of *a, b, c* or *d* is below 5 (some, e.g. Downie & Heath 1974, p.196, say < 10), most statistical texts suggest the χ^2 is a serious overestimate and the result highly doubtful. Even for larger cell-numbers, some statisticians advocate **Yates' correction for continuity**, which changes the formula to the following:

$$\chi^2 = \frac{N(|ad - bc| - N/2)^2}{(a+b)(c+d)(a+c)(b+d)} \qquad [13.12b]$$

Different views on whether or not this correction should be used are stated by, for example, Conover (1974), Crow(1952) and Plackett (1981). Pirie & Hamden(1972) suggest replacing *N/2* simply by $\frac{1}{2}$ (a much smaller correction), and Downie & Heath (1974) support this idea — at least for 2 x 2 tables.

χ^2 can be used to analyze nominal data with *many* rather than just 2 categories, using contingency tables larger than 2 x 2. The only difference is that calculation of observed and expected frequencies has to be done the 'long' way, as there is no summary formula corresponding to [13.12a]. The *df* for the resulting χ^2 is $(r-1)(c-1)$, where *r* and *c* are the numbers of rows and columns, respectively. This reduces to *df* = 1 for a 2x 2 table, as in Tables 13.5-6.

Table 13.6. Example calculations for a 2 x 2 contingency table

	In Cambrian limestone	Below Cambrian limestone	Totals
Pyroxene-bearing dykes	*a*: 39	*b*: 12	*a*+*b*: 51
Pyroxene-free dykes	*c*: 54½*	*d*: 35½*	*c*+*d*: 90
Totals	*a*+*c*: 93½	*b*+*d*: 47½	*a*+*b*+*c*+*d*=*N*: 141

* ½ relates to a borderline case

Characteristic analysis (Botbol et al. 1978) is an application of contingency table analysis, designed to analyze relationships among *many* binary variables, where a '1' indicates a favourable factor for mineralization and a '0' indicates one whose favourability is uncertain or indeterminate. A target exploration region is divided into cells, and selected model cells with definite or probable mineralization are studied to determine which geological variables associate most clearly with the mineralization. Other cells can then be compared with the model cells to suggest their prospectivity.

13d2. Fisher's Exact Probability Test

Worked geological examples: *Palaeontology* (Till 1974, p.119); *Petrography* (Cheeney 1983, p.54).
Worked non-geological examples: Lewis(1977,p.81).
Statistical tables: Lewis(1977, table 4); Powell(1982,p.52).
Stand-alone computer programs: Romesburg et al.(1985).
Mainframe software: covered in most statistics packages (e.g. SPSS).
Microcomputer software: covered in SYSTAT.

The χ^2 analysis of contingency tables in the previous section is, as indicated, a continuous approximation to discontinuous data. For 2 x 2 contingency tables, the *exact* probability *p* can be calculated via the following formula:

$$p = \frac{(a+b)!(c+d)!(a+c)(b+d)!}{N!a!b!c!d!} \qquad [13.13]$$

where ! indicates a factorial (4! = 4x3x2x1, $N! = N(N-1)(N-2)...1$). Manual calculation is only viable for very small N, although judicious cancellation of overlapping factorials helps. Pocket calculators and microcomputers will also quickly balk at the enormous factorials involved, as N rises. Nevertheless, this test usefully complements χ^2 because it is practicable in precisely those circumstances where χ^2 is unreliable. Mainframe computer programs are available to perform the test for larger N, but may involve approximations which annul the intrinsic advantages of the test over χ^2. SPSS^X (1986, p.346), for example, calculates the Fisher exact test only for $N < 20$, and reverts to χ^2 for larger N.

13d3. Correlation coefficients for dichotomous and nominal data: Contingency Coefficients

Further non-geological discussion: Coxon (1982), SPSS (1986).
Worked non-geological examples: Lewis(1977,p.93).
Statistical tables: nearly all contingency coefficients use χ^2 tables (see §8b3).
Stand-alone computer programs: Press et al.(1986,p.479).
Mainframe software: covered in most statistics packages (e.g. SPSS).
Microcomputer software: programmed in many statistics packages (e.g. Statview 512+™).

It is obvious from formula [13.12a] that χ^2 from a contingency table increases with *N, r and c.* Hence, although the more *ad* diverges from *bc,* the more significant the association, the *absolute* value of χ^2 itself does not actually give any clear indication of the *strength* of the association. Ideally, a coefficient independent of N is required, which yields a value of +1 or –1 for perfect positive and negative associations, and of 0 for no association. This is difficult to achieve in practice, and a plethora of measures based on χ^2 have been proposed, which are not only dependent to variable (though lesser) extents on N, but also take differing account of the situation where one cell-number vanishes (for example, does $a = 0$ mean perfect association, or not?). To compound the number of measures, several have received more than one name, or are differently described in different texts, sometimes being confused with **matching coefficients** (§21a1). For simplicity and convenience, therefore, full details and formulae are given in the Glossary (see initially under **contingency coefficients**), while Table 13.1 attempts a rationalised list of all the commonly encountered coefficients. Most of those listed will be encountered in later Topics.

Among those listed in Table 13.1 are two **probabilistic contingency coefficients**, which are not based on χ^2, but measure the extent to which one variable can be predicted, given the other. Since predicting X given Y may not be identical to predicting Y given X, these coefficients are asymmetric, in general having *two* values for any *one* contingency table. Because these measures have so far been little used in geology, they are not discussed further here, but as they are occasionally encountered in computer printout, sufficient details are given in the Glossary for readers to follow them up.

One final contingency coefficient, the **tetrachoric coefficient**, is more suitable whether *both* variables represent continuous data forced into a dichotomy, rather than true dichotomies. Its exact calculation involves solving a quadratic equation, but Downie & Heath(1974, appendix H) provide tables giving r_t from the ratio *ad/bc* or *bc/ad* of Table 13.5 (whichever ratio is ≥ 1). For example, the following results gives one group of students' performances on two geological tests marked between 0 and 100 (pass in both cases means mark ≥ 50; fail < 50, so these are forced dichotomies):

	Number passing Test 1	Number failing Test 1
Number passing Test 2	18	20
Number failing Test 2	11	32

The value $ad/bc = 2.62$ yields $r_{tet} \approx 0.36$ from the tables, which in fact work best if the original continuous data have been dichotomized by a 50–50 split like this.

The coefficient might have application in economic geology: for example the 'yes/no' of the Pb-Zn deposits in the previous section is really a dichotomy forced by ephemeral economic conditions. There is really a continuum of host areas from those with *no* sulphides to those with a prodigious resource; the current economic climate determines the cutoff where 'resource' becomes 'reserve' and 'no deposit' becomes 'deposit exists', and this cutoff will of course fluctuate with time. Where both variables are *true* dichotomies, however, r_{tet} has no advantages over the many contingency coefficients. It also has the disadvantage that tests of significance are complicated (Downie & Heath 1974, p.230).

13e. Comparing Pearson's Correlation Coefficient with itself: FISHER'S Z TRANSFORMATION

Worked geological examples: *Geomorphology* (Till 1974, p.87); *Palaeontology* (Miller & Kahn 1962,p.224).
Selected geological applications: *Igneous petrology* (Rock 1979); *Metamorphic stratigraphy* (Rock et al. 1984).
Statistical tables: Downie & Heath(1974); Powell(1982,p.79).
Computer software: none known.

Standard tables based on [13.2] are good only for testing $H_0: r = 0$. To test r against a *non-zero* value, we use the fact that the function:

$$Z = \frac{1}{2}\ln\frac{(1+r)}{(1-r)} = \tanh^{-1}(r) \qquad [13.14]$$

rather than r itself, is normally distributed (except for very small N), with mean Z_r and variance $1/\sqrt{(N-3)}$. This relationship can be used in several ways:

(1) *Confidence intervals for r:* intervals for Z become $\pm Z_{\alpha/2}/\sqrt{(N-3)}$, where $Z_{\alpha/2}$ is the chosen level of a standard Normal deviate (1.96 for 95% etc). These can then be converted back to corresponding intervals for r, via the inverse of [13.14], i.e. $r = [e^{2Z}-1]/[e^{2Z}+1]$. Unlike confidence intervals for a mean (§9e1), these will be *highly asymmetric*. For example, with the r value from Table 13.2 of 0.8906, $Z = 1.42$, so 95% limits on Z are $1.42 \pm 1.96/\sqrt{6} = 1.42 \pm 0.80$, which translates into 95% confidence *limits* for r of 0.89{+0.08,–0.34}, or confidence *intervals* of $0.55 \leq r \leq 0.977$.

(2) *Testing r against a non-zero value:* the test $H_0: r = r_0$, where $r_0 \neq 0$ is some fixed (e.g. population correlation) coefficient, can be carried out either by determining whether r_0 lies within the confidence interval for r, or equivalently by comparing the value $(r - r_0)/\sqrt{(N-3)}$ with a standard Normal deviate. In the above example, we would accept $H_0: r = r_0$ for any value of r_0 between 0.55 and 0.977, at the 95% level.

(3) *Comparing two sample correlation coefficients:* comparing r_1 and r_2, based respectively on n_1 and n_2 data-pairs, can be tested by comparing the following with a standard Normal deviate:

$$Z = \frac{Z_{r_1} - Z_{r_2}}{\sqrt{\frac{1}{(n_1 - 3)} + \frac{1}{(n_2 - 3)}}} \qquad [13.15]$$

The denominator is the pooled estimate of the standard deviation s_d for the difference between the Zs. Let us supposed that correlation of K versus Rb in one igneous rock-suite yields $r_1 = 0.85$ (based on $n_1 = 103$) and in another suite yields $r_2 = 0.75$ ($n_2 = 147$), then $Z_{r1} = 1.256$, $Z_{r2} = 0.973$, $s_d = 0.13$ and $Z = 2.18$, which is greater than the 5% two-tailed level for the standard Normal deviate (1.96), and thus indicates $r_1 > r_2$.

Equation [13.15] can be used if we are comparing the same *two* variables between *different* data-sets (suites of samples), as above, or *four* variables within the *same* data-set (e.g. to compare the K–Rb correlation with that of, say, Ca-Ba). However, if we are comparing coefficients based on *three* variables (e.g. K-Rb with K–Ba) in the *same* data-set, an inherent correlation between r_{ab} and r_{ac} is introduced by their sharing of one variable (in this case, a = K). The revised test then becomes:

$$t_{N-3} = \frac{(r_{ab} - r_{ac})\sqrt{(N-3)(1 + r_{bc})}}{\sqrt{2(1 - r_{ab}^2 - r_{ac}^2 - r_{bc}^2 + 2r_{ab}r_{ac}r_{bc})}} \qquad [13.16]$$

For example, if $r_{\text{K-Rb}} = 0.85$, $r_{\text{K-Ba}} = 0.75$ and $r_{\text{Rb-Ba}} = 0.60$ for one suite with $N = 50$, t calculates to 1.77. The critical 90% and 95% values of Student's t_{47} are ≈1.68 and 2.01, so once again $r_{\text{K-Rb}}$ is accepted as larger than $r_{\text{K-Ba}}$ at just above the 90% significance level.

(4) *Comparing three or more coefficients*: $r_1, r_2 ... r_k$, can be compared using the following formula:

$$\chi^2_{q-1} \approx \sum_{i=1}^{i=q}(n_i - 3)Z_i^2 - \frac{\left\{\sum_{i=1}^{i=q}(n_i - 3)Z_i\right\}^2}{\sum_{i=1}^{i=q}(n_i - 3)} \qquad [13.17]$$

where Z_i and n_i correspond to each r_i. This is sometimes known as **Rao's test.** For example, r coefficients of 0.3, 0.4, and 0.5, respectively based on 6, 15 and 20 data-values, translate into a χ^2 value of 4.85; this is significant at between the 90 and 95% levels, implying inhomogeneity.

(5) *Averaging coefficients:* values of r cannot be added, subtracted or averaged like raw data-values. They can however, be averaged after conversion to Z values, and the average Z converted back to an r coefficient. In example (4), the averages of the 3 rs is ≈ 0.404 (Z = 0.428), slightly larger than the conventional mean of 0.400. A better method is to weight each r, multiplying each Z_i by $(N_i - 3)$ and dividing the sum by $(N - 3q)$, where q is the number of r coefficients being compared; this would here yield a still higher r ≈ 0.447 (Z = 0.480).

13f. Measuring Agreement: Tests of Reliability and Concordance

Suppose that an exploration manager has received word of the imminent release of a number of exploration lease areas, and wants his senior geologists to judge which areas are most worth staking. Given that their judgments are unlikely to be identical, he can either base his decision on the consensus verdict (if the verdicts basically agree), or, if they do not agree, he may need to seek further opinions. But how does he determine whether or not they do agree? **Concordance** is a way of *quantifying* agreement in such cases, and hence of determining how reliable (reproducible) they are. It can be regarded as an extension of the correlation concept to 3 or more variables, as a different way of looking at contingency problems, or as a special case of the randomized complete block design in ANOVA (§12). It can be used to determine agreement between any b objects (blocks) as classified by c criteria (treatments). Objects and criteria might both be human (e.g. the verdicts of several examiners on students' abilities), or abstract (e.g. verdicts of viewers on TV programmes). As usual, there are different tests for nominal and ordinal variables.

13f1. Concordance between Several Dichotomous Variables: Cochran's Q test

Worked non-geological examples: Conover(1980,p.201).
Statistical tables: uses χ^2 tables (§8b3).
Mainframe software: programmed in some large statistics packages (e.g. SPSS[X]) .

This amounts to a two-way ANOVA (§12e) on dichotomous data, in which c treatments are applied to b blocks. It is also an extension of the McNemar test (§10c2) to more than two groups. Suppose our exploration manager has 5 leases (b = 5, notated A–E in Table 13.7) and 4 geologists' verdicts on them (c = 4; notated 1–4) ; the geologists supplied verdicts of 0 for 'no go' and 1 for 'area worth staking':

Table 13.7. Calculations for Cochran's Q test

Geologist I Lease area	A	B	C	D	E	Total (C_i)
1	0	1	1	0	1	3
2	1	0	0	0	1	2
3	1	0	1	1	0	3
4	1	1	1	0	1	4
Total (B_i)	3	2	3	1	3	12 = N

The statistic for testing H_o (that the judgements are consistent) is then defined as:

$$\chi^2_{c-1} \approx Q = c(c-1)\frac{\sum_{i=1}^{i=c}\left[C_i - \frac{N}{c}\right]^2}{\sum_{i=1}^{i=b} B_i(c - B_i)} \qquad [13.18]$$

Where B_i and C_i refer to the total scores for each object (lease area) and criterion (geologist), respectively. For Table 13.7, this yields Q = 4x3x[0^2+$(-1)^2$+0^2+1^2]/[3+4+3+3+3] = 1.5, which is not significant. We conclude that the judgements are consistent and thus, by implication, reliable.

13f2. Concordance between ordinal & dichotomous variables: Kendall's coefficient of concordance, W

Worked non-geological examples: Downie & Heath(1974, p.121); Sprent(1981,p.223).
Statistical tables: Downie & Heath(1974,p.322); Sprent (1981,p.262). For values of *b* and *c* not covered, the value $(c-1)W/(1-W)$ is distributed as Fisher's *F* (Powell 1982,p.75)
Mainframe software: in some large packages (e.g SPSS^X) and libraries (e.g. NAG routine G08DAF).
Microcomputer software: covered by a few statistics packages (e.g. Stat80 for Macintosh).

Suppose that the geologists are now asked to *rank* the lease areas in order of their mineral potential from 1–5 (1-*q*) as in Table 13.8, rather than just supplying 'yes/no' verdicts:

Table 13.8. Calculation of Kendall's coefficients of concordance, *W*

Geologist | Area	A	B	C	D	E	
1	1	2	3	4	5	
2	2	1	4	3	5	
3	3	1	5	2	4	
4	3	5	1	2	4	$c = 4$ (1–4); $b = 5$ (A–E)
Observed rank sum, S_i	9	9	13	11	18	ΣSum of ranks = $R = 60$ [= $bc(b+1)/2$ as check].
$D = S - E$	3	3	1	1	6	E = expected average rank sum = $R/b = 60/5 = 12$.
D^2	9	9	1	1	36	$\Sigma D^2 = 56$

Kendalls' coefficient of concordance, *W*, is then defined as follows:

$$W = \frac{12\sum_{i=1}^{i=b} D_i^2}{c^2 b(b^2-1)} = \frac{12\sum_{i=1}^{i=b} S_i^2 - 3c^2 b(b+1)^2}{c^2 b(b^2-1)} \qquad [13.19]$$

Unlike *Q*, whose value is unbounded, *W* is like a correlation coefficient, taking values between 0 (indicating total disagreement) and +1 (indicating complete agreement). It cannot of course take negative values, since one cannot have 'positive' and 'negative agreement' for >2 variables. Table 13.8 yields $W = 0.35$, which is below the 90% critical value (0.47), and thus this time indicates that the geologists' judgements are *in*consistent. More opinions are therefore needed!

Note the similarity of [13.19] to [13.5a] defining Spearman's ρ, which arises because *W* is similarly based on differences between expected and observed ranks. Also, calculating *W* is exactly equivalent to performing a two-way ANOVA (§12e) on related, *ordinal* data, and hence *W* is directly related to Friedmans' test statistic *T* (§12e2) by: $T = c(b-1)W$.

Note that *b* (the number of individuals or objects being judged) and *c* (the number of judges or criteria whose verdicts are being compared), in both this and the previous section, are *not* reversible: *Q*,*W* and their critical values will *both* be wrong if they are interchanged. To confuse matters further, available tables use different symbols for *b* and *c*: e.g. Conover (1980) respectively uses *r* and *c*, Downie & Heath(1974) *N* (or *n*) and *m*, Sprent(1981) ν and *n*, and so on.

Other coefficients of concordance are mentioned in the statistical literature (Table 13.1), but appear to offer little more than Kendall's *W*, and so are not considered here.

13g. Testing *X-Y* plots Graphically for Association, with or without Raw Data

Sometimes, we may wish to test for association between two variables of which we only have an *X-Y* scatter plot (e.g. in some previously published paper which does not give the raw data). Pearson's *r* cannot be calculated without the raw data at all, and although it is sometimes possible to work out the respective paired ranks for *X* and *Y*, and so calculate Spearman's ρ or Kendall's τ, this becomes increasingly difficult as *N* rises, or as the points become more closely spaced. Fortunately, alternative **medial tests** are available which count up numbers of points in various sectors of the plot; these are extremely quick and simple to carry out even on the smallest and scruffiest scatter plot. Two tests are covered here: the first uses mainly the outer values on the scatter plot, which makes it even easier to carry out, albeit sensitive to erroneous data. The second is especially useful , for it can detect the presence of *non-linear* (curvilinear) trends, as in Fig.13.1g. These tests can equally well be used where we *do* have the raw data, but then have few (if any) advantages over the coefficients covered above. Nevertheless, to make a 'once only' test on data it is not intended to enter into a computer, or to test where no computer program is available, they can still be quick and effective.

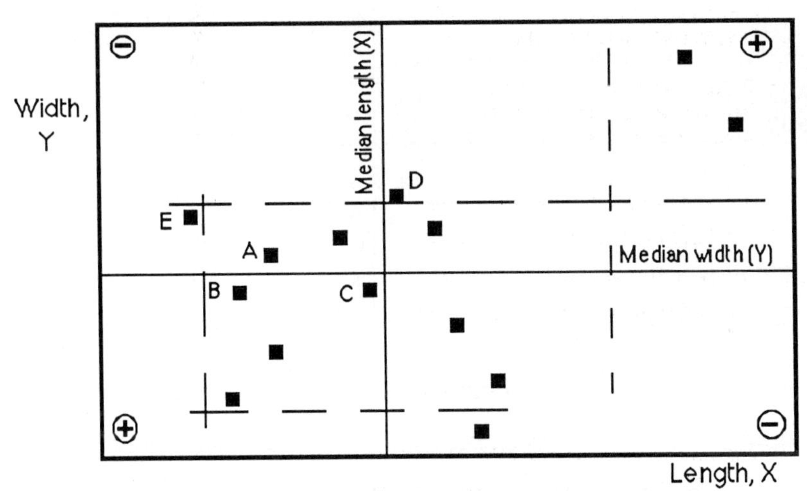

Fig.13.2. Plot of length versus width for a hypothetical set of fossils, illustrating method of calculating the Corner test for association (§13g1)

13g1. The Corner (Olmstead-Tukey Quadrant Sum) Test for Association

Further non-geological discussion and worked example: Dixon & Massey(1969,p.352); Steel & Torrie(1980, p.551).
Statistical tables: Dixon & Massey(1969, table A-30d); Steel & Torrie(1980, table A.20).
Computer programs: calculations too simple (and graphically based) for software be required.

We start by dividing a plot such as Fig.13.2 into 4 quadrants, at the *median* values for *X* (length) and *Y* (width), and label the quadrants + or – as shown. This can still be done by eye even if the raw data-values, or even the scales of the *X* and *Y* axes, are unknown: in Fig.13.2, for example, it is obvious, merely from their positions on the plot, that the median *Y* line comes between the 7th and 8th points moving from bottom to top (or vice-versa), namely A and B, and similarly the median X line lies between C and D. Now, moving from top to bottom, we sum the number of points we encounter in the direction of decreasing Y, before the values cross the median X line. In this case the crossover occurs after point D, when we have encountered 3 points higher up in the + quadrant, so we score S = +3. We repeat the procedure moving from west to east (S = –1); from bottom to top (S = –1 again); and finally from right to left (S = +2). The inner box shows the points at which all the crossovers occur. The sum of these 4 numbers, ΣS = +3–1–1+2 = +3 is the test statistic, which for N = 14 is not significant ($\Sigma S \geq 9$ to be significant at > 90% level). Thus we accept H_0: no correlation, in this case.

If N is odd, the median lines will actually pass through two points $\{X,Y_{med}\}$ and $\{X_{med}, Y\}$. In this case, replace these two points by the single point $\{X,Y\}$ to leave an even number of pairs, then proceed as above. If ties occur (for example if points D and E had identical, rather than slightly different, Y values) there are two alternative values of S for this quadrant, one (S_1) including the extra point, the other (S_2) excluding it (in this case, 4 or 5). The value of S to be incorporated into ΣS in this case should be $S_2/[1 + S_1]$.

It can be seen that ΣS is based primarily on the outlying points, giving them special weight.

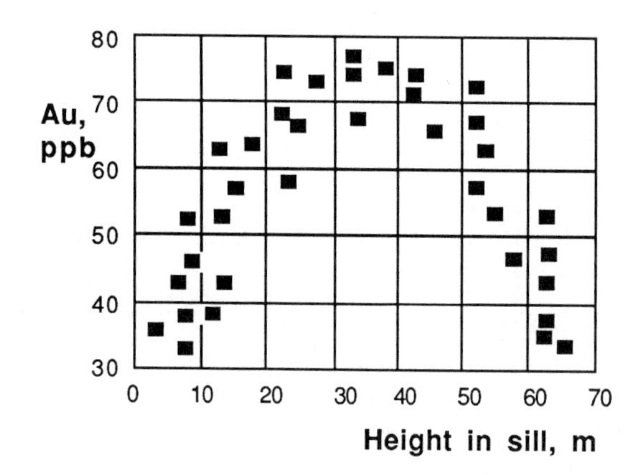

Fig.13.3. Division of an X-Y scatterplot into cells for calculation of the correlation ratio, eta

13g2.A test for curved trends: the Correlation Ratio, eta (η)

Worked non-geological examples: Downie & Heath(1974, p.111); Powell(1982,p.50).
Statistical tables: uses F tables (§8b2).
Proprietary and published software: none known.

If a scatter plot of X versus Y shows even the slightest indication of a non-linear trend, r and derived coefficients are inapplicable. Non-linear correlations are instead detected and tested in one of 3 ways:

(1) *Appropriate transformation:* e.g. if the relationship is believed to be $Y = kX^n$, perform a correlation calculation on the log-transformed data, for which the relationship will be linear and r can be used.

(2) *Use rank correlation coefficients:* as mentioned above, Spearman's ρ and Kendall's τ can detect all *monotonic* relationships (i.e. as in Fig.13.1d,e though *not* g),*without* prior transformation.

(3) *Use the correlation ratio,* η, which requires a plot of the type in Fig.13.3.

Eta has 2 different values for Y versus X and X versus Y; η_{xy} is defined analogously to η_{yx}:

$$\eta_{yx} = \sqrt{\frac{\text{Y "between" SS}}{\text{Y total SS}}} = \sqrt{\frac{\Sigma y_b^2}{\Sigma y_t^2}} = \sqrt{\frac{\Sigma\left[\frac{(\Sigma y)^2}{f_x}\right] - \frac{[\Sigma(\Sigma y)]^2}{N}}{\Sigma f(y)^2 - \frac{(\Sigma f y)^2}{N}}} \qquad [13.22]$$

where all summations are from 1 to N and 'SS' as usual denotes a sum of squares (§12). Fig.13.3 depicts Au contents in the same sill as discussed earlier. The plot is first divided into boxes, as shown, and then calculations proceed as in Table 13.9. This yields:

$$\eta_{\text{Au-height}} = \sqrt{\frac{118378.3 - \frac{(2020)^2}{36}}{120700 - \frac{(2020)^2}{36}}} = \sqrt{\frac{5033.9\ (\Sigma y_b^2)}{7355.6\ (\Sigma y_t^2)}} = 0.827$$

Since relationships such as Fig.13.1g or 13.3 vary from +ve to –ve across the plot, η *has no sign.*

From the above calculations, it should be apparent that η is closely modelled on ANOVA (§12), and it is not therefore surprising that its significance is tested by conversion to an F ratio:

$$F_{(c-1, N-c)} = \frac{\text{"between" mean square}}{\text{"within" mean-square}} = \frac{\Sigma y_b^2}{c-1} \times \frac{N-c}{[\Sigma y_t^2 - \Sigma y_b^2]} \qquad [13.23]$$

where c is the number of X columns into which the scatter plot is divided. For the above example, where $N = 36$, $c = 7$, this yields $F_{6,29} = (5033.9 \times 29)/(6 \times [7355.6 - 5033.9]) = 10.48$, which compares with a 99% critical $F_{6,29}$ value of 3.50. Hence η is significant at well above the 99% level, and a curvilinear correlation between Au content and height in the sill is confirmed for Fig.13.3.

Table 13.9. Calculation of the correlation ratio η from the data plotted in Fig.13.3

y (Au) \ x(Ht)	0–10(=5)	10–20(=15)	20–30(=25)	30–40(=35)	40–50(=45)	50–60(=55)	60–70(=65)	f_y	fy	$f(y)^2$
70–80 (=75)			2	3	2	1		8	600	45000
60–70 (=65)		2	2	1	1	2		8	520	33800
50–60 (=55)	1	2	1			2	1	7	385	21175
40–50 (=45)	2	1				1	2	6	270	12150
30–40 (=35)	3	1					3	7	245	8575
f_x	6	6	5	4	3	6	6	N =	36	Σfy=2020
	Σ120700									
Σy	250	320	335	290	215	360	250	$\Sigma(\Sigma y)$ =	2020	
$(\Sigma y)^2/f_x$	10416.7	17066.7	22445	21025	15408.3	21600	10416.7	$\Sigma\{(\Sigma y)^2/f_x\}$=118378.3		

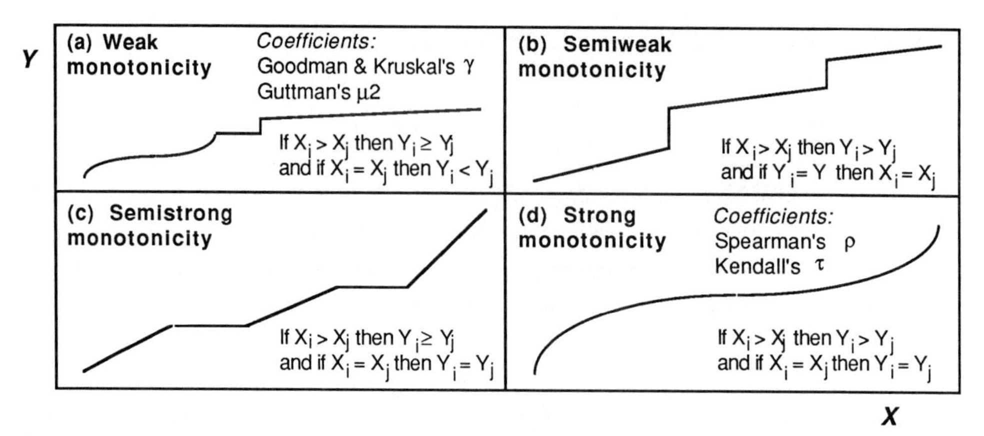

Fig.13.4. Definitions of the various strengths of monotonicity and the coefficients sensitive to each
Note that the various inequalities are defined for *positive* relationships; they are reversed for negative relationships.

13h. Measures of weak monotonicity

Further non-geological discussion: Kotz & Johnson(1982-8, vol.8, p.80).
Proprietary software: programmed in SPSS and Systat.

We contrasted Pearson's r against Spearman's ρ and Kendall's τ above according to their applicability to interval/ratio versus ordinal data, and also indicated that r only measured strictly *linear* trends, whereas ρ and τ detect *monotonic* trends as well. In fact, these coefficients are just the best-known examples of an infinity of possible coefficients which could be employed to detect different strengths of *monotonicity* (Fig.13.4): specifically, ρ and τ detect **strong monotonicity.** There are other coefficients capable of detecting weaker, but still monotonic relationships, which differ in the way they treat tied values. Of these, Guttman's μ_2 and Goodman & Kruskal's γ are mentioned here, solely because they are already implemented in some proprietary packages. Table 21.4 contrasts matrices of r, ρ, τ, μ_2 and γ for one particular set of data. Whereas all these coefficients have a value of 1 for a perfect linear relationship, the rate at which they fall towards zero as the relationship departs from strict linearity is in the order $r \gg \rho, \tau > \mu, \gamma$.

Guttman's μ_2 is defined as follows (note the close relationship to r in [13.1]):

$$\mu 2 = \frac{\sum_{i=1}^{i=N}\sum_{j=1}^{j=N}(X_i - X_j)(Y_i - Y_j)}{\sum_{i=1}^{i=N}\sum_{j=1}^{j=N}|X_i - X_j|\,|Y_i - Y_j|} \qquad [13.20]$$

Goodman and Kruskal's γ is defined as follows (note the close relationship to τ in [13.9]):

$$\gamma = \frac{\sum_{i=1}^{i=N}\sum_{i=1}^{i=N}\alpha_{ij}\beta_{ij}}{\sum_{i=1}^{i=N}\sum_{j=1}^{j=N}|\alpha_{ij}\beta_{ij}|} \qquad [13.21]$$

13i. Spurious and illusory correlations

Further geological discussion: Mann(1987); Mogarovskii (1962).

After obtaining a *statistically* significant correlation coefficient, the user must always ask whether the correlation is *geologically* meaningful. Two types of correlation represent pitfalls:

(1) **Spurious correlations** are those where a false (non-reproducible) correlation has been induced in data by some error: the correlation is statistically, but maybe not geologically, significant. Fig.13.1i is a typical example (all too frequently found on isochron plots a few years ago!) The one outlying point has produced a highly significant value of r, but if it is deleted, all the other points scatter randomly and give a non-significant r. The geologist must then ask himself whether there are *geological* grounds for excluding the outlier (in which case the correlation becomes spurious), or whether its inclusion can be supported (in which case the correlation is at best unconfirmed until further data corroborate the outlier). Because many sophisticated statistical methods rely on arrays of r values (e.g. factor analysis: §23), Fig.13.1i underlines a fundamental pitfall with 'black box' use of computer packages. A computer printout of r values listed to 16 decimal places may look very impressive, but if all the corresponding X-Y plots are like this plot, subsequent analysis may be garbage. Correlation coefficients complement, but in no way substitute for careful examination of *plots*.

(2) **Illusory correlations** are those which are both statistically and geologically significant, but do not tell the whole story: two variables are not primarily correlated with each other, but with a third, underlying, variable. For example, geologists' waist measurements may well correlate with their salary, but the correlation arises not because the two variables are related directly, but because both are more fundamentally related to, say, beer consumption!

TOPIC 14. QUANTIFYING RELATIONSHIPS BETWEEN TWO VARIABLES: Regression

Dedicated monographs and reviews: Acton(1966); Belsley et al.(1980); Daniel & Wood(1980); Draper & Smith(1966); Furnival (1971); Graybill(1961); Mann(1981); Montgomery(1982); Mosteller & Tukey(1977); Open University (1974a); Troutman & Williams(1987); Vinod & Ullah(1981); Weisberg(1985); E.T.Williams(1959); G.P.Williams(1983); Williams & Troutman(1987).

Regression proceeds a major step beyond correlation (§13). Having determined that a significant relationship between two variables (X and Y) exists, it attempts to fit a *line* (linear regression: of the form $Y = \beta_0 + \beta_1 X$) or a *curve* (non-linear regression: of such form as $Y = \beta_0 + \beta_1 X^2$) which best describes that relationship, and so can be used to predict the value of Y given X. Geologists are familiar with 'eyeballing' trends on X-Y plots, but regression does the fitting objectively, and tests the significance of the results. Correlation and regression are intimately intertwined (if there is no correlation, regression is pointless), but they are separated into two Topics here because: (a) whereas correlation methods cover the whole range of data-types, from dichotomous to ratio, full regression can only be carried out on ratio data; (b) the two methods can stand on their own. For example, correlation alone is enough in some geological applications: the *existence* of relationships between incompatible trace elements like Nb and Zr in igneous rock suites is more genetically important (in indicating a fractionation relationship) than the actual form of a predictive regression line, say Nb = 4.63 + 0.225(Zr). Conversely, in geochronology, regression is usually used without reference to correlation, for isochrons, pseudo-isochrons, scatterchrons and errorchrons are *all* characterized by extremely high correlation coefficients, but are distinguishable as regression lines.

Regression is perhaps one of the most widespread but mis-used techniques in geology; only a few pointers to available methods and their correct usage can be given here. Table 14.1 and Fig.14.1 compare the results of various methods (used for illustrative purposes only) on one small data-set.

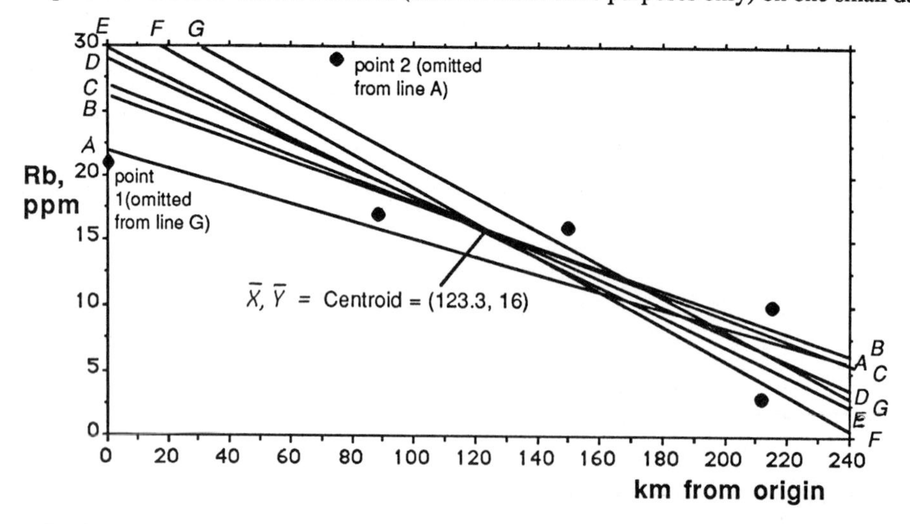

Fig.14.1. Comparative results of various regression methods on the raw data in Table 14.1 (see this table for key)

14a. Estimating Lines to Predict one Dependent (Response) Variable from another Independent (Explanatory) Variable: CLASSICAL PARAMETRIC REGRESSION

Dedicated monographs and reviews: Acton(1966); Mullet & Murray(1971).
Further geological discussion: Hey(1977); Miller & Kahn (1962,ch.8); Davis (1986, p.176ff).
Worked geological examples: *Mining* (Koch & Link 1971, p.8); *Sedimentology* (Till 1974,p. 91); *Tectonics* (Le Maitre 1982, p.49ff).
Selected geological applications: *Geomorphology* (Doornkamp & King 1971; Till 1973); *Igneous petrology* (Bailey & Macdonald 1975; Flinter 1974; Weaver et al. 1972; Weaver & Gibson 1975); *Metamorphic petrology* (Lang & Rice 1985); *Mineralogy* (Taylor et al.1985); *Mineral exploration* (Austria & Chork 1976; Chung & Agterberg 1980; Taylor et al. 1984); *Sedimentology* (Krumbein & Graybill 1965, p.238); *Structural geology* (James 1970); *Tectonics* (Hutchison 1975; Whitford & Nicholls 1976).
Statistical tables: uses *F* and *t* tables (§8b2–3).
Stand-alone computer programs: Davis(1973,p.199);Olhoeft(1978);Press et al.(1986,p.508ff); Till(1977).
Mainframe software: Well covered in packages (BMDP, GENSTAT, MINITAB, SAS, SPSS etc) and libraries (e.g. NAG provides several routines G02C*n*F where n is A,B,C or D, to cover various options). Also implemented in *GLIM* (Generalized Linear Interactive Modelling), a specialized and highly sophisticated regression package.
Microcomputer software: programmed in many statistics packages (e.g. Statview 512+™).

14a1. Introduction: important concepts

Classical regression studies the variation of one **dependent (response) variable,** X (subject to error) with another **independent (explanatory) variable,** Y (theoretically not subject to error). Examples would include the dependence of (1) specific gravity (Y) on whole-rock sulphide content (X); or (2) extinction angle of plagioclase twin lamellae (Y) on anorthite content (X). It is intuitively obvious that extinction angle depends on anorthite content, and not the reverse. The model for a classical Y on X regression takes the form:

$$\textbf{Population: } Y = \delta + \gamma X + \varepsilon \qquad \textbf{Sample: } Y = \beta_0 + \beta_1 X \qquad [14.1]$$

where β_1 and β_0 are *sample* estimates of the the *population* slope γ and Y intercept δ, and ε is a random error term generating the spread of Y values about the line. The '=' sign is somewhat misleading in this conventional equation, and it does *not* follow that regression of [14.1] will always produce equivalent results to some conventional algebraic manipulation of this equation, such as:

$$\{Y - \beta_0\}/\beta_1 = X \qquad [14.1a]$$

The discussion below focuses on explaining the various parameters output by a typical linear regression program (Table 14.2).

14a2. Calculating the regression line: Least-squares

The traditional method for estimating a regression line involves minimising the squares of distances of the points from the line, called **least-squares.** In classical Y on X regression, since X is assumed to be without error, the error must be taken up parallel to the Y axis, that is, the sum of squared *vertical* distances from the points to the regression line is minimised (Fig.14.2). A corollary of this is that the calculated line must pass through the **centroid** or *centre of gravity* $(\overline{X},\overline{Y})$ (Fig.14.1).

Least-squares is a parametric method which makes the following assumptions:

(1) Errors (ε) about the regression line are Normally distributed with zero mean and variance σ^2.

(2) The **residuals** are uncorrelated and **homoscedastic**: that is, the differences between the *observed* and *predicted* Y values (the residuals), or equivalently the spread of the Y values about the regression line, do not increase, decrease or vary cyclically with the value of X. Fig.14.3 contrasts residuals which obey this assumption (Fig.14.3a) with those which do not (Fig.14.3b).

The quantity to be minimized in least-squares regression of Y (the dependent variable) on X is:

$$\sum\varepsilon^2 = \sum(Y - \delta - \gamma X)^2 \qquad [14.2]$$

The slope $ß_1$, intercept $ß_o$, and variance s^2 of ε are then estimated by the following formulae:

$$\beta_1 = \frac{\sum XY - \frac{\sum X \sum Y}{N}}{\sum X^2 - \frac{(\sum X)^2}{N}} = \frac{SP_{XY}}{SS_X}; \quad \beta_o = \bar{Y} - \beta_1 \bar{X}; \quad s^2 = \frac{1}{N-2}\left[SS_Y - \frac{(SP_{XY})^2}{SS_X}\right] \qquad [14.3]$$

where, as usual, $SS_y = \sum(Y - \bar{Y})^2$, $SS_x = \sum(X - \bar{X})^2$, $SP_{xy} = \sum(Y - \bar{Y})(X - \bar{X})$, and all summations are from 1 to N. The corresponding values $ß_1'$ and $ß_o'$ for the regression of X on Y are obtained by appropriate substitution; it can readily be shown than $ß_1ß_1' = r^2$, so that the two slopes are only equal when $r = 1$. Correspondingly, when $r = 0$ the two lines are orthogonal.

Table 14.1. Data and comparative results of regression worked examples in Topic 14

Raw data		Comparative regressions on Fig.14.1						Regression *B* on Fig.14.1		
Distance from origin,km(X)	Rb content, ppm (Y)	Line label	Intercept, $ß_o$	Slope, $ß_1$	*N*	Regression method	*r*	Residual residual	Standard	Y-calc
0	21	*A*	22.39	-.068	5	Y on X,outlier 1 rejected	.764	-5.285	-.859	26.28
75	29	*B*	26.29	-.083	6	Y on X, all data-pairs	.662	8.969	1.459	20.031
88	17	*C*	26.32	-.084	6	Major axis line	.662	-1.946	-.317	18.946
150	16	*D*	29.04	-.105	6	Reduced major axis line	.662	2.224	.362	13.776
212	3	*E*	29.94	-.113	6	Theil's complete method	.784*	-5.606	-.912	8.606
215	10	*F*	32.54	-.134	6	X on Y, all data-pairs	.662	1.644	.267	8.356
		G	33.95	-.128	5	Y on X, outlier 2 rejected	.776			

* Spearman's ρ **Note:** not all of the above regressions are *statistically*, let alone geologically, meaningful.

14a3. Assessing the significance of the regression: Coefficient of determination, ANOVA

The following parametric tests on a regression are conventionally made:

(a) Correlation coefficient, r (Table 14.2a). As fully explained in (§13), r indicates whether a significant relationship exists between X and Y. Additionally in regression, r^2, termed the **coefficient of determination**, indicates the proportion of total variance in X and Y which is explained by the regression line. In Table 14.2a, 62.2% of the variance is explained in this way.

(b) ANOVA (Table 14.2b). This is an overall test that the regression is meaningful: that X accounts for a significant amount of the variation in Y. Formally, it is a test of H_o: $ß_1 = 0$ (i.e. that the regression line parallels the X axis); if this is accepted, the regression is meaningless. Note that this is *not* the same as a test of $r = 0$, for, if $ß_1$ is indeed 0, r is undefined (Fig.13.1c). Davis (1973, fig.5.9) gives a useful comparison of cases where these two tests will give significant or insignificant results, which can together help to tie down the type of relationship present. In Table 14.2, both r *and* F are significant, so we have a real ability to predict 'Rb' from 'km'.The ANOVA operates as usual (§11–12) by separating the total sum of squares about the mean into a component explained by the regression and an error component; the larger their ratio, estimated by F, the more significant the regression.

(c) Beta coefficient table t tests (Table 14.3d). The t statistics here test the *individual* ß coefficients (other than $ß_0$) in the same way that the ANOVA tests the *overall* regression; in this case of a simple linear regression, therefore, the t statistic of 2.565 is testing *exactly the same hypothesis* (H_0: $ß_1 = 0$) as the F statistic of 6.579 in the ANOVA table, and we find that $t = \sqrt{F}$ (cf. §8b4). Although computer packages almost invariably provide an overall ANOVA table, some use the t test for the individual ßs, others a series of further F tests (cf. Table 14.2 with Le Maitre 1982, p.86). The user must thus be aware of their equivalence for, when we proceed to more complicated forms of regression (§14e & §19), where there are *several* ß coefficients ($ß_0$, $ß_1$, $ß_2$... etc.), the purposes of the ANOVA and Beta coefficient tables diverge: the ANOVA still tests the *overall* regression (where X can explain Y), but the beta coefficient tables allow the *individual* ßs to be assessed one by one.

Table 14.2. Statistics for classical Y on X regression (line B of Fig.14.1, data of Table 14.1)

(a) Correlation coefficients

DF:	R:	R-squared:	Adj. R-squared:	Std. Error:
5	.789	.622	.527	6.149

(b) ANOVA table

Source	DF:	Sum Squares:	Mean Square:	F-test:
REGRESSION	1	248.755	248.755	6.579
RESIDUAL	4	151.245	37.811	p = .0623
TOTAL	5	400		

(c) Durbin-Watson test for autocorrelation of residuals

SS[e(i)-e(i-1)]:	e ≥ 0:	e < 0:	DW test:
453.604	3	3	2.999

(d) t tests of individual ß coefficients

Parameter:	Value:	Std. Err.:	Std. Value:	t-Value:	Probability:
INTERCEPT	26.285				
SLOPE	-.083	.033	-.789	2.565	.0623

(e) Confidence intervals on mean and slope $ß_1$

Parameter:	95% Lower:	95% Upper:	90% Lower:	90% Upper:
MEAN (X,Y)	9.029	22.971	10.648	21.352
SLOPE	-.174	.007	-.153	-.014

14a4. Testing the regression model for defects: Autocorrelation and Heteroscedasticity

Further non-geological discussion: Mather(1976,p.77ff).
Worked geological examples: *Sedimentology*: Mather (1976, p.91).
Statistical tables: Powell(1982,p.86).
Mainframe software: covered in large statistics packages (e.g. $SPSS^X$).
Microcomputer software: programmed in some statistics packages (e.g. Statview 512+™).

The values of r and F in Table 14.2 are not enough to confirm a regression model as fully adequate. It is important to examine the *residuals* (differences between observed Y values and those predicted from a regression equation) for two possible defects in the model: **autocorrelation** (the residuals themselves or, in effect, their *average*, shows either a trend or a periodicity) and **heteroscedasticity** (the *variance* of the residuals shows a trend). The presence of either invalidates the model (Fig.14.3).

A well-established test is available for autocorrelated residuals: the **Durbin-Watson statistic.** This compares residuals with the preceding ones in the manner of a χ^2 statistic (cf. [8.27]):

$$d = \frac{\sum_{i=2}^{i=N}(e_i - e_{i-1})^2}{\sum_{i=1}^{i=N} e_i^2} \qquad [14.4]$$

If the residuals show strong autocorrelation, d is near zero; if uncorrelated, d is near 2. The null hypothesis H_0 that the residuals are uncorrelated is tested by comparison against lower *and* upper critical values of d at the chosen significance level, termed d_L and d_U. If $d_{calc} > d_U$, H_0 is accepted; if $d_{calc} < d_L$, H_0 is rejected, but if $d_L < d_{calc} < d_U$, no decision can be made (a "case of indecision"). Critical values of d are unfortunately unavailable for $N = 6$ as in Table 14.1, but let us assume $N = 15$; the calculated value of 3.678 in this case would then lie *outside* the 99% critical limits 0.81–1.07, indicating autocorrelation and invalidating the model.

Heteroscedasticity is probably most easily tested by visual checks of scatter plots (Fig.14.3c,d). In ambiguous cases, however, Spearman's ρ (§13b1) can be calculated between the *absolute* values of the residuals and X. If ρ is significant, the residuals are heteroscedastic and the regression invalid.

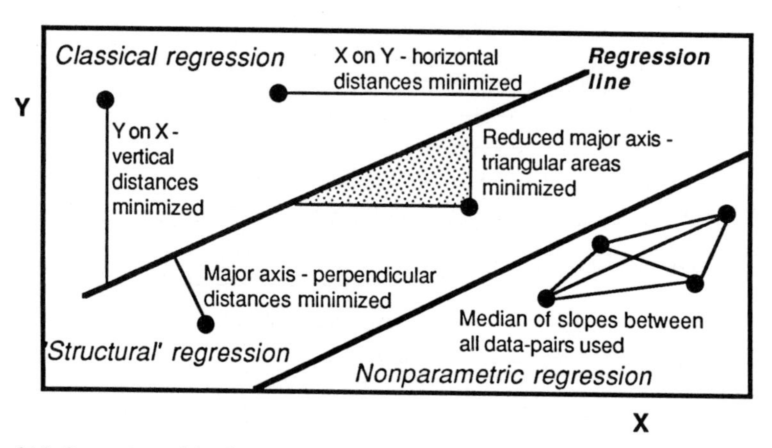

Fig.14.2. Comparison of the distances between data-points and regression line which are minimized in the different types of regression summarised in Fig.14.1 and Table 14.1

14a5. Assessing the influence of outliers on a regression

Classical regression is sensitive to outliers (§9f). Table 14.1 and Fig.14.1 show two lines (*A* and *G*) with just 1 or 2 outliers eliminated: these move outside the range of the other 5 lines depicted, while the r and F statistics increase substantially (F rises from 6.6 to 10.4 when outlier 1 is eliminated). Many computer packages provide detailed statistics for those points most influencing the exact position of a regression line. The user can then decide whether these points in fact represent true, false or bizarre outliers, and accordingly retain or reject them in his regression model.

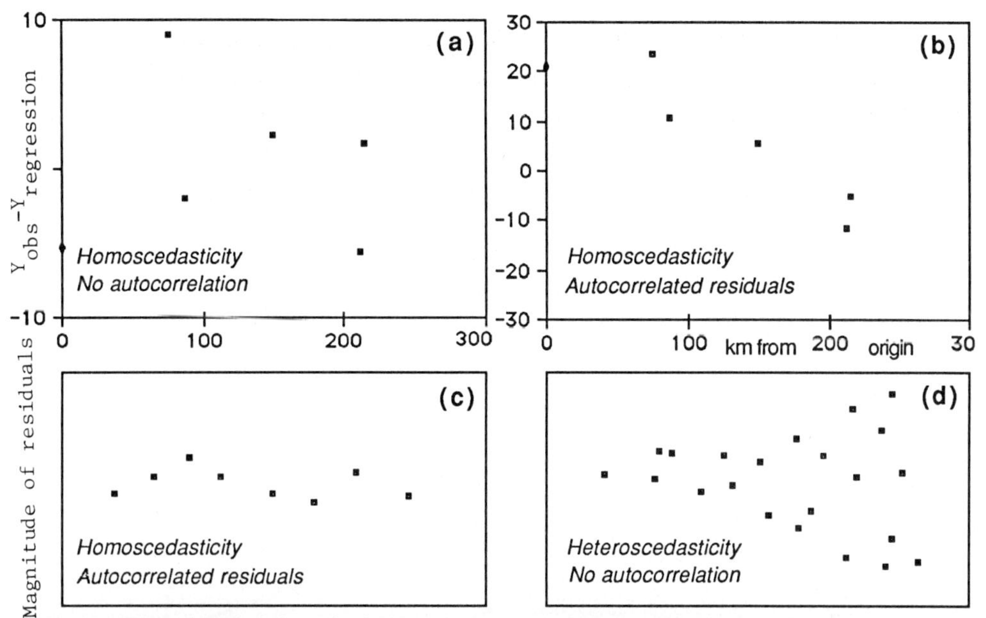

Fig.14.3. Plots of residuals from classical Y on X regression of Fig.14.1 (line B), to test for defects in the model.
Plots (**a**) and (**b**) test for trend (autocorrelation) in the residuals, and illustrate what happens when a 'silly' model is attempted. In (**a**), the regression was calculated from the data in Table 14.1 in the usual form $Y = \beta_1 X + \beta_0$; in (**b**), however, β_0 was constrained to be zero, i.e. the equation was made to be $Y = \beta_1 X$, even though we know from Table 14.1 that β_0 should be far from zero. Plot (**b**) reveals this model as statistically invalid because the residuals decrease with X (i.e. show a negative trend or autocorrelation). In less obvious cases, a plot like this can be vital.
Plots (**c**) and (**d**) test for heteroscedasticity in the residuals using imaginary data. In (**c**), the residuals show a constant variance with X (homoscedasticity), but they now show a cyclicity so the corresponding regression model is still invalid. In (**d**), the residuals show increasing variance with X; the corresponding model is again invalid. Overall, only plot (**a**) satisfies the assumptions of homoscedasticity and no autocorrelation in classical regression.

14a6. Confidence bands on regression lines

The confidence interval for a Y value predicted from a given X value (i.e. the interval on the mean value of Y that would be obtained by repeated random sampling at the particular X value) is given by:

$$Y \pm t_{(\alpha/2;N-2)} s \sqrt{\frac{1}{N} + \frac{(X-\overline{X})^2}{SS_X}} \qquad [14.5]$$

If this interval is calculated for several X values a **confidence band** for the regression line begins to emerge (Fig.14.4). This band is narrower near the centroid because there are (by definition) more data-pairs confining it here than at the extremes. The band is obviously wider as a whole as r decreases, since there is then less overall confidence in the position of the line.

14a7. Comparing regressions between data-sets or data-sets and populations: Confidence Intervals

Further non-geological discussion: Potthoff(1974).
Selected geological applications: *Palaeontology* (Miller & Kahn 1962, p.206).

Confidence limits at the α% level (i.e. $\alpha = 0.05$ for 95%) on β_1 and β_0 are estimated by:

$$\beta_1 \pm t_{(\alpha/2;\,N-2)}\sqrt{\frac{s^2}{SS_X}}; \qquad \beta_0 \pm t_{(\alpha/2;\,N-2)}\sqrt{\frac{s^2\Sigma X^2}{N\mathrm{x}SS_X}} \qquad [14.6]$$

To decide whether a lines has an assumed (population) slope or *Y*-intercept, therefore, we merely determine whether the population value lies within the confidence interval of the sample value. In Table 14.2, the confidence interval (often called the 'standard error of estimate' on computer output) on the slope is 26.285 ± 4.731 and on the *Y*-intercept it is –0.083 ± 0.033.

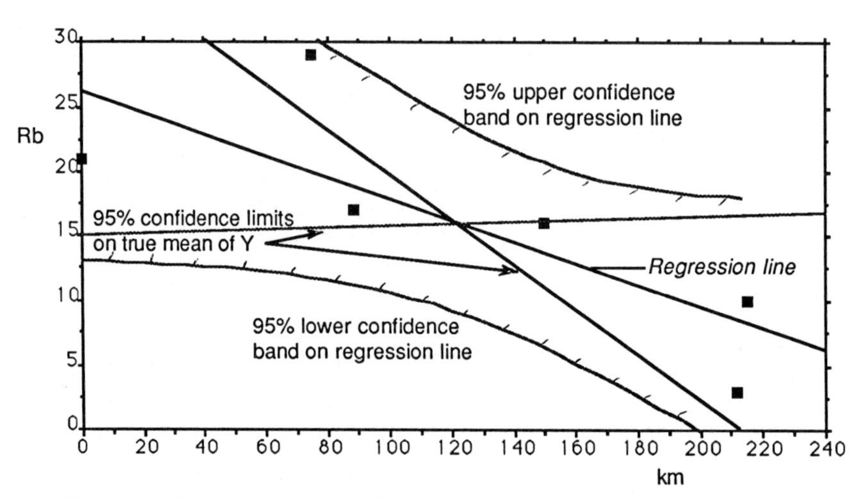

Fig.14.4. Confidence bands and *Y* confidence limits on the *Rb* on *km* regression line from Table 14.1
Note how the confidence bands are not continued beyond the actual data-points (right of plot)

14b. Calculating Linear Relationships where Both Variables are Subject to Error: 'STRUCTURAL REGRESSION'

Further discussion: Mann(1987); Moran(1971); Troutman & Williams (1987); York(1966).
Worked geological examples: *Igneous petrology* (Le Maitre 1982, p.54); *Palaeontology* (Till 1974, p.99; Miller & Kahn 1962,p.204; Davis 1986, p.200).
Geochronological discussion: Brooks & Hart(1969), Brooks et al.(1968,1972,1975), T.A.Jones(1979), Mark & Church(1977), McIntyre et al. (1966); Nicolaysen(1971); Titterington & Halliday(1979); York(1967,1969).
Selected geological applications: *Geomorphology* (Till 1973); *Igneous petrology* (Yoder & Chayes 1982); *Mining* (Taylor 1973); *Palaeontology* (Parkinson 1954).
Stand-alone computer programs: Faure(1977), Rock & Duffy (1986), York(1969).
Proprietary software: could be programmed in GENSTAT or MINITAB.

In many geological situations, there is no obvious "dependent" or "independent" variable, and classical regression does not strictly apply. The statistical assessment of Rb-Sr and Nd-Sm isochrons, for example, is a form of 'structural' regression in which *both* variables (*X* and *Y*) are subject to error. Regression between length and breadth of fossil tests, or between two elements (say K and Rb) in a rock-suite, would also be structural, since neither obviously depends directly on the other, but on some more fundamental, organic or petrogenetic control. Some statisticians maintain that the term "regression" itself is inappropriate, in the absence of an independent (error-free) variable but, as opinions differ, it is retained here.

'Structural' regression solutions again use least-squares methods, and many of the statistical results (r, F ratio from the ANOVA of Table 14.1) do not change, but different distances between the points and the regression line are minimised (Fig.14.2). The revised slope β_1 is given by:

$$\beta_1 = \frac{SS_y - \lambda SS_x + \sqrt{\{(SS_y - \lambda SS_x)^2 + 4\lambda SP_{xy}^2\}}}{2SP_{xy}} \qquad [14.7]$$

where λ is the ratio of the *random error* variances in Y and X, and SS, SP are defined as usual. Four special cases can be defined (Fig.14.2): two for classical and two for 'structural' regression:

(1) $\lambda = \infty$ (X is without error): classical Y on X regression already dealt with, that is $Y = \beta_0' + \beta_1' X$.

(2) $\lambda = 0$ (Y is known without error), classical X on Y regression, that is $X = \beta_0' + \beta_1 Y$; the sum of squared *horizontal* distances to the regression line is minimized.

(3) $\lambda = 1$: **Pearson's Major Axis solution** — also known as **Least Normal Squares (LNS)** — the same as the first eigenvector of Principal Components Analysis (§23a); here, *perpendicular* distances to the regression line are minimized.

(4) $\lambda = [s_y / s_x]$, i.e. the ratio of *total* variances in Y and X: the **Reduced Major Axis solution (RMA)**; in this case, [14.7] reduces to the simplest form $\beta_1 = \lambda = [s_y / s_x]$, the ratio of standard deviations; here, the *triangular areas* between the points and regression line are minimized.

All 4 of these solutions pass through the point $(\bar{X}, \bar{Y})$, the 'structural' solutions lying *between* the classical Y on X and X on Y lines (Fig.14.1). Of course, as $r_{X,Y}$ approaches 1, the separation between all four lines diminishes, and the difference between 'classical' and 'structural' regression becomes less and less important.

'Structural' regression should always be used where *both* X and Y have associated errors, and where neither is intuitively dependent on the other. But what about those geological situations where there *is* (in theory) a dependent variable, yet both X and Y still have associated errors? For example, in regressing specific gravity of plagioclase against anorthite content, specific gravity is obviously the dependent variable, yet neither can be measured without error. Statisticians may well give divergent answers here, but one plausible approach is to use classical Y on X regression *whenever we are trying to predict Y from X*, even if X has associated errors (and likewise vice-versa). On the other hand, if we are merely interested in the slope of the regression line, use 'structural' solutions based on some reasonable estimate of λ. If it is intuitively obvious (even without quantitative justification) that X can be measured far more accurately than Y, a high value of λ will be appropriate, with a line near to the Y on X classical regression. This would apply, for example, in measuring chemistry or other parameters down a borehole, where distance along the core material can be measured far more precisely than the chemistry. Even taking a line midway between the Y on X and X on Y regressions may be better than nothing, provided r is reasonably high!

Two 'structural' slopes β_1 and β_2 can be compared by conversion to a standard Normal deviate:

$$Z = \frac{\beta_1 - \beta_2}{\sqrt{s_{\beta_1}^2 + s_{\beta_2}^2}} \text{ where } s_m = \beta\sqrt{\frac{1 - r^2}{N}} \qquad [14.8]$$

14c. Avoiding sensitivity to outliers: ROBUST REGRESSION

Further geological discussion: Garrett(1983); Hawkins(1981b); Troutman & Williams(1987).
Further non-geological discussion: Bornstedt & Carter(1971); Chen & Dixon(1972); Hogg(1979); Huber(1973,1981); Huynh(1982); Press et al. (1986,p.539); Yale & Forsythe(1976).
Selected geological applications: *Metamorphic petrology* (Powell 1985).
Stand-alone computer programs: Press et al.(1986,p.544).
Mainframe software: can be implemented in GENSTAT and MINITAB.

Robust regression imitates what robust estimates (§8c1) did in estimating location and scale: namely, it reduces the deleterious effect of outliers and of departures from assumptions in classical regression. The method is attracting a growing literature; only an outline of available approaches is therefore attempted here. Readers are referred to the appropriate sections in §8 and the Glossary for more details:

(1) **Winsorizing**: as with Winsorized location estimates, Winsorized regression replaces extreme values with the next values 'in' towards the bulk of the data.

(2) **Jackknife:** here, many possible subsets of regressions are calculated with various points omitted, and the final result is a type of compromise.

(3) ***M,L,R*- estimates.** Regressions based on these location estimates have been widely advocated. One popular technique is **Least Absolute Deviations**, based on absolute rather than squared distances from the regression line, which gives extremely deviant points less weight.

(4) **Weighted least-squares (WLS).** An alternative (often used in geochronology) is to weight each value such that least precise determinations receive least weight. This, however, requires that replicate determinations are available for each and every data-point (or no estimates of precision are possible), and that some theoretical model for assigning the weights can be invoked. In many senses, WLS does manually what the other robust regression methods do automatically, and thus allows the user more control over his technique in circumstances where his data really do allow him to exercise such control knowledgeably.

14d. Regression with few assumptions: NONPARAMETRIC REGRESSION

14d1. A method based on median slopes: Theil's Complete Method

Further discussion: Conover(1980,p.265); Hogg & Randles(1975); Sen(1968); Sprent(1981).
Worked geological examples: *Geochronology* (Vugrinovich 1981).
Stand-alone computer programs: Rock & Duffy(1986).
Proprietary software: none known.

Nonparametric regression can model geological data where the assumptions behind none of the parametric methods above are met. It is robust, "resistant", that is insensitive to localized misbehaviour in data, and also structural, that is applicable where both X and Y are subject to error.

Any two points, i and j, on a scatter point will define an estimate of the slope $S_{ij} = (Y_i - Y_j)/(X_i - X_j)$, where $1 \leq i < j \leq N$, and there will be $N(N-1)/2$ such estimates for N data-pairs (Fig.14.2). For example, in the data of Table 14.1, the first two listed points yield a slope of $(21 - 29)/(75 - 0) = -0.11$, and there are 14 other such estimates. In Theil's complete method, the overall slope, β_1, is taken as the median of all S_{ij}. The Y-intercept $ß_0$ is then estimated as the median of N intercepts calculated using $ß_1$ with all pairs (i.e. $ß_0$ = median of $[Y_i - ß_1X_i]$ for i = 1 to N). Confidence intervals for $ß_1$ are based on Kendall's τ (§13b2), or, more specifically, Kendall's S statistic which is related simply to τ by [13.7a] — a completely different concept from confidence interval estimation in least-squares regression. Confidence limits for $ß_0$ are based on the same principle as for the median (§9e4). Unlike parametric confidence limits, neither set is usually symmetric about $ß_1$ or $ß_0$, being calculated from the actual set of estimators generated from the data themselves.

Although Teil's complete method (based on medians) is resistant to outliers which actually form part of the data set in question, it will still be affected by "bizarre outliers" (§9f1). A nonparametric procedure for detecting these is therefore available, based on the median M_d of absolute deviations from the median. An "outlier" in this sense is defined as any point (X,Y) whose residual falls outside the limits $ß_1 \pm \kappa M_d$, where M_d is the median of $|Y_i - ß_1X_i - ß_0|$, for $1 \le i \le N$, where κ is a constant. The value of κ is selected arbitrarily, but a value between 3 and 4 is suggested by Vugrinovich (1981, p.447) as appropriate, $\kappa = 3$ being "somewhat more stringent than the customary 5% significance level".

14d2. A quicker nonparametric method: Theil's incomplete method

Worked non-geological examples: Sprent(1981,p.227).
Computer software: none known.

With Theil's complete method (previous section), the number of slope estimates escalates rapidly with N, from only 3 for N = 3 to ≈ 5,000 for N = 100! Theil's incomplete method offers a feasible alternative for hand calculation. It uses the same basic procedure, but uses only $N/2$ of the $N(N-1)/2$ possible slope estimates between all N data-pairs. The procedure is as follows:

(1) Rank the N data-pairs in increasing order of X.

(2) *Only* if N is odd, delete the point with the median X value.

(3) Work out $S_{1,(N+2)/2}$: that is, the slope between the point with the lowest X value, X_1, and the X value immediately above X_{med}, namely $X_{(N+1)/2}$. Similarly, work out $S_{2,(N+4)/2}$, $S_{3,(N+6)/2}$ and so on as far as $S_{N/2,N}$: that is, $N/2$ slopes in all.

(4) As in the complete method, arrange all the S_{ij} values in order and take $ß_1$ as the median S_{ij}

For the data in Table 14.1, we use only 3 slopes: $S_{1,4} = (16-21)/(150-0) = -0.0333$, and similarly $S_{2,5} = -0.189$,and $S_{3,6} = -0.0551$, the median of which is −0.0551, rather lower than the other slope estimates in Table 14.1. Clearly, much of the robustness of nonparametric methods is here sacrificed in the interests of rapid calculation.

14e. Fitting curves: POLYNOMIAL (CURVILINEAR, NONLINEAR) REGRESSION

14e1. The parametric approach

Further discussion: Harrell(1987); Mather(1976,p.94ff); McCammon (1973).
Worked geological examples: *Geomorphology and hydrology* (Mather 1976, p.102 & 109); *Sedimentology* (Davis 1986, p.186ff).
Selected geological applications: *Igneous Petrology* (Fuster 1975; Vasilenko & Kholodova 1977; Verma 1972); *Metamorphic petrology* (Lang & Rice 1985); *Mineralogy* (Kahma & Mikkola 1946); *Sedimentology* (Ribeiro & Merriam 1979); *Structural geology* (Attoh & Whitten 1979; James 1970).
Stand-alone computer programs: Davis(1973, p.212); Mather(1976,p.110); McCammon(1973); Press et al.(1986,p.526ff).
Microcomputer software: covered in SYSTAT.

Polynomial regression allows *powers* of the independent X variable, in an equation of the form:

$$Y = \beta_0 + \beta_1 X + \beta_2 X^2 + ...\beta_n X^n \qquad [14.9]$$

It can be used with one of two aims: (1) simple 'curve-fitting', i.e. approximating the form of the curve which best fits the data, and hence interpolating Y from X; (2) testing the hypothesis that the dependent Y and independent X variables have some theoretical population relationship. In case (1), no statistical assumptions are necessarily required, and other methods (e.g. **splines**, §20) may be preferable. In case (2), however, the usual assumptions of least-squares methods are made (§14a).

This is still a bivariate problem, since we involve only X and Y variables. Furthermore, although [14.9] defines a curve, the underlying model is still said to be *mathematically* linear because X, X^2 and X^n could alternatively be regarded as three separate variables X, Y and Z; the polynomial regression would then become a form of **multiple regression** (§19). These two are often in fact handled together in computer programs. Any equation such as [14.9] can moreover be reduced to a linear form by taking logarithms of both sides; it can then be handled just like a linear regression:

$$\log Y = \log\beta_0 + [\log\beta_1 + 2\log\beta_2 + ...n\log\beta_n]\log X = \beta_0' + \beta_1'\log X \qquad [14.10]$$

If a simple X-Y plot shows obvious non-linearity, and taking logarithms is not acceptable for some reason, polynomial regression must be used. It must also be used for models of the form:

$$Y = \beta_0 X^{\beta_1} \text{ or } Y = \beta_0 + X^{\beta_1} + \beta_2^X \qquad [14.11]$$

which are not reducible to linear forms via logarithms. Unfortunately, the presence of a curvilinear rather than linear trend will become less and less obvious as X-Y scatter increases in geological data. However, plots of residuals can also help to reveal non-linearity: groups of mostly +ve residuals alternating with groups of mostly –ve residuals in a linear model clearly indicate that a curved trend is nearer the truth, as the curve departs from the assumed line in first one direction, then another.

Allowing higher powers of X in a polynomial model does introduce its own problems. Firstly, unless we have some *theoretical* basis for determining the highest power n to be introduced (such as an experimental model), an infinity of answers can be generated from one data-set. This is because a straight line can always be drawn through two points, a parabola through 3, a cubic curve through 4, and a curve of degree $[n-1]$ through n points (Fig.14.5e): hence, *any* given array of data-points can have a curve fitted *exactly* to them, simply by introducing high enough power terms. In the absence of a model (as is the case with probably the majority of geological data), therefore, polynomial regression

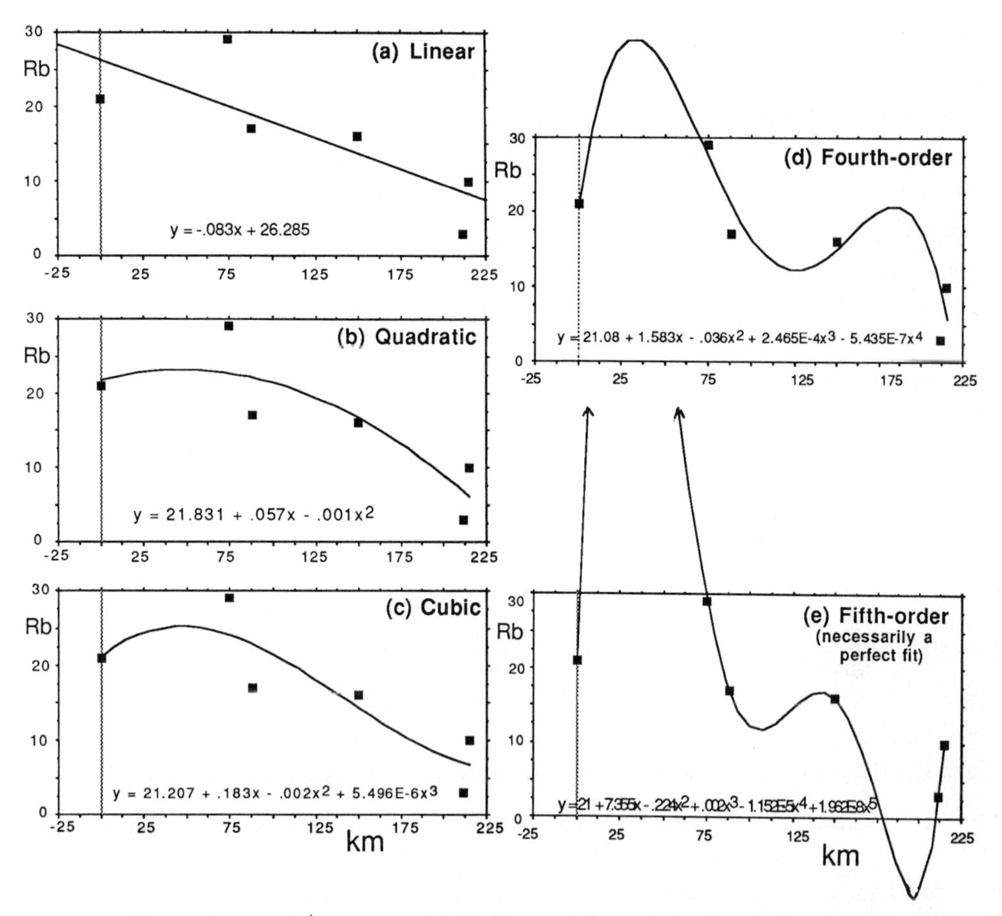

Fig.14.5. Linear and four possible polynomial regression lines using the 6 Rb-km data-pairs of Table 14.1. Note in plot (e), that it is always possible to find a polynomial of degree $n-1$ to fit through n points, just as it is always possible to draw a straight line through 2 points

requires caution and a great deal of common sense.

Secondly, the successive terms $ß_1X$, $ß_2X^2$... $ß_nX^n$ are obviously not independent, but *correlated.* This leads to the phenomenon of **ill-conditioned matrices** in the actual computations: the results become less and less accurate as n increases (see e.g. final column in Table 14.3).

Otherwise, however, the method of estimating a polynomial regression is very similar to linear regression, and generates much the same statistics. Table 14.3 and Fig.14.5 compare the linear and 4 possible polynomial models fitted to the 6 data-pairs in Table 14.1, while Table 14.4 details the statistics from the *quadratic* model (i.e. where one more term in km^2 has been added). An ANOVA table is generated, as before, which in Table 14.4 indicates that the overall regression barely reaches significance at the 90%, even though the percentage of total variation explained has risen to 75.9% (r^2). In the 'beta coefficient' table, the 'value' column indicates the coefficients of the calculated equation, namely:

$$Rb = 21.831 + 0.057km - 0.001km^2$$

In this case, t statistics have been generated for *each* of the X terms, but note that neither t-value is now the square of the F ratio in the ANOVA table (cf. Table 14.1). Note also that the t values are both non-significant: i.e. *neither* ß coefficient is statistically distinguishable from zero. The same conclusion applies here (with even more force) to the cubic and higher-order regressions, and of course, the 5th-order polynomial fits the points exactly (Fig.14.5e), whereupon the F ratio vanishes entirely.

Table 14.3. Comparison of the linear and 4 possible polynomial regression models for the Rb-km data in Table 14.1

km	Rb	Linear	Quadratic	Cubic	Fourth-order	Fifth-order	
0	21	-5.285	-0.831	-0.207	-0.08	-1.816E-13†	R
75	29	8.969	6.33	4.966	2.56	4.236E-12†	E
88	17	-1.946	-5.124	-5.864	-3.214	-5.059E-12†	S
150	16	2.224	-0.706	1.625	1.277	1.613E-12†	I
212	3	-5.606	-3.637	-3.861	-5.359	-5.580E-12†	D
215	10	1.644	3.968	3.341	4.816	4.969E-12†	S.
r^2		0.622	0.759	0.780	0.824	1.000	
F ratio		6.58	4.72	2.37	1.17	0	
F probability		0.062	0.119	0.311	0.59	∞	
$ß_0$		+26.285	+21.831	+21.207	+21.08	+21.0	
$ß_1$(X term)		–0.083*	+0.057	+0.183	+1.583	+7.355	
$ß_2$(X^2 term)		—	–0.001	–0.002	+0.036	–0.224	
$ß_3$(X^3 term)		—	—	+0.000005496	+0.0002465	+0.002	
$ß_4$(X^4 term)		—	—	—	–0.0000005435	–0.00001152	
$ß_5$(X^5 term)		—	—	—	—	+0.00000001962	

* This is the only ß coefficient in the table which has a significant t statistic

† These terms should all be exactly zero: the tiny error reflects computer imprecision due to the ill-conditioned problem.

In general, the user would begin with a quadratic model and then add higher powers successively, examining the 'usefulness' of each addition. Although r^2 tends to rise as higher-order terms are added, reaching 1 automatically for a regression of order n–1, F falls antithetically towards 0 at the same point. The balance between these two statistics should determine where the user should stop: ideally, this would be at the point where the overall F or the t for the added term cease to be significant, or (more subjectively) where the unexplained variation $(1 - r^2)$ is considered to be negligible. Despite the increase in r^2, we should probably conclude in the present case that adding even the quadratic term to our original linear model has contributed little or nothing to the regression: statistically, we have been using a polynomial sledge-hammer to crack a nut!

If the object is to isolate a *trend* rather than to interpolate Y from X, knowing where to stop can become a circular argument. We should then stop only where this trend is isolated — but this is itself the object of the exercise! Fortunately, with geological data, we can usually add the constraint that the curve should be 'smooth'; hence adding higher terms would stop where the curve began to 'break up' — i.e. involve inflections which bore no relation to actual data-points. This is clearly the case in Fig.14.4d–e, so that the 4th and 5th order curves would be dismissed. The quadratic and cubic curves might now however be regarded as superior to the simple linear model, given their higher r^2 values.

Table 14.4. Statistics from quadratic regression on the 6 *Rb-km* data-pairs in Table 14.1

Polynomial Regression X_1 : km Y_1 : Rb

DF:	R:	R-squared:	Adj. R-squared:	Std. Error:
5	.871	.759	.598	5.671

Analysis of Variance Table

Source	DF:	Sum Squares:	Mean Square:	F-test:
REGRESSION	2	303.519	151.76	4.719
RESIDUAL	3	96.481	32.16	p = .1185
TOTAL	5	400		

Residual Information Table

SS[e(i)-e(i-1)]:	e ≥ 0:	e < 0:	DW test:
268.408	2	4	2.782

Beta Coefficient Table

Parameter:	Value:	Std. Err.:	Std. Value:	t-Value:	Probability:
INTERCEPT	21.831				
x	.057	.111	.535	.508	.6465
x^2	-.001	4.635E-4	-1.374	1.305	.283

We should finally mention in this topic that a nonparametric approach to *monotonic* non-linear regression is also available (Conover 1980, p.272ff). This is based on the fact that if data display a monotone relationship, their *ranks* will display a linear relationship (which is why rank correlation coefficients attain values of 1 for perfectly monotone trends: Fig.13.1d). This method produces neither a straight line nor a smooth curve, but a regression fit comprising a zigzag series of line segments. It can be used to predict Y from X in the usual way, however, and is almost certainly preferable to parametric polynomial regression in cases where no theoretical non-linear model can be assumed, or where the usual Normality assumptions are not met. As it has not yet been applied in geology, the reader is referred to Conover's ample description and to several further references he cites.

TOPIC 15. SOME PROBLEMATICAL DATA-TYPES IN GEOLOGY

Throughout this topic, X refers to absolute quantities and P to **P**roportions, **P**ercentages, etc.

15a. Geological Ratios

Dedicated monographs and reviews: Chayes (1971).
Further geological discussion: Koch & Link (1971, p.154ff); Le Maitre (1982, p.77).
Stand-alone computer programs: Chayes (1975).

Ratios of the form X_1/X_2, $X_1/[X_1+X_2]$, $[X_1-X_2]/X_2$ are ubiquitous in the geological literature, whether explicit (e.g. $^{87}Sr/^{86}Sr$; K/Rb) or implicit (e.g. ε_{Nd}, ε_{Sr}, μ, ∂O^{18} or ∂C^{13} values in isotope geology; points in ternary diagrams, where $P_1 = 100X_1/[X_1 + X_2 + X_3]$).

Plots of ratios have a number of problematical properties, for example:

(1) A linear plot of X_1 versus X_2 will transform into a convex or concave ratio plot of P_1 versus P_2.

(2) The lever rule does not normally apply; i.e. equal amounts of two rocks will not plot halfway between their respective compositions in a ratio plot. In triangular diagrams, it will apply to the experimental phase diagram Di-Ab-An, but *not* to a normative plot of Di, Ab and An constituents from a series of rock analyses, since in the latter case the sum [Di + Ab + An] will differ between analyses.

MacCaskie(1984) and many others have advocated plots of various trace element ratios (e.g. A/B versus A/C, or B/A versus C/A) to aid interpretation of igneous petrogenesis: different genetic processes (fractionation, mixing, partial melting, etc.) variously yield straight lines, hyperbolic or exponential curves on such plots, allowing their role in a particular suite to be assessed (§18). Although *visual* inspection of such plots is perfectly valid, their use of ratios creates certain problems for *statistical* interpretation. It should, for example, be intuitively obvious that, because A has a perfect correlation with itself, the ratios A/B, A/C (or their inverses) will be correlated to some extent even if B and C are completely uncorrelated, simply because they share the common numerator or denominator, namely A. Hence if a tightly clustered, sub-linear plot is obtained on a plot of A/B vs. A/C, the normal method of testing for linearity (comparing the correlation coefficient for significance with a null coefficient of zero: §13a) is no longer valid, for the null correlation, r_o is now a positive value.

The following are approximate null correlations for common combinations of ratios used in geology (Chayes 1971); in each case, as usual, $\overline{X}_1$ and $s^2{}_1$ are the mean and standard deviation for X_1:

$$\text{For } \frac{X_1}{X_2} \text{ versus } X_1\text{, null correlation} \approx \frac{\overline{X}_2 s_1 - \overline{X}_1 s_2 \rho_{12}}{\sqrt{\overline{X}_2^2 s_2^2 + \overline{X}_1^2 s_2^2 - 2\overline{X}_1 \overline{X}_2 s_2 s_1 \rho_{12}}} \quad [15.1a]$$

where $\rho_{12} \neq 0$ is the correlation between X_1 and X_2. Where $\rho_{12} = 0$ this reduces to:

$$\text{For } \frac{X_1}{X_2} \text{ versus } X_1\text{, null correlation} \approx \frac{\overline{X_2}s_1}{\sqrt{\overline{X_2^2}s_1^2 + \overline{X_1^2}s_2^2}} \quad [15.1b]$$

$$\text{and } \frac{X_1}{X_2} \text{ versus } X_2\text{, null correlation} \approx -\frac{\overline{X_1}s_2}{\sqrt{\overline{X_2^2}s_1^2 + \overline{X_1^2}s_2^2}} \quad [15.1c]$$

$$\text{For } \frac{X_1}{X_2} \text{ versus } \frac{X_3}{X_2}\text{, null correlation} \approx \frac{\overline{X_1X_3}s_2^2}{\sqrt{\left[\overline{X_2^2}s_1^2 + \overline{X_1^2}s_2^2\right]\left[\overline{X_2^2}s_3^2 + \overline{X_3^2}s_2^2\right]}} \quad [15.1d]$$

$$\text{For } \frac{X_1}{X_2} \text{ versus } \frac{X_2}{X_3}\text{, null correlation} \approx -\frac{\overline{X_1X_3}s_2^2}{\sqrt{\left[\overline{X_2^2}s_1^2 + \overline{X_1^2}s_2^2\right]\left[\overline{X_2^2}s_3^2 + \overline{X_3^2}s_2^2\right]}} \quad [15.1e]$$

$$\text{For } [X_1 + X_2] \text{ versus } X_1\text{, null correlation} \approx \frac{s_1}{\sqrt{s_1^2 + s_2^2}} \quad [15.1f]$$

$$\text{For } [X_1 + X_2] \text{ versus } [X_1 + X_3]\text{, null correlation} \approx \frac{s_1^2}{s_{[X_1+X_2]}s_{[X_1+X_3]}} \quad [15.1g]$$

$$\text{For } [X_1 + X_2] \text{ versus } [X_1/(X_1 + X_2)]\text{, null corr.} \approx \frac{\overline{X_2}s_1^2 - \overline{X_1}s_2^2}{\sqrt{\{s_1^2 + s_2^2\}\left[\overline{X_2^2}s_1^2 + \overline{X_1^2}s_2^2\right]}} \quad [15.1h]$$

$$\text{For } [X_1/(X_1+X_2)] \text{ versus } [X_3/(X_2+X_3)]\text{, null corr.} \approx \frac{\overline{X_1X_3}s_2^2}{\sqrt{\left[\overline{X_1^2}s_2^2 + \overline{X_2^2}s_1^2\right]\left[\overline{X_2^2}s_3^2 + \overline{X_3^2}s_2^2\right]}} \quad [15.1i]$$

Equation [15.1c] also applies for X_2/X_1 versus X_2/X_3 (i.e. common numerator). Once obtained, r_{ij} is tested for significance against the null correlation using Fisher's Z transformation (§13e).

Butler (1982b) and Chayes et al. (1977) commented that $^{87}Sr/^{86}Sr$ versus $^{87}Rb/^{86}Sr$ (and similarly Sm-Nd, Hf-Lu, Re-Os, Pb-Pb) isochron diagrams involve a common denominator, so that some of the observed 'trend' could be a closure effect. Dodson(1982) disagreed. In fact, the difference between an isochron and an 'errorchron' is determined by MSWD, not by correlation; with today's extremely high-precision isotopic analysis, even a (geochronologically speaking) highly scattered errorchron may have a correlation coefficient > 0.999; the closure effect is thus likely to be small.

15b. Geological Percentages and Proportions with Constant Sum: CLOSED DATA

Dedicated monograph: Aitchison(1986).
Further geological discussion: Butler (1975,1976,1977,1978,1979abcd,1980,1981,1982a,b), Butler & Woronow (1986a,b), Chayes (1960,1962,1964a,1968c, 1980,1981,1983cd), Chayes & Kruskal (1966), Chayes & Trochimczyk (1977,1978), Darroch (1969); Darroch & Ratcliff (1970,1978), Darroch & Ratcliff(1978); Davis (1986, p.42); Hohn & Nuhfer (1980), Krumbein & Watson(1972), Koch & Link (1971, p.168ff); Le Maitre(1982,p.45); Miesch (1969a,1976d,1979), Miesch et al.(1966), Rosenblatt et al.(1978), Pearce(1970); Skala(1979a,b); Smith(1969); Snow (1975), Vistelius & Sarmanov (1961).
Worked geological examples: *Petrology* (Koch & Link 1971, p.170);
Mathematical modelling and significance testing: Chayes & Kruskal (1966); Koch & Link (1970, p.180); Kork(1977); Skala (1977); Smith(1969; 1972); Smith & Watson (1972); Watson (1987); Zodrow (1976).

Closed data include all compositional data expressed as %, ‰ or ppm, which sum to a constant value (100, 1000, or 10^6); this means that, given values for $N-1$ variables, the Nth value is automatically determined. Mineral formulae (cations per unit cell) are also closed. **Open data** show no such constraint, and can take any values.

Because a large proportion of its data are compositional (e.g. rock and mineral analyses, modal data), and therefore closed, the problem looms particularly large in geology. Unfortunately closure is unavoidable with compositional data: the absolute amount of SiO_2 (in kg) within a particular rock specimen may be open, but it is quite meaningless: only the *percentage* of SiO_2 constitutes usable data. Furthermore, many open geological data which have more intrinsic value still have to be recast as proportions or percentages for meaningful presentation: for example, numbers of individual specimens of different fossil species collected from a particular outcrop are theoretically unlimited, and therefore open; but though more meaningful than absolute amounts of SiO_2 in rocks, they still become useful for comparative purposes only if recast as *percentages* of the total number of fossils each species represents within the collection. Meanwhile, geology's ubiquitous ternary diagrams often plot *double-closed* data: original percentages A, B, C have been closed again via dividing by $[A+B+C]$!

Data-sets related by closure are termed **basis** (X_i, the parent, open data-set) and **composition** (P_i, the closed data-set formed by percentage or proportion formation), where:

$$P_j = \frac{X_j}{\sum_{i=1}^{i=v} X_i} \qquad [15.2]$$

In a whole-rock analysis, increasing SiO_2 must decrease the sum of the other oxides by virtue of the smaller 'space ' remaining for them, within the nominally 100% total. In fewer-component systems, the closure effect becomes more obvious: in a 2-component mixture, say, SiO_2-MgO, there *must* be a perfect negative correlation between the two oxides, since increasing one automatically decreases the other by the same amount. Adding one more component (say, SiO_2-MgO-FeO, as in olivines), means that increasing MgO no longer decreases FeO or SiO_2 by exactly the same amount, but there must still be a proportionate decrease in both.

In closed data, the covariances of one variable with all others sum to its own negative variance:

$$\sum_{i=1}^{i=v} \text{Covariance}_{ij} = -\text{Variance}_j, \text{ with } i \neq j \qquad [15.3]$$

Thus Pearson's r correlation coefficients (i.e. covariance/variance ratios [13.1]) will tend to be more negative than in open data. In fact, if all v means and variances are homogeneous, all correlation coefficients will take the value $-(v-1)^{-1}$, i.e. -0.1 for $v = 11$, etc. Furthermore, the component with the largest variance will shown the highest –ve correlations. Harker diagrams (Fig.15.1), long used to illustrate chemical trends in rocks, are a case in point. Because SiO_2 has the highest variance, negative trends (correlations) will arise between SiO_2 and other elements *purely* because of closure. Put another way, the *null correlation* between any two elements in whole-rock analyses is no longer zero, but some unknown, negative value. When testing correlation coefficients (or variation trends) for statistical significance, we cannot therefore use standard tables (§13a), but need to know the null correlation and use Fisher's Z transformation (§13e). A negative correlation is *less* significant in a closed array, and a positive correlation *more* significant, than the same value in an open array.

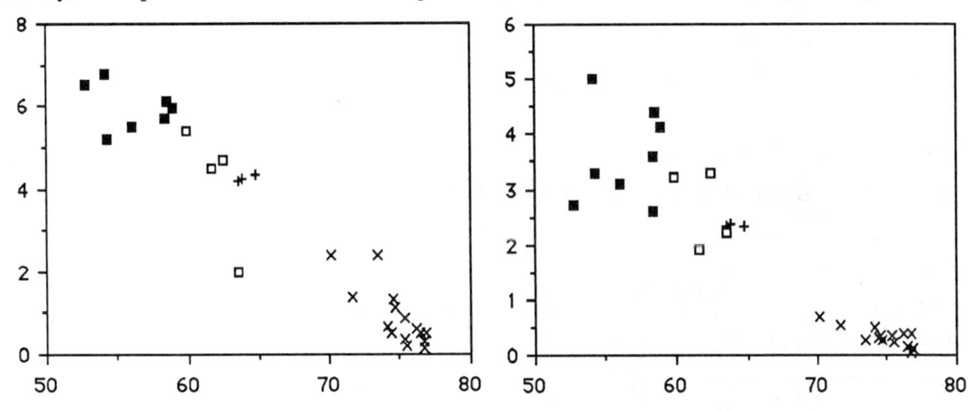

Fig.15.1. Typical Harker diagrams, such as appear routinely in the geological literature.
Though 'smooth trends' have traditionally been imbued with mystic petrological significance (and taken to imply cogeneticity, comagmaticity, etc.), the tendency of most oxides to decrease with increasing SiO_2 is to a large extent a consequence of closure, due to the large variance of SiO_2 (i.e. what is added to SiO_2 must come away from the other oxides, to preserve the 100% total). Removing this purely statistical effect to determine whether any remaining trend is still 'smooth' is extremely difficult.

The effect of closure on Spearman's ρ or Kendall's τ (§13b1–2) has barely been discussed in the literature, compared to its effect on Pearson's r. Whereas closure *must* affect r, because the individual values automatically change in relation to each other, their *ranks* may or may not change. In the imaginary data-set of Table 15.1, for example, closure changes the rank orders for variables 3 and 4 but not for variables 1 and 2; hence, τ_{12} or ρ_{12} are unchanged by closure, whereas between any other pair of variables, τ or ρ are different. Meanwhile, r_{12} changes from 0.96 to 0.98 after closure. The most that can be said at present, therefore, is that τ and ρ are *less* affected by closure than r, but that the complete absence of a closure effect *cannot* be assumed.

Some other geological data-types and procedures which can be seriously affected by (or arise via) closure, and which consequently need to be interpreted with extreme care, include the following:

(1) 'Trends' and 'clusters' on petrological ternary and principal component diagrams (§23), which can have little or no geological significance. For example, trace element fields used to discriminate tectonic settings of igneous rocks (e.g. Pearce & Cann 1971) can be closure-induced.

(2) Skew and leptokurtosis (§8c3), which can be induced by closure even in Normal data distributions.

(3) Dendrograms from cluster analysis (§21b), which can be severely biased.

(4) Some results from discriminant analysis (§22a), which can be illusory.

(5) Modal data from sediments, where erroneous genetic interpretations can be imposed.

(6)Trace element correlations used to infer petrogenetic processes, such as high correlations between incompatible elements (Nb, Zr, etc.) used to imply fractionation (e.g. Weaver et al. 1972; MacCaskie 1984). Null correlations between traces are almost certainly *positive* (Chayes 1983d), not zero, so that high correlations are *less* significant than they appear (cf. majors, considered above).

(7) Multivariate methods, where ill-conditioned matrices resulting from high internal correlations between variables can lead to unstable or sometimes erroneous results.

(8) Confidence limits, numbers of values lying so many standard deviations from the mean, or values within a certain range, which can all differ from those expected from Normally distributed data, even for large data-sets believed intrinsically to be Normal.

(9) Contouring, which can be dangerous — especially in 3-component systems (Smith 1969).

Table 15.1. Illustration of the effect of percentage formation on ranks

	Open data				Closed data (percentages)		
	Rock 1	Rock 2	Rock 3		Rock 1	Rock 2	Rock 3
X_1	2	4	14	P_1	8	12	14
X_2	3	11	23	P_2	16	22	23
X_3	5	17	30	P_3	20	34	30
X_4	15	18	33	P_4	60	36	33
Total	25	50	100		100	100	100
®(X_1)	1	2	3		1	2	3 (NB ranking is across *rows*)
®(X_2)	1	2	3		1	2	3
®(X_3)	1	2	3		1	3	2
®(X_4)	1	2	3		3	2	1

15c. Methods for reducing or overcoming the Closure Problem

The geological literature is unfortunately replete with examples where significance tests have been made on percentage or 'ppm' data without regard to the closure effect. This may be partly because those concerned were unaware of the problem, or because the mathematics involved in reducing closure are complex, leading them to ignore the problem.

15c1. Data transformations and recalculations

No amount of conventional data transformation will eliminate closure (Butler 1980, 1981). Recalculation of whole-rock analyses to 100% is meaningless, since the difference between actual totals and 100% is merely an error factor. Neither the transformations applied to individual variables in §8f2 (log-transformation, etc.), nor standard petrological methods which treat all variables at once (e.g.

norms, cations, molecular proportions, formula units in minerals, ternary diagrams) will 'open' data which are originally closed.

Perhaps the only transformation which does afford partial escape from closure is principal components analysis (§23), in which successive components are defined to be at right angles, and thus by definition, uncorrelated. Even here, however, there are restrictions (Butler, 1976; Chayes & Trochimczyk 1978).

15c2. Ratio normalising

Russian petrologists favour correlation diagrams in which major elements are expressed as ratios to the total O content of the rock (Si/O etc.: Podol'sky 1962; Ivanov 1963), while Miesch et al.(1966) use ratios such as MgO/SiO_2. Such ratios are unconstrained although, of course, their common denominators introduce other correlation problems (§15a)! Russell (1986) advocates plotting ratios such as Mg/K vs. Fe/K, where the denominator K is chosen for its independence of the geological processes involved.

15c3. Hypothetical open arrays

Chayes & Kruskal (1966) postulated the existence of a *theoretical* open array (the basis) corresponding to any given set of closed data (the composition). Such a basis would have zero covariances between the variables but, when closed, would generate the observed variances and covariances in the composition. From the basis, the expected values for correlation coefficients could be calculated, and used as null correlations to be compared with those in the composition via Fisher's Z transformation (§13e). Their method involved testing the null hypothesis that there is no correlation in the closed data, other than that induced purely by closure — a condition known as **complete subcompositional independence**. If this null hypothesis is accepted, none of the correlations observed in the closed array are deemed statistically significant, and further analysis is precluded.

Chayes & Kruskal's method has unfortunately since been found to have the following problems:

(1) As indicated earlier, the basis itself has little physical meaning for compositional data.

(2) An infinite number of closed arrays can generated any one closed array. For example, the open pairs 2,3; 6,9; 600,900; 2000, 3000; all generate percentages of 40%, 60%. The particular type of open array generated by Chayes & Kruskal's method may not be any more relevant than any other.

(3) For some data-sets, the procedure paradoxically generates open arrays with negative variances (Chayes 1971; Butler 1975), which tends to undermine its fundamental assumptions. (A variance is a *squared* number and cannot therefore be negative!)

(4) If the null hypothesis above is rejected, the method is incapable of deciding *which* correlations are significant, so that one of the main objects of the exercise cannot be achieved (Kork 1977).

(5) The power of the method is commonly as low as 91% (i.e. there is a 9% probability of α or ß error, §7g), due partly to its inherent approximations, and partly to a failure to meet the Normality assumptions of the Fisher Z transformation (Zodrow 1976).

15c4.Remaining space variables

If we have two associated proportions P_1 and P_2, where $P_1 > 0$, $P_2 > 0$, $P_1 + P_2 < 1$ (or 100%) — such as major oxide or mineral percentages in a rock — remaining space variables assess one component in relation to the space in the whole rock which is *not* occupied by another (the 'remaining space'). For example, $\psi_1 = P_1/(1 - P_2)$ and $\psi_2 = P_2/(1 - P_1)$ are *not* closed, and correlation between P_1 and P_2 can thus be tested via median tests (§10d1) of ψ_1 versus P_2 or ψ_2 versus P_1 (Darroch & Ratcliff 1978). Such tests will not necessarily yield identical results: P_1 may for example inhibit P_2 whereas P_2 may be unaffected by P_1.

Niggli numbers in petrology (*si, mg,k, al* etc.) are remaining space variables, representing the amount of space in a rock which is *not* occupied by SiO_2; they can also therefore be used to test correlations between proportions (Chayes 1983c).

15c5.A recent breakthrough: log-ratio transformations

Further discussion: Aitchison (1981,1982,1983,1984a,1986); Woronow (1986).
Stand-alone computer programs: Aitchison (1986); Woronow & Butler (1985); — but see Woronow 1988 for details of bugs in former program.

All experts appear now to accept that most methods in §15c1–4 are of limited value. One of the major problems is that closed data comprising v variables lie in a **simplex** (a restricted region) in $(v-1)$-dimensional space, and this greatly complicates statistical testing. Aitchison's (*op.cit*) method makes the entire $(v-1)$-dimensional real space available, via a log-ratio transformation, in which each closed data-value P_i (in the composition) is divided by a particular component P_v (e.g. other major oxides by SiO_2) and logarithms are taken, to yield the unrestricted (open) values in the basis, X_i:

$$X_i = \ln(P_i/P_v), \qquad i = 1,2,........(v-1) \qquad [15.4]$$

The method of testing the basis is complex, involving interactive maximization of a **likelihood** function. The eventual result is a **Wilks'** Λ or (for large data-sets) a χ^2 statistic, which tests the same null hypothesis (complete subcompositional independence) as in Chayes & Kruskal's method (§15c3). Only if this is *rejected* (i.e. correlation is present *other than* that caused purely by closure), is it worth testing the individual correlations. The method appears to be both robust and powerful, but its enormous potential in geology has only just begun to be realized.

15d. The Problem of Missing Data

Dedicated monographs and reviews: Beale & Little (1975); Dodge (1985); Little (1987).
Further geological discussion: Rock (1988).

Files with intermittently missing values are extremely common in geological work. As in other sciences, experiments may fail, data may become corrupted, determinations may lie below machine detection limits, etc., etc., but there are many other reasons peculiar to geology why critical data may not be obtainable in the first place: outcrops may not exist in appropriate locations, limited sample sizes may preclude analysis for the whole range of elements determined on other samples, and so on.

Missing data have little effect on the *univariate* statistical techniques of §8–12, except that they reduce N by the number of individually missing values for each calculation. In *bivariate* techniques (§13-14), they can be dealt with by either **casewise** or **pairwise** omission, the latter maximising information in the data but producing correlation coefficients based on different N for different pairs of variables. Unfortunately, professional statisticians disagree strongly as to the relative merits of pairwise and casewise omission: Wilkinson (1986), for example, refuses even to output significance levels for pairwise correlation coefficients in the Systat package (Table 2.2), but his approach has been criticized by several reviewers (e.g. Berk 1987b). Fortunately, the choice is not too critical if coefficients are an end in themselves (for example, if a geologist simply wishes to identify those elements which are most strongly correlated in an igneous rock-suite, for subsequent plotting on X-Y variation diagrams). Pairwise and casewise coefficients will differ more and more numerically, as the proportion of missing data rises, but it would be most unusual to find a casewise coefficient being highly significant where a pairwise coefficient was not significant, or vice-versa.

When we come to more sophisticated multivariate techniques (§18-23), missing data become more crucial. Many of the calculations simply cannot be performed unless a value is provided for *every* variable of interest, in *every* specimen. This leaves two alternatives: (1) do not use any variables in statistical analysis which include missing values; (2) interpolate values for the missing data. Option (1) becomes more and more the *only* viable option as the proportion of missing to 'real' data for a particular variable increases; for example, if 1,000 specimens were analyzed for Au, but 995 gave values below detection limits, Au should *under no circumstances* be included in any multivariate statistical treatment. Unfortunately, it is often the elements of greatest interest which include most missing data (e.g. Au, PGE and other rare elements may be analyzed for all samples from an exploration programme, but only the rare mineralized zones will yield values above detection limits!)

Option (2) — interpolation — is therefore the most commonly adopted, but the problem then of course arises of how to interpolate. It is vital at this point to distinguish two types of missing data: (a) values for which *no information whatever* is available (not analyzed at all, or analyzed but rejected as 'bad data' for some reason); (b) values for which *some* information is available (most commonly, values below machine detection limits (dl), which can be specified as, say, '< 5 ppm'). Values of type (a) can only be interpolated by analogy with values determined for similar objects (e.g. another specimen of the same formation, fossil, dyke, etc.), but the process is obviously highly dangerous. Because the '<' cannot be handled by a computer for statistical analysis (it is a character, not a number), values below dl must be replaced with processable numbers, but there are at least 4 alternative replacement values: (1) the dl itself; (2) an arbitrary value of zero; (3) an arbitrary value of some fixed proportion of the dl (e.g. 2.5 or 5 for a dl of 10 ppm); (4) a random number between zero and the dl. Arguments can be proferred in favour of each of these alternatives; as the dl itself approaches zero, there is less and less to choose between them.

In short, there are no hard-and-fast rules for dealing with missing data, but the treatment adopted may change the results of statistical analysis substantially, and must therefore be chosen very carefully.

15e. The Problem of Major, Minor and Trace elements

The elements which comprise a complete rock analysis are conventionally divided into *major elements* (Si, Al, etc.), measured in % or tens of %; *minor elements* (MnO etc.), usually one or two %; and *trace elements,* measured in ppm or ppb. This means that, *in absolute magnitude*, geochemical values can range by a factor of some 10^8, from SiO_2 (say, 50% of the rock) to Au (say, 1-5 ppb). This has not been a problem with the techniques so far discussed, since these have either considered one element at a time (§8–12), or two elements in such a way that their relative magnitudes are irrelevant (as in the calculation of correlation coefficients in §13, where the data are effectively **standardized**).

The problem arises when we turn to *multivariate* analysis (§18–23), which considers *all* variables *simultaneously*. In many multivariate techniques, the variable which shows greatest absolute (as well as relative) variance will have greatest influence on the outcome, and variance is obviously related to absolute magnitude (the variance of SiO_2 is, once again, measured in $\%^2$, but that of Au is in ppb^2). We *cannot* therefore use the data in the form that they are conventionally quoted in tables of rock analyses. For example, consider typical contents for SiO_2 and Ba (the most abundant major and trace elements) in basalts: 500 ppm and 50%. Even though Ba is actually 3 orders of magnitude *lower* than SiO_2 in absolute content, a computer will interpret an input data-value of '500' as 10 times *higher* than a figure of '50', *unless* the user tells it not only which data-value represents % and which ppm, but *also* tells it to recalculate all values to the same measurement unit. Similarly, an MnO (%) value of '0.05' will be treated by a computer as 4 orders of magnitude *lower* than a Ba (ppm) value of '500', though in fact the two contents are exactly equal in absolute terms!

Geomathematicians have suggested three ways around this problem, none of which, however, entirely overcomes the difficulties; some individual solutions are proferred in the Topics which follow:

(1) *Re-express all data in one fixed measurement unit (%, ppm or ppb)* This option unfortunately means that variations in trace elements are often completely swamped by those in minor elements, and these in turn by variations in major elements. In other words, much multivariate analysis will effectively be a study of SiO_2 — the most variable element — and little else.

(2) *Standardize all data.* This option re-expresses all data distributions to zero mean and unit variance. It unfortunately has the opposite effect to option (1): variation for elements such as Mn is blown up out of all proportion to its real geological importance. In equating all the variances, moreover, it can obscure many of the important relationships between geological variables.

(3) *Convert the data to ranks:* Because ranking reduces all raw variables to the same set of ranks (between 1 and *N*), it has much the same effect as standardization. Although use of the multivariate techniques in §18–23 on ranked data is perfectly legitimate, a better alternative would be to use properly defined nonparametric multivariate techniques, which unfortunately are still in their infancy.

(4) *Log-transform the data.* Log_{10} values of 100%, 1%, 1000 ppm, 1 ppm and 1 ppb (using 100% = 1) are 0, –2, –3, –6 and –9. Decreasing 100% to 50%, 1 ppb to 0.5 ppb, etc., involves *the same* logarithmic decrease of 0.3 in each case (i.e. 0 to –0.3, –2 to –2.3, –3 to –3.3, –6 to 6.3, or –9 to –9.3). Hence larger *relative* log-variances receive larger *relative* weights, but small *absolute* values, and log-variances for trace elements are no longer completely submerged by those for majors.

TOPIC 16. ANALYSING ONE-DIMENSIONAL GEOLOGICAL SEQUENCES IN SPACE OR TIME

This Topic concerns bivariate data, characterized by X as *position* (in space or time) along some sequence, and Y as a measured variable. The X axis for sequences in *space* can be as concrete as a borehole (for Y measurements such as lithology, resistivity, porosity), as irregular as a water-course, (for Y such as pH, salinity), or as abstract as a vertical 'sequence' in a layered intrusion. Sequences where X represents *time* (e.g. seismic or volcanic records) are known as **time-series** (Cryer 1985; Hannan 1976; Williams & Gottmann 1982), and now command a substantial literature of their own. The techniques used below, however, are applicable where X represents *either* space *or* time, and no special treatment can be afforded to time-series in the space available. In geology, furthermore, space and time sequences are often inextricable: a stratigraphical succession represents not only a sequence of lithologies, bed thicknesses etc. in space, but also a record of sedimentation rates in time.

16a. Testing whether a single Series is Random or exhibits Trend or Periodicity

A 'trend' here is simple a *correlation* or *association* within a sequence. Thus the correlation coefficients covered in §13a–c can all be used to test for trend, with the same provisos. Pearson's r would tend to be used when both X and Y are ratio variables, where no outliers are present, and where the points are fairly regularly spaced. For example, if Cd values along a stream were measured from its source to a confluence, r could be used to check whether Cd (the Y variable) increased linearly downstream, decreased linearly, or stayed constant, provided that distances of all sampling points from the source were measured quantitatively (the X variable). The same would apply to Au analyses down a borehole, provided exact distances along the core were measured (as X). Drift in, say, a mass spectrometer could also be checked by measuring the isotopic ratio (Y) of some standard solution over a period, and using the time from day 1 as X. However, where either X or Y is an *ordinal* (ranked) variable, where a monotonic (rather than strictly linear) trend is of interest, or where there are outliers, Spearman's ρ or Kendall's τ would be more appropriate. This would be the case if the downstream or down-borehole distances, or the dates of the isotopic measurements, were not known exactly, but the *order* (in space or time) in which these measurements were taken *was* known. The X values are then taken as ranks of 1,2,3...N, corresponding to the 1st, 2nd, 3rd...Nth measurements in this known order. These uses of ρ and τ are known by the special names of **Daniel's** and the **Mann-Kendall** tests for trend; the special names arise because the X variable is inferred from the Y values, rather than being a separate measured variable. Miller & Kahn(1962, p.334) give a worked application of Daniel's test to the petrology of layered gabbros.

The following three, more specialized, tests are less efficient than r, ρ or τ in detecting linear or monotonic trends, but more efficient in detecting periodic (cyclic), U- or $\cap$-shaped trends. In each case, H_0 {no trend} (i.e. randomness), is tested against H_1{monotonic or periodic trend}. All three tests are nonparametric, equally applicable to ranked ordinal as to ratio data.

16a1. Testing for trend in ordinal or ratio data: Edgington's nonparametric test

Description and worked example: Edgington(1961); Gibbons(1971); *Hydrology* (Lewis 1977,p.191).
Statistical tables: Lewis(1977,table 11); *Z* approximation for large *N* following equation (16.1).
Computer implementations: none known.

This is a combination of the Sign and Runs tests (§10c1 & 10f1). The sign of the change between adjacent values is noted, and the number of runs of the same sign summed. For example:

```
13  16  22  21  20  24  26  18  13  10  9  6     yields signs
  +   +   -   -   +   +   -   -   -   -  -
```

Clearly, the *lower* the number of runs, the more likely a trend is present. Here, the total number of runs, R, is 4, which has an associated probability of 0.008. The test therefore indicates a ∩-shaped trend at >99% confidence. If tied values occur (yielding no change rather than + or –), the test can be made conservative by minimising the number of runs (and hence the chance of accepting H_0), or the converse, as appropriate. For $N > 25$, the following is compared with a standard Normal deviate:

$$Z = \frac{9.487(R - 0.333[2N - 1])}{\sqrt{(16N - 29)}} \qquad [16.1]$$

16a2. Testing for cycles in ordinal or ratio data: Noether's nonparametric test

Description and worked example: Noether(1956); *Geography* (Lewis 1977,p.194)
Statistical tables: uses *cumulative binomial probabilities* (e.g. Lewis 1977, table 3).
Computer implementations: none known.

This test is *specifically designed* to detect cyclical trends. It is based on splitting the data into triplets, ranking each triplet from 1 to 3 separately, and counting the number of triplets in which the *ranks* are monotonic (i.e. form the sequence 1,2,3 or 3,2,1). Tied values within a triplet are omitted, and the triplets moved up or down the data-set. Such sequences have an individual probability of 1/3, since they represent 2 of 6 possible arrangements (the others being 1,3,2; 2,1,3; 2,3,1; 3,1,2). As the number of monotonic triplets, m, increases, a trend becomes more likely. From n possible triplets, m is a binomial random variable (§8a3) with associated probability as follows [cf. 8.7a]:

$$\text{probability} = {}^nC_m p^m q^{n-m}, \qquad \text{here with } p = 0.333,\ q = 0.667.$$

The overall probability of having $m >$ some value is given by *cumulative* binomial tables. For the example above, $m = 3$, as underlined below, and the associated probability ($m \geq 3$) is about 0.0001.

<u>13 16 22</u> 21 20 24 <u>26 18 13</u> <u>10 9 6</u>

Clearly, sinusoidal data which reverse trend every 3 values would give $m = n = N/3$ and always lead to rejection of H_0.

16a3. Testing for specified trends: Cox & Stuart's nonparametric test

Description/worked examples: *Geochemistry* (Conover 1980,p.135); *Hydrology*(Lewis 1977,p.196)
Statistical tables: uses Binomial Distribution tables (see §8a3).
Computer implementations: none known.

This versatile adaptation of the Sign Test (§10c1) can be used to detect any nonrandom trend, including periodic ones (e.g. sine curve), or even longer-term trends in data known to follow short-term trends (e.g. annual rainfall or temperature data). The idea behind the test is shown in Fig.16.1.

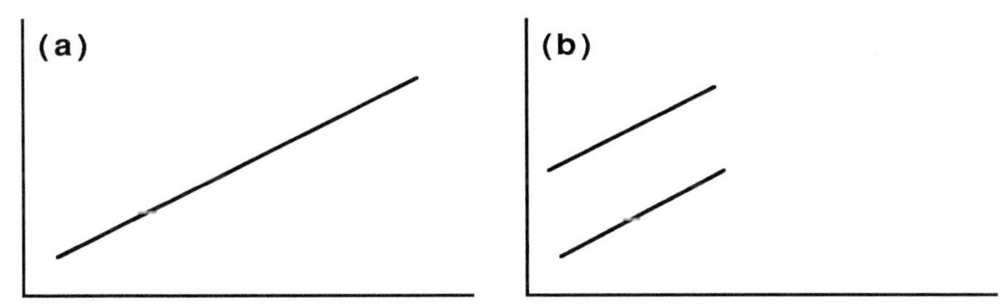

Fig.16.1. The idea behind the Cox-and-Stuart test for trend (see text below for explanation)

If we take a linear trend as in Fig.16.1a, bisect it at the median, and slide one half under the other as in Fig.16.1b, all values in one half will necessarily exceed all those in the other, scoring a '+'. However, if there is no trend, values in one half are as likely to be less as greater than those in the other: + or – should occur randomly along the compared halves. H_0: {no trend} can in this way be tested against H_1 of: (1) an unspecified trend (two-tailed); (2) an upward trend, or (3) a downward trend (both one-tailed). The test statistic, *m,* is simply the number of + in the predicted direction, when we compare the two halves on either side of the median value (small m favours H_0). If N is odd the median value itself is omitted. Tied values yield zero differences and are also omitted, with a concomitant reduction of N. For the data above we obtain:

1st half:	13	16	22	21	20	24	
2nd half:	26	18	13	10	9	6	
yielding	–	–	+	+	+	+	or $m = 4$.

The statistic is a binomial random variable (§8a3) and the associated probability (cf.[8.7a]) is:

probability = ${}^nC_m p^m q^{n-m}$, here with $p = q = 0.5$ and n = *number of untied pairs* = 6

For testing H_0: {no trend} against H_1: {*downward* trend}, we require the probability of having $m \le 4$ for $n = 6, p = 0.5$, which is 0.8906, so H_0 is firmly accepted. For testing H_0 against H_1:{*upward* trend}, we require the probability of having $m' \le (6 - 4) = 2$, which is 0.3438, so H_0 is again accepted. For testing H_0 against H_1: {*any* trend}, $m = 0$ has a two-tailed probability of $2 \times 0.0156 = 0.0312$, so m would have to be 0 or $m' = 6$ to reject H_0 at ≈97% level. (The same tables show $m = 1$ has probability 0.1094, so $m = 1$ or $m' = 5$ would not achieve even 90% significance).

This test is already efficient, but can sometimes be still further improved by removing the middle third of the data, and comparing the 1st and 3rd thirds remaining, as below:

1st third:	13	16	22	21	
3rd third:	13	10	9	6	
yielding	0	+	+	+	or $m = 3$

which has associated probability 0.125. H_0 is much nearer to being rejected than above.

To detect a nonlinear trend (e.g. sine curve), we merely rearrange the data such that the smaller values *as predicted by the trend* are on the left and the larger ones on the right, and then proceed as above. Accepting H_1 for the rearranged sequence implies the predicted trend is present in the original sequence. For example, to test whether the following rainfall data show a minimum in July and maximum in January, we rearrange the months in successive pairs away from July on the left; thus:

January	February	March	April	May	June	July	August	September	October	November	December
189	154	145	159	123	109	78	89	102	95	146	167

is rearranged into the following, on which the test is performed:

July	June	August	May	September	April	October	March	November	February	December	January
78	109	89	123	102	159	95	145	146	154	167	189

Testing for longer-term trends in data known to shown short-term periodicity is even easier. To compare the above rainfall data for one year with data for another year, we would merely pair the 12 months; m becomes the number of months in which the second year has higher rainfall than the first.

16a4.Testing for trend in dichotomous, nominal or ratio data: the one-group Runs Test

Further geological discussion: Miller & Kahn(1962,p.327ff).
Worked geological examples: *Exploration/stratigraphy* (Miller & Kahn 1962, p.341, 351); *Igneous petrology* (Miller & Kahn 1962, p.334); *Petrography* (Davis 1986, p.170).
Statistical tables: see under §10f1.
Mainframe software: covered in most statistics packages (e.g. SPSS) or libraries (e.g. NAG).
Microcomputer software: programmed in many statistics packages (e.g. Statview 512+).

A one-group form of the Runs Test (§10f1) can be applied to any sequence of data in the form of a series of + and –. Dichotomous data are already in this form, so we simply look up the number of runs, R, in tables against n_1 = no. of +, n_2 = no.of –, or use [10.13]. Ratio data can be dichotomized by giving + and – respectively to values above and below the median. For the same sequence above:

13	16	22	21	20	24	26	18	13	10	9	6	have a median of 17, giving
–	–	+	+	+	+	+	+	–	–	–	–	or R = 3 runs.

$R = 3$ is significant at the 95% level (of course $n_1 = n_2$ for dichotomization about the median).

The test can also be adapted to the case of nominal data with *three* possible states. As above, we count the number of runs, R, and the number of times each state occurs, $A,B,C,$ and then use:

$$Z = \frac{R - \dfrac{2(AB+AC+BC)}{(A+B+C)} - 1}{\sqrt{\left\{\dfrac{4(AB+AC+BC)^2}{(A+B+C)^2(A+B+C-1)} - \dfrac{2(AB+AC+BC)+6ABC}{(A+B+C)(A+B+C-1)}\right\}}} \qquad [16.2]$$

In this case, $A, B, C,$ could be *dominant* lithologies in a sequence, pebble varieties in a till, actual lithologies every fixed interval in a succession, as well as actual sedimentary beds, etc. For example, with the sequence $ABBCBCABAA$ we have $R = 8$, $A = 4$, $B = 4$, $C = 2$, giving $Z = 0.46$ from [16.2]; this is non-significant, indicating that the sequence is random.

Table 16.1. Results of auto-association (forward matches only) on the sequence 12323132134, with sliding step 1
(Probability of a match for this sequence, p = 0.2)

Match position	Low α position	High α position	No.of matches	No.of compare	Matches/ comparisons	Standard deviations	Chi-square (uncorrected)	Chi-square (Yates corrected)
1	1	1	0	1	0.0000	-0.9273	0.2500	0.5625
2	1	2	0	2	0.0000	-1.3114	0.5000	0.0313
3	1	3	1	3	0.3333	0.5260	0.3333	0.0208
4	*1*	*4*	*1*	*4*	*0.2500*	*0.2398*	*0.0625*	*0.1406*[16.4]
5	1	5	1	5	0.2000	0.0000	0.0000	0.3125
6	1	6	2	6	0.3333	0.7438	0.6667	0.0937
7	1	7	2	7	0.2857	0.5307	0.3214	0.0089
8	1	8	1	8	0.1250	-0.5786	0.2812	0.0078
9	1	9	3	9	0.3333	0.9110	1.0000	0.3403
10	1	10	0	10	0.0000	-2.9324	2.5000	1.4062
11	1	11	11	11	1.0000	7.3440	44.0000	39.1420
12	2	11	0	10	0.0000	-2.9324	2.5000	1.4062
13	3	11	3	9	0.3333	0.9110	1.0000	0.3403
14	4	11	1	8	0.1250	-0.5786	0.2812	0.0078
15	5	11	2	7	0.2857	0.5307	0.3214	0.0089
16	6	11	2	6	0.3333	0.7438	0.6667	0.0937
17	7	11	1	5	0.2000	0.0000	0.0000	0.3125
18	8	11	1	4	0.2500	0.2398	0.0625	0.1406
19	9	11	1	3	0.3333	0.5260	0.3333	0.0208
20	10	11	0	2	0.0000	-1.3114	0.5000	0.0313
21	11	11	0	1	0.0000	-0.9273	0.2500	0.5625

Sum of χ^2 for match positions 1–10 = 5.9152, corresponding to Z = –0.9194

16a5. Testing parametrically for cyclicity in nominal data-sequences: AUTO-ASSOCIATION

Further geological discussion: Sackin & Merriam(1969).
Worked geological examples: *Stratigraphy* (Davis 1986, p.238).
Stand-alone computer programs: Srivastava & Sackin(1971); the cross-association program in Davis (1973, p.252) can also be used, by comparing the sequence with itself.
Proprietary software: Not implemented in the well-known statistical packages, but available in some of the innumerable packages output by oil companies and associated software houses to analyze borehole data (*see* program sources in §2).

Autoassociation (self-association) copes with a sequence of qualitative data in states — e.g. lithologies in a bore log, coded perhaps in the form 123234... (1 = limestone, 2 = sandstone, 3 = shale, 4 = coal). The issue of concern is whether the states are arranged randomly along the log or show any periodicity (e.g. the same sequence is repeated at intervals). The method thus involves comparing the sequence with itself, and has obvious applications to investigating cyclothems. For example, the sequence 12323132134 is moved past itself at successive match positions (numbers of overlapping points), ν, and the numbers of matches (*m*) and the ratio *m*/ν (the auto-association) are obtained thus:

```
ν  Match position 1 .....Match position 4 ....Match position 9
   12323432134          12323432134         12323432134
             12323432134         12323432134   12323432134
m            0                   1                 3
```

A **auto-associatogram** of ν against auto-association *m*/ν can then be made. The probability that a match will occur when a *random* sequence is compared with itself at any position is given by [16.3]:

$$p = \frac{\sum_{i=1}^{i=\mu} X_i^2 - n}{n(n-1)} \qquad [16.3]$$

where μ is the number of different states, n is the number of positions in the sequence, and X_i is the total number of times state i occurs in the sequence ($\Sigma X_i = n$). Here, we have $\mu = 4$, $n = 11$, $X_1 = 2$, $X_2 = 3$, $X_3 = 4$, $X_4 = 2$, so $p = 0.2$. The expected numbers of matches and mismatches in position ν are thus νp and $\nu(1-p)$; that is, in match position 5 (5 overlapping points) we expect 1 match and 4 mismatches. We then calculate a χ^2 statistic for each match position. With O, E signifying the observed and expected numbers of matches (O*, E* correspondingly for mismatches), for position 4 above we have O = 1, E = 0.8, O* = 4, E* = 3.2, yielding the result in [16.4] overleaf:

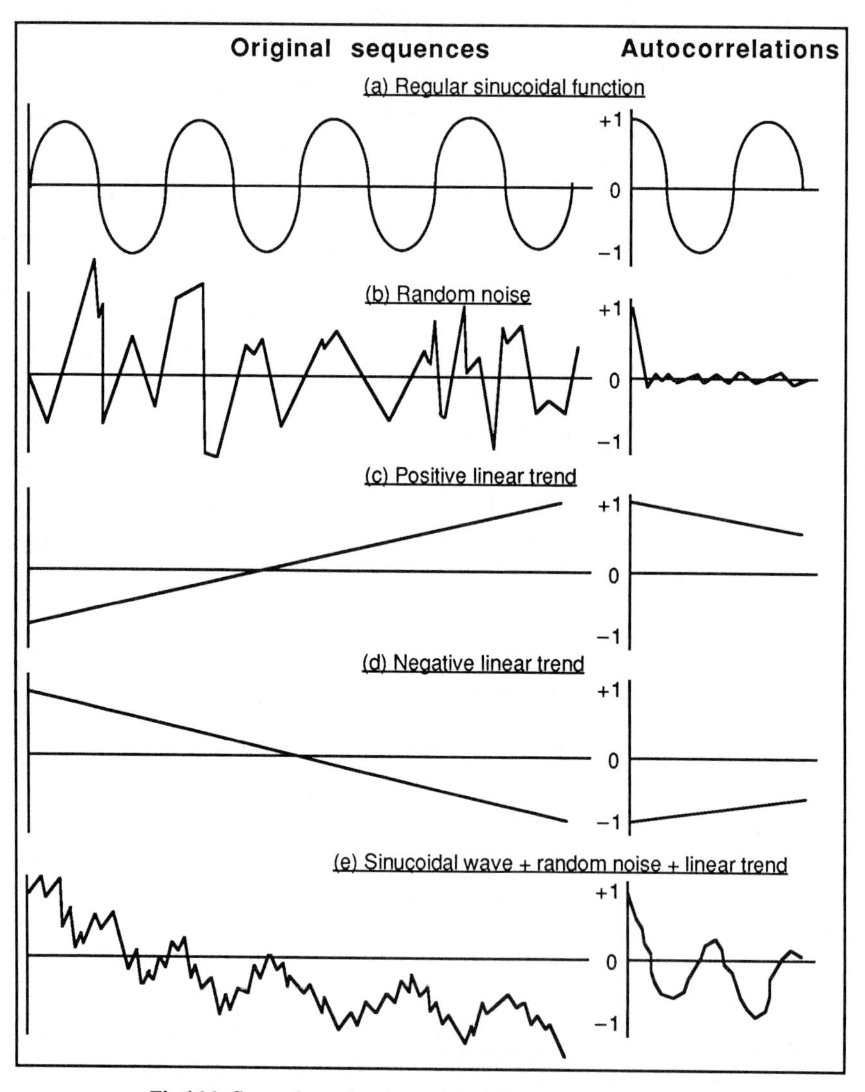

Fig.16.2. Comparison of various original sequences and their autocorrelograms

$$\chi_1^2 = \frac{(O - E - 0.5)^2}{E} + \frac{(O* - E* - 0.5)^5}{E*} = \frac{(1 - 0.8 - 0.5)^2}{0.8} + \frac{(4 - 3.2 - 0.5)^2}{3.2} = 0.14 \quad [16.4]$$

Yates' correction (the – 0.5 in [16.4]) is usually applied for $\nu \leq 5$ (§13d1). Our χ^2 is non-significant, so we conclude that the number of matches in this position is no larger than that expected from matching a *random* sequence with itself. Furthermore if, as here, χ^2 for *none* of the 11 non-trivial match positions is significant, evidence for periodicity in the sequence as a whole is lacking.

16a6. Looking for periodicity in a sequence of ratio data: AUTO-CORRELATION

Further geological discussion: Miller & Kahn(1962, p.349).
Worked geological examples: *Stratigraphy* (Davis 1986, p.217).
Selected geological applications: *Sedimentology* (Parsley1971).
Stand-alone computer programs: Davis (1973, p.238).
Proprietary software: SYSTAT and specialist packages; also in NAG(routines G13ABF, G13ACF etc.)

Autocorrelation (self-correlation) copes with the same situation as auto-association, except that the data are now quantitative (interval or ratio scales of measurement). It is thus suitable for seeking periodicity of, say, resistivity measurements, porosities, thicknesses of sediment beds, etc. in borehole logs. It can also be used where measurements are spread along a *time* rather than distance sequence, and can thus be useful in assessing annual numbers of earthquakes or eruptions in a particular area (or globally), so as to predict periods when seismic or volcanic activity may be particularly intense.

Repetition is found by comparing the sequence with itself at all successive positions, and computing a 'self-similarity coefficient' for each comparison. Autocorrelation requires measurements to be equally spaced along the sequence (in space or time), so that the *d*th measurement can be notated as Y_d, its actual position along the sequence then being automatically implied as $d\eta$, where η is the spacing. If the entire sequence contains N observations, its total length is similarly $\eta(N-1)$.

The separation between two (non-adjacent) points Y_d and Y_{d+l} is termed the **lag**, where l is the number of spacing intervals. If we chop a chain into two halves, and lay the two halves exactly alongside, we are performing auto-correlation at *lag 0*. If we move one half a single link to the left or right, we are performing auto-correlation at *lag 1*, and so on.

A sequence is usually analyzed by calculating the **autocorrelation coefficient**, a ratio of covariance to variance which is a direct analogy of Pearson's r linear correlation coefficient (§13a):

$$r_l = \frac{\text{autocov}_l}{\text{var}_Y} = \frac{\sum_{i=1+l}^{i=N} Y_i Y_{i-l} - \overline{Y_i Y_{i-l}}}{\sum_{i=1}^{i=N} (Y_i - \overline{Y})^2} = \frac{\sum_{i=1+l}^{i=N} Y_i Y_{i-l} - \sum_{i=1+l}^{i=N} Y_i \sum_{i=1+l}^{i=N} Y_{i-l}}{\sqrt{\left[\sum_{i=1+l}^{i=N} Y_i^2 - \left\{\sum_{i=1+l}^{i=N} Y_i\right\}^2\right]\left[\sum_{i=1+l}^{i=N} Y_{i-l}^2 - \left\{\sum_{i=1+l}^{i=N} Y_{i-l}\right\}^2\right]}} \quad [16.5]$$

As its name suggests, r_l measures the correlation of a sequence with itself, at lag l. A plot of r_l against l, known as a **autocorrelogram** (Fig.16.2 & 3b), is usually then calculated for $0 < l \leq N/4$. Autocovariance (the numerator in [16.5]) can be plotted in the same way, to generate an **autocovariogram**, but the units are then squares of the original measurement units, which makes it more difficult to compare different plots. The autocorrelation coefficient, by analogy with Pearson's r,

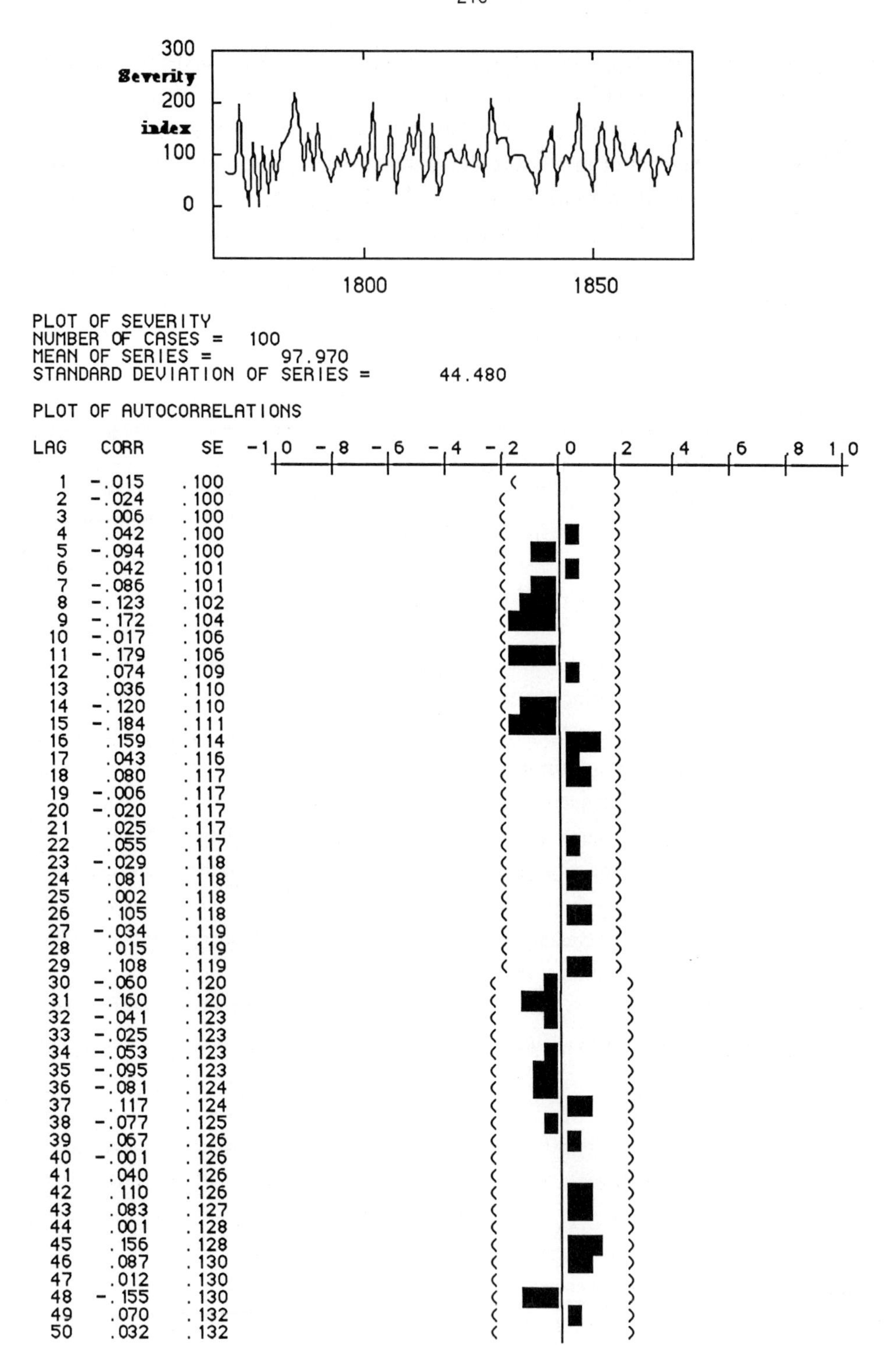

Fig.16.3. Auto-correlation example: index of worldwide earthquake severity, 1770–1869 (Davis 1986, p.225). Upper figure gives original time-series; lower gives correlogram [NB: ( ) give positions of 2 standard errors around zero]

represents the autocovariance on *standardized* data, and so has the same range (–1 to +1) for all data. Because $r_l = r_{-l}$, correlograms are usually given only for $l > 0$.

Autocorrelograms are assessed by comparison with theoretical models (Fig.16.2). If there is no periodicity (no relationship between Y_i and Y_{i+l} for any l), then $r_l = 0$ for all l except near 0 (Fig.16.2b), while a sine wave will generate a cosine curve (Fig.16.2a), and so on. It is usual to distinguish **stationary** signals (no significant trend) from *non-stationary*. Note that all auto-correlograms start at $r = 1$ for $l = 0$ ($r_0 = 1$), since the sequence at zero lag is being compared exactly with itself. Again by analogy with significance testing of Pearson's r, any individual r_l value can be significance tested using the fact that its expected variance in a random sequence is $(N - l + 3)$. In fact:

$$Z = r_l\sqrt{N - l + 3} \qquad [16.6]$$

is a *conservative* test of H_0: $r_l = 0$, provided N is 'large' (say > 50) and l 'small' (say < $N/4$). These limitations apply because the theory assumes the autocorrelation is a sample from an infinitely long sequence but, as l becomes large relative to N, r_l is in fact based on fewer and fewer measurements.

Fig.16.3 works Davis' (1986) unsolved exercise on periodicities in worldwide seismic activity. The original time-series suggests slight periodicity with much random noise; the correlogram indicates periodicity a little more clearly, although the auto-correlation coefficients are barely significant.

16b. Comparing/correlating two sequences with one another

One dream of many geologists in the mining and exploration industries is for an computerized way of correlating borehole sequences. While in principle perfectly possible, there are in practice major obstacles to the realization of this dream (Mann 1979): e.g. gaps and breaks (faults, unconformities, disconformities, nonconformities), which can easily lead the computer 'astray'. As usual, therefore, we here deal only with the very basic principles for statistical correlation of two sequences.

16b1. Comparing two sequences of nominal (multistate) data: CROSS-ASSOCIATION

Further geological discussion: May & Jones (1982); Nosal & Vrbik(1982); Vrbik(1985).
Worked geological examples: *Stratigraphy* (Davis 1986, p.236).
Selected geological applications: *Sedimentology* (Read 1976; read & Sackin 1971).
Stand-alone computer programs: Davis (1973, p.252).
Proprietary software: see under auto-association (§16a5).

Cross-association performs the same task as auto-association (§16a5), but between *two different* sequences (e.g. boreholes). It requires the *order* in which the successive observations occur, but does *not* require equal spacing. Let us compare the same sequence as in §16a5 (12323432134) with another sequence of these 4 lithologies (213424312431). As before, we move the sequences past one another to obtain cross-associations (and a cross-associatogram if required):

```
v Match position 1.......Match position 4.....Match position 7
  12323432134            12323432134          12323432134
            213424312431           213424312431     213424312431
m       0                     4                    1
```

If μ, by analogy with §16a5, represents the number of different states (here, $\mu = 4$), X_{Ai} the number of occurrences of state i in borehole A (here, $X_{A1} = 2, X_{A2} = 3, X_{A3} = 4, X_{A4} = 2$), n_1 the total number of positions in sequence A ($\Sigma X_{Ai} = n_1 = 11$ here), and similarly X_{Bi}, n_2, then the probability of a match at any position in *random* sequences with the same lengths as A and B is given by:

$$p = \frac{\sum_{i=1}^{i=\mu} X_{Ai}X_{Bi}}{n_1 n_2} \qquad [16.7]$$

As in auto-association, the expected number of matches and mismatches between the two random sequences for v overlapping points are as usual pv and $v(1-p)$.

Lithology	X_{Ai}	X_{Bi}	$X_{Ai}X_{Bi}$
Limestone	2	3	6
Sandstone	3	3	9
Shale	4	3	12
Coal	2	3	6
	$n_1 = 11$	$n_2 = 12$	$\Sigma X_{Ai}X_{Bi} = 33$

For the above example, we obtain $p = 33/11\text{x}12 = 0.25$. Calculating χ^2 as in [16.4], for match position 4 we have O = 4, E = 1, O* = 0, E* = 3, which yields:

$$\chi_1^2 = \frac{(O - E - 0.5)^2}{E} + \frac{(O^* - E^* - 0.5)^5}{E^*} = \frac{(4 - 1 - 0.5)^2}{1} + \frac{(0 - 3 - 0.5)^2}{3} = 10.3$$

which is significant at the 99.5% level, so we conclude that the two sequences match better, at position 4, than can be expected to occur in random sequences of equivalent length. The position of 'best' match among all 12 possible match positions is identified as that with the highest (significant) value of χ^2. Of course, if none of the 12 χ^2 values is significant, the two sequences are taken to be wholly dissimilar.

Variants of the coefficient of cross-association have been advocated: for example, May & Jones(1982), recommend a *dissimilarity coefficient* which allows data measured on an ordinal scale to be compared. Vrbik(1985) favours a statistic based on numbers of *runs* of matches between the two sequences; this seems to detect nonrandom similarities better.

16b2. Comparing two sequences of ratio data: CROSS-CORRELATION

Further geological discussion: Southam & Hay(1978).
Worked geological examples: *Earthquake prediction* (Davis (1986, p.228).
Selected geological applications: *Sedimentology* (Shaw & Cubitt 1979).
Stand-alone computer programs: Davis (1973, p.245); Gordon(1980).
Proprietary software: in SYSTAT and various specialist packages; also in NAG(routines G13BCF etc).

Cross-correlation performs the same task as auto-correlation (§16a6), but between *two different* sequences (e.g. two boreholes). This enables the degree of similarity of the two sequences at a given lag (offset) between them, at their position of maximum similarity, to be evaluated. In this case, 'zero lag' is undefined, unless a definite marker band or point in time exists in both sequences, and $r_l \neq r_{-l}$ (the cross-correlogram is *not* symmetrical about $r_l = 0$), so plots are usually given for $-N/4 < r_l <$

$N/4$ (Fig.16.4). When sequences A and B are of different lengths, the procedure often consists of moving a distinctive subsection of A past the whole of B, to determine the position of best match.

The cross-correlation coefficient for match position m has an equivalent formula to Pearson's r:

$$r_m = \frac{N^*\sum Y_A Y_B - \sum Y_A \sum Y_B}{\sqrt{\left[N^*\sum Y_A^2 - \left\{\sum Y_A\right\}^2\right]\left[N^*\sum Y_B^2 - \left\{\sum Y_B\right\}^2\right]}} = \frac{\text{cov}_{A,B}}{s_A s_B} \quad [16.8]$$

where N^* is the number of overlapping positions between sequences A and B, and the summations extend only over this overlapping section. Note that in this formula the denominator changes with m, whereas in the auto-correlation coefficient formula [16.5] it is constant, being based on the variance of the (one) entire sequence. Testing r_m for significance is based on the same transformation to Student's t as for Pearson's r in [13.2], or we can merely enter standard tables of Pearson's r for $[N^*-2]$ df.

The main *statistical* difficulties in achieving the oil geologists' dream of using cross-correlation to perform automated borehole correlations are these: (1) measurements in boreholes can rarely be taken at discrete, uniformly spaced points; (2) measurement intervals in different boreholes are rarely the same; (3) we cannot sample at equally spaced points in geological time; (4) even if we are able instead to sample at equally spaced points along a bore, the intervals do not correspond to equal intervals in time because sedimentation rates are not constant. In effect, comparing two boreholes by cross-correlation is like comparing two chart recorders which were running at different, unknown speeds, and were switched off for a large but equally unknown portion of the time.

Nevertheless, there are circumstances in which cross-correlation offers the only hope for a solution. Fig.16.4 works Davis' (1986) unsolved problem concerning measurements of varve thicknesses in an Eocene lake deposit (the Green River dolomitic oil shales). The shorter sequence is believed to be equivalent to some part of the larger, logged some 15 km away, but there are no marker beds or distinctive features to identify the position of 'zero lag', so the 'best' correlation must be based on the most 'perfect' match of varve thicknesses in the two sequences. The three highest cross-correlations on Fig.16.4 (all > 2 *SE*, and so highly significant) correspond to aligning peaks **D,E,F** for sequence 1 with **X,Y,Z** for sequence 2; peaks **A,B,C** with **V,W,Z;** and finally, peaks **B,C,D** with **V,W,Z**.

16b3. Comparing two ordinal or ratio sequences nonparametrically: Burnaby's χ^2 procedure

Further geological discussion: Burnaby(1953).
Worked geological examples: *Exploration* (Miller & Kahn 1962,p.362).
Statistical tables: uses χ^2 tables.
Software: not necessary; suitable for simple manual calculation.

This procedure performs the same task as cross-correlation, but is easy enough to be calculated manually, for a 'quick check', where data are not already computerized. It can also be used where the sequences comprise only ordinal data. It is a kind of combination between contingency tables (§13d1) and the Sign test (§10c1). Let us assume that some marker band in fact exists in the data used for Fig.16.4, such that the short fragments as below of the two sequences can be directly compared at zero

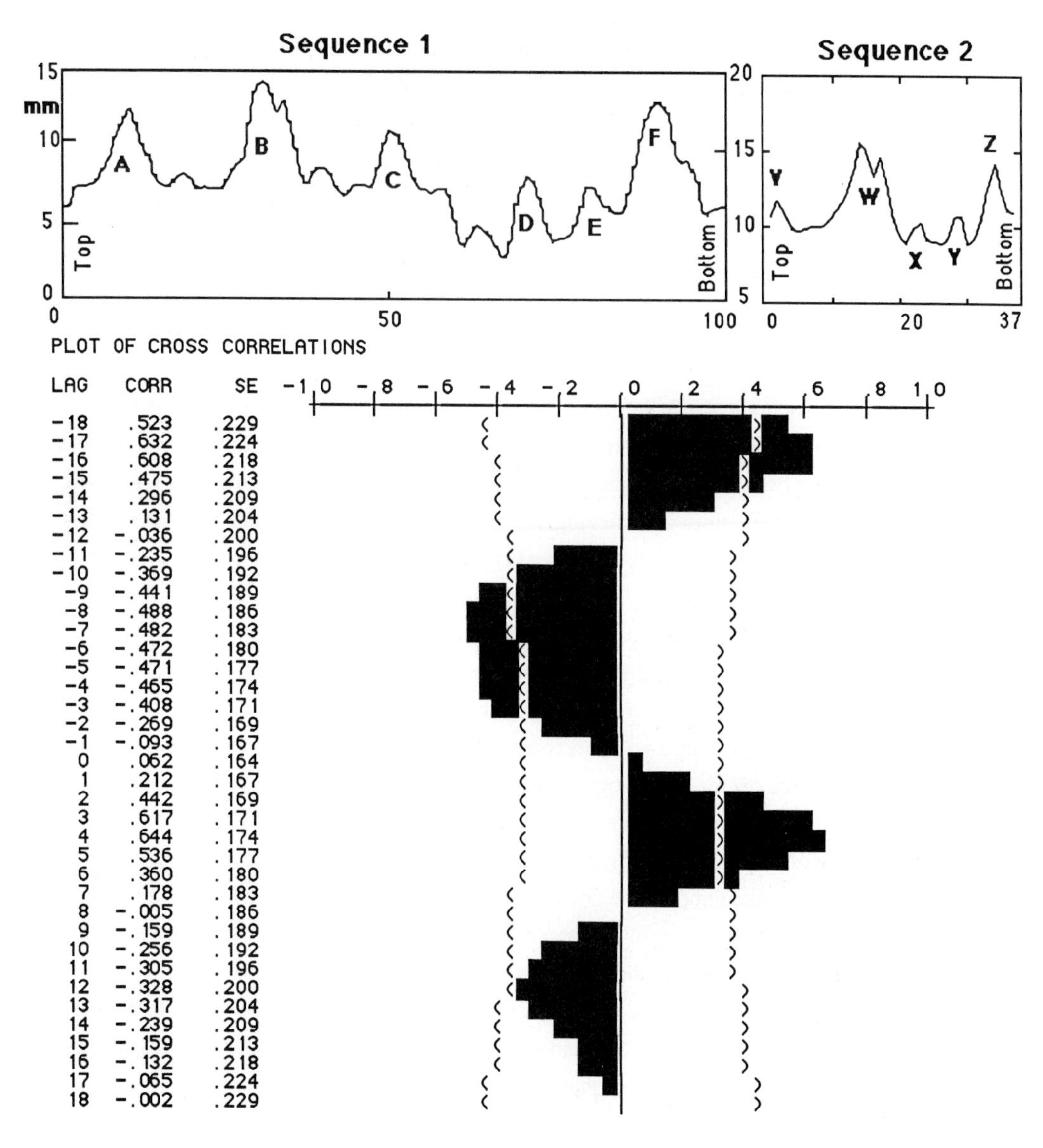

LAG	CORR	SE
-18	.523	.229
-17	.632	.224
-16	.608	.218
-15	.475	.213
-14	.296	.209
-13	.131	.204
-12	-.036	.200
-11	-.235	.196
-10	-.369	.192
-9	-.441	.189
-8	-.488	.186
-7	-.482	.183
-6	-.472	.180
-5	-.471	.177
-4	-.465	.174
-3	-.408	.171
-2	-.269	.169
-1	-.093	.167
0	.062	.164
1	.212	.167
2	.442	.169
3	.617	.171
4	.644	.174
5	.536	.177
6	.360	.180
7	.178	.183
8	-.005	.186
9	-.159	.189
10	-.256	.192
11	-.305	.196
12	-.328	.200
13	-.317	.204
14	-.239	.209
15	-.159	.213
16	-.132	.218
17	-.065	.224
18	-.002	.229

Fig.16.4. Cross-correlogram example: upper figures give varve thicknesses for two Eocene lake deposits (Davis 1986, p.224 & 233); in lower cross-correlogram, the ( ) give positions of 2 standard errors around zero

lag. Calculations then proceed as in Table 16.2, by comparing signs in quartets of values.

If there is only chance agreement between the two sequences, numbers of + and – should both equal $[N/4 - m]$ — here, 2 — where N is the overall number of points in each sequence, and m is the number of quartets giving zero difference. The more the observed numbers of + and – diverge from this expected value (and consequently, the larger χ^2), the less the similarity between the two series can be due to chance. Here, the resulting χ^2_1 is non-significant, so no significant match is assumed.

Table 16.2. Calculations for Burnaby's χ^2 on part of the data used for Fig.16.4

A	B.	Signs of cross differences		Product sign
a	b	Sign(a–d)	Sign(b–c)	Sign(a–d)(b–c)
c	d			
6.0	11.7	—	+	—
7.2	11.0			
7.1	9.9	—	+	—
7.1	9.8			
7.2	**9.9 marker band(say)**	0	+	0
7.4	7.2			
8.0	7.8	—	—	+
8.6	8.1			
10.0	7.8	+	—	—
11.4	7.1			

Summary: 3(—), 1(+), 1(0).
Expected nos. of +, – = 2 each
$$\chi^2_1 = \Sigma\frac{(O-E)^2}{E} = \frac{(3-2)^2}{2} + \frac{(1-2)^2}{2} = 1$$
This χ^2 is non-significant, with $p \approx 0.2$ indicating no significant match.

If one particular match has probability p as inferred from χ^2, its true probability is $p\,[n_A - n_B + 1]$ — there are $[n_A - n_B + 1]$ ways of comparing 2 sequences — i.e. considerably higher (less significant).

Although quick and simple for comparing two sequences of equal length, Burnaby's procedure becomes more tedious as the lengths diverge. For the original data used for Fig.16.4, where $n_A = 101$, $n_B = 37$ and there is in reality no marker band, Burnaby's procedure would have to be repeated 65 times, for all 65 possible ways of comparing the two sequences. It then has no advantages over analyzing the data via a full cross-correlation computer program.

16c. Assessing the control of geological events by past events

16c1. Quantifying the tendency of one state to follow another: transition frequency/probability matrices

Worked geological examples: *Stratigraphy* (Till 1974, p.11; Davis 1986, p.150ff).
Selected geological applications: *Sedimentology* (Allen 1970; Selley 1969).
Stand-alone computer programs: Godinho(1976a).
Proprietary software: none known.

Another way a sedimentologist might quantify a borehole sequence such as Fig.16.5, is by looking at *transitions* from one facies to another (rather than in the relative *positions* of facies) — that is, determining whether one lithology tends to follow another. One way of analysis begins by merely counting the number of different types of transition (Table 16.3a), and then dividing by the total number of transitions for a given facies (i.e. the row totals), to generate an upward-transition probability matrix (TPM: Table 16.3b). If the core in Fig.16.5 is fully representative of the parent stratigraphy, this tells us that A always follows B (p_{BA} or, strictly, $p_{A|B}$, the probability of B *given* A = 1.0), and that from facies D we will most probably pass to C ($p_{D|C} = 0.75$). Note that both matrices are generally asymmetric, i.e. $p_{ij} \neq p_{ji}$.

Another way (which may be preferable when there are disconformities or other clear breaks within particular facies), is to construct the TPM from facies transitions *at a fixed interval* up the succession (say every 10cm); this allows like→like transitions wherever the fixed interval occurs within a bed. Using this approach, if A is the facies of greatest thickness, the most likely transition from A will of course be $A \rightarrow A$. The choice of fixed interval can thus be critical: if it is too large, some transitions will be overlooked, while if too small, a surfeit of $A \rightarrow A$ transitions will be recorded and the overall TPM will merely reflect the thickest facies. Hence numerous TPMs based on fixed intervals can be generated, but only one based on facies transitions.

Table 16.3. Upward-transition frequencies (a) and probabilities (b) for the sequence in Fig.16.5

Upper / Lower\	(a) Transition frequency matrix A	B	C	D	E	Total	(b) Transition probability matrix A	B	C	D	E
A	–	0(n_{AB})	1	2	3	6	–	0.0(p_{AB})	0.167	0.333	0.5
B	5(n_{BA})	–	0	0	0	5	1.0(p_{BA})	–	0.0	0.0	0.0
C	0	1	–	1	2	4	0.0	0.25	–	0.25	0.5
D	0	1	3	–	0	4	0.0	0.25	0.75	–	0.0
E	1	2	1	1	–	5	0.2	0.4	0.2	0.2	–
Total	6	4	5	4	5	24	0.25 (p_A)	0.167(p_B)	0.208	0.167	0.208

16c2. Assessing whether sequences have 'memory': MARKOV CHAINS and PROCESSES

Dedicated monographs and reviews: Collins(1975), Cox & Lewis(1966); Cox & Miller(1965,ch.3-5); Kemery & Snell (1960); Kulback et al. (1962).
Further geological discussion: Dacey & Krumbein(1970); Harbaugh & Bonham-Carter(1970); Harbaugh & Merriam(1968); Hattori(1976), Krumbein(1975), Krumbein & Dacey (1969), Potter & Blakely (1968); Schwarzacher (1969,1975); Selley (1969); Türk(1979).
Worked geological examples: *Stratigraphy* (Till 1974,p.74; Davis (1986, p.150ff).
Selected geological applications: *Petrology* (Agterberg 1966; Whitten & Dacey 1975); *Stratigraphy* (Doveton 1971; Doveton & Skipper 1974; Ethier 1975; Hattori 1973; Lumsden 1971; Miall 1973; Osborne 1971; Read 1969); *Volcanology* (Reyment 1976; Wickman 1976).
Stand-alone computer programs: Agterberg & Cameron(1967); Godinho(1976a); Krumbein(1967); Rohrlich et al.(1985).

Processes in which the occurrence of a given state depends partly on a previous state, and partly on random fluctuations — that is, the sequence displays 'memory' — are called **Markov processes. Markov chains** are physical sequences (such a sedimentary successions) which are governed by Markov processes. **Stationary** chains display constant transition probabilities through the sequence. In **first-order chains**, the 'memory' extends back only to the previous state, whereas in ***n*th-order**, it extends n transitions back. Markov chains and processes, which are midway between **deterministic** and **probabilistic** processes, are common in earth science: for example, the weather (today's weather is often like yesterday's), and sedimentation. Examples like Fig.16.5, where we do not count transitions from one state to itself, are called **embedded Markov chains.**

The Markov property can be detected in a sequence using a χ^2 test:

$$\chi^2_{(\mu-1)^2} \approx 2\sum_{i=1}^{i=\mu}\sum_{j=1}^{j=\mu} n_{ij}\ln\left[\frac{p_{ij}}{p_j}\right] \qquad [16.9]$$

where μ = number of states (lithologies, etc., as above), and n_{ij}, p_{ij} and p_j are defined via Table 16.3. H_o is that the successive states are independent, H_1 that they are not (and hence Markovian). Strictly speaking, calculating the test for Table 16.3 requires *either* that estimates of p_{AA}, p_{BB}.... be obtained, *or* a TPM based on a fixed time interval approach be used. It also requires that all $n_{ij} > 5$. As the former is rather complicated (Davis 1986, p.157ff), we will assume for the sake of illustration that the TPM in Table 16.3b was derived from a fixed interval approach, with diagonal probabilities actually zero, and ignore the latter, whereupon we obtain (with $\mu = 5$):

$$\chi^2_{16} \approx 2\left\{0 + 0 + 1\text{xln}\left[\frac{0.167}{0.208}\right] + 2\text{xln}\left[\frac{0.333}{0.167}\right] + ...+ 1\text{x}\left[\frac{0.2}{0.208}\right] + 1\text{xln}\left[\frac{0.1}{0.167}\right] + 0\right\} \approx 38.4$$

which lies between the 99.5 and 99.9% values of χ^2_{16} (34.3 and 39.3), so we accept H_1. This is reasonable, when $p_{BA} = 1.0$ clearly indicates dependency of at least one lithology on another, although the result will be conservative because we have ignored the $n_{ij} > 5$ requirement. If p_{AA}, p_{BB}.... were actually estimated, χ^2 would lose μ further *df* in [16.9], i.e. have $\{(\mu-1)^2 - \mu\}$ *df*.

To test for a *n*th-order Markov property, we raise the first-order TPM in Table 16.3b to the *n*th power (i.e. square it for second-order), and then perform exactly the same χ^2 test. Eventually, the matrix ceases to change when multiplied one more time by itself, and the probabilities become equal to the fixed totals in Table 16.3b.

16c3. Analyzing the tendency of states to occur together: SUBSTITUTABILITY ANALYSIS

Geological discussion and worked examples: Davis (1986, p.276).
Selected geological applications: *Remote sensing* (Rosenfeld & Huang 1968); *Sedimentology* (Davis & Cocke 1972); Doveton & Skipper 1974).
Software: none known.

We showed that in Fig.16.5 lithology A always follows B. This suggests the two lithologies are related, and can substitute (proxy) for one another in a preferred succession. Analyzing this tendency towards substitutability can sometimes reveal groupings of states (lithologies) which are not otherwise obvious. The tendency is quantified via a **substitutability coefficient.** This takes several forms, of which the *left* substitutability coefficient, L_{AB}, between states A and B refers to the substitutability of the left-hand state (lower, in stratigraphy) for the right-hand (upper), and is defined as:

$$L_{AB} = \frac{\sum_{i=1}^{i=\mu} p_{Ai}p_{Bi}}{\sqrt{\sum_{i=1}^{i=\mu} p_{Ai}^2 \sum_{i=1}^{i=\mu} p_{Bi}^2}} \qquad [16.10]$$

This lies between 0 and 1 (for L_{AA}, the substitutability of a state with itself). From Table 16.3b:

$$L_{DE} = \frac{(0\text{x}0.2 + 0.25\text{x}0.4 + 0.75\text{x}0.2 + 0\text{x}0.2 + 0\text{x}0)}{\sqrt{(0.2^2+0.4^2+0.2^2+0.2^2+0)(0+0.25^2+0.75^2+0+0)}} = \frac{0.25}{0.4183} \approx 0.6$$

whereas $L_{AB} = 0$. Hence despite the fact that *A* always follows *B*, the *two* states are followed by a different two states, whereas *D* and *E* tend to be followed by the same two states (i.e. they occur in similar contexts). A left substitutability matrix can be built up from all the *L* values. **Right** and **mutual** substitutability coefficients/matrices, respectively indicating the tendency for states to be *preceded* (underlain) by, or *enclosed* by, the same states, are defined analogously. The resulting coefficients can even be used to generate a dendrogram (§21), which shows the relationship of all states to all others most clearly (Davis 1986, p.480).

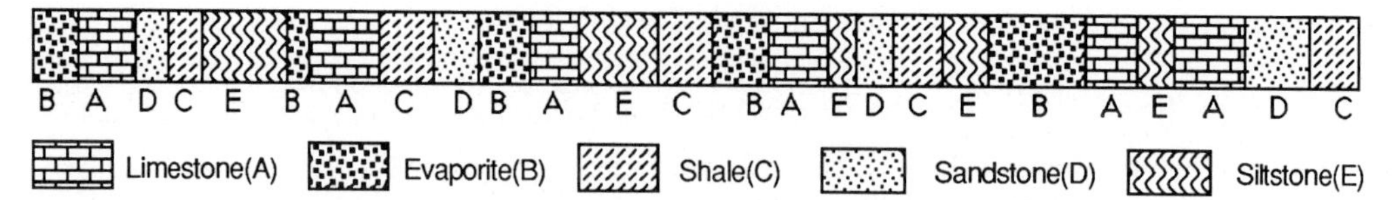

Fig.16.5. An imaginary borehole sequence for testing the Markov property

16d. Sequences as combinations of waves: SPECTRAL (FOURIER) ANALYSIS

Dedicated monographs and reviews: Bloomfield(1976); Jenkins & Watts(1968); Rayner(1971).
Further geological discussion: Haan(1977).
Further non-geological discussion: Bendat & Piersol(1971); Koch & Link (1971,p.54); Panofsky & Brier(1965);.
Worked geological examples: *Hydrology* (Davis 1986, p.248ff).
Selected geological applications: *Hydrology* (Yevjevich1972); *Palaeontology* (Kaesler & Waters 1972); *Sedimentology* (Anderson & Koopmans 1963); also very widely used in geophysics and remote sensing.
Stand-alone programs: Davis (1973, p.372).
Proprietary software: numerous packages available (see sources in §3).

Fourier(1968–1830) first showed that any continuous, single-valued function could be represented as the sum of several sinusoidal (sine, cosine) waves of different wavelengths and amplitudes (see Davis 1986, figs.4.48 & 4.54). This applies as much to complex, non-periodic sequences (such as X-ray diffractometer traces) as to intrinsically periodic ones (e.g. tidal fluctuations, seasonal variations):

$$Y = \sum_{k=0}^{k=\infty} \alpha_k \cos(k\theta) + \beta_k \sin(k\theta) \qquad [16.11]$$

k is called the **harmonic number**. Because [16.11] resembles a regression equation, the coefficients α and $ß$ can be found by least-squares (§14a), although the mathematics can be very complex. However, if an intrinsically periodic time-series is sampled at N points, the α and $ß$ are more simply given by:

$$\alpha_0 = \frac{1}{N}\sum_{i=0}^{i=N-1} Y_i; \quad \beta_0 = 0; \quad \alpha_i = \frac{2}{k}\sum_{i=0}^{i=N-1} Y_i \cos\left[\frac{2\pi ik}{N}\right]; \quad ß_i = \frac{2}{k}\sum_{i=0}^{i=N-1} Y_i \sin\left[\frac{2\pi ik}{N}\right] \qquad [16.12]$$

(the square brackets merely convert the ith position into radians). From this, we can express the proportion of the total variance in the original time-series which is contributed by each component wave-form in a **periodogram**. This can be tested statistically, and usually gives a clearer indication of the long-term periodicity of the variation than does than original time-series.

Although most geological phenomena are not intrinsically periodic, but **stochastic**, a form of spectral analysis can still be applied to them. Such analysis involves the concepts of samples and populations, but has evolved its own exquisite terminology (see **continuous spectrum, ensemble, ergodic, levelling, stationary**, in Glossary). Because the mathematics involves complex numbers ($i = \sqrt{-1}$ etc.) it will not be covered here, except to mention that two methods are available, based on **fast Fourier transforms** and on autocorrelation (§16a5).

16e. Separating 'noise' from 'signal': FILTERING, SPLINES, TIME-TRENDS

Further geological discussion & worked examples: Miller & Kahn (1962,p.355); Davis (1986, p.204ff,263ff & 378).
Selected geological applications: *Sedimentology* (Vistelius 1961); fundamental in seismology.
Proprietary software: widely available in geophysical laboratories.

Filters smooth out time-series such that short-term fluctuations ('noise') are removed and longer-term 'signals' become more evident. They necessarily attenuate the signal in the process. Filters can also search for a particular wave-form in a time-series and, having found it (in digital form), make it more (or less) conspicuous. The Earth filters in reverse the signal from a dynamite shot in reflection seismology, making it noisier, and the seismologist's task is often to remove the resulting unwanted frequencies, to pass others preferentially, and to sharpen diffuse reflections. 'Perfect' filters which remove only the unwanted noise are, of course, unattainable in practice; achieving the right blend of smoothing, without attenuating the signal itself out of existence, is a complex art to which justice cannot be done within the present scope.

Smoothing (see Davis 1986, figs.4.60-63) can be achieved in 4 main ways:

(1) **Derivatives.** Using conventional differential calculus, the *slope* dY/dX of the original time-series is obtained; a 'spike' on the original is then mirrored as a high derivative, whereas a constant portion becomes a zero derivative.

(2) **Splines.** Short segments of the original sequence can be approximated with smooth curves, called splines. These, too, involve differential calculus but are simple empirical functions — nothing more than mathematical analogues of the flexible curves used by draftsman to draw smoothly through series of points. They are commonly cubic:

$$Y = ß_1 + ß_2X + ß_3X^2 + ß_4X^3 \qquad [16.13]$$

Since cubic functions can pass exactly through up to 4 points, sequences of more than 4 points are fitted with a *series* of splines (a **spline function**), in such a way that there are no abrupt changes in slope or curvature where the individual splines meet (i.e. the first and second derivatives dY/dX and d^2Y/dX^2 are constrained to be zero). This is actually done by fitting the splines only to 2 rather than 4 points — i.e. a different spline is fitted between every pair of points on the original time-series.

(3) **Polynomials** can achieve the same end as splines, but are formally fitted by least squares, as in regression (§14a).

(4) **Moving averages** can be calculated, such as **Shepard's 5-term filter**:

$$Y_{smooth} = \frac{1}{35}\{17Y_i + 12(Y_{i+1} + Y_{i-1}) - 3(Y_{i+2} + Y_{i-2})\} \qquad [16.14]$$

These are effective because the signal varies far less between adjacent points than the noise, i.e. an average of this kind will tend to converge on the signal. This procedure is commonly known as **time-trend analysis.**

TOPIC 17. ASSESSING GEOLOGICAL ORIENTATION DATA: AZIMUTHS, DIPS AND STRIKES

17a. Special Properties of Orientation Data

Dedicated monographs & reviews: Gaile & Burt(1980); Mardia(1972,1981a,b); Pincus(1953: includes useful bibliography for 1938–52); Sugranes(1977); Watson(1966,1970,1983).
Further geological discussion: Koch & Link (1971, p.135); Cheeney (1983, ch.8); Davis (1986, p.314).

Orientation (angular) data are rarely treated in statistical texts, but are extremely important in geology in general, and, of course, fundamental in mapping, palaeomagnetic, structural and (often) metamorphic studies. The coverage below attempts to be considerably more comprehensive than that available in any of the geological texts just listed, but to be more accessible than Mardia's(1972) excellent, but highly mathematical treatise. (Readers should note that complete copies of Mardia(1972) include an extremely long *erratum*, and that many errors in the book are not included even in this *erratum*. Discrepancies between Mardia, the other listed books and the formulae below are not necessarily therefore real!)

Two classes of orientation data give rise to four statistically distinct data-types in all:

(1) *2-D (circular) versus 3-D (spherical):* Whilst orientation data are commonly 3-*D* (e.g. dips and strikes of sediments, plunges of folds, etc.), some are either intrinsically 2-*D* (e.g.azimuths of glacial striae, palaeocurrent directions, photogeological lineaments, mineral orientations within foliation planes), or can be reduced for convenience to two dimensions (e.g. azimuths of near-vertical dykes).

(2) *Axial versus directional.* **Directional** orientation data have both direction and sense, i.e. they should be represented by an arrow, and can range around the full circle of 360° (e.g. palaeocurrent, palaeomagnetic and glacial striae directions). **Axial** data have direction only, and can be represented adequately by a line within an angular range of 180° (e.g. dyke and lineament azimuths). Statisticall axial data present a difficult problem. For the present purposes, however, they can be treated exactly as directional data, if the angles are first doubled; results such as actual mean directions (rather than test statistics) are correspondingly halved (Koch & Link, 1971, p. 135; Mardia, 1972, p. 169, etc.)

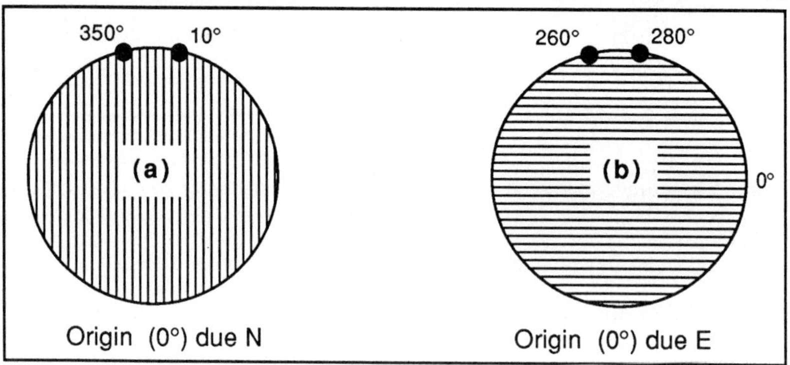

Fig.17.1. Illustrating dependence of the arithmetic mean on choice of origin for circular data

A fundamental statistical problem posed by orientation data is that calculations are sensitive to the choice of origin (Chayes 1954a). As a simple example, the arithmetic mean of 10° and 350°, expresse relative to the conventional north zero point as on Fig.17.1a, is 180°, but this is clearly *not* the logical average of these two directions. The correct answer is obtained only if 350° is rewritten as –10°. Furthermore, resetting the zero point as east, the angles become 260° and 280°, yielding the correct mean of 270°. Hence the arithmetic mean is *not* a suitable way of averaging orientation data. Otherwis expressed, orientation data are *vectors*, and cannot be summarised by *linear* statistics. All of the tests in this Topic take their vectorial properties into account (mostly via use of sines and cosines), and thereby become insensitive to the choice of origin.

An important concept with orientation data, having no linear equivalent, is that of **uniformity**. This arises because angular location and dispersion are *not* independent. If we talk geologically of angular data showing a very strong *preferred orientation*, we mean that they show a small *spread* around a well-defined *location* (mean direction). As the spread increases over the circle or sphere, the preferred orientation becomes less and less meaningful, and eventually vanishes when dispersion reaches 360°; the distribution is then **uniform** (continuous, random, with no preferred orientation).

It is worth mentioning *en passant* that orientation statistics can also be used to analyze *cyclical* (e.g. annual, seasonal) data: for example, 0–90° = Spring; 90–180° = Summer, etc.

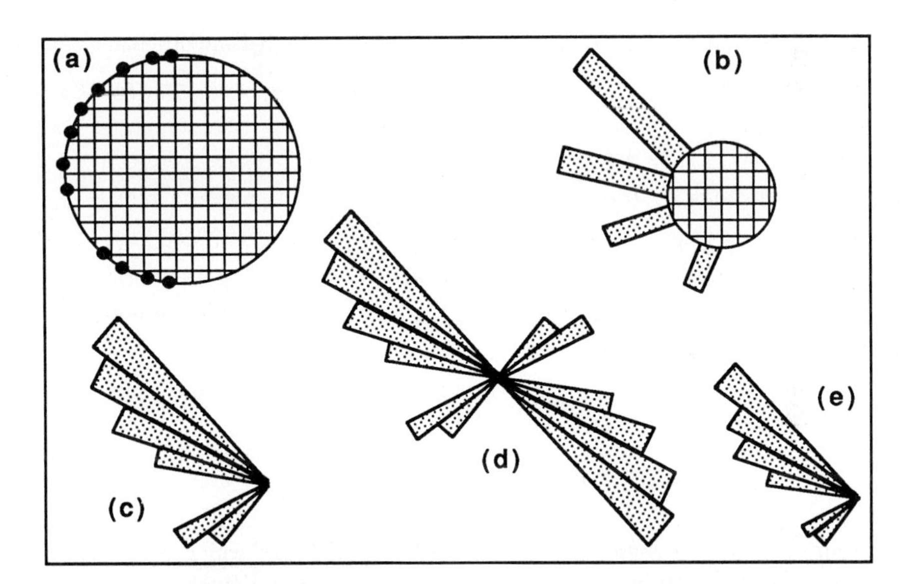

Fig.17.2. Graphical means for displaying circular (2-*D*) orientation data
(**a**) circular plot; (**b**) circular histogram; (**c**) rose diagram for directional data (radius of segments proportional to class frequencies); (**d**) as (c) but as conventionally mirrored for axial data; (**e**) as (c) but with radius proportional to frequencies (generates 'stubbier' and probably preferable diagram)

17b.Describing distributions of 2-dimensional (circular) orientation data

17b1.Graphical display

Further discussion: Cheeney(1983,p. 22); Fields & Egemeter (1978); Mardia(1972, p.1ff).
Stand-alone computer program: Srivastava & Merriam (1976).

Conventional means of displaying circular data are compared in Fig.17.2. Opinions differ as to whether Fig.17.2c is a reasonable or absurd representation, as compared with Fig.17.2e.

17b2. Circular summary statistics

Further discussion: Mardia(1972, ch.2)

Circular analogues can be derived for all the linear statistics (§8): mean, median, mode, variance, skew, kurtosis, etc. Only widely used statistics are covered below. The **circular mean** is defined as:

$$\bar{\theta} = \cos^{-1}\left\{\frac{C}{R}\right\} = \sin^{-1}\left\{\frac{S}{R}\right\}, \text{ where } C = \frac{\sum_{i=1}^{i=N}\cos\theta_i}{N},\ S = \frac{\sum_{i=1}^{i=N}\sin\theta_i}{N},\ R = \sqrt{C^2 + S^2} \quad [17.1]$$

It has similar properties to the linear mean — the sum of deviations about it equals zero, and it minimizes a measure of dispersion — but is not affected by the choice of origin, as the arithmetic mean is with angular data. R in [17.1] is called the **mean resultant length.** It is itself a measure of dispersion, lying between 0 and 1, but takes *high* values for *small* dispersions (i.e. tight clustering about the mean). The **circular variance** can therefore be simply defined as $[1 - R]$, but it is not widely used.

The **circular median**, which divides the data in two, is best obtained via a circular plot: the linear median of the numbers 33, 45, 57, 75, 88, 180, 353 is 75, but the circular median is 57° (Fig.17.3):

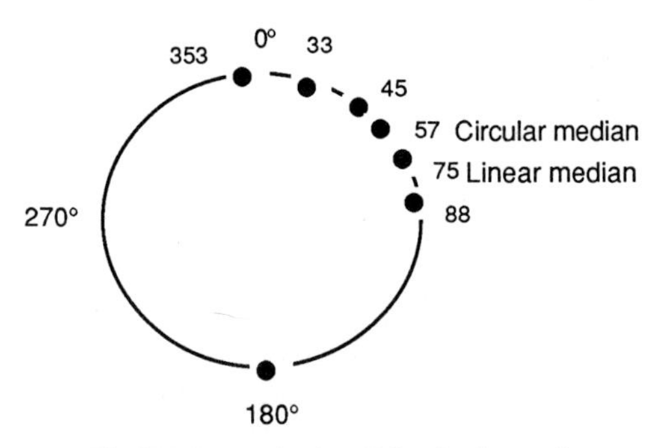

Fig.17.3. Determination of the circular median

The **circular range** is simply the length of the smallest arc which encompasses all the observations, and can vary from 0 to 360° (for a uniform distribution). In Fig.17.3 it is 187°.

17b3. Circular data distributions

Further discussion: Cheeney(1983,p.98ff); Jones (1968); Mardia(1972, ch.3).
Statistical tables: Mardia(1972, p.290ff).

The Normal Distribution on the line does not have a single exact equivalent on the circle. Rather, there are *two* distributions, which jointly (but not individually) possess the statistical properties of the linear Normal. If we 'wrap' a Normal Distribution curve around the circumference of a circle, we obtain the **Wrapped Normal Distribution,** which has the following equation:

$$f(\theta) = \frac{1}{\sigma\sqrt{2\pi}} \sum_{j=-\infty}^{j=+\infty} \exp\left\{ \frac{(\theta + 2\pi j^2)}{2\sigma^2} \right\}, 0 \le \theta \le 2\pi \qquad [17.2]$$

The **Von Mises distribution** has the following equation, and the form in Fig.17.4:

$$f(\theta) = \frac{1}{2\pi I_o(k)} e^{k\cos(\theta-\theta_o)} \text{ where } I_o(k) = \frac{1}{2\pi}\int_0^{2\pi} e^{k\cos\theta} d\theta \qquad [17.3]$$

θ_0 is the mean direction, k is called the **concentration,** and $I_o(k)$ is a complicated constant dependent on the concentration called a **Bessel function,** which has the following formula:

$$I_0(k) = \sum_{j=0}^{j=\infty} \frac{(\frac{1}{2}k)^{2j}}{j!^2} \qquad [17.4]$$

The concentration k acts for the Von Mises distribution like the standard deviation for the Normal Distribution, but in reverse: it is an *inverse* measure of spread, with *high* concentration indicating, as the name is of course meant to imply, *low* spread (Fig.17.4b). It is sometimes called "precision" or "estimate of precision" (e.g. Koch & Link 1971, p.137). We indicated above that the resultant length of a sample is also an inverse measure of spread, so it is not therefore surprising that k and R are directly related, for the Von Mises case. In fact, their relationship is mathematically complex, involving **maximum likelihood** estimation, but tables are available (e.g. Davis 1986, p.323; Mardia, 1972, p.298), together with approximations (to 2 sig.figs.) such as (Cheeney, 1983, p.100):

$$\text{For } R < 0.65,\ k = \frac{R}{6}(12 + 6R^2 + 5R^4); \quad \text{for } R \ge 0.65,\ k = \frac{1}{2(1-R) - (1-R)^2 - (1-R)^3}, \qquad [17.5]$$

Whereas the wrapped Normal distribution is the more natural adaptation of the linear Normal Distribution onto the circle (a variant form of the **central limit theorem** on the circle leads to it, and it possess such properties as additivity), the Von Mises is governed by θ_0 and k in more exactly the way that μ and σ govern the linear Normal. Because it also leads to more tractable statistical tests, 'parametric' tests for circular data in this Topic assume that the parent population is Von Mises.

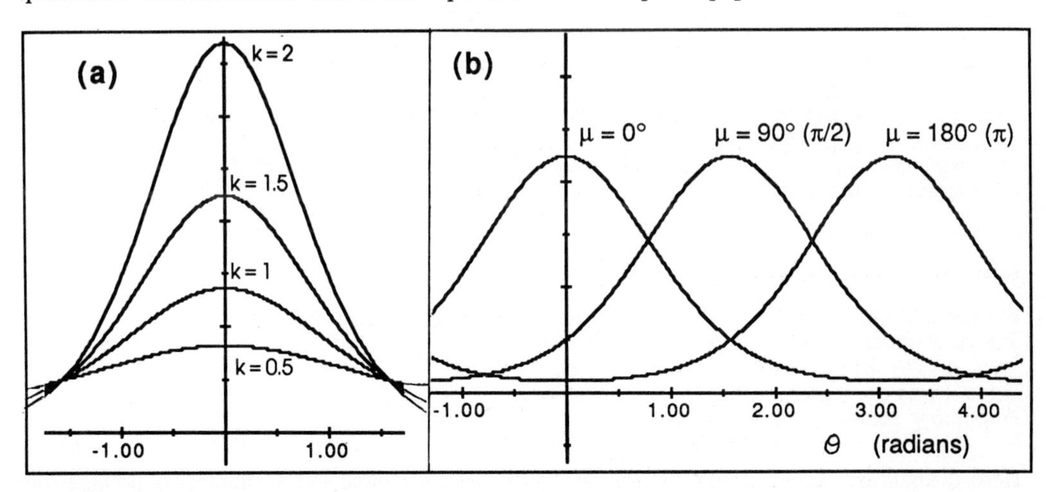

Fig.17.4. The Von Mises distribution: variation with (a) concentration k; (b) with mean direction μ (cf. Fig. 8.1).

Tests and computer programs analogous to Normality tests for linear data (§8e3) are available for assessing goodness-of-fit to a Von Mises or wrapped Normal distribution (Harvey & Ferguson 1976; Upton 1973). Analogous methods for separating mixed or multi-modal distributions (cf. §8f3) are also known (Hsu et al.1986; Jones & James 1969; Spurr 1981).

17c.Testing for uniformity versus preferred orientation in 2-D orientation data

Dedicated review: Dudley et al. (1975).

In analyzing angular data, it is important to test for uniformity, since a uniform distribution has no meaningful location, and further statistical analysis is then generally meaningless. Many tests for uniformity have been suggested. Three popular nonparametric tests and one parametric test specifically developed for angular data are covered below. Because these examine different properties of the group distribution, their results need not be the same: for example, Cheeney (1983) found 36 fossils to be uniformly oriented (at the 95% level) using Hodges-Ajne's and Kuiper's tests, but non-uniformly using Watson's and Rayleigh's tests. In extensive checks (summarised by Mardia 1972, p.194-5), **Kuiper's test** (§17c2) proved preferable for small group sizes, while the **Rayleigh test** (§17c4) proved most powerful where the parent population distribution was Von Mises.

Other uniformity tests, such as the **Equal spacings**, and **Range** tests (Mardia 1972) have no particular advantages over those covered below. The χ^2 test used by some geologists, and included by Davis(1986,p.326), is specifically criticised by Ballantyne and Cornish(1979) and Bayer (1985) as being sensitive to the choice of origin, and therefore wholly unsuitable for angular data.

Table 17.1. Quick guide to statistical tests for orientation data covered in Topic 17

	2-dimensional (circular) data		*3-dimensional (spherical) data*
	Parametric tests	Nonparametric tests	Parametric tests¶
Uniformity (one-sample) tests	**Rayleigh test** (§17c4)	**Hodges-Ajne test** (§17c1) **Watson's U^2 test** (§17c2) **Kuiper's test** (§17c3)	**Rayleigh test** (§17i)
2-sample tests	**Watson-Williams test** (§17e4)	**Runs,Watson's U^2 tests** (§17e2-3) **Mardia Uniform Scores test**(§17e1)	**Watson-Williams test** (§17j)
q-sample tests	**Watson-Williams test** (§17f2)	**Mardia's Uniform Scores test**(§17f1)	**Watson-Williams test** (§17k)

¶Virtually no nonparametric tests for 3-dimensional data are available; Mardia(1972,p.282) mentions a single nonparametric test for uniformity, which is omitted here because of its obscurity.

17c1. A simple nonparametric test: the Hodges-Ajne Test

Worked examples: Mardia (1972, p.186); *Palaeontology:* Cheeney (1983, p.94).
Statistical tables: $10 < N < 50$: Mardia (1972, p.309); Cheeney (1983, p.94); $N > 50$ approximate critical value = $(N-2M_{min})/\sqrt{N}$.
Stand-alone computer programs: Rock (1987b).
Proprietary software: none known.

Here, the data are examined on a circular plot. A diameter is plotted and the number of points to one side of it, M, counted. Different diameters are a chosen until the *smallest* value of M (M_{min}) is found;

low M_{min} indicates a preferred orientation (Fig.17.5a). The test is not very sensitive, and for $N < 10$ in particular, critical values are indistinguishable from zero, so no significance level can be determined. Like other graphically-based tests, this one does have the advantage that it can be carried out if the data are available only as a circular plot or histogram, and not as raw data.

The dyke-swarm data in Table 17.2 all give significant results, indicating preferred orientations.

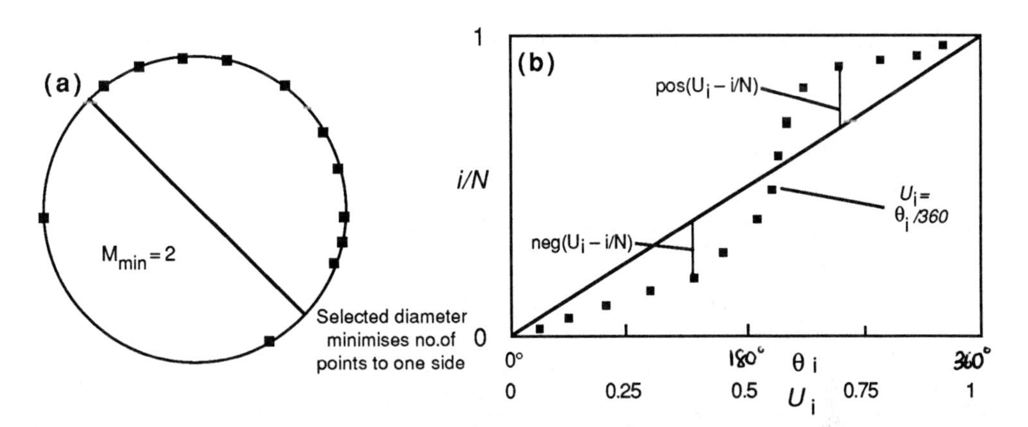

Fig.17.5. Method of calculating test statistics for the (a) Hodges-Ajne; and (b) Kuiper tests for uniformity

17c2. A more powerful nonparametric EDF test: Kuiper's Test

Worked examples: Lewis (1977, p.229); Mardia (1972, p.178); *Palaeontology* (Cheeney 1983, p.95).
Statistical tables: $5 \le N \le 17$: Mardia (1972, p.308); Cheeney (1983, p.97). $N > 17$ use (17.2). See also Abrahamson (1967).
Stand-alone computer programs: Rock (1987b).
Proprietary software: none known.

This, an angular analogue of the linear Kolmogorov test (§8e2), is based on the difference between the maximum, +ve *and* –ve deviations of the data distribution from a uniform distribution (Fig.17.5b):

$$V_N = \text{pos}\left[U_i - \frac{i}{N}\right] - \text{neg}\left[U_i - \frac{i}{N}\right] + \frac{1}{N} \qquad [17.6]$$

where $U_i = \theta_i /360$ for each angular measurement θ_i, arranged in numerical order. For $N > 8$, the critical value V_c is calculated using the following formula (Cheeney 1983, p.97):

$$V_N^* = V_c\left[\sqrt{N} + 0.155 + \frac{0.24}{\sqrt{N}}\right] \qquad [17.7]$$

where V^*_N = 1.620, 1.747, 1.862 and 2.001 for the 90, 95, 97.5 and 99% levels (Mardia 1972).

All 3 dyke-swarms in Table 17.2 again give significant results, indicating preferred orientations.

17c3. A powerful nonparametric test: Watson U^2 Test

Worked examples: Mardia (1972, p.182); *Palaeontology* (Cheeney 1983, p.97).
Statistical tables: Burr (1964); $N > 7$: Cheeney (1983, p.98); Mardia (1972, p.182); Stephens(1970).
Stand-alone computer programs: Rock (1987b).

This is an angular analogue of the linear **Cramér-von Mises** test. The test statistic U^{*2} is calculated from the formula:

$$U^{*2} = \left[1.0 + \frac{0.8}{N}\right]\left[\sum_{i=1}^{i=N}\{U_i^2\} - \frac{2}{N}\sum_{i=1}^{i=N}\{iU_i\} + \overline{U} + N\left\{\overline{U} - \overline{U}^2 + \frac{1}{12}\right\} - \frac{0.1}{N} + \frac{0.1}{N^2}\right] \quad [17.8]$$

where U_i is defined in the previous section, and $\overline{U}$ is the mean U_i, i.e. $\{\Sigma U_i\}/N$.

All 3 dyke-swarms in Table 17.2 again give significant results, indicating preferred orientations.

Table 17.2. Data for 2-D examples in Topic 17a-f (azimuths for vertical dykes from three swarms)

	Swarm 1		Swarm 2		Swarm 3	
Raw data	Azimuth	Rank	Azimuth	Rank	Azimuth	Rank
	113	6	065	5	102	4
	141	11	74	7	113	5
	126	9	088	9	125	7
	095	3	090	10	096	2
	105	5	039	2	138	8
	114	7	061	4	175	12
	123	8	070	6	101	3
	149	12	036	1	091	1
	088	2	052	3	147	11
	081	1	095	11	139	9
	102	4	076	8	140	10
	156	13			118	6
	135	10				
Summary statistics						
n_i	13		11		12	
Mean azimuth	117.5°		67.9°		123.4°	
k (estimated)	6.9366		9.7077		6.2433	
Uniformity tests on individual swarms	Statistic	% sig.level	Statistic	% sig.level	Statistic	% sig.level
Hodges-Ajne	0	>95	0	>90	0	>95
Kuiper's test	0.7917	>99	0.8361	>99	0.7667	>99
Watson's U^2	0.7018	>99	0.6567	>99	0.6400	>99
Rayleigh test	0.9249	>99.9999	0.9470	>99.9999	0.9161	>99.9999

17c4. The standard parametric test: Rayleigh's Test

Further geological discussion: Curray(1956); Durand & Greenwood(1958); Pincus(1956).
Worked examples: Mardia (1972, p.133); *Glaciology* (Davis 1986, p.325); *Palaeontology* (Cheeney 1983, p.100).
Statistical tables: $4 \leq N \leq 100$: Mardia (1972, p.300); Davis(1986, p.324). For larger N, $2NR^2$ is distribut as χ^2; exact probability of R for any N given by (17.6); critical values can also be approximated by $\sqrt{\{3/N\}}$ and $\sqrt{\{4.61/N\}}$ at the 95% and 99% levels, where $N > 15$ (Cheeney 1983, p.101).
Stand-alone computer programs: Rock (1987b).
Mainframe and microcomputer software: none known.

This parametric test (which assumes that the parent population is Von Mises) is simply based on the **mean resultant length** of a group of data, R, as defined in [17.1]. (**NB:** some texts such as Davis (1986) use both R for resultant length, i.e. [17.1] without the N divisors, and $\overline{R}$ for the *mean* resultant length, but as this can cause confusion, only R, R_1, R_2, referring throughout to *means*, i.e.

values between 0 and 1, are used throughout this Topic). In addition to looking it up in tables, the probability of a value $\kappa \geq 2NR^2$ can be calculated more exactly as follows (Mardia 1972, p.135):

$$p = e^{-\kappa}\left[1 + \frac{2\kappa - \kappa^2}{4N} - \frac{24\kappa - 132\kappa^2 + 76\kappa^3 - 9\kappa^4}{288N^2}\right] \qquad [17.9]$$

This of course gives the probability for rejection of the null hypothesis of uniformity.

The Rayleigh test can also be used with directional data to test whether an *axis* exists (i.e. that there are *two* preferred orientations at 180° to one another). This could obviously be useful in palaeocurrent or glaciological studies, to check for a palaeotrough direction. This is done (Mardia 1972, p.168) simply by doubling the angles (i.e. treating the data as if they were axial) and adjusting so that all measurements are in the 0–180° range. Thus 20° becomes 40°, and 90° becomes 180°, as 120° becomes (2x120 – 180) = 60°. The test is then performed as usual.

All 3 dyke-swarms in Table 17.2 again give significant results, indicating preferred orientations.

17d. One-group tests for mean directions and concentrations

Further discussion and worked examples: Davis (1986, p.325); Durand & Greenwood(1958); Mardia(1972,p.137ff).
Stand-alone programs: Rock(1987b).

If the tests of the previous section establish that data-sets are uniformly distributed, this is equivalent to establishing that $k = 0$ and that, consequently, there are *no* mean directions. There is then no point in further statistical testing. If, on the other hand, data-sets prove to be non-uniform ($k \neq 0$), we can proceed to one-, two- and multi-group tests which are analogues of those in §8-11 for linear data.

The most obvious use for one-group tests would be to determine whether the mean direction $\bar{\theta}$ of an angular data-set has a pre-specified value: for example, whether some acicular crystals display preferred orientations parallel to a *known* regional strike direction, μ. The simplest method is to use R which, being measure of dispersion, can be employed to estimate confidence limits for a circular mean. This can be done either using **Batschelet charts** (e.g. Mardia, 1972, p. 302-3), which give a ±° figure directly for given $\bar{\theta}$ and N, or by estimating k from R using [17.5], and then calculating the standard error of $\bar{\theta}$: $s_e = 1/\sqrt{\{NRk\}}$. If the pre-specified strike direction μ lies within the confidence limits $[\bar{\theta} \pm t_{\alpha,N-1}s_e]$, i.e. for moderate sized data-sets $[\bar{\theta} \pm 1.96/\sqrt{NRk}]$ at the 95% level, $\bar{\theta}$ and μ can be equated. This is analogous to one-group *t*-tests and parametric confidence limits for linear data (§8g & §9e).

Tests are also available whether a given angular data-set could come from a population with specified μ *and* κ (the same problem but with κ now known, and analogous to the *zM* test of §8g1), and for hypothesis tests about k and κ themselves (analogous to the F tests of §10a1). Confidence limits for k, for example, can also be read off further Batschelet charts (e.g. Mardia 1972, p.304-5). These tests use tables and distribution functions of the Von Mises Distribution in the same way that the linear tests use tables and distribution functions of the Normal Distribution. See Mardia(1972).

17e. Comparing two groups of 2-dimensional data

Here, the object is to determine whether two groups of angular data might have come from the same population. Once again, uniformity tests on each group (§17c) are a necessary prerequisite. If all distributions prove to be uniform, for example, identity of parent populations is already suggested. Conversely, if one or more distributions prove uniform, but others non-uniform, parent populations cannot be the same. In either case, two- or multi-group tests may be redundant. The assumptions of *parametric* tests in particular are unlikely to have been met in such cases, though nonparametric tests may still have some value in comparing overall shapes of the group distributions. All parametric two- (and multi-)group tests for orientation data assume that the data are drawn from Von Mises distributions, and have equal concentrations (just as two-group *t*-tests and multi-group ANOVA assume Normal distributions with equal standard deviations for linear data). The circular tests are also based on approximations, however, and have different procedures depending on the value of k or R.

Table 17.3. Intermediate calculations for Mardia's 2-sample uniform scores test on data of Table 17.2

Dyke-swarm 1				Dyke-swarm 2		
Azimuth	Azimuth–50°	*Combined* rank,®	Uniform score, υ	Azimuth	*Combined* rank,®	Uniform score, υ
113	63.	10.0	2.618	65.	12.0	3.141
141	91.	21.0	5.498	74.	15.0	3.927
125	76.	16.5	4.320	88.	19.0	4.974
95	45.	5.0	1.309	90.	20.0	5.235
105	55.	8.0	2.094	39.	4.0	1.047
114	64.	11.0	2.880	61.	9.0	2.356
123	73.	14.0	3.665	70.	13.0	3.403
149	99.	23.0	6.021	36.	2.0	0.524
88	38.	3.0	0.785	52.	6.5	1.702
81	31.	1.0	0.262	95.	22.0	5.760
102	52.	6.5	1.702	76.	16.5	4.320
156	106.	24.0	6.283 = 2π			
135	85.	18.0	4.712			

17e1. A nonparametric test: Mardia's uniform scores

Worked geological examples: Mardia (1972, p.200); *Glaciology* (Lewis 1977, p.231).
Statistical tables: $N \leq 20$: Lewis (1977, table 18); Mardia (1972, p. 312). For $N>20$, use (17.10).
Stand-alone computer programs: Rock (1987b).
Mainframe and microcomputer software: none known.

This nonparametric test is broadly analogous to the Mann-Whitney test for linear data (§10d3). It involves converting each angular measurement to a **uniform score**, $\upsilon_i = 2\pi®_i/N$, where $®_i$ is the rank of the measurement in the two groups *combined*. These uniform scores behave like angles in radians, ranging up to 2π (Table 17.3). The statistic is the resultant length of the uniform scores for the *smaller* group, R_1 (that is, replace θ_i in [17.1] with υ_i and sum only to n_1), and is tested via:

$$\chi_2^2 = \frac{2(N-1)R_1^2}{n_1 n_2} \qquad [17.10]$$

A variant is available to test the null hypothesis of a *bimodal* distribution (Mardia, 1972).

We might wish to test that the preferred orientations of dyke-swarms 1 and 2 in Table 17.2, differed by 50°; that is, test H_o: $\theta_1 - 50 = \theta_2$ against H_1: $\theta_1 - 50 \neq \theta_2$. To do so, we subtract 50° from each azimuth for swarm 1, combine the two sets of data and rank the combined data, as in Table 17.3. This yields $C_2 = -0.9025$, $S_2 = -1.1572$, $R^2{}_2 = 2.1535$, $\chi^2{}_2 = 0.6927$, which is not significant, so H_o is accepted. Had the hypothesis been that the two preferred orientations were *equal,* rather than differing by a pre-set amount, H_o would have been strongly rejected.

17e2. An alternative nonparametric test: Watson's U^2

Theoretical discussion: Beran(1969); Maag(1966); Stephens(1965); Wheeler & Watson(1964).
Worked non-geological example: Mardia (1972, p. 202).
Statistical tables: $8 < N < 17$: Mardia(1972,p.314). Asymptotic values used for larger *N.*
Stand-alone computer programs: Rock (1987b).
Mainframe and microcomputer software: none known.

This is an extension of the one-group test (§17c3). Let i, j, ®_i and ©_j be the rank orders of measurements respectively within group 1 alone (i) , within group 2 alone (j), and within groups 1 (®_i) and 2 (©_j) for the two groups *combined*, then (Stephens 1965):

$$U^2 = \frac{1}{N}\left[\frac{Q}{n_1 n_2} - \frac{4n_1 n_2 - 1}{6}\right] \qquad [17.11]$$

$$\text{where } Q = n_1\sum_{i=1}^{i=n_1}(\text{®}_i - \text{i})^2 + n_2\sum_{j=1}^{j=n_2}(\text{©}_j - \text{j})^2 - \left[\sum_{j=1}^{j=N}(\text{©}_j - \text{j}) - \frac{n_1 n_2}{2}\right]^2 \qquad [17.12]$$

(note that supposedly equivalent formula 7.4.13 in Mardia 1972 is ambiguous).

The same test as in §17e1 yields $U^2{}_{13,11} = 0.0344$, which is not significant; H_o is again accepted. Once again, if H_o were that the azimuths were *equal,* it would have been strongly rejected.

17e3. A linear nonparametric test applicable to angular data: the Wald-Wolfowitz Runs test

This is almost the *only* linear test (§10f1) which works with angular data. A run is an uninterrupted sequence of azimuths belonging to one of the two data-sets, and the test statistic is the total number of runs. Testing for H_o: {equal azimuths} with swarms 1 and 2 in Table 17.2, we obtain:

36 39 52 61 65 70 74 76 81 88 88 90 95 95 102 105 113 114 123 125 135 141 149 156

that is, 4 runs, which is below the critical 99% value of 6 for $n_1 = 13$, $n_2 = 11$; H_o is once again strongly rejected. The only condition for the test to work with angular data, clearly, is that the entry point for counting the number of runs (here, 36° but effectively 0°) should be chosen at the *start* of a run. Hence any angle between 76-81°, 88° or 95° would also have sufficed in this case, for example:

81 88 88 90 95 95 102 105 113 114 123 125 135 141 149 156 36 39 52 61 65 70 74 76

17e4. A parametric test for equal concentrations and mean directions: the Watson-Williams Test

Worked geological examples: *Geophysics* (Mardia 1972, p. 170); *Glaciology* (Lewis 1977, p.231); *Palaeontology* (Cheeney 1983, p.101); *Remote sensing/exploration* (Davis 1986, p.329).
Statistical tables: uses *t* and *F* tables (§8b1 & 2).
Stand-alone computer programs: Rock (1987b).
Mainframe and microcomputer software: none known.

This parametric test is based on the resultant lengths from [17.1] for the two groups, R_1 and R_2, and is similar (though not strictly equivalent) to a linear *t*-test with preparatory analysis of variance (§10a1). Like the Rayleigh test, it assumed both groups are drawn from Von Mises populations with equal concentrations. The underlying assumption of equal concentrations is first tested using one of three procedures, depending on the value of R, the mean resultant length of the *combined* groups:

1) For $R < 0.45$, the following statistic is compared with a standard Normal deviate (Cheeney 1983, p.102), where $\phi(R) = \sin^{-1}(1.225R)$:

$$Z = \frac{1.155\{\phi(R_1) - \phi(R_2)\}}{\sqrt{\dfrac{1}{(n_1 - 4)} + \dfrac{1}{(n_2 - 4)}}} \qquad [17.13]$$

2) For $0.45 < R \leq 0.70$, however, the following is compared with a standard Normal deviate:

$$Z = \frac{f(R_1) - f(R_2)}{0.893\sqrt{\dfrac{1}{(n_1 - 3)} + \dfrac{1}{(n_2 - 3)}}} \qquad [17.14]$$

where $f(R) = \ln\,[x + \sqrt{(1+x^2)}]$ and $x = (R - 1.0894)/0.25789$.

3) For $R > 0.7$, Fisher's F provides critical values:

$$F_{(n_1-1,\, n_2-1)} = \frac{n_1\,(1-R_1)\,(n_2-1)}{n_2\,(1-R_2)\,(n_1-1)} \qquad [17.15]$$

In cases where the calculated F ratio or Z value is *not* significant at the chosen level, the null hypothesis that the two groups have equal concentrations is accepted, and calculations proceed to a *t*-test for equality of mean directions, where (Cheeney 1983, p.103):

$$t^2 = \frac{\left[1 + \dfrac{0.375}{k}\right](N-2)\,(n_1R_1 - n_2R_2 - NR)}{(N - n_1R_1 - n_2R_2)} \qquad [17.16]$$

and here, k is the estimated concentration parameter for the *combined* groups, calculated from [17.5].

Comparing dyke-swarms 1 (angles –50°) with 2, as in Table 17.3, we obtain $R = 0.9350$, $k = 7.9707$, and $F_{10,12} = 0.7162$, which is non-significant and thus indicates equal concentrations. This yields $t^2 = 0.0492$ which is also non-significant, so the two mean azimuths are taken to differ by 50°.

17f.Comparing three or more groups of 2-dimensional data

For > 2 groups, q-group versions of the Watson-Williams and uniform scores tests are available. With both tests, a significant result indicates that one *or more* preferred orientations are distinct, while a non-significant result suggests that all the preferred orientations are homogeneous.

17f1.Nonparametric testing: multigroup extension of Mardia's Uniform Scores Test

Worked non-geological examples: Mardia (1972, p. 208);
Statistical tables: $q = 3$, $N > 14$, all $n_i < 6$: Mardia(1972, p.316). For larger data-sets compare W from (17.15) with χ^2 for $2(q-1)$*df.*
Stand-alone computer programs: Rock (1987b).
Mainframe and microcomputer software: none known.

Here, the test statistic (W) is the sum of resultant lengths of uniform scores for all groups:

$$W = 2\sum_{i=1}^{i=q}\frac{(C_i^2 + S_i^2)}{n_i}, \text{ where } C_i = \sum_{j=1}^{j=n_i}\frac{\cos(2\pi®_{ij})}{N} \text{ and } S_i = \sum_{j=1}^{j=n_i}\frac{\sin(2\pi®_{ij})}{N} \qquad [17.17]$$

and $®_{ij}$ is the rank of the jth measurement in the ith group in a ranked set of N total measurements for all groups combined (Mardia 1972, p.208).

For the 3 swarms in Table 17.2, we obtain $W = 18.81$, which exceeds the critical 99% value of 13.28, indicating inhomogeneity of median azimuths; intermediate results are:

	Dyke-swarm 1	Dyke-swarm 2	Dyke-swarm 3
C_i	–1.9536	+3.6371	–1.7055
S_i	–3.3123	+7.6206	–4.3016

17f2.Parametric test for equal concentrations and mean directions: multi-group Watson-Williams Test

Worked non-geological example: Mardia (1972, p. 164).
Stand-alone computer programs: Rock (1987b).
Mainframe and microcomputer software: none known.

This is analogous to a one-way ANOVA for linear data (§11b) and, like ANOVA, assumes equality of dispersions (concentrations). This assumption is therefore first tested via a χ^2 statistic in one of three ways, depending on the mean resultant length of the q samples combined, R (Mardia 1972, p.165):

If $R < 0.45$:

$$\chi^2_{q-1} \approx \sum_{i=1}^{i=q}\frac{4(n_i-4)}{3}\left[\sin^{-1}(1.225R)\right]^2 - \frac{3\left\{\sum_{i=1}^{i=q}\frac{4(n_i-4)}{3}[\sin^{-1}(1.225R)]\right\}^2}{4(N-q)} \qquad [17.18]$$

If $0.45 \le R \le 0.70$:

$$\chi^2_{q-1} \approx \sum_{i=1}^{i=q}\frac{(n_i-3)}{0.7979}\left[\sinh^{-1}\left\{\frac{R_i-1.0894}{0.25789}\right\}\right]^2 - \frac{0.7979\left[\sum_{i=1}^{i=q}\frac{(n_i-3)}{0.7979}\sinh^{-1}\left\{\frac{R_i-1.0894}{0.25789}\right\}\right]^2}{(N-3q)} \qquad [17.19]$$

Finally, if $R > 0.70$, the test below is equivalent (really!!) to Bartlett's test for linear data (§11a3):

$$\chi^2_{q-1} \approx \frac{(N-q)\ln\left[\dfrac{N-\sum_{i=1}^{i=q} R_i}{N-q}\right] - \sum_{i=1}^{i=q}(n_i - q)\ln\left[\dfrac{n_i - R_i}{n_i - q}\right]}{1 + \dfrac{\sum_{i=1}^{i=q}\dfrac{1}{n_i - 1} - \dfrac{1}{N-q}}{3(q-1)}} \qquad [17.20]$$

If the concentrations are accepted as equal, the test statistic for equal mean directions again varies, depending on R. For $R > 0.45$, the following is treated as an F ratio (Mardia 1972, p.163):

$$F_{(q\text{-}1,N\text{-}q)} = \left[1 + \frac{0.375}{k}\right] \frac{(N-q)\,(\sum_{i=1}^{i=q} R_i - R)}{(N - \sum_{i=1}^{i=q} Ri)\,(q-1)} \qquad [17.21]$$

For $R < 0.45$, the following value is instead compared with χ^2 for $(q-1)$ *df*:

$$\chi^2 = \frac{\dfrac{2}{N}\left[(\sum_{i=1}^{i=q} R_i)^2 - R^2\right]}{\left[1 - 0.375k^2 + \dfrac{q}{2Nk^2}\right]} \qquad [17.22]$$

For the 3 groups in Table 17.2, we obtain $R = 0.8467$, $k = 3.5778$, and $F_{2,33} = 20.976$, which greatly exceeds the critical 99.5% value of ≈6.3, confirming inhomogeneity of the mean azimuths.

17g. Introduction to 3-dimensional Orientation Data

Dedicated monographs and reviews: Mardia (1972); Watson(1983).
Further geological discussion: Koch & Link(1971,p.132ff), Cheeney(1983,ch.9); Davis(1986,p.330ff).

Many geological orientation data (dips and strikes) are of course *spherical* (3-D) rather than circular (2-D), and require stereographic projection to be plotted in 2-D. Unfortunately, spherical data are far harder to handle mathematically than circular data, requiring knowledge of matrix algebra, vector manipulations, etc. Furthermore, many of the statistical tests below, despite their formidable-looking formulae, are only approximations, and very few have yet been implemented in computer programs. Only the briefest of outlines is therefore attempted here.

As with 2-D data, we must distinguish *directional* 3-D data, which must be plotted on the full sphere (magnetization directions, palaeocurrent directions in folded strata, orientations of *c*-axes of quartz crystals in thin sections, etc.) from *axial* data, which only require the hemisphere (poles to bedding or foliations; plunges and lineations, etc.) It is precisely because most geological 3-D data are axial that stereographic projection works. It is also intuitively obvious that the doubling of angles we employed with 2-D data is no answer here (if we double the azimuth, what do we do with the dip, for example?) Fortunately, some (though not all) of the statistical methods below work reasonably well with axial as well as directional 3-D data.

Table 17.4. Transforming 3-dimensional angular data to direction cosines

Direction	Azimuth,° (from North)	Dip,° (from horiz)	Direction cosines: *L cosine*	*M cosine*	*N cosine*
General	θ	ϕ	$cos\phi cos\theta$	$cos\phi sin\theta$	$sin\phi$
Down	000	+90	0	0	+1
Up	000	–90	0	0	–1
North	000	0	1	0	0
East	090	0	0	1	0
South	180	0	–1	0	0
West	270	0	0	–1	0
{Data	075	-61	0.1255	0.4683	-0.8746
{for	098	-73	-0.0407	0.2895	-0.9563
{dyke-	125	-71	-0.1867	0.2667	-0.9455
{swarm 1	219	-78	-0.1616	-0.1308	-0.9781
{in	150	-71	-0.2820	0.1628	-0.9455
{Table	117	-77	-0.1021	0.2004	-0.9744
{17.5	082	-74	0.0384	0.2730	-0.9613

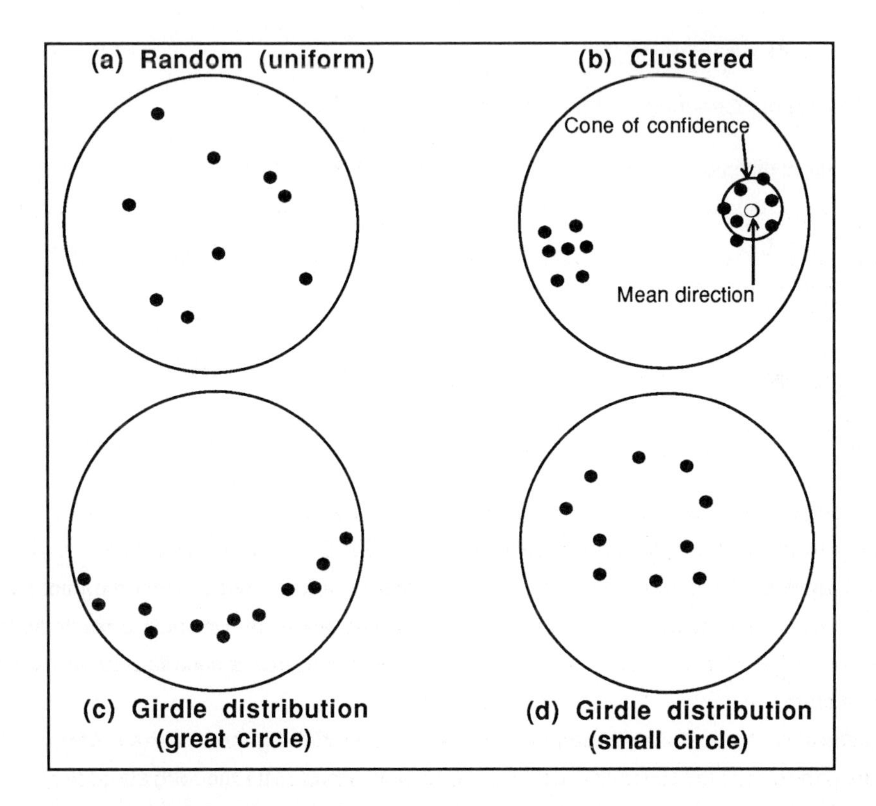

Fig.17.6. Important ways in which 3-D (spherical) data may be distributed, as represented stereographically
Case (b) might also be called *bimodal* (a single cluster would have been *unimodal*), and both a mean direction and cone of confidence can be usefully calculated for it.

17h. Describing distributions of 3-dimensional orientation data

Further geological discussion: Alldredge et al.(1974); Bailey(1975); Kiraly(1969), Mardia (1972,1981a,b).

Spherical data are analyzed statistically by first converting strikes θ and ϕ and dips to **direction cosines** *L, M, N,* as in Table 17.4.

17h1. Displaying and interpreting spherical data graphically

Further geological discussion: Cheeney(1983,p.108); Koch & Link(1971, p.134); Mardia(1972,p.222).
Selected geological applications: *Glaciology* (Kamb 1959).

More complicated methods are required here than those of Fig.17.2. Simple descriptive display is almost inextricable from geological interpretation. Apart from stereographic projection, which has its own substantial literature, methods include *petrofabric diagrams* (Braun 1969; Griffis *et al.*1985), *contouring* (Diggle & Fisher 1975; Starkey 1974,1976), *fitting great and small circles* (Mancktelow 1981), *reorientation*, for example to remove folding (Cooper & Marshall 1981), *cluster analysis* and *contour methods* (Schaeben 1984; Shanley & Mahtab 1973,1975,1976; Wallbrecher 1978, 1979).

17h2. Spherical data distributions

Further discussion: Cheeney(1983,p.112ff); Mardia(1972, ch.8).

It is important to distinguish several categories of spherical data-set (Fig.17.6), which must be treated in different ways statistically. Although many geologists would of course determine which category on Fig.17.6 applied to a particular data-set by inspecting a stereographic projection by eye, statistical methods are invaluable in less clear-cut cases than those on Fig.17.6.

Because 3-D data are usually plotted stereographically, distributions in Fig.17.6 can be analyzed like point distributions in flat 2-D space, i.e. like maps (§20). For example, a test for the random distribution in Fig.17.6a, for example, can in theory be undertaken by determining whether the projected points form a Poisson distribution across the stereographic net (cf. §20b1). Determining whether data are more towards *clusters* or girdles involves inspection of the **eigenvalues** of the spherical data-matrix; 3 cases may be distinguished:

(a) If 2 eigenvalues are large and roughly equal, a girdle (Fig.17.6c or d) is indicated.
(b) If 1 eigenvalue is far larger than the others, a cluster is indicated (or two antipodal clusters; Fig.17.6b). This is true for all 4 data-sets in Table 17.5, where eigenvalue (3) is much the largest.
(c) If all 3 eigenvalues are subequal, the data form neither a cluster nor a girdle (nor necessarily are they random).

A number of theoretical distributions are available for modelling these various alternatives. The **Fisher distribution** is the spherical extension of the Von Mises, with the following formula:

$$f(\mathbf{c}) = \frac{k}{2\sinh(k)}e^{k[\mathbf{c}.\mathbf{l}]} \qquad [17.23]$$

where **c** is a random unit vector (spherical direction), **l** a unit vector specifying the mean direction, **c.l** is the 'scalar product' of these two vectors, and k , as usual, specifies concentration. Data-sets can be tested for goodness-of-fit to Fisher distributions (Kulatilake 1985), or the Fisher parameters estimated from them (Mardia 1972, p.252). Fisher-distributed data are clustered, as in Fig.17.6b (which could represent samples from *two* Fisher distributions with different mean directions). Parametric tests in all remaining sections of this Topic assume the data concerned are sampled from a Fisher distribution, and consequently apply to clustered data. Few statistical tests have yet been devised for girdled data, although purely descriptive procedures are available (Koch & Link 1971, p.144).

The **Bingham distribution** is a hemispherical (axial) distribution, with the following formula:

$$f(\mathbf{c}) = \frac{1}{4\pi d(\mathbf{K})} e^{[K_1(\mathbf{c}.\mathbf{t}_1)^2 + K_2(\mathbf{c}.\mathbf{t}_2)^2 + K_3(\mathbf{c}.\mathbf{t}_3)^2]} \qquad [17.24]$$

where **c** is a random unit axis, **K** is a matrix of concentrations with diagonal elements K_1, K_2, K_3, and $\mathbf{t}_1 \leq \mathbf{t}_2 \leq \mathbf{t}_3$ are three vectors representing mutually perpendicular *principal axes*, which have associated **eigenvalues** $\tau_1 \leq \tau_2 \leq \tau_3$. Three main Bingham forms are usefully distinguished: (1) **uniform** ($\tau_1 \approx \tau_2 \approx \tau_3$); (2) **point ('bipolar')** $\tau_1 \approx \tau_2 \ll \tau_3$; and (3) **girdle** ($\tau_1 \ll \tau_2 \approx \tau_3$).

Dimroth-Watson's is another hemispherical, symmetric girdle distribution, with the formula:

$$f(l,m,n) = \frac{1}{4\pi \int_0^1 e^{-kx^2} dx} e^{-(l\lambda + m\mu + n\nu)^2}, \quad -\infty < k < \infty, \qquad [17.25]$$

17h3. Summarising real distributions of spherical data: Vector Means and Confidence Cones

Further geological discussion: Cheeney(1983,p113ff); Koch & Link(1971,p.140); Scheidegger (1965).

Spherical means, concentrations k and resultant lengths R are analogous to those for 2-D. Thus:

$$R^2 = (\Sigma L)^2 + (\Sigma M)^2 + (\Sigma N)^2 \qquad [17.25b]$$

$$\overline{L} = \Sigma L/R, \ \overline{M} = \Sigma M/R, \ \overline{N} = \Sigma N/R \qquad [17.25c]$$

Estimate of mean azimuth $\overline{\theta} = \tan^{-1}(\overline{M}/\overline{L})$; estimate of mean inclination $\overline{\phi} = \sin^{-1}(\overline{N})$.

Instead of confidence intervals, 3-D azimuths have **cones of confidence** (Fig.17.6b). The semivertical angle of such a cone, A, is given at the 90% confidence level by the formula:

$$\cos(A) = 1 - \frac{n-R}{R}\left\{\left[\frac{1}{0.9}\right]^{\frac{1}{(n-1)}} - 1\right\} \qquad [17.25d]$$

If the product $nRk > 3$, this can be simplified to: $\cos(A) = [1 + \ln(\alpha/nRk)]$, where α is the significance level (0.01 or 0.05) and, as usual, n is the number of data-values.

17i.Testing for uniformity versus preferred orientation in 3-D orientation data

Further geological discussion: Dudley et al. (1972); Flinn(1958); Stauffer(1966); Steinmetz(1962).
Worked geological examples: *Palaeomagnetism* (Koch & Link 1971, p.142); Mardia(1972,p.258)
Statistical tables: Davis(1986,p.342); Mardia(1972,p.325); Koch & Link(1971,p.420 — NB gives critical values for nR rather than R as in the others).
Stand-alone programs: Peron & Nemec(1977).

An extensive literature has built up on this topic, and readers are advised to consult the above references before embarking on a particular method. The test mostly commonly used is a direct analogue of the 2-*D* **Rayleigh test**, using the spherical resultant length, R. For the data in Table 17.5, R for all 4 swarms exceeds the 5% critical values, so each swarm is taken to have a preferred magnetization direction.

This is one test which does *not* work with axial data (Koch & Link 1971, p.148), since points restricted to one hemisphere clearly cannot be random on the full sphere (the doubling of angles made 2-D axial data behave as if they *were* distributed on the whole circle). The best way to test uniformity for axial 3-D data (which also works with directional data) is to determine whether the points as plotted on a stereographic projection are randomly distributed, as described earlier (§17h2, cf. §20b1).

Table 17.5. Data for 3-D examples (§17g–i): magnetization directions in dolerite dykes from 4 swarms

(a) Raw data

Swarm 1		Swarm 2		Swarm 3		Swarm 4	
Azimuth	Dip	Azimuth	Dip	Azimuth	Dip	Azimuth	Dip
075	-61	290	-72	250	-85	339	-70
098	-73	284	-70	005	-59	311	-86
125	-71	281	-75	271	-72	048	-87
219	-78	275	-81	337	-77	215	-89
150	-71	271	-76	261	-69		
117	-77						
082	-74						

(b) Summary statistics

	Swarm 1	Swarm 2	Swarm 3	Swarm 4
n_i	7	5	5	4
R_i	0.977	0.997	0.964	0.989
R_{crit} (5%)	4.18	3.50	3.50	3.10
Mean azimuth	111.71°	281.35°	309.07°	339.23°
Mean dip	-76.06°	-74.90°	-77.85°	-84.02°
Concentration, k	36.81	306.02	22.27	70.26
Cone of confidence, A^*	8.56°	3.66°	13.83°	9.05°
Eigenvalue (1)	0.05259	0.00245	0.31107	0.08078
Eigenvectors	-0.83152	-0.86886	-0.85434	-0.92332
	0.51931	-0.48805	-0.51925	0.37005
	0.19722	0.08302	-0.02207	-0.10270
Azimuth,°	148.01°	209.32°	211.29°	158.16°
Dip,°	11.37°	4.76°	-1.26°	-5.89°
Eigenvalue (2)	0.26522	0.02363	0.03561	0.00365
Eigenvectors	-0.54821	0.49240	0.50416	-0.37183
	-0.82446	-0.83458	-0.83834	-0.92830
	-0.14048	0.24704	0.20738	-0.00189
Azimuth,°	236.38°	300.54°	301.02°	248.17°
Dip,°	-8.08°	14.30°	11.97°	-0.11°
Eigenvalue (3)	6.68219	4.97391	4.65332	3.91557
Eigenvectors	-0.08964	0.05128	0.12618	0.09603
	0.22493	-0.25552	-0.16605	-0.03644
	-0.97024	-0.96544	-0.97801	-0.99471
Azimuth,°	111.73°	281.35°	307.23°	339.22°
Dip,°	-75.99°	-74.89°	-77.96°	-84.10°

* Semivertical angle of cone of confidence at 90% significance level, as given by [17.25d].

(c) Sum of cross product matrices

	Swarm 1			Swarm 2			Swarm 3			Swarm 4		
	L	M	N	L	M	N	L	M	N	L	M	N
L	0.16978			0.02066			0.31019			0.10548		
M	-0.03758	0.53254		-0.07384	0.34179		0.02545	0.23719		-0.04004	0.01941	
N	0.59300	-1.42222	6.29769	-0.24354	1.22204	4.63755	-0.56465	0.75305	4.45262	-0.36638	0.13886	3.87511

17j. Comparing two groups of 3-dimensional orientation data

17j1.Testing for equality of two concentrations

Worked geological examples: *Palaeomagnetism* (Koch & Link 1971,p.142); *Sedimentology &structural geology* (Mardia 1972, p.267); *Structural geology* (Cheeney 1983,p.118, but note his formula 9.8 is incomplete).
Statistical tables: uses Normal Distribution or *F* tables (Topics 8a1, 8b2).
Stand-alone and proprietary software: none known.

As with the equivalent 2-*D* test, this (in addition to a uniformity test) is prerequisite to testing equality of mean directions. There are, as in earlier cases, 3 alternative formulae, depending on the value of *R*. For $R < 0.44$, the following is compared with a standard Normal deviate (Mardia 1972, p.266):

$$Z = \frac{\left[\sin^{-1}\left(\frac{3R_1}{\sqrt{5}}\right) - \sin^{-1}\left(\frac{3R_2}{\sqrt{5}}\right)\right]}{\sqrt{\frac{3}{5}\left(\frac{1}{(n_1-5)} + \frac{1}{(n_2-5)}\right)}} \qquad [17.26]$$

For $0.44 \leq R \leq 0.67$, the following is again treated as a standard Normal deviate:

$$Z = \frac{\left[\sin^{-1}\left(\frac{R_1 + 0.17595}{1.02903}\right) - \sin^{-1}\left(\frac{R_2 + 0.17595}{1.02903}\right)\right]}{0.62734\sqrt{\frac{1}{(n_1-4)} + \frac{1}{(n_2-4)}}} \qquad [17.27]$$

For $R > 0.67$, however, the following is treated as an *F* ratio:

$$F_{[2(n_1-1),\, 2(n_2-1)]} = \frac{n_1(1-R_2)(n_2-1)}{n_2(1-R_1)(n_1-1)} \qquad [17.28]$$

Comparing swarms 1 and 2 from Table 17.5 gives $F_{8,12} \approx 8.2$ via [17.28], which exceeds the 99.5% *F* value of 5.35, so these concentrations are taken as unequal. Alternatively, and rather more simply in this case, the ratio of two concentrations k_1/k_2 (where $k_1 > k_2$) can also be treated approximately as an *F* ratio, with the same *df* as in [17.28] — by more direct analogue with the ratio of variances for linear data (§10a1). Here, this yields $F_{8,12} = 306.02/36.81 = 8.31$, almost the same as via [17.28].

17j2. Testing for equality of two mean directions: Watson-Williams test

Worked geological examples: *Metamorphic petrology* (Cheeney 1983,p.118); *Palaeomagnetism* (Koch & Link 1971, p.141); *Structural geology* (Mardia 1972, p.264).
Statistical tables: uses *F* tables (§8b2), exact formula for F_{2n} in (17.29), or Mardia (1972, p.329-30).
Stand-alone computer programs Schuenemeyer et al.(1972).
Proprietary software: none known.

If the test of the previous section gives a non-significant *F* ratio, another *F* ratio becomes the test for equality of mean directions, provided $R > 0.7$ (Mardia 1972, p.263):

$$F_{[2,2(N-2)]} = (N-2)\left[\frac{1}{\alpha^{1/(N-2)}} - 1\right] \approx \frac{(N-1)(n_1R_1 + n_2R_2 - NR)}{N(1-R)} \qquad [17.29]$$

where α is the chosen probability level (numerically expressed as 0.1, 0.05, 0.01, etc.) For $R < 0.7$, we calculate $R' = (R_1 + R_2)/N$, and compare it with critical values (Mardia 1972, p.329-30).

Comparing swarms 1 and 2 in Table 17.5, we have $n_1 = 7$, $n_2 = 5$, $N = 12$, $F_{2,20} = 20.74$ compared with critical value of 3.49, so the two mean directions are statistically different. Similar tests show that all four mean directions are in fact distinct in Table 17.5.

17k.Comparing three or more groups of 3-dimensional data

17k1. Testing whether three or more concentrations are homogeneous

Worked geological examples: *Sedimentology* (Mardia 1972,p.270; Steinmetz 1962).
Statistical tables: uses χ^2 tables (§8b3).
Stand-alone and proprietary software: none known.

This extension of the two-group test again yields 3 alternative statistics, all here compared with χ^2.

For $R < 0.44$ (Mardia 1972, p.270):

$$\chi^2_{q-1} \sim \sum_{i=1}^{i=q} \frac{5(n_i-5)}{3}\left[\sin^{-1}\left(\frac{3R_i}{\sqrt{5}}\right)\right]^2 - \frac{3\left[\left(\sum_{i=1}^{i=q}\frac{5(n_i-5)}{3}\sin^{-1}\left(\frac{3R_i}{\sqrt{5}}\right)\right]^2}{5(N-5q)} \quad [17.30]$$

For $0.44 \le R \le 0.67$:

$$\chi^2_{q-1} \approx \sum_{i=1}^{i=q}\left\{\frac{n_i - 4}{0.39356}\right\}\left[\sin^{-1}\left\{\frac{R_i + 0.17595}{1.02903}\right\}\right] - \frac{0.39356\left[\sum_{i=1}^{i=q}\left\{\frac{n_i - 4}{0.39356}\right\}\sin^{-1}\left\{\frac{R_i + 0.17595}{1.02903}\right\}\right]}{(N-4q)} \quad [17.31]$$

For $R \ge 0.67$:

$$\chi^2_{2(q-1)} = \frac{2(N-q)\ln\left[\frac{N - \sum_{i=1}^{i=q} n_i R_i}{2(N-q)}\right] - \sum_{i=1}^{i=q} 2(n_i-1)\ln\left[\frac{n_i(1-R_i)}{2(n_i-1)}\right]}{1 + 3(q-1)\left[\sum_{i=1}^{i=q}\frac{1}{2(n_i-1)} - \frac{1}{2(N-q)}\right]} \quad [17.32]$$

17k2. Testing whether three or more mean directions are homogeneous

Worked geological examples: *Geophysics* (Koch & Link 1971,p.141); *Sedimentology* (Mardia 1972, p.269; Steinmetz 1962).
Statistical tables: uses F tables (§8b2).
Stand-alone and proprietary software: none known.

This is exactly analogous to the multi-group test for 2-D data. Two alternatives again arise. For $R < 0.32$, small N, and k not near 0 or 1, the test statistic is given by (Mardia 1972, p.268):

$$\chi^2_{2(q-1)} \approx \frac{3\left[\left\{\sum_{i=1}^{i=q} n_i R_i\right\}^2 - \{NR\}^2\right]}{N\left[1 - \frac{k^2}{15} + \frac{3q}{2Nk^2}\right]} \qquad [17.33]$$

For $R \geq 0.32$, the test statistic is an F ratio approximated by:

$$F_{[2(q-1),2(N-q)]} = \left[1 - \frac{1}{5k^2}\right]\left[\frac{\{N-q\}\left\{\sum_{i=1}^{i=q} n_i R_i - NR\right\}}{\{q-1\}\left\{N - \sum_{i=1}^{i=q} n_i R_i\right\}}\right] \qquad [17.34]$$

For $R \geq 0.67$, [17.34] again applies, but the correction term $[1 - 1/(5k^2)]$ can be omitted.

From the data in Table 17.5, we obtain the following F ratios and consequent conclusions:

$F_{6,34}$	= 6.83 comparing swarms	1, 2, 3, 4,	which indicates inhomogeneous means at >99.5% level
$F_{4,70}$	= 22.09 comparing swarms	1, 2, 3,	which indicates inhomogeneous means at »99.9% level
$F_{4,102}$	= 35.61 comparing swarms	1, 2, 4,	which indicates inhomogeneous means at »99.9% level
$F_{4,134}$	= 41.79 comparing swarms	1, 3, 4,	which indicates inhomogeneous means at »99.9% level
$F_{4,186}$	= 61.06 comparing swarms	2, 3, 4,	which indicates inhomogeneous means at »99.9% level.

SECTION VI: ADVANCED TECHNIQUES

Introduction

Dedicated monographs & reviews on Multivariate Statistical Methods: Anderson(1958); Anderson et al. (1972); Dempster (1969); Flury (1988); Fornell (1982a); Giri (1977); Gnanadesikan (1977); Hope (1968); Lunneborg & Abbott (1983); Manly (1986); Mather (1986); Morrison (1967); Srivastava & Carter (1983); Tatsuoka (1971).

A Note on Matrix Algebra

Dedicated monographs and reviews: P.J.Davis (1973); Gere & Weaver (1982); Jennings (1977); Maron (1982); Pettofrezzo (1978).
Further geological discussion: Davis (1986); Ferguson (1987); Joreskog et al.(1976); Le Maitre (1982); Mather (1976). David(1986) includes a floppy disk containing a series of lessons on matrix algebra, and an entire course on the subject is available in PLATO (§5).

This final Section deals with techniques beyond the capabilities of pocket calculators, and requiring computer programs. Some readers will doubtless bemoan the lack of a section on matrix algebra at this point, although many of the important terms involved are explained in the Glossary. The reasons are two: (1) matrix algebra is covered by innumerable excellent books (listed above); (2) mere mention of matrix algebra is sufficient in the writer's experience to discourage many geology students completely from the techniques which follow, even where these techniques offer the only solution to their particular problems. He would nevertheless council any aspiring multivariate geomathematicians to acquaint themselves with matrix algebra, at least *before* they try to cope with the output of 'everyman' computer packages (SPSS etc.): geological data are replete with such pitfalls for the unwary as **ill-conditioned** and **singular matrices**, which can render computer output meaningless.

A note on Scales of Measurement and Multivariate Nonparametric Methods

Dedicated monographs and reviews: Bradley et al.(1971); Brofitt et al.(1976), Conover & Iman(1980), Kendall(1966), Puri & Sen(1971), Puri et al. (1970), Randles et al.(1978), Richards(1972), Simon(1977), Zimmermann & Brown(1972).
Further discussion: Burnaby(1966); Gerig (1969).
Selected geological applications: *Exploration geochemistry* (Howarth 1971c, 1973a); *Sedimentary petrology* (Howarth 1971a).

Most of the techniques in Section VI assume data measured on a **ratio scale.** However, *any* technique based on a correlation matrix (e.g. Cluster Analysis in §21, Factor Analysis in §23) can also be applied to **ordinal** data, since such a matrix can equally well be built up from Spearman's ρ or Kendall's τ, as from the admittedly more conventional Pearson's r (requiring ratio data). Geological examples of such usage are given by Johnson(1960) and Miller & Kahn(1962,p.292). A few further techniques (e.g. nonmetric multidimensional scaling) are specifically designed for ranked data.

Given our earlier emphasis on nonparametric methods, the existence of nonparametric equivalents for multivariate techniques in this Section must also be mentioned. Unfortunately, their calculation by hand is extremely difficult, and few computer programs are yet available. The reader is therefore referred once again to the references above, which no more than provide an *entrée* to the literature.

TOPIC 18. MODELLING GEOLOGICAL PROCESSES NUMERICALLY

Selected geological applications of numerical modelling, simulation and experimentation:
Economic geology: Raikar (1987);
Geochemistry: Ruan et al.(1985); Tsitsiashvili et al.(1977);
Igneous petrology: used to model the emplacement and solidification of igneous bodies (e.g. Frenkel 1979; Frenkel & Arishkin 1975; Kawabe 1979; Ribando et al. 1978);
Lunar petrology: Aggarwal & Oberbeck(1979); Comstock (1979);
Metamorphic petrology: Ferguson & Harvey (1980); Harvey & Ferguson (1978); Keylman & Panyak (1977);
Palaeontology: Savazzi (1985);
Sedimentology: Bitzer & Harbaugh(1987); Horowitz(1976); Merriam(1972);
Structural geology: now widely used to model fracturing, deformation and emplacement processes (e.g. Dieterich 1970; Fletcher 1972; Hattori & Mizutani 1971; Howard 1971; Nidd & Ambrose 1971; Wray 1973), to complement more traditional physical deformation experiments in the laboratory, where the inherent complexity of the problems precludes formal analytical solutions. Techniques include *finite difference, finite element* and *boundary element* methods, which have important ramifications in the fields of geomechanics, petroleum exploration (e.g. deformation related to formation of structural traps, hydrocarbon migration, thermal maturation) and mineral exploration (e.g. fluid migration); for exhaustive compilations and reviews of such applications, see Whitten (1969,1981);
Tectonics: Belbin & Rickard(1973); Mareschal & West(1980).

As explained in the Introduction to this book, numerical modelling (often called 'simulation' or 'experiments') has now spread to most branches of geology, but the forms it takes are too diverse and too specialized to be covered individually in the space available here. Selected references and comments are merely therefore given above, as an indication of the rapidly growing literature. Specific mention should, however, be made of **Monte Carlo modelling** (Ferguson & Harvey 1982; Hamersley & Handscomb 1964; Shrieder 1966), a technique of increasingly recognized power, which uses random numbers to imitate natural processes.

The remainder of this Topic briefly overviews modelling methods for *igneous* processes.

18a. Univariate Modelling of Magmatic Processes

Dedicated monographs & reviews: Trustrum(1971).
Further theoretical discussion & equations: Allègre & Minster (1978); Arth(1976); Bardsley & Briggs(1984); Consolmagno & Drake(1976); De Paolo (1980); Gray(1973); Greenland(1970); Hertogen & Gijbels(1976); Maaløe(1976); MacCaskie(1984); Minster et al.(1977); Neumann et al.(1984); Ottonello(1983); Powell(1984); Provost & Allègre(1979); Shaw(1970); Smith(1983).
Selected geological applications: *Igneous petrology:* (Gast 1968; McCarthy & Hasty 1976; Robb 1983; Rock 1984a); *Lunar petrology* (Weil et al.1974); *Mantle evolution* (Allègre et al.1984).
Stand-alone computer programs: Conrad(1987); Duke(1979); Haimes & Dowsett(1976); Morris(1984); Nielsen(1985); Pearce(1983); Petitpierre & Boivin(1983); Stormer & Nicholls(1978); Wright et al. (1983); Woussen & Côté (1987).
Proprietary software: none known.

Here, we are concerned with explaining the 'smooth trends' on conventional X-Y scatterplots so beloved by petrologists (e.g. Fig.15.1). Does a given trend indicated it was controlled for example by crystal fractionation, partial melting or partial fusion? Fundamental in answering this question is the *bulk distribution coefficient* for a given element, defined as:

$$D = \frac{\text{Concentration of element in solid phase}}{\text{Concentration of element in liquid phase}} \quad [18.1]$$

which separates 'incompatible' elements (F, Cl, K, Rb, Zr, Nb, Th, U, etc. in most magmatic systems), which have $D \ll 0$, from 'compatible' elements, which have $D > 0$ (Cr, Ni, etc.)

The main processes usually considered in univariate modelling are as follows:

Fractional (Rayleigh) crystallisation: crystals are removed rapidly from the melt as they form.
Partial fusion or Equilibrium crystallisation: crystals and melt remain in equilibrium as one forms.
Batch partial melting: source-rock is melted, the liquid remaining in equilibrium with the solid residue.
Fractional partial melting: liquid is drawn off from the source rock as soon as it is formed.
Zone refining: liquid moves upward through solid, reacting continuously with it by solution-stoping.
Mixing/contamination: of one magma with another, or with wall-rocks.

Equations governing trace element behaviour during these processes are as below. In each case, X is the concentration when the melt fraction in the system is f (when $f = 0$ or 1, $X = X_o$); D_x and D_{ox} refer to the bulk distribution coefficients for the element and for an initial source rock, respectively:

Rayleigh crystallization: $X = X_o f^{(D_x - 1)}$ [18.2a]

Equilm.crystallisation: + partial fusion $X = \dfrac{X_o}{[D_x(1-f)+f]}$ [18.2b]

Batch partial melting: $X = \dfrac{X_o}{[D_{ox} + f(1 - D_x)]}$ [18.2c]

Fractional partl,melting: $X = \dfrac{X_o[1 - f(D_x/D_{ox})]^{(1/D_x - 1)}}{D_{ox}}$ [18.2d]

Zone refining: $X = X_o\left\{\dfrac{1}{D_o} - \left[\dfrac{1}{D_o} - 1\right]e^{-D_x N}\right\}$ [18.2e]

Mixing: $X = \dfrac{[X_1 + kX_2]}{[1+k]}$ [18.2f]

If phases in the source melt in proportion to their modal abundances in [18.2b–d], D_{ox} can be replaced by D_x, which simplifies these equations somewhat. In [18.2e], N is the number of zone lengths processed (the volume of solid reacted with a given volume of liquid). In [18.2f], one part of magma of concentration X_1 mixes with k parts of X_2. Fig.18.1 illustrates trends resulting from [18.2a,b].

A decision to reject a particular model as inconsistent with a given set of data can be made *qualitatively*, by comparing the plot of an actual suite of rocks with the following theoretical trends:

Table 18.1. Trends resulting on scatterplots from 6 major igneous differentiation processes (MacCaskie 1984)

Process(es)	X/Y vs. X/V*	X/Y vs. V/Y*	Element-element(X-Y,Y-V etc.)
Rayleigh crystallisation/fractional melting (modal)	Exponential§	Curved	Exponential
Equilibrium crystallisation/batch partial melting	Hyperbolic¶	Curved	Hyperbolic
Collection melting	Curves resemble hyperbolae		
Magma mixing/contamination	Hyperbolic	Linear	Linear

*X, Y, V are 3 trace elements, of which D_Y and D_V are as different as possible, and $D_X \approx D_V$.
§ Equation of the form $A = B^k$, where k is a constant (a function of D_x, D_y, D_v).
¶ Equation of the form $aA + bAB + cB + d = 0$.

Goodness-of-fit tests (§8) can be used to quantify which model works best: for example, Pearson's r (§13) can easily be calculated to check that a trend is linear. However, such comparisons can only be used to eliminate *implausible* models; they cannot unequivocally identify the *correct* model.

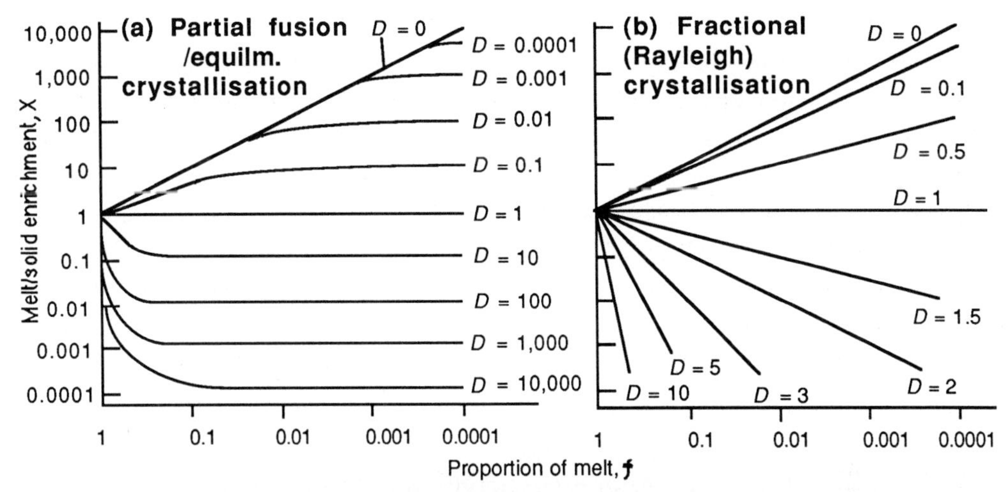

Fig.18.1. Dependence of melt enrichment factor X on melt proportion f and bulk distribution coefficient D for two important igneous differentiation processes; (curves defined by [18.1] and [18.2a,b]

18b. Multivariate Modelling of Mixing, Reactions, & Parent-Daughter Problems

Further geological discussions Albarède & Provost (1977); Bryan et al.(1969a,b); Chayes(1968b); Harbaugh & Bonham-Carter(1970); LeMaitre(1982,p.89ff); Miesch(1981a); Trustrum(1971).
Worked geological examples: *Igneous petrology* (Le Maitre 1982, p.94,101).
Selected geological applications: *Igneous petrology* (Le Roex et al.1981; Rock & Hunter 1987; Wright 1974); *Lunar petrology* (Schonfeld 1973); *Metamorphic petrology:* (Rock et al. 1986a).
Stand-alone computer programs: Albarède & Provost(1977); Banks (1979); Le Maitre (1979,1981a); Wright & Docherty(1970).
Proprietary software: none known, although multiple regression routines in some packages can be adapted to modelling problems.

In what follows, $\boldsymbol{X, Y, A, B}$ (bold italics) indicate multivariate objects (rocks or minerals defined by ν variables from SiO_2 to whatever) — and generally described in actual computation as one-column matrices — while x, y, a, b,.. represent scalar coefficients (proportions or reactants or products), and X indicates the proportion of component C in rock X (i.e. a single variable from object $\boldsymbol{X}$).

Among major applications of multivariate modelling in petrology are the following related problems:

(1) Estimating the modal composition of a rock $\boldsymbol{X}$, given the compositions of its minerals $\boldsymbol{A,B}$,..., i.e. estimating the proportions a,b,... in the matrix equation $\boldsymbol{X} = a\boldsymbol{A} + b\boldsymbol{B} + \ldots\ldots$

(2) Estimating the composition of a daughter $\boldsymbol{Y}$ derived by fractionation of predetermined amounts of given minerals from a parent magma $\boldsymbol{X}$, i.e. solving the equation $\boldsymbol{Y} = x\boldsymbol{X} - a\boldsymbol{A} - b\boldsymbol{B} - \ldots$.for $\boldsymbol{Y}$.

(3) Estimating whether or not magmas $\boldsymbol{X,Y}$ can in fact have a parent-daughter relationship, given the compositions of minerals in $\boldsymbol{X}$, i.e. solving $\boldsymbol{Y} = x\boldsymbol{X} - a\boldsymbol{A} - b\boldsymbol{B} - \ldots$.for x, a, b, c.

(4) Estimating whether or not a reaction, e.g. $a\boldsymbol{A} + b\boldsymbol{B} + \ldots = x\boldsymbol{X} + y\boldsymbol{Y}\ldots$ is chemically feasible.

Problems (1) to (3), in which only one object $\boldsymbol{Y}$ appears on the left of the equations, are known as **simple mixing models**, and problem (4) as a **generalized mixing model.**

These problems have wider applications than to igneous processes alone. Problem (1) is of course applicable to *any* geological object formed of various components. Problem (4) could equally well represent a metamorphic mineral reaction, or a metasomatic reaction of the type "parent + added components = metasomatic rocks + removed components". Problems (2) and (3) also have obvious applications, to see how or whether one rock could be related metasomatically to another.

These equations are essentially related by moving components between the left and right-hand sides of a more general equation. Modelling and **multiple regression** (§19) are in effect Q- and R-mode variants of the same procedures (though modelling does not usually involve a constant term in the equation; cf. [19.2]). Modelling assumes that the number of participating phases ($\boldsymbol{A}$,$\boldsymbol{B}$,...) is known; **Q-mode factor analysis** (§23c2) can be used to much the same effect when this number has not yet been determined (i.e. when it is uncertain which phases are in fact participating).

If N, the number of unknown coefficients, equalled v, the number of variables, then the above problems could be solved *exactly* from a set of v simultaneous equations of the form:

$$X_{MgO} = aA_{MgO} + bB_{MgO} + ...$$
$$X_{CaO} = aA_{CaO} + bB_{CaO} + ...$$

In fact, however, Gibbs' Phase rule dictates that $v > N$, so the set of equations is overdetermined, and is usually solved by **least-squares**, i.e. by minimising SS (the sum of squares) where:

$$SS = \sum_{i=1}^{i=v}(X_i - aA_i - bB_i - ...)^2 \qquad [18.3]$$

To do this requires calculation of the **inverse** (reciprocal) of a matrix, and can cause problems with geological data. Matrices of closed data (§15), for example, are **singular**, that is, they do not have an inverse. They *can* be inverted, however, if one variable is omitted.

One difference between the four examples above is that in problem (1), only +ve values for a,b...are physically meaningful (a rock cannot have a negative modal % of a given mineral). This is called the **constrained mixing model.** A least-squares result giving a –ve result for a coefficient in such a model provides a clear indication that the result (or model itself) is wrong. In problems (2) to (4), by contrast, a,b,... can be either +ve or –ve (obtaining a –ve a in problem 2, for example, means that mineral $\boldsymbol{A}$ is *added* rather than subtracted as assumed in the equation).

Another constraint in many petrological applications is that the coefficients must sum to unity, i.e.

$$\Sigma(a + b + ...) = 1 \qquad [18.4]$$

In problem (1) above, for example, the modal composition must total 100%.

Some petrologists have recast generalized mixing models (problem 4 above) as simple models, by arbitrarily choosing one component to be $\boldsymbol{Y}$. For example, they might recast the metamorphic reaction:

{chloritoid + quartz = staurolite + almandine + water} as

{chloritoid = staurolite + almandine + water – quartz}

Unfortunately, this process can produce different results depending on $\boldsymbol{Y}$, and is thus unsatisfactory.

Mixing models are least satisfactory in solving problems of the form:

{clinopyroxene$_1$ + orthopyroxene$_1$ + spinel = clinopyroxene$_2$ + orthopyroxene$_2$ + garnet}

Mathematically, this is equivalent to finding the point of intersection of two nearly parallel lines: the uncertainty is large, and the position of intersection may be in a region of extreme composition.

Table 18.2 shows the results of a typical simple mixing calculation. The model being tested is:

{Daughter rock = Parent rock – phenocrysts in parent rock}

Table 18.2. Results of a typical simple mixing model

	Parent rock	Phenocrysts in parent rock: Mg-biotite	Fe-biotite	Amphibole	Magnetite	Plagioclase	Calculated daughter rock	Observed daughter rock
SiO_2	61.30	36.47	36.50	41.23	-	56.31	69.60	69.40
Al_2O_3	17.74	13.89	15.07	12.29	0.07	26.46	16.26	15.79
FeO_t	5.10	19.33	17.68	12.13	92.37	0.39	2.65	2.43
MgO	2.94	11.03	11.72	13.77	0.02	0.07	1.41	1.00
CaO	4.03	0.04	0.99	10.82	-	8.18	1.61	1.83
Na_2O	3.92	0.07	0.09	2.61	-	6.26	3.71	3.85
K_2O	3.23	9.82	9.29	0.87	-	1.11	4.13	4.47
TiO_2	0.92	3.61	2.52	3.42	0.06	-	0.86	0.45
MnO	0.11	0.36	0.35	0.16	0.09	-	0.12	0.06
Reaction %	159.60	+14.08	-20.33	-17.65	-2.54	-33.16	100.00	

Residual SS $[\Sigma(\text{obs} - \text{calc})^2] = 0.8310$ Distance between calculated and expected compositions = **0.9116**
Calculations (from Rock & Hunter 1987) carried out via the GENMIX package (Le Maitre1981a).

Mixing models involving trace as well as major elements should ideally take into account the differing magnitudes of the numbers and errors involved (§15e), via **weighting.** This merely multiplies all values by a weighting factor w_1, which need only be determined relatively (rather than absolutely) for the different oxides. Such weighting does, however, assume that each oxide has a constant standard deviation, whereas in practice standard deviations tend to increase with oxide concentrations.

An alternative to least-squares in mixing models is **linear programming**, which has the dubious advantage of more easily accommodating the requirement for all-positive coefficients in problem (1) above. Unfortunately, all-positive coefficients will then be generated from an unsound model as well!

Finally, in this fleeting overview, how do we assess a given model? In general, a model is taken to be acceptable if $SS < 1$, as in Table 18.2. The SS is a measure of distance in ν-dimensional space (cf. Fig.21.1), and hence a small value implies the observed and calculated compositions are close together. If $SS \gg 1$, however, this implies that additional participating phases need to be considered, or that some of those incorporated do not in fact participate, or that the envisaged reaction is quite wrong. Statistical tests of the kind outlined in other Topics are *not* applicable to mixing problems (cf. Le Maitre 1982, p.104 with Banks 1979), because the uncertainties cannot be calculated from the data themselves. It is not, for example, possible to put confidence limits on the determined coefficients *a,b,....* This is a major difference between modelling and regression (§14 & 19): in modelling, we attempt, say, to express a rock composition *exactly* in terms of its constituent minerals, and any misfit must be due to *measurement* errors, or to an incorrect choice of constituents. In multiple regression, the proposed model is known to be only approximate, and the independent variable(s) will never express the dependent variable exactly, because of the *random* errors. Methods are available for assessing the overall significance of a fit, once the measurement errors associated with the participating oxides are taken into account, but these are rather complicated to be detailed here.

TOPIC 19. ANALYZING RELATIONSHIPS BETWEEN MORE THAN TWO VARIABLES

Fig.19.1 and Table 19.1 summarise relationships between the techniques covered in §13-14 (correlation and bivariate regression), and those now covered in §19-20.

19a. Homing in on the Correlation of Two Variables among Many: PARTIAL CORRELATION

Further non-geological discussion: Conover(1980, p.260).
Worked geological examples: *Sedimentology* (Krumbein & Graybill 1965, p.298)
Selected geological applications: *Igneous petrology* (Vistelius 1958).
Statistical tables: uses same tables as for appropriate coefficients.
Mainframe software: covered in large statistics packages (e.g. SPSS 1986, p.649ff).

Sometimes it is of interest to determine the correlation between only two variables of a multivariate data-set, say, X and Y, when we already know that both X and Y are also correlated with Z, so that the X-Z and Y-Z correlations indirectly influence that between X and Y. This is the situation encountered in closed major-element data-sets (§15), where the correlation between, say, MgO and CaO is heavily influenced by the strong –ve correlation between both oxides and SiO_2. We may wish to determine the 'true' or **partial** correlation between MgO and CaO after eliminating the effect of SiO_2.

As another example, this time in data not so affected by closure, the K content of lavas in classical island-arcs increases with distance to the trench, reflecting increasing depth to the Benioff zone, h (Hatherton & Dickson 1969). However, K content also varies due to igneous processes (fractionation, etc.), and it is therefore customary to calculate K *at a particular* SiO_2 *content* (usually 55%), termed K_{55}, for differentiated suites from a number of volcanoes, and then test for a relationship between K_{55} and h (e.g. Whitford & Nicholls 1976). Le Maitre (1982, p.83) treats this problem by full **multiple regression**, which we come to in the next section, but a third way is to calculate a *partial* correlation between K and h, namely that which is *independent of* the K–SiO_2 correlation, occurring at constant SiO_2. Rather than calculating K_{55} for each volcano and correlating it with h, therefore, we can use all the data from different volcanoes at once, and determine *partial r* between K and h, as below.

The partial Pearson coefficient between variables 1 and 2, at constant variable 3, is defined as:

$$r_{12.3} = \frac{r_{12} - r_{13}r_{23}}{\sqrt{(1 - r_{13}^2)(1 - r_{23}^2)}} \qquad [19.1]$$

For Whitford & Nicholls' (1976) Javanese data, let us guestimate r_{K-h}, r_{K-Si}, r_{Si-h} as 0.5, 0.7 and 0.3 respectively. This yields $r_{[K-h].Si} \approx 0.425$, slightly less than r_{K-h} alone. Some further examples of partial correlation coefficients are displayed in Table 19.4.

Kendall's partial *rank* correlation coefficient, $\tau_{12.3}$, is defined analogously to [19.1]. According to Conover(1980, p.260), programs for calculating Pearson's partial r can also be used on the ranks of the data (rather than the raw data), to yield Spearman's partial ρ — in the same way that a normal ρ can be calculated from a program for r by performing the calculations on the ranks rather than the raw data.

Table 19.1. Forms of regression and contingency problems as related to the numbers and types of variables involved

		1 dependent (response) variable		2 or more dependent variables	
	Type of variables	Interval/ratio	Dichotomous/nominal	All interval/ratio	Mixed
1 independent (explanatory) variable	Interval/ratio	SIMPLE BIVARIATE REGRESSION (§14)	(Biserial correlation coefficients — §13c)	(No methods — logically impossible!	
	Dichotomous/ nominal	Logically impossible	2-way CONTINGENCY TABLES (§13d1)		
2 or more independent (explanatory) variables	All interval/ratio	MULTIPLE REGRESSION (§19a)	LOGISTIC/GROUPED REGRESSION(§19c)	CANONICAL COR -RELATION (§19e)	LOGIT MODELS(§19d)
	Dichotomous/ nominal	Logically impossible	MANY-WAY CONTIN- -GENCY TABLES(§13d)	Logically impossible	LOGIT MODELS(§19d)

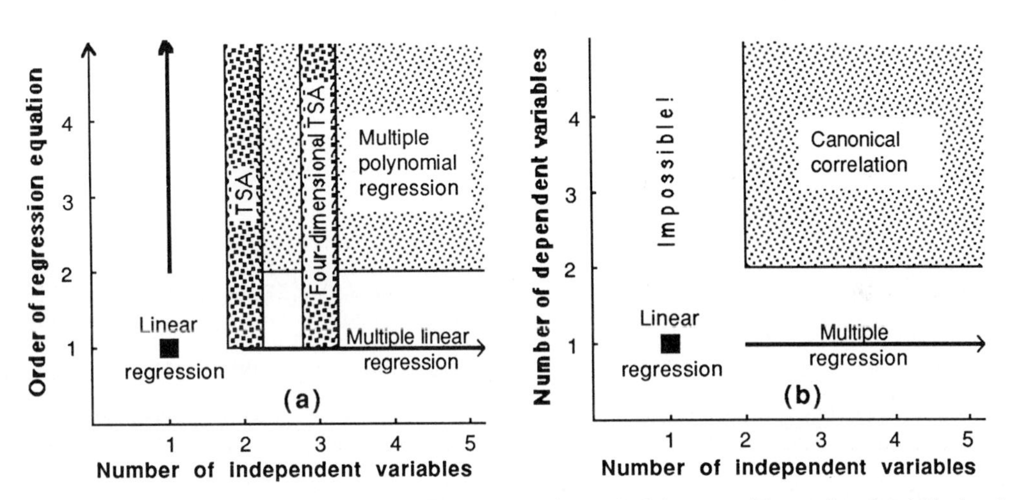

Fig.19.1. Relationship between various correlation and regression methods in terms of the number of variables involved and the powers (X, X^2, X^3, etc.) to which those variables are raised. (TSA = trend surface analysis).

19b. Assessing the Effect of Several Explanatory Ratio Variables on One Dependent Ratio Variable: MULTIPLE REGRESSION

19b1. Multiple Linear Regression

Dedicated monographs and reviews: Geary & Leser(1968); Hoerl & Kennard(1980); Kerlinger & Pedhazur(1973).

Further geological discussion: Jones(1972); Karlinger & Troutman(1985); Whitten(1966).

Worked geological examples: *Geomorphology and oil exploration* (Davis 1986, p.469ff); *Hydrogeology:* (Mather 1976, p.60 & 188); *Igneous Petrology* (Koch & Link 1971, p.24 & 91; Le Maitre 1982, p.83); *Meteorites* (Ma & Schmitt 1976); *Sedimentology* (Miller & Kahn 1962,p.242); *Weathering* (Sarazin 1978).

Selected geological applications: *Igneous petrology* (Rock 1978); *Volcanology* (Fuster 1975).

Stand-alone computer programs: Borkon & Boles(1970); Brookes et al.(1978); Davis (1973, p. 415, 423ff); Esler et al.(1976); Ghiorso(1983); Mather (1976,p.54 & 179).

Mainframe software: covered in most large statistics packages (SPSS, BMDP,etc.) and libraries (e.g. NAG routines G02C*n*F, where *n* is G,H or J, cover various options).

Microcomputer software: programmed in many statistics packages (e.g. Statview 512+™).

Multiple regression differs from simple bivariate regression in that *more than one* independent (X) variable is involved in explaining the variation in Y, i.e. we are now considering the equation:

$$Y = \text{ß}_0 + \text{ß}_1X_1 + \text{ß}_2X_2 + ... \text{ß}_nX_n \qquad [19.2]$$

where X_1, X_2... are now *completely different and (usually) unrelated variables.* As with simple bivariate regression (§14), polynomial forms can be envisaged for [19.2], but we will here deal only with linear forms, in order to keep the available options within reasonable limits. The sedimentological data and results used as an example are shown in Table 19.2-3. The equation being tested is as follows:

$$\text{Porosity} = \text{ß}_0 + \text{ß}_1\{\%\text{ clay}\} + \text{ß}_2\{\text{roundness index}\} + \text{ß}_3\{\text{sorting index}\} + \text{ß}_4\{\%\text{silt}\} \qquad [19.3]$$

The reader should note that so many independent variables should not normally be employed in a multiple regression with N only 8; N should preferably be 20 or 30 for such a problem. However, space restrictions dictate the use of small data-sets in this text for the purposes of illustration only.

Table 19.2. Raw sandstone data for multiple, stepwise and logistic regression examples, with residuals and predicted *Y* values from multiple and stepwise treatments

Sst. spec. no.	Fossil†	Dependent variable, *Y* %porosity	Independent (independent) variables, *X* %clay fraction	Roundness index	Sorting index	%silt fraction	Multiple regression Residual	Predicted *Y* porosity	Stepwise regression Residual	Predicted *Y* porosity
1	1	5	2	0.15	16	4	3.921	1.079	1.227	3.773
2	0	12	10	0.37	25	2	-3.726	15.726	-0.706	12.706
3	1	23	19	0.54	49	6	-4.666	27.666	-6.113	29.113
4	1	36	55	0.25	19	17	-1.207	37.207	-1.010	37.010
5	0	49	41	0.06	53	15	3.931	45.069	4.828	44.172
6	1	65	73	0.19	78	24	-10.198	75.198	-9.865	74.865
7	1	78	62	0.46	72	18	10.939	67.061	12.498	65.502
8	0	85	89	0.28	81	29	1.005	83.995	-0.858	85.858

† 1 indicates presence of fossil in particular sandstone specimen; 0 indicates absence (only used in §19d).

In many respects, the results of a multiple regression differ little from a simple bivariate regression (cf. Tables 14.2; 19.3), except that there are now more of them! Instead of a single *X-Y* correlation coefficient, we now have *two sets* of coefficients: a matrix of correlation coefficients between pairs of variables (Table 19.3a), and a **multiple correlation coefficient**, which expresses the total variation in *Y* explained by *all the X variables put together* (Table 19.3b). Here, Table 19.3a–b indicate that 95.4% of the variation in *Y* (porosity) can be explained by the other 4 (*X*) variables combined, but that roundness is uncorrelated with porosity, so its individual contribution is likely to be insignificant. The ANOVA table (Table 19.3c) has identical form to that for a bivariate regression, and again tests the hypothesis that the slope of the regression line is zero, except that this is now a *multiple* hypothesis:

H_0: $\{\text{ß}_0 = \text{ß}_1 = \text{ß}_2 = ... \text{ß}_n = 0\}$; against H_1: {at least one of the ß coefficients $\neq$ 0}.

In this case, the overall regression *is* significant. Table 19.3d–e give estimated ßs with their confidence limits, and test whether each individually is zero via t tests. The statistics are the same as in Table 14.2, but are now calculated for *more than one* ß. Here, the results indicate that %clay and sorting index have more effect on porosity (*Y*) than do roundness or % silt. This is *only partly* what one might expect from Table 19.3a: the low porosity–roundness correlation is certainly reflected in an insignificant ß in Table 19.3d; however, the %silt correlation with porosity is *higher* than that of sorting, yet this variable has a *lower* ß. In other words, one cannot extract the results intuitively simply by looking at the correlation matrix: the *interplay* of the variables, and their relative variabilities, all play a part.

Table 19.3. Multiple regression of porosity (Y) on % clay, roundness, sorting and % silt (X) from Table 19.2

(a) Individual correlation coefficient matrix between pairs of variables

	Porosity	%clay	Roundness	Sorting i...	%silt
Porosity	1				
%clay	.941	1			
Roundness	.003	-.089	1		
Sorting index	.893	.779	.122	1	
%silt	.921	.986	-.217	.77	1

(b) Multiple correlation coefficient R and coefficient of multiple determination, R^2

Count:	R:	R-squared:	Adj. R-squared:	RMS Residual:
8	.977	.954	.893	9.876

(c) ANOVA table for overall regression and Durbin-Watson test for autocorrelated residuals

Source	DF:	Sum Squares:	Mean Square:	F-test:
REGRESSION	4	6060.254	1515.064	15.533
RESIDUAL	3	292.621	97.54	p = .024
TOTAL	7	6352.875		

Residual Information Table

SS[e(i)-e(i-1)]:	e ≥ 0:	e < 0:	DW test:
842.838	4	4	2.88

(d) Significance tests for individual ß coefficients in [19.2]

Parameter:	Value:	Std. Err.:	Std. Value:	t-Value:	Probability:
INTERCEPT	1.202				
%clay	1.188	1.172	1.235	1.014	.3854
Roundness	-16.133	43.128	-.086	.374	.7332
Sorting index	.509	.26	.451	1.959	.145
%silt	-2.056	4.078	-.662	.504	.6488

(e) Confidence limits on ß coefficients and partial F table

Parameter:	95% Lower:	95% Upper:	90% Lower:	90% Upper:	Partial F:
INTERCEPT					
%clay	-2.541	4.916	-1.57	3.945	1.028
Roundness	-153.386	121.121	-117.63	85.365	.14
Sorting index	-.318	1.336	-.103	1.121	3.836
%silt	-15.034	10.922	-11.653	7.541	.254

In common with many geological examples, Table 19.3 shows 3 of the 4 independent (X) variables to be highly correlated: a problem known as **multicollinearity.** The larger multicollinearity is, the more **ill-conditioned** the calculations — that is, the more sensitive the results are to small changes in the data, to the algorithm used, or to the number of decimal places to which the computer stores numbers. Perfect multicollinearity can arise not only where two (or more) variables have a mutual correlation of ± 1.0, but even where all variables are mutually correlated to only a moderate extent. Where this happens, the correlation matrix becomes **singular**, and the regression cannot be performed at all. Because badly ill-conditioned data-sets may yield different results if different computers and/or different programs are used, the results in Table 19.3 have been confirmed by two different packages.

Note that *both* computer *and* software are critical here. Even highly ill-conditioned problems are usually soluble via software which uses double precision arithmetic on machines with 32-bit accuracy (§1), but if *either* the program supports only single precision, *or* the machine stores data to fewer decimal places, even moderate collinearity may yield 'wrong' results. Multicollinearity also makes it difficult to sort out the separate contributions of the various independent variables to the regression.

Table 19.4. Stepwise multiple regression on the data of Table 19.1

(a) Step 1. *X* variable entered: %clay

R:	R-squared:	Adj. R-squared:	Std. Error:
.941	.885	.866	11.014

Analysis of Variance Table

Source	DF:	Sum Squares:	Mean Square:	F-test:
REGRESSION	1	5625.062	5625.062	46.372
RESIDUAL	6	727.813	121.302	
TOTAL	7	6352.875		

Variables in Equation

Parameter:	Value:	Std. Err.:	Std. Value:	F to Remove:
INTERCEPT	4.409			
%clay	.905	.133	.941	46.372

Variables Not in Equation

Parameter:	Par. Corr:	F to Enter:
Roundness	.255	.349
Sorting index	.75	6.446
%silt	-.114	.065

(b) Step 2 (last step). *X* variable entered: Sorting Index

R:	R-squared:	Adj. R-squared:	Std. Error:
.975	.95	.93	7.974

Analysis of Variance Table

Source	DF:	Sum Squares:	Mean Square:	F-test:
REGRESSION	2	6034.934	3017.467	47.453
RESIDUAL	5	317.941	63.588	
TOTAL	7	6352.875		

Variables in Equation

Parameter:	Value:	Std. Err.:	Std. Value:	F to Remove:
INTERCEPT	-4.76			
%clay	.601	.154	.625	15.312
Sorting index	.458	.18	.405	6.446

Variables Not in Equation

Parameter:	Par. Corr:	F to Enter:
Roundness	.041	.007
%silt	-.192	.152

Note: `Par.Corr.` = partial correlation of variables with the dependent variable (porosity) - see §19a.

19b2. Stepwise Multiple Regression

Worked geological examples: *Igneous petrology* (Koch & Link 1971, p.89ff).
Selected geological applications: *Applied geochemistry* (Malmqvist 1978).
Mainframe software: covered in some large statistics packages (e.g. SPSS^X, GENSTAT).
Microcomputer software: programmed in many statistics packages (e.g. Statview 512+™).

Having performed a multiple regression using *all* the measured variables in Table 19.2, and determined that some variables are more useful in predicting porosity than others, the next question is: can we simplify the regression, and predict porosity as successfully using a *smaller subset* of the measured variables? Answering this question requires a number of subjective judgments. For example, unless any variable has a perfect correlation with one of the independent variables (r = 1), and a null correlation with the response variable (r = 0), each variable added *must* increase the total percentage of variance explained, so we are left to decide whether the percentage increase is significant.

Various methods of selecting subsets of variables are available, including **backward elimination**, **stagewise**, and **forward selection.** Table 19.4 illustrates **stepwise regression**, which starts (step 1) with a simple bivariate regression (as in §14), and adds variables successively in an order determined by the user, testing at each step to see whether each new variable has added significantly to the regression. Readers should note that professional statisticians take widely varying attitudes towards this procedure, ranging from tacit acceptance to outright hostility!

Step 1 naturally begins with the variable with which the dependent variable (porosity) shows highest correlation, namely % clay (Table 19.4a), but also calculates a F ratios for the other three variables. The user sets predetermined levels of F above or below which these variables will be added to or removed from the regression. Here, these values were taken as 4 (to add) and 3.996 (to remove), so that the only other variable to be added is sorting index, in step 2 (Table 19.4b). All the statistics are then recalculated with both % clay and sorting index as the independent variables, but the F ratios here remain below the pre-set limit of 4, so no further steps are calculated and the regression ends here.

19b3. Overcoming problems: Ridge Regression and Generalized Least Squares (GLS)

Dedicated monographs and reviews: Draper & Smith(1966); Hoerl & Kennard(1970).
Further non-geological discussion: Mather (1976, p.73).
Selected geological applications: *Sedimentology* (Jones 1972).
Proprietary software: programmable in GENSTAT.

As with bivariate regression, both *autocorrelation* of residuals and *heteroscedasticity* invalidate the assumptions of classical multiple regression. They yield ß coefficients which are no longer the 'best' (minimum variance) values, and statistical tests whose significance levels may be misleading. They can be identified by much the same techniques as outlined in §14 for bivariate regression — Durbin-Watson tests, simple scatter plots of residuals versus the X variables, etc. (Notice that a 'DW' (Durbin-Watson) statistic appears in Table 19.3c just as in Table 14.2). Additional methods have, however, been used in the multivariate case: for example, Mather (1976, p.81) recommends calculating Spearman's ρ between absolute residuals and each X variable.

As in the bivariate case, autocorrelation in geological multiple regression probably arises most commonly because the model has been incorrectly specified (e.g. linear regression is being used to model a curved trend), or because some important X variable has been inadvertently omitted.

However, if either autocorrelation or heteroscedasticity prove resistant both to log-transformation (which normally deals with non-linear trends: §14e1) and to the addition of all reasonable X variables, the method of **generalized least squares** can be used to overcome them. This will not be described here because of its involved matrix algebra, but basically it involves weighting the X variables.

The problem of *multicollinearity* is an additional danger with multiple regression, which did not of course arise with bivariate regression. It can often be dealt with by eliminating all but one of the most highly correlated variables (the others are counterproductive, not merely redundant). There may, however, be cases where throwing away variables leads to undesirable loss of information. Multicollinearity can then be overcome using **ridge regression**. This requires performing some matrix additions to the input correlation matrix, which involve a non-negative parameter, k, and then calculating the various ß coefficients as k increases from 0 to around 1 ($k = 0$ for ordinary multiple regression). A plot of the ß values against k is known as a **ridge trace**, and might look like Fig. 19.2.

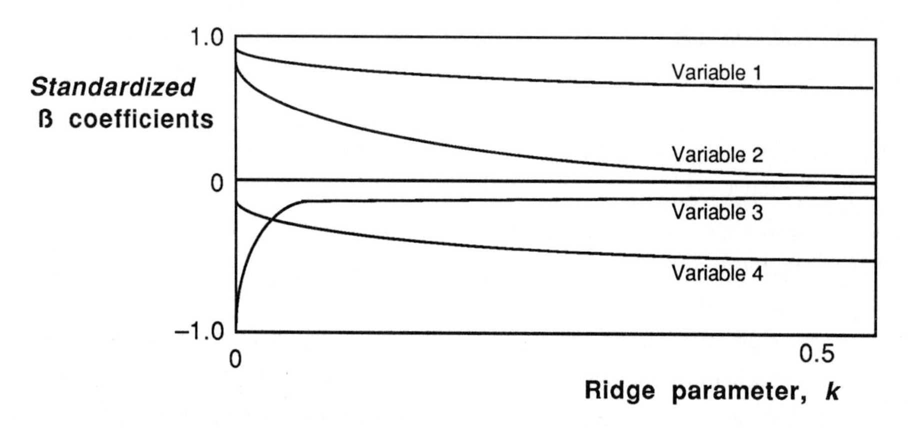

Fig.19.2.Hypothetical ridge trace from a ridge regression designed to deal with high multicollinearity.
Variable 3 quickly loses its predictive power as k increases, even though its coefficient starts off as the largest; it would normally therefore be eliminated from the regression. Variable 1 (whose influence barely changes), and variable 4 (whose influence actually rises), would both be unequivocally retained. Variable 2 is more equivocal, since its influence falls gradually, but would probably also be retained. Note that it is the *absolute* values of the ß coefficients (not their signs) which should be compared.

19b4. Nonparametric multiple regression

Nonparametric techniques corresponding broadly to the nonparametric bivariate regression discussed in §14d, are also available. Subroutines G08R in the NAG (1987) library, for example, perform nonparametric analysis by simply replacing the original data by their ranks. Little appears to have been published on them, however, and they have not so found their way into the geological literature.

19c. Relationships involving Dichotomous or Nominal Explanatory Variables: GROUPED REGRESSION

Further discussion: Mather(1976, p. 193).
Selected geological applications: *Engineering geology* (Simonett 1967); *Sedimentology* (Ethridge & Davies 1973).
Stand-alone computer programs: none known.

Grouped regression allows us to determine how a regression relationship between two continuous variables X and Y varies between several groups of data defined on a third (dichotomous or nominal) grouping variable, Z; in other words, it examines how Z affects the X-Y regression. With sandstone data of the type in Table 19.2, for example, we might wish to know whether the regression between porosity (Y) and the other physical parameters (Xs) varies between different sandstone horizons, or even between different sampling points within the same horizon (Z). Using bivariate methods, we must then: (a) assume one regression model for all horizons; (b) assume a common slope but different Y intercept for each horizon; or (c) calculate a separate regression for all horizon. Using the grouped regression method, which combines regression with ANOVA (§12), the effect of Z can be isolated.

A variant of grouped regression uses ranges of values within one of the independent variables to define the groups. Again with the sandstone data, for example, we might wish to see how the regression differs for *high* and *low* values of porosity. It is possible to formulate a model either by specifying a breakpoint between what constitutes 'high' and 'low, or to allow the regression to determine the breakpoint itself, statistically.

19d. Relationships involving Dichotomous or Nominal Dependent Variables or Ratios: LOGISTIC REGRESSION

Dedicated monographs and reviews: Wrigley (1985).
Further discussion and selected geological applications: Chung(1978), Chung & Agterberg(1980).
Stand-alone computer programs: Chung(1978).
Mainframe software: available in large statistical packages (BMDP, GLIM, SAS) and libraries (NAG).

Logistic regression, sometimes called the **logit model**, can be viewed in several ways whose mutual equivalence are unfortunately rather difficult to appreciate at first, without going into the details of the General Linear Model (see §12a). The explanation here is necessarily therefore rather contrived.

We begin by considering a situation where the response variable is a *ratio* (proportion), lying in the range $-1 \leq p \leq +1$. Because p cannot vary outside these limits, i.e. cannot vary freely with the independent variables (which are not constrained), it is not difficult to appreciate intuitively that some special calculations might have to be applied. In fact, what is termed a **logit transformation** of the ratio, $\lambda = \ln[p/(1-p)]$ is necessary, after which the constraint on p disappears because $-\infty \leq \lambda \leq +\infty$.

Now the inverse of the logit transformation $L = \exp(\lambda)/[1 + \exp(\lambda)]$ is known as the *logistic distribution,* and therefore a regression model of the following form is known as a logistic regression:

$$Y = \exp(\beta_0 + \beta_1 X_1 + \beta_2 X_2 + \ldots \beta_n X_n)/\{1 + \exp(\beta_0 + \beta_1 X_1 + \beta_2 X_2 + \ldots \beta_n X_n)\} \quad [19.4]$$

This model can of course be bivariate ($\beta_2 = \ldots = \beta_n = 0$) or multivariate ($\beta_2$ or ... or $\beta_n \neq 0$). Such a model can be applied to a ratio, in a way which a standard multiple regression cannot. The reason this might be useful in geology can be seen if we further restrict our Y variable not merely to being a ratio *between* 0 and 1, but to being a *dichotomous* variable which can *only take the values 0 or 1* (or yes/no, etc.) It turns out that the same method can be applied in this case too, or to cases where the dependent variable is nominal, or even where the independent variables are mixtures of dichotomous, nominal, and/or ratio. One obvious application is therefore to exploration, where the probability of occurrence of a mineral deposit (*yes* or *no*) is determined by numerous interlocking variables of the geological environment (e.g. nominal variables such as rock-type; ordinal variables such as intensity of metamorphism; interval/ratio variables such as *P*/*T*). To ask whether the occurrence of mineral deposits can be predicted statistically from this mixture of variables is thus to demand a logistic regression.

Logistic regression can be regarded in 3 other ways: (1) as a kind of converse of Grouped Regression (§19c), in which the *dependent* variable is now categorical; (2) as the regression model related to the biserial correlation coefficient (§13c) in the same way that classical regression is related to Pearson's *r*; (3) as a special case of what statisticians term the **loglinear model,** the simplest case of which is the analysis of 2 categorical variables in a contingency table (§13d1). ('Loglinear' here has nothing to do with plots such as 'spidergrams', where one axis is linear but the other logarithmic). Overall, it is perhaps easiest to view logistic regression as an *extension* of contingency tables to a problem where some of the variables are measured on higher scales than nominal, or as a *degeneration* of multiple regression to a problem where at least the dependent variable is dichotomous.

A number of mathematical techniques are available for solving logistic models, but most involve iterative calculations, because solutions are not generally exact. As an example, Table 19.2 includes data on the occurrence of a particular fossil in the 8 sandstone specimens (5 contain the fossil, 3 do not). Table 19.5 shows the results of a logistic regression which investigates to what extent the presence of the fossil is determined by the roundness and %clay characteristics of the sandstones. The first two treatments in Table 19.5 are bivariate logistic regressions with *one* continuous independent variable and *one* binary dependent variable (Y = fossil; X_1 = clay *or* roundness). The third is a multiple logistic regression with *two* continuous independent variables (Y = fossil; X_1 = clay; X_2 = roundness). Calculations continued in each example until changes in the estimated ß parameter(s) between successive iterations were < 0.001% of the absolute ß value. The 'loss function' is a measure of the fit of the model, here based on least-squares estimation, and hence equal to $\Sigma(X_{observed} - X_{estimated})^2$. Other methods of estimation are also commonly used (e.g. **maximum likelihood).**

The ß coefficients in Table 19.5 suggest that roundness has more effect than %clay (the geological implications are immaterial in these purely artificial data!) This particular problem also proves to be extremely sensitive to changes in the data: if %silt is used as X_1, for example, the iterations fail to converge, and if small changes are made to the '0' and '1' allocations in the fossil column of Table 19.2, two different computer programs (SYSTAT and Chung 1978) both 'crash'. This is just one illustration of how full of booby traps logistic regression can be: computational instability (overflows, underflows, round-off errors), and other factors which may cause results to differ wildly depending on the algorithm and computer used, the estimation method, the criterion for convergence, etc.

All in all, logistic regression is an 'art' which is not to be recommended for the faint-hearted or inexperienced. Yet, if used knowledgeably, it has tremendous and largely unrealised potential in solving geological problems, because binary and categorical data are so commonly generated in routine geological work, and there is simply no other way to assess their combined effect on a range of other, continuous variables in a typical multivariate geological data-set, than via logistic regression. The reader is therefore urged to study the selected applications listed above and to apply the technique himself!

Table 19.5. Results of logistic regression on data from Table 19.2
(Dependent variable = binary variable 'fossil' in all three examples; independent variables = clay and/or roundness)

Explanatory	ROUNDNESS			CLAY			ROUNDNESS & CLAY			
Iteration	Loss†	β_0	β_1	Loss†	β_0	β_1	Loss†	β_0	β_1	β_2
0	2.009374	0.1000	0.1000	3.772497	0.1000	0.10000	3.731272	0.1000	0.1000	0.10000
1	2.000628	-0.0279	0.0510	2.991001	0.0959	-0.06713	2.896698	0.0959	-0.0968	-0.06718
2	2.000040	-0.0047	0.0368	1.962357	0.1333	-0.01058	1.953247	0.1319	-0.0970	-0.01036
3	1.999893	0.0654	-0.1557	1.944510	0.1305	-0.00688	1.952731	0.1337	-0.0960	-0.00428
4	1.999783	0.1274	-0.4085	1.908263	0.4433	-0.01161	1.993584	0.1739	-0.0646	-0.00817
5	1.999776	0.1313	-0.4528	1.901079	0.6510	-0.01489	1.906935	0.5184	-0.2061	-0.01294
6	1.999776	0.1299	-0.4520	1.901052	0.6639	-0.01507	1.898284	0.7715	-0.4058	-0.01515
7	1.999776	0.1297	-0.4515	1.901052	0.6646	-0.01508	1.898137	0.7963	-0.4265	-0.01514
8	1.999776	0.1297	-0.4514	1.901052	0.6646	-0.01508	1.898132	0.7997	-0.4314	-0.01513
9	1.999776	0.1297	-0.4514	1.901052	0.6646	-0.01508	1.898111	0.8107	-0.4596	-0.01511
10	1.999776	0.1297	-0.4514	**1.90105**	**0.6646**	**-0.01508**	1.898058	0.8172	-0.5116	-0.01503
11	**1.999776**	**0.1297**	**-0.4514**	.Iterations terminated*			1.898057	0.8147	-0.5076	-0.01503
12	.Iterations terminated*						1.898057	0.8146	-0.5074	-0.01504
13							1.898057	0.8146	-0.5074	-0.01504
14							1.898057	0.8146	-0.5074	-0.01504
15							1.898057	0.8146	-0.5074	-0.01504
16							**1.898057**	**0.8146**	**-0.5074**	**-0.01504**
							.Iterations terminated*			
Std.error on ßs		4.2424	16.1451		1.6802	0.02997		3.219	7.060	0.034
Corr.‡	β_0/β_1 = -0.991383			β_0/β_1 = -0.861739			β_0/β_1 = -0.865 β_1/β_2 =0.372; β_0/β_2 =-0.725			

† Least-squares loss function. ‡ Correlation coefficients. * Termination at 0.0001 of abs. magnitude for each parameter
Emboldened values are final estimates of loss function and ß coefficients at convergence/termination of iterations

19e. Correlating two sets of several variables each: CANONICAL CORRELATION

Dedicated monographs and reviews: D.Clark (1975); Fornell(1982b); Gittins(1979); Hotelling(1936); Knapp(1982); Stewart & Love(1982); Van der Wollenberg(1977, 1982).
Further non-geological discussion: Cooley & Lohnes(1971); Fornell(1982a), Gower(1978), Lee (1969).
Selected geological applications: *Hydrology* (Reyment 1968); *Palaeontology* (Ivert 1980; Reyment 1975a,b; Reyment & Banfield 1976); *Sedimentology* (Reyment 1972).
Stand-alone computer programs: none known.
Proprietary software: available in SPSS,SAS,SYSTAT (mainly via General Linear Model).

Multiple regression extends bivariate regression to the case where there are *several* independent variables, X_1, X_2, etc. but only *one* dependent variable, Y. Canonical correlation extends it further to the case where there are *several dependent* variables Y_1, Y_2, etc. as well. It calculates linear functions of the two sets of variables, in such a way that the correlation between X' and Y' is a maximum, i.e.:

$$Y' = \alpha_1 Y_1 + \alpha_2 Y_2 + ... \alpha_n Y_n \quad \text{and} \quad X' = \beta_1 X_1 + \beta_1 X_2 + ...\beta_n X_n \qquad [19.5]$$

It thus generates two results: (a) a *canonical correlation coefficient* between X' and Y' (exactly analogous to a simple Pearson's r); and (b) the α and ß scores in each of the above linear functions.

Most published uses of canonical correlation in geology have concerned relationships between physical and chemical parameters: for example, correlations between the dimensions of fossils and the chemistry of their host-rocks have indicated that some fossils grow larger in arenaceous than argillaceous environments. Canonical correlation can also compare two sets of chemical parameters: for example, the chemistry of pore-water in sediments as compared with that of superimcumbent water. One so far unused application would be to compare the chemistry of minerals and their igneous host-rocks; this could help our understanding of a rock such as kimberlite, the origins of whose major mineral phases (especially olivine and ilmenite) remain contentious (Mitchell 1986). The assumption would be that a mineral showing a clear chemical relationship with its host-rock is more likely to be cognate than xenocrystic, whereas the converse would apply to one showing no such relationship.

Table 19.6 adds some accessory mineral abundances to the sediment data of Table 19.2, to illustrate canonical correlation between the physical parameters and the accessory mineral data. (Readers should note that such an analysis would normally require more cases, i.e. higher *N*, to be useful and effective).

Table 19.6. Raw data and additional correlation matrix for canonical correlation analysis (expanded from Table 19.2)

Sediment	*X* variables (physical parameters)				*Y* variables (% accessory mineral abundances)		
specimen	%clay	Roundness	Sorting	%Silt			
number	fraction	index	index	fraction	Apatite	Zircon	Garnet
1	2	0.15	16	4	0.32	1.65	0.09
2	10	0.37	25	2	1.95	0.54	0.61
3	19	0.54	49	6	1.03	0.71	0.34
4	55	0.25	19	17	0.65	1.04	0.97
5	41	0.06	53	15	1.45	0.96	1.56
6	73	0.19	78	24	0.43	1.35	0.26
7	62	0.46	72	18	1.86	1.27	0.57
8	89	0.28	81	29	3.02	1.31	0.49

	%clay	Roundness	Sorting	%silt	Apatite	Zircon	Garnet
Apatite	.399	.269	.463	.346	1		
Zircon	.316	-.445	.212	.419	-.231	1	
Garnet	.132	-.378	-.055	.124	.189	-.384	1

The linear combinations of *X* and *Y* below prove to show a maximum canonical correlation of +0.829:

$$2.127\ \%\text{clay} + 0.586\ \text{roundness} + 0.336\ \text{sorting} - 2.876\ \%\text{silt} \quad \textit{with}$$

$$1.214\ \text{zircon} + 0.999\ \text{garnet} - 0.040\ \text{apatite}$$

Note that this correlation is substantially higher than that between any of the physical parameters and accessory minerals *individually* (the maximum is 0.463 between sorting and apatite — Table 19.6). Canonical correlation has thus revealed a strong *collective* relationship between accessory abundances and physical parameters which is by no means so obvious between the original variables.

Canonical correlation analysis will always output a number of correlation coefficients (with corresponding sets of ß coefficients) equal to the number of variables in the *smaller* set. Here, therefore, the maximum correlation of 0.829 is just the largest of *three* correlations. Two smaller canonical correlations of 0.656 and 0.530 are between linear combinations of *X* and *Y* which are mutually orthogonal to the combinations above. Generally, however, only the maximum correlation is of practical interest, so the corresponding ß coefficients for these lower correlations are not listed here.

Although canonical correlations behave like Pearson's *r* insofar as they lie between 0 and ± 1, their significance levels *cannot* be looked up in tables as if they actually were Pearson's *r* (cf. §13a). In fact, the maximum correlation has a highly skewed, extreme-value distribution, precisely because it *is* a maximum value. Determination of its significance therefore requires special calculations.

TOPIC 20. ANALYZING SPATIALLY DISTRIBUTED DATA: Thin Sections, Maps, Mineral deposits and the like

20a. Analyzing Thin Sections: Petrographic Modal Analysis

Dedicated monographs and reviews: Chayes(1956b).
Further geological discussion: Bankier(1955); Bardsley(1983); Belov(1963); Chayes(1954b,1956a); Chayes & Fairbairn(1951); Demirmen(1971); Neilson & Brockman(1977); Saha et al.(1974); Sahu(1974); Shaw & Harrison(1955); Solomon(1963); Watson & Nguyen(1985).
Stand-alone computer programs: Davaud & Strasser(1982); Korsch(1977); Root(1978); Svane Petersen(1977); Tocher(1978a,b); Usdansky(1985). See also Dunn et al. (1985).

The estimation of the mode of a three-dimensional rock specimen (with its errors), from point-counts on a two-dimensional thin section, is a complex statistical sampling problem which has long occupied the minds of petrologists. With the advent of rapid machine-based chemical analysis, however, some petrologists now seem to regard modal analysis as *passée*, and all too many rock analyses are accompanied by few or no data on the rock's mineralogical composition, let alone its quantitative mode. It is, unfortunately, also true that modal data have 4 main disadvantages with respect to, say, chemical data: (1) they are primarily an end in themselves, and are amenable to comparatively little further statistical manipulation; (2) the set of variables (mineral species) required to describe one rock fully is not the same as the set for another, whereas nearly all rocks are expressible in terms of the same 9 major chemical components — SiO_2 etc.; (3) modes give virtually no information about the trace element geochemistry of a rock; (4) different modes can arise from a single chemical composition (one rock — the quintessence of metamorphic petrology), whereas different chemical compositions *must* relate to different rocks. The recent appearance of several programs to calculate chemical analyses from modes (e.g. Usdansky 1985) is in this respect surprising, since nowadays few petrologists would determine a mode where they could obtain a chemical analysis first! Such programs may nevertheless find use in adding chemical data to databases for occurrences which for some reason can no longer be sampled, and for which only old modal data are available.

For these reasons, and because an extensive literature already exists, it was regretfully decided not to review modal analysis here. Readers interested in determining modes are nevertheless urged to familiarise themselves with some of the benchmark literature cited above.

20b. Analyzing Spatial Distributions of Points on Maps

Dedicated monographs and reviews: Agterberg(1978); Bartels & Ketellapper(1979); Bouille(1976); Berry & Marble(1968); Burns et al.(1969); Davis & McCullagh(1975); Fukunuga (1972); Lewis (1977); Ripley(1981); Rogers (1974); Unwin(1981).
Further geological discussion: Agterberg & Fabbri(1978); Cadigan (1962); Chork & Cruikshank (1984); Henley (1976c); Lundstrom & Bjork (1979); Tinkler(1971).
Worked geological examples: Miller & Kahn (1962,ch.16 & 17); Davis (1986, ch.5).
Stand-alone computer programs: Burns & Remfry(1976); Davis(1973, p.317ff); Howarth(1971b).
Mainframe software: numerous proprietary packages available (e.g. SURFACE II); see catalogue provided by Calkins & Marble(1980).
Microcomputer software: numerous proprietary packages available.

Maps are the 'bread and butter' of many geologists, but although their skills in displaying and interpreting 3-dimensional data on a map are probably unsurpassed, ways of quantifying maps, or of comparing one map with another, are not so highly developed as among, for example, geographers. The methods for quantifying map interpretation in this (and the next) section are applicable to *any* scale of 'map', from microscopic to global.

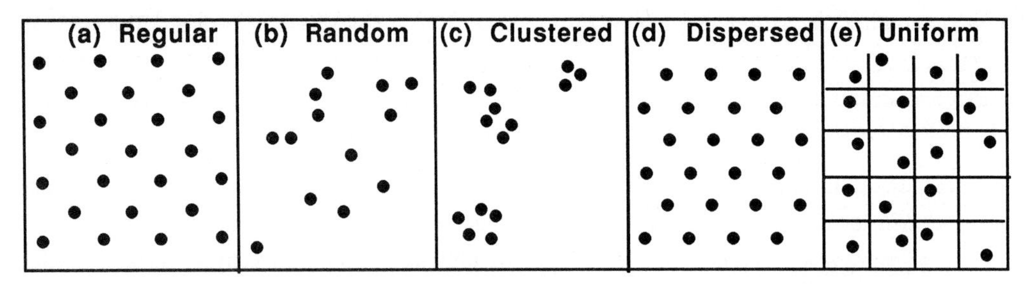

Fig.20.1. Types of point distributions: note how the uniform distribution looks remarkably 'random'

20b1. Testing for random, uniform and clustered point distributions: QUADRAT ANALYSIS

Dedicated monographs and reviews: Cole & King(1968,ch.4); Haggett et al.(1977, ch.4,8,13); King(1969,ch.5).
Further geological discussion and worked examples: Davis(1986,p.297ff); Griffiths(1962,1966a); Miller & Kahn(1962, p.375).

Many geologists are concerned with point distributions on maps (outcrops, oil wells, volcanoes, etc.) . Any point distribution can be classified as some combination of these extreme distributions (Fig.20.1):

Regular and **dispersed** (Fig.20.1a &d): points are disposed on some kind of grid; the distance between any pair of points in some specified direction is constant. A hexagonal distribution (Fig.20.1d) has the property that its points are at the *maximum possible* separation (most dispersed).

Random (Fig.20.1b): any subarea is as likely to have a point as any other equal subarea, and positioning of one point has no influence over any other point; the number of points in any subarea of constant size and shape is governed by a Binomial or (for an infinite map) Poisson distribution.

Clustered (Fig.20.1c): some areas have far more points within them than others.

Uniform (Fig.20.1e): the number of points in any constant size and shape subarea is itself constant. 'Uniform' is the vaguest of these terms, for both random and regular distributions are also uniform.

Because the human eye is not good at distinguishing these extremes (most people would classify Fig.20.1e as random, whereas it is in fact highly uniform) mathematical methods are advisable. For these, a map is divided into equal-sized subareas called **quadrats** or 'tracts'. The hypothesis of equal points per quadrat is then a test for H_0: {uniform point distribution} versus H_1: {non-uniform}. This is clearly a problem for χ^2, comparing expected *(E)* and observed (*O*)numbers of points, where:

$$E = \frac{\text{total number of points}}{\text{number of quadrats}}, \text{ and } \chi^2_{N-2} = \sum_{i=1}^{i=N} \frac{(O-E)^2}{E} \text{ as usual} \quad [20.1]$$

and *N* is here the number of quadrats.

If the distribution proves uniform (H_o is accepted), we can proceed to test for randomness. Suppose the map has area a, there are m points in all, we divide the map into N quadrats of area A, and then divide each quadrat into n extremely small subareas. The overall density of points, λ, and the probability, p, of having a point within a subarea are given by [cf. 8.5a]:

$$\lambda = \frac{m}{a} \quad \text{and } p = \lambda \frac{A}{n} \qquad [20.2]$$

The probability of obtaining ν points on u subareas is then given by a Binomial expression [cf. 8.5a]:

$$p(\nu) = {}^{u}C_{\nu} \left\{ \lambda \frac{A}{u} \right\}^{\nu} \left\{ 1 - \lambda \frac{A}{u} \right\}^{u-\nu} \qquad [20.3]$$

As u becomes very large, the Binomial distribution approaches its limiting form, the Poisson distribution (§8a7), and [20.3] simplifies to:

$$p(\nu) = e^{-\lambda A} \frac{(\lambda A)^{\nu}}{\nu!} \approx e^{-(m/N)} \frac{(m/N)^{\nu}}{\nu!} \qquad [20.4]$$

because λA, the average number of points per quadrat, can then be estimated by m/N, the ratio of total points to total quadrats. The number of quadrats with exactly ν points is then $mp(\nu)$, and this expected number E can then be built into a χ^2 test against the observed numbers O, as in [20.1], except that N is replaced by c degrees of freedom, the number of categories the results are divided into. Rejection of H_0 implies that the distribution is not random (not Poisson-determined).

The *variance* in the number of points per quadrat is calculated as:

$$s^2 = \frac{\sum_{i=1}^{i=N} (\nu_i - m/N)^2}{N - 1} \qquad [20.5]$$

Comparing s^2 with its mean m/N can also give a useful idea of the type of distribution as follows:

(1) if $m/N < s^2$ the distribution is more uniform than random;

(2) if $m/N \approx s^2$ the distribution is random;

(3) if $m/N > s^2$ the distribution is more clustered than random.

We can further test which of the above holds using a t test (§10a2) that $m/N = 1$:

$$t_{N-1} = \left[\frac{m}{Ns^2} - 1 \right] \sqrt{\frac{(N-1)}{2}} \qquad [20.6]$$

where the $\sqrt{}$ term is the standard error in the mean number of points per quadrat expected from randoı fluctuations. Significant positive t values indicate relative uniformity, while significant negative values indicate relative clustering, and non-significant values suggest randomness.

Should this test indicate *clustering*, we can proceed further. Clustered distributions are most easily described by the interaction of *two* distributions: one to control the positions of cluster centres, and one to control the positions of points within clusters. The **negative binomial distribution** (§8a8) is commonly taken to control the distributions of geological points (oil fields, ore bodies, etc.) in space. The probability $p(\nu)$ of having ν points within a quadrat then becomes:

$$p(\nu) = {}^{k+\nu-1}C_{\nu}\left[\frac{p}{1+p}\right]^{\nu}\left[\frac{1}{1+p}\right]^{k} \qquad [20.7]$$

where p is the probability that a given *subarea* contains a point, estimated as $m/(Nk)$, and k is a cluster statistic estimated as:

$$k = \frac{\left[\frac{m}{N}\right]^2}{s^2 - \frac{m}{N}} \qquad [20.8]$$

As $k \rightarrow 0$, clustering becomes stronger; as $k \rightarrow \infty$, the distribution becomes random (Poisson). If k is not an integer, [20.7] cannot be solved exactly, and the following approximations must be used:

$$p(0) = \frac{1}{(1+p)^k}, \text{ and } p(\nu) = p(\nu-1)\frac{(k+\nu-1)}{\nu}\frac{(1+p)}{p} \qquad [20.9]$$

We then proceed to a χ^2 test exactly as in the random case. Significant χ^2 indicates that the points are *not* clustered in the way expected from a negative binomial, while non-significant χ^2 indicates the negative binomial as *one probable* model for the point distribution.

20b2. Analyzing distributions via distances between points: NEAREST NEIGHBOUR ANALYSIS

Dedicated monographs and reviews: Cliff & Ord(1981); Ebdon (1977); Getis & Boots(1978); Ripley(1981,p.121ff).
Further geological discussion and worked examples: Davis (1986, p.308); Ripley(1981,p.175).
Statistical tables: for nearest-neighbour statistic see Ripley(1981, p.185).

Nearest neighbour analysis has two advantages over quadrat analysis (previous section): (1) it is based on *distances* between closest pairs of points, which usually means there are many more data-values on which to base an analysis (i.e. greater sensitivity); (2) it requires no choice of quadrat size, and so avoids the problem that the statistical analysis of a pattern may sometimes vary with the chosen size.

A simple nearest-neighbour test compares the observed mean distance $\bar{d}$ between points with means expected from particular distributions, δ, determined by the Poisson distribution; for example:

$$\overline{\delta_{ran}} = \frac{1}{2}\sqrt{\frac{A}{N}} \text{ for a random distribution; } \overline{\delta_{hex}} = 1.0743\sqrt{\frac{A}{N}} \text{ for a hexagonal distribution} \qquad [20.10]$$

where A is the map area and N the number of points. The distribution of δ is Normal, provided $N > 6$, so a crude Z test can be made of the hypothesis that $\bar{d}$ equals the value δ_{ran} by way of the sampling variance of δ_{ran}, s_{δ}:

$$Z = \frac{\bar{d} - \bar{\delta}}{s_{\bar{\delta}}} = \frac{\bar{d} - \bar{\delta}}{\frac{1}{N}\sqrt{\frac{(4-\pi)A}{4\pi}}} = \left[\bar{d} - \frac{1}{2}\sqrt{\frac{A}{N}}\right]\left[\frac{3.826N}{\sqrt{A}}\right] \qquad [20.11]$$

This test unfortunately ignores edge effects (i.e. it assumes the point pattern extends to ∞ in all directions). Nearest neighbours of points on the edges are of course within the map, which biases $\bar{d}$ towards higher values. This can be corrected in 3 ways: (1) Surround the area being analyzed with a *buffer region* (this is no problem if the map extends beyond the area of immediate interest); d for the edge points then uses points in the buffer region, although the latter are not considered further. (2) Consider the map as being spherical (or on a torus) rather than planar (cf. stereonets), so that the nearest neighbours for points on the right-hand edge are those adjacent to the left-hand edge, and similarly for top and bottom. (3) Use modifications for δ and s_δ in [20.10], based on numerical simulation, which allow for edges:

$$\bar{\delta} = \frac{1}{2}\sqrt{\frac{A}{N}} + \left\{0.514 + \frac{0.412}{\sqrt{N}}\right\}\frac{p}{N}; \quad s_{\bar{\delta}} = \sqrt{\frac{0.070A}{N^2} + 0.035p\sqrt{\frac{A}{N^5}}} \quad [20.12]$$

where p is the perimeter of the map. If $p = 0$, [20.12] reduces to [20.10].

The **nearest-neighbour statistic** G is a useful general index to the spatial pattern:

$$G = \frac{\bar{d}}{\bar{\delta}} \quad [20.13]$$

G varies from zero (where all points are superimposed), to a maximum of 2.15 for a hexagonal distribution (Fig.20.1d). A random distribution has $G = 1.0$, a clustered distribution has $G < 1$, and a dispersed (uniform) distribution $G > 1$. Critical values of G have been tabulated for testing a given distribution directly as random, clustered or uniform, rather than going through these procedures. Thus a one-tailed test H_0: $G = 1$ against H_1: $G < 1$ would distinguish a random from a clustered distribution.

Analogous tests are available for *line* (as opposed to point) distributions on maps (Davis 1986, p.312), but are not so well developed. They could be useful for analyzing lineaments seen on remote sensing images, or for studying fault, joint or microfracture patterns.

20b3.Contouring methods

Dedicated monographs and reviews: Jones et al. (1986).
Further geological discussion: Agterberg (1974b); Chayes & Suzuki(1963); Dahlberg(1975); Merriam & Sneath(1966); Philip & Watson (1986a); Smith (1969); Walters (1969).
Worked geological examples: Koch & Link (1971, p.59); Davis (1986, p.353ff).
Selected geological applications: methods used routinely in oil and mineral exploration; rarer applications in other disciplines include: *Igneous petrology* (Rock 1978,1983).
Stand-alone computer programs: Davis (1973, p.182, 319 etc.); Koelling & Whitten (1973); Sowerbutts (1983); Yates (1987). See also Walters (1969).
Proprietary software: several widely distributed packages available, e.g. SURFACE II, GeoMap™. See catalogue provided by Calkins & Marble (1980).

The two previous sections have been concerned with 2-dimensional data — point distributions in X, Y space. Contour maps portray the form of a *3-D* surface in 2-D, based on points which not only have associated X,Y coordinates, but also values of a third variable, Z, which might represent topographic elevation, thickness, composition, number of veins or any other geological measurement. Contour lines ('isolines') are drawn to connect points of equal Z, while the space between two successive contours with associated Z values z_1 and z_2 ideally contains only points whose Z values lie in the range z_1 to z_2.

Contour maps are still commonly drawn by hand, which on the one hand introduces the geologist's subjective biases and preconceptions, but on the other his experienced concepts and ideas about how a real geological surface should behave. Although computer-drawn contour maps are based on 'precise' mathematical methods, and are hence in theory more objective, many different 'precise' (but in essence *ad hoc*) methods have in fact been employed, with little theoretical basis to choose between them. In a fascinating experiment, Dahlberg (1975) presented a team of geologists and one particular computer program with a subset of well elevations, to see how well the 'correct' contour map (as derived from the complete set) could be approximated by both men and machines, from the subset they were given. He found that the geologists varied very widely in their 'correctness' while the program behaved exactly like an 'average' geologist, giving grounds for confidence in computer-drawn contours!

As well as assuming that Z is known without error (other than measurement errors), most contouring programs assume that map surfaces obey the following 3 conditions, which may not hold for certain common types of geological data and therefore need comment:

(a) *They are single-valued*, that is, Z can have only one value at a specific X, Y location. While many geological surfaces (e.g. depths to some gently folded stratigraphical horizon, surface values of say Na_2O in a granite batholith), indeed intrinsically obey this condition, *multi-valued* surfaces can also arise for both statistical *and* geological reasons. For example, repeated sampling of the granite batholith at each locality would yield a suite of varying Na_2O values, due to the various error sources (§9); multi-valued data would thus arise for an intrinsically single-valued surface. In areas of overturned folding, a stratigraphical horizon might in reality occur at more than one depth for a given X,Y, also yielding an intrinsically multi-valued surface. The depth to a balloon-like, diapiric structure (e.g. salt dome), even in areas of simple folding, could also be multi-valued. Few computer programs can cope with multi-valued data, and it is necessary to reduce the multiple observations for each X,Y location to a single (usually, average) value, for contouring purposes. The difference between single- and multi-valued data can be quite subtle — see Fig.20.2, which treats the increasingly important alkali-silica type plots used for petrological classification purposes in two slightly different ways.

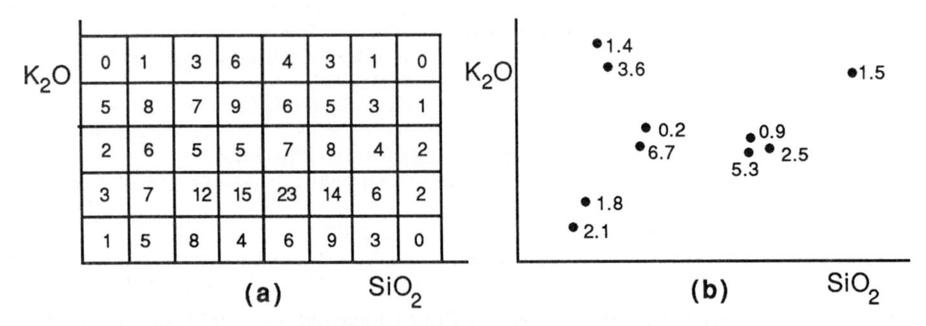

Fig.20.2. Illustrating the subtle difference between single- and multi-valued data for contouring purposes. We are here interested in contouring a diagram of potash-versus-silica for igneous rocks. In **(a)** the numbers of analyses within cells have been calculated; this is *single-valued* contouring, because there can only one such number for any cell. In **(b)**,we wish to contour the plot for the associated value of Na_2O (figures shown); this is *multi-valued*, because rocks of very different composition can be superimposed at or around the same point

(b) *They have continuity within the map area*, that is, Z varies *smoothly* from place to place. The common occurrence of faults or other discontinuities in geological surfaces can still be accommodated by separating the map into two parts, along the fault line. However, no contouring program so far available can detect such discontinuities and perform the separation automatically. The most that can be expected is for previously unanticipated faults to appear as areas of very steep gradient on the contoured map. Hence areas of complex faulting are rarely suitable for machine-based contouring.

(c) *They show autocorrelation* (§16a6) over a distance greater than the typical spacing of X, Y points; that is, the Z value at any point is closely related to that at nearby points. This is built into most contouring procedures by estimating the Z value of the surface at any position as the average Z for a number of nearby points. The reliability of the resultant contour map decreases with the degree of autocorrelation (i.e. as the relationship between nearby points becomes more random).

Two main methods for generating actual contours are used (Fig.20.3). **Triangulation** uses the actual known points to model the surface as a mesh of tilted triangles (with the known points as their apices), and then draws horizontal lines (the contours) across the triangles. **Gridding** converts the irregular distribution of known points into a regular (e.g. rectangular) grid of hypothetical points ('grid points' or 'node points') with interpolated Z.

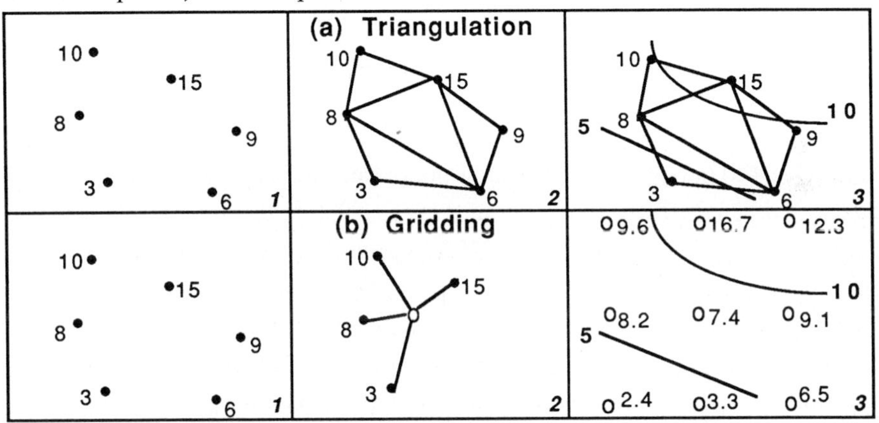

Fig.20.3. The three stages of the two main methods of contouring
In *triangulation*, values are interpolated along the sides of the triangles joining the pairs of points; in *gridding*, the original irregular point distribution is recast into a regular network whose values are interpolated.

20c. Fitting Planes and Surfaces to Spatial Data: TREND SURFACE ANALYSIS (TSA) AND TWO-DIMENSIONAL REGRESSION

Dedicated monographs and reviews: Watson(1971,1972).

Further geological discussion: Chayes(1970b); Chayes & Suzuki(1963); Doveton & Parsley(1970); Ferguson (1982); Hall (1973); Howarth(1967); Lahiri & Rao(1978); Lee(1969); Miller & Kahn (1962,p.394ff); Rao(1971); Rao & Rao(1970); Shepherd & Gaskell(1977); Sturgul & Aiken(1970); Sutterlin & Hastings(1986); Trapp & Rockaway(1977); Whitten (1970, 1972, 1975); Whitten & Koelling (1972).

Worked geological examples: *Hydrology* (Mather 1976, p.161ff); *Mining* (Koch & Link (1971, p.18ff); *Oil exploration* (Davis 1986, p.405ff).

Selected geological applications: *Geochemistry* (Nichol & Webb 1967; Parsley & Doveton 1969); *Igneous petrology* (Baird et al.1967; Dawson & Whitten 1962; Waard et al.1976; Whitten 1961,1962a,b;1963; Yeap & Chan 1981); *Metallogenesis and exploration* (Agterberg & Cabilio 1969; Burkov et al.1978); *Sedimentology* (Cameron 1968; Read et al. 1971; Thrivikramaji & Merriam 1976).

Statistical tables: covered in many statistics texts and most compilations of statistical tables.

Stand-alone computer programs: Attoh & Whitten (1979); I.Clark(1977c); Davis (1973, p. 332); Mather(1976,p.133).

Mainframe software: programmed in most statistics packages (e.g. SPSS) or libraries (e.g. NAG).

Microcomputer software: programmed in many statistics packages (e.g. Statview 512+™).

TSA is concerned with separating spatial data into two components: 'regional trends' (structural dip, metamorphic grain, etc.) and 'local anomalies'. These components are relative to 3 factors: (1) *scale*: the 'regional trend' on a map of large-scale can be the 'local anomaly' on one of small-scale'; (2) *spacing of data:* 'local features' can only be detected if their magnitude significantly exceeds the sample spacing; (3) *model*: the direction and magnitude of 'local anomalies' differ according to the shape of the assumed 'regional trend'; in the extreme, a complex enough model will fit the observed spatial pattern perfectly, so that the 'anomalies' disappear altogether (as in Fig.14.5e).

The simplest TSA model fits a planar (first-degree) surface in X,Y to Z:

$$Z = \beta_0 + \beta_1 X + \beta_2 Y \qquad [20.14a]$$

This differs from multiple linear regression (§19; cf. especially [19.1]) only in that both X and Y are now perpendicular, spatial (geographic) coordinates. Solving the equation for a given data-set proceeds in the same way, by least-squares minimization of the residuals, and yields similar significance tests: a **coefficient of multiple correlation** assesses the extent of fit between plane and data (R = 1.0 for data which do in fact define a plane), and an ANOVA assesses the hypothesis $H_o: \beta_1 = \beta_2 = 0$ (i.e. that there is no regression). The same assumptions are also made, namely that (i) Z is Normally distributed about the regression surface; (ii) the variance of Z is independent of changes in X or Y (homoscedasticity); (iii) the samples are randomly drawn from the parent population; (iv) there is no autocorrelation of residuals. The manner of testing is so similar to that for multiple regression that repetition here is hardly necessary (see Mather 1976 for a particularly detailed treatment of TSA). Note, however, that the sampling randomness condition (iii) may be particularly difficult to meet in geology, where the X,Y data-points (especially drillhole locations) are often chosen very carefully!

Of course, geological surfaces (such as stratigraphic horizons) rarely form planes. If a first-degree surface fails to fit the data, therefore, we can add terms to [20.14] to form higher-degree surfaces:

$$\textit{2nd-degree: } Z = \beta_0 + \beta_1 X + \beta_2 Y + \beta_3 X^2 + \beta_4 Y^2 + \beta_5 XY \qquad [20.14b]$$

$$\textit{3rd degree: } Z = \beta_0 + \beta_1 X + \beta_2 Y + \beta_3 X^2 + \beta_4 Y^2 + \beta_5 XY + \beta_6 X^3 + \beta_7 Y^3 + \beta_8 X^2 Y + \beta_9 XY^2 \qquad [20.14c]$$

and so on. These represent variants of polynomial regression (§14 & §19) and, as in that method, the significance of the higher-order terms can be assessed via stepwise ANOVA. For example, ANOVA assessing whether a 3rd order surface fitted data better than a 2nd order surface would be performing the test H_o: $\{\beta_6 = \beta_7 = \beta_8 = \beta_9 = 0\}$ against H_1: $\{\beta_6 \text{ or } \beta_7 \text{ or } \beta_8 \text{ or } \beta_9 \neq 0\}$. Individual terms can also be singled out for testing. These tests are important with geological TSA, where the exact form of the surface can rarely be predicted from theoretical considerations, and a 'hit-and-miss' approach is often required to find the best fit. The number of data-points must always exceed the number of coefficients β_n, preferably by a large margin, if the power of the tests is to be reasonable.

In some geological applications of TSA, particularly in mineralogy and petrology, attention is essentially focussed on the 'regional trend' (i.e. the order of the surface and the values of β_o...) By contrast, in exploration, it is the 'local anomalies' which are of interest, since these may reveal mineral, oil or gas deposits. Although the trend surfaces for these contrasting applications are obtained in the same way, the statistical models being used are different and significance tests are only valid for the former type. Fortunately, the regression itself is only of incidental interest in an exploration strategy.

Two further possible pitfalls can either distort or totally invalidate a TSA:

(1) *Edge effects:* As with nearest-neighbour analysis, contouring, etc. (§20b), these can have a major effect. It is wise to have data for a larger area than is actually analyzed, so that a buffer region is available to define the slope along the edges of the area of interest, in the same way as near its centre.

(2) *Point distribution*: in general, trend surfaces vary as the arrangement of data-points on a map diverges from random (Mather 1976, fig.3.5): in particular, trend surfaces will be elongated parallel to an elongated band of points (such as those resulting from a series of parallel geological traverses). Distortions are more severe if local 'noise' is also present. Trend surfaces appear to be more robust to clustered point distributions (such as almost inevitably result from borehole sites, which tend to be concentrated in oilfields, goldcamps, etc.), but do not give grounds for complacency. Certain point distributions can give rise to badly ill-conditioned matrices, and consequent problems in computer calculation (Mather 1976, p.125ff).

Four-dimensional trend surfaces (sometimes called 'hypersurfaces' or 'isopleth surfaces'), concerning the variation of some parameter V with *three* spatial coordinates X,Y,Z (N-S, E-W and vertical) can also be modelled, by extension of the above methods (Clark 1977c). They have obvious application to modelling the grade of ore deposits in 3-D. The final result is a three-dimensional solid (e.g. Davis 1986, p.432), enclosing contour envelopes (instead of contour lines), which enclose *volumes* of equal composition.

20d. Spatial Statistics with Confidence Estimation: GEOSTATISTICS and KRIGING

Short introductions: Brooker(1979); Froidevaux(1982); Royle(1978,1980); Steffens(1985).

Longer monographs, course notes & reviews: Armstrong (1983); Blais & Carller(1967); Brooker(1975); Clark(1979); David(1977); Guarascio et al. (1976); Henley(1981); Journel(1983,1985,1986); Journel & Huijbregts(1981); King et al.(1982); Knudsen & Kim(1977); Krige(1951,1960,1981,1984,1985); Matheron (1963,1971); Matheron & Armstrong (1987); McCammon(1975a); Merriam(1970); Pauncz(1978); Rendu(1978); Royle et al.(1980); Verly et al.(1984).

Polemics — for and against geostatistics: Henley(1987); Journel(1985,1986,1987); Krige(1986); Matheron (1967,1986); D.E.Myers(1986); Philip & Watson(1986b,1987b); Serra(1987); Shurtz(1985); Srivastava (1986).

Further discussion of specific theoretical aspects: Benest & Winter 1984; Brooker (1977,1986); Clark (1983); Cressie (1985); Cressie & Hawkins(1980); Dowd(1982); Dubrule(1983); Freund (1986); Hawkins & Cressie(1984); Heikkila & Koistinen(1984); Krige(1984,1985); Krige & Magri(1982a,b); Lemmer(1985); Magri & Longstaff(1979,1982); Magri & Mostert(1985); Matheron(1981); Olea(1975,1977); Philip & Watson (1985); Rendu(1979,1984a,b); Sichel(1966); Starks & Fang(1982); Starks et al.(1982); Vallee et al. (1982).

Worked examples & geological applications (mainly in mineral exploration): (Agterberg et al.1972; Camisani-Calzolari et al.1985; Davis 1986, p.239 & 383; Holm & Neal 1986; Magri & Mostert 1985; Matheron & Armstrong 1987; Miller & Morrison 1985; Wood 1985).

Statistical tables and graphs: *Means & confidence limits for logNormal populations* (Sichel 1966); *Means, paylimits, payvalues & % pay* (Krige 1981).

Stand-alone computer programs: Carr et al.(1985); Carr(1987); Clark(1977a); Rendu(1984b); Tough & Leyshon (1985); Yates et al.(1986).

Mainframe and microcomputer software: numerous proprietary packages are now available, both from software houses and the large computer manufacturers (e.g. MINEVAL, DATAMINE™).

Geostatistics (see Glossary definition) is one of few techniques in this book which has been developed primarily by and for geoscientists (most notably in France), and only then attracted the attention of statisticians, or been exported to other sciences. It is also the technique which has attracted most polemic in the literature (see above)!

Geostatistics is concerned with **regionalized variables**, which have continuity from point to point but vary in a manner too complex for simple mathematical description. Although it is beginning to find applications in other fields, attention will be confined here to mining, where it has found its main use to date. The word "grade" (of ore) will thus substitute for "value of a regionalized variable at a point in space". Geostatistics attempts (a) to model the form of a regionalized variable (such as the Au content of an ore-body) in 2- or 3-*D*; (b) to allow unbiased estimates of the grade to be made at a point *A* from known grades at points *B*,*C*... (in a similar manner to the 2nd stage of gridding in Fig.20.3); (c) to provide *confidence limits* on those estimates; and (d) to relate grades obtained from samples of different sizes (core samples, grab samples, stope blocks, truck-loads, etc.) Although other techniques (such as moving averages, triangulation, etc.) are capable of tasks (a) and (b), capability to perform (c) and (d) is unique to geostatistics.

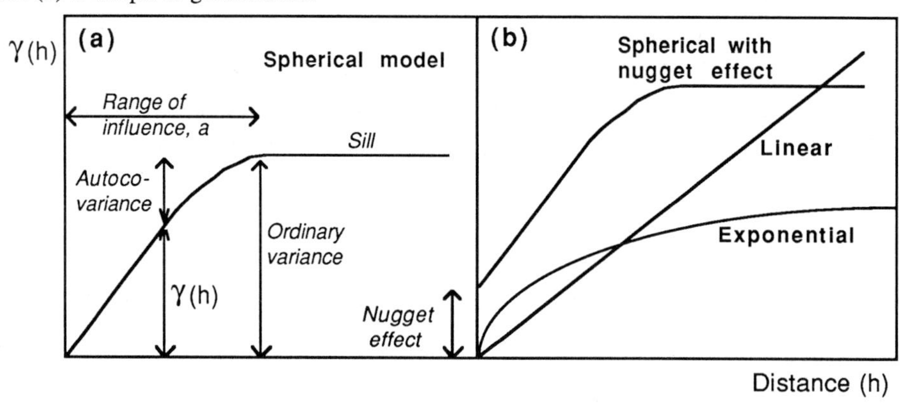

Fig.20.4. Idealized semivariogram models (for equations see [20.16])

20d1. Estimating rates of change of regionalized variables along specific trends: Semi-variograms

Fundamental to geostatistics is **semivariance**, notated γ_h or $\gamma(h)$, a measure of spatial dependence between samples in a specified direction, which may be vertical (i.e. depth), horizontal (N–S, E–W), etc. Semivariance is defined for a constant sampling spacing (interval between samples), Δ, as follows:

$$\gamma_h = \sum_{i=1}^{i=N-h} \frac{(X_i - X_{i+h})^2}{2N} \qquad [20.15]$$

where X_i is the grade at location *i*, and X_{i+h} the grade *h* intervals (i.e. Δh in absolute distance) away. **Semi-variograms** (sometimes incorrectly called "variograms", but "semi-" should be used because of the factor 2 in the denominator), are plots of γ_h against *h* (Fig.20.4). A different semi-variogram should be plotted for several different directions, to obtain a feel for the anisotropy in a deposit. Conventionally, N–S, NW–SE, E–W are used in surface analysis, while for boreholes *h* is clearly measured along the hole.

The semi-variogram indicates over what distance, and to what extent, grade at a given point influences grade at adjacent points; and conversely, therefore, how close together points must be for known grade at one point to be capable of predicting unknown grade at another. For $h = 0$, for example, grade is being compared with itself, so normally $\gamma_o = 0$, except where a so-called **nugget effect** is present (Fig.20.4b), which implies that grade is highly variable at distances less than the sampling interval. As h increases, grades become less and less closely related and γ_h also therefore increases. In most deposits, a **sill** (Fig.20.4a) is eventually reached, where γ_h equals the overall variance (i.e. that around the average grade). This corresponds to a point separation called the **range of influence**, or **span,** a, beyond which grades can be considered unrelated (Fig.20.4a). We can only usefully predict grades between points less than a apart, but if we do use points within this separation, the semi-variogram provides the proper *weightings* which should apply to grades at different separations.

When a semivariogram is calculated for an ore-body, an irregular spread of points is usually obtained, and one of the most subjective tasks of the geostatistician is to identify a *continuous,* theoretical, semi-variogram model which most closely follows the real data. Davis(1986,p.245) describes this as "to a certain extent an art, requiring experience, patience and sometimes luck. The process is not altogether satisfying, because the conclusions are not unique". Fitting is done essentially by trial-and-error, rather than by formal statistics. A 'good' fit allows γ_h to be confidently estimated for *any* value of h, rather than only at the sampled points. The main theoretical models (Fig.20.4) are:

Spherical model: $$\gamma_h = \sigma^2\left[\frac{3h}{2a} - \frac{h^3}{2a^3}\right] \qquad [20.16a]$$

Exponential: $$\gamma_h = \sigma^2(1 - e^{-h/a}) \qquad [20.16b]$$

Linear: $$\gamma_h = \alpha h \qquad [20.16c]$$

De Wijsian: $$\gamma_h = 3\alpha \ln(h) \qquad [20.16d]$$

Generalized linear: $$\gamma_h = \alpha h^p \qquad [20.16e]$$

where a is the range of influence, as above, and α is a constant (the slope). The above equations apply where $h < a$; for $h > a$, $\gamma_h = \sigma^2$, the ordinary variance. Hence the sill can also be defined for a stationary regionalized variable as the point at which the semivariance reaches the ordinary variance. At smaller values of h, the sum of the **autocovariance** (§16a6) and the semivariance also equals the ordinary variance (Fig.20.4a). Because of this, the semivariance is also a mirror image of the autocorrelation for a *standardized* stationary variable (Davis 1986, p.243).

<u>20d2. Kriging</u>

To estimate the grade X_A at point A from points $B,C...$, we have much the same problem as in the gridding method for contouring (Fig.20.3b), and need to use some kind of weighted average:

$$E_A = \Sigma w_i X_i = w_B X_B + w_C X_C + ... \qquad [20.17]$$

where E indicates the estimate. The weights w can be estimated in an infinite number of ways, but it is intuitively obvious that w should vary inversely with d. the separation between the known and unknown points (i.e. points farther away should have less influence). Consequently, by analogy with many physical phenomena such as gravity and electric charge, an inverse square relationship was popular in the past, i.e. $w \propto 1/d^2$.

An alternative approach is based on the estimation error between the true value X_A and its estimate, $\varepsilon_A = [E_A - X_A]$. Although, if $\Sigma w_i = 1$, E_A is **unbiased** (i.e. average $\varepsilon_A = 0$, provided there is no drift), ε_A values may scatter widely about zero, and their scatter can be expressed as an *error variance*:

$$s_\varepsilon^2 = \frac{1}{N}\sum \varepsilon^2 = \frac{1}{N}\sum (E - X)^2 \qquad [20.18]$$

This bears a clear relationship to [20.15], and it turns out that kriging, which bases w on γ_h, provides a unique combination of weights which not only give the *minimum possible* error variance, but also indicates the magnitude of the error. Neither of these is obtainable from the inverse-square method. Kriging has the further advantage that it is an "exact interpolator" — it predicts the actual values at the known points exactly. This means that if kriging is used for contouring, the estimate surface will pass *exactly* through with the control points, where its confidence bands will be of zero width.

Numerous different kinds of kriging are in use: for example, **punctual** (simple) **kriging** is used where the measured variable is **stationary**, and **universal** (general) **kriging** where it is not. The formal equivalence of kriging with **splines** is shown by Matheron (1981): any curve fitted by a spline function can also be identified using kriging, and vice-versa.

Since deriving even simple kriging equations requires calculus, and solving them requires matrix algebra and esoteric mathematical devices (Lagrangian multipliers), neither is attempted here. In practice, geostatistics is virtually unmanageable without a computer program (which in itself introduces the "little knowledge is a dangerous thing" syndrome). Suffice it to say that if N known points are being used to estimate grade at unknown point A, the ws are obtained by solving N simultaneous equations of the form below, together with the equation $\Sigma w_i = 1$:

$$w_1\,\gamma(h_{11}) + w_2\,\gamma(h_{i2}) + \ldots\ldots w_N\,\gamma(h_{iN}) = \gamma(h_{iA}) \qquad [20.19]$$

where $\gamma(h_{ij})$ is the semivariance corresponding to the separation h_{ij} between points i and j, and is read off the semi-variogram previously modelled. The Lagrangian multipliers enter because the set of equations is overdetermined ([N + 1] equations in N unknowns). Once the weights have actually been determined, the variance of the estimate E_A can also be obtained using the semivariances $\gamma(h_{i1})$.

In *universal kriging*, a regionalized variable is divided into **drift** (its nonstationary component, a kind of trend-surface) and **residual** (the difference between the actual values and the drift). The drift is first estimated and stripped away, the stationary residual is then kriged to obtain the estimates required, and finally the two are recombined to estimate the actual surface. The mathematics are excessively complicated for incorporation into the present Notes.

One legitimate objection to kriging is that it is **parametric**: confidence intervals and the like can only be derived by assuming something about the underlying distribution. In practice, geostatisticians assume Normality or logNormality, and the dangers of this were emphasised earlier (§8-9). However, robust and even nonparametric forms of kriging are now emerging (Henley 1976).

TOPIC 21. CLASSIFYING OBJECTS FROM FIRST PRINCIPLES

It is useful to distinguish **classification** from **assignment,** covered in §22. If we take a series of measurements on, say, an assemblage of conodonts, and, without pre-conceived palaeontological criteria, attempt to assess relationships *purely* from these measurements, we are *classifying*. If, by contrast, we assign the conodonts to known species by comparing them with established species, we are *assigning*. Classification is a 'suck-it-and-see', exploratory approach with no *a priori* framework; assignment is a follow-up approach where some established framework already exists.

A set of 52 analyses of metalimestones in the Scottish Highlands is chosen to illustrate §21-23, because they lie midway between these two techniques. These limestones come from 6 isolated outcrops in structurally complex terrain, cut by numerous faults and major slides. Country-rocks are Proterozoic, psammitic metasediments of the 'Moine' assemblage. This assemblage is elsewhere free of calcareous rocks, which raises the strong possibility that these limestones are actually tectonic slices of a limestone-rich, younger (Vendian–Cambrian) 'Dalradian' Supergroup, whose main outcrop lies to the SE. There does therefore exist an *a priori* framework for assignment in this case, namely the 6 outcrops, but it is not known whether the outcrops actually correspond to geologically meaningful units (formations, etc.). Hence this particular problem can be attacked from both directions: classification to see whether the limestone chemistry groups the analyzed specimens according to their source outcrops, and assignment to see whether the chemistries of each source outcrop are in fact distinct. Further details can be found in Rock(1986c,d,e), Rock & Waterhouse (1986) and Rock et al.(1984, 1986b). The data are displayed in Table 21.3.

Before tackling this problem, however, we briefly consider how similarity is measured between multivariate objects measured on all the various scales of measurement.

21a. Measuring similarity and dissimilarity between multivariate objects: MATCHING AND DISTANCE COEFFICIENTS

Further geological discussion: Burnaby(1970); Cheetham & Hazel(1969); De Capeariis(1976); Gower (1970,1971); Hay & Southam(1978); Hazel(1970); Hohn(1976); Peschel & Zezel(1977); Sepkuski (1974).
Worked geological examples: Till (1974, p.134); Davis (1986, p.503).
Selected geological applications: *Sedimentology*: (Bonham-Carter 1965; Hennebert & Lees 1985).
Stand-alone computer programs: Millendorf et al.(1979) — used for results in Table 21.1.
Mainframe software: covered in most large statistics packages (e.g. $SPSS^X$).
Microcomputer software: none known.

Similarities between objects (Q-mode) or variables (R-mode) can be expressed by a wide variety of coefficients (Table 21.1). **Correlation coefficients** (§13a–c) are sometimes used Q-mode, but the intrinsic meaning of this procedure is unclear (e.g. how can the *X-Y* plots in Fig.13.1 be interpreted or generated with axes defined by 'rocks' rather than chemical elements?) On the other hand, **distance coefficients**, though conventionally Q-mode, are fairly commonly used R-mode. The main difference is that correlation coefficients lie between –1 and +1, with zero indicating total dissimilarity, whereas distance coefficients lie between 0 and ∞, with zero indicating complete identity.

Table 21.1. Conventional usage of similarity/dissimilarity coefficients according to mode and data-type

Data-type	Q-mode	R-mode
Dichotomous/nominal	**Matching coefficients**(Table 13.1 & §21a1)	Matching coefficients might be used
Ordinal/interval/ratio	**Distance coefficients** (§21a2)	**Correlation coefficients** (§13a-c)

21a1. Measuring similarity between dichotomous/nominal data: MATCHING COEFFICIENTS

At least 20 coefficients have been proposed (Table 13.1 and Glossary). Although all have the same aim of converting originally dichotomous/nominal data into decimal numbers between 0 and 1, different coefficients place varying emphasis on **positive** and **negative matches** and **mis-matches**. Table 21.2 compares 5 matching coefficient matrices for the same set of data. Although the *absolute* values of the coefficients vary considerably (e.g. the similarity between rocks 4 & 5 varies between 0.1250 and 0.5333), *relative* values within each matrix (which would actually control the results of a cluster analysis based on these coefficients) are much less variable. Thus if the 10 coefficients < 1 are *ranked* down each column, or over the whole of each matrix, the results for matrices (c) – (f) are almost identical to those for matrix (b) in Table 21.2 (see g,h). Hence the choice of matching coefficient is not so critical (at least in this example) as it might first appear. Nevertheless, if the user wishes to stress negative matches (attributes absent in both specimens), he should obviously choose a coefficient which takes these into account (e.g. Sokal & Sneath) rather than one which ignores them (e.g. Jaccard).

Table 21.2. Comparison of 5 different Q-mode matching coefficient matrices for the same data*

(a) Raw data†

Rock	Variables
1	000000110110110100110000100100
2	100000111110000110000100100100
3	000000100100111110100000100100
4	000000000100110100100000111000
5	111110110100000100000000000100

(b) Dice coefficient

Rock	1	2	3	4	5
1	1.0000				
2	0.6364	1.0000			
3	0.7619	0.5714	1.0000		
4	0.6316	0.3158	0.6667	1.0000	
5	0.4762	0.5714	0.4000	0.2222	1.0000

(c) Jaccard coefficient ‡

Rock	1	2	3	4	5
1	1.0000				
2	0.4667	1.0000			
3	0.6154	0.4000	1.0000		
4	0.4615	0.1875	0.5000	1.0000	
5	0.3125	0.4000	0.2500	0.1250	1.0000

(d) Simpson coefficient

Rock	1	2	3	4	5
1	1.0000				
2	0.6364	1.0000			
3	0.8000	0.6000	1.0000		
4	0.7500	0.3750	0.7500	1.0000	
5	0.5000	0.6000	0.4000	0.2500	1.0000

(e) Sokal & Sneath simple matching coefficient

Rock	1	2	3	4	5
1	1.0000				
2	0.7333	1.0000			
3	0.8333	0.7000	1.0000		
4	0.7667	0.5667	0.8000	1.0000	
5	0.6333	0.7000	0.6000	0.5333	1.0000

(f) Otsuka coefficient

Rock	1	2	3	4	5
1	1.0000				
2	0.6364	1.0000			
3	0.7628	0.5721	1.0000		
4	0.6396	0.3198	0.6708	1.0000	
5	0.4767	0.5721	0.4000	0.2236	1.0000

(g) Ranks of coefficients for matrix (b) by column¶

Rock	1	2	3	4
2	3			
3	4	2.5		
4	2	1	2	
5	1	2.5	1	(1)

(h) Ranks of coefficients for matrix (b) overall¶

Rock	1	2	3	4
2	8			
3	10	5.5		
4	7	2	9	
5	4	5.5	3	1

* Matrices generated by program of Millendorf et al.(1969).

† Data indicate the presence or absence of 30 fossil types in 5 rock specimens, adapted from Till(1974, p.136) with Till's '2' here changed to '1', indicating 'present', and his '1' to '0', indicating 'absent'.

‡ Agrees with values in Till(1974, table 7.13).

¶ Coefficients of 1.0 ignored (leaving 10 values to compare).

Table 21.3. Limestone data used to illustrate techniques in Topics 21-23

	Label	Outcrop	La	Ce	Cr	Ni	Cu	Zn	Rb	Sr	Zr	Ba	SiO2	Al2O3	Fe2O3	MgO
1	S70859 G	Findhorn	7	21	0	4	0	32	13	1660	21	5	16.32	1.57	1.06	4.09
2	S70860 G	Findhorn	29	32	0	5	0	35	20	1587	38	21	15.07	2.05	1.54	3.09
3	KYA G	Findhorn	0	0	3	11	0	11	1	1367	0	5	2.78	.27	.19	.70
4	KYC G	Findhorn	0	0	6	10	0	16	4	1671	0	8	4.41	.31	.22	1.45
5	KYB G	Findhorn	0	2	4	2	0	5	9	1368	14	12	5.08	.53	.23	1.34
6	KYD G	Findhorn	0	7	4	2	0	4	3	1247	6	3	1.69	.24	.12	.58
7	KYE G	Findhorn	1	14	3	0	0	6	3	1783	11	5	4.80	.54	.22	1.51
8	S70820 G	Glen Lui	12	0	5	2	3	9	37	1738	7	22	12.25	2.74	.89	.55
9	S70816 G	Glen Lui	21	0	8	1	0	8	45	1562	29	20	14.22	2.82	.97	.60
10	S70818 G	Glen Lui	6	0	0	1	0	9	23	1845	0	14	6.88	1.08	.42	.34
11	S70822	Glen Lui	5	21	0	6	5	8	15	2133	7	1	7.50	1.40	.70	.60
12	GRe G	Grantown	2	75	0	0	0	4	5	2012	0	107	3.49	.65	.24	1.04
13	GRb G	Grantown	14	77	12	7	8	51	1	1160	52	81	19.53	3.31	1.77	4.61
14	GRC G	Grantown	3	84	0	0	2	15	1	1676	2	42	4.78	.52	.31	1.72
15	GRD G	Grantown	1	97	0	0	0	7	0	1683	0	22	2.73	.38	.34	1.39
16	GRE G	Grantown	4	84	1	0	2	4	7	2007	1	100	3.90	.70	.26	1.07
17	GRG G	Grantown	26	28	17	7	0	20	33	1045	20	16	16.61	2.22	1.05	8.11
18	S70799 G	Grantown	0	11	0	0	0	2	4	1105	12	7	3.82	.70	.22	.77
19	S70844 G	Kincraig	0	1	0	6	14	10	27	2196	2	14	9.19	1.88	.66	.38
20	AS4 G	Kincraig	7	77	14	3	3	10	19	1855	20	170	9.90	2.48	1.04	.85
21	AS3 G	Kincraig	9	93	12	2	9	12	12	1561	20	119	9.88	2.61	1.45	.96
22	ASII G	Kincraig	11	75	11	3	5	19	18	1961	9	107	7.63	2.01	.73	.72
23	ASI G	Kincraig	7	95	7	0	1	5	9	2189	0	64	5.97	1.53	.73	.73
24	S34502 C	Loch an Eilean	0	10	0	1	4	7	11	2763	21	80	6.80	1.59	.64	.56
25	OB G	Loch an Eilean	11	82	10	2	6	9	19	2906	210	127	10.46	2.53	1.04	1.48
26	OBa G	Loch an Eilean	7	88	8	0	3	4	13	2072	10	120	7.20	1.55	.69	.68
27	OBL G	Loch an Eilean	12	86	18	3	2	7	16	2052	2	126	7.98	2.06	.93	.77
28	OBM G	Loch an Eilean	2	82	16	7	0	14	4	2254	9	87	9.01	2.46	1.21	1.10
29	OBK G	Loch an Eilean	9	83	25	1	2	5	23	1994	28	148	12.62	2.69	.92	.78
30	OBg G	Loch an Eilean	7	86	1	5	0	4	20	2191	0	144	9.70	2.46	.83	.75
31	OBe G	Loch an Eilean	7	91	6	0	1	3	11	2226	0	72	4.54	1.06	.61	.69
32	OBh G	Loch an Eilean	8	84	11	0	1	5	20	2174	0	96	8.26	1.95	.72	.72
33	OBb G	Loch an Eilean	9	81	11	1	2	9	15	2175	19	89	8.90	2.01	.71	.73
34	OBd G	Loch an Eilean	0	83	15	2	4	7	17	2175	17	109	4.82	.53	.31	1.66
35	OBJ G	Loch an Eilean	11	91	14	3	4	7	16	2010	9	122	8.66	2.37	1.00	.81
36	OBC G	Loch an Eilean	9	85	15	1	3	5	17	2180	24	125	10.00	1.99	.77	.76
37	OBF G	Loch an Eilean	0	0	20	1	0	7	25	2107	36	0	13.94	3.00	.90	.80
38	OBN G	Loch an Eilean	8	92	18	6	8	8	18	2291	1	117	8.40	2.14	.90	.85
39	S34501 C	Kinlochlaggan	0	0	0	0	11	40	0	1698	0	20	1.68	.40	.64	2.38
40	KIN4782P	Kinlochlaggan	2	27	17	0	17	3	12	2090	121	71	4.40	1.04	.56	.73
41	KIN4788P	Kinlochlaggan	4	40	12	2	19	13	23	1529	92	26	7.12	1.64	.56	.95
42	LS1 G	Kinlochlaggan	2	77	0	1	1	13	10	2018	109	0	5.80	.89	.41	.90
43	GSQ1 G	Kinlochlaggan	12	84	5	2	18	7	23	1830	27	62	12.48	2.00	.50	.62
44	GSQ2 G	Kinlochlaggan	0	89	0	0	0	6	2	1527	0	18	10.47	2.34	.78	.91
45	AN94d G	Kinlochlaggan	8	88	0	0	0	22	17	1858	0	43	5.56	1.39	.69	.84
46	AN94e G	Kinlochlaggan	7	82	0	0	5	23	5	1899	0	16	2.33	.51	.60	2.15
47	AN42VI G	Kinlochlaggan	16	100	0	0	1	3	3	893	0	27	9.66	.48	.86	.76
48	AN42a2 G	Kinlochlaggan	7	95	9	1	5	5	21	2041	0	73	9.48	2.11	.64	.84
49	AN94a G	Kinlochlaggan	2	91	0	0	0	15	9	1717	0	16	.97	.34	.60	2.08
50	AN42a1 G	Kinlochlaggan	9	80	10	2	1	4	17	2207	2	90	7.60	1.90	.68	.66
51	GS49a G	Kinlochlaggan	1	97	8	0	6	3	6	2014	0	64	4.12	.95	.59	.70
52	GSq3	Kinlochlaggan	0	87	6	1	4	6	15	2147	0	120	2.10	.30	1.10	.80

The analyses represent systematically collected samples covering the full range of macroscopic variation in 6 isolated clusters of limestone (marble) outcrops within otherwise limestone-free metamorphic terranes of the Grampian Highlands of Scotland. The rocks are at least late Proterozoic in age and may be rather older, but have suffered Caledonian metamorphism and disruption at c. 500 Ma.

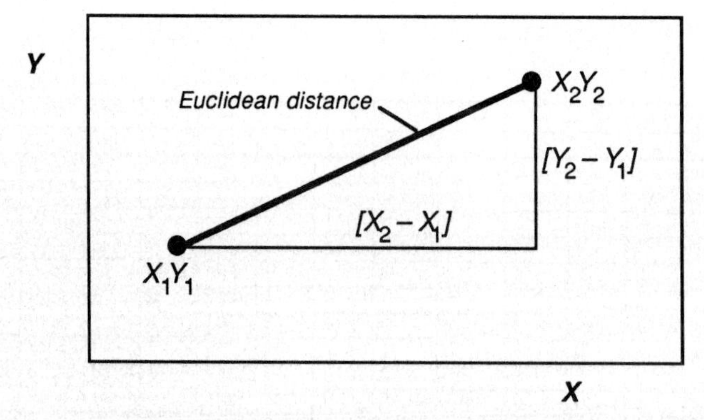

Fig.21.1. Simple geometrical interpretation of the Euclidean distance in 2-D.
The full multivariate Euclidean distance defined via [21.1] is simply an extension of this concept to *v*-dimensions

21a2. Measuring similarity between ordinal to ratio data: DISTANCE COEFFICIENTS

Distances between two multivariate objects X and Y defined on v variables can be expressed as:

$$d = \left[\sum_{i=1}^{i=v} (Y_i - X_i)^p \right]^{\frac{1}{p}} \qquad [21.1]$$

For $p = 2$, [21.1] defines **Euclidean distance**, the *v*-dimensional equivalent of Pythagorean distance (Fig.21.1). For $p = 1$, using the *absolute* differences $|Y_i - X_i|$, d is the **City Block (Manhattan) distance**, while the general form in [21.1] itself is called the **Minkowski distance.**

Table 21.4 compares matrices of various similarity/distance coefficients obtainable from the data of Table 21.3; *any* of these matrices could be used as input to the techniques outlined in §21-23.

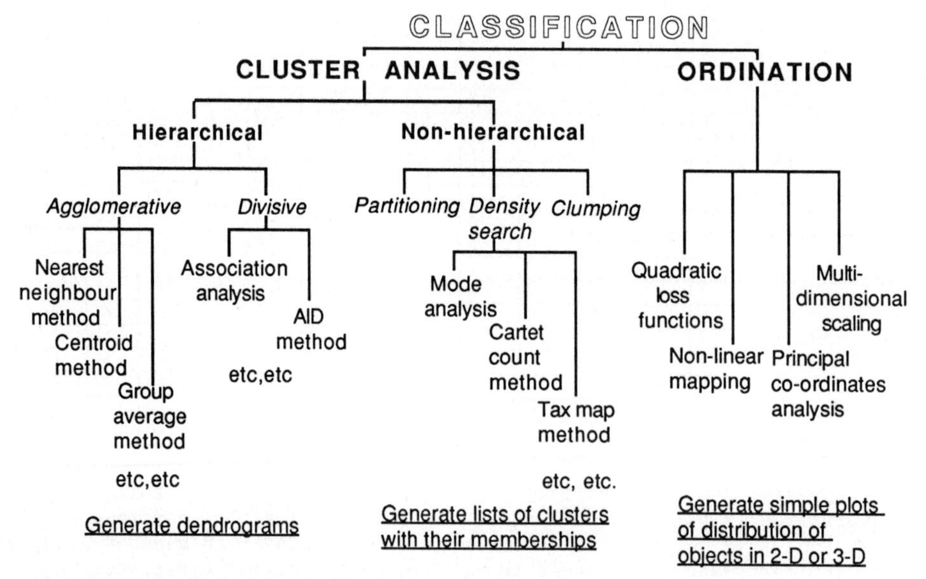

Fig.21.2. Relationship between classification methods covered in Topic 21 (only a selection of techniques is given).

Table 21.4. Alternative R-mode input matrices for techniques in §21-23, all derived from the limestone data in Table 21.3

(a) VARIANCE-COVARIANCE MATRIX (*Variances* on diagonal; covariances off diagonal)

	La	Ce	Cr	Ni	Cu	Zn	Rb	Sr	Zr	Ba	Si	Al	Fe	Mg
La	*42.369*													
Ce	30.211	*1411.7*												
Cr	9.802	51.892	*46.195*											
Ni	2.854	-31.04	2.870	*7.335*										
Cu	-1.620	-0.098	6.273	0.125	*23.038*									
Zn	19.562	-58.85	-8.671	8.964	5.427	*98.453*								
Rb	33.703	-68.85	24.129	1.747	6.747	-12.29	*95.072*							
Sr	-530.3	4030.3	572.37	-153.7	243.06	-1199.	715.10	*156053*						
Zr	29.591	-57.92	59.035	-2.605	65.878	30.300	58.926	3498.4	*1404.0*					
Ba	43.383	1160.8	181.15	-10.04	28.030	-112.3	62.455	10081.	142.62	*2436.8*				
Si	18.743	-7.378	10.591	3.565	0.891	14.230	22.407	-198.1	34.924	21.836	*17.82*			
Al	2.985	3.982	3.278	0.478	0.562	1.122	5.117	70.830	7.663	16.886	3.076	*0.766*		
Fe	1.421	3.043	0.922	0.230	0.165	1.635	1.111	4.500	2.429	5.974	1.087	0.228	*0.125*	
Mg	3.657	-5.123	0.522	1.092	-0.564	8.131	0.151	-207.5	4.034	-13.34	2.254	0.093	0.156	*1.638*

(b) MATRIX OF PEARSON's *r* (§13a; a variance-covariance matrix (a) on the *standardized* data would be identical)

Ce	0.124	1.000												
Cr	0.222	0.203	1.000											
Ni	0.162	-0.305	0.156	1.000										
Cu	-0.052	-0.001	0.192	0.010	1.000									
Zn	0.303	-0.158	-0.129	0.334	0.114	1.000								
Rb	0.531	-0.188	0.364	0.066	0.144	-0.127	1.000							
Sr	-0.206	0.272	0.213	-0.144	0.128	-0.306	0.186	1.000						
Zr	0.121	-0.041	0.232	-0.026	0.366	0.081	0.161	0.236	1.000					
Ba	0.135	0.626	0.540	-0.075	0.118	-0.229	0.130	0.517	0.077	1.000				
Si	0.682	-0.047	0.369	0.312	0.044	0.340	0.544	-0.119	0.221	0.105	1.000			
Al	0.524	0.121	0.551	0.202	0.134	0.129	0.600	0.205	0.234	0.391	0.832	1.000		
Fe	0.617	0.229	0.383	0.240	0.097	0.466	0.322	0.032	0.183	0.342	0.727	0.737	1.000	
Mg	0.439	-0.107	0.060	0.315	-0.092	0.640	0.012	-0.411	0.084	-0.211	0.417	0.083	0.344	1.0

(c) MATRIX OF NORMALIZED EUCLIDEAN DISTANCES (§21a2) ON **RAW** DATA-MATRIX

Ce	64.05	0.000												
Cr	8.240	63.326	0.000											
Ni	7.824	68.202	8.344	0.000										
Cu	8.730	66.815	8.237	5.582	0.000									
Zn	10.931	62.352	13.201	12.717	12.791	0.000								
Rb	11.075	60.286	11.740	15.214	14.465	14.912	0.000							
Sr	1908.3	1855.1	1907.3	1912.3	1910.9	1904.9	1900.5	0.000						
Zr	39.091	66.520	38.222	41.048	39.038	38.555	37.230	1894.0	0.000					
Ba	72.973	38.766	70.726	76.665	75.303	72.252	67.760	1850.0	72.255	0.000				
Si	4.900	63.377	6.531	6.977	7.537	9.793	10.155	1906.9	38.226	72.120	0.000			
Al	7.788	68.209	8.319	2.730	5.099	13.518	15.310	1912.8	41.010	76.794	7.148	0.000		
Fe	8.473	68.904	9.096	3.012	5.476	14.038	16.196	1913.7	41.489	77.563	8.076	1.045	0.000	
Mg	7.926	68.564	8.852	2.743	5.490	13.221	15.857	1913.2	41.221	77.386	7.542	1.498	1.316	0.000

(d) MATRIX OF NORMALIZED EUCLIDEAN DISTANCES (§21a2) ON **STANDARDIZED** DATA-MATRIX

Ce	1.311	0.000												
Cr	1.236	1.250	0.000											
Ni	1.282	1.600	1.287	0.000										
Cu	1.436	1.401	1.259	1.394	0.000									
Zn	1.169	1.507	1.488	1.143	1.318	0.000								
Rb	0.959	1.527	1.117	1.353	1.296	1.487	0.000							
Sr	1.538	1.195	1.242	1.498	1.308	1.601	1.264	0.000						
Zr	1.313	1.429	1.228	1.418	1.115	1.342	1.283	1.224	0.000					
Ba	1.215	1.437	1.236	1.245	1.679	1.553	1.307	0.973	1.345	0.000				
Si	0.997	1.125	1.197	1.286	1.325	1.138	0.945	1.481	1.236	1.325	0.000			
Al	1.543	1.376	1.184	1.736	1.746	1.307	0.886	1.249	1.226	1.093	0.573	0.000		
Fe	1.657	1.475	1.364	1.394	0.897	1.024	1.153	1.378	1.266	1.136	0.732	0.718	0.000	
Mg	1.049	1.473	1.358	1.159	1.463	0.840	1.392	1.663	1.340	1.541	1.069	1.341	1.134	0.0

Table 21.4 (contd.)

(e) MATRIX OF SPEARMAN'S ρ RANK CORRELATION COEFFICIENT (§13b1)

	La	Ce	Cr	Ni	Cu	Zn	Rb	Sr	Zr	Ba	Si	Al	Fe	Mg
Ce	0.261	1.000												
Cr	0.324	0.131	1.000											
Ni	0.206	-0.333	0.328	1.000										
Cu	0.165	0.185	0.299	0.071	1.000									
Zn	0.163	-0.302	-0.104	0.436	0.064	1.000								
Rb	0.467	-0.128	0.400	0.325	0.184	0.058	1.000							
Sr	-0.025	0.239	0.280	-0.011	0.298	-0.262	0.287	1.000						
Zr	0.270	-0.339	0.400	0.313	0.232	0.182	0.347	-0.036	1.000					
Ba	0.408	0.534	0.529	0.026	0.395	-0.240	0.236	0.483	0.047	1.000				
Si	0.645	-0.022	0.380	0.457	0.063	0.199	0.592	-0.006	0.519	0.202	1.000			
Al	0.594	0.036	0.546	0.381	0.184	0.133	0.613	0.200	0.479	0.402	0.866	1.000		
Fe	0.608	0.213	0.410	0.363	0.149	0.285	0.404	0.144	0.286	0.425	0.733	0.740	1.000	
Mg	-0.032	0.100	-0.025	0.026	-0.125	0.476	-0.328	-0.317	0.103	-0.042	-0.018	-0.095	0.117	1.0

(f) MATRIX OF KENDALL'S τ RANK CORRELATION COEFFICIENT (§13b2)

Ce	0.187	1.000												
Cr	0.237	0.073	1.000											
Ni	0.165	-0.259	0.256	1.000										
Cu	0.137	0.136	0.233	0.070	1.000									
Zn	0.118	-0.218	-0.068	0.358	0.059	1.000								
Rb	0.367	-0.099	0.293	0.255	0.146	0.055	1.000							
Sr	-0.006	0.160	0.194	0.005	0.204	-0.189	0.191	1.000						
Zr	0.199	-0.244	0.299	0.219	0.178	0.143	0.252	-0.020	1.000					
Ba	0.279	0.342	0.401	0.029	0.273	-0.149	0.179	0.331	0.036	1.000				
Si	0.508	-0.012	0.282	0.342	0.054	0.143	0.462	0.019	0.387	0.151	1.000			
Al	0.485	0.036	0.385	0.284	0.140	0.095	0.490	0.155	0.352	0.300	0.718	1.000		
Fe	0.486	0.147	0.284	0.287	0.099	0.221	0.296	0.129	0.221	0.319	0.575	0.584	1.000	
Mg	-0.042	0.068	0.002	0.017	-0.085	0.349	-0.234	-0.235	0.074	-0.029	-0.017	-0.068	0.119	1.0

(g) MATRIX OF GOODMAN & KRUSKAL'S γ COEFFICIENT OF WEAK MONOTONICITY (§13h)

Ce	0.199	1.000												
Cr	0.265	0.080	1.000											
Ni	0.192	-0.292	0.294	1.000										
Cu	0.156	0.150	0.267	0.086	1.000									
Zn	0.128	-0.228	-0.075	0.405	0.067	1.000								
Rb	0.392	-0.102	0.317	0.286	0.161	0.057	1.000							
Sr	-0.007	0.162	0.207	0.006	0.223	-0.194	0.193	1.000						
Zr	0.223	-0.265	0.334	0.250	0.205	0.157	0.274	-0.022	1.000					
Ba	0.295	0.349	0.429	0.033	0.299	-0.154	0.183	0.332	0.039	1.000				
Si	0.535	-0.012	0.300	0.379	0.059	0.147	0.469	0.019	0.415	0.152	1.000			
Al	0.510	0.037	0.410	0.315	0.153	0.098	0.498	0.155	0.378	0.302	0.719	1.000		
Fe	0.513	0.150	0.304	0.320	0.109	0.228	0.301	0.130	0.238	0.322	0.578	0.587	1.000	
Mg	-0.045	0.069	0.003	0.019	-0.094	0.360	-0.239	-0.236	0.079	-0.029	-0.017	-0.068	0.120	1.0

(h) MATRIX OF GUTTMAN'S μ_2 COEFFICIENT OF WEAK MONOTONICITY (§13h)

Ce	0.199	1.000												
Cr	0.362	0.339	1.000											
Ni	0.273	-0.510	0.269	1.000										
Cu	-0.107	-0.001	0.347	0.020	1.000									
Zn	0.501	-0.311	-0.252	0.582	0.234	1.000								
Rb	0.720	-0.282	0.560	0.109	0.264	-0.238	1.000							
Sr	-0.309	0.422	0.344	-0.240	0.245	-0.536	0.281	1.000						
Zr	0.293	-0.105	0.505	-0.073	0.674	0.219	0.393	0.427	1.000					
Ba	0.225	0.861	0.717	-0.130	0.220	-0.449	0.204	0.750	0.170	1.000				
Si	0.873	-0.074	0.550	0.480	0.082	0.465	0.712	-0.177	0.482	0.163	1.000			
Al	0.755	0.190	0.754	0.313	0.250	0.220	0.771	0.306	0.500	0.556	0.952	1.000		
Fe	0.841	0.375	0.588	0.358	0.184	0.675	0.485	0.048	0.402	0.518	0.883	0.897	1.000	
Mg	0.664	-0.247	0.130	0.593	-0.260	0.921	0.023	-0.730	0.286	-0.509	0.603	0.176	0.614	1.0

21b. Producing Dendrograms: HIERARCHICAL CLUSTER ANALYSIS

Fig.21.2 shows the relationship between the various classification methods covered in this Topic, in the form of a dendrogram ('tree diagram') — the diagnostic output of hierarchical cluster analysis.

21b1. An Overview of Available Methods

Dedicated monographs and reviews: Aldenderfer & Blashfield(1984); Anderberg (1973); Bijnen (1973); Blashfield & Aldenderfer (1978); Duran & Odell(1974); Everitt(1974,1979); Gordon(1981); Jambu & Lobeaux(1983); Klastorin (1983); Massart & Kaufman (1983); Tryon & Bailey(1970); Zupan(1982).
Further geological discussion: Gill & Tipper (1978); Griffiths (1983); Le Maitre(1982, ch.10); Mancey (1982); McCammon(1968); Michie(1978); Parks(1966); Rickwood (1983); Warshauer & Smosna (1981); *effects of data-closure* (Butler1978,1979c).
Further non-geological discussion: Gower (1967); Mather(1976,p.316ff).
Worked geological examples: *glaciology* (Mather 1976,p.342); *igneous petrology* (Le Maitre 1982, p.168); *palaeontology* (Miller & Kahn 1962,p.295); *sedimentology* (Till 1974, p.137; Davis 1986, p.504).
Selected geological applications: *geochemistry* (Obial 1970); *glaciology* (Miller 1979); *igneous petrology* (Ferguson 1973; Godinho 1976b; Joyce 1973; Lenthall 1972; Rickwood et al.1983; Stephens & Dawson 1977; Stone & Exley 1978; Upadhyay et al. 1986; Whitten 1985; Whitten et al.1987); *mineralogy* (Jago & Mitchell 1988); *mineral exploration* (Collyer & Merriam 1973; Pirkle et al.1984); *palaeontology* (Oltz 1971; Valentine & Peddicord 1967); *sedimentology* (Bonham-Carter 1965; Buchbinder & Gill 1976; Rao & Naqvi 1977); *stratigraphy* (Rock et al.1984).
Statistical tests for significance of clusters: Sneath(1977,1979a,1985); Sokal & Rohlf (1962).
Stand-alone computer programs: Coradini et al.(1977), Davis(1973,p.467ff); Grimm (1987), Hill(1979), Jacob (1975), Jambu (1981), Jones & Baker (1975), Lance & Williams(1967), Mather(1976,p.381), Sneath(1979b,1985), Späth(1980), Tipper(1979a), Wishart(1969,1978); also CLUSTER, available from H-J.Mucha, Inst.Mathematics, Acad.Sci.DDR, Mohrenstraße 39, Postfach 1304, Berlin, DDR-1086. *Libraries of algorithms*: Anderberg(1973), Hartigan(1974), Jambu & Lebeaux(1983), Zupan(1982).
Mainframe software: covered in most statistics packages (e.g. SPSS) or libraries (e.g. NAG).
Microcomputer software: programmed in some large statistics packages (e.g. SYSTAT).

Dendrograms (Fig.21.4) are an appealing method of displaying relationships between multivariate objects. Closest relationships are between objects nearest together. Davis (1986, p.502) gives a blow-by-blow explanation of how dendrograms are generated. They do, however, have 3 disadvantages.
(1) *A surfeit of options.* A plethora of alternative algorithms is available, based on 5 main divisions of choice (Table 21.5)— see Glossary for further details. Choices of input data, matrix and similarity measure have been dealt with above. (Note that, once any of the matrices as in Tables 21.1 or 21.2 have been generated, the original scale of measurement of the raw data — nominal or whatever — is no longer material). Perhaps the broadest choice is that of *method*, where **agglomerative** algorithms start with the objects and successively amalgamate them into the final dendrogram, whereas **divisive** algorithms start with the complete data-set and successively divide it. **Linkage** refers to how the distance between an *object* and a *cluster*, or between *two clusters* (as opposed to two individual objects) is measured, and hence how a decision is made whether or not to link the object to that cluster, or to fuse the two clusters. Such a decision may reflect the similarity of their two closest or farthest individuals (**single linkage**), the average similarity of all pairs of individuals within the clusters (**average linkage**), or the distances between their centroids (**centroid linkage**). Finally, pooling or separation of individuals can be made using *one* measured attribute/variable (**monothetic**), *several* attributes (**polythetic**) or *all available* attributes (**omnithetic**).

Fig.21.4 shows only a selection of many possible dendrograms from data of Table 21.3. Even though all were derived by the same agglomerative, polythetic algorithm — differing merely in the choice of input matrix, similarity measure and linkage method— they appear very different. How, then, is the bemused user to choose between them? Probably only by following convention— the most widely used distance measures being correlation coefficients (especially Pearson's r) and the Euclidean distance. Many authors recommend the Euclidean because, as already mentioned, the "correlation" between two limestones does not have any obvious geological meaning. Furthermore, a plot of one limestone versus another in the present example looks like Fig.21.3 — the value for Pearson's r between *any* pair of limestones is > 0.999 because it is controlled by the much higher absolute value for Sr relative to all other elements. Even ρ and τ, though as usual less affected, are still essentially only measuring the absolute magnitude of element contents (e.g. that Si tends to have higher numerical values than Al), rather than any intrinsic measure of affinity between the two limestones. Correlation coefficient matrices are thus ruled out as the starting point for *any* Q-mode classification method based on major and trace element data, unless the data are standardized first.

(2) *Interpretation.* Even assuming a 'best' dendrogram can be selected (hopefully on a more objective basis than just proving the users' preconceptions!) it is by no means always obvious what a dendrogram is saying. Where, for example, is it logical to draw boundaries in what is essentially a continuous structure? Although significance tests are being developed (see above), the plethora of options makes even these difficult to apply in practice.

(3) *Structure.* The rigid hierarchical structure of a dendrogram is only appropriate for data where probable groupings are themselves intrinsically hierarchical. This rigidity causes another problem in that an object which happens to come near the top of the input file may get 'stuck' in an inappropriate place in the dendrogram hierarchy, simply because its place is assigned early in the computation, if the computer algorithm does not allow for iterative passes through the data to refine the structure. Different results can sometimes therefore be obtained from certain algorithms, even by inputting the objects in different orders — at which point the surfeit of permutations might tempt most users to give up!

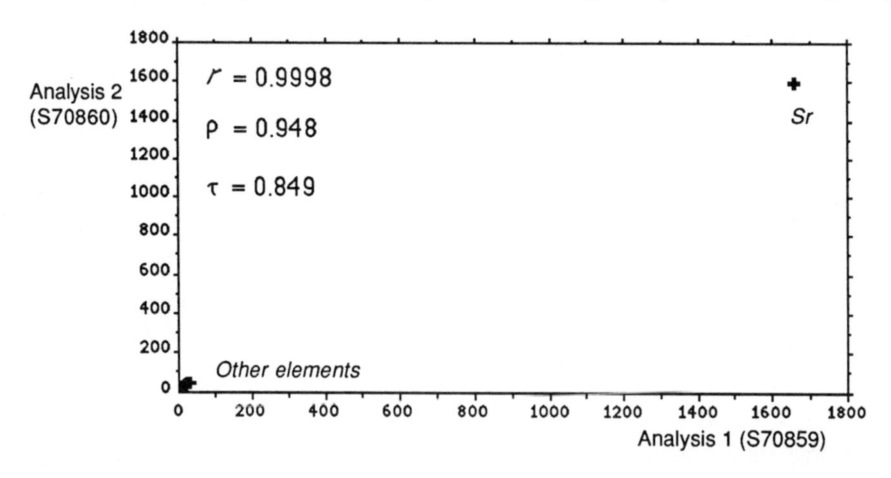

Fig.21.3. Illustrating the futility of correlation coefficients for measuring similarity between two major+trace element analyses where one element has a much higher absolute value than the others (data from Table 21.3)

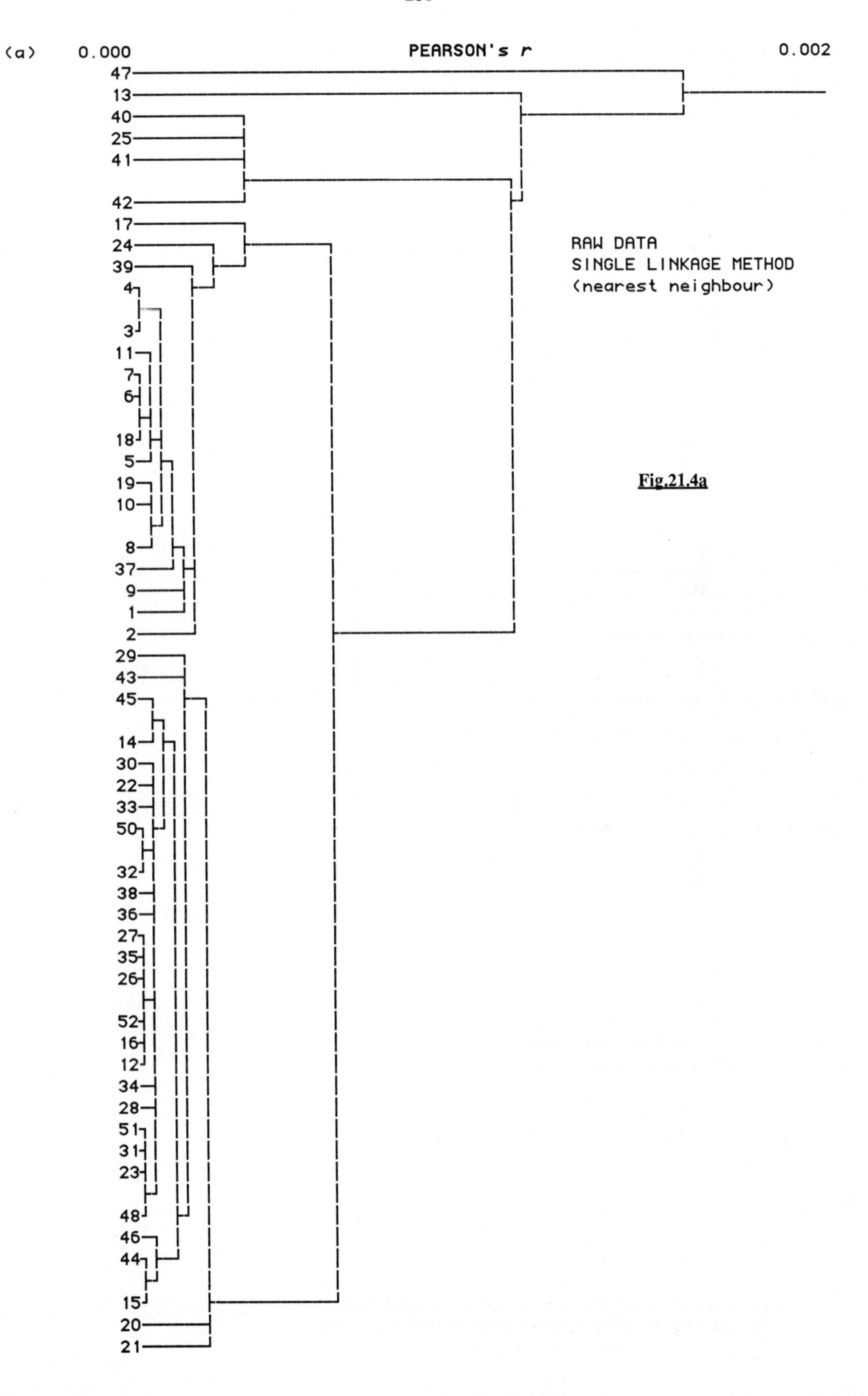

(a)
0.000
PEARSON's r
0.002
47
13
40
25
41
42
17
24
39
4
3
11
7
6
18
5
19
10
8
37
9
1
2
29
43
45
14
30
22
33
50
32
38
36
27
35
26
52
16
12
34
28
51
31
23
48
46
44
15
20
21
RAW DATA
SINGLE LINKAGE METHOD
(nearest neighbour)

Fig.21.4a

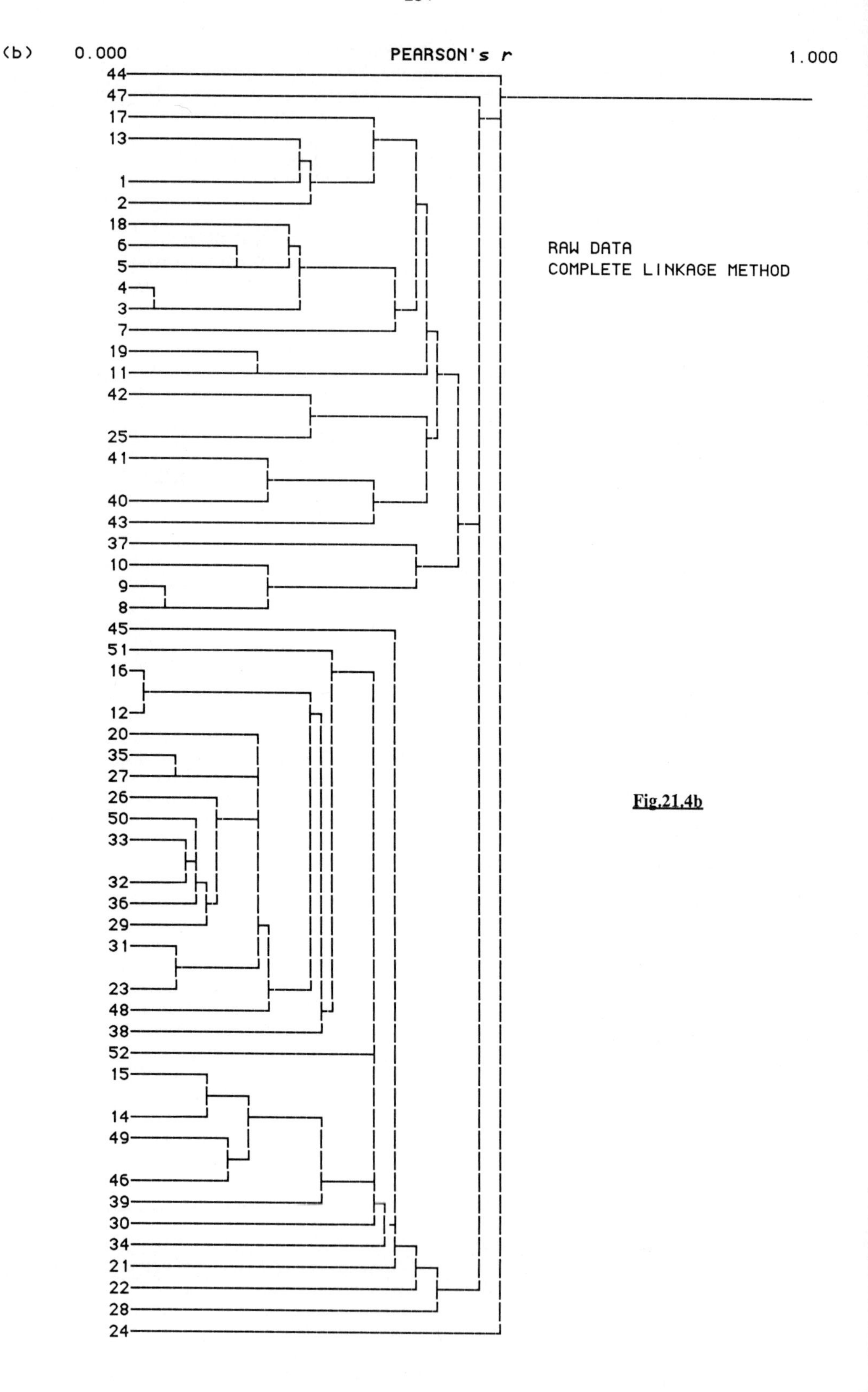
(b)
0.000
PEARSON's r
1.000
44
47
17
13
1
2
18
6
5
4
3
7
19
11
42
25
41
40
43
37
10
9
8
45
51
16
12
20
35
27
26
50
33
32
36
29
31
23
48
38
52
15
14
49
46
39
30
34
21
22
28
24
RAW DATA
COMPLETE LINKAGE METHOD

Fig.21.4b

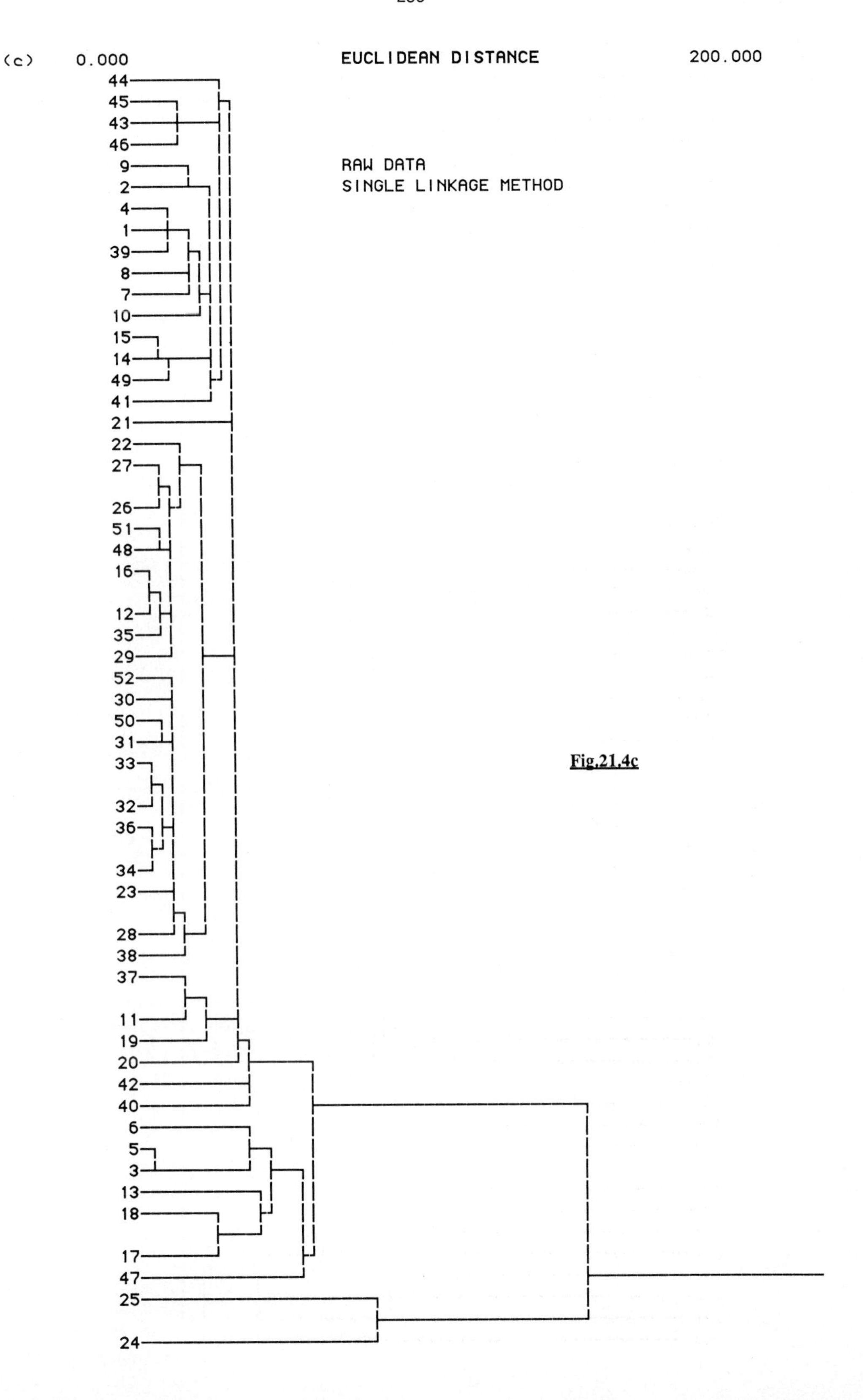
(c)
0.000
EUCLIDEAN DISTANCE
200.000
44
45
43
46
9
2
4
1
39
8
7
10
15
14
49
41
21
22
27
26
51
48
16
12
35
29
52
30
50
31
33
32
36
34
23
28
38
37
11
19
20
42
40
6
5
3
13
18
17
47
25
24
RAW DATA
SINGLE LINKAGE METHOD

Fig.21.4c

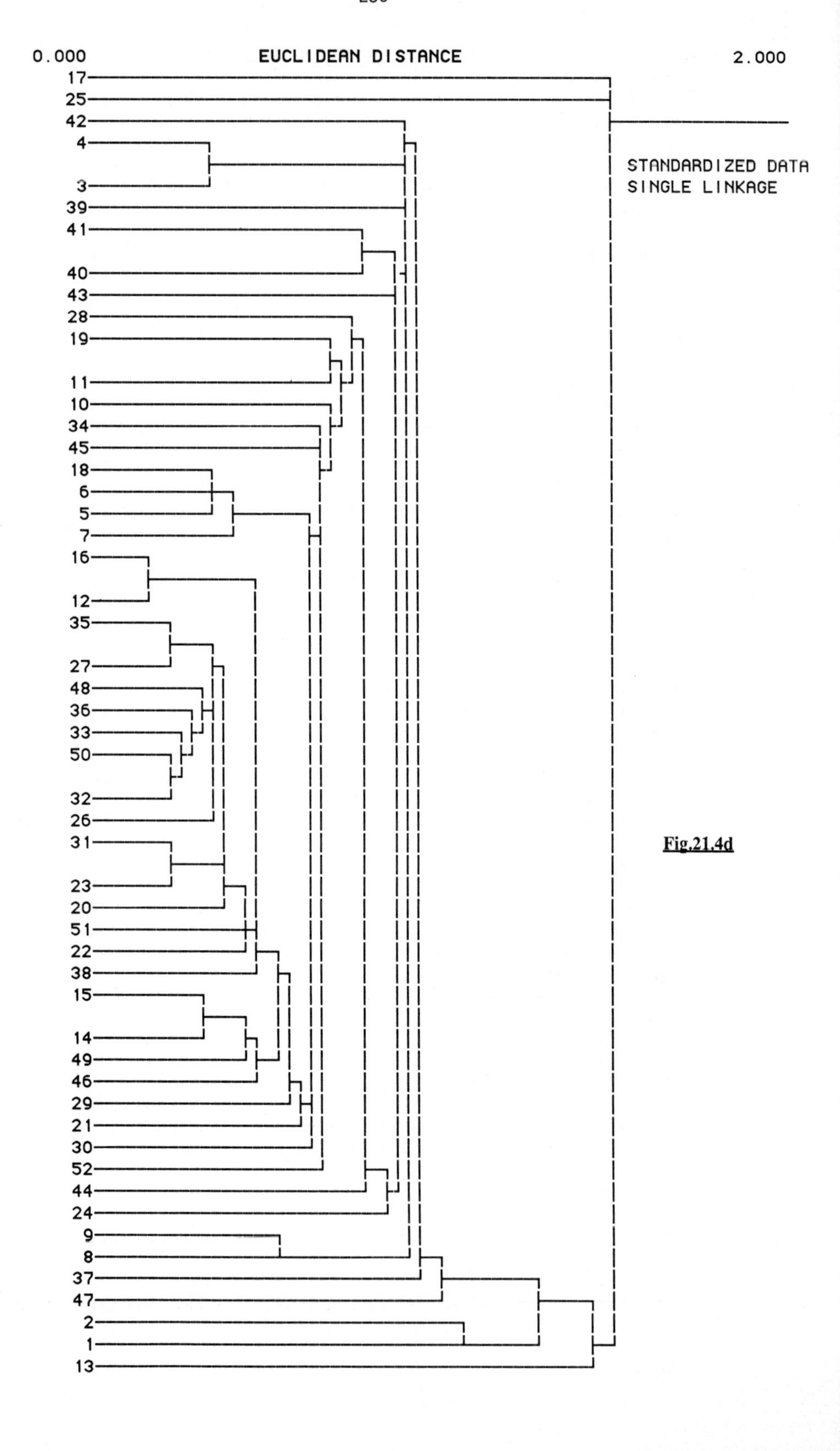
(d)
0.000
EUCLIDEAN DISTANCE
2.000
17
25
42
4
3
39
41
40
43
28
19
11
10
34
45
18
6
5
7
16
12
35
27
48
36
33
50
32
26
31
23
20
51
22
38
15
14
49
46
29
21
30
52
44
24
9
8
37
47
2
1
13
STANDARDIZED DATA
SINGLE LINKAGE

Fig.21.4d

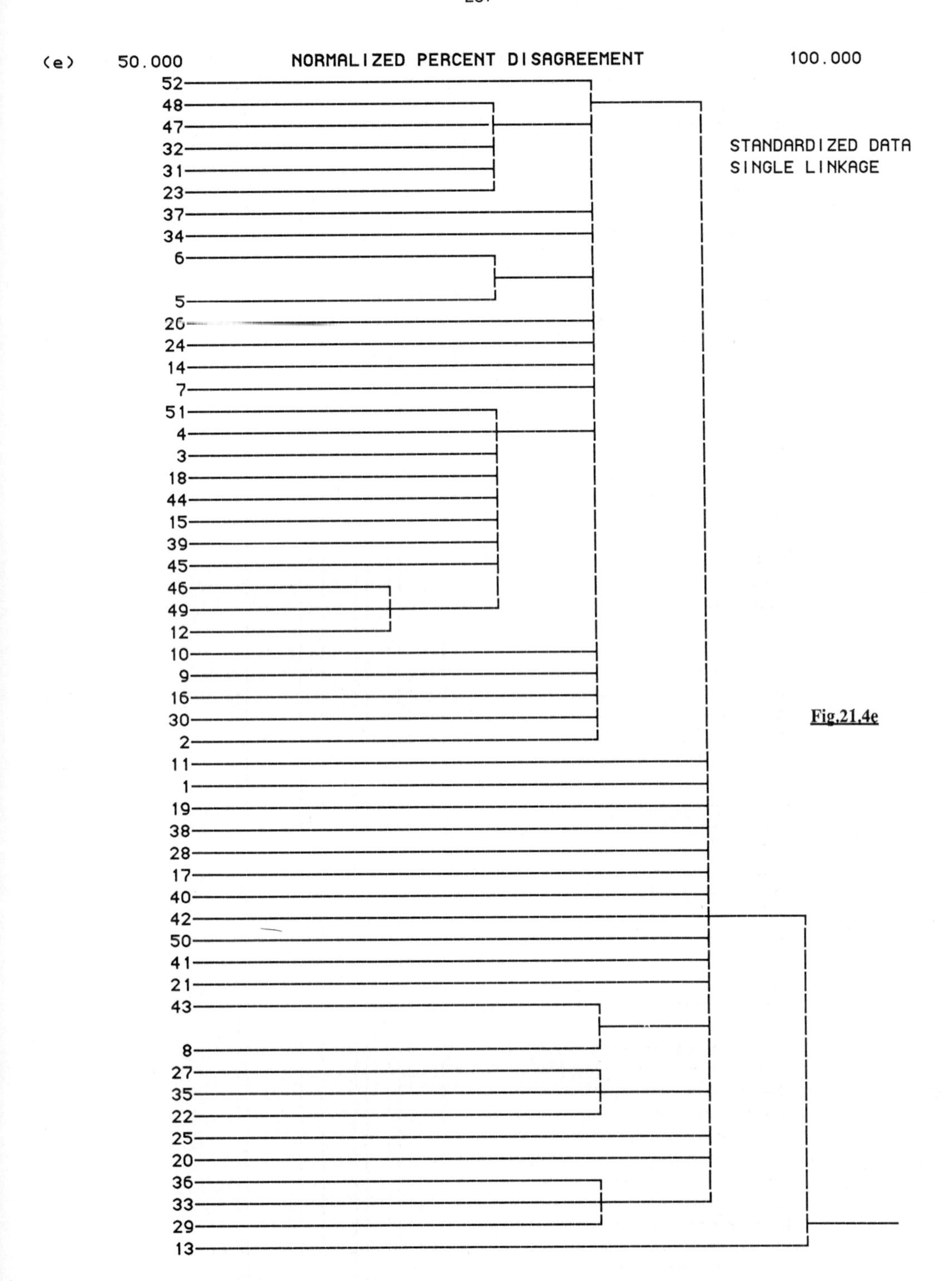

Fig.21.4. Representative dendrograms from 9 variants of hierarchical cluster analysis on the data of Table 21.3

(f)

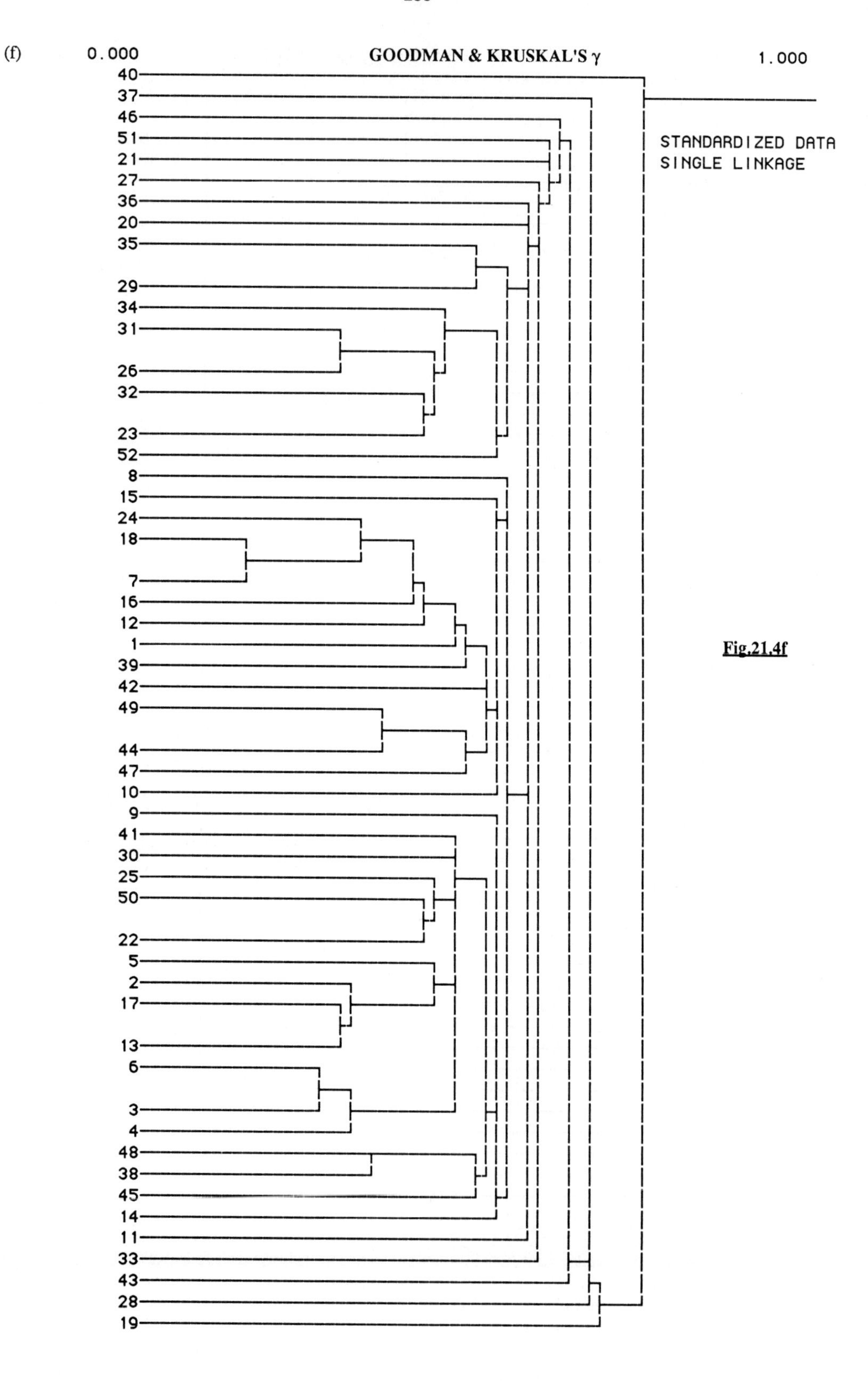
0.000
GOODMAN & KRUSKAL'S γ
1.000
40
37
46
51
21
27
36
20
35
29
34
31
26
32
23
52
8
15
24
18
7
16
12
1
39
42
49
44
47
10
9
41
30
25
50
22
5
2
17
13
6
3
4
48
38
45
14
11
33
43
28
19
STANDARDIZED DATA
SINGLE LINKAGE

Fig.21.4f

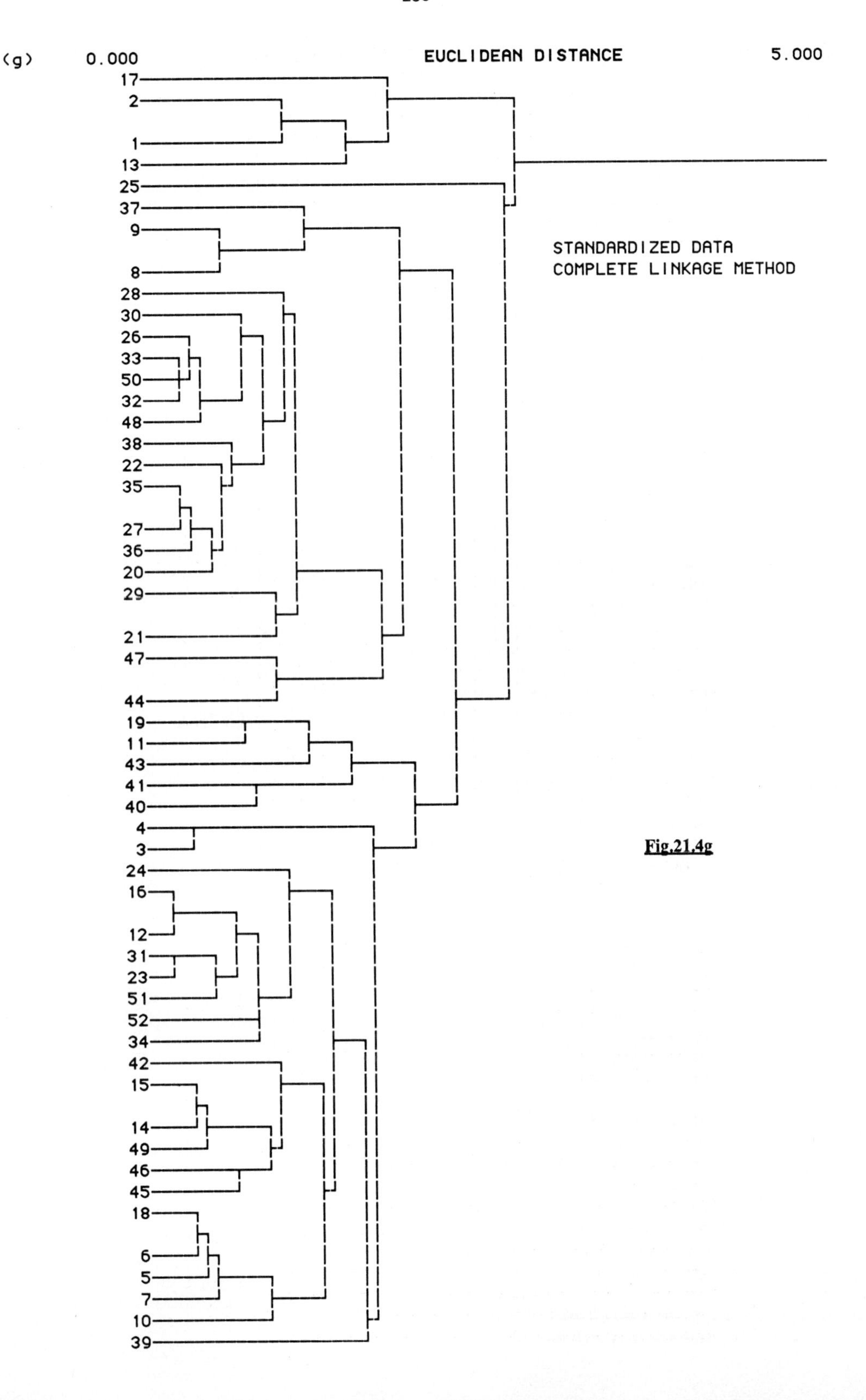

(g)
0.000
EUCLIDEAN DISTANCE
5.000
STANDARDIZED DATA
COMPLETE LINKAGE METHOD
17
2
1
13
25
37
9
8
28
30
26
33
50
32
48
38
22
35
27
36
20
29
21
47
44
19
11
43
41
40
4
3
24
16
12
31
23
51
52
34
42
15
14
49
46
45
18
6
5
7
10
39

Fig.21.4g

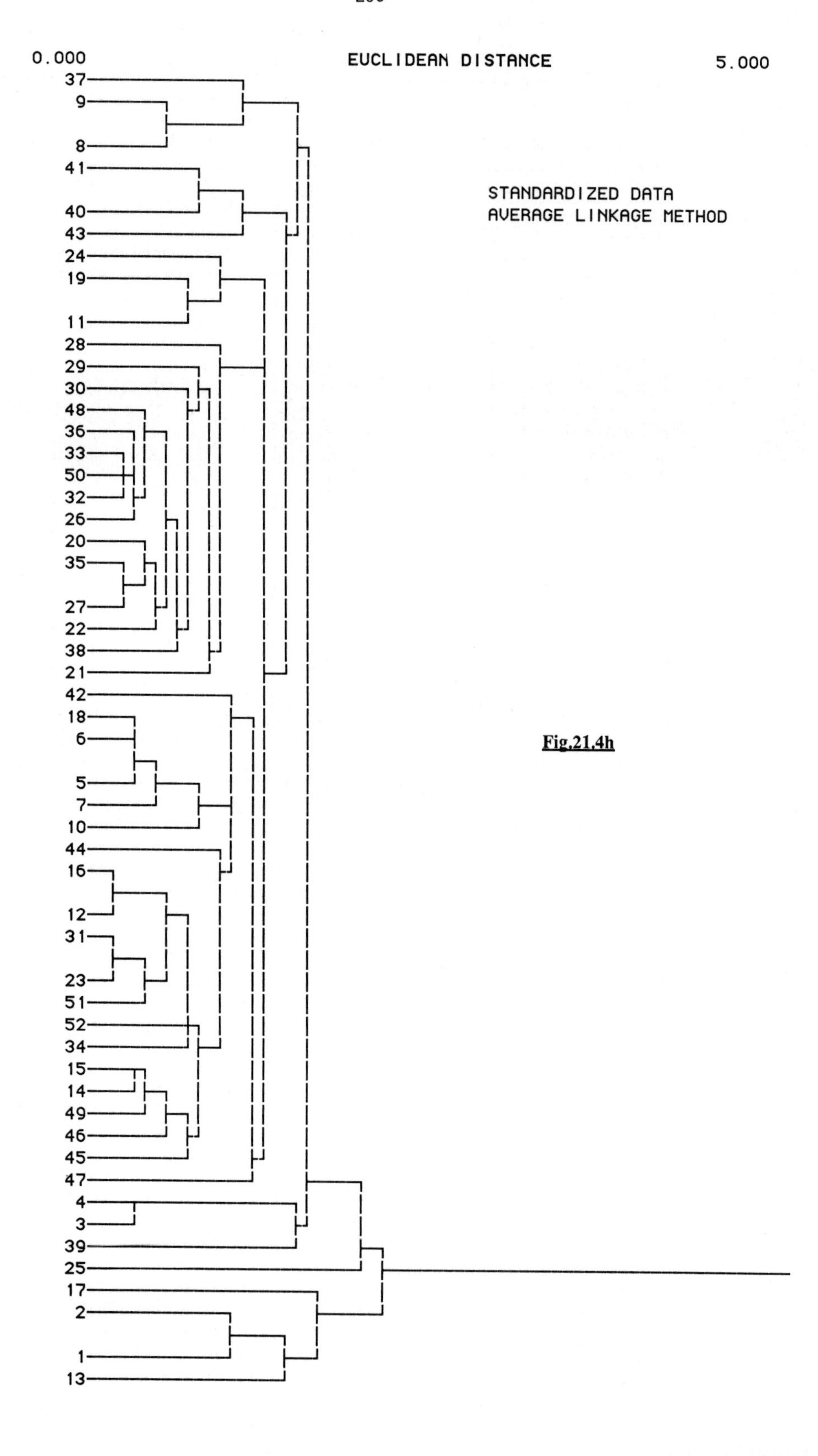

(h)
0.000
EUCLIDEAN DISTANCE
5.000
STANDARDIZED DATA
AVERAGE LINKAGE METHOD
37
9
8
41
40
43
24
19
11
28
29
30
48
36
33
50
32
26
20
35
27
22
38
21
42
18
6
5
7
10
44
16
12
31
23
51
52
34
15
14
49
46
45
47
4
3
39
25
17
2
1
13

Fig.21.4h

(i)

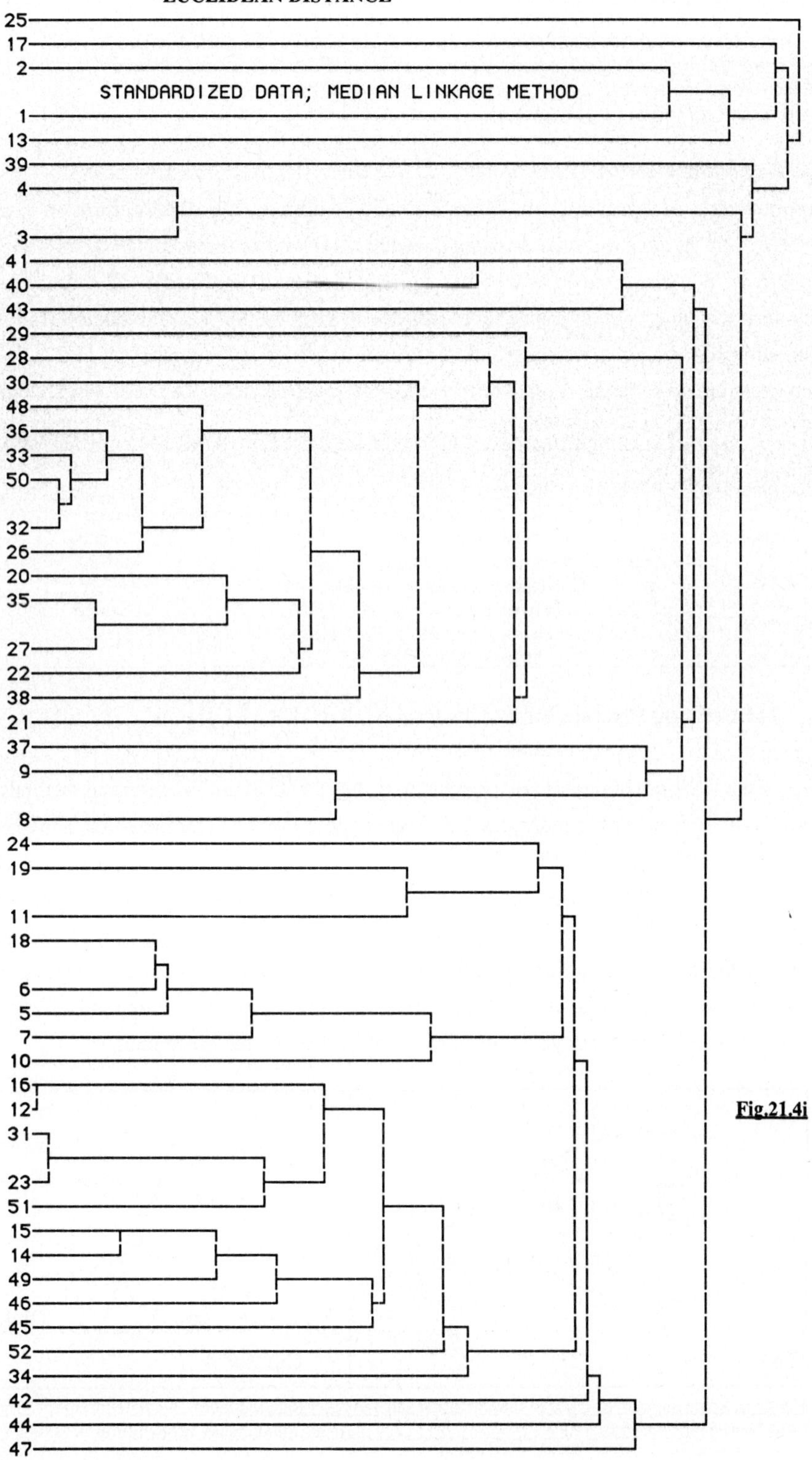
EUCLIDEAN DISTANCE
STANDARDIZED DATA; MEDIAN LINKAGE METHOD
25
17
2
1
13
39
4
3
41
40
43
29
28
30
48
36
33
50
32
26
20
35
27
22
38
21
37
9
8
24
19
11
18
6
5
7
10
16
12
31
23
51
15
14
49
46
45
52
34
42
44
47

Fig.21.4i

21b2. Divisive cluster analysis on dichotomous/nominal data: ASSOCIATION ANALYSIS

Further non-geological discussion: Williams & Lambert(1959,1960,1961).
Selected geological applications: *Sedimentology* (Buchbinder & Gill 1976; Gundu Rao 1980).
Stand-alone computer programs: Gill et al.(1976); Lance & Williams(1965).
Proprietary software: none known.

All the common forms of hierarchical cluster analysis are defined by stringing together different combinations of options from Table 21.5. For example, the dendrograms in Fig.21.4 were all generated by an *agglomerative polythetic* method, but with different linkages, similarity measures, and input data. For some reason best known to its progenitors, one form of *divisive monothetic* cluster analysis, which uses χ^2 as a matching coefficient, goes by the special name of *association analysis*. The end-product, however, is still a dendrogram. χ^2 defined as usual by [8.27], according to the nomenclature of Table 13.5; quite different formulae appear in the literature, but all reduce to [8.27].

Table 21.5. Possible permutations of hierarchical cluster analysis based on 5 major divisions of choice

(1) Input data	(2) Similarity/distance measure	(3) Method	(4) Linkage	(5) Attributes used
Raw	*Dichotomous/nominal data:* χ^2,matching	Agglomerative	Single	Monothetic
Standardized	coefficients(Table 13.1)	Divisive	Complete	Polythetic
	Ordinal data: Spearman's ρ, Kendall's τ,		Median	Omnithetic
	Goodman & Kruskal's γ (§13b)		Average	
	Ratio data (similarity): Pearson's r (§13a)		Centroid	
	Ratio data (dissimilarity): City Block, Euclidean,			
	Minkowski distances; PCT			

21c.Defining Discrete/Fuzzy Clusters: NON-HIERARCHICAL CLUSTER ANALYSIS

Where the structure of data is not expected to be hierarchical, non-hierarchical methods are intrinsically preferable. Instead of a dendrogram, they generate a table of memberships, showing which objects belong to which of a limited number of groups (clusters). Usually, one individual is allowed to belong to one cluster only (i.e. the clusters are discrete), so the result is truly non-hierarchical. Certain algorithms, however, can generate nested clusters ('clusters within clusters'), which produces another form of hierarchical structure, while others allow one individual to belong to several clusters — so-called **fuzzy** cluster analysis (Fig.21.5). The latter is probably more realistic with real geological data, although it introduces yet more elements of choice into the procedure!

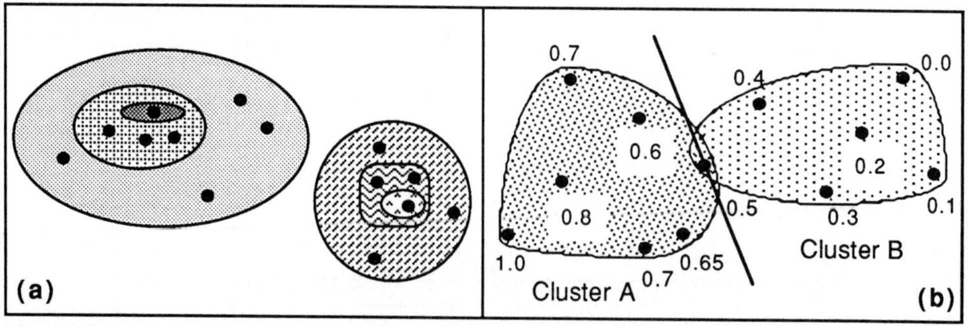

Fig.21.5. Some alternative types of classification scheme other than the hierarchical dendrogram structure in Fig.21.3: **(a)** Nested clusters (dots indicate objects). **(b)** Fuzzy clusters (numbers are probabilities of belonging to cluster A)

21c1. *K*-means Cluster Analysis

Further discussion: Davis (1986, p.513).
Stand-alone computer programs : none known.
Proprietary software: covered in a few large statistics packages (e.g. Systat).

The user must here predetermine a number of clusters, *K*, into which the data-set is to be divided (*K* can be up to the number of objects, but more than a few will merely reveal noise rather than structure). The technique usually starts by taking the first *K* points in the data-set as the initial estimates of the future cluster means. The remaining objects are then assigned to the nearest of these clusters. For example, if $K = 2$, the first two objects initially represent the 2 clusters, and the 3rd object is assigned to cluster 1 if its mean is the nearer. The mean of cluster 1 is then recalculated as the mean of objects 1 and 3, and object 4 is then assigned to the (revised) cluster 1 or (the original) cluster 2, whichever mean is nearer. Once all objects have been assigned to the continually evolving clusters in this way, the process is repeated iteratively with the entire data-set, each object being reassigned as necessary, wherever this will increase the ratio of between/within cluster dispersion (ANOVA is thus used as the decision rule). Dozens of passes through the data-set may be needed until the solution stabilises.

Table 21.6 (which greatly condenses the output from a typical computer program!) illustrates the technique on the data of Table 21.3, using both the raw and standardized data as input. Since there are 6 limestone occurrences, $K = 6$ is the obvious choice, to determine whether chemical clusters reproduce geographical clusters. Note that *K*-means analysis produces a wholly numerical, tabular output, which in its bare essence merely lists which objects lie within which cluster; this contrasts with the graphical, visually more easily interpreted output from hierarchical (§21a) or ordination methods (§21d). However, for certain purposes *K*-means can lead to quicker and less subjective classification, *provided* there exists a reasonably objective way of choosing *K* in the first place.

In Table 21.6a, the *F* ratios indicate that, for the standardized data at least, the between/within cluster spread greatly exceeds 1 for all variables except Ni — i.e. 6 *non-overlapping* clusters have been generated. For the raw data, however, the clusters are significantly overlapping for 8 of the 14 elements at the 90% level — a much less successful result. Consequently, the cluster memberships in Table 21.6c are rather different from the two treatments. For reasons discussed in §15e, standardized data are probably preferable here, in view of the orders of magnitude differences in variable values.

Table 21.6b gives representative results for a single cluster; the full results show 12 such tables, and have been summarised in Table 21.6c & d. Note that the numbers of the clusters (1-6) are **nominal**, i.e. cluster 6 is in no sense '6 times', or even merely larger than, cluster 1 (§6).

Note that in *both* treatments the limestone with highest Sr content, no. 25, forms either a cluster on its own, or a cluster of 2 analyses. Conversely, no. 47 forms its own cluster in the standardized treatment because it has the *lowest* Sr and *highest* Ce contents. This form of cluster analysis can thus be used empirically to identify multivariate outliers: in fact, the data-set in Table 21.3 has already been screened by the *K*-means method from a larger set of 57 analyses, in which setting $K = 6$ identified 5 single-analysis clusters (i.e. 5 outliers) and a 6th cluster with all the 52 analyses used here. Each of these 5 outliers had one or more element values much higher than the others, but it is more objective to use *K*-means to screen such data out than to do so by eye alone.

The conclusion apparent from Table 21.6 is that the 6 geographically distinct limestone occurrences overlap one another very considerably in chemistry: no single occurrence forms its own cluster.

Table 21.6. *K*-means cluster analysis for the data of Table 21.3 ($K = 6$)

(a) Overall success of cluster analysis

(1) Raw data

VARIABLE	BETWEEN SS	DF	WITHIN SS	DF	F-RATIO	PROB
La	134.373	5	2026.454	46	0.610	0.693
Ce	20358.073	5	51641.677	46	3.627	0.008
Cr	366.921	5	1989.002	46	1.697	0.154
Ni	60.136	5	313.940	46	1.762	0.140
Cu	44.156	5	1130.767	46	0.359	0.874
Zn	951.113	5	4070.195	46	2.150	0.076
Rb	387.307	5	4461.366	46	0.799	0.556
Sr	7639010.265	5	319718.716	46	219.815	0.000
Zr	19512.337	5	52095.971	46	3.446	0.010
Ba	43733.122	5	80545.859	46	4.995	0.001
Si	12.486	5	896.750	46	0.128	0.985
Al	2.953	5	36.101	46	0.752	0.589
Fe	0.242	5	6.151	46	0.362	0.872
Mg	20.183	5	63.347	46	2.931	0.022

(2) Standardized data

VARIABLE	BETWEEN SS	DF	WITHIN SS	DF	F-RATIO	PROB
La	22.737	5	28.263	46	7.401	0.000
Ce	18.019	5	32.981	46	5.026	0.001
Cr	23.839	5	27.161	46	8.075	0.000
Ni	8.135	5	42.865	46	1.746	0.143
Cu	24.007	5	26.993	46	8.183	0.000
Zn	26.278	5	24.722	46	9.779	0.000
Rb	28.235	5	22.765	46	11.410	0.000
Sr	27.630	5	23.370	46	10.877	0.000
Zr	40.417	5	10.583	46	35.135	0.000
Ba	31.448	5	19.552	46	14.797	0.000
Si	35.442	5	15.558	46	20.958	0.000
Al	30.698	5	20.302	46	13.911	0.000
Fe	29.409	5	21.591	46	12.531	0.000
Mg	37.827	5	13.173	46	26.419	0.000

(b) Full output statistics for a single cluster (cluster 6 for raw data)

MEMBERS OF CLUSTER		CLUSTER STATISTICS				
CASE	DISTANCE	VARIABLE	MINIMUM	MEAN	MAXIMUM	ST.DEV.
10	41.09	La	1.00	7.07	12.00	3.61
12	17.16	Ce	0.00	78.14	97.00	22.68
16	15.15	Cr	0.00	7.50	25.00	7.77
20	36.03	Ni	0.00	1.29	3.00	1.22
22	7.54	Cu	0.00	3.79	18.00	4.41
27	28.45	Zn	3.00	9.86	23.00	6.56
29	21.27	Rb	5.00	14.93	23.00	6.70
35	18.27	Sr	1830.00	1956.86	2052.00	77.92
42	37.36	Zr	0.00	14.64	109.00	27.98
43	34.87	Ba	0.00	82.29	170.00	50.17
45	29.10	Si	2.33	7.20	12.62	3.11
46	24.39	Al	0.51	1.56	2.69	0.73
48	23.50	Fe	0.24	0.64	1.04	0.25
51	17.63	Mg	0.34	0.89	2.15	0.39

*(c) Cluster membership (numbers in **bold** are common to both raw and standardized treatments)*

	(1) Raw data	(2) Standardized data
	Cluster 1	*Cluster 1*
Analyses nos. in Table 21.3	11,**19,23,26,28,30,31,32,** **33,34,36**,37,38,40,**50,52**	**19**,20,21,22,**23**,24,**26**,27,**28**,29,**30,31,32,** **33,34**,35,**36,38**,43,48,**50**,51,**52**
	Cluster 2	*Cluster 2*
Analyses nos.	47	40,41
	Cluster 3	*Cluster 3*
Analyses nos.	**1,2**,4,7,8,9,14,15,21, 39,41,44,49	**1,2**,13,17
	Cluster 4	*Cluster 4*
Analyses nos.	24,**25**	**25**
	Cluster 5	*Cluster 5*
Analyses nos.	10,12,16,20,22,27,29,35, 42,43,45,46,48,51	8,9,37
	Cluster 6	*Cluster 6*
Analyses nos.	**3,5,6**,13,17,**18**	**3**,4,**5,6**,7,10,11,12,14,15,16,**18**,39,42,44,45 46,47,49

(d) Multivariate cluster mean-values

(1) Raw data						
La	5	16*	7	5.5	7	7
Ce	68	100	43	46	78	21
Cr	10	0	4	5	8	7
Ni	2	0	2	1.5	1	5
Cu	4	1	3	5	4	1
Zn	6	3	16	8	10	16
Rb	16	3	13	15	15	9
Sr	2175	893	1645	2834	1957	1215
Zr	16	0	17	115.5	15	17
Ba	82	27	26	103.5	82	21
SiO_2	7.6	9.7	8.1	8.63	7.2	8.2
Al_2O_3	1.7	0.48	1.4	2.06	1.6	1.2
Fe_2O_3	0.75	0.86	0.74	0.84	0.64	0.60
MgO	0.79	0.76	1.7	1.02	0.89	2.7

(2) Standardized data						
La	+0.03	-0.53	1.93	0.70*	0.70	-0.53
Ce	0.55	-0.67	-0.51	0.62	-1.57	-0.27
Cr	0.51	1.11	0.04	0.45	0.59	-0.86
Ni	-0.01	-0.44	1.31	-0.07	-0.32	-0.17
Cu	0.18	3.03	-0.30	0.53	-0.51	-0.43
Zn	-0.37	-0.29	2.38	-0.19	-0.29	0.03
Rb	0.23	0.38	0.30	0.53	2.24	-0.76
Sr	0.60	-0.16	-1.29	2.61	-0.18	-0.55
Zr	-0.26	2.32	0.36	5.09	0.12	-0.29
Ba	0.81	-0.25	-0.61	1.34	-0.95	-0.74
SiO_2	0.07	-0.47	2.16	0.64	1.35	-0.74
Al_2O_3	0.38	-0.21	0.87	1.15	1.52	-0.92
Fe_2O_3	0.26	-0.41	1.83	0.94	0.60	-0.80
MgO	-0.37	-0.33	2.90	0.17	-0.48	-0.06

* Actual values for the single specimen in the cluster.

21c2. The Refined, Iterative *K*-means Method

Further discussion & worked geological examples: *glaciology* (Mather 1976,p.327 & 366).
Selected geological applications: *petrology* (Brotzen 1975).
Stand-alone computer programs: Froidevaux et al.(1977); Mather(1976,p.393); McRae(1971).
Proprietary software: none known.

Here, the user specifies both a *maximum* and a *minimum* number of clusters he would expect, K_{max} and K_{min}. The algorithm starts as in the K-means method, by selecting K_{max} cluster centroids either by inspection of the data or by generating random numbers, and iteratively searching for the optimum group membership. An extra stage is then introduced, however, when K is reduced by 1 and the whole process repeated successively until K is reduced to K_{min}. For each reduction stage, an F ratio can be calculated which allows a rough statistical test of the improvement (or otherwise) of the fit with K clusters relative to $K+1$ or $K-1$. A plot of K against this ratio (cf. Fig.23.3) can give a more objective choice of the 'correct' number of clusters to choose than the guestimate required fo the simple K-means method — especially if there is a substantial jump in the ratio at some value of K (as in Fig.23.3b). Of course, the basic K-means method could be repeated for various values of K, but this would not give the F ratio or consequent objectivity.

21d. Displaying groupings in as few dimensions as possible: ORDINATION

Cluster analysis assumes *a priori* that the objects being examined fall into several *discrete* classes, which can be arranged as a dendrogram or as non-overlapping or nested clusters. Ordination methods cover a more exploratory situation, where either there is no knowledge about possible classes, or it is likely that classes are *gradational* (i.e. groupings are continuous rather than discrete). Ordination essentially aims to reduce the **dimensionality** of data (**data reduction**) — that is, to allow relationships between multivariate objects to be displayed as accurately as possible in as few dimensions as possible (ideally in the 2-*D* of a sheet of paper). Relationships and possible groupings can then be sought using the remarkable natural abilities of the human eye. Principal components analysis can also be used for ordination, but is discussed in §23a because of its close relationship to factor analysis. Variants of techniques discussed below appear in Friedman & Rafsky (1981).

21d1.Metric and Nonmetric MultiDimensional Scaling (MDS)

Dedicated monographs and reviews: Coxon (1982); Davies & Coxon (1982); Davison(1983); Kruskal (1964); Kruskal & Wish (1978); Schiffman et al.(1981);
Worked geological examples: Mather (1976,p.379)
Selected geological applications: *Palaeontology* (Whittington & Hughes 1972); *Sedimentology* (Feldhausen 1970; Reyment et al.1976); *Sedimentary petrology* (Doveton 1976; Smosna & Warschauer 1979)
Stand-alone computer programs: Mather(1976,p.407); Young (1968).
Mainframe software: covered in some recent statistics packages (e.g. $SPSS^X$); also MDS(1980).
Microcomputer software: covered in a few larger packages (e.g. SYSTAT).

This is probably the most powerful (and widely used) ordination technique. Though various algorithms are available, all start with a matrix of similarities (e.g. correlation coefficients) or dissimilarities (e.g. Euclidean distances), and have the same aim — to generate coordinates for a set of points in 2-D space such that distances between pairs of points as closely as possible mirror the input similarities or dissimilarities. MDS does not require that distances in reduced space and similarities are linearly related, and it can work equally on ratio (metric MDS) or ranked (non-metric MDS) data.

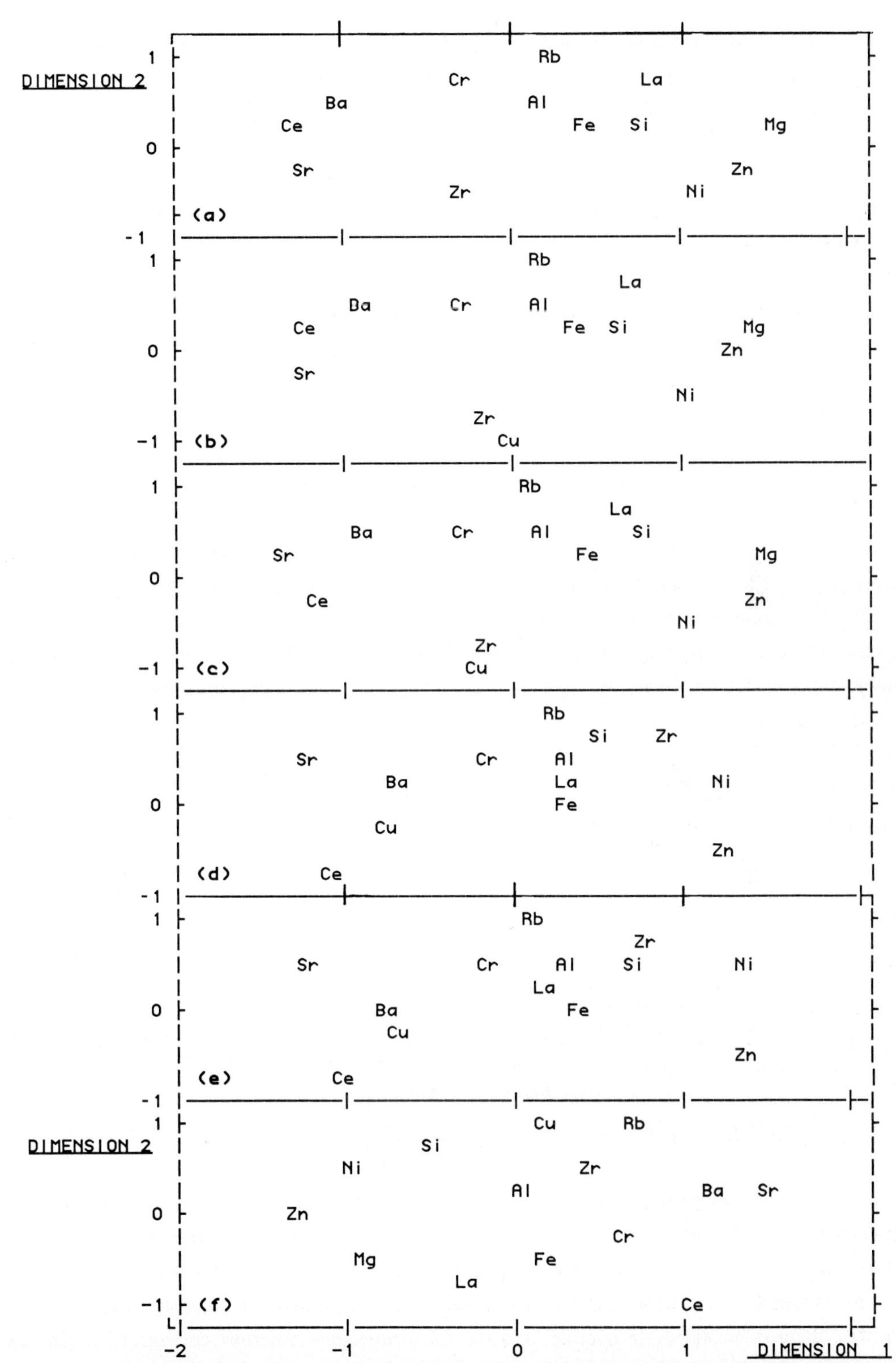

Fig.21.6. Alternative R-mode multidimensional scaling solutions for the limestone data of Table 21.3
See Table 21.7 for exact details of input parameters and final stress results

Apart from the usual choice of similarity/dissimilarity measure in the input matrix (Table 21.5), the investigator usually also has the following MDS options available:

(1) *Number of dimensions*: MDS is often capable of squeezing multivariate data into fewer dimensions than, say, principal components or factor analysis (§23); 2 dimensions would normally be chosen, dimensionality only being increased if the final 2-D solution were unsatisfactory.

(2) *Type of scaling*. MDS measures the approach of the scaled solution to the input matrix via a factor called **stress:** the larger the stress, the poorer the fit. Among measures of stress are **Kruskal's formula 1** and **Guttman-Lingoes' stress.**

(3) *Distance measure:* A measure of the *output* dissimilarities in the final scaled solution is required, to compare with that in the input matrix. Most commonly, Euclidean or City Block is used.

(4) *Iterations:* most MDS implementations continue until (a) a predetermined number of iterations has been completed, (b) the stress reaches a predetermined minimum value (e.g. 0.01), or (c) stress decrements on successive iterations reach a predetermined, vanishingly small, value (e.g. 0.001).

Figs.21.6–7 illustrate a selection of MDS solutions on the limestone data of Table 21.3. Of the R-mode variants, Fig.21.6f illustrates that a variance-covariance input matrix is of little use for rock analyses, since this solution merely reflects the relative size of the input variables (absolute magnitude increasing from left, Zn, to right, Sr). The remaining solutions, however, appear consistent and equally valid: notice that elements have been grouped in a geochemically sensible way (e.g. Rb with Al, presumably reflecting association in feldspar). The exception is that La is far from Ce, but this in fact faithfully mirrors the rather low input correlation, rather than any fault in the solutions themselves.

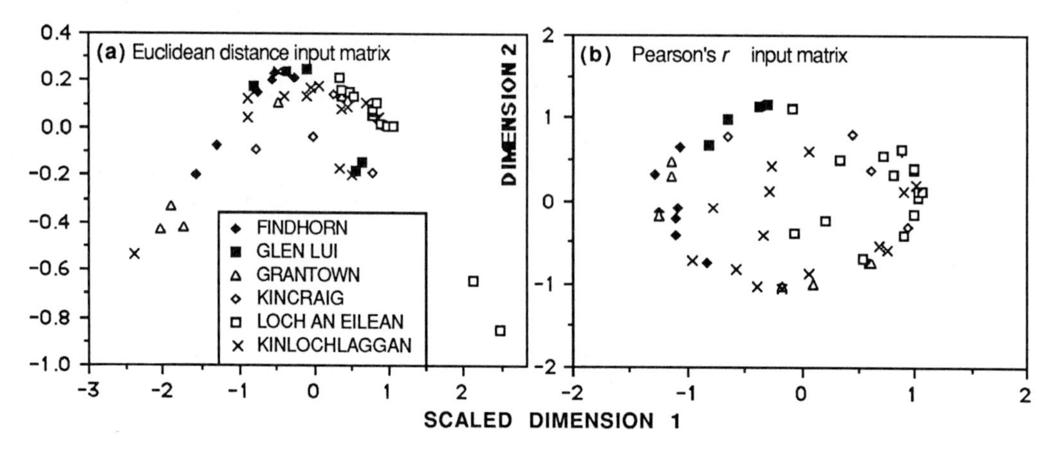

Fig.21.7. Alternative Q-mode MDS representations of the limestone data of Table 21.3

The Q-mode variants (Fig.21.7) illustrate how successfully MDS can reduce a 52-dimensional input matrix to a 2-dimensional plot. Thus the 52 x 14 analysis matrix in Table 21.3 can be *very satisfactorily* displayed by MDS on a sheet of paper, so the method clearly offers the geologist a powerful alternative to conventional two-element scatter diagrams! We see that, apart from two outliers, there is considerable overlap between all 6 limestone outcrops on Fig.21.7a; the less satisfactory (much higher stress — Table 21.7b) solution in Fig.21.7b shows rather less overlap.

Table 21.7 compares stresses resulting from selected MDS permutations; as can be seen, the input similarity/dissimilarity measure affects the result very much more than the MDS method or distance measure chosen. The choice of raw versus standardized input data is also critical for both Q-mode and R-mode MDS, but their efficacies are reversed: a Q-mode matrix of Pearson's r consists entirely of values > 0.99, because all plots of 'Analysis 1' versus 'Analysis 2' look like Fig.21.3; r is merely tracing a line between the Sr content at high absolute values and the other trace elements at low values.

Shepard diagrams (Fig.21.8) — here shown for the Q-mode variants but equally applicable to the R-mode — show how closely the output similarity/dissimilarity measures reflect those input. As shown, the spread becomes tighter as stress decreases. Generally, the diagram conveys little more information to the general user than the simple stress statistic itself.

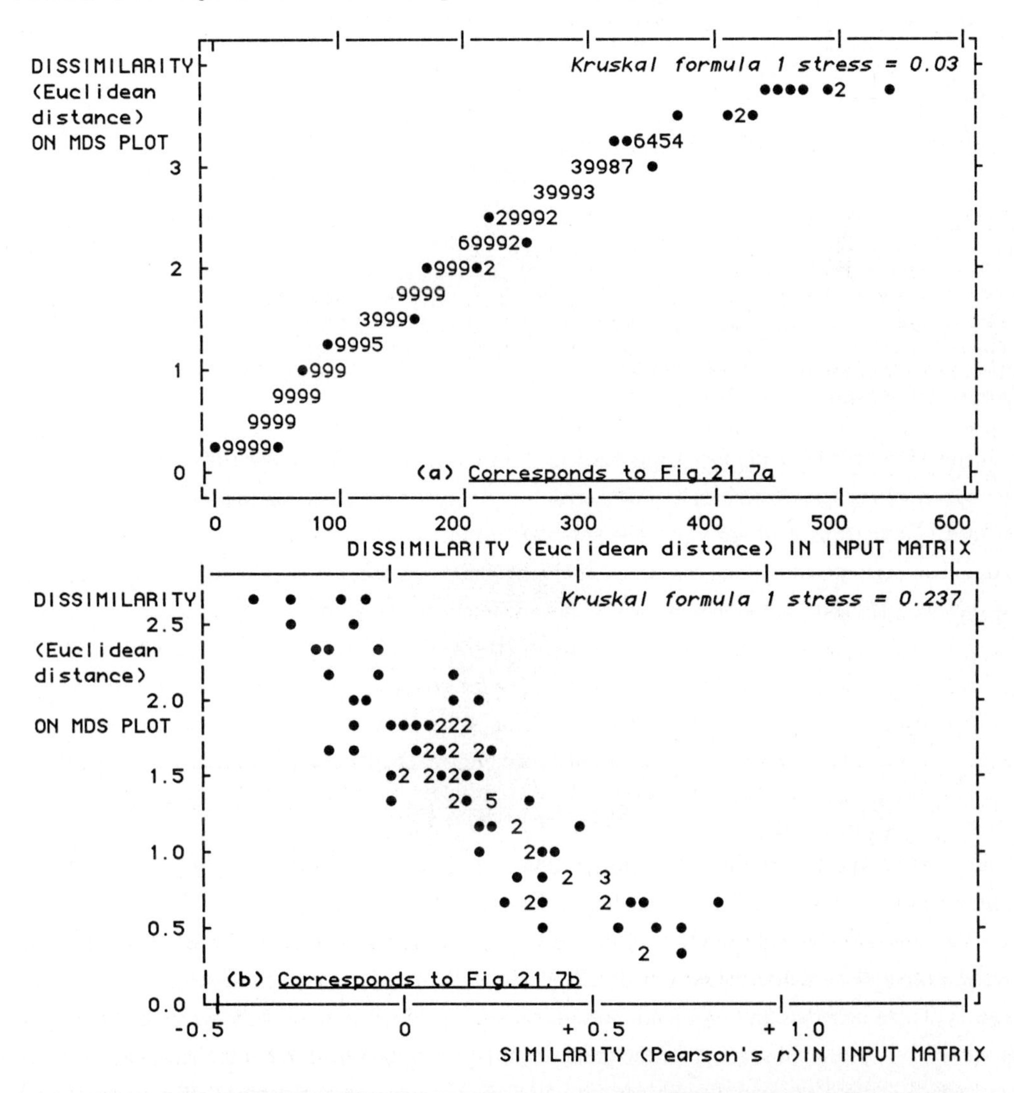

Fig.21.8. Shepard diagrams illustrating correlation between input and MDS-output similarity/dissimilarity measures for various stresses, corresponding to Q-mode plots on Fig.21.7.

(● indicates one data-point, numbers indicate superposition of the given number of data-points, >1).

Table 21.7. Comparison of stresses resulting from various MDS options on data of Table 21.3

Input distance/similarity measure	MDS distance measure	Stress measure	Final stress	See
	(a) R-mode			
Pearson's r	City Block distance	Kruskal formula 1	0.140	Fig.21.6a
Pearson's r	Euclidean distance	Kruskal formula 1	0.130	Fig.21.6b
Pearson's r	Euclidean distance	Guttman-Lingoes	0.182	Fig.21.6c
Kendall's τ	Euclidean distance	Kruskal formula 1	0.140	Fig.21.6d
Spearman's ρ	Euclidean distance	Kruskal formula 1	0.141	Fig.21.6e
Variance-covariance	Euclidean distance	Kruskal formula 1	0.237	Fig.21.6f
	(b) Q-mode			
Pearson's r	Euclidean distance	Kruskal formula 1	0.237	Fig.21.7b/21.8b
Pearson's r	City Block distance	Kruskal formula 1	0.258	—
Pearson's r	Euclidean distance	Guttman-Lingoes	0.256	—
Variance-covariance	Euclidean distance	Kruskal formula 1	0.243	—
Sum of squares/cross products	Euclidean distance	Kruskal formula 1	0.247	—
Euclidean distance	Euclidean distance	Kruskal formula 1	0.030	Fig.21.7a/21.8a
Euclidean distance	Euclidean distance	Guttman-Lingoes	0.032	—
Euclidean distance	City Block distance	Kruskal formula 1	0.036	—

21d2. Principal Coordinates Analysis

Further geological discussion: Davis (1986, p.574ff); Joreskog et al.(1976,p.100ff); Mather(1976, p.330).
Worked geological examples: *Glaciology/mineralogy* (Mather 1976,p.400) .
Selected geological applications: *Palaeontology* (Reyment & Van Valen 1969); *Sedimentology* (Baer 1969; Imbrie & Purdy 1962); *Structural geology* (Fisher et al. 1985).
Stand-alone computer programs: Mather(1976,p.402).
Proprietary software: none known.

This can be regarded as a Q-mode application of Principal Components Analysis (§23a), but it can also be applied where some variables are measured only on a dichotomous, nominal or ordinal scale, via the use of measures such as **Gower's general similarity coefficient,** g. It is somewhat less heavy than MDS on computing resources. It begins like the methods discussed earlier, by computing similarity coefficients, but variables measured on different scales are dealt with separately at first. For example, if there are v_d dichotomous variables, v_n nominal variables and v_r ratio variables, we first generate *three* matrices of similarity coefficients, one for each class of variable. A combined matrix is then computed, in which each final similarity coefficient is derived from the coefficients in the three separate matrices by weighting them according to the numbers of variables involved in each matrix:

$$g_{final} = \frac{g_d}{v_d} + \frac{g_n}{v_n} + \frac{g_r}{v_r} \qquad [21.2]$$

After a minor further manipulation of this matrix, the method then proceeds by a form of Principal Components Analysis. The *X-Y* coordinates of the individual objects on the final 2-D plot in Fig.21.9 are determined by the **eigenvectors** of this matrix, while the first two **eigenvalues** determine how well the two displayed dimensions actually display the full, multidimensional relationships between the objects. These concepts are explained more fully in §23a. Usually, these first two eigenvalues (and hence the *X-Y* plot) will account for the great majority of the total variance in the data. Occasionally, however, 3-D plots, or several combinations of 2-D plots such as *X-Y, Y-Z* and *X-Z*, may be required to display the full relationships adequately.

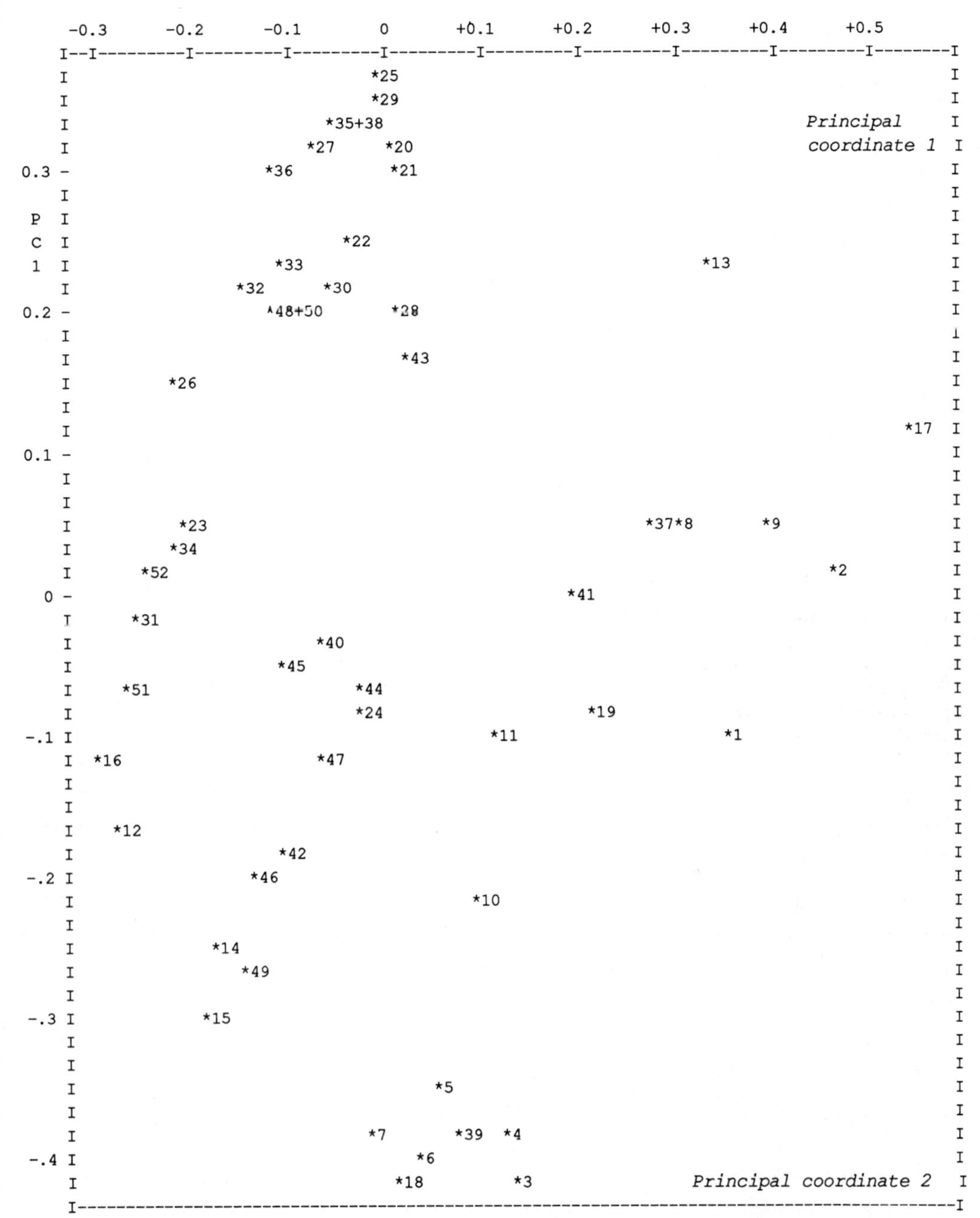

Fig.21.9. Plot of first 2 principal co-ordinates for limestone data of Table 21.3 using Gower's general coefficient as the similarity measure; generated using program in Mather(1976)

Table 21.8. Progress of mapping error in non-linear mapping applied to limestone data of Table 21.3

Iteration	Mapping error	Iteration	Mapping error	Iteration	Mapping error	Iteration	Mapping error	Iteration	Mapping error
1	0.004351245	4	0.0388806318	7	0.0055147414	10	0.0022581662	13	0.0016597634
2	0.241506675	5	0.0180966787	8	0.0036743969	11	0.0019663235	14	0.0015727116
3	0.090427000	6	0.0094133378	9	0.0027560161	12	0.0017827373	15	Termination

21d3. Non-linear mapping (NLM)

Further geological discussion: Howarth(1973b); Mather(1976, p.339).
Worked geological examples: *Glaciology & mineralogy* (Mather 1976,p.369);
Selected geological applications: *Palaeontology* (Temple 1980); *Sedimentology* (Clark & Clark 1976); *Stratigraphy* (Read 1976).
Stand-alone computer programs: Henley (1976b); Mather (1976,p.416).
Proprietary software: none known.

This is similar in both concept and execution to MDS, but is generally used only on input matrices of Euclidean distance coefficients. NLM differs from MDS in aiming not to ensure monotonicity between the input dissimilarities and distances on the 2-*D* (or, much more rarely, 3-*D*) output plot, but to minimise the distortion of interpoint distances introduced by mapping from v dimensions into 2. The goodness-of-fit measure which is minimized (equivalent to MDS *stress*) is called **mapping error**, E:

$$E = \frac{1}{\sum_{i=1}^{i=N}\sum_{j=1}^{j=N} d_{ij}} \sum_{i=1}^{i=N}\sum_{j=1}^{i=N} \frac{(\partial_{ij} - d_{ij})^2}{\partial_{ij}^2} \qquad [21.3]$$

d_{ij} being the distance on the output 2-*D* plot, and ∂_{ij} the original input (Euclidean) distance in v-dimensions. If compared with [8.27], it can be seen that E is a kind of multidimensional χ^2.

As with MDS, the user has considerable control over both 'stopping' and 'starting', but this can be both a blessing and a curse. The method proceeds iteratively by choosing 2 initial variables (normally those with maximum variance), and then altering the projection iteratively until E minimizes or until a maximum number of iterations is reached. Hence the user must specify both the minimal acceptable E and the maximum number of iterations. Secondly, in some computer implementations, he chooses whether to supply initial coordinates for each of the objects, or allow these to be computed. Thirdly, he must specify a value for a 'mafic factor' involved in one of the equations. As its name suggests, this is particularly difficult to do objectively, but its value is usually taken to be between 0.3–0.4. Fig.21.10 shows the results of NLM on the limestone data of Table 21.3; it should be compared with Fig.21.9.

21d4. Quadratic Loss Functions

Further geological discussion: Mather(1976, p.340).
Further non-geological discussion: Anderson(1971).
Selected geological applications: *Exploration and pedology* (Anderson 1971).
Software: none known.

This so far little-used technique represents another attempt to minimize a goodness-of-fit function,L:

$$L = \sum_{i=1}^{i=N}\sum_{j=1}^{j=N} w_{ij}(d_{ij} - \partial_{ij})^2 \qquad [21.4]$$

where d and ∂ are defined as in [21.3] and w_{ij} is a weight which merely increases or decreases the importance given to large distances. Execution proceeds similarly to the two ordination techniques above, and (from what limited examples have been published), appears to give similar results, but it is claimed to be computationally simpler.

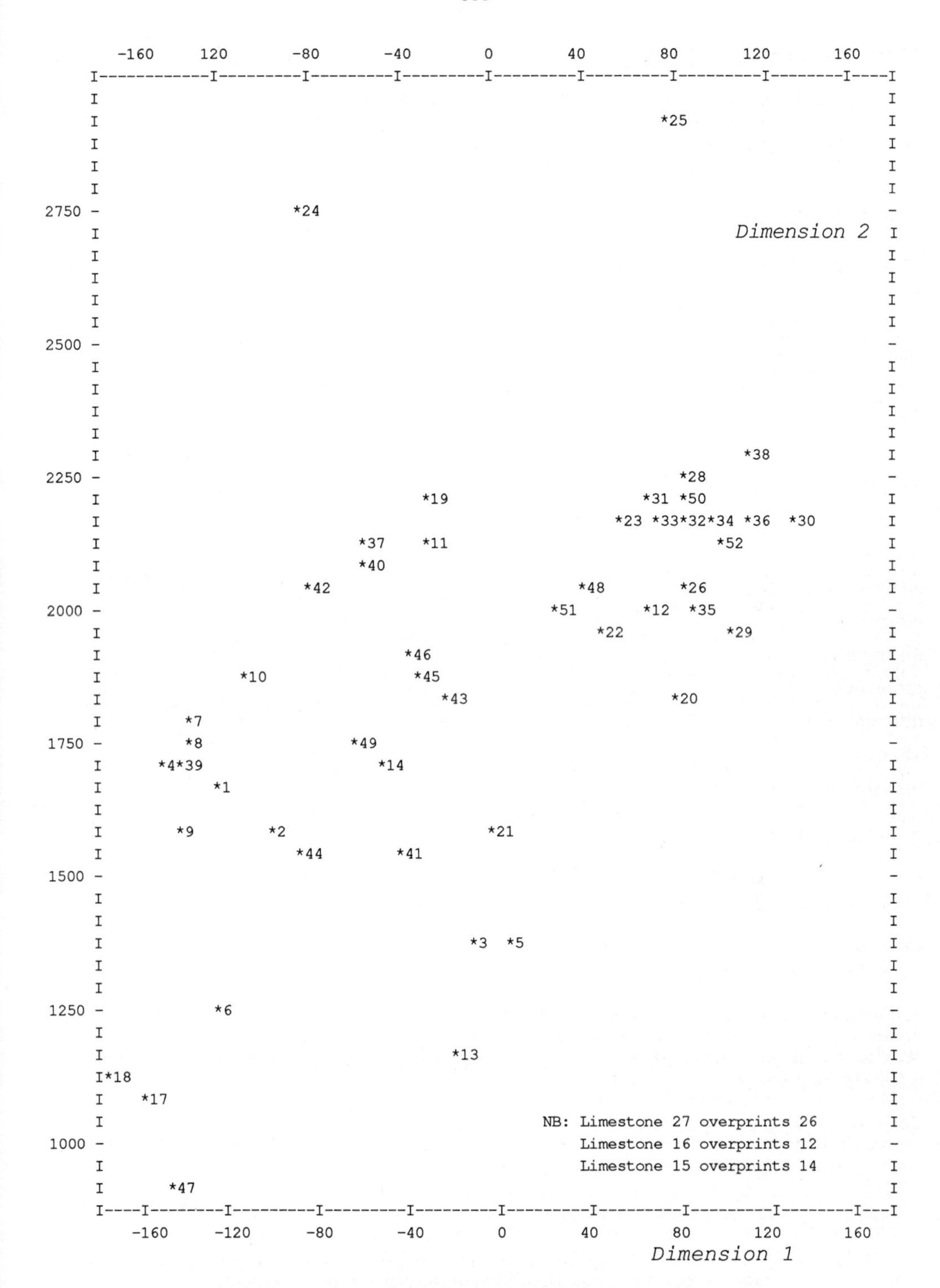

Fig.21.10. Output from Nonlinear Mapping of the limestone data in Table 21.3
(magic factor = 0.35; final mapping error 0.001; initial configuration computed).
The initial plot configuration was determined by the 2 variables of maximum variance (Sr and Ba);
Table 21.8 shows the progress of a mapping error through the 14 calculated iterations.
Calculations performed using program in Mather(1976,p.416)

TOPIC 22. COMPARING 'KNOWNS' & ASSIGNING 'UNKNOWNS': Discriminant Analysis and related methods

22a. Introduction

The major difference between this Topic and the previous one is that we now have some *geological* means of distinguishing groups, independent of the variables which we are going to treat statistically. We wish to compare groups which we *already* have some basis for separating. For example, we might wish to compare the chemistry of two granite batholiths, or of two sedimentary formations, *which we have already proven to be geologically separable* by, say, field mapping or microscope work. Two questions can be answered by the discriminant methods outlined here: (1) given that we have *already* distinguished two (or more) groups *A, B*.... how efficiently can they be distinguished on available criteria, i.e. how different are they overall? (2) Assuming that *A, B*... do prove to be different, to which of these groups is another, 'unknown', single object *X* most likely to belong?

As Le Maitre (1982, p.141) points out, it is important that the criterion for defining the groups is *independent* of the variables which are to be used in the analysis. For example, if discriminant analysis is to be used to see how whether two ammonite species can be separated by simple morphological measurements (number of whorls, size, etc.), the ammonite species *must* have been distinguished in the first place on some quite different criterion (e.g. suture pattern). There is thus no problem in using discriminant analysis to investigate *chemical* differences between, say, basalts in different *tectonic settings*, since the tectonic settings would be identified by criteria totally independent of chemistry (rock associations, evidence for fossil sutures, etc.); but if we were to use chemistry to discriminate I-, S- and A-type granites, there would be a strong danger of a circular argument, since these granite types are already in part defined by their chemistry (e.g. S-type have higher Al/[Na+K]).

22b. Comparing Two or more Groups of Multivariate Data: DISCRIMINANT ANALYSIS

Dedicated monographs and reviews: Cacoullos(1973), Cacoullos & Styan(1973), Hand(1981); Kendall(1966); Klecka(1980); Kleeman (1980); Lachenbruch (1975), Le Maitre(1982,ch.9); Lohnes(1961); Mather(1976,ch.7).

Further discussion: Butler (1982a); Campbell (1984); Cochran (1964); Griffiths(1966b); Melton (1963); Press & Wilson (1978); Saprykin et al. (1977); Williams (1961).

Worked geological examples: *Economic Geology* (Koch & Link 1971, p.102ff); *Igneous petrology* (Le Maitre 1982, pp.148,151,158); *Palaeontology*: (Miller & Kahn 1962, p.258ff); *Sedimentology* (Mather 1976,p.430 & 449; Davis 1986, p.482).

Selected geological applications: *Exploration geochemistry* (Armour-Brown & Olesen 1984; Beauchamp et al.1980; Castillo-Munoz & Howarth 1976; Clausen & Harpoth 1983; Divi et al.1979; Kalogeropoulous 1985; Marcotte & David 1981; Prelat 1977; Reidel & Long 1980; Smith et al. 1982, 1984); *Igneous petrology* (Barnes et al.1986; Carr 1981; Chayes 1965, 1966, 1968a; 1976, 1979a,b, 1987; Chayes & Velde 1965; Defant & Ragland 1982; Embey-Isztin 1983; Gruza 1966; Le Maitre 1976b; Métais & Chayes 1964; Pearce 1976; Pouclet et al.1981; Rhodes 1969; Rock 1977,1984b,1986a,1987e; Saha & Rao 1971); *Lunar petrology* (Gleadow et al.1974); *Metamorphic petrology* (Duncan & Watkeys 1986; Hickman & Wright 1983; Hutchison et al.1976; Pavlova & Agukina 1967; Piboule 1977; Rock et al. 1984,1985,1986a,b; Shaw 1964); *Mineralogy* (Alberti & Brigati 1985); *Palaeontology* (Campbell & Reyment 1978; Reyment 1978; Wright & Switzer 1971); *Pedology* (Digby & Gower 1981); *Sedimentology & sed.petrology* (Chyi et al. 1978; Ghosh et al.1981; Greenwood 1969; Hails 1967; Hawkins & Rasmussen 1973; Middleton 1962; Moiola & Spencer 1979; Potter et al. 1963).

Stand-alone computer programs: Davis (1973,p.452), Mather (1976,p.426,443,etc.)
Proprietary software: covered in most large statistics packages (e.g. BMDP, SAS, SPSS,SYSTAT).

Discriminant analysis, as the above list shows, is one of the most popular multivariate techniques in geomathematics. Thus, two- and q-group discriminant analysis can be envisaged simply as multivariate extensions of the t-test (§10a2) and of one-way ANOVA (§11b), determining whether or not *multivariate* group means differ. A multivariate analogue of Student's t (**Hotelling's T^2**) is in fact generated by many two-group discriminant analysis programs, while q-group discriminant analysis is performed in some programs (e.g. Systat) as a **MANOVA** (Multivariate ANOVA).

Discriminant analysis proceeds be calculating **linear discriminant functions:** linear combinations of the original variables which maximise the differences between the groups (Fig.22.1):

$$Z = \beta_1 X_1 + \beta_2 X_2 + \beta_3 X_3 + \ldots\ldots \beta_n X_n \qquad [22.1]$$

The near-identity between [22.1] and the multiple linear regression equation [19.1] should be noted: discriminant analysis is merely another facet of the General Linear Model (§12a). Coefficients β_1, β_2, β_3 are now called **discriminant (canonical) *coefficients***, while the value of Z for an object is called its **discriminant (canonical) *score*.** The number of discriminant functions calculated is one fewer than the number of groups. Thus there is only *one* such function for 2 groups, but for $q > 2$ groups, there are several, which are mutually uncorrelated and conventionally shown as being at right angles. As with principal components analysis (§23a), it is unusual for more than the first 2–3 discriminant functions to be useful in separating groups, so that even for 4 or more groups, it is usually possible to generate a 2-dimensional X-Y plot to show differences clearly (as on Fig. 22.2).

If a discriminant analysis is to be performed using v variables on q groups, there should ideally by at least v measured objects (analyses, etc.) in each group. However, the technique can still be executed if some data-sets are smaller, provided that the total number of objects measured exceeds vq. (If if is less than this, the discriminant analysis amounts to drawing a straight line through two points!)

Like the t test and ANOVA, discriminant analysis and MANOVA are parametric techniques with analogous assumptions; the effects of violating these have so far barely been aired in the geological literature, but are briefly considered in the discussion which follows:

(1) *The data are distributed multivariate Normally*: This can be tested by multivariate equivalents of the Normality tests in §8e3: for example, multivariate kurtosis and skew (Mardia 1970; Malmgren 1979). Some texts recommend the conventional log-transformation (§8f) for non-Normal data, but this introduces a common and thorny problem, for the distributions of the same variable in the different groups are often different, and a single transformation will not then induce Normality in both together. For example, if MgO in one group is nearly Normal, but logNormal in the other, neither leaving the data alone nor log-transformation will satisfy the Normality assumption.

(2) *The dispersion matrices are equal:* This is equivalent to the equal variance assumption in the t test and one-way ANOVA, and can be tested using **Box's M** — a multivariate F extension. If the dispersion matrices are unequal, the **Browne-Forsythe statistic** can proxy for the statistics in Table 22.2a to test for equal means (it is equivalent to the t-test with unequal variances in §10a2).

To overcome more serious violations of these two assumptions, polynomial and nonparametric forms of discriminant analysis are available (e.g. Bhapkhar 1966; Bhaphkar & Patterson 1977; Bickel 1965, 1969; Chatterjee & Sen 1966; Specht 1967), but they are not considered here because computer implementations are as yet rare. The effect of non-Normality is further considered below.

Many geologists use discriminant analysis empirically, choosing the variables most likely to separate their chosen groups intuitively via a combination of geological reasoning and inspection of the data, based on which variables show the most dissimilar averages. They then calculate the discriminant functions and eliminate individual variables which prove to be ineffective from subsequent reruns, until an optimal separation is achieved using the smallest possible subset of variables. Unfortunately, this can lead the user astray, for it is by no means always the variables which show the most dissimilar *univariate* averages which contribute most to discriminant functions: in Fig.22.1, for example, variable 1 shows similar means for the two groups, but when combined with variable 2 provides an effective discriminant function.

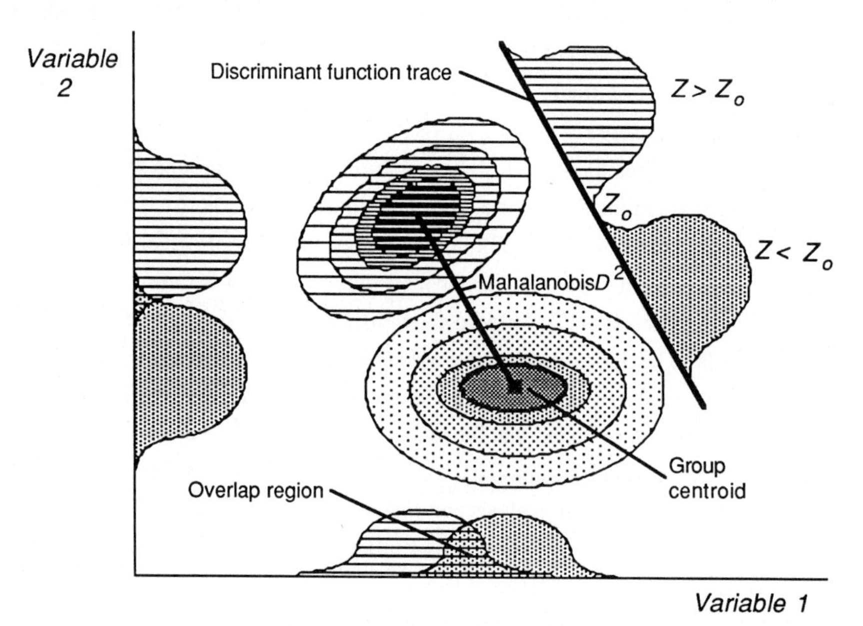

Fig.22.1. Geometric representation of a discriminant function in two-dimensional space.
The two ellipses represent the density distributions of the two groups of data. Discriminant analysis requires that the two ellipses be not only of the same shape but also of the same orientation; only the former condition is here obeyed. Whereas the distributions of the two groups of data overlap substantially along the axes of variable 2 and (especially) variable 1, there is virtually no overlap when projected along the trace of the discriminant function.

For many geological applications, determining whether the groups can in fact be separated is more important than determining which variables do the separation best. Rigorous choice of the 'best' subset of discriminatory variables may not therefore be too critical (except to the extent that if a poor subset is chosen, a misleadingly bad impression of the possible discrimination will result). If delineating the 'best' subset of discriminatory variables does however become important, rigorous statistical procedures are available, analogous to the stepwise regression technique of §19b2. They are briefly considered in §22b2.

Butler(1982a) warns of the dangers of discriminant analysis with **closed data** (§15b), suggesting that some of the structure within the result can be induced by closure itself. In fact, discriminant analysis cannot be performed *at all* on completely closed data, because the matrix manipulations either become impossible or yield an infinity of solutions (**singular matrices** are involved). If all the oxides from major element analyses, for example, were input into a discriminant analysis, many computer programs would yield a solution, but it would be **ill-conditioned** and largely reflect inaccuracies rather than real structure in the data (it would only work at all to the extent that the analyses do *not* total 100%). It is therefore best to input a subset of closed data by eliminating at least one of every pair of highly correlated variables.

22b1. Two-group Parametric Discriminant Analysis

Two-group discriminant analysis is conceptually one of the simplest multivariate techniques. It merely recasts the data via [22.1] in a form such that the overlap between the two groups is minimized (Fig.22.1). Computer implementations will generate a selection of the following data, as in Table 22.1:

(1) *Hotelling's* T^2, *Mahalanobis* D^2. As mentioned above, T^2 is the multivariate analogue of Student's t, and tests whether the multivariate means of the two groups are statistically identical. D^2 is simply related mathematically to T^2 (see Glossary), but has a simple graphical explanation, namely as the *distance* in v-dimensional space between the multivariate means (centroids) of the two groups (Fig.22.1). Both of these statistics are convertible into an F statistic, just as Student's t and F were interchangeable in the univariate case (Table 11.3-4).

(2) *Discriminant scores and resulting assignments.* Because a single discriminant function is generated, each object can be given a single Z score, resulting from application of equation [22.1] to the object's X values. Combining all the Z scores for each group yields mean scores, $\bar{Z}_1$ and $\bar{Z}_2$ and a value halfway between these, Z_o, is conventionally taken as the 'cutoff' criterion between the two groups (Fig.22.1). In a 100% efficient discriminant analysis, all Z values for group 1 are $> Z_o$, and for group 2 they are $< Z_o$. The efficiency of the analysis can thus be measured by the total proportion of objects which obey these relationships: as with classical hypothesis testing, a 90% efficiency would conventionally be regarded as a minimum (meaning that 1 object in 10 is misclassified), but 95% efficiency or more would be preferable.

(3) *Contingency table analysis.* Application of the Z scores in (2) yields a 2 x 2 contingency table (§13d1) as below. Cells a and d represent objects correctly assigned to their original group by the Z scores (and should ideally equal n_1 and n_2), whereas cells b and c represent incorrect assignments (and should both ideally equal zero). A χ^2 statistic can then be calculated to see whether a and d diverge significantly from b and c; the more efficient the separation, the more significant χ^2 will be.

		Original group	
		Group 1	Group 2
Assigned group	Group 1	a	b
	Group 2	c	d

In this author's experience with geological data, discriminant Z scores offer a more sensitive index of discrimination efficiency than T^2, D^2, F or especially χ^2 tests: the latter are commonly significant at > 99% level where the efficiency based on Z scores has barely reached 90%. Discriminant scores are also useful in revealing severe violations of the method's assumptions: asymmetry of dispersion matrices or non-Normality, for example, are commonly revealed by unequal classification efficiencies of the two groups (e.g. group 1 may have all values > Z_o, i.e. 100% efficiency, but group 2 may have only 70% of scores < Z_o). In such cases, T^2, D^2 and F are unlikely to give reliable significance levels anyway because their fundamental assumptions are not met.

This leads to the question: how meaningful or reliable is a result where the two groups are *asymmetrically* discriminated, and hence where the usual assumptions of discriminant analysis are violated? This has barely been discussed in the geological literature, but there is little reason to doubt the usefulness of such results. A discriminant function is merely a linear combination of original variables [22.1]. If calculation of a such a function on the X values for two groups produces a different result for, say, 95% of the *combined* sample, then the Normality or otherwise of the distributions would seem largely immaterial. What is sacrificed with asymmetry or non-Normality is the ability to seek out the *maximum possible* discrimination; however, if 95% efficiency is achieved even without Normality, this too may be irrelevant, for the 'real' efficiency may be even higher.

With geological data, it thus often pays to try moving the cutoff value Z_o along the line joining the centroids, and recalculate the discrimination efficiencies. The efficiency for group 1 will of course rise towards 100% (and for group 2 fall towards 0%), as Z_o moves towards $\bar{Z}_1$, but the *combined* efficiency may rise and fall irregularly, and reach a maximum at some other percentage of the distance from $\bar{Z}_1$ to $\bar{Z}_2$ than at 50%. Since taking Z_o at the mid-point is based on the parametric assumptions of discriminant analysis, there is no reason why an *empirical* discriminant function based on a cutoff at, say, 40% or 70% should not be used — particularly if these assumptions are known to be suspect.

Note that Table 22.1 only gives an ordinal assessment of the contribution of the individual variables to the final discrimination, without any formal statistical tests.

Table 22.1. An example of two-group discriminant analysis

Compositional differences between alnöites and aillikites (ultramafic lamprophyres), from Rock (1986)

Number of variables used in discriminant function [1]	10
Number of analyses of alnöites, aillikites	49,75
Variables in descending order of contribution to discriminant function [2]	Al, Na, Mg, Si, Fe^3, Fe^2, Ca, K, Ti, P
Limiting value of discriminant function Z_o	30.3
Average Z value for alnöites, $\bar{Z}_1$	33.6
Average Z value for aillikites, $\bar{Z}_2$	27.0
Mahalanobis D^2 [3]	6.6
F ratio (degrees of freedom)	18.1 (10, 113)
% significance level of F ratio [3]	» 99.9
% of alnöites correctly classified (i.e. Z value > Z_o)	92
% of aillikites correctly classified (i.e. Z value < Z_o)	93
χ^2 from contingency table analysis	89.3
% significance level of χ^2	» 99.99

Notes:

[1] Program used is listed in Davis (1973, p. 442).

[2] Listed in order of the value $|\lambda_j D_j / D^2|$ which measures contribution to the discriminant (see Davis, 1973, p. 452).

[3] The precise significance levels of these statistics depend on data Normality and equal dispersion matrices (see text). As these assumptions are only partly valid in this case, the % correct classification is probably a better guide to the discrimination efficiency, and the quoted significance levels should only be taken as a rough guide.

22b2. Multi-group Discriminant (Canonical Variate) Analysis (MDA)

MDA tests the differences between *several* multivariate means, and is a special case of the General Linear Model — a multivariate analysis of variance (MANOVA). As mentioned above, though some computer packages (e.g. SPSS) implement MDA as a separate "canned" program or module, others (e.g. MacSS, Systat) actually perform MDA step-by-step as if it were a one-way MANOVA.

Table 22.2 and Fig.22.2 show critical results from MDA on the limestone data of Table 21.3; most packages will output a far wider range of other statistics, including means, sum of product and covariance matrices. In interpreting these results, we shall assume multivariate Normality for the purposes of illustration. The statistics in Table 22.2a test the overall significance of the discriminant analysis — here, the null hypothesis that the 6 limestone occurrences have the same multivariate mean composition, against the alternative that *at least one mean* is different. All these statistics are equivalent to F tests in univariate ANOVA (cf. §11b). In this case, all have vanishingly low associated probabilities, so the alternative hypothesis is accepted. Another statistic sometimes output by some MDA programs and testing the same hypothesis is **Rao's *R***. Note that although these statistics are different from those used to test the same hypothesis in two-group discriminant analysis (§22b1), *all* of them are actually tested by conversion to F. Readers intimidated by them can therefore ignore them!

Table 22.2b-c test the contribution to this overall result of the *five* discriminant functions we obtain for 6 groups. (This test was unnecessary in §22a1 with only one discriminant function, for the overall F test was necessarily there a test of this function). If the 'residual root' in Table 22.2b is significant, then insufficient of the total variance has been extracted. In this case, roots 4 and 5 add little to the discriminant, as can also be seen by their small contributions to the total variance in Table 22.2c, so we conclude that 3 discriminant functions are enough to separate the 6 limestone groups.

Table 22.2d conversely tests the contribution of the 15 *variables* to the discrimination: the higher F (and lower the associated probability), the more the variable contributes. In this case, we would expect elimination of La, Zn, Zr, SiO_2 and Fe_2O_3 from the original data-set to have little effect on the separation efficiency. Some packages would then perform *stepwise* MDA, in which such variables are deleted until the 'best subset' is located. However, dangers with stepwise MDA are denounced by some statisticians as "fishing expeditions", more dangerous even than those with stepwise regression analysis (§19b2). 'Best subsets' may not be reproducible between different sample data-sets, and the probabilities of some of the test statistics are invalid if more than one model is being tested.

Table 22.2d also shows the ß coefficients [see 22.1] for each of the 5 discriminant functions. The *raw* values (only shown for function 1) yield the discriminant scores illustrated by a subset in Table 22.2e. Thus for limestone no.1, we obtain, for discriminant function 1 (and similarly for Z_2 to Z_5):

$$Z_1 = 0.0531\ \text{La} - 0.0227\ \text{Ce} \ldots + 0.2028\ \text{MgO} = 3.44 \quad \text{(from Table 22.2e)}$$

Assignment to groups is then based on the Mahalanobis distances of the individual object from the mean score for the original 6 groups (a similar procedure to that used in cluster analysis). The scores represent locations in discriminant space. Thus the nearer the object's score (location) is to the mean score for a particular group, the more probable is its membership within that group. Table 22.2e also shows these distances, together with resultant probabilities for the membership of 4 limestone analyses

Table 22.2. Multi-group linear Discriminant Analysis (MDA) on the limestone data of Table 21.3

(a) MULTIVARIATE TESTS FOR OVERALL SIGNIFICANCE OF THE DISCRIMINATION

Statistic	Value	*F equivalent*	*df*	Prob.
WILKS' Λ (lambda)	0.017	3.125	[70,161]	0.00
PILLAI TRACE	2.475	2.590	[70,185]	0.00

(b) TESTS OF RESIDUAL ROOTS

	χ^2	*df*	Assoc.prob
ROOTS 1–5	167.364	70	0.00
ROOTS 2–5	106.142	52	0.00
ROOTS 3–5	56.636	36	0.01
ROOTS 4–5	17.798	22	0.71
ROOT 5	5.938	10	0.82

(c) CANONICAL CORRELATIONS, LATENT ROOTS AND PERCENTAGES OF VARIANCE

	1	2	3	4	5
Canonical correlations	0.881	0.837	0.782	0.501	0.367
Latent roots	3.451	2.345	1.578	0.335	0.123
% of total variance	43.88	29.81	20.07	4.26	1.98

(d) TESTS FOR THE CONTRIBUTION OF EACH VARIABLE TO THE DISCRIMINATION, AND DISCRIMINANT COEFFICIENTS 1–5 (standardized by conditional [within-groups] standard deviations)

	F-test	Prob.	1(raw)	1	2	3	4	5
La	0.5716	0.7236	0.0531	0.353	0.541	0.167	0.006	0.261
Ce	8.9245	0.0000	-0.0227	-0.639	-0.880	-0.327	0.295	0.204
Cr	4.4840	0.0024	-0.0484	-0.284	0.317	0.427	0.825	0.146
Ni	2.7384	0.0297	0.1340	0.335	0.369	0.094	0.130	-0.497
Cu	2.8529	0.0249	-0.0945	-0.417	-0.633	-0.426	-0.063	-0.332
Zn	1.1229	0.3616	0.0053	0.053	0.080	-0.230	0.329	-0.123
Rb	5.5821	0.0006	-0.0670	-0.543	0.173	-0.990	-0.199	0.188
Sr	8.2502	0.0000	-0.0015	-0.445	0.524	0.430	0.481	0.358
Zr	0.3210	0.8976	0.0063	0.246	-0.162	-0.034	0.241	0.171
Ba	10.148	0.0000	0.0074	0.264	0.000	0.708	-0.611	-0.417
SiO_2	0.9906	0.4349	0.2066	0.872	0.576	0.678	1.013	0.389
Al_2O_3	4.4321	0.0025	-0.9114	-0.690	-0.566	0.215	-1.647	-0.083
Fe_2O_3	1.2899	0.2840	-2.0506	-0.716	0.349	-0.824	0.401	-0.553
MgO	3.3666	0.0113	0.2028	0.234	-1.066	0.620	-0.786	0.337

(e) ASSIGNMENT DETAILS (example limestones only)

	Discriminant scores					Mahalanobis D^2 from group means						Group membership probabilities						Group	
No	1	2	3	4	5	1	2	3	4	5	6	1	2	3	4	5	6	Actual	Predicted
1	3.44	0.25	0.61	0.54	0.65	4.30	6.70	5.12	6.83	6.47	6.27	0.98	0.00	0.02	0.00	0.00	0.00	Find	Find
2	2.49	2.70	-1.38	0.26	0.10	4.81	4.91	6.72	6.55	6.59	6.57	0.61	0.38	0.00	0.00	0.00	0.00	Find	Find
10	1.38	2.73	-1.66	-0.27	0.69	3.74	1.99	5.56	4.68	4.79	4.86	0.01	0.99	0.00	0.00	0.00	0.00	Lui	Lui
52	-2.29	-0.20	-1.55	1.27	-1.30	7.84	6.82	7.39	5.24	5.88	5.33	0.00	0.00	0.00	0.61	0.02	0.37	Laggan	Kincraig

(f) SUMMARY OF ASSIGNMENT EFFICIENCY (actual versus predicted group membership)

		Findhorn	Glen Lui	Grantown	Kincraig	Lochan Eilean	Kinlochlaggan	TOTAL
	Findhorn	**7** (100%)	0	0	0	0	0	7
Actual	Glen Lui	0	**4** (100%)	0	0	0	0	4
limestone	Grantown	0	0	**6** (86%)	0	0	1	7
occurrence	Kincraig	0	1	0	**3** (60%)	1	0	5
	L.Eilean	0	0	0	2	**13** (87%)	0	15
	KinlochL	1	0	0	2	1	**10** (71%)	14
	TOTAL	8	5	6	7	15	11	52

Numbers in **bold** are totals assigned to correct occurrences; numbers in brackets are % efficiencies for each occurrence within each of the 6 groups, and their consequent assignment to the most probable group. For limestone no.1, for example, the probability is 98% that it lies within group 1 (Findhorn), 2% within group 3 (Grantown) and effectively nil for the others, so it is unequivocally assigned to Findhorn. These are of course the assignments resulting from chemistry alone, so the final test of the MDA is to compare these chemical assignments with the original *a priori* classification (Table 22.2f). All of the Findhorn and Glen Lui limestones in fact prove to be assigned to their original groups, an assignment efficiency of 100%, but the efficiency is down to 60% for Kincraig. The final row for limestone 52 in Table 22.2d illustrates one of the *misclassified* examples, for its group membership probability within its original (Kinlochlaggan) group (37%) proves less than its probability within the Kincraig group (61%). The overall classification efficiency is 43 out of 52 analyzed limestones (83% success).

Here, the classification efficiency (Table 22.2f) is more useful *geologically* than the *statistical* tests in Tables 22.2a–b. Although the means are clearly unequal, the 83% success rate implies an error 17 times out of 100; if this were a significance level, an 'informal null hypothesis' that we can separate *all* these limestones on a chemical basis would be rejected. This is more easily seen on Fig.22.2, which plots the 1st and 2nd discriminant scores for all 52 limestones. Although the separations here are necessarily less than optimal (the ≈ 26% contribution from discriminant functions 3–5 being ignored) the Findhorn and Glen Lui limestones are clearly distinct (hence their 100% assignment efficiencies in Table 22.2f), while the remainder (especially Loch an Eilean and Kinlochlaggan) overlap considerably.

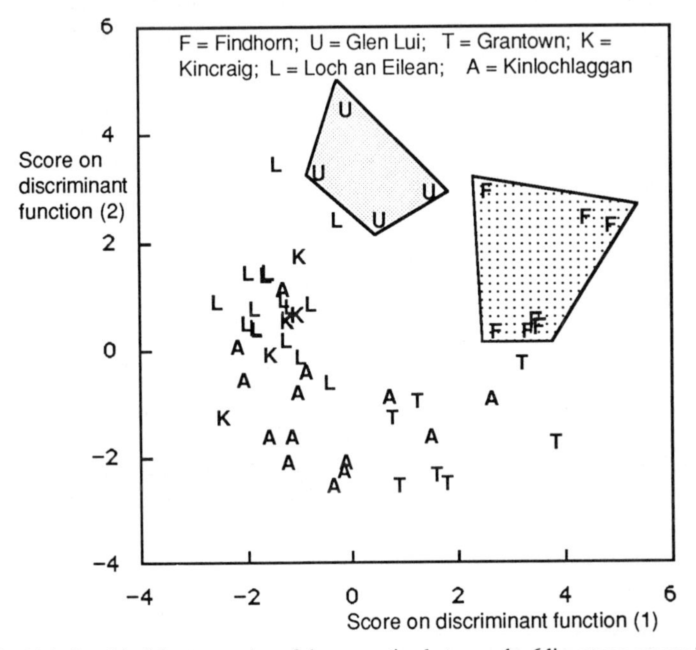

Fig.22.2. Graphical demonstration of the separation between the 6 limestone groups in Table 21.3 effected by multi-group discriminant analysis in Table 22.2

22c. Comparing Groups of Multivariate Dichotomous Data: ADAPTIVE PATTERN RECOGNITION

Further non-geological descriptions: Aitchison & Aitken(1976); Mather (1976,p.453).
Selected geological applications: *Sedimentology/palaeontology* (Ramsayer & Bonham-Carter 1974).
Stand-alone computer programs: Mather (1976,p.458).
Proprietary software: none known.

In many geological situations, we may only have measurements of several *dichotomous* (yes/no) variables for two groups of objects. Topic 21a has already shown how *similarities* between such objects can be measured; we now consider how to *differentiate* groups of objects. Table 22.3 lists the occurrence of 8 accessory minerals in 9 samples of quartzite from 2 separate horizons. Can the quartzites be distinguished from these data, and can we assign an unknown sample to either quartzite *A* or *B* using its accessory mineral assemblage alone? Exactly the same technique could be applied if, for example, we wished to distinguish two stratigraphical units on their fossil assemblages (replacing minerals with fossils in Table 22.3), to distinguish two types of mineralization on the presence or absence of certain geological features, and so on.

Table 22.3. Raw data for Adaptive Pattern Recognition: accessory minerals in two quartzite units

	Apatite	Baryte	Corundum	Garnet	Kyanite	Monazite	Pyrite	Zircon	*Dummy*
Quartzite A,sample 1	1	1	1	0	1	1	1	0	1
Quartzite A,sample 2	1	0	0	0	1	0	1	1	1
Quartzite A,sample 3	1	1	1	1	1	1	1	1	1
Quartzite A,sample 4	1	0	0	0	1	1	1	0	1
Quartzite A,sample 5	0	1	0	0	0	1	1	0	1
Quartzite B,sample 6	0	0	1	1	0	0	0	1	1
Quartzite B,sample 7	0	1	1	1	1	0	0	1	1
Quartzite B,sample 8	0	0	1	1	0	0	0	0	1
Quartzite B,sample 9	0	0	0	1	0	0	0	1	1
ß coefficients (1)	2	0	−1	−2	1	1	2	−1	0
ß coefficients (2)	3	0	−1	−2	0	3	3	−2	−1
Unknown quartzite 1	0	0	1	1	0	1	1	1	1
Unknown quartzite 2	1	1	1	1	1	0	1	1	1

Note: 1 = present, 0 = absent for each mineral

Adaptive Pattern Recognition, which can be thought of as binary discriminant analysis, works on the same equation as [22.1], except that the *X*s now denote *dichotomous* variables. A little thought will also show that the ßs can only take whole number values (+ve or −ve) — multiplying a '0' or '1' by say 3.5731 is not very logical! — and that the method *must* be iterative, beginning with a trial set of ßs which are modified until maximum separation of the two groups is achieved.

The method starts by adding a dummy variable (Table 22.3), whose purpose is to ensure that the average discriminant Z score for the combined groups is zero. The ßs are calculated in such a way that $Z > 0$ if the object belongs to group A, and $Z < 0$ if it belongs to group B. If any object is wrongly allocated, the ßs are adjusted and calculations repeated. Iterations are terminated as soon as perfect discrimination is achieved, or after many iterations (say 1,000) for problems which do not converge (i.e. where the two groups are not in fact linearly separable).

For the data of Table 22.3, a solution is quickly achieved after only 3 iterations, with the ß coefficient set (1) shown. In other words, the Z value:

$$Z = 2 \text{ apatite} - \text{corundum} - 2 \text{ garnet} + \text{kyanite} + \text{monazite} + 2 \text{ pyrite} - \text{zircon}$$

is +ve for all samples from Quartzite *A* (the actual scores for the 5 samples are 5, 4, 2, 6, 3), and –ve for all those from Quartzite *B* (Z = –4, –3, –3, –3). Hence apatite, garnet and pyrite are the most consistently different in the two horizons, while baryte has no discriminatory power, and therefore has a zero ß coefficient. This is intuitively reasonable from inspection of the raw data, where it can be seen for example that 4 of the 5 *A* horizon quartzites have apatite whereas none of horizon *B* have it, while the reverse is true for garnet, and whereas baryte is evenly distributed between the two horizons. On this basis, we would assign unknown samples 1 and 2 in Table 22.3 to Quartzite *B* and *A* respectively (their Z scores are respectively –1 and +1).

A limitation of this technique is that there can be several solutions which give perfect discrimination between the two groups: different solutions can even be obtained by inputting the data into the computer program in different orders. For example, ß coefficient set (2) in Table 22.3 was derived merely by interspersing the samples in the fashion *ABABABABA* in the input file, rather than *AAAAABBBB* for ß set (1). The two sets of coefficients are quite different, yet the discrimination is perfect with both (with set 2, horizon *A* samples yield Z scores of 7, 3, 3, 8, 5 and horizon *B* samples yield scores of –6, –6, –4, –5). This indicates that there may be no such thing as a 'best' solution to a given data-set, which can severely limit the possibilities of giving the ß coefficients any geological interpretation. For example, kyanite is a contributing variable to ß set (1) but not to set (2) in Table 22.3, so we cannot say overall whether it is important or not! This also of course implies that computing the most efficient discriminatory function may require several attempts, and with a large data-set the user may never be sure that he has obtained all possible solutions. Finally, it implies that an unknown may be classified inconsistently using different possible solutions — particularly if it is a borderline case. For example, the two unknown quartzites in Table 22.3 are both classified as 'undefined' (Z scores = 0) on ß coefficient set (2), whereas they are classified as quartzites *B* and *A* respectively using ß coefficient set (1).

Despite these obstacles, adaptive pattern recognition has considerable potential in geology for assigning groups of multivariate data of the type which come out particularly from studies of fossil or mineral assemblages in rocks. Applications in other fields could also easily be envisaged.

TOPIC 23. EXAMINING STRUCTURE, RECOGNIZING PATTERNS & REDUCING THE DIMENSIONALITY OF MULTIVARIATE DATA

23a. Graphical Methods of Displaying Multivariate Data in Two-Dimensions

Dedicated monographs and reviews: Barnett(1982, p.254ff); Everitt(1978); Wang(1978).
Further geological discussion: Chernoff (1973); Kleiner & Hartigan (1981); McDonald & Ayers(1978).
Selected geological applications: *Geochemistry* (Garretrt 1983; Lepeltier 1969; Howarth & Garrett 1986).
Stand-alone computer programs: unpublished programs are known to be available.
Proprietary software: none known.

Graphics are used extensively in geology — more so than in some other sciences (Howarth & Turner 1987). Because of the inherent difficulty of grasping purely numerical analyses of multivariate data, a recent trend has been towards representing such data in cartoon or symbol form (Fig.23.1). Such symbols have the considerable additional advantage that they can be displayed on geological maps at the point from which the analyzed sample was taken (e.g. Fig.23.1b), thus providing an immediate visual comparison of data-values for the user of the map (the basic aim of all geological maps).

Of the symbols in Fig.23.1, **Chernoff faces** (Fig.23.1a) require further explanation, since it is sometimes difficult to get geologists to take them seriously at first sight! These stylised faces can in fact be used to represent the interplay of up to 16 different variables. Their principal advantages are that: (a) the human brain is specially equipped to analyze and interpret facial expressions: *very subtle* differences can thus be instantaneously detected between different faces, which are quite impossible using the other abstract graphics on Fig.23.1; (b) because facial features are linked (e.g. tightly knit eyebrows usually go with pursed lips), inter-relationships between different variables can also be readily depicted. The obvious disadvantage is that only data with some emotive quality can be successfully represented (it would hardly be appropriate, for example, to distinguish say an I-type granite by a sad face and an S-type by a happy face!) However, series of faces have been used effectively in depicting the progress of a company's finances (the mouth would presumably be set to 'happy' for profits and 'sad' for losses!) Fig.23.1a also suggests how faces might also be used to depict the probability of diamond deposits in a given area. Primary diamonds occur almost exclusively in intrusions of kimberlites, lamproites and other lamprophyres, which are usually discovered by following up stream sediments carrying certain indicator minerals (Cr-pyrope, Cr-diopside, Mg-ilmenite, etc.) Certain factors are already known to influence the diamond content of such pipes (e.g. the most prospective occur in the centres of ancient stable cratons), but other possible controls (e.g. kimberlites as pipes or sills) are less well known. A Chernoff face could be designed with care which would reflect many of these subtleties. For example, the most obvious, indicative facial features (the mouth and eyes) could be linked to the most important known controls (presence of indicator minerals and ancient cratons), and the secondary features (nose, ears, etc.) to the less well known controls. The three faces in Fig.23.1a immediately suggest 'bad', 'thank heaven' and 'pull the other one': they represent exactly how an exploration manager reacts to potential exploration stimuli, and thus have enormous potential for simplification of complex geological data.

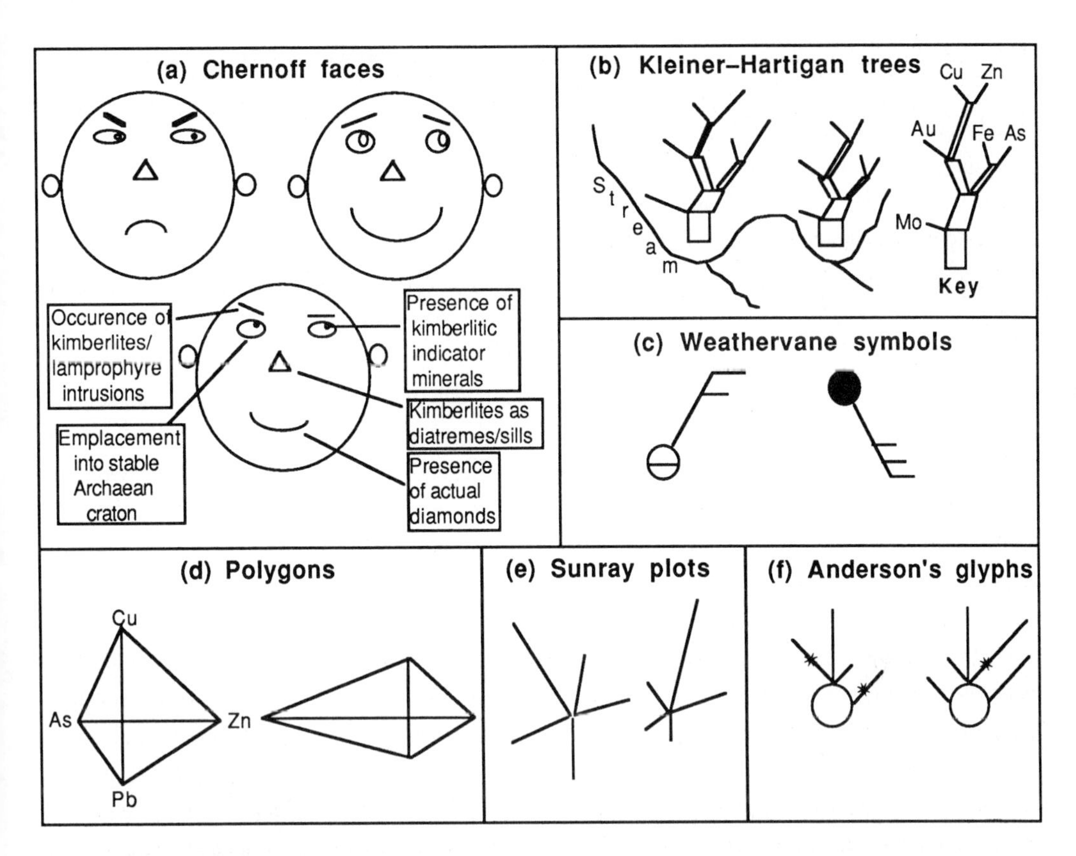

Fig.23.1. Some compound characters (symbols) capable of displaying multivariate data in 2 dimensions
In **(a)**, up to 16 different variables can be represented, which control the shape of the head, the slant of the eyes and mouth, the size of the pupils, the position of the ears, etc. In **(b)**, the positions of different elements are constant but their different tenures are indicated by changes of shape. In **(c)** and **(f)**, the 'whiskers' represent different variables. In **(d)** and **(e)**, the lengths of the sides or sunrays here indicate the tenure of 4 variables, but further variables can be depicted by adding more sides or rays. Any of these symbols can be placed at the correct geographical position, as on the map of a stream course in (b). In this way, multi-element anomalies and other multivariate patterns can be detected by eye.

23b.Finding Structure in Multivariate Data: PRINCIPAL COMPONENTS ANALYSIS

Dedicated monographs and reviews: Jolliffe(1986).
Further geological discussion: Aitchison(1984b); Butler(1976); Hohn(1978,1979); Le Maitre(1982, ch.7 — gives full mathematical treatment, including matrix equations); Miesch(1980); Rao(1964); Stephens (1976); Till & Colley (1973); Trochimczyk & Chayes (1977a,b;1978);
Worked geological examples: *geomorphology* (Mather 1976, p.233); *igneous petrology & mineralogy* (Le Maitre 1982, p.114 & 147); *sedimentology* (Davis 1986, p.540).
Selected geological applications: *economic geology* (Krauchenko 1982); *geochemistry* (Armands 1972; Garrett et al. 1980; Gilirab & Churchman 1977; Howarth 1970); *hydrology* (Lahiri & Rao 1974); *igneous petrology* (Cann 1971; Coffrant et al.1978; Hohn & Friberg 1979; Le Maitre 1968; Podol'sky 1975; Rock 1978; Ruegg 1976; Saxena 1970; Vistelius et al.1970); *mineralogy* (Atzori et al.1973; Joreskog et al.1976, p.147; Mitchell 1986,p.222; Saxena 1969; Saxena & Ekström 1970; Saxena & Walter 1974; Webb & Briggs 1966); *palaeontology* (Berthou et al. 1975; Malmgren 1972; Reyment 1973,1979; Spicer & Hill 1979); *sedimentology* (Brower et al.1979; Read & Dean 1972; Yamamoto et al.1979); *stratigraphy*(Hohn 1978; McCammon 1966).
Stand-alone computer programs: Davis (1973,p.494); Fatti & Hawkins(1976); Mather(1976,p.228).
Mainframe software: covered in most large statistics packages (e.g. SPSS,SAS); Hohn(1985).
Microcomputer software: covered in some larger statistics packages (e.g. Systat, MacSS).

PCA, at its simplest, can be regarded simply as another **ordination** technique (§21d: e.g. Brower et al.1979), for reducing multivariate data into fewer dimensions. However, it has further possibilities, notably as the first stage of the final and most complex of the techniques covered in this book, **factor analysis** (§23c), and has a rather different mathematical basis — hence its deferral to this final Topic.

PCA transforms an original set of *N* variables into a new set of *N* **principal components** which are at right angles to each other (uncorrelated). Fig.23.2 illustrates what this means in 3-dimensional space. Although there are as many principal components as variables, the transformation is such that the first 1 or 2 components almost invariably account for a far larger proportion of the total variance than an equivalent combination of the original variables. Thus if we plot the first two components against one another, we obtain a far more representative idea of the total data variation than in, say, a traditional Harker diagram (Fig.15.1). In a typical igneous rock suite (9 major elements + many minors and traces), 2 components will usually be adequate, and the remainder can be discarded, so we have conveniently reduced *N*-dimensional data to 2 dimensions for viewing on flat paper. PCA is perfectly general, makes no assumptions, tests no hypotheses and has no underlying model. It is merely another form of data transformation — more complicated than straightforward logarithmic transformation, no doubt, but no more than that.

PCA is actually performed by extracting **eigenvectors** and **eigenvalues** from the original data. These frightful-sounding terms are in fact simple to visualise geometrically (Fig.23.2). There are, however, two ways of extracting them, which use a different input matrix formed from the raw data:

(1) *Variance-covariance matrix* (e.g. Table 21.4a). Here, the original data are unscaled, so that the variable with maximum absolute variation contributes most to the 1st eigenvector. With many geological data, this method leads to a highly satisfactory solution in which the first 2 eigenvectors account for 80-90% of total variance (or even more), allowing 2-D plots to represent *N-D* data quite adequately. As indicated in §15e & §21, this method is best suited to data where all the variables have similar variances and are measured in the same units. Variance-covariance matrices used on raw rock analyses invariably result in elements like Si and Ba contributing most to the eigenvectors, simply because they have the highest numerical values; this says little about the data of interest.

(2) *Correlation matrix* (e.g. Tables 21.4b,e,f). This is equivalent to a variance-covariance matrix on standardized data (§13). Hence using the correlation matrix means that all variables are scaled to unit variance and zero mean. This means that variables with small *absolute* but high *relative* variations (e.g. a trace element varying from 5 to 100 ppm, a factor of 20) would receive more weight in the eigenvectors than most major oxides, which vary by much smaller relative factors (say 50–70% for SiO_2, or a factor of 1.4). Geometrically, this method stretches the axes in Fig. 23.2 corresponding to the variables with small variance, and compresses those corresponding to a large variance. In principle, it should be used where data are expressed in different units, or there is any reason to 'upgrade' the contribution of the minor variables. It has the disadvantage, however, that the first few eigenvectors will almost invariable contribute *far less* to the total variance than is the case from method (1), so in a sense nullifying the basic purpose of PCA: that is, it is no longer possible to plot just 2 principal components, knowing that they will adequate represent the *N*-dimensional data.

Different authors have made quite contradictory recommendations as to whether raw or standardized data should be used in different circumstances. This is because, in truth, neither method is wholly satisfactory with geological data, where we usually want a procedure somewhere in between — something which gives greatest weight to variables we 'know' to be most significant geologically (e.g. SiO_2), but still does not wholly ignore the minor and trace elements. One alternative is to log-transform the data first, and then use the variance-covariance matrix. This means that *changes* by the same order of magnitude will be given equal weight (§15e).

Table 23.1a-b show a correlation matrix between 7 chemical variables measured on 25 water samples, and the composition of the first 4 eigenvectors (principal components) extracted from this. Note that there are actually 7 eigenvectors — as many as variables. Table 23.1d reveals that the first 2 associated eigenvalues account for 79.5% (0.568 + 0.227) of the total variance. Hence a plot of these two eigenvectors would be a fairly reasonable display of the chemical variation between the water samples. To produce this, we multiply the data-values by the weights in Table 23.1b to give scores:

$$\text{Score on PCA}_1 = -.414Cl^- - .405Mg^{++} - .398\text{dissolved solids} - .351SO_4^{=} - .383Ca^{++} - .362HCO_3^- - .325pH$$
$$\text{Score on PCA}_2 = .308Cl^- + .391Mg^{++} + .389\text{dissolved solids} + .097SO_4^{=} - .45Ca^{++} - .43HCO_3^- - .452pH$$

and then plot these as *X-Y* coordinates on a 2-D variation diagram, instead of the original variables. If the eigenvectors were extracted from the variance-covariance matrix, the values to use for 'Cl' etc. in these equations are *raw* data-values, but if (as here) they were extracted from the correlation matrix, *standardized* values must be used. This is an important distinction (Le Maitre 1982, p.118).

Note that weights of the 7 variables on PCA 1 are here subequal, so that no one variable appears to dominate the variation. On the other hand, there are three *groups* of variables in the scores for PCA 2 — 3 subequal and –ve, 3 subequal and +ve, and one between. *X-Y* plots of the original raw variables against one another would therefore have been highly ineffective in displaying the variation.

As with all multivariate techniques, care is needed in interpreting principal components with closed data (§15b), since eigenvalues and eigenvectors may in part be controlled by closure rather than by real geological variations. Unfortunately, no simple way has been suggested of separating these controls.

Should it be necessary to display more than 3 principal components on a 2-*D* diagram, Stephens(1976) suggests a way of doing this using different sized circles to provide perspective.

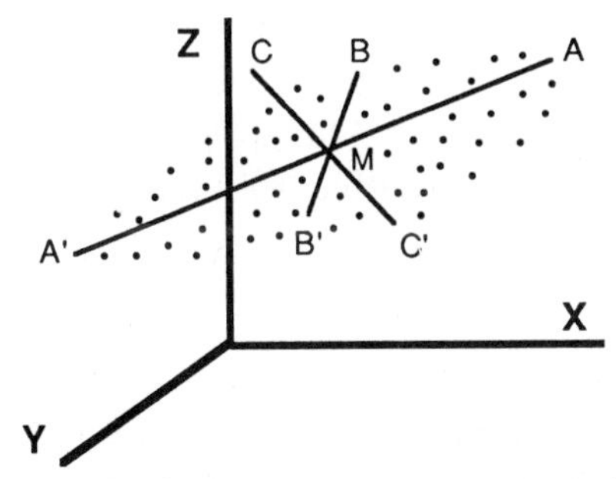

Fig.23.2.Geometric interpretation of Principal Components Analysis (in 3-D)
In the 3-D space defined by the axes X,Y,Z, the cluster of data-points forms an ellipsoid with three **eigenvectors** in the directions MA, MB, MC, where M is their centroid. MA, the 1st eigenvector, defines the *direction* of maximum variability (spread) of the data cluster (the longest axis of the ellipsoid), and is also the direction of 'best fit' to a straight line in 3-D space. MB, the 2nd eigenvector, defines the maximum spread at right angles to MA, and so on when translated into *N*-dimensional space. The 3 **eigenvalues** are the *lengths* of the axes AA' > BB'> CC'.

23c. Fitting Multivariate Data to a Model: FACTOR ANALYSIS

Dedicated monographs and reviews: Blackrith & Reyment(1971); Cattell(1978); Child (1970); Claudon et al. (1977); Francis (1974); Gorsuch(1974); Harman (1967); Joreskog et al. (1976); Klovan(1975); Koch & Link(1971,p.119ff); Lawley & Maxwell (1963,1971); Le Maitre(1982,ch.8); Mather(1976,ch.5); Mulaik(1972); Thurstone (1947); Tripathi (1979).
Polemics — for and against factor analysis in geology: Francis(1974); Kufs(1979); Miesch(1969b); Temple (1979).
Further geological discussion: Dumitriu et al.(1980); Vistelius & Ruiz Fuller(1969); Zhou et al.(1983).
Stand-alone computer programs: Cameron(1967); Clarke(1978); Davis(1973,p.519ff); Mather (1976, p.258,264,276,etc.)
Mainframe software: covered in most large statistics packages (e.g. BMDP,SAS,SPSS).
Microcomputer software: programmed in some statistics packages (e.g. Statview 512+™).

After geostatistics (§20e), this is probably the most controversial statistical technique, in terms of its geological efficacy. Factor analysis was developed initially by psychologists, to handle data dealing with 'attitudes', 'social groups' and the like. To some 'pure' scientists, it is therefore a suspect technique from the start! Its other principle problem is that it not only makes wide-ranging assumptions about the data, but also requires the geologist to make numerous decisions about which of a medley of possible techniques to use at each stage of the analysis. Time and again, this author has seen students who decide Factor Analysis is for them, but then find they are confronted with pages of alternative but equally incomprehensible methods of execution. They have no alternative but to opt for the program's 'default' methods in each case, but are then faced with still more pages of computer output, detailing a host of statistical parameters and tests which mean equally little to them. The whole exercise is then not merely fruitless but a clear case of "a little knowledge is a dangerous thing".

Altogether, a single data-set can probably be 'solved' in 100 or more different ways; the 'best' solution is an inevitably subjective choice, even for experienced geomathematicians, and some (e.g. Temple 1979) claim it is far *too* subjective to be geologically meaningful. Certainly, with such a choice, it is all too tempting to repeat the analysis until the desired result is obtained (cf. Fig.21.4!) Factor analysis is placed last of all in this book precisely because it is believed to be the technique requiring the most expert handling; in experienced hands, it can nevertheless produce valuable results.

The literature is confusing as to the relationship between Factor analysis and Principal Components Analysis (§23b). PCA involves no assumptions, and is merely a method of mathematically manipulating multivariate data so as to represent it in fewer dimensions, and in this respect can be treated as another ordination technique (§21d). It can actually be used to extract the initial factors in a factor analysis, as shown below, but factor analysis then goes several stages further, involving rotation, significance testing and geological interpretation: a far more subjective process.

The aim of a factor analysis is normally to explain the variation in a multivariate data-set in terms of as few 'factors' as possible (3-4 at most). Factors are similar to discriminant functions in being weighted linear combinations of the original variables. However, unlike discriminant analysis, two quite different methods can be distinguished. **R-mode analysis** is used to look for geological control over observed variations. **Q-mode analysis**, by contrast, is mainly used to identify end-members. Some authors have found Q-mode to be relatively unproductive (e.g. Shaw et al.1974) or to yield information only on anomalous samples (Saager & Sinclair 1974). Others (e.g. Carusi et al.1974; Cosgrove 1972) have used Q-mode analysis to identify groups prior to R-mode analysis.

The stages in factor analysis can be summarised as follows:

(1) Determine how many variables (or objects) to input.

(2) Decide which measure of similarity or dissimilarity should form the input matrix: (r, ρ, τ, dissimilarity coefficient, variance-covariance matrix etc; see §14, §21a–b & §23b).

(3) Extract the initial factors. Numerous alternative methods are available, including PCA (§23b), **Harris** or **Kaiser Image analysis** and other forms, and **Iterated Principal Axis** methods. Statisticians tend to favour another alternative, the **maximum likelihood** method, because it has desirable properties such as efficiency, consistency, and generation of confidence limits and hypothesis tests on the factor solution. However, as this method is mathematically rather obscure to geologists, it is not treated here.

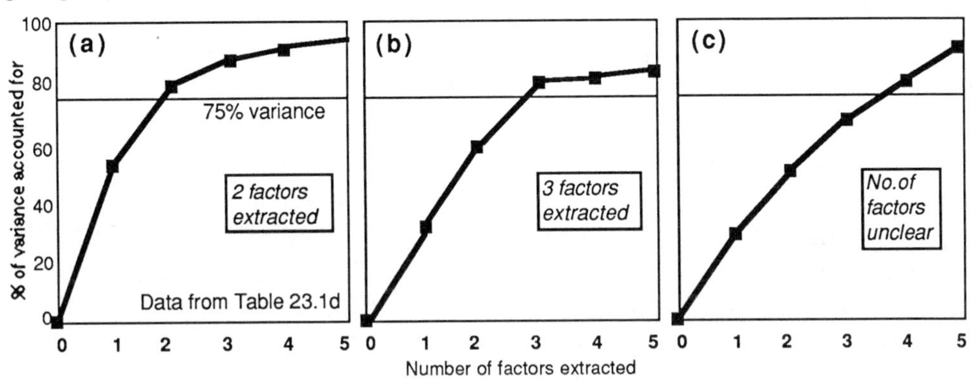

Fig.23.3. Factor-variance diagram for determining the number of relevant factors in a factor analysis
The proportion of total variance accounted for is plotted against the number of factors extracted. In **(a)**, the 75% cutoff rule leads to 2 factors being extracted. In **(b)**, a more clear-cut case, 3 factors are indicated either by the 75% rule or by the flattening out of the curve thereafter. Case **(c)** is much less clear cut. Note that 100% of variance is accounted for when the number of factors equals the number of variables. It is also possible to generate a factor-variance diagram for *each variable*, based on the **coefficient of determination** (Le Maitre 1982, p.138), which allows the user to choose the number of factors according to the most geologically significant variables.

(4) Determine how many of the factors are significant: The cutoff can be determined using: (a) the **variance rule**, when a certain percent of total variance has been reached (commonly 75%), on a factor–variance diagram (e.g. Fig.23.3a); (b) the **root curve criterion**, namely when the same curve markedly flattens out (Fig.23.3b), indicating that additional factors account for an insignificant further amount of variation; (c) the **roots >1 criterion**, i.e. as many factors are retained as there are eigenvalues ≥ 1. Fig.23.3c shows an ambiguous case where factor analysis may be of little value.

(5) Rotate the factors into a meaningful orientation: The initially extracted factors are usually nothing more than mathematical abstractions, impossible to interpret in 'real' terms. Meaningful factors are generally only obtained by rotating the solution. Here again, there are numerous alternatives, including *orthogonal solutions*, which maintain the factors at right angles (e.g. **varimax, equamax, quartimax**), and *oblique solutions*, which allow the factors to be correlated, i.e. at oblique angles to one another (e.g. **orthotran**).

(6) Interpret the results in geological terms. This is often the most subjective and contentious stage, and no rigid rules can be imposed. It is best illustrated by example, as in the succeeding sections.

Even the relatively simple computer implementation of factor analysis used below (Statview 512+) provides 4 alternative factor extraction methods, 5 alternative criteria for the number of factors, and 3 alternative rotation methods. This, when added to the 4 or more alternative input correlation matrices (see 1 above) can generate 240 possible factor solutions from a single set of data! More expansive implementations in large mainframe statistics packages (e.g. SPSS) may provide an even more bewildering (and in practice totally unmanageable) range of alternatives.

23c1. Looking for Geological Control over Data Variations: R-MODE FACTOR ANALYSIS

Further geological discussion: Le Maitre (1982, p.127); Davis (1986, p.546).
Selected geological applications: *Exploration geochemistry* (Armands 1972; Garrett et al.1980; Saager & Sinclair 1974; Santos Oliveira 1978); *Hydrogeochemistry* (Hitchon et al. 1971); *Igneous petrology*: (Cosgrove 1972; Manson 1967; Piispanen & Alapieti 1977; Potenza 1973; Prinz 1967; Shaw et al.1974; Size 1973); *Lunar petrology* (Carusi et al.1974); *Metamorphic petrology*: (Das Gupta 1978; Holland & Winchester 1983); *Meteorites* (Harmon & Shaw 1975; Shaw 1974; Shaw & Harmon 1975; Shaw et al.1973); *Mineralogy* (Mottana et al.1971; Schweitzer et al.1979); *Sedimentology*: (Cameron 1968; Dean & Gorham 1976; Lonka 1967; Merriam & Pena Daza 1978; Osborne 1967); *Stratigraphy*: (Harbaugh & Demirmen 1964; Krumbein & Graybill 1965, p.372).

Tables 23.1–2 and Fig.23.4 show the results of an R-mode factor analysis starting with a correlation matrix (of Pearson's r) between 7 variables (Table 23.1a) measured on 25 objects. The extraction of the eigenvectors and eigenvalues by PCA was discussed in §23b. The researcher is perhaps interested in determining whether he can identify *components* of the borewater (e.g. connate, meteoric, different aquifers) from their chemistry, and is hoping that the factors will be geologically equivalent to these components. Two tests should first, however, be performed on the input matrix to determine whether it is in principle worth even proceeding to extract factors in the first place.

The first test examines whether or not the variables have a statistical 'logic'. In the present example, we have measured 7 variables which might on *geological* reasoning be logically expected to reflect the components of the borewater. Other variables could have been measured, but some (e.g. the *temperature* of each water sample), would not be relevant to identifying water *sources*. Statistically speaking, temperature should show very low *correlations* with the 7 measured variables. The full statistical test examines whether the partial correlation coefficients (§19a) between each variable and any other *single* variable are small, but the multiple correlation coefficients (§19b1) of each variable with *all* other variables are large. The measures of **matrix** and **variable sampling adequacy** in Table 23.1e quantify the extent to which the matrix of Table 23.1b actually approximates this ideal of small partial and large multiple correlations. Zero partial correlations lead to total matrix sampling adequacy (TMSA) = 1, but Kaiser (1970) suggests that TMSA > 0.5 is adequate, so the test is well met by the matrix in Table 23.1e (TMSA = 0.779). We gain an indication of why this is so by looking at the individual correlations in Table 23.1b: for example, the squared partial coefficient between Ca^{2+} and Cl^- (0.402^2= 0.162) indicates that only ≈16% of the variance of Ca^{2+} can be explained in a linear regression by Cl^- (or vice-versa) *alone*, whereas the squared multiple correlation (SMC) for Ca^{2+} (0.802^2 = 0.643) indicates that fully four times as much of its variation can be explained by the other *six* variables combined. Had any of the individual variables shown sampling adequacy < 0.5, we would have regarded them as statistically unrelated to the others, and would have had to consider rerunning the factor analysis without them. In this case, however, all show sampling adequacy > 0.7 except for $SO_4^=$, which is still (just) > 0.5, so all 7 variables are statistically acceptable as ingredients to the analysis. Of course, this test only identifies any variables *among those measured* which are *not* suitable for the analysis; it cannot tell whether some *other* variables should have been included.

The other test, **Bartlett's test for sphericity,** determines whether the correlation matrix in Table 23.1a is statistically significant *as a whole* (i.e. whether, in general, the coefficients differ from zero). This is performed because it is possible to obtain interpretable factors from a matrix of non-significant correlations (derived from unrelated, random measurements). In this case the test is passed because a highly significant χ^2_{27} value ($\approx$135) results.

It is therefore acceptable to proceed to a factor analysis. Full results (Table 23.1) are now discussed for factor extraction via PCA (§23b) — to show how factor analysis can be merely an extension and elaboration of PCA discussed above — and for varimax plus orthotran-varimax rotation. Plots for some other extraction algorithms and rotations are compared in Fig.23.4.

Table 23.1d shows that $\approx$80% of the total variance can be accounted for by the first two eigenvalues and eigenvectors (§23b) from the PCA; the third adding only 8%. Given Table 23.1a and Table 23.1d we therefore choose to extract only 2 factors (the first two principal components). Table 23.1f then shows the **loadings** (correlations) of each variable on these. Loadings have a similar meaning to correlation coefficients: that is, the square of the loading for Cl^- on Factor 1 (0.826^2 = 0.682) represents the proportion of its variance ($\approx$68%) which can be predicted by Factor 1 alone. The sums of the two squared loadings, the **communality estimates**, similarly represent the proportion of variance in each variable which is explicable by *both* factors combined; these communalities are compared in Table 23.11 with the squared multiple correlations (SMC) from the original data-matrix (diagonals in Table 23.1b). For example, the final estimate for Cl^-, 0.832 (= $0.826^2 + 0.388^2$ from Table 23.1f) indicates an improvement over the SMC (0.722) of 11%, while for the other variables, improvements reach as high as 20%. It is possible to obtain communality estimates >1 (the so-called **Heywood case**), without entirely invalidating the factor analysis.

We can see from Tables 23.1f and 23.2a1 that the three variables Cl, Mg and Dissolved Solids have mutually comparable negative loadings on Factor 2, whilst Ca, HCO_3 and pH share positive loadings, with SO_4 intermediate. The spectrum of 7 original variables, in other words, is *beginning* to split into 2 or 3 groups, which we ultimately want to interpret in a geological sense. On factor 1, however, the groupings are not so clear, for there is a more-or-less continuous drop in loadings from Cl to pH. We therefore now proceed to *factor rotation* , to see how much 'cleaner' this separation can be made, and also to see whether SO_4 can in fact be bracketed with one or other group, or forms a group of its own. Table 23.1g and 23.2a2 show the revised loadings after **varimax** (orthogonal) rotation, which keeps the two factors at right angles (i.e. uncorrelated). The two groups of variables are now more completely separated — i.e. rotated Factor 2' is considerably less numerically loaded with Cl, Mg and dissolved solids, and rotated factor 1' with Ca, HCO_3 and pH than the unrotated factors, which translates graphically into the two groups lying nearer the *X* and *Y* axes. Yet these remnant loadings are still not negligible, so in Table 23.1h, we finally relax the orthogonality restraint in the **orthotran-varimax** (oblique) solution, which takes the rotated varimax factors but allows

(a) Correlation matrix

	Cl-	SO4-	Dissolve...	Mg	Ca	HCO3-	pH
Cl-	1						
Mg	.825	1					
Dissolved solids	.769	.883	1				
SO4-	.552	.498	.513	1			
Ca	.402	.323	.34	.466	1		
HCO3-	.367	.348	.329	.359	.842	1	
pH	.345	.267	.239	.317	.73	.633	1

(b) Eigenvectors

Vector 1	Vector 2	Vector 3	Vector 4
-.414	.308	.112	.196
-.405	.391	.255	-.028
-.398	.389	.195	-.117
-.351	.097	-.913	.039
-.383	-.45	-.039	-.205
-.362	-.43	.153	-.569
-.325	-.452	.162	.761

(c) Partials in off-diagonals and Squared Multiple R in diagonal

	Cl-	SO4-	Dissolve...	Mg	Ca	HCO3-	pH
Cl-	**.722**						
Mg	.475	**.839**					
Dissolved solids	.104	.686	**.793**				
SO4-	.196	.014	.105	**.402**			
Ca	.084	-.189	.122	.275	**.802**		
HCO3-	-.072	.177	-.061	-.117	.71	**.725**	
pH	.103	.057	-.113	-.064	.457	.034	**.546**

(d) Eigenvalues and Proportion of Original Variance

	Magnitude	Variance P
Value 1	3.975	.568
Value 2	1.588	.227
Value 3	.586	.084
Value 4	.389	.056

FACTOR LOADINGS

(e) Sampling Adequacy

Total matrix sampling adequacy : .779

Cl-	.87
Mg	.723
Dissolved solids	.785
SO4-	.898
Ca	.685
HCO3-	.741
pH	.842

(f) Unrotated factor matrix

	Factor 1	Factor 2
Cl-	.826	-.388
Mg	.807	-.493
Dissolved solids	.793	-.491
SO4-	.701	-.122
Ca	.763	.567
HCO3-	.722	.542
pH	.647	.57

(g) Varimax Orthogonal Solution

	Factor 1	Factor 2
Cl-	.884	.226
Mg	.936	.132
Dissolved solids	.924	.125
SO4-	.618	.351
Ca	.229	.922
HCO3-	.213	.877
pH	.137	.851

(h) Orthotran/Varimax Oblique Solution

	Primary Pattern Matrix		Reference Structure	
	Factor 1	Factor 2	Factor 1	Factor 2
Cl-	.898	.032	.806	.029
Mg	.979	-.082	.879	-.074
Dissolved solids	.967	-.087	.868	-.078
SO4-	.577	.233	.518	.209
Ca	.005	.948	.004	.851
HCO3-	-1.13E-4	.903	-1.02E-4	.81
pH	-.075	.893	-.067	.801

(i) Proportionate Variance Contributions

	Orthogonal	Oblique		
	Direct	Direct	Joint	Total
Factor 1	.541	.451	.182	.632
Factor 2	.459	.383	-.015	.368

(j) Factor intercorrelations -Orthotran/Varimax Oblique Solution

	Factor 1	Factor 2
Factor 1	1	
Factor 2	.441	1

(k) VARIABLE COMPLEXITY

	Varimax Orthogonal Solution	Orthotran /Varimax Oblique Solution
Cl-	1.13	1.003
Mg	1.04	1.014
Dissolved solids	1.037	1.016
SO4-	1.584	1.316
Ca	1.122	1
HCO3-	1.118	1
pH	1.052	1.014
Average	1.155	1.052

(l) COMMUNALITY SUMMARY

	SMC	Final Estimate
Cl-	.722	.832
Mg	.839	.894
Dissolved solids	.793	.869
SO4-	.402	.506
Ca	.802	.903
HCO3-	.725	.815
pH	.546	.744

(m) FACTOR SCORE WEIGHTS

	Varimax Orthogonal Solution		Orthotran/Varimax Oblique Solution	
	Factor 1	Factor 2	Factor 1	Factor 2
Cl-	.315	-.056	.356	-.137
Mg	.354	-.111	.412	-.206
Dissolved solids	.35	-.112	.409	-.206
SO4-	.185	.053	.186	.013
Ca	-.079	.397	-.19	.451
HCO3-	-.077	.379	-.183	.431
pH	-.102	.38	-.211	.439

Bartlett's chi-squared statistic = 135; df = 27; p < 0.0001

Table 23.1. Input correlation matrix (a) and R-mode factor analysis solution (b-m) for factor analysis example discussed in text

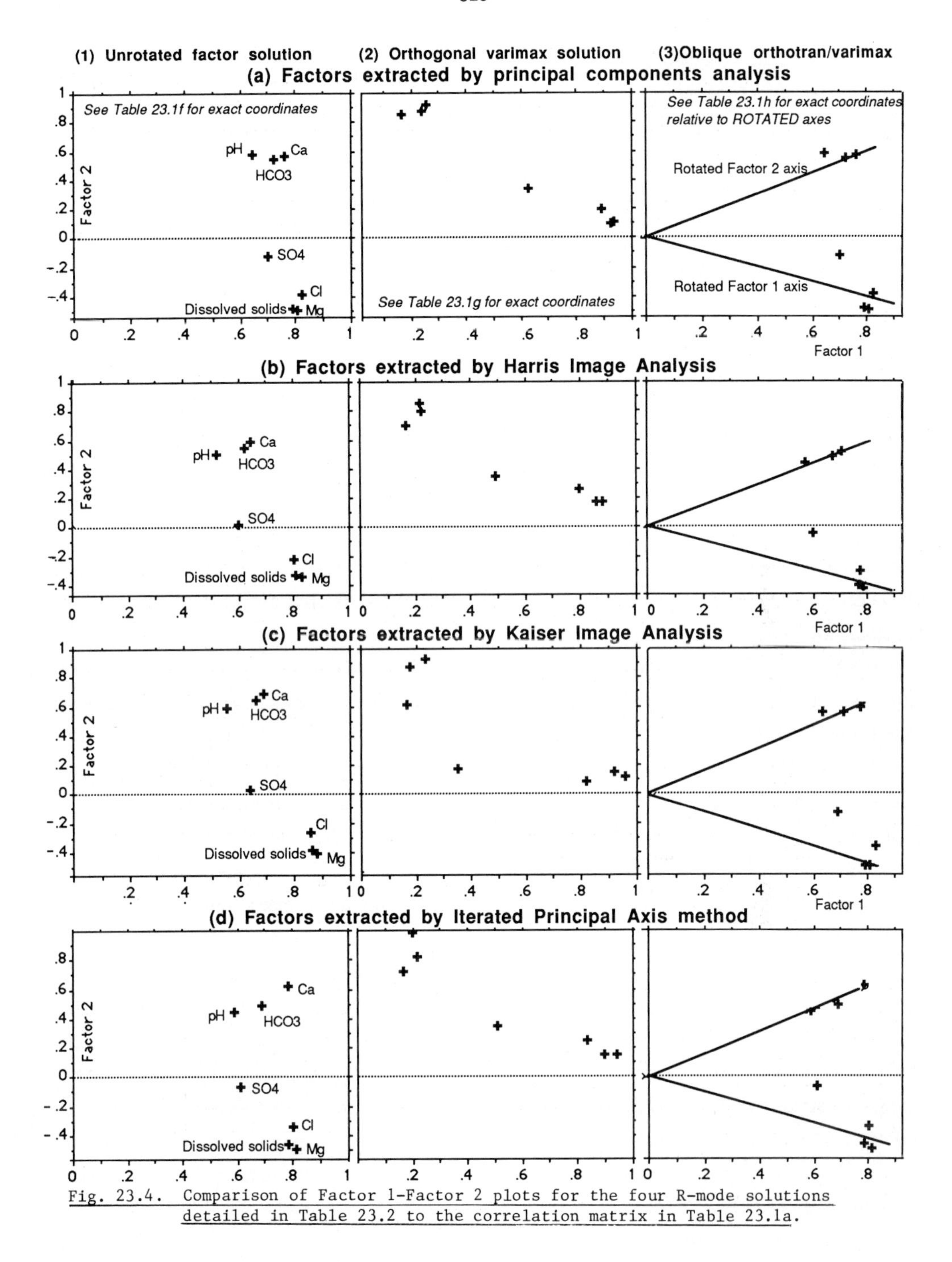

Fig. 23.4. Comparison of Factor 1-Factor 2 plots for the four R-mode solutions detailed in Table 23.2 to the correlation matrix in Table 23.1a.

them to be correlated (i.e. oblique to one another, Table 23.2a3). The alternative **primary pattern** and **reference structure** matrices for this solution are merely column rescalings of one another; the former tends to give larger loadings (sometimes > 1), but their interpretation should be the same.

From their loadings, we can now (at last!) attempt to name the factors geologically. The Cl–Mg–dissolved solids group of variables can probably be interpreted as a *seawater* component in the borewater, and the Ca–HCO_3–pH group as a *connate* component from a carbonate aquifer. The SO_4 can probably be interpreted as a *pollution* (?acid rain) component; this would obviously have been incorporated into *both* the other components, so accounting for its intermediate position on the plots.

In this relatively simple case, the interpretation for both the varimax and orthotran solutions would be the same. In the orthotran-varimax solution, however, the seawater and carbonate components (factors) have a mutual correlation of 0.441 (Table 23.1j). Which rotation we prefer depends on whether we have geological reason to believe the two interpreted factors might be related. Thus if seawater addition to the borewater is wholly independent of connate addition, the varimax solution might be preferred. We might, on the other hand, infer from the orthotran solution that the two components tend to be added together. This could be explained if the borewater is actually a three component mixture, mainly of acid rain (which yields only SO_4), with lesser connate water and seawater; since these are 'closed' data (§15b), the less acid rain is present, the more there will be of the other two components.

Two final checks on these various solutions are now usefully made. (1) **Variable complexity.** After an ideal factor analysis, each variable is wholly accounted for by one factor *alone;* otherwise stated, the average variable complexity is 1. To the extent that variable complexity exceeds 1, a variable is accounted for by more than one variable, and thus the factor analysis has failed to simplify (reduce the dimensionality of) the original data-matrix. In this case, however, Table 23.1k shows that 6 of the 7 variables (except SO_4) are indeed accounted for by one factor alone. (2) **Factor intercorrelation.** On an orthotran solution, intercorrelation > 0.5 would suggest that insufficient factors had been recognized at the extraction stage, but here the factor intercorrelation is 0.441 (Table 23.1j).

Table 23.1i indicates how important each factor is in explaining the original correlation matrix (Table 23.1a). The *direct* proportionate variance contribution, for both varimax and orthotran-varimax solutions, represents the proportion of common variance that each factor accounts for, independent of the other. Thus for both solutions, Factor 1 accounts on its own for slightly more (54 or 45%) of the common variance than Factor 2 (46 or 38%). The *joint* contribution measures shared variance (common to more than one factor), and is thus only defined for the orthotran-varimax solution, where the two factors are correlated. Here, 18% of the common variance is attributable to covariation of oblique factors 1 & 2.

Table 23.1m finally gives weights for converting original data-values (Cl^-, etc.) to standardized factor scores (here, a seawater and a connate score). The intercorrelations of these factor scores for the varimax solution are zero, while for the orthotran solution they equal the factor intercorrelation, 0.441.

Fig.23.4 and Table 23.2 compare 4 alternative factor extraction methods using the same rotation algorithm. Although the graphical differences in Fig.23.4 are apparently small, the Harris and Kaiser

(a) VARIABLE COMPLEXITY

	Varimax orthogonal solution	Orthotran / Varimax oblique solution	Varimax orthogonal solution	Orthotran / Varimax oblique solution	Varimax orthogonal solution	Orthotran / Varimax oblique solution	Varimax orthogonal solution	Orthotran / Varimax oblique solution
Cl-	1.13	1.003	1.201	1.016	1.66	1.419	1.177	1.016
Mg	1.04	1.014	1.076	1.006	1.067	1.01	1.048	1.008
Dissolved solids	1.037	1.016	1.081	1.005	1.184	1.067	1.057	1.005
S04-	1.584	1.316	1.882	1.646	1.407	1.035	1.738	1.527
Ca	1.122	1	1.138	1	1.567	1.236	1.086	1.001
HC03-	1.118	1	1.158	1.007	1.161	1.012	1.145	1.003
pH	1.052	1.014	1.117	1.004	2.099	1.866	1.108	1

(reproduces Table 23.1k)

(b) COMMUNALITY SUMMARY

	SMC	Final Estimate	SMC	Final Estimate	SMC	Final Estimate	SMC	Final Estimate
Cl-	.722	.832	.722	.702	.722	.888	.722	.762
Mg	.839	.894	.839	.807	.839	.958	.839	.909
Dissolved solids	.793	.869	.793	.766	.793	.921	.793	.828
S04-	.402	.506	.402	.37	.402	.93	.402	.377
Ca	.802	.903	.802	.758	.802	.952	.802	1
HC03-	.725	.815	.725	.683	.725	.927	.725	.708
pH	.546	.744	.546	.517	.546	.893	.546	.546

(reproduces Table 23.1l)

(c) FACTOR INTERCORRELATIONS

	Factor 1	Factor 2
Factor 1	1	
Factor 2	.425	1

(reproduces Table 23.1j)

	1	2	3	4
Factor 1	1			
Factor 2	.364	1		
Factor 3	.471	.352	1	
Factor 4	.223	.292	.252	1

	1	2	3	4
Factor 1	1			
Factor 2	.47	1		
Factor 3	.092	.158	1	
Factor 4	-.006	-.013	.107	1

	Factor 1	Factor 2
Factor 1	1	
Factor 2	.425	1

FACTOR EXTRACTION METHOD

(1) PRINCIPAL COMPONENTS ANALYSIS (yields 2 factors)

(2) HARRIS IMAGE ANALYSIS (yields 4 factors)

(3) KAISER IMAGE ANALYSIS (yields 4 factors)

(4) ITERATED PRINCIPAL AXIS (yields 2 factors)

Table 23.2. Comparison of four R-mode factor analysis solutions to the correlation matrix in Table 23.la

methods both extract 4 factors (i.e. Fig.23.4 shows only 3 out of 18 plots generated for these 2 methods), as against 2 for the PCA and Iterated methods. Inspection of Table 23.2 is needed to see which method is 'best'. Fortunately, in the present relatively simple case, the conclusion can be made fairly objectively: the Harris and Kaiser methods generate substantially higher variable complexities than the others, and the final communalities for the Harris are all lower than the initial estimates (a *reductio ad absurdum).* The PCA yields slightly better overall figures (higher communalities, lower variable complexities) than the Iterated method — which is why it was chosen in the first place for full illustration in Table 23.1! The 'best' solution is, unfortunately, not always so obvious.

23c2.Analyzing Series and Mixtures in Terms of End-members: Q-MODE FACTOR ANALYSIS

Further geological discussion: Ehrlich & Full(1987); Full et al.(1981,1982); Klovan(1981); Miesch(1976b,d).
Worked geological examples: *Igneous Petrology* (Le Maitre 1982, p.133); Davis (1986).
Selected geological applications: *Exploration geochemistry* (Saager & Sinclair 1974); *Hydrogeochemistry* (Hitchon et al. 1971); *Igneous Petrology* (Cosgrove 1972; Stuckless et al.1979; Miesch 1979,1981a; Miesch & Reed 1979); *Mineralogy* (Alberti 1979; Middleton 1964); *Palaeontology* (Nigam 1987; Sepkoski 1981). *Sedimentology* (Dean & Gorham 1976; Drapeau 1973; Klovan 1966; Mazzullo & Ehrlich 1983; Merriam & Pena Daza 1978).
Stand-alone computer programs: Klovan & Imbrie(1971); Klovan & Miesch(1976); Miesch(1976c); Full et al.(1981-no code).

The objectives of Q-mode factor analysis can be summarised in 3 stages as follows:
(a) determine the minimum number of end-members (factors) required to explain a set of data;
(b) determine the composition of these end-members *A,B*... in terms of the original variables;
(c) describe the original objects as proportions of *A,B*.. ($aA + bB + ...$), *a,b* being the factor loadings.

There are in fact an infinity of possible solutions to any given problem, and each of these stages requires subjective input. Stage (a) has to be decided on some independent geological criterion. Stage (b) requires either the most divergent possible end-members, or theoretical compositions, to be chosen.

As with R- versus Q-mode classification (§21), Q-mode factor analysis is more appropriately based on a matrix of **cosine theta** similarity coefficients, than a matrix of Pearson's *r* or a variance-covariance matrix. Otherwise, Q-mode analysis initially proceeds much as R-mode analysis (§23c1). An important difference does, however, emerge once the factor loadings and scores have been obtained. Because the factor loadings for one object do not sum to unity, they cannot be considered as proportions, and because the factor scores are dimensionless, they cannot be considered as compositions. Geological applications therefore almost invariably employ a variant method known as **extended Q-mode factor analysis**, in which the loadings and scores are scaled such that they sum to unity and represent end-member compositions respectively. The extended method is in fact applicable not only to percentage data with constant sums, but also to open data (§15).

Q-mode factor analysis is the logical alternative to mixing models (§18) where the number of end-members is uncertain, or where the end-members themselves are unknown, for both uncertainties are actually tested during the analysis, rather than being pre-requisite as with mixing models.

23c3.Pattern Recognition by Combining Q- and R-modes: CORRESPONDENCE ANALYSIS

Dedicated monographs and reviews: Benzécri(1973); Hill(1974); Lebart et al.(1984).
Further geological discussion: David et al.(1974), Joreskog et al.(1976, p.107 — includes full exposition of mathematical methodology); Miesch(1983),Teil(1975),Valenchon(1982,1983).
Worked geological examples: Joreskog et al.(1976, p.167).
Selected geological applications: *Economic geology* (Teil 1976); *igneous petrology* (Dagbert & David 1974; Dagbert et al.1975; Dumitriu et al.1979; Karche & Mahé 1977; Ploquin & Royer 1977; Teil & Cheminée 1975); *metamorphic petrology:* (David et al. 1974); *palaeontology* (Mahé 1971; Bonham Carter et al.1986; Spicer & Hill 1979).
Stand-alone computer programs: Available from M.David and associated workers.
Mainframe and microcomputer software: none known.

Readers may well be confused by the variety of guises under which correspondence analysis is described (always assuming that they have no additional problem with the French in which much of the literature is cast!) Indeed, some recent French literature uses the term 'factor analysis' for what we here term correspondence analysis, after Benzécri (1973). Sometimes the method is cast as one for displaying *both* variables *and* objects on the same graph (i.e. combined Q- and R-mode factor analysis), elsewhere as factor analysis applied to binary or nominal data. In fact, both descriptions are correct. Originally developed to apply to contingency table problems (§13d1) with nominal data, it was then extended to cover interval/ratio data by a complicated scaling procedure which involves dividing up the ranges of continuously distributed variables into discrete 'chunks'. The adopted similarity measures are equally arcane, being based variously on χ^2 or on multidimensional scaling (§21d1).

Fig.23.5 is a hypothetical plot to illustrate what one might expect from a correspondence analysis of igneous host-rocks from a 'typical' Archaean mesothermal greenstone-hosted gold deposit (e.g. from the Yilgarn Province, W.Australia or the Superior Province, Canada). Both rock samples (objects) and some chemical variables are plotted together. Factor 2 is weighted at one end with Mg, Cr, Co and Ni, and at the other with Na, Al and Si, and is clearly reflecting primary igneous variations within the host-rocks (from ultramafic/komatiitic to more felsic hosts). Factor 1, by contrast, is loaded

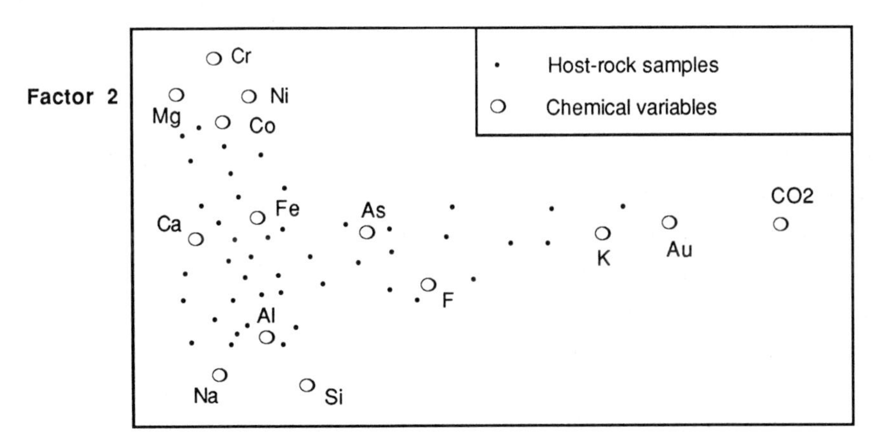

Fig.23.5. An imaginary correspondence analysis (see §23c3 for explanation)

with CO_2, F, As, Au and K, which are the elements known to be added during gold alteration (Groves & Phillips 1987). We can thus identify Factors 1 and 2 as related to primary and alteration chemistry, and those rock samples which plot more towards CO_2, K etc. are those which have suffered greater alteration. Whilst this may be intuitively obvious in the present example, correspondence analysis would have obvious application where sets of elements related to alteration, metamorphism, metasomatism or primary controls were not so well-defined.

23d.Epilogue: the Statistical 'Zap' versus the 'Shotgun' Approach

Further geological discussion: Brower & Veinus (1974); Reyment(1974).
Selected geological applications of the 'statistical 'zap': see under individual methods above.
Selected geological applications of the 'shotgun' approach: *exploration geochemistry* (Chapman 1978); *igneous petrology* (Frangipane 1977; Naney et al.1976); *mineralogy* (Hawkins 1974; Le Maitre 1982, ch.7–8; Schweitzer et al.1979); *palaeontology* (Reyment 1971b); *sedimentology* (Clark & Clark 1976; Feldhausen & Ali 1976).

We have now touched on the vast majority of statistical tests and techniques which have been used to interpret real geological data — together with a number whose geological potential has not yet been realised. The final question must be — which technique should the geologist use, when faced with a mountain of data? Two possible approaches have now been given populist codenames, which presumably underline their respectability: the **statistical zap** involves blasting the data to pieces via one particular technique, and is most appropriate when theoretical notions of how the data may behave are already available. This book has attempted all along to guide readers to the correct method on this basis. However, the alternative **shotgun approach** is also available as a fallback, both for the relative novice or for data whose behaviour is indeterminate. This involves spraying the data with a panoply of statistical techniques and inspecting the results: a 'suck-it-and-see' philosophy.

SELECTIVE BIBLIOGRAPHY OF NUMERICAL GEOLOGY

The following principles have governed compilation of this bibliography, which (despite unavoidable selectivity in a subject so vast), is believed to be the largest (> 2,000 references) so far published:

Subject coverage: The bibliography emphasises applications of computers and statistical techniques to such branches of geology as exploration, geochemistry, mineralogy, palaeontology, petrology, sedimentology and stratigraphy, because individual techniques are usually portable between these different disciplines (e.g. cluster analysis can be applied in much the same way to rocks as to fossils). Applications to geophysics and structural geology are less comprehensively covered, partly because these are far more specialized (and commensurately less portable), and partly because bibliographies to these disciplines already exist (see below).

Textbook coverage: Geomathematical textbooks are comprehensively compiled, together with a substantial selection of the huge number of statistical textbooks. However, only selected examples from the probably even huger range of texts on computers and computing could be included.

Journal coverage: Most articles from *Computers & Geosciences, Mathematical Geology* and *Journal of Geological Education* are included, except for a few issues which were unavailable to the author, and subject to the general omissions below.

General omissions: the following are not claimed as being covered to any significant extent:

—Publications prior to 1960 (except for a smattering of germinal, benchmark papers).

—Publications in languages other than English and some European languages (e.g. French, Portuguese, Spanish), in journals to which the author has access.

—Abstracts, theses and other unpublished, obscure, proprietary or commercial materials.

—Publications based on (to geologists) the most esoteric branches of mathematics, such as set, information and group theory.

—Modelling (e.g.finite element modelling in structural geology), other than as specified in §18.

—Individual references from pre-existing bibliographies: for example, Whitten(1969,1981) gives two comprehensive bibliographies of computer applications to structural geology, covering the years 1959–1969 and 1969–1979; little purpose would be served by adding these to the present list.

—Special lists of references extracted from commercial databases; for example, AMF (1984a,b) publishes two lists (of 334 and 589 references), dealing with statistical and computer applications to *Australian* geology. It is clearly neither possible nor necessary to incorporate these here.

Notes on listing conventions:

1.Titles in other than West European languages are translated into English and quoted in [....].

2. Order is strictly alphabetical, ignoring spaces, hyphens, apostrophes and other diacritical marks; Mac and Mc are treated as such, *not* at start of M's. Names starting with *De, Le, La, Van, Von* etc. are listed under the *prefix.*

3. For ease of location, authors' names for useful, general-purposes statistical reference texts are **emboldened**, while important geomathematical tests are underlined *in toto.*

Abbey,S.(1983).Studies in "standard samples" of silicate rocks and minerals,1969-82. *Pap. Geol. Surv. Can.* **83-15**.

Abrahamson,I.G.(1967).Exact Bahadur efficiencies for the Kolmogorov-Smirnov and Kuiper one- and two-sample statistics.*Ann.Math.Stat.***38**:1775-1790.

Acton,F.S.(1966).*Analysis of straight-line data.*Dover,New York.

Adamis,E.(1985).*Multiplan on the Apple Macintosh: the Microsoft desktop dictionary and cross reference guide.* Microsoft Press,Bellevue,Washington.

Afifi,A.A. & Azen,S.P.(1972).*Statistical analysis,a computer oriented approach.* Academic Press,New York,366pp.

Aggarwal,H.R. & Oberbeck,V.R.(1979).Monte Carlo simulation of lunar regolith and implications. In: *Abstr.10th Lunar Planet.Sci.Conf.Part I.***12-3/19:3/23**:289-291.

AGIA (Australian geoscience information association, 1984). Information needs of the Exploration Geoscientist. (ed. Campbell, G.E.) Proceedings of Seminar, December 1983, Adelaide. AGIA, Parkside, South Australia.

Agresti,A.(1984).*Analysis of ordinal categorical data.*Wiley,New York,287pp.

Agterberg,F.P.(1966).The use of multivariate Markov schemes in petrology. *J.Geol.***74**,764-85.

Agterberg,F.P.(1967a).Computer techniques in geology.*Earth-Sci.Rev.***3**,47-77.

Agterberg,F.P.(1967b).Mathematical models in ore evaluation.*J.Can.Res.Soc.***5**,144-158.

Agterberg,F.P.(1974a).*Geomathematics: mathematical background and geoscience applications. (Developments in Geomathematics Vol.1).*Elsevier,Amsterdam,596pp.

Agterberg,F.P.(1974b).Automatic contouring of geological maps to detect target areas for mineral exploration.*J.Math.Geol.***6**,373-394.

Agterberg,F.P.(1977).Quantification and statistical analysis of geological variables for mineral resource evaluation.*Mem BRGM (Paris)* **91**:399-406.

Agterberg,F.P.(1978).*Analysis of spatial patterns in the Earth Sciences.* In: Merriam(1978b), qv, 7-18.

Agterberg,F.P.(1979).Mineral resource estimation and statistical evaluation Pp.301-318 in *Mem.Can.Soc.Petrol.Geol.* **6**.

Agterberg,F.P.(1981).*Computers as an aid in mineral-resource evaluation.* In: Merriam(1981b), qv, 43-62.

Agterberg,F.P. & Cabilio,P.(1969).Two-stage least-squares model for the relationship between mappable geological variables.*J.Math. Geol.***1**,137-154.

Agterberg,F.P. & Cameron,G.D.(1967).Computer program for the analysis of multivariate series and eigenvalue routines for asymmetrical matrices. *Geol.Surv.Can.Pap.***67-14**,54pp.

Agterberg,F.P. & David,M.(1979).Statistical exploration.In: *Computer methods for the 1980s*, 90-115. Soc.Ming.Eng. AIME, New York.

Agterberg,F.P. & Divi,S.R.(1978).A statistical model for the distribution of copper,lead and zinc in the Canadian Appalachian region.*Econ.Geol.* **73**,230-245.

Agterberg,F.P. & Fabbri,A.G.(1978).Spatial correlation of stratigraphical units quantified from geological maps. *Comput.Geosci.***4**,285-294.

Agterberg,F.P.,Chun,G.C.F.,Fabbri,A.G.,Kelly,A.M. & Springe,J.S.(1972). Geomathematical evaluation of Cu and Zn potential of the Abitibi area, Ontario and Quebec.*Pap.Geol.Surv.Can.* **71-41**, 55pp.

Ahrens,L.H.(1954).The log-normal distribution of the elements.*Geochim Cosmochim.Acta* **5**,49-73; **6**,121-31;

Ahrens,L.H.(1957).Lognormal distributions. *Geochim.Cosmochim.Acta* **11**,205-212.

Ahrens,L.H.(1963).Lognormal-type distributions in igneous rocks.*Geochim.Cosmochim.Acta* **27**,877-90.

Aitchison,J.(1981).A new approach to null correlations of proportions. *J.Math.Geol.***13**,175-90.

Aitchison,J.(1982).The statistical analysis of compositional data.*J.R.Stat.Soc.***B94**,139-77.

Aitchison,J.(1983).Principal components analysis of compositional data. *Biometrika* **70**,57-67.

Aitchison,J.(1984a).The statistical analysis of geochemical compositions. *J.Math.Geol.***16**,531-564.

Aitchison,J.(1984b).Reducing the dimensionality of compositional data-sets. *J.Math.Geol.***16**, 617-636.

Aitchison,J.(1986).*The statistical analysis of compositional data.* Wiley, New York, 416pp.

Aitchison,J. & Aitken,C.G.E.(1976).Multivariate binary discrimination by the kernel method.*Biometrika* **63**,413-420.

Aitchison,J. & Brown,J.A.C.(1957).*The lognormal distribution,with special reference to its use in economics.* Cambridge Univ.Press.

Aitchison,J. & Shen,S.M.(1984).Measurement error in compositional data. *J.Math.Geol.***16**,637-650.

Ajne,B.(1968).A simple test for uniformity of a circular distribution. *Biometrika* **55**,343-354.

Albarède,F. & Provost,A.(1977).Petrological and geochemical mass-balance equations: an algorithm for least-squares fitting and general error analysis. *Comput.Geosci.***3**,307-26.

Alberti,A.(1979).Chemistry of zeolites—Q-mode multivariate factor analysis. *Chem.Erde* **38**,64-82.

Alberti,A. & Brigati,M.F.(1985).Dependence of chemistry on genesis in zeolites: multivariate analysis of variance and discriminant analysis. *Am.Mineral.***70**,805-13.

Aldenderfer,M.S. & Blashfield,R.K.(1984).*Cluster analysis.*Sage,Beverly Hills,88p.

Aleksandrov,I.V.(1985).Paragenetic rare and ore-element associations in granitoids and the use of correlation analysis in studying element paragenesis in multicomponent systems.*Geochem.Int.***22**, 39-48.

Alldredge,J.R.,Aahbt,M.A. & Panek,C.A.(1974).Statistical analysis of axial data*J.Geol.***82**, 519-524.

Allègre,C.J. & Minster,J.F.(1978).Quantitative models of trace element behaviour in magmatic processes.*Earth Plan.*

Atzori,P.,Lo Giudice,A.,Pezzino,A. & Rittman,L.(1973).Analisi fattoriale della correlazione fra le variabili chimiche di biotiti di diversi ambienti genetici. *Rivista Mineraria Siciliana* **142**:171-189

Auble,D.(1953).Extended tables for the Mann-Whitney statistic.*Bull.Inst.Educ.Res.Indiana Univ.***1** **(2)**.

Aubrey,M.C.(1986). The distribution of mineralization: an analytical procedure using a geographic information system. *Proc.3rd Int.Conf.Geosci.Info.(Adelaide,S.Australia, June '86)*, Vol.1, 143-158.

Augenstein, K. & Wolf, F. (1987). Pilot study to assist environmental management decision in the Elizabeth River Basin. *Proc. 7th Ann. ESRI User Conf.* ESRI, California.

Augustithis,S.S.(1983).*The significance of trace elements in solving petrogenetic problems and controversies.* Theophrastus,Athens.

AusIMM (Australian Institute for Mining and Metallurgy, 1980). *Computers in Mining.* Seminar at 8th World Computer Congress, Melbourne, 1980. AusIMM, 95pp.

Austria,V. & Chork,C.Y.(1976).A study of the application of regression analysis for trace element data from stream sediments in New Brunswick.*J.Geochem. Explor.***6**,211-232.

Ayora,C.(1984).A program in BASIC for creating and using a regional mineralisation minicomputer data file. *Comput.Geosci.***10**,251-262.

Bachu,S.,Savveplane,C.M.,Lytviak,A.T. & Hitchon,B.(1987).Analysis of fluid and heat regimes in sedimentary basins: techniques for use with large databases.*Bull.Am.Assoc.Pet.Geol.***71**,822-843.

Baer,J.L. (1969). Palaeoecology of cyclic sediments of the Lower Green River Formation, central Utah. *Brigham Young Univ.Geol.Studies* **16**, 3-95.

Bagnold,R.A. (1983). The nature and correlation of random distributions. *Proc.R.Soc.Lond. Ser.A* **388**, 273-291.

Bahadur,R.R.(1960).Asymptotic efficiency of tests and estimates.*Sankhya* **22**: 229-252.

Bailey,A.I.(1975).A method for analyzing polymodal distributions in orientation data.*J.Math.Geol.***7**, 285-294.

Bailey,D.K. & Macdonald,R.(1975).Fluorine and chlorine in peralkaline liquids and the need for magma generation in an open system. *Mineral.Mag.***40**,405-414.

Baird,A.K.,McIntyre,D.B.,Welday,E.E. & Madlem,K.W.(1964).Chemical variation in a granitic pluton and its surrounding rocks.*Science* **146**,258-259.

Baird,A.K.,McIntyre,D.B. & Welday,E.E.(1967).Geochemical and structural studies in batholithic rocks of southern California.Part II.Sampling of the Rattlesnake Mountain pluton for chemical composition,variability and trend analysis. *Bull.Geol.Soc.Am.***78**,191-222.

Ballantyne,C.K. & Cornish,R.(1979).Use of the chi-square test for the analysis of orientation data. *J.Sed. Petrol.* **49(3)**,773-776.

Bankier,J.D.(1955).The theory of thin section analysis: a discussion.*J.Geol.* **63**,287-288.

Banks,R.(1979).The use of linear programming in the analysis of petrological mixing models. *Contrib.Mineral.Petrol.***70**,227-44.

Bardsley,W.E.(1983).Random error in point counting.*J.Math.Geol.***15**, 469-476.

Bardsley,W.E. & Bridges,R.M.(1984).Note on fitting quantitative models of magmatic processes to trace-element data. *Comput.Geosci.***13**,445-8.

Barnes,R.P.,Rock,N.M.S. & Gaskarth,J.W.(1986).Late Caledonian dyke-swarms in southern Scotland: new field, petrographical and geochemical data for the Wigtown Peninsula,Galloway. *Geol.J.* **21**,101-125.

Barnett,V.(1982).*Interpreting multivariate data.* Wiley, New York, 370 pp.

Barnett,V. & Lewis,T.(1984).*Outliers in statistical data.*2nd Edn.Wiley,Chichester,463pp.

Barr,D.L.,Mutschler,F.E. & Lavin,O.P.(1977).KEYBAM,a system of interactive computer programs for use with the PETROS petrochemical data-bank. *Comput.Geosci.***3**,489-496.

Bartels,C.P.A. & Ketallapper,R.H.(1979).*Exploratory and explanatory statistical analysis of spatial data.* M.Nijhoff, Boston,268pp.

Bartholemew,R.B. & Jonas,E.C.(1974).The computer! A supplementary teaching device for undergraduate geological classes. *J.Geol.Educ.***22**,157-161.

Bartlett,M.S.(1937).Properties of sufficiency and statistical tests. *Proc.Roy.Soc.Lond.* **A168**, 268-282.

Bary,C.de,Marchal,M.,Crehange,M. & Grandclaude,P.(1977).Main logical and practical features of the GISCO system: data structures and retrieval language applied to geochemical data bank.*Sci.Terre Inf.Geol.*(Nancy),10:83-93.

Basu,A.P.(1967).On the large sample properties of a generalized Wilcoxon-Mann-Whitney statistic. *Ann.Math.Stat.* **38**,905-915.

Bates,R.C.(1959).An application of statistic analysis to exploration for uranium on the Colorado Plateau. *Econ.Geol.***54**,449-466.

Bates,R.L. & Jackson,J.A.(1980). *Glossary of geology.* 2nd Edn. Am.Geol.Inst.,Falls Church,VA, 749 pp.

Batten,D.J.(1973).Use of palynological association types in Wealden correlation.*Palaeontol.***16**,1-40.

Baxter,J.(1986a).*Macintosh desktop typography.* Baxter Group,Sunnyvale,CA.

Baxter,J.(1986b).*Macintosh desktop design.* Baxter Group,Sunnyvale,CA.

Bayer,U.(1985).*Pattern recognition problems in geology and palaeontology.* Springer-Verlag,Berlin(Lecture Notes in Earth Sciences,**2**), 229 pp.

*Sci.Lett.***38**,1-25.

Allègre,C.J.,Hamelin,B. & Dupré,B.(1984).Statistical analysis of isotopic ratios in MORB: the mantle blob cluster model and the convective regime of the mantle. *Earth Planet.Sci.Lett.***71**,71-84.

Allen,J.R.L.(1970).Studies in fluviatile sedimentation: a comparison of fining-upwards cyclotherms with special reference to course-member composition and interpretation.*J.Sed.Petrol.***40**,298-323.

Al Mehaldi,H.M.,Al Amiri,H.M.,Al Dewachi,B.M. & Levandowski,D.(1978).An evaluation of computer-aided analysis of LANDSAT MSS data for regional mapping: folded mountain zone,North Iraq.*J.Geol.Soc.Iraq* **11**:1-24.

Al-Turki,K.I. & Stone,M.(1978).Petrographic and chemical distinction between the megacrystic members of the Carnmenellis granite,Cornwall. *Proc.Ussher Soc.***4**,182-189.

Alvey,N.,Galwey,N. & Lane,P.(1982).*An introduction to GENSTAT*. Academic Press,London.

AMF (Australian Mineral Foundation, 1984a). *Geostatistical and statistical methods: a special list of references from AESIS 1976-1984*. Special list no. 13A. AMF, Adelaide (334 references).

AMF (Australian Mineral Foundation, 1984b). *Computer applications: a special list of references from AESIS 1976-1984*. Special list no. 14A. AMF, Adelaide (589 references).

AMF (Australian Mineral Foundation, 1984c). *Geoscience numeric and bibliographic data: papers and recommendations.*. Project seminar no. 154/81. AMF, Adelaide.

AMF (Australian Mineral Foundation, 1987). *Statistical methods for exploration and mining*. Workshop course 525/87. AMF, Adelaide, 123 pp.

AMIRA (Australian Mineral Industries Research Association, 1985). *Computer Applications in the Australian Mining Industry*. Vol.1:*Introduction and Summary*; Vol.2: *Hardware and systems software*; Vol.3: *Current Australian Practice;* Vol.4:*Application software and services*; Vol.5:*Management aspects*. AMIRA, Melbourne.

Ammar,A.A.,Meleik,M.L. & Fouad,K.M.(1982).The aerial radiometric identification of granitic plutons by statistical analysis. *Comput.Geosci.***8**,209-220.

Andar,P.(1976).[The automatic computation of Rittmann's norms.]In Romanian. *Dari Seama Sedintelor* **62(1)**:59-62.

Anderberg,M.R.(1973).*Cluster analysis for applications*.Academic Press,New York.

Anderson,E.(1960). A semigraphical method for the analysis of complex problems. *Technometrics* **2**, 387-391.

Anderson,J.B.(1971).Numeric examination of multivariate soil samples. *J.Math.Geol.***3**,1-14.

Anderson,P.Y. & Koopmans,L.H.(1963).Harmonic analysis of varve time series. *J.Geophys.Res.* **68**,872-893.

Anderson,T.W.(1958).*Introduction to multivariate statistical analysis*.Wiley,374p

Anderson,T.W.,Das Gupta,S. & Styan,G.P.H.(1972).*A bibliography of multivariate statistical analysis*. Oliver & Boyd, Edinburgh,642pp.

Anderssen,R.S. & Osborne,M.R.(1969).*Least-squares methods in data analysis*. Australian National Univ.Computer Centre Publication CC2/69.Univ.Queensland Press.

Anderssen,R.S. & Osborne,M.R.(1970).*Data representation*.Univ.Queensland Press.

Andrew,A.S. & Linde,J.(1981).PRP: a FORTRAN IV interactive plotting program. *Comput.Geosci.* **7**,3-20.

Andrews,D.F.,Bickel,P.J.,Hampel,F.R.,Huber,P.J.,Roger,W.H. & Tukey,J.W.(1972). *Robust estimates of location: survey and advances*. Princeton Univ.Press,373pp.

Angelis,G.de,Farinato,R.,Febraro,F. & Loreto,L.(1977).[Computerised data bank of volcanic rocks analyses: preliminary note].*Mem.Soc.Geol.Ital.***13**(-suppl.2) 407-414.

Anon (1978a).Numerical databases for Australian science and technology. *Search (Sydney)* **9(12)**: 448-452.

Anon (1978b).Computer-based files on mineral deposits: guidelines and recommended standards for data content. *Pap.Geol.Surv.Can* **78/26**,72pp.

Anon (1979).Standards for computer applications in mineral resource studies (IGCP Project 98). *Nat.Resourc. (Paris)*,**15(1)**,p.34.

Ansari,A.R. & Bradley,R.A.(1960).Rank-sum tests for dispersion.*Ann.Math.Stat.* **31**,1174-1189.

Anscombe,F.J.(1960).Rejection of outliers.*Technometrics* **2**,123-147.

Armands,G.(1972).Geochemical studies of uranium,molybdenum and vanadium in a Swedish alum shale.*Stockholm Contrib.Geol.* **27**,1-148.

Armour-Brown,A. & Olesen,B.L.(1984).Condensing multi-element reconnaissance geochemical data from S.Greenland using empirical discriminant analysis. *J.Geochem.Explor.***21**,395-404.

Armstrong,M. (1983). *Coal geostatistics*. Australian Mineral Foundation Course Notes 245/83, Adelaide, 98 pp.

Artandi,S.(1972).*An introduction to computers in information science*. (2nd Ed). Scarecrow Press, Methuchen,New Jersey, 190pp.

Arth,J.G.(1976).Behaviour of trace elements during magmatic processes — a summary of theoretical models and their applications.*J.Res.U.S.G.S.* **4**,41-47.

Arthur,L.J.(1987). *Software evaluation: a software maintenance challenge*. Wiley, New York, 272 pp.

ASTM (American Society for testing and materials, 1969).Recommended practice for dealing with outlying observations. In: *Annual Book of ASTM standards* **30**,429-445.

Attoh,K. & Whitten,E.H.T.(1979).Computer program for regional model for discontinuous structural surfaces. *Comput.Geosci.* **5**,47-71.

Beale,E.M.L. & Little,R.J.A.(1975).Missing values in multivariate analysis. *J.R.Stat.Soc.***37B**, 129-145.

Beauchamp,J.,Begovich,C.L.Kane,V.E. & Wolf,D.A.(1980).Application of discriminant analysis and generalised distance measures to uranium exploration *J.Math.Geol.***12**,539-558.

Beckman,R.J. & Cook,R.D.(1983).Outlier.......s.*Technometrics* **25**,119-149.

Belbin,L. & Rickard,M.(1973).Computer program for continental reconstructions: CONTPLOT. *Tectonics Struct. Newsletter (Sydney)* 2:13-21.

Belov,A.N.(1963).Statistical analysis of the paragenetic sequence in igneous rocks.*Dokl.Acad. Sci.USSR* **151**:152-154.

Belsley,D.A.,Kuh,E. & Welsh,R.E.(1980).*Regression diagnostics: identifying data and sources of collinearity*.Wiley,New York.

Bendat, J.S. & Piersol,A.G. (1971). *Random data: analysis and measurement procedures.* Wiley,New York, 407pp.

Benest,J. & Winter,P.E.(1984).Ore-reserve estimation using geologically-controlled geostatistics. *Tr.Inst. Ming. Metall. Sect. B*,B173-180.

Benzécri,J.P.(1973). *L'analyse des données. 2.L'analyse des correspondances.* Dunod,Paris, 619 pp.

Beran,R.J.(1969).The derivation of non-parametric two-sample tests for uniformity of a circular distribution.*Biometrika* **56**,561-70.

Beran,R.J.(1977).Robust location estimates.*Ann.Stat.***5**,431-444.

Berger,G.W. & York,D.(1970).Precision of the $^{40}Ar/^{39}Ar$ dating technique. *Earth Planet.Sci.Lett.* **9**, 39-44.

Bergeron,R.,Robinson,S.C. & Burk,C.F.(1972).Computer-based storage, retrieval and processing of geological information. *24th Int.Geol.Congr.(Montreal), Session* **16**,222pp.

Bergerud,M. & Keller,T.(1987). *Computers for managing information.* Wiley, New York, 480 pp.

Berk,K.N. (1987a). Effective microcomputer statistical software. *The American Statistician* **41**, 222-228.

Berk, K. N. (1987b). Statistical software computing reviews. *The American Statistician* **41 (4)**, 314-324. Includes reviews of DataDesk (by P.E.Shrout), Systat (by D.Morgenstein) and Statview 512+ (by A.L.Best).

Bernhardt,H.J.(1979).[Computer-supported ore mineral identification using spectral reflectance in visible light]. *N.Jb.Mineral.Monatch.***79(9)**:403-407.

Berry,B.J.L. & Marble,D.F.(1968).*Spatial analysis,a reader in statistical geography*.Prentice-Hall, New Jersey,512pp.

Berthou,P.Y., Brower,J.C. & Reyment,R.A. (1975). Morphometrical study of Choffat's vascoceratids from Portugal. *Bull.Geol.Inst.Univ.Uppsala N.Ser.* **6**, 73-83.

Bertrand,H. & Coffrant,H.(1976).Application of multivariate statistical methods to the geochemical study of tholeiites of the Eastern Shelf of North America. *Réun.Ann.Sci.de la Terre* **4**,51.

Beyer,W.H.(1968).*Handbook of tables for probability & statistics*.(2nd Ed). Chemical Rubber Co.,Cleveland,Ohio,642pp.

Bhapkhar,V.P.(1966).Some nonparametric tests for the multivariate several sample location problem.*Proc.1st Int.Symp.Multivzr.Anal.* (ed.P.R.Krishnaiah):29-42.

Bhapkhar,V.P. & Patterson,K.W.(1977).On some non-parametric tests for profile analysis of several multivariate samples. *J.Multivar.Anal.***7**,265-277.

Bhattacharyya,G.K. & Johnson,R.A.(1977).*Statistical concepts and methods.* Wiley,New York, 639pp.

Bichteler,J. (1986). Expert systems for geoscience information processing. *Proc.3rd Int.Conf. Geosci.Info. (Adelaide,S.Australia, June '86)*, Vol.1, 179-191.

Bickel,C.E.(1979).The CIPW normative calculation.*J.Geol.Educ.***27**,80-82.

Bickel,P.J.(1965).On some asymptotically nonparametric competitors of Hotelling's T^2.*Ann.Math. Stat.* **36**:160-173.

Bickel,P.J.(1969).A distribution-free version of the Smirnov two sample test in the p-variate case. *Ann.Math.Stat.***40**,1-23.

Bier,R.A. (1987). The use and geoscience information. *Proc.Geosci.Info.Soc.* **17**, 113 pp.

Bièvre, P.de. (1978). Accurate isotope ratio mass spectrometry: some problems and possibilities. In: *Advances in mass spectrometry* , N.R.Daly (ed.), **7A**, 395-447.

Bijnen,E.J.(1973).*Cluster analysis—survey and evaluation of techniques.* Tilburg Univ.Press, Groningen, 112pp.

Binsler,E.C.,Trexler,D.T.,Kemp,W.R. & Bonham,H.F.(1976).PETCAL: a BASIC language computer program for petrologic calculations.*Rep.Nevada Bur.Mines Geol.***28**:27pp.

Birnbaum,Z.W.(1952).Numerical tabulation of the distribution of Kolmogorov's statistic for finite sample sizes. *J.Amer.Stat.Assoc.***47**,425-441.

Birnbaum,Z.W. & Hall,R.A.(1960).Small-sample distribution for multi-sample statistics of the Smirnov type. *Ann.Math.Stat.***31**,710-720.

Bitzer, K. & Harbaugh, J.W. (1987). DEPOSIM: a Macintosh computer model for two-dimensional simulation of transport, erosion and compaction of clastic sediments. *Comput.Geosci.* **13**, 611-638.

Bjorklund,A.(1983).Effects of exponentially decaying patterns on the probability distribution of anomalous values.*J.Geochem.Explor.***19**,349-359.

Bkouche-Palen,E. (1986). Système expert dans la classification des roches. In: Royer (1986), qv.

Blackrith,R.E. & Reyment,R.A.(1971).*Factor analysis and multivariate morphometrics.* Academic Press, New York.

Blais,R.A. & Carller,P.A.(1967).Application of geostatistics in ore evaluation. *Ore Reserve Estimation & Grade Contol* **9**,41-68.

Blalock,H.M.(1972).*Social statistics*. McGraw-Hill,New York,583pp.

Blanchard,F.N.(1984).A FORTRAN program for computing refractive indices using the double refraction method.*J.Geol.Educ.***32**,17-19.

Blashfield,R.K. & Aldenderfer,M.S.(1978).The literature on cluster analysis. *Multivar.Behav.Res.* **13**,271-295.

Blasi,A.(1979).Mineralogical applications of the lattice constant: variance-covariance matrices. *Tschermaks Mineral. Petrogr.Mitt.***26(3)**:139-148.

Blattner,P. & Hulston,J.R.(1978).Proportional variations of geochemical δO^{18} scales — an interlaboratory comparison. *Geochim.Cosmochim.Acta* **42**, 59-62.

Blencoe,J.G.(1976).RECPLT and TRIPLT: Fortran printer-plotting routines for rectangular and triangular plots. *Comput.Geosci.***2**,171-94.

Bliss,J.D.(1986a).Management of earth science databases — a small matter of data quality.In: Hanley & Merriam(1986), qv,113-120.

Bliss, J.D.(1986b). Management of the life and death of an earth-science database: some examples from GEOTHERM. *Comput.Geosci.***12**, 199-205.

Bliss, J.D. & Rapport, A.(1983). GEOTHERM: the US Geological Survey geothermal information system. *Comput.Geosci.* **9**, 35-39.

Bloomfield,P.(1976).*Fourier analysis of time-series.*Wiley,New York,258pp.

Bohrnstedt,G.W. & Carter,T.M.(1971).Robustness in regression analysis. In: *Sociological Methodology*, Costner,H.L.(ed.), 118-145.Jossey-Bass,San Fransisco.

Bonham-Carter,G.F.(1965).A numerical method of classification using quantitative and semi-quantitative data as applied to the facies analysis of limestones. *Bull.Can.Assoc.Pet.Geol.* **13**,481-502.

Bonham-Carter,G.F.,Gradstein,F.M. & D'iorio,M.A.(1986).Distribution of Cenozoic foraminifera from the northwestern Atlantic margin analyzed by correspondence analysis.*Comput.Geosci.***12**,621-636.

Bonyum,D. & Stevens,G.(1971).A general purpose computer program to produce geological stereonet diagrams.In: *Data processing in biology and geology,* Cutbill,J.et al.(eds.).Academic Press,New York,165-188.

Borkon,E. & Boles,J.N.(1970). *The 1130 multiple linear regression system.* Giannini Foundation of Agricultural Economics, California, Berkeley.

Borovkov,A.A.Markova,N.P. & Sycheva,M.M.(1964).[Tables of N.V.Smirnov's two-sample test of equality].In Russian. Akad.Nauk.SSSR Novosibirsk.

Botbol,J.M.(1986).An overview of the state of computing in the earth sciences. In: Hanley & Merriam(1986),qv,1-8.

Botbol,J.M.,Sinding Larsen,R.,McCammon,R.B. & Gott,G.B.(1978).A regionalized multivariate approach to target selection in geochemical exploration. *Econ.Geol.***73(4)**:534-546.

Botbol,R. & Botbol,J.M.(1975).The geological retrieval and synopsis program (GRASP). *U.S.G.S. Prof.Pap.***66**.

Bouille,F.(1976).Graphic theory and digitization of geological maps. *J.Math.Geol.* **8**,375-393.

Bowen,R.W.(1986).Micro-grasp,a microcomputer data system.In: Hanley & Merriam(1986), qv, 121-134.

Bowers,T.S. & Helgeson,H.C.(1985).FORTRAN programs for generating fluid isochores and fugacity coefficients for the system H_2O-CO_2-NaCl at high pressures and temperatures. *Comput. Geosci.* **11**:203-214.

Box,G.E.P. & Cox,D.R.(1964). An analysis of transformations. *J.Roy.Stat.Soc.***B26**, 211-252.

Boyer,P.S.(1986).Customized geological map patterns for the Macintosh computer. *J.Geol.Educ.* **34**,265-267.

Boyle,R.W.(1979).The geochemistry of gold and its deposits. *Mem.Geol.Surv.Can.* **280**, 584 pp.

Bozdar,L.B. & Kitchenham,B.A.(1972).Statistical approach of the occurrence of lead mines in the northern Pennines.*Tr.Inst.Ming.Metall.***81**,B183-8.

Bradley,J.V.(1968).*Distribution-free statistical tests.* Prentice-Hall,New Jersey,388pp.

Bradley,R.A.,Patel,K.M. & Wackerly,D.D.(1971).Approximate small sample distribution for multivariate two sample nonparametric tests.*Biometrics* **27**, 515-30.

Braun,G.(1969).Computer calculated counting rates for petrofabric and structural analysis.*Neues Jb. Mineral. Monatsch.* **10**,469-476.

Bridges,N.J. & McCammon,R.B.(1980).DISCRIM: a computer program using an interactive approach to dissect a mixture of normal or log-normal distributions.*Comput.Geosci.* **6**,361-96.

Brodlie, K.W. (1985). GKS — the standard for computer graphics. *Comput. Geosci.* **11**, 339-344.

Brofitt,J.P.,Randles,R.H. & Hogg,R.V.(1976).Distribution-free partial discriminant analysis. *J.Am. Stat. Assoc.* **71**,934-939.

Brooker,P.I.(1975).Optimal block estimation by kriging.*Tr.Aust.Inst.Ming.Metall.* **253**,15-69.

Brooker,P.I.(1977).Robustness of geostatistical calculations: a case study. *Proc.Aust.Inst.Ming. Metall.* **264**, 61-68.

Brooker,P.I.(1979).Geostatistics,part 2:Kriging.*Eng.Ming.J.***180(9)**:148-153.

Brooker,P.I.(1986).A parametric study of robustness of kriging variance as a function of range and relative nugget effect for a spherical semivariogram.*J.Math.Geol.***18**,477-488.

Brookes,S.M., Rees,P.H. & Stillwell,J.H. (1978). A manual for using MULREQ: a multiple regression program on the DEC system-10 and the 1906A. *Computer Manual School of Geography, Univ.Leeds* **3**.

Brooks,C. & Hart,S.R.(1969).Discussion of the use of Rb-Sr isochron regression treatment. *Carn.Inst. Wash. Yearbk.*

68,413-7.

Brooks,C.,Hart,S.R. & Wendt,I.(1972).Realistic use of two-error regression treatment as applied to Rubidium-Strontium data.*Rev.Geophys.Space Phys.* **10**,561-77.

Brooks,C.,Wendt,I. & Harre,W.(1968).A two-error regression treatment and its application to Rb-Sr and initial $^{87}Sr/^{86}Sr$ ratios of younger Variscan granitic rocks from the Schwarzwald massif,SW Germany.*J.Geophys.Res.* **73**,6071-6084.

Brooks,C.,James,D.E.,Hart,S.R. & Hoffman,W.(1975).Rb-Sr mantle isochrons. *Carn.Int.Wash. Yearbk.***75**,176-207.

Brooks,R.R.,Boswell,C.R. & Reeves,R.D.(1978).Elemental abundance data for standard rocks as a measure of the quality of analytical methods. *Chem.Geol.***21(1-2)**:25-38.

Brotzen,O.(1967).Geochemical ranking of rocks.*Sver.Geol.Unders.Ser.C.***617**,57pp.

Brotzen,O.(1972).A graph for the chemical diagnostics of rocks.*Sver.Geol.Unders.Ser.C.***678**,30pp.

Brotzen,O.(1975).Analysis of multivariate point distributions and chemical groupings of rocks. *J.Math.Geol.* **7**,191-214.

Brower,J.C.(1981).Quantitative biostratigraphy.In: Merriam(1981b),qv,63-104.

Brower,J.C. & Veinus,J.(1974).The statistical zap versus the shotgun approach. *J.Math.Geol.* **6**, 311-332.

Brower,J.C.,Cubitt,J.M.,Veinus,J. & Morton,M.(1979).Principal components analysis, factor analysis and point coordinates in the study of multivariate allometry.In: Gill & Merriam(1979), qv,245-266.

Brown,I.F.(1984).Three techniques to help students teach themselves concepts in environmental geochemistry. *J.Geol.Educ.* **32**, 177-178.

Brown,I.F.(1985).The use of simple statistics to show the uncertainty inherent in geochemical information. *J.Geol.Educ.***33(5)**:271-273.

Browne,M.B. & Forsythe,A.B.(1974a).Robust tests for the equality of variances. *J.Am.Stat.Assoc.* **69**,364-367.

Browne,M.B. & Forsythe,A.B.(1974b).The small sample behaviour of some statistics which test the equality of several means.*Technometrics* **16**,129-132.

Brownstein,I. & Kerner,N.B.(1982).*Guidelines for evaluating and selecting software packages.* Elsevier,278pp.

Bryan,W.,Finger,L.W. & Chayes,F.(1969a).A least-squares approximation for estimating the composition of a mixture.*Carn.Int.Wash.Yearbk.***67**,243-244.

Bryan,W.,Finger,L.W. & Chayes,F.(1969b).Estimating proportions in petrographic mixing equations by least-squares approximation.*Science* **163**,926-927.

Buchanan,B.G. & Shortliffe,E.H.(1984).*Rule-based expert systems.*Addison-Wesley,Reading, 748pp.

Buchbinder,B. & Gill,D.(1976).Numerical classification of multivariate petrographic presence-absence data by association analysis of Miocene Ziglag reef complex.In: Merriam (1976c),qv,23-32.

Burch,C.R.(1972).Tests of normality of frequency distributions of Na,K,SiO_2 and Cl in the Dartmoor granite.*Tr.R.Geol.Soc.Cornwall* 20,179-198.

Burger,H.R.(1983-4).Catalog of computer programs used in undergraduate geology education. (In 4 parts). *J.Geol.Educ.* **31**,219-229, 315-321 & 379-386; **32**,20-28.

Burger,H.R.(1986).Personal computer software for geological education.*US Nat.Assoc.Geol.Teachers Spec.Publ.No.***1.**

Burk,C.F.(1973).Computer-based storage and retrieval of geoscience information: bibliography,1970-2. *Geol.Surv. Can.Pap.* **73-14**,38pp.

Burk,C.F.(1978).The national data referral system for Canadian geoscience. *Proc.Geosci.Inf.Soc.* **8**:31-41.

Burk,C.F. (1986). Managing geoscience information for success and profit. *Proc.3rd Int.Conf.Geosci.Info. (Adelaide,S.Australia, June '86)*, Vol.1, 94-98.

Burkov,Y.K.,Darasansuscho,Y.I.,Yegorov,V.N. & Tuganova,Y.V.(1978).Geochemical zoning of the East Siberian nickel province based on trend analysis. *Dokl.Acad.Sci.USSR Earth Sci.Sect.***232**:82-84.

Burnaby,T.P.(1953).A suggested alternative to the correlation coefficient for testing the significance of agreement between pairs of time series and its application to geological data. *Nature* **172**, 210.

Burnaby,T.P.(1966).Distribution-free quadratic discriminant functions in palaeontology.In: Merriam(1969),qv.

Burnaby,T.P.(1970).On a method for characterising a similarity coefficient employing the concept of information. *J.Math.Geol.***2**,25-38.

Burns,K.L.(1979).Electronic-survey computing in Australia.*Comput.Geosci.* **5**,277-8.

Burns,K.L.(1981).Computers in geological photointerpretation.In: Merriam (1981b),qv,105-124.

Burns,K.L. & Remfry,J.G.(1976).EVENT: a computer program for analysis of geological maps. *Comput.Geosci.* **2**,141-162.

Burns,K.L.,Marshall,B. & Gee,R.D.(1969).Computer-assisted geological mapping. *Proc.Aust.Inst.Ming.Metall.* **232**, 41-47.

Burns,K.L.,Shepherd,J. & Berman,M.(1976).Reproducibility of geological lineaments and other discrete features interpreted from imagery: measurement by a coefficient of association.*Remote Sensing Env.***5**,267-301.

Burns,K.L.,Shepherd,J. & Marshall,B.(1978).Analysis of relational data from metamorphic tectonites: derivation of deformation sequences from overprinting relations.In: Merriam(1978a), qv, 171-200.

Burr,E.J.(1964).Small-sample distributions of the two sample Cramér-Von Mises and Watson's U^2. *Ann.Math.Stat.* **35**,1091-1098.

Burrows,D.R.,Wood,P.C. & Spooner,E.T.C.(1986).Carbon isotope evidence for a magmatic origin for Archaean gold-quartz vein ore deposits. *Nature* 321, 851-854.

Burwell,A.D.M. & Topley,C.G.(1982a).GEOFILE: an interactive program in BASIC for creating and editing data-files. *Comput.Geosci.* **8**,285-322.

Burwell,A.D.M. & Topley,C.G.(1982b).Programs in BASIC for plotting graphs and triangular diagrams. *Geol.Soc.Lond. Misc.Pap.***15**,124-134.

Busby,J.P. (1987). An interactive FORTRAN 77 program using GKS graphics for 2.5D modelling of gravity and magnetic data. *Comput.Geosci.* **13**, 639-644.

Butler,J.C.(1972).A FORTRAN IV plotting program utilizing an online printer. *Mineral.Mag.* **38**,899-900.

Butler,J.C.(1975).Occurrence of negative open variances in ternary systems. *J.Math.Geol.* **7**,31-45.

Butler,J.C.(1976).Principal components analysis using the hypothetical closed array*J.Math.Geol.* **8**,25-36.

Butler,J.C.(1977).Non-zero skewness — a geochemical or numerical consequence.? *Abstr.Progr. (Boulder)* **9(7)**:917-918.

Butler,J.C.(1978).Visual bias in R-mode dendrograms due to the effects of closure*J.Math.Geol.* **10**,243-252.

Butler,J.C.(1979a).Trends in ternary petrologic variation diagrams — fact or fantasy?*Am.Mineral.* **64**,1115-21.

Butler,J.C.(1979b).Numerical consequences of changing the units in which chemical analyses of igneous rocks are analysed. *Lithos* **12**,33-40.

Butler,J.C.(1979c).Effect of closure on the measures of similarity between samples *J.Math.Geol.***11**,431-440.

Butler,J.C.(1979d).The effects of closure on the moments of a distribution. *J.Math.Geol.***11**:75-84.

Butler,J.C.(1980).Numerical consequences of computing structural formulas. *Lithos* **13**,55-59.

Butler,J.C.(1981).Effect of various transformations on the analysis of percentage data*J.Math.Geol.* **13**,53-68.

Butler,J.C.(1982a).The closure problem as reflect in discriminant function analysis.*Chem.Geol.***37**, 367-375.

Butler,J.C.(1982b).Artificial isochrons*Lithos* **15**,207-14.

Butler,J.C.(1986).An interactive program for creating and manipulating datafiles using an Apple microcomputer system. In: Hanley & Merriam(1986),qv,135-40.

Butler,J.C. & Woronow,A.(1986a).Discrimination among tectonic settings using trace element abundances of basalts. *J.Geophys.Res.***91**,10289-10301.

Butler,J.C. & Woronow,A.(1986b).Extracting information from coarse clastic modes.*Comput. Geosci.***12**,643-652.

Cable,D. & Rowe,B.(1983).Software for statistical and survey analysis. *Study Group on Computers in Statistical Analysis (British Computer Soc.)*

Cacoullos,T.(1973).*Discriminant analysis and applications*.Academic Press,New Jersey.

Cacoullos,T. & Styan,P.H.(1973).A bibliography of discriminant analysis. In: Cacoullos(1973), qv,375-434.

Cadigan,R.A.(1962).A method for determining the randomness of regionally distributed quantitative data. *J.Sed.Petrol.***32**,813-818.

Cain,J.A. & Cain,L.S.(1968).Non-parametric statistics: largely ignored for analysing geologic data (abstr.) *Geol.Soc.Am.Abs.Spec.Pap.***115**,p.32.

Cairncross,B. & McCarthy,T.S.(1986).A FORTRAN-IV computer program for displaying coal seam data. *Tr.Geol.Soc.S.Afr.***89**,409-414.

Calkins,H.W. & Marble,D.F.(1980).*Computer software for spatial data handling. Vol.1.Full geographic information systems. Vol.2.Data manipulation programs. Vol.3.Cartography and graphics*.Int.Geogr.Union/U.S.G.S.

Calkins,H.W. & Tomlinson,R.F.(1977).*Geographic information systems, methods and equipment for land use planning,1977.* Int.Geogr.Union,Ottawa,394pp.

Cameron,E.M.(1967).A computer program for factor analysis of geochemical and other data. *Geol.Surv.Can.Pap.* **67-34**,42pp.

Cameron,E.M.(1968).A geochemical profile of the Swan Hills Reef.*Can.J.Earth Sci.***5**,287-309.

Camina,A.R. & Janacek,G.J. (1984). *Mathematics for seismic data processing and interpretation.* Int.Human Resour.Dev.Corp., Boston, Mass., 255 pp.

Camisani-Calzolari,F.A.G.M.,de Klerk,W.J. & van der Merwe,P.J.(1985).Assessment of South African uranium resources: methods and results.*Tr.Geol.Soc.S.Afr.***88**,83-98.

Campbell,A.N.,Hollister,V.F.,Duda,R.O. & Hart,P.E.(1982).Recognition of a hidden mineral deposit by an artificial intelligence program.*Nature* **217**,927-929.

Campbell,N.A.(1984).Canonical variate analysis with unequal covariance matrices: generalization of the usual solution. *J.Math.Geol.***16**, 109-124.

Campbell,N.A. & Reyment,R.A.(1978).Discriminant analysis of a Cretaceous foraminifera using shrunken estimation. *J.Math.Geol.***10**,347-359.

Cann,J.R.(1971).Major element variations in ocean floor basalts.*Phil.Tr.R.Soc.Lond.*, **A**495-505.

Cargill,S.M. & Clark,A.L.(1977).Standards for computer application in resource studies*J.Math. Geol.***9**,205-237 & **10**,405-442.

Carlile,R.E. & Gillett,B.E.(1973).*FORTRAN and computer mathematics for the engineer and scientist*.Petroleum Publ.Co.,Tulsa,520p.

Carnahan,B.,Luther,H.A. & Wilkes,J.O.(1969).*Applied numerical methods*.Wiley, New York.

Carpenter,J.,Deloria,D. & Morganstein,D.(1984).*Statistical software for micros*.Byte (April '84),234-264.

Carr,J.D.,Myers,D.E. & Glass,C.E.(1985).Cokriging — a computer program. *Comput. Geosci.* **11**:111-128.

Carr,J.R. (1987). A comparison of FORTRAN, Pascal and C for variogram computation on a microcomputer. *Comput.Geosci.* **13**, 645-654.

Carr,M.J. & Rose, W.I.(1987). CENTAM — a database of Central American volcanic rocks. *J. Volcanol. Geotherm. Res.* **33**, 239-240.

Carr,P.F.(1981).Distinction between Permian and post-Permian igneous rocks in the southern Sydney Basin,NSW,on the basis of major element geochemistry. *J.Math.Geol.***13**,193-200.

Carraro,C.C. & Gamermann,N.(1978).Programa para o cálculo da norma pelo método CIPW em computador.*Pesquisas***9**, 159-179.

Carras,S.(1987).Gold sampling — the importance of getting it right.*Bull.Aust.Inst.Geoscientists* **7**, 1-25.

Carrasco,C.(1977).[Some techniques of multivariate statistic analysis in geology].In Spanish. *Minerales (Santiago),* **33(140)**:35-41.

Carroll,J.B.(1971).The nature of the data — or,how to choose a correlation coefficient.*Psychometrica* **26**,347-372.

Carter,C.S. & Chorley,R.J.(1961).Early stage development in an expanding stream system.*Geol. Mag.***98**,117-130.

Carter,J.D.(1986).*The Remote Sensing Handbook: a guide to remote sensing products,services,facilities,publications and other materials.*Kogan Page, London.

Carter,G.R. (1987). *FERROS — a computer hosted database for geochemical analysis of rocks related to auriferous iron formations.* Eastern Washington University, M.S. thesis, Cheney, Washington.

Carusi,A.,Cavarretta,G.,Coradini,A., Fanucci,O., Funiciello,R.,Salomone,M.,Fulchignoni,M. & Teddeucci,A. (1974). Lunar rocks and glasses: a unifactorial and multivariate analysis of chemical data.*Geochim.Cosmochim. Acta Suppl.(Proc.5th.Lunar Sci.Conf.)*,**5**:605-620.

Cary,L.(1985).*Mac: the Apple Macintosh Book.* (2nd Ed).Microsoft Press,Washington.

Cary, M.S.(1987). Establishing a global database. *Outlook* (Prime computers Inc., Mass.,USA), Feb '87, 3-5.

Castillo-Munoz,R. & Howarth,R.J.(1976).Application of the empirical discriminant function to regional geochemical data from the United Kingdom.*Bull.Geol.Soc. Am.***87**,1567-1581.

Cattell,R.B.(1978).*The scientific use of factor analysis.* Plenum,New York.

Cawthorn,R.G. & Fraser,D.G.(1979).Element partitioning in immiscible liquids: a statistical model.*Chem.Geol.* **27**:99-113.

CEED (Comparison,Evaluation,Exhibition,Demonstration, 1986).Mapping systems compared and evaluated. *Geobyte,* Winter '86, 25-40.

Cerven,J.F.,Robinson,P.D. & Fang,J.H.(1968).Fortran IV program for molecular norm calculations. *Mineral.Mag.* **36**:1175-1176.

Chambers,J.M.,Pregibon,D. & Zayas,E.R.(1981).Expert software for data analysis — an initial experiment.*43rd Ser. Int. Stat. Inst.Buenos Aires,*9pp.

Chandor,A.,Graham,G. & Williamson,R.(1985).*The Penguin dictionary of computers.* 3rd Ed.Penguin, Harmondsworth, 488 pp.

Chapman,R.P.(1976).Limitations on correlation and regression analysis in geochemical exploration. *Tr.Inst. Ming. Metall. Ser.B.***85**,279-283.

Chapman,R.P.(1976-7).Some consequences of applying lognormal theory to pseudo-lognormal distributions. *J.Math.Geol.***8**,209-14 & **9**,191-194.

Chapman,R.P.(1977).Log transformations in geochemistry.*J.Math.Geol.* **9**,194-198.

Chapman,R.P.(1978).Evaluation of some statistical methods of interpreting multi-element geochemical drainage data from New Brunswick.*J.Math. Geol.***10**,195-224.

Charlesworth,H.A.K.,Langenberg,C.W. & Ramsden,J.(1975).Determining axes, axial planes and profiles of microscopic folds using computer-based methods.*Can.J.Earth Sci.***13**,54-65.

Chatfield,C. & Collins,A.J.(1980)*Introduction to multivariate analysis.* Chapman & Hall.

Chatterjee,A.K.,Sharma,K.K. & Nagasubramanian,S.(1977).Development of a statistical oriented software system for iron ore exploration.*Indian Ming. Eng.J.***16(7)**:25-33.

Chatterjee,S.K. & Sen,P.K.(1966).Nonparametric tests for the multivariate multi-sample location problem.In: *Essays in probability and statistics in Memory of S.N.Roy,* Bose et al.,(eds.),197-228. Univ.N.Carolina Press.

Chattic,A.G.(1977).Guide to computer program directories.*Spec.Publ.Natl.Bur. Standards USA* **500(22)**:168pp.

Chayes,F.(1949).On correlation in petrography.*J.Geol.***57**,239-254.

Chayes,F.(1950).Composition of the granites of Westerly and Bradford,Rhode Island. *Am.J.Sci.* **248**, 378-407.

Chayes,F.(1953).In defense of the second decimal.*Am.Mineral.***38**,784-793.

Chayes,F.(1954a).Effect on change of origin on mean and variance of two-dimensional fabrics. *Am.J.Sci.* **252**,567-570.

Chayes,F.(1954b).The theory of thin section analysis.*J.Geol.***62**,92-101,& **63**,288-290.

Chayes,F.(1954c).The log-normal distribution of the elements: a discussion. *Geochim. Cosmochim. Acta* **6**,119-120.

Chayes,F.(1956a).The Holmes effect and the lower limit of modal analysis. *Mineral.Mag.* **31**, 276-281.

Chayes,F.(1956b).*Petrographic modal analysis.*Wiley,113pp.

Chayes,F.(1960).On correlation between variables of constant sum.*J.Geophys. Res.***65**,4185-4193.

Chayes,F.(1962).Numerical correlation and petrographic variation.*J.Geol.* **70**,440-452.

Chayes,F.(1964a).Variance-covariance relations in some published Harker diagrams of volcanic suites. *J.Petrol.***5**,219-237.

Chayes,F.(1964b).A petrographic distinction between Cenozoic volcanics in and around the open oceans. *J.Geophys.Res.***69**,1573-1588.

Chayes,F.(1965).Classification in a ternary diagram by means of discriminant functions. *Am.Mineral.* **50**,1618-1633.

Chayes,F.(1966).Alkaline and subalkaline basalts. *Am.J.Sci.* **264**,128-145.

Chayes,F.(1968a).On locating field boundaries in simple phase diagrams by means of discriminant functions. *Am.Mineral.* **53**,359-371.

Chayes,F.(1968b).A least-squares approximation for estimating the amounts of petrographic partition products. *Miner.Petrogr.Acta* **14**,111-114.

Chayes,F.(1968c).On the graphical appraisal of the strength of associations in petrographic variation diagrams.In: P.H.Abelson (ed.) *Researches in Geochemistry*,vol.II,322-339. Wiley,New York.

Chayes,F.(1969).A last look at G1-W1. *Carn.Inst.Wash.Yearbk.***68**,239-41.

Chayes,F.(1970a).Another last look at G1-W1. *J.Math.Geol.***2**,207-10.

Chayes,F.(1970b).On deciding whether trend surfaces of progressively higher order are significant. *Bull.Geol.Soc.Am.* **81**,1273-1278.

Chayes,F.(1971).*Ratio Correlation*.Univ.Chicago Press.

Chayes,F.(1972).Effect of the proportion transformation on central tendency. *J.Math.Geol.* **4**,269-270.

Chayes,F.(1975).A priori and experimental approximation of simple ratio correlation.In: McCammon(1975a), qv,106-137.

Chayes,F.(1976).Characterising the consistency of current usage of rock-names by means of discriminant functions.*Carn.Int.Wash.Yearbk.***75**,782-784.

Chayes,F.(1977).A measure of consistency in current usage of the names of common volcanic rocks. *Fortschr.Mineral.***55(1)**:24-25.

Chayes,F.(1978).On translating petrographic information into bits and bytes. *Geol.Contrib.Syracuse Univ.***5**:19-22.

Chayes,F.(1979a).A comparison of two methods for classifying basalts. *Carn.Inst.Wash.Yearbk.* **1978-9**,481-484.

Chayes,F.(1979b).Partitioning by discriminant analysis: a measure of consistency in the nomenclature and classification of volcanic rocks. In: Yoder,H.S.(ed),*Evolution of the Igneous Rocks: 50th anniversary perspectives*,521-532. Princeton Univ.Press.

Chayes,F.(1979-80).A world data base for igneous petrology.*Carn.Inst.Wash.Yearbk.* **78**,484-485 & **79**,288.

Chayes,F.(1980).An effect of closure correlation on classification by ranges. *Carn.Inst.Wash.Yearbk.* **79**,289-290.

Chayes,F.(1981).Characterizing associations between proportions by means of remaining space variables. *Carn.Inst.Wash.Yearbk.***80**,345-347.

Chayes,F.(1982).The rock information system RKNFSYS.*Carn.Inst.Wash.Yearbk.* **81**,315-316.

Chayes,F.(1983a).Inversion of variance relations by the logarithmic transformation. *Carn.Inst.Wash.Yearbk.* **82**,281-284.

Chayes,F.(1983b).A FORTRAN decoder and evaluator for use at operation time.*Comput.Geosci.***9**,537-550.

Chayes,F.(1983c).Detecting nonrandom associations between proportions by tests of remaining-space variables. *J.Math.Geol.***15**,197-206.

Chayes,F.(1983d).*On the possible significance of strong positive correlations between trace elements*. Unpubl.typescript, Geophys.Lab.Washington.

Chayes,F.(1987).Consistency of the two group discriminant function in repartitioning rocks by name. In: Size(1987a),qv, 47-54.

Chayes,F. & Fairbairn,H.W.(1951).A test of the precision of thin-section analysis by point counter. *Am.Mineral.* **36**,704-712.

Chayes,F. & Kruskal,W.(1966).An approximate statistical test for correlation between proportions. *J.Geol.***74**,692-702.

Chayes,F. & Le Maitre,R.W.(1972a).The number of published analyses of igneous rocks. *Carn.Inst.Wash.Yearbk.* **71**,493-495.

Chayes,F. & Le Maitre,R.W.(1972b).Published analyses of igneous rocks.*Nature* **236**,449-450.

Chayes,F. & Suzuki,Y.(1963).Geological contours and trend surfaces; a discussion. *J.Petrol.***4**,307-312.

Chayes,F. & Trochimczyk,J.(1977).The sampling variances of eigenvalues and of the coefficients of eigenvectors of closed sets. *Carn.Inst.Wash.Yearbk.***77**, 640-642.

Chayes,F. & Trochimczyk,J.(1978).An effect of closure on the structure of principal components analysis. *J.Math.Geol.***10**,323-334.

Chayes,F. & Velde,D.(1965).On distinguishing basaltic lavas of circumoceanic and oceanic-island type by means of discriminant functions.*Am.J.Sci.***263**, 206-222.

Chayes,F.,Groetsch,G.,Li,S-Z. & Possolo,A.(1982).The IGBA data base for igneous petrology. *Carn.Inst.Wash.Yearbk.* **81**,316-319.

Chayes,F.,Li,S-Z. & Stewart,D.(1983).Data verification procedures in IGBA. *Carn.Inst.Wash.Yearbk.* **82**,280-281.

Chayes,F., Brooks,C.,Strobel,M.,James,D.E. & Hart,S.(1977).Use of correlation statistics with Rubidium-Strontium systematics.*Science* **196**, 1234-1235.

Cheeney,R.F.(1983).*Statistical methods in geology for field and laboratory decisions.* Allen & Unwin,London,169pp.
Cheetham,A.H. & Hazel,J.E.(1969).Binary (presence-absence) similarity coefficients. *J.Palaeontol.***43**,1130-1136.
Chen,E.H. & Dixon,W.J.(1972).Estimates of parameters of a censored regression sample*J.Am.Stat.Assoc.***67**, 664-671.
Chernicoff,S.(1985).*Macintosh revealed.*2 vols.Hayden,Hasbrouck Heights,New Jersey.
Chernoff,H.(1973).Using faces to represent *k*-dimensional space graphically. *J.Am.Stat.Assoc.***68**,361-368.
Child,D.(1970).*The essentials of factor analysis.* Holt,Reinhart & Winston,London.
Chinadi,V.C., Gokhale,N.W. & Kulkarni,S.R.(1972). Geostatistical investigation of gabbro occurrences near Savantadi, Riri district, India. *J.Math.Geol.* **4**, 147-153.
Chork,C.Y. & Cruikshank,B.I.(1984).Statistical map analysis of regional stream sediment data from Australia. *J.Geochem.Explor.* **21**, 405-419.
Choubert,B. & Zanone,L.(1976).On the distribution of atoms in magmatic systems: normal geochemical law. *J.Math.Geol.***8**,349-374.
Christie,O.H.J. & Alfsen,K.N.(1977).Data transformation as a means to obtain reliable consensus values for reference materials.*Geostandards Newsletter* **1(1)**:47-49.
Chung,C.F.(1978).Computer program for the logistic model to estimate the probability of occurrence of discrete events.*Geol.Surv.Can.Pap.***78-11**.
Chung,C.F. & Agterberg,F.P.(1980).Regression models for estimating mineral resources from geological map data. *J.Math.Geol.***12**,473-488.
Chung,C.F.,Divi,S.R. & Fabbri,A.G.(1977).User's guide for SIMGRA 1: an interactive program for simulating the distribution of transformations of several independent random variables.*Open File Geol.Surv.Can.***493**:9pp.
Chyi,L.L.,Elizalde,L.,Smith,G.E. & Ehmann,W.D.(1978).Multivariate analysis in characterisation of limestone units based on minor and trace element contents.In: Merriam(1978a),qv,35-45.
Clanton,U.S. & Fletcher,C.R.(1976).Sample size and sampling errors as the source of dispersion in chemical analyses. *Geochim.Cosmochim.Acta* **7(2)**:1413-1428.
Clark,D.(1975).Understanding canonical correlation analysis (Concepts and techniques in modern Geography,3).*Geo Abstracts (Norwich)*,36pp.
Clark,I.(1977a).Regularization of a semivariogram.*Comput.Geosci.***3**,341-346.
Clark,I.(1977b).ROKE,a computer program for nonlinear least-squares decomposition of mixtures of distributions. *Comput.Geosci.***3**,245-256.
Clark,I.(1977c).SNARK — a four-dimensional trend-surface analysis computer program. *Comput.Geosci.***3**,283-308.
Clark,I.(1979).*Practical geostatistics.*Applied Science Publishers,London,129pp.
Clark,I.(1983).Regression revisited.*J.Math.Geol.***15**,517-536.
Clark,I. & Garnett,R.H.T.(1974).Identification of multiple mineralization phases by statistical methods. *Tr.Inst. Ming. Metall.* **83**,A43-52.
Clark,M.W.(1976).Some methods for statistical analysis of multimodal distributions and their application to grain-size data. *J.Math.Geol.* **8**,267-282.
Clark,M.W.(1977).GETHEN: a computer program for the decomposition of mixtures of two normal distributions by the method of moments.*Comput.Geosci.***3**,257-267.
Clark,M.W.(1984).Mix'n'match: proportional parts of univariate normal mixtures. *Comput.Geosci.***10**,245-50.
Clark,M.W. & Clark,I.(1976).A sedimentological pattern recognition problem. In: Merriam(1976c), qv,121-142.
Clarke,T.L.(1978).An oblique factor analysis solution for the analysis of mixtures. *J.Math.Geol.***10**,225-242.
Claudon,G.,Pharisat,A. & Thomas,A.(1977).Examples of application of factor analysis in geology (palaeontology, geochemistry and hydrochemistry). *Ann.Sci.Univ.Besançon Ser.3.Geol.***28**:43-63.
Clausen,F.L. & Harpoth,O.(1983).On the use of discriminant analysis for classifying chemical data from panned heavy-mineral concentrates in central E.Greenland*J.Geochem.Explor.***18**,1-24.
Clemency,C.V. & Borden,D.M.(1978).The precision of rapid rock analysis and the homogeneity of new U.S.G.S. standard rock samples.*Geostandards Newsletter* **2**:147-56.
Cliff,A.D. & Ord,J.K.(1981).*Spatial processes, models and applications.* Pion, London, 266 pp.
Clifford,A.A.(1973).*Multivariate error analysis.*Applied Science Publ.,Barking.
Clocksin,W.F. & Mellish,C.S.(1981).*Programming in Prolog.*Springer-Verlag,279pp.
Cobb,D.F.(1985).*Excel in business: number-crunching power on the Apple Macintosh.*Microsoft Press,New York.
Cobb,D.A. (1986). Access to cartographic information: considerations and limits. *Proc.3rd Int.Conf.Geosci.Info. (Adelaide,S.Australia, June '86)*, Vol.1, 69-82.
Cochran,W.G.(1964).On the performance of the linear discriminant function. *Technometrics* **6**,179-190.
Cochran,W.G.(1977).*Sampling techniques.*Wiley,New York.
Coffrant,D.,Girod,M. & Roca,J.L.(1978).Application of principle component analysis to the petrological study of lavas from the New Hebrides. *Bull.Soc.Géol.Fr.Sér.7*,**20(4)**:503-509.
COGS (Computer-Oriented Geological Society, 1986). *Geology programs for microcomputers.* 5th Ed. COGS, Denver.
Cole,J.P. & King,C.A.M.(1968).*Quantitative geography.* Wiley,New York, 692 pp.
Coleman, D. & Naiman, A. (1987). *The Macintosh Bible.* Goldstein-Blair, Berkeley, CA, 418 pp.

Collins,L.(1975).An introduction to Markov chain analysis (concepts and techniques in Modern Geography,1).*Geo Abstracts (Norwich)*,36pp.
Collyer,P.L. & Merriam,D.F.(1973).An application of cluster analysis in mineral exploration.*J.Math.Geol.***5**,213-224.
Comstock,G.M.(1979).Miniregoliths and Monte-Carlo modelling.In: *Abstr.10th Lunar Plan.Sci.Conf.*Part 1. **12-3/19:3/23**: 229-231.
Condit,C.D. & Chavez,P.S.(1979).Basic concepts of computerized digital image processing for geologists. *Bull.U.S.G.S.***1462**:16pp.
Connors,M.(1987).*Computers and computing information resources directory.* Gale Research Co.,Detroit,1271pp.
Conover,W.J.(1974).Some reasons for not using the Yates continuity correction on 2x2 contingency tables. *J.Am.Stat.Assoc.***69**,374-376.
Conover,W.J.(1980).*Practical nonparametric statistics.* 2nd Edn. Wiley,462pp.
Conover,W.J. & Iman,R.L.(1980).The rank transformation as a method of discrimination with some examples. *Communications in Statistics* **A9**.
Conover,W.J. & Iman,R.L.(1981).Rank transformation as a bridge between parametric and nonparametric statistics.*The American Statistician* **35**,124-129.
Conrad,W.K.(1987).A FORTRAN program for simulating major and trace element variations during Rayleigh fraction with melt replenishment and assimilation. *Comput.Geosci.***13**,1-12.
Consolmagno,G.J. & Drake,N.J.(1976).Equivalence of equations describing trace element distribution during equilibrium partial melting.*Geochim.Cosmochim. Acta* **40**,1421-1422.
Cook,R.D.(1977).Detection of influential observations in linear regression. *Technometrics* **19**,15-18.
Cook,R.D.(1979).Influential observations in linear regression.*J.Am.Stat. Assoc.***74**,169-174.
Cooley,W.W. & Lohnes,P.R.(1971).*Multivariate data analysis.*Wiley,364pp.
Cooper,M.A. & Marshall,I.D.(1981).ORIENT: a computer program for the resolution and rotation of palaeocurrent data. *Comput.Geosci.***7**,153-166.
Coppock,J.T. & Anderson,E.K.(1987). Editorial review. *Int.J.Geogr.Info.Systems* **1**,3-11.
Coradini,A.,Fulchignoni,M.,Fanucci,O. & Gavrishin,A.I.(1977).A FORTRAN IV program for a new classification technique: the G-mode central method. *Comput.Geosci.***3**,85-106.
Corbett,R.G.,Manner,B.M. & Tompkins,F.G.(1985).Using a FORTRAN program and piper diagrams to facilitate understanding of water chemistry. *J.Geol.Educ.***37**,171-174.
Cosgrove,M.E.(1972).The geochemistry of the potassium-rich Permian volcanic rocks of Devonshire. *Contrib.Mineral. Petrol.* **36**,155-170.
Cox,A. & Dalrymple,G.B.(1967). Statistical analysis of geomagnetic reversal data and the precision of potassium-argon dating. *J.Geophys.Res.* **72**, 2603-2614.
Cox,D.R. & Hinkley,D.V.(1974).*Theoretical statistics.*Chapman & Hall,London.
Cox,D.R. & Lewis,P.A.W.(1966).*The statistical analysis of series of events.* Methuen,London, 285pp.
Cox,D.R. & Miller,H.D.(1965).*The theory of stochastic processes.*Wiley,New York,398pp.
Cox,D.R. & Stuart,A.(1955).Some quick tests for trend in location and dispersion. *Biometrika* **42**,80-95.
Coxon,A.P.M.(1982).*The user's guide to multidimensional scaling.*Heinemann, London,271pp.
Crain,I.K.(1976).Statistical analysis of geotectonics.In: Merriam(1976a),qv,3-15.
Crandall,R.E. & Colgrove,M.M.(1986).*Scientific programming with Macintosh Pascal.* Wiley,New York,288pp.
Cressie,N.(1985).Fitting variogram models by weighted least squares. *J.Math.Geol.***17**,563-586.
Cressie,N. & Hawkins,D.M.(1980).Robust estimation of the variogram;I: *J.Math.Geol.***12**,115-125.
Crow,E.L.(1952).Some cases in which Yates' correction should not be applied. *J.Am.Stat.Assoc.***47**,303-4.
Cryer,J.D. (1985). *Time-series analysis.* Duxbury Press, Boston, 286 pp.
Cubitt,J.M.(1976).An analysis and management system suitable for sedimentological data.In: Merriam(1976c),qv,1-10.
Cubitt,J.M. & Celenk,O.(1976).FORTRAN program for producing stereograms in geology.*Comput.Geosci.***1**,207-211.
Cubitt,J.M. & Henley,S.(1978).*Statistics in geology.*Benchmarks Papers in Geology,vol.37. Dowden, Hutchison & Ross,Stroudsburg,Pa,340pp.
Cuneo,M. & Feldman,D.S.(1986). *Statview 512+— the professional, graphic, statistics utility.* Brainpower, Calabasas, CA, 180pp.
Curray,J.R.(1956).The analysis of two-dimensional orientation data.*J.Geol.* **64**,117-131.
Currie,K.L.(1980).A modified norm calculation.*Can.J.Earth Sci.***17**,1342-1350.
Dacey,M.F. & Krumbein,W.C.(1970).Markovian chains in statigraphic analysis. *J.Math.Geol.***2**,175-192.
Dagbert,M. & David,M.(1974).Pattern recognition and geochemical data: an application to the Monteregian Hills. *Can.J.Earth Sci.***11**,1577-1585.
Dagbert,M.,Pertsowsky,R.,David,H. & Perrault,G.(1975).Agpaicity revisited — pattern recognition in the geochemistry of nepheline syenite rocks. *Geochim.Cosmochim.Acta* **39**,1499-1504.
D'Agostino,R.B. & Tietjen,G.L.(1971).Simulation probability points of b_2 for small samples.Biometrika **58**,669-672.
Dahlberg,E.C.(1975).Relative effectiveness of geologists and computers in mapping potential hydrocarbon exploration targets.*J.Math.Geol.* **7**,373-394.

Dampney,C.N.G., Pilkington,G. & Pratt,D.A. (1986). ASEG-GDF: the Australian Society of Exploration Geophysicists standard for digital transfer of data. *Proc.3rd Int.Conf.Geosci.Info.(Adelaide,S.Australia, June '86)*, Vol.1, 21-49.

Daniel,C. & Wood,F.(1980).*Fitting equations to data.*(Revised Ed).Wiley,New York.

Daost,G. & Gelinas,L.(1981).TELECINO: an interactive petrological and geochemical diagram generator. *Comput. Geosci.* **7**,21-26.

Darroch,J.N.(1969).Null correlation for proportions.*J.Math.Geol.***1**, 221-227.

Darroch,J.N. & Ratcliff,D.(1970).Null correlation for proportions.II.*J.Math.Geol.***2**,307-312.

Darroch,J.N. & Ratcliff,D.(1978).No-association of proportions.*J. Math.Geol.***10**,361-368.

Das Gupta,H.C.(1978).An application of R-mode factor analysis to the study of metamorphic petrogenesis.*Lithos* **11**:121-131.

Date,C.J.(1981).*An introduction to database systems.* 3rd Edn.Addison-Wesley,574pp.

Davaud,E. & Strasser,A.(1982).GEOMAN: a FORTRAN program for the management of geological thin section data. *Comput.Geosci.***8**,61-68.

David,H.(1982).*The computer package STATCAT.*North-Holland,Amsterdam 779pp.

David,H.A.(1981).*Order statistics.*Wiley,New York,360pp.

David,H.A.,Hartley,H.O. & Pearson,E.S.(1954).The distribution of the ratio in a single sample of range to standard deviation.*Biometrica* **41**,482-493.

David,M.(1977).*Geostatistical ore reserve estimation.*Elsevier Developments in Geomathematics vol.2, Amsterdam, 364pp.

David,M.,Campiglio,C. & Darling,R.(1974).Progress in R- and Q-mode analysis and its application to the study of geological processes.*Can.J.Earth Sci.* **11**,131-166.

David,P.P. & Lebuis,J.(1976).LEDA: a flexible codification system for computer-based files of geological field data. *Comput.Geosci.***1**,265-278.

Davies,D.M. & Coxon,A.P.M. (1982). *Key texts in multidimensional scaling.* Heinemann,London.

Davies,O.(1984). *Omni complete catalog of computer software.* Macmillan, New York, 471 pp.

Davies,O.L. & Goldsmith,P.L.(1972).*Statistical methods in research and production.*Hafner,New York,478pp.

Davis,J.C.(1973).*Statistics and data analysis in geology.*1st Edn.Wiley,New York,550pp.

Davis,J.C.(1986).*Statistics and data analysis in geology.*2nd Edn.Wiley,New York,646pp.

Davis,J.C.(1975).Training geologists in geomathematics and use of computers. *Bull.Am.Assoc.Pet.Geol.***59**,159-160.

Davis,J.C.(1981).Looking harder and finding less — use of the computer in petroleum exploration.In: Merriam(1981b), qv,125-144.

Davis,J.C. & Cocke,J.M.(1972). Interpretation of complex lithologic successions by substitutability analysis. In: Merriam(1972),qv,27-52.

Davis,J.C. & McCullagh,M.J.(1975).*Display and analysis of spatial data.*Wiley,London.

Davis,P.J.(1973).*The mathematics of matrices.*(2nd Ed).Wiley,New York,348pp.

Davison,M.L.(1983).*Multidimensional scaling.*Wiley,New York.

Dawson,J.B. & Stephems,W.E. (1975). Statistical analysis of garnets from kimberlites and associated xenoliths. *J.Geol.* **83**, 589-607.

Dawson,K.R.(1958).An application of multivariate variance analysis to mineralogical variation, Preissac-Lacorne Batholith,Abitibi County,Quebec, Canada.*Can.Mineral.* **6**,222-233.

Dawson,K.R. & Whitten,E.H.T.(1962).The quantitative mineralogical composition and variation of the Lacome, LaMotte and Preissac granitic complex,Quebec, Canada.*J.Petrol.***3**,1-37.

Dean,R.B. & Dixon,W.J.(1951).Simplified statistics for small numbers of observations. *Anal.Chem.***23**,636-638.

Dean,W.E. & Gorham,E.(1976).Classification of Minnesota lakes by Q- and R-mode factor analysis of sediment mineralogy and geochemistry.In: Merriam(1976c),qv,61-72.

Deb,M. & Saxena,S.K.(1978).Multivariate statistical analysis — an application to lunar minerals.*Indian J.Earth Sci.* **5(1)**:77-94.

De Backer,L.W. & Kuypers, J.P. (1986). User experiencewith data collection networks. *Proc.3rd Int.Conf.Geosci.Info. (Adelaide,S.Australia, June '86)*, Vol.1, 1-19.

De Capeariis,P. (1976). Errors in the calculation of similarity coefficients. *J.Math.Geol.* **8**, 499-505.

Deer,W.A., Howie,R.A. & Zussman,J.(1962). *An introduction to the rock-forming minerals.* Longmans, London.

Defant,M.J. & Ragland,P.C.(1982).Classification of trondhjemites into four tectonically unique environments by discriminant factor analysis. *Abstr.Progr.(Boulder)* **14:**1-2,p.14.

De la Roche,H., Stussi,J.M., Marchal,M. & Ploquin,A. (1986). Communication and exchange of computerized chemical data. Problems, solutions, scientific results. In: Royer (1986), qv.

Demirmen,F.(1971).Counting error in petrographic point-count analysis: a theoretical and experimental study. *J.Math.Geol.***3**,15-42.

Demirmen,F.(1976).RANK: a FORTRAN IV program for computation of rank correlations. *Comput.Geosci.***1**,221-229.

Dempster,A.P.(1969).*Elements of continuous multivariate analysis.*Addison-Wesley, Reading, Mass.,388pp.

Dennison,J.M.(1962).Graphical aids for determining reliability of sample means and an adequate sample size. *J.Sed.Petrol.*

32,743-750.
Dennison,J.M.(1972).Statistical meaning in geologic fieldwork. *Geol.Soc.Am.Spec.Pap.* **146**, 25-38.
De Paulo, D.J.(1980). Trace element and isotopic effects of combined wallrock assimilation and fractional crystallization. *Earth Planet.Sci.Lett.* **53**, 189-202.
Devore,J.L.(1982).*Probability and statistics for engineering and the sciences.* Brooks & Cole, Monterey,640pp.
Dhanoa,M.S.(1982).A BASIC computer program which tests for normality and the presence of outliers: some modifications. *Lab.Practice* **31**,32-33.
Dieterich,J.H.(1970).Computer experiments in the mechanics of finite amplitude folds.*Can.J.Earth Sci.***7**,467-476.
Digby,P.G.N. & Gower,J.C.(1981).Ordination between- and within-groups applied to soil classification. In: Merriam(1981d), qv,63-76.
Diggle,P.J. & Fisher,N.I.(1985).SPHERE: a contouring program for spherical data. *Comput.Geosci.***11**:725-766.
Dinneen,L.C. & Bradley,B.C.(1976). *Appl.Stat.* **25**, 75-77.
Divi,S.R.,Thorpe,R.I. & Franklin,J.M.(1979).Application of discriminant analysis to evaluate compositional controls of stratiform massive sulphide deposits in Canada.*J.Math.Geol.***11**,391-406.
Dixon,W.J.(1953).Processing data for outliers.*Biometrica* **9**,74-89.
Dixon,W.J.(1975).*BMDP-Biomedical computer programs.*Univ.Calif.Press,Berkeley,792pp.
Dixon,W.J.(1981).*BMDP Statistical software.*Univ.Calif.Press,Berkeley.
Dixon,W.J. & Brown,M.B.(1979).*Biomedical computer programs,P series.* Univ.Calif.Press, Berkeley, 880pp.
Dixon,W.J. & Massey,F.J.(1969).*Introduction to statistical analysis.*McGraw-Hill, New York,638pp.
Dodd,J.R. & Ilmmega,N.T.(1974).The portable computer as an aid: examples from a geobiology course. *J.Geol.Educ.***23**,109-114.
Dodge,Y.(1985). *Analysis of experiments with missing data.* Wiley, New York, 499 pp.
Dodson,M.H.(1982).On 'spurious' correlations in Rb-Sr isochron diagrams. *Lithos* **15**,215-20
Doe,B.R. & Rohrbough,R.(1979).*Lead isotope data bank: 3459 samples and analyses cited.* U.S.G.S.Open File Rept.**79(661)**:152pp.
Dolan,N.(1984).*Cracking the Mac: tutorials,tips and traps.University of Western Australia Use of Microcomputers Project.*Nedlands,Western Australia.
Doornkamp,J.C. & King,C.A.M.(1971).*Numerical analysis in geomorphology.*Arnold,London,372pp.
Dorn,W.S. & McCracken,D.D.(1972).*Numerical methods with Fortran IV case studies.* Wiley,New York.
Dowd,P.A.(1982).Lognormal kriging — the general case.*J.Math.Geol.***14**,475-498.
Doveton,J.H.(1971).An application of Markov chain analysis to the Ayrshire coal measures. *Scott.J.Geol.***7**,11-27.
Doveton,J.H.(1976).Multidimensional scaling of sedimentary rock descriptors. In: Merriam (1976c),qv,143-156.
Doveton,J.H. & Cable,H.W.(1979).Fast matrix methods for the lithological interpretation of geophysical logs.In: Gill & Merriam(1979),qv,101-116.
Doveton,J.H. & Parsley,A.J.(1970).Experimental evaluation of trend surface distortions indicated by inadequate data point distributions.*Tr.Inst. Ming.Metall.***B79**,197-207.
Doveton,J.H. & Skipper,K.(1974).Markov chains and substitutability analysis of turbidite successions,Cloridome Formation,Gaspe,Canada.*Can.J.Earth Sci.***11**,472-488.
Downie,N.M. & Heath,R.W.(1974).*Basic statistical methods.*Harper & Row, New York.
Drapeau,G.(1973).Factor analysis — how it copes with complex geological problems.*J.Math.Geol.* **5**,351-364.
Draper,N.R. & Smith,H.(1966).*Applied regression analysis.*Wiley,New York,407pp.
Dubrule,O.(1983).Two methods with different objectives:splines and kriging. *J.Math.Geol.* **15**,245-258.
Duckworth,W.E.(1968).*Statistical techniques in technological research.* Methuen,London,303pp.
Dudley,R.M.,Perkins,P.C. & Gine,M.E.(1975).Statistical tests for preferred orientation.*J.Geol.* **83**,685-705.
Duke,J.M.(1979).Computer simulation of the fractionation of olivine and sulfide from mafic and ultramafic magmas.*Can.Mineral.***17(2)**:507-514.
Dumitriu,C.,Webber,R. & David,M.(1979).Correspondence analysis applied to a comparison of some rhyolitic zones in the Noranda area,Quebec,Canada. *J.Math.Geol.***11**:299-307.
Dumitriu,M.,Dumitriu,C. & David,M.(1980).Typological factor analysis: a new classification method applied to geology. *J.Math.Geol.***12**,69-78.
Duncan,A.C.(1985a).PLANE: an interactive program for calculating intersection lineations from planes,planes from lines and plunges from pitches. *Comput.Geosci.***11**:183-202.
Duncan,A.C.(1985b).DRAFT: an interactive map plotting program for structural geologists.*Comput. Geosci.* **11**:149-182.
Duncan,A.R. & Watkeys,M.K.(1986).Geochemical correlation techniques for quartzo-feldspathic gneisses and their application in the Proterozoic Namaqualand metamorphic complex,S.Africa. *Geocongress '86, Geol.Soc.S.Afr. Ext.Abs.* 777-780.
Duncan,C.B.(1955).Multiple range and multiple *F* tests.*Biometrika* **11**,1-42.
Dunn,T.L.,Hessing,R.B. & Sandkuhl,D.L.(1985).Application of voice recognition in computer-assisted point counting. *J.Sed.Petrol.***55(4)**:602-603.

Duran,B.S. & Odell,P.L.(1974).*Cluster analysis.*Springer-Verlag,New York.

Durand,D. & Greenwood,J.A.(1958).Modification of the Rayleigh test for uniformity of two-dimensional orientation data. *J.Geol.***66**,229-238.

Durbin,J.(1973).*Distribution theory for tests based on the sample distribution function.* Society for Industrial & Applied Mathematics, Philadelphia, PA, 64 pp.

Durbin,J. & Watson,G.S.(1950).Testing for serial autocorrelation in least-squares regression. *Biometrika* **37**,409-428 & **38**,159-178.

Durovic,S.(1959).Contribution to the lognormal distribution of the elements. *Geochim.Cosmochim. Acta* **15**,330-336.

Dybczynski,R.(1980).Comparison of the effectiveness of various procedures for the rejection of outlying results and assigning consensus values in interlaboratory programs involving determination of trace elements or radionuclides.*Anal.Chim.Acta* **117**,53-70.

Ebdon,D.(1977).*Statistics in geography: a practical approach.*Blackwell, Oxford,195pp.

Edgington,E.S.(1961).Probability table for number of runs of signs of first differences in ordered series. *J.Am.Stat.Assoc.***56**,156-159.

Efron,B.(1982).*The jackknife,bootstrap and other resampling plans.*Society for Industrial and Applied Mathematics, Philadelphia,Penn.,92pp.

Ehier,V.G.(1975).Application of Markov analysis to the Banff Formation, Alberta.*J.Math.Geol.* **7**,47-61.

Ehrenberg,A.S.C.(1982).*A primer in data reduction (an introductory statistics textbook).* Wiley,305pp.

Ehrlich,R.Vogel,T.A.,Weinberg,B.,Kamilli,D.C.,Byerly,G. & Richter,H.(1972).Textural variation in petrogenetic analysis. *Bull.Geol. Soc.Am.***83**,665-676.

Ehrlich,R. & Full,W.E.(1987). Sorting out geology — unmixing mixtures. In: Size(1987a),qv,33-46.

Einot,I. & Gabriel,K.R. (1975). A study of the power of several methods of multiple comparisons. *J.Am.Stat.Assoc.* **70**, 574-583.

Ellis,A.J.(1976).The I.A.G.C.interlaboratory water analysis programme. *Geochem.Cosmochim. Acta* **40**,1359-74.

Ellis,P.J.(1981).Confidence limits for the Gastwirth median.*Geostandards Newsletter* **5**,161-166.

Ellis,P.J. & Steele,T.W.(1982).Five robust indicators of central value. *Geostandards Newsletter* **6:2**,207-216.

Ellis,P.J.,Copelowitz,I. & Steele,T.W.(1977).Estimation of the mean by the dominant cluster method.*Geostandards Newsletter* **1**,123-130.

Embey-Isztin,A.(1983).Major element patterns in Hungarian basaltic rocks.An approach to determine their tectonic settings.*Elsevier Developments in Solid Earth Geophysics* (eds.E.Bisztricsany & G.Szeidovitz) **15**:601-609.

Engels,J.C. & Ingamells,C.O.(1970).Effect of sample inhomogeneity in K-Ar dating. *Cheochim.Cosmochim.Acta* **34**, 1007-1017.

Enslein,K.,Ralston,A. & Wilf,H.(1977).*Statistical methods for digital computers.*Wiley,New York.

Erskine,J.C. & Smith,E.L.(1978).The logic that underlies a feasibility study and calculations for mining projects. *J.Aust.Bur.Miner.Resour.Geol.Geophys.***3(3)**:257-258.

Esimai,G.O.(1982).Teaching of statistics to non-majors and preservation of statistics as a mathematical science.In: Rustagi & Wolfe(1982),qv,p.431-442.

Esler,J.E.,Smith,P.F. & Davis,J.C.(1976).KWIKR8,a FORTRAN IV program for multiple regression and geologic trend analysis.*Kansas.Geol.Surv.Computer Contrib.***28**,31pp.

ESRI (Environmental Systems Research Institute, 1985a). *ARC/INFO user's manual.* ESRI, Redlands, California.

ESRI (Environmental Systems Research Institute, 1985b). *ARC/INFO study guide & notes.* ESRI, Redlands, California.

ESRI (Environmental Systems Research Institute, 1985c). *ARC/INFO maps.* ESRI, Redlands, California.

ESRI (Environmental Systems Research Institute, 1986). *ARC/INFO maps.* ESRI, Redlands, California.

Ethier,V.G.(1975).Application of Markov analysis to the Banff formation, Alberta. *J.Math.Geol.* **7**,47-61.

Ethridge,F.G. & Davies,D.K.(1973).Grouped regression analysis — a sedimentological example.*J.Math.Geol.***5**,377-388.

Everitt,B.(1974).*Cluster analysis.*Heinemann,London,122 pp. 2nd Edn. (1980).

Everitt,B.(1978).*Graphical techniques for multivariate data.*North-Holland, New York,117pp.

Everitt,B.(1979).Unresolved problems in cluster analysis.*Biometrics* **35**,169-181.

Eyre,J.M.(1983).*The application of computers to mine surveying,planning and geology in the USA.*The Mining Engineer, Feb.(1983,463-472.

Fabbri,A.G.(1986).Promising aspects of geological image analysis.In: Hanley & Merriam(1986),qv,199-214.

Fabbri,A.G.,Divi,S.R. & Wong,A.S.(1975).A database for mineral potential evaluation in the Appalachian Region of Canada. *Pap.Geol.Surv.Can.* **75-1C**,123-32.

Fabbri,A.G. & Kasvand,T.(1978).Picture processing and geological images. *Pap.Geol.Surv.Can.* **78-1B**,169-174.

Fabregat,G.(1977).[Computer program for the identification of minerals from their diffraction pattern].In Spanish.*Rev.Inst.Geol.Univ.Nac.Mexico* **1(1)**:85-91.

Fairbairn,H.W.(1953).Precision and accuracy of chemical analysis in silicate rocks.*Geochim. Cosmochim.Acta* **4**,143-156.

Farris,J.S.(1969).On the cophenetic correlation coefficient.*Zoology* **18**,279-285.

Fatti,L.P. & Hawkins,D.M.(1976). FORTRAN IV program for canonical variate and principal component analysis.

*Comput.Geosci.***1**, 335-338.
Fears,D.(1985).A corrected CIPW norm program for interactive use.*Comput.Geosci.***11**:787-798.
Feldhausen,P.H.(1970).Ordination of sediments from the Cape Hatteron continental margin. *J.Math.Geol.***2**,113-129.
Feldhausen,P.H. & Ali,S.A.(1976).Sedimentary environment analysis of Long Island Sound,UWA, with multivariate statistics. In: Merriam(1976c),qv,73-98.
Feller,W.(1968).*An introduction to probability theory and its applications.* Vol.I.Wiley,New York.
Fenner,P.(1969). Models of geological processes: an introduction to mathematical geology. *AGI short course lecture notes.* American Geological Institute, Washington, 359 pp.
Fenner,P.(1972).Symposium on quantitative geology.*Geol.Soc.Am.Spec.Rep.***143.**
Ferguson,C.C. & Harvey,P.K.(1980).Applications of Monte Carlo simulation in petrography.*J.Geol. Soc.Lond.* **137**:109-110.
Ferguson,C.C. & Harvey,P.K.(1982).The Monte Carlo method: a brief review and some applications in metamorphic petrology. *Geol.Soc.Lond.Misc.Pap.***14**,262-284.
Ferguson,J.(1973).The Pilanesberg alkaline province,southern Africa.*Tr. Geol.Soc.S.Afr.* **76**:249-270.
Ferguson,J.(1982).The application of information theory to trend surface analysis.*Geol.Soc.Lond. Misc.Pap.***15**,3-44.
Ferguson,J.(1987).*Mathematics in geology.* Allen & Unwin, Boston,Mass., 256 pp.
Ferguson,J. & Currie,K.L.(1972). Geology and petrology of Callender Bay. *Bull.Geol.Surv.Canada* **217.**
Ferguson,J. & Winer,P.(1980).Pine Creek geosyncline: statistical treatment of whole-rock chemical data.In: J.Ferguson & A.B.Goleby (eds.) *Uranium in the Pine Creek Geosyncline.* Int.Atomic Energy Authority,Vienna.
Ferguson,T.S.(1961).Rules for the rejection of outliers.*Rev.Inst.Inst.* Stat.**3**,29-43.
Fidel, R.(1987). *Database design for information retrieval: a conceptual approach.* Wiley, New Yor, 256 pp.
Fields,R.W. & Egemeter,S.J.(1978).Statistical significance of orientation data — a new approach with applications to rose diagrams.*Abstr.Progr.***10(2)**,p.42.
Fieller,E.C.,Hartley,H.O. & Pearson,E.S.(1957).Tests for rank correlation coefficients:1.*Biometrika* **44**,470-81.
Filliben,J.A.(1975).The probability plot correlation coefficient test for normality.*Technometrics* **17**,111-117.
Fisher,N.I.(1983).Graphical methods in nonparametric statistics: a review and annotated bibliography. *Int.Stat.Rev.* **51**,25-58.
Fisher,R.A.(1970).*Statistical methods for research workers.*(14th Ed).Hafner,New York,362pp.
Fisher,N.H.(1975).Multidisciplinary study of earth resources imagery of Australia,Antarctica and Papua New Guinea. *Gov.Rep.Announce Springfield* VA E75-10352,31pp.
Fisher,N.I.,Huntington,J.F.,Jacket,D.R.,Willcox,M.E. & Creasey,J.W.(1985). Spatial analysis of two-dimensional orientation data.*J.Math.Geol.* **17**,177-194.
Fitzgerald,J.D. & Mackinnon,I.D.R.(1977).PETPAK — a computing package for the petrologist. *Comput.Geosci.* **3**,637-638.
Flast,R. & Flast,L.(1986).*Macintosh spreadsheets: using Microsoft multiplan Chart and File.* McGraw-Hill,New York.
Flanagan,F.J.(1976a).G1-W1: requiescant in pace!*U.S.G.S.Prof.Pap.***840**,189-192.
Flanagan,F.J.(1976b).Description and analysis of 8 new U.S.G.S.rock standards. *U.S.G.S. Prof.Pap.***840.**
Fletcher,R.C.(1972).Application of a mathematical model to the emplacement of mantled gneiss domes. *Am.J.Sci.* **272**,197-216.
Flinn,D.(1958).On tests of significance of preferred orientation in three-dimensional fabric diagrams.*J.Geol.***66**,526-539.
Flinn,D.(1959).An application of statistical analysis to petrochemical data. *Geochim.Cosmochim.Acta* **17**,161-175.
Flinter,B.H.(1974).The differentiation index applied to the New England igneous complex,New South Wales,Australia: a preliminary study. *Pacific Geol.* **7**:45-63.
Flury, B.(1988). *Multivariate statistics — a practical approach.* Chapman & Hall, London.
Foley,J.D. & Van Dam,A.(1982).*Fundamental of interactive computer graphics.* Addison-Wesley, Reading,Mass.,684pp.
Forester,R.M.(1977).Abundance coefficients: a new method for measuring microorganism relative abundance. *J.Math.Geol.* **9**:619-633.
Fornell,C.(1982a).*A second generation of multivariate analysis.*Vol.1 *Methods*, 392pp; Vol.2. *Measurement and evaluation*,430pp.Praeger.
Fornell,C.(1982b).Three approaches to canonical correlation analysis.In: Fornell(1982a), qv,Vol.I,pp.36-54.
Forsyth,R.(1984).*Expert systems.*Chapman & Hall,London.
Francis,I.(1974).Factor analysis:fact or fabrication?*Math.Chronicle* **3**,9-44.
Francis,I.(1981).*Statistical software: a comparative review.*North-Holland,542pp.
Frangipane,F.(1977).Analysis of multvariate data: an application to some recent volcanics of the central Andes. *Schweiz.Mineral.Petrogr.Mitt.***57(1)**,115-134.
Frenkel,M.Y.(1979).Computer solution of the heat and mass transport equations for the emplacement of a sill. *Geochem.Int.***15**:162-170.
Frenkel,M.Y. & Arishkin,A.A.(1985).Computer simulation of basalt magma equilibrium and fractional crystallisation. *Geochem.Int.***22**,73-84.
Freund,J.E. & Williams,F.J.(1966).*Dictionary/outline of basic statistics.* McGraw-Hill,New York,195pp.

Freund,M.J.(1986).Cokriging: multivariate analysis in petroleum exploration. *Comput.Geosci.* **12**,485-492.

Fridlund,A.J.(1986).Statistics software.*Infoworld* (Sept.1,'86),31-38.

Friedman,H.(1937).The use of ranks to avoid the assumptions of normality implicit in the analysis of variance. *J.Am.Stat.Assoc.***32**,675-701.

Friedman,J.H. & Rafsky,L.C.(1981).Graphics for the multivariate two-sample problem.**76**,277-291.

Frizado,J.(1985).A microcomputer-based X-ray diffractogram simulation program.*J.Geol.Educ.* **33**,277-280.

Froidevaux,R.(1982).Geostatistics and ore reserve estimation.*Can.Inst.Ming.Metall.Bull.* July 1982,77-83.

Froidevaux,R.,Jaquet,J.M. & Thomas,R.L.(1977).AGCL,a FORTRAN IV program for agglomerative,non-hierarchical Q-mode classification of large data-sets. *Comput.Geosci.***3**,31-48.

Frost,M.J.(1977a).Two computer programs for teaching igneous petrology.*J.Geol. Educ.* **25(5)**:148-149.

Frost,M.J.(1977b).A new interactive computer program to process electron microprobe data.*Mineral. Mag.***41**:414-416.

Fryer,H.C.(1966).*Concepts and methods of experimental statistics*.Allyn & Bacon, New Jersey.

Fukunaga,K.(1972).*Introduction to statistical pattern analysis*.Academic Press, New York.

Full,W.E.,Ehrlich,R. & Klovan,J.E.(1981).EXTENDED Q-MODEL — objective definition of external end-members in the analysis of mixtures.*J.Math.Geol.* **13**,331-344.

Full,W.E.,Ehrlich,R. & Beldek,J.C.(1982).FUZZY QMODEL: a new approach for linear unmixing. *J.Math.Geol.***14**, 259-270.

Furnival,G.M.(1971).All possible regressions with less computation.*Applied Stat.***14**,196-200.

Fuster,J.M.(1975).Las Islas Canarias: un ejemplo de evolución espacial y temporal del vulcanismo oceánico. *Estud.Geol.* **31**:439-463.

Gabriel,K.P. & Sen,P.K.(1968).Simultaneous test procedures for one-way ANOVA and MANOVA based on rank scores.*Sankya,Ser.A*:303-312.

Gaile,G.L. & Burt,J.E.(1980).*Directional statistics. concepts and techniques in modern geography No.25*,Univ.East Anglia,Norwich,39pp.

Gani,J.(1982).Consulting and research in the CSIRO division of mathematics and statistics.In: Rustagi & Wolfe(1982),qv,p.215-232.

Garrett,R.G.(1969).The determination of sampling and analytical errors in exploration geochemistry. *Econ.Geol.* **64**,568-71 & **68**,282-3.

Garrett,R.G.(1983).Opportunities for the 80s.*J.Math.Geol.***15**,385-398.

Garrett,R.G.,Kane,V.E. & Zeigler,R.K.(1980).The management and analysis of regional geochemical data. *J.Geochem.Explor.***13**,115-152.

Gast,P.W.(1968).Trace element fractionation and the origin of tholeiitic and alkaline magma-types. *Geochim.Cosmochim. Acta* **32**,1057-1086.

Gastwirth,J.(1966).On robust procedures.*J.Am.Stat.Assoc.***61**,929-948.

Gateau,C.(1978).Analyse quantitative des images: applications au minéralogie. *Bull.Minéral.***101** **(2)**:305-314.

Gates,C.E. & Ethridge,F.G.(1973).A generalized set of discrete frequency distributions with FORTRAN program. *J.Math.Geol.***5**,1-24.

Geary,R.C.(1935).The ratio of the mean deviation to the standard deviation as a test of normality. *Biometrika* **27**,310-332.

Geary,R.C.(1936).Moments of the ratio of the mean deviation to the standard deviation for normal samples.*Biometrika* **28**,295-307.

Geary,R.C. & Leser,C.E.V.(1968).Significance tests in multiple regression. *The American Statistician* **22**,20-21.

GEC (General Electric Company, 1962). *Tables of the individual and cumulative terms of the Poisson distribution.* Van Nostrand, Princeton, 202 pp.

Geoscience Information Society (1987). Microcomputers, Minis and Geoscience Information. *Proc.Geosci.Inf.Soc.* **16**, 176 pp.

Gerasimovsky,V.I.,Roschina,I.A. & Shevaleyevskiy,I.D.(1981).Chemical composition of basalts in the Baykal rift zone. *Geochem.Int.***17**,1-10.

Gere,J.M. & Weaver,W.(1982).*Matrix algebra for engineers*.(2nd Ed).Brooks-Cole,Monterey,175pp.

Gerig,T.M.(1969).A multivariate extension of Friedman's test.*J.Am.Stat.Assoc.* **64:**1595-1608.

Getis,A. & Boots,B.(1978).*Models of spatial processes, an approach to the study of point, line and area patterns.* Cambridge Univ. Press, Cambridge, 198 pp.

Ghiorso,M.S.(1983).LSEQIEQ: a FORTRAN IV subroutine package for the analysis of multivariate linear regression problems with possibly deficient pseudorank and linear equality and inequality constraints.*Comput.Geosci.***9**,391-416.

Ghiorso,M.S. & Carmichael,I.S.E.(1981).A FORTRAN IV computer program for evaluating temperatures and oxygen fugacities from the composition of coexisting iron-titanium oxides.*Comput.Geosci.***7**,123-129.

Ghosh,A.K.(1965).A statistical approach to the exploration of Cu in the Singhbhum shear zone,Bihar.*Econ.Geol.* **60,**1422-30.

Ghosh,J.K.,Saha,M.R. & Sen Gupta,S.(1981).Gondwana stratigraphic classification by statistical method.In: Merriam(1976c),qv,47-62.

Ghosh,S.B. & Rajan,T.N. (1986). Integrated geoscience information system for India: a perspective. *Proc.3rd Int.Conf.Geosci.Info. (Adelaide,S.Australia, June '86)*, Vol.1, 112-127.
Gibbons,J.D.(1971).*Nonparametric statistical inference*.McGraw-Hill,Tokyo.
Gilirab,D.J. & Churchman,G.J.(1977).The use of principal components analysis for the detection of equilibria and the evaluation of equilibrium constants. *Geochim.Cosmochim.Acta* **41**:387-392.
Gill,D. & Merriam,D.F.(1979).*Geomathematical and petrophysical studies in sedimentology*. Pergamon (Computers & Geology series,vol.3),Oxford,267pp.
Gill,D. & Tipper,J.C.(1978).The adequacy of non-metric data in geology: tests using a divisive omnithetic clustering technique*J.Geol.* **86**:241-260.
Gill,D.,Beylin,J.,Boehm,S.,Frendel,Y. & Rosenthal,E.(1977).Design of geological data systems for developing nations. *J.Math.Geol.***9**,145-158.
Gill,D.,Boehm,S. & Erez,Y.(1976).ASSOCA: FORTRAN IV program for Williams and Lambert association analysis with printed dendrograms.*Comput.Geosci.* **2**,219-248.
Gill,E.M.(1975).Feasibility studies for a petrological databank*Rept. Inst.Geol.Sci.(UK)*,**75/3.**
Gilliland,J.A. & Grove,G.(1973).Some principles of data storage and information retrieval and their implications for information exchange*J.Math.Geol.***5**,1-10.
Giri,N.C.(1977).*Multivariate statistical inference*.Academic Press,New York.
Gittins,R.(1979). Ecological appolications of canonical analysis. In: *Multivariate methods in ecological work*, L. Orlóci et al.(eds.), 309-535. International Cooperative publishing house, Fairlna,d Maryland, USA.
Glazner,A.F.(1984).A short CIPW norm program.*Comput.Geosci.***10**,449-50.
Glazner,A.F. & McIntyre,D.B.(1979).Computer-aided X-ray diffraction identification of minerals in mixtures. *Am.Mineral.***64**:902-905.
Gleadow,A.J.W.,Le Maitre,R.W.,Sewell,D.K.B. & Lovering,J.F.(1974).Chemical discrimination of petrographically defined cluster groups in Apollo 14 and 15 lunar breccias.*Chem.Geol.***14**,39-61.
Gnanadesikan,R.(1977)*Methods for statistical data analysis of multivariate observations*.Wiley, 311pp.
Gnanadesikan,R. & Kettering,J.R.(1972).Robust estimates,residuals,and outlier detection with multiresponse data. *Biometrics* **28**,81-124.
Godinho,M.M.(1976a).Programas FORTRAN IV pelo analise de sequências geológicas. *Mem.Not.Publ. Mus.Mineral. Geol. Univ.Coimbra* **81:**29-50.
Godinho,M.M.(1976b).Uma tentativa a uma classificação química e mineralógica dos granitoides da região de Guardão. *Mem.Not.Publ.Mus.Mineral.Geol.Univ. Coimbra* 81:1-27.
Goldfarb,R.J.,Folger,P.F.,Smaglik,S.M. & Tripp,R.B.(1984).A statistical interpretation of geochemical data from Chugach National Forest.*U.S.G.S. Circ.***939**,47-50.
Golding,S.D.,Groves,D.I.,McNaughton,N.J.,Barley,M.E. & Rock,N.M.S.(1987). Carbon isotopic composition of carbonates from contrasting alteration styles in supracrustal rocks of the Norseman-Wiluna Belt, Yilgarn Block, Western Australia: their significance fo the source of Archaean auriferous fluids. *Univ.W.Aust.Geology Dept. & Extension Publ.* **11**, 215-238. See also *Nature* **331**, 254-257.
Goodman,L.A. & Kruskal,W.H.(1954).Measures of association for cross-classification.*J.Am.Stat. Assoc.***99**,732-764.
Gordon,A.D.(1980).SLOTSEQ: a FORTRAN IV program for comparing two sequences of data. *Comput.Geosci.***6**,7-20.
Gordon,A.D.(1981).*Classification: methods for the exploratory analysis of multivariate data*.Chapman & Hall,193pp.,
Gordon,A.D. & Reyment,R.A.(1979).Slotting of borehole sequences*J.Math. Geol.***11**,309-327.
Gorsuch,R.L.(1974).*Factor analysis*.Saunders,Philadelphia.
Goubin,N.(1978).Some examples of management and processing of geological and geochemical data. *Comput. Geosci.***4**,37-52.
Govett,G.I.S.(1972).Interpretation of rock geochemical exploration survey in Cyprus — statistical and graphical techniques. *J.Geochem.Explor.* **1**,77-102.
Govett,G.I.S.,Goodfellow,W.D.,Chapman,R.P. & Chork,C.Y.(1975).Exploration geochemistry: distribution of elements and recognition of anomalies. *J.Math.Geol.***7**,415-446.
Gower,J.C.(1967).A comparison of some methods of cluster analysis.*Biometrics* **23**,623-637.
Gower,J.C.(1970).A note on Burnaby's character-weighted similarity coefficient. *J.Math.Geol.* **2**,39-46.
Gower,J.C.(1971).A general coefficient of similarity and some of its properties. *Biometrics* **27**, 857-872.
Gower,J.C.(1978).Growth-free canonical variates and generalized inverses. *Bull.Geol.Inst.Univ. Uppsala New Ser.***7**:1-10.
Gradstein,F.M.,Agterberg,F.P.,Brower,J.C. & Schwarzacher,W.(1985). *Quantitative stratigraphy*. Kluwer Academic Press,Hingham,Mass., 632pp.
Graffenreid,J.A.de & Cable,H.W.(1978).A computer aid to the distribution of geologic publications. *Proc.Geosci.Inf.Soc.* **8**:57-65.
Grandclaude,P.(1976).Design and use of a geochemical data bank.*Comput. Geosci.***2**,163-170.
Grandclaude,P.(1978).About some categories of problems and standards in communication and processing of computerised geochemical and added data. *Sci.Terre Inf.Geol.(Nancy)* **11**:11-12.
Grandclaude,P. & Stussy,J.M.(1978).[Problems and standards in exchange and processing of geochemical data.

Comparative geochemistry of two-mica granites from various orogens].*Sci.Terre Inf.Geol.(Nancy)* **11**,94pp.
Grandclaude,P.,Marchal,M. & De la Roche,H.(1976).[The geochemical card indexes of the petrographical and geochemical research centre (CRPG): their content and the means of their utilisation as a data bank].*I ndust.Mineral.(St.Etienne)* **58(10)**:457-464.
Gray,H.L. & Schucany,W.R.(1972).*The generalized jackknife statistic.* Marcel Dekker,New York, 308 pp.
Gray,N.H.(1973).Estimation of parameters in petrologic materials balance equations.*J.Math. Geol.***5**,225-236.
Graybill,F.A.(1961).*An introduction to linear statistical models.*McGraw-Hill, New York,463pp.
Green, N.P., Finch, S. & Wiggins, J.C. (1985). The state of the art in Geographic Information Systems. *Area* **17**, 295-301.
Green,W.R.(1985).*Computer-aided data analysis: a practical guide.*Wiley,268pp.
Greenfield,A.A. & Siday,S. (1980). Statistical computation for business and industry. *The Statistician* **29**, 33-55.
Greenland,L.P.(1970).An equation governing trace element distribution during magmatic crystallisation. *Am.Mineral.***55**,455-465.
Greenwood,B.(1969).Sediment parameters and environmental discrimination: an application of multivariate statistics. *Can.J.Earth Sci.***6**,1347-1358.
Griffin,M.E.,Mutschler,F.E. & Stevens,D.S.(1985).ALKY — a databank of chemical analyses for alkaline and related igneous rocks from western North America. *U.S.G.S.Open File report* **85-XX**,4329pp.
Griffis,R.A.,Gustafson,S.K. & Adams,H.G.(1985).PETFAB: user-considerate FORTRAN-77 program for the generation and statistical evaluation of fabric diagrams.*Comput.Geosci.***11**:369-408.
Griffiths,J.C.(1960).Frequency distributions in accessory mineral analysis. *J.Geol.***68**,353-365.
Griffiths,J.C.(1962).Frequency distributions of some natural resource materials. *Penn. StateUniv.Mineral Indust. Experiment Station Circ.***63**, 174-198.
Griffiths,J.C.(1966a).Exploration for natural resources. *J.Am.Oper.Res.Soc.***14(2)**, 189-209.
Griffiths,J.C.(1966b).Application of discriminant functions as a classification tool in the geosciences. *Comput.Contrib. Kansas Geol.Surv.***7**,48-52.
Griffiths,J.C.(1967).*Scientific method in the analysis of sediments.* McGraw-Hill,New York,508pp.
Griffiths,J.C.(1970).Current trends in geomathematics.*Earth-Sci.Rev.***6**,121-140.
Griffiths,J.C.(1971).Problems of sampling in geoscience.*Tr.Inst.Ming.Metall.* **80**,B346-356.
Griffiths,J.C.(1974).Quantification and the future of geoscience.In: Merriam(1974),qv,83-102.
Griffiths,J.C.(1978).Some alternate exploration strategies.In: Merriam (1978b),qv,23-36.
Griffiths,J.C.(1981).Systems behaviour and geoscience problem-solving. In: Merriam(1981b), qv,1-13.
Griffiths,J.C.(1983).Geologic similarity by Q analysis.*J.Math.Geol.* **15**,85-108.
Griffiths,P. & Hill,I.D.(1985).*Applied statistics algorithms.*Ellis Horwood, London,307pp.
Grimm,E.C.(1987).CONISS: a FORTRAN 77 program for stratigraphically constrained cluster analysis by the method of incremental sum of squares. *Comput.Geosci.***13**,13-36.
Gross, A.M.(1976). Confidence interval robustness with long-tailed symmetric distributionas. *J.Am.Stat.Assoc.* **71**, 409-416.
Grout,J.C.(1983).*Fundamental computer programming using FORTRAN 77.* Prentice-Hall,New Jersey,384pp.
Groves,D.I. & Phillips,G.N.(1987). The genesis and tectonic control on Archaean gold deposits of the Western Australian shield — a metamorphic replacement model.*Ore-Geol.Rev.***2**, 287-322.
Grubbs,F.E.(1950).Sample criteria for testing outlying observations. *Ann.Math.Stat.***21**,27-58.
Grubbs,F.E.(1969).Procedures for detecting outlying observations in samples. *Technometrics* **11**,1-21.
Grubbs,F.E. & Beck,G.(1972).Extension of sample sizes and percentage points for significance tests of outlying observations. *Technometrics* **14**,847-854.
Gruza,V.V.(1966).Differentiation criteria for nepheline rocks in the Altai-Sayan region.*Dokl.Acad. Sci.USSR* **167**,129-131.
Guarascio,M.,David,M. & Huijbregts,C.(1976).*Advanced geostatistics in the mining industry.*Reidel, Dordrecht,491pp.
Gubac,J. (1986). On the characteristics of the distribution of chemical elements in Nature. *J.Math.Geol.* **18**, 429-432.
Guenther,W.C.(1965).*Fundamentals of statistical inference.*McGraw-Hill,New York,353pp.
Guenther,W.C.(1966).*Analysis of variance.*Prentice-Hall,Englewood Cliffs,199pp.
Gumbel,E.J.(1958).*Statistics of extremes.*Columbia Univ.Press,New York,375pp.
Gundu Rao,C.(1980).Numerical classification by association analysis of multivariate mineralogic, petrographic, presence-absence data from some sub-Himalyan carbonates,northwest India.*J.Math.Geol.***12**,607-614.
Gy,P.M.(1979).*Sampling of particulate materials:theory and practice.*Elsevier Developments in geomathematics **4**,431pp.
Haan,C.T.(1977).*Statistical methods in hydrology.*Iowa State Univ.Press,378pp.
Hage,G.L.(1983).KRS: a fast,special-purpose database system.*Comput.Geosci.* **9**,41-52.
Haggett,P., Cliff,A.D. & Frey,A.(1977). *Locational analysis in human geography.* 2nd Ed.Wiley,New York, 605 pp.
Haight,F.A.(1967). *Handbook of the Poisson distribution.* Wiley, New York, 168 pp.
Hails,J.R.(1967).Significance of statistical parameters for distinguishing sedimentary environments. *J.Sed.Petrol.* **37**,1059-1069.

Haimes,R. & Dowsett,F.R.(1976).COOLIT: a FORTRAN IV program that simulates fractional crystallisation in the formation of layered intrusions. *Comput.Geosci.***2**,377-406.

Hall,A.(1973).The median surface: a new type of trend surface.*Geol.Mag.***100**, 467-472.

Hall,A.(1983).The application of non-parametric statistical methods to studies of trace element distribution in igneous rocks. In:Augustithis(1983), qv,161-174.

Hamersley,J.M. & Handscomb,D.C.(1964).*Monte Carlo methods.*Methuen,London,178pp.

Hampel,F.R.,Rousseeuw,P.J.,Ronchetti,F.M. & Stahel,W.A.(1986).*Robust statistics.* Wiley, New York, 502 pp.

Hand,D.J.(1981).*Discrimination and classification.*Wiley,New York.

Hanley,J.T. & Merriam,D.F.(1986).*Microcomputer applications in geology.* Pergamon, 258 pp.

Hannan, E.J.(1976). *Time-series analysis.* Methuen, London, 152 pp.

Harbaugh,J.W.(1981).Regional mineral and fuel resources forecasting — a major challenge and opportunity for mathematical geologists. In: Merriam (1981b),qv,169-178.

Harbaugh,J.W. & Bonham-Carter,G.F.(1970).*Computer simulation in geology.* Wiley,New York,575pp.

Harbaugh,J.W. & Demirmen,F.(1964).Application of factor analysis to petrological variation in Americus Limestone (Lower Permian),Kansas and Oklahoma. *Kansas State Geol.Surv.Lawrence Spec.Dist.Publ.***15.**

Harbaugh,J.W. & Merriam,D.F.(1968).*Computer applications in stratigraphic analysis.*Wiley,New York,282pp.

Hardyck,C.D. & Petrinovich,L.F.(1969).*Introduction to statistics for the behavioural sciences.* Saunders,Philadelphia.

Harman,H.H.(1967).*Modern factor analysis.*(2nd Ed).Univ.Chicago Press.

Harmon,R.S. & Shaw,D.M.(1975).Factor analysis of chondrite and achondrite elemental abundances.*Meteoritics* **10(4)**:412-413.

Harrell,J.B.(1987).The analysis of bivariate association. In: Size(1987a),qv,142-166.

Harris,D.P. (1984). *Mineral resources appraisal.* Clarendon Press, Oxford, 445 pp.

Harris,R.J.(1975).*A primer of multivariate statistics.*Academic Press,New York.

Harrison,R.K. & Sabine,P.A. (1970). A petrological-mineralogical code for computer use. *Rep.Inst.Geol.Sci.* **70/6.**

Harter,H.L.(1980).Modified asymptotic formulas for critical values of the Kolmogorov test statistic.*The American Statistician* **34**:1-11.

Harter,H.L. & Owen,D.B.(1970).*Selected tables in mathematical statistics.*Vol.1. Markham,Chicago.

Hartigan,J.(1974).*Clustering algorithms.*Wiley, New York,351pp.

Hartnell,T.(1986).*Desktop publishing: the book.* Interface, London, 160 pp.

Harvey,P.K.(1974).The detection and correction of outlying determinations that may occur in geochemical analysis.*Geochim.Cosmochim.Acta* **38**,435-451.

Harvey,P.K. & Ferguson,C.C.(1976).On testing orientation data for goodness-of-fit to a Von Mises distribution. *Comput.Geosci.***2**,261-268.

Harvey,P.K. & Ferguson,C.C.(1978).A computer simulation approach to textural interpretation in crystalline rocks.In: Merriam(1978a),qv,201-232.

Hastings,G.P.(1986).*Computer-aided design on the Macintosh.* Prentice-Hall,Englewood Cliffs,New Jersey.

Hastings,N.A.J. & Peacock,J.B.(1975).*Statistical distributions.* Butterworth,London.

Hatherton,T. & Dickson,W.R.(1969).The relationship between andesitic volcanism and seismicity in Indonesia, the Lesser Antilles and other island arcs. *J.Geophys.Res.* **74**, 5301-5310.

Hattori,I.(1973).Mathematical analysis to discriminate two types of sandstone-shale alternation. *Sedimentology* **20**,331-345.

Hattori,I.(1976).Entropy in Markov chains and discrimination of cyclic patterns.*J.Math.Geol.* **8**,477-497.

Hattori,I. & Mizutani,S.(1971).Computer simulation of fracturing of layered rocks.*Eng.Geol.* **5**,253-269.

Hawkes,D.D.(1985).INTAL — an expert system for the identification of igneous rocks in the hand specimen. *Geol.J.***20**,367-76.

Hawkins,D.B.(1974).Statistical analysis of the zeolites clinoptilite and heulandite.*Contrib.Mineral. Petrol.***45**,27-36.

Hawkins,D.M.(1981a).*Identification of outliers.*Chapman & Hall,London.

Hawkins,D.M.(1981b).Robust statistics in the geosciences.In: Merriam(1981d), qv,29-38.

Hawkins,D.M. & Cressie,N.(1984).Robust kriging — a proposal.*J.Math. Geol.***16**,3-18.

Hawkins,D.M. & Kew,D.S.(1974).Use of multivariate statistical methods in data-base maintenance. *J.Math.Geol.* **6**,47-58.

Hawkins,D.M. & Rasmussen,S.E.(1973).Use of discriminant analysis for classification of strata in sedimentary successions.*J.Math.Geol.* **5**,163-177.

Hay,W.W. & Southam,J.R.(1978).Binary similarity coefficients and confidence intervals.*Bull.Am. Assoc.Pet.Geol.* **63(2)**,p.520.

Hayami,I.(1971).Discontinuous variation in an evolutionary species.*Cryptopecten vesiculosus* from Japan. *J.Palaeontol.* **47**,401-20.

Hayes,W.R. & Allard,G.O.(1981).A mineral-exploration exercise based on an interactive computer-simulated drilling program. *J.Geol.Educ.***29**,247-250.

Hayes-Roth,F.,Waterman,D.A. & Lenat,D.B.(1983).*Building expert systems.* Addison-Wesley, London,444pp.

Hazel,J.E.(1970).Binary coefficients and clustering in biostratigraphy. *Bull.Geol.Soc.Am.***81,** 3237-3252.
Hazen,S.W. (1967). Some statistical techniques for analyzing mines and mineral deposits sample and assay data. *Bull. US Bureau Mines* **621** 223pp.
Heikkila,M. & Koistinen,E.(1984).Kriging made easy,or assessing the linear and unbiased minimum variance estimate by adapting the projection theorem.*Mineral.Deposita* **19**,2-6.
Hein,P.(1977).La géochimie des nodules du Pacifique nord-est: étude statistique. *Rapp.Sci.Tech.Cent.Natl.Exploit. Ocean (Paris)* **35**,74pp.
Hellewell,E.G. & Myers,J.O.(1973).Measurement and analysis of in situ rock densities in British Carboniferous and Permo-Triassic.*Tr.Inst.Ming.Metall.***82,**B51-60.
Helms,H.L.(1980).*Computer language reference guide*.Sams & Co.,Indianapolis.
Henderson,W.G.(1986).Implementation of geological knowledge in a relational database: another task for the earth science information centre? *Proc.3rd Int.Geosci.Info.Conf.,Adelaide*,Vol.1,38-51.
Henkes,L. & Roettger,B.(1980).MODAL: a program to calculate compositions of ternary systems within the Qz-Ab-Or-An tetrahedron from modal analysis data. *Comput.Geosci.* **6(1)**:69-85.
Henley,S.(1976a).Catastrophe theory models in geology.*J.Math.Geol.* **8**,649-656.
Henley,S.(1976b).An R-mode nonlinear mapping technique.*Comput.Geosci.* **1**,247-254.
Henley,S.(1976c).The identification of discontinuities from areally distributed data.In: Merriam (1976c),qv,157-168.
Henley,S.(1977).Communication of geological information among different machines.*Comput. Geosci.***3(3)**:465-468.
Henley,S.(1981).*Nonparametric geostatistics*.Applied Science,London.
Henley,S.(1987).Kriging: blue or pink? *J.Math.Geol.***19**,55-58.
Hennebert,M. & Lees,A.(1985).Optimized similarity matrices applied to the study of carbonate rocks. *Geol.J.***20**,123-132.
Herbst,N.F.(1977).Computerised literature searches in Australia.*Met.Australas (Parkville)* **9(2)**:33+1.
Hertogen,J. & Gijbels,R.(1976).Calculation of trace element fractionation during partial melting. *Geochim.Cosmochim. Acta* **40**,313-322.
Hexagon Software (Australia) Ltd. (1987). Mining data interchange format: draft specification V1.10, October 1987.
Hey,M.H.(1977).Regression lines in mineralogy are usually bunkum.*Mineral. Mag*.**37**:p.5.
Hickman,A.H. & Wright,A.E.(1983).Geochemistry and chemostratigraphical correlation of slates, marbles and quartzites of the Appin Group, Argyll, Scotland. *Tr.R.Soc.Edinburgh: Earth-Sci.* **73**, 251-278.
Hicks,C.R.(1973).*Fundamental concepts in the design of experiments*.(2nd Ed). Holt,Rinehart & Winston,New York, 293pp.
Hill,M.(1974). Correspondence analysis — a neglected multivariate method. *J.R.Stat.Soc.* **C23**, 340-354.
Hill,M.O.(1979). TWINSPAN — a FORTRAN program for arranging multivariate data in an ordered 2-way table by classification of the individuals and attributes. *Ecol.Systematics Section, Cornell Univ, New York*, 90pp.
Hitchon,B.,Billings,K.G. & Kolvan,J.E.(1971).Geochemistry and origin of formation waters in the western Canada sedimentary basin.III.Factors controlling chemical composition.*Geochim.Cosmochim.Acta*.**35**,567-598.
Ho,S.E. & Groves,D.I.(1987).Recent advances in understanding Archaean gold deposits. *Univ.W.Aust.Geology Dept. & Univ.Extension Publ.* **11**, 368 pp
Hoaglin,D.C.,Mosteller,F. & Tukey,J.W.(1983).*Understanding robust and exploratory data analysis*.Wiley,New York, 447 pp.
Hoaglin,D.C.,Iglewicz,B. & Tukey,J.W.(1986).Performance of some resistant rules for outlier labelling. *J.Am.Stat. Assoc.***81**, 991-999.
Hochberg,A.C.(1987). *Multiple comparison procedures*. Wiley, New York, 368 pp.
Hodell,D.A. & Estep,K.W.(1985).Using graphics to illustrate geology. *Geotimes October* 1985,10-13.
Hoel,P.G.(1962).*Introduction to mathematical statistics*.Wiley,New York.
Hoerl,A. & Kennard,R.W.(1980).Ridge regression — advances,algorithms and applications.*Am. J.Math.Management Sci.* **1**,5-83.
Hoff,D.(1973).*How to lie with statistics*.Penguin,Hamondsworth,124p. Videocassettes: Pt 1.*The gee-whiz graph*. Pt.2.*The average chap.* (Video Arts,London)
Hoffman,P.(1986).*Macintosh paperwork: integrating Microsoft products.* Osborne/McGraw-Hill, Berkeley,CA.
Hogg,R.V.(1974).Adaptive robust procedures: a partial review and some suggestions for future applications and theory. *J.Am.Stat.Assoc.***69**,909-923.
Hogg,R.V.(1977).Robustness.Communications in Statistics-Theory and Methods. *Special Issue* **A6**,789-894.
Hogg,R.V.(1979). Statistical robustness: one view of its use in applications today. *Am.Statistician* **33(3)**, 108-115.
Hogg,R.V. & Randles,R.H.(1975).Adaptive distribution-free regression methods and their applications.*Technometrics* **17**,399-407.
Hohn,M.E.(1976).Binary coefficients: a theoretical and empirical study. *J.Math.Geol.* **8**,137-150.
Hohn,M.E.(1978).Stratigraphic correlation by principal components: effects of missing data.*J.Geol.* **86**,524-532.
Hohn,M.E.(1979).Principal components analysis of three-way tables.*J. Math.Geol.***11(6)**:611-626.
Hohn,M.E.(1985).SAS program for quantitative stratigraphic correlation by principal components. *Comput.Geosci.* **11**:471-478.

Hohn,M.E. & Friberg,L.M.(1979).A generalized principal components model in petrology.*Lithos* **12(4)**:317-324.

Hohn,M.E. & Nuhfer,E.R.(1980).Asymmetric measures of association,closed data and multivariate analysis. *J.Math.Geol.* **12**,235-246.

Holland,J.G. & Winchester,J.A.(1983).The use of geochemistry in solving problems in highly deformed metamorphic complexes. In: Augustithis(1983),qv,389-405.

Hollander,M. & Wolfe,D.A.(1973).*Nonparametric statistical methods*.Wiley, New York.

Holm,M.E. & Neal,D.W.(1986).Geostatistical analysis of gas potential in Devonian shales of West Virginia. *Comput.Geosci*.**12**,611-618.

Hope,K.(1968).*Methods of multivariate analysis*.University Press,London.

Horder,M.F.(1981).The use of databanks and databases within the Institute of Geological Sciences. *J.Geol.Soc.Lond.* **138**,575-582.

Horder,M.F.(1982).Computer applications in geology III. *Geol.Soc. Lond.Misc.Pap.***15**.

Horder,M.F. & Howarth,R.J.(1982).Computer applications in geology I & II. *Geol.Soc.Lond. Misc.Pap.***14**,284pp.

Hore,M.K. & Harpavat,C.L.(1975).Application of statistical analysis in exploration for copper in Satkui,Khetri copper belt,Jhunjhunu district, Rajasthan.*Indian Mineral.* **29(3)**:25-34.

Horowitz,D.H.(1976).Mathematical modelling of sediment accumulation in prograding deltaic systems.In: Merriam(1976c),qv,105-120.

Horton,I.F.,Russell,J.S. & Moore,A.W.(1968).Multivariate-covariance and canonical analysis: a method for selecting the most effective discrimination in a multivariate situation.*Biometrics* **24**,845-58.

Horton,R.L.(1978). *The general linear model.* McGraw-Hill, New York, 274 pp.

Hotelling,H.(1936).Relationships between two sets of variables.*Biometrika* **28**,321-377.

Hotelling,H. & Pabst,M.R.(1936).Rank correlation and tests of significance involving no assumption of normality. *Ann.Math.Stat.***7**:29-43.

Howard,J.C.(1971).Computer simulation models of salt domes.*Bull.Am.Assoc.Pet.Geol.* **55**,495-513.

Howarth,R.J.(1967).Trend surface fitting to random data — an experimental test. *Am.J.Sci.* **265**,619-625.

Howarth,R.J.(1970).Principal components analysis of the geochemistry and mineralogy of the Portaskaig Tillite and Kiltyfannad Schist (Dalradian) of Co.Donegal,Eire.*J.Math.Geol.* **2**,285-302.

Howarth,R.J.(1971a).An empirical discriminant method applied to sedimentary rock classification from major element geochemistry.*J.Math.Geol.* **3**,51-60.

Howarth,R.J.(1971b).FORTRAN IV program for grey-level mapping of spatial data. *J.Math.Geol.* **3**,95-122.

Howarth,R.J.(1971c).Empirical discriminant classification of regional stream-sediment geochemistry in Devon and east Cornwall.*Tr.Inst.Ming.Metall.***80**,B142-149.

Howarth,R.J.(1973a).FORTRAN IV program for empirical discriminant classification of spatial data. *Geocom.Bull.* **6**,1-31.

Howarth,R.J.(1973b).Preliminary assessment of a nonlinear mapping algorithm in a geological context. *J.Math.Geol.* **5**,39-58.

Howarth,R.J.(1982).The Box-Cox lambda transform as a geological tool. *Geol.Soc.Lond. Misc.Pap.***15**,88.

Howarth,R.J.(1983a).*Statistics and data-analysis in geochemical prospecting*. Elsevier,Amsterdam,438pp.

Howarth,R.J.(1983b).Graphical aids for some statistical tests.In: Howarth(1983a), qv,393-401.

Howarth,R.J.(1984).Statistical applications in geochemical prospecting: a survey of recent developments. *J.Geochem.Explor.***21**,41-51.

Howarth,R.J. & Earle,S.A.M.(1979).Application of a generalised power transformation to geochemical data. *J.Math.Geol.* **11**,45-62.

Howarth,R.J. & Garrett,R.G.(1986).The role of computing in applied geochemistry.In: Thornton & Howarth(1986), qv,163-184.

Howarth,R.J. & Leake,B.E.(1980).The role of data processing in the geological sciences. *Sci.Progr.Oxford* **66**,295-329.

Howarth,R.J. & Lowenstein,P.L.(1971).Sampling variability of stream sediments in broad-scale regional geochemical reconnaissance. *Tr.Inst.Ming.Metall.***B**,363-372.

Howarth,R.J. & Martin,L.(1979).Computer-based techniques in the compilation, mapping and interpretation of exploration geochemical data.*Econ.Geol.Rep. Geol.Surv.Can.***31**:545-574.

Howarth,R.J. & Turner,M.St.J.(1987).Statistical graphics in geochemical journals. *J.Math.Geol.***19**,1-24.

Howell,J.A.(1983).A FORTRAN-77 program for computing percentiles of large data-sets. *Comput.Geosci.* **9**,281-296.

Hruska,J.(1976).Current data-management systems:problems of application in economic geology. *Comput.Geosci.* **2**:299-304.

Hruska,J. & Burk,C.F.(1971).Computer-based storage and retrieval of geoscience information: bibliography 1946-1969. *Geol.Surv.Can.Pap.***71-40**,52pp.

Hsu,Y-S.,Walker,J.J. & Ogren,D.E.(1986).A stepwise method for determining the number of component distributions in a mixture.*J.Math.Geol.***18**,153-161.

Hubaux,A.(1969).Archival files of geological data.*J.Math.Geol.***1**,41-52.

Hubaux,A.(1970).Description of geological objects.*J.Math.Geol.***2**,89-95.

Hubaux,A.(1971).Scheme for a quick description of rocks*J.Math.Geol.* **3**,317-322.
Hubaux,A.(1973).A new geological tool — the data.*Earth-Sci.Rev.***9,**139-196.
Hubbert,M.K.(1974).Is being quantitative sufficient? In: Merriam(1974),qv, 27-50.
Huber,P.J.(1972).Robust statistics:a review*Ann.Math.Stat.***43**,1041-1067.
Huber,P.J.(1973).Robust regression: asymptotics,conjectures and Monte Carlo. *Ann.Stat.***1**,799-821.
Huber,P.J.(1977).*Robust statistical procedures.* Society for Industrial & Applied Mathematics, Philadelphia, PA, 56 pp.
Huber,P.J.(1981).*Robust statistics.*Wiley,New York.
Hudson,T.,Askevold,G. & Plafker,G.(1975).A computer-assisted procedure for information processing of geologic field data.*J.Res.U.S.G.S.***3:3**,369-75.
Huitson,A.(1966).*Analysis of variance.*Griffin,London,83pp.
Hull,C.H. & Nie,N.H.(1981).*SPSS update 7-9.*McGraw-Hill,New York.
Huntsberger,D.V.(1961).*Elements of statistical inference.*Allyn & Bacon,Boston.
Hutchison,C.S. (1975). Correlation of Indonesian active volcano geochemistry with Benioff zone depth. *Geol.Mijnbouw* **54**, 157-168.
Hutchison,R.I.,Skinner,D.L. & Bowes,D.R.(1976).Discriminant trace element analysis of strata in the Witwatersrand system. *J.Math.Geol.* **8**,413-428.
Hutchison,W.W.(1975).Computer-based systems for geological field data. *Geol.Surv.Can.Pap.* **74-63.**
Huxham,F.A.(1986).*Using the Macintosh toolbox with C.*Sybex,Berkeley,CA.
Huynh,H.(1982). A comparison of four approaches to robust regression. *Psych.Bull.* **92(2),** 505-512.
Ikeda,K.(1979).Calculating procedure of end-members of clinopyroxene,computer program and estimation of ferric and ferrous in EPMA analyses*J.Jap.Assoc. Mineral.Petrol.Econ,Geol.* **74(4)**: 135-149.
Iman,R.L.(1976).An approximation to the exact distribution of the Wilcoxon-Mann-Whitney rank sum statistic. *Communications in Statistics-Theory and Methods*, **A5**,587-598.
Iman,R.L. & Davenport,J.M.(1976).New approximations to the exact distribution of the Kruskal-Wallis test statistic. *Communications in Statistics-Theory and Methods* **A5**,1335-1348.
Iman,R.L.,Quade,D. & Alexander,D.A.(1975).Exact probability levels for the Kruskal-Wallis test.*Selected tables in Math.Stats.***3**,329-384.
Imbrie,J. & Purdy,E.G.(1962).Classification of modern Bahaman carbonate sediments.*Am.Assoc. Pet. Geol.Mem.* **1**,253-272.
Imbrie,J. & Van Andel,T.H.(1964).Vector analysis of heavy mineral data. *Bull.Geol.Soc.Am.* **75**,1131-1156.
IMM (Institute of Mining & Metallurgy, 1984). *Application of computers and mathematics in the mineral industries.*
Ingamells,C.O.(1974).Control of geochemical error through sampling and subsampling diagram. *Geochim. Cosmochim. Acta* **38**,1225-1237.
International Study Group.(1982).An inter-laboratory comparison of radiocarbon measurement in tree rings. *Nature* **298**, 619-623.
Iregui,H.G.(1972).[Computations applied to the chemical classification of the igneous rocks (CIPW & Niggli methods)] *Bol.Geol.Inst.Nac.Invest.Geol.Mineral. (Bogota)*,**20(3)**:7-44.
ISO (International Standards Organization, 1983). Information processing — Graphics Kernel System (GKS) — functionl description. Draft International Standard ISO/DIS 7942.
Ivanov,D.N.(1963).Linear parageneses of the chief rock-forming minerals in the granites of central Kazakhstan. *Dokl.Acad.Sci.USSR* **150**:134-136.
Ivert,H. (1980). Relationship between stratigraphic variation in the morphology of *Gabonella elongata* and geochemical composition of the host sediment. *Cretaceous Research* **1**, 223-233.
Iversen,G.R. & Norpoth,H.(1976).*Analysis of variance.*Sage,Beverly Hills, 95p.
Jaccard,P.(1908).Nouvelles récherches sur la distribution florale.*Bull.Soc.Vaud. Sci.Nat.***44**, 223-270.
Jacob,A.F.(1975).FOLKSS: a FORTRAN program for petrographic classification of sandstones.*Comput.Geosci.* **1**,97-104.
Jaech,J.L.(1985). *Statistical analysis of measurement errors.* Wiley, New York, 293 pp.
Jaffrezic,H.,Villemant,B. & Joron,J.L.(1978).Concentration of trace elements in rocks:estimation of the error due to their statistical distribution in the powdered sample.*Geostandards Newsletter* **2(1)**:57-60.
Jago,B.C. & Mitchell,R.H.(1988).A new garnet classification technique: divisive cluster analysis applied to garnet populations from Somerset Island kimberlites.*Spec.Publ.Geol.Soc.Aust.* (in press).
Jambu,M.(1981).FORTRAN IV program for rapid hierarchical classification of large data-sets.*Comput.Geosci.* **7**,297-310.
Jambu,M. & Lebeaux,M-O.(1983).*Cluster analysis and data analysis.* North-Holland, 898 pp.
James,C.H. & Radford,N.W.(1980).Apportionment of error in geochemical prospecting.*Proc.Geol.Assoc.***91**,69-70.
James,C.J.(1970).A rapid method for calculating the statistical precision of geochemical prospecting analyses.*Tr.Inst.Ming.Metall..***79**,B88-9.
James,W.R.(1970).Regression models for faulted structural surfaces.*Bull.Am. Assoc.Pet.Geol.* **54**,638-646.
Jeffrey,K.G. & Gill,E.M.(1976a).The design philosophy of the G-EXEC system. *Comput.Geosci.* **2**:345-346.

Jeffrey,K.G. & Gill,E.M.(1976b).The geological computer.*Comput.Geosci.* **2**:347-9.

Jenkins,G.M. & Watts, D.G.(1968). *Spectral analysis and its applications.* Holden Day, San Fransisco, 525pp.

Jennings,A.(1977).*Matrix computation for engineers and scientists.*Wiley, Chichester,330pp.

Jeran,P.W. & Mashey,J.R.(1970).A computer program for the stereographic analysis of coal fractures and cleats. *Info.Circ.U.S. Bur.Mines* **8454**,34pp.

Jizba,Z.V.(1959).Frequency distribution of elements in rocks.*Geochim.Cosmochim. Acta* **16**,79-82.

John,P.(1971).*Statistical design and analysis of experiments.*Macmillan,New York.

Johns,M.V.(1974).Nonparametric estimation of location. *J.Am.Stat.Assoc.***69**, 453-460.

Johnson,R.G.(1960).Models and methods for the analysis of the mode of formation of fossil assemblages. *Bull.Geol.Soc.Am.***76**,1075-1086.

Jolliffe,I.T.(1986).*Principle components analysis.*New York,Springer, 271pp.

Jones,A.D.(1978).Computers and the teaching of airphoto interpretation. *Photogramm.Eng.Remote Sensing* **44(10)**:1267-1272.

Jones,B.G. & Facer,R.A.(1982).CORRMAT/PROB,a program to create and test a correlation coefficient matrix for data with missing values. *Comput.Geosci.***8**,191-198.

Jones,R.L. (1986). Geoscience data gathering — are we getting value from the harvest? *Proc.3rd Int.Conf.Geosci.Info. (Adelaide,S.Australia, June '86)*, Vol.1, 31-37.

Jones,T.A.(1968).Statistical analysis of orientation data.*J.Sed.Petrol.***38**,61-7.

Jones,T.A.(1972).Multiple regression with correlated independent variables. *J.Math.Geol.***4**,203-218.

Jones,T.A.(1979).Fitting straight lines when both variables are subject to error:I.Maximum likelihood and least-squares estimation.*J.Math.Geol.***11**,1-26.

Jones,T.A. & Baker,R.A.(1975).PIP1 and PIP2: FORTRAN IV programs to aid in the determination of important parameters in a classification scheme. *Comput.Geosci.***1**,3-26.

Jones,T.A. & James,W.R.(1969).Analysis of bimodal orientation data.*J.Math.Geol.***1**,129-136.

Jones,T.A.,Hamilton,D.G. & Johnson,C.R.(1986).*Contouring geological surfaces with the computer.*Van Nostrand,New York,320pp.

Joreskog,K.G.,Klovan,J.E. & Reyment,R.A.(1976).*Geological factor analysis.* Elsevier,New York,178pp.

Jorgensen,B.(1982). *Statistical properties of the generalized inverse Gaussian distribution.* Springer, New York, 188 pp.

Jorstad,R.B.(1986).Down-to-earth software reviews.*J.Geol.Educ.***34**,132-133.

Journel,A.G.(1983).Nonparametric estimation of spatial distributions.*J. Math.Geol.***15**,445-468.

Journel,A.G.(1985).The deterministic side of geostatistics.*J.Math.Geol.* **17**,1-15.

Journel,A.G.(1986).Geostatistics: models and tools for the earth sciences. *J.Math.Geol.***18**,119-139.

Journel,A.G.(1987).Reply to "comments" by J.Serra.*J.Math.Geol.***19**,357-60.

Journel,A.G. & Huijbregts,C.J.(1981).*Mining geostatistics.*Academic Press,New York,600pp.

Joyce,A.S.(1973).Application of cluster analysis to detection of subtle variation in a granitic intrusion. *Chem.Geol.* **11**,297-306.

Kaesler,R.L. & Waters,J.A.(1972).Fourier analysis of the ostracode margin. *Bull.Geol.Soc.Am.* **83**,1169-1178.

Kahma,A. & Mikkola,T. (1946). A statistical method for the quantitative refractive index analysis of minerals in rocks. *CR Soc.Géol.Finlande* **19.**

Kaiser,H.F. (1970).A second generation little jiffy.*Psychometrika* **35**,401.

Kalogeropoulos,S.I.(1985).Discriminant analysis for evaluating the use of lithogeochemistry along the Tetsusekiei horizon as an exploration tool in the search for Kuroko type deposits.*Mineral. Deposita* **20**:135-142.

Kamb,W.B.(1959).Ice petrofabric observations from the Blue glacier, Washington, in relation to theory and experiment. *J.Geophys.Res.* **64**, 1891-1909.

Karche,J.R. & Mahé,J.(1977).Étude pétrochimique et petrogénétique des associations volcaniques pour analyse correspondence des facteurs appliqué au Madagascar du norte.*Rev.Geogr.Phys. Geol.Dyn.Ser.2(Paris)* **19(2)**:125-136.

Karlinger,M.R. & Troutman,B.M.(1985).Error bounds in cascading regression. *J.Math.Geol.***17**, 287-296.

Kasvand,T.(1983).Computerised vision for the geologist.*J.Math.Geol.* **15**,3-24.

Kater,D.A.(1985).*The printed word: the Microsoft guide to advanced word processing on the Apple Macintosh.* Microsoft Press,Washington.

Kawabe,I.(1979).A stochastic model of fractional crystallisation and its application to the Japanese geosynclinal basalt.*J.Math.Geol.* **9**,39-54.

Kellaway,F.W.(1968).*Penguin-Honeywell book of tables.*Penguin, Harmondsworth.

Kelly,P.C. & Phillips,M.J. (1986).Better data for decision-making: implications for the geosciences. *Proc.3rd Int.Conf. Geosci.Info. (Adelaide,S.Australia, June '86)*, Vol.1, 51-62.

Kemery,J.G. & Snell,J.L.(1960).*Finite Markov chains.*Van Nostrand,Princeton.

Kendall,M.G.(1955).*Rank correlation methods.*(2nd Ed).Hafner,New York,196pp.

Kendall,M.G.(1957).*A course in multivariate analysis.*Griffin,London.

Kendall,M.G.(1966).Discrimination and classification.In: P.R.Krishnaiah (ed.)*Multivariate analysis.*Academic,New York.

Kendall,M.G. & Buckland,W.R.(1982).*A dictionary of statistical terms.* (4th Ed).Longmans,London.

Kendall,M.G. & Stuart,A.(1965).*The advanced theory of statistics*.Hafner,New York,678pp.
Kennedy,W.J.(1982).The statistical computing portion of a graduate education program in statistics.In: Rustagi & Wolfe(1982), qv,p.233-247.
Kennedy,W.J. & Gentle,J.E.(1980).*Statistical computing*.Dekker,New York,591pp.
Kerlinger,F.N. & Pedhazur,E.J.(1973).*Multiple regression in behavioral research.* Holt,Rinehart & Wilson,New York.
Kern,E.L. & Carpenter,J.R.(1986).Enhancement of student values, interests and attitudes in earth science through a field-oriented approach. *J.Geol.Educ.* **32**, 299-305.
Keylman,G.A. & Panyak,S.G.(1977).Mathematical modelling of metasomatic processes and their geologic interpretation.*Dokl.Acad.Sci.USSR Earth Sci. Sect.***227**:139-140.
Kim,P.J.(1969).On the exact and approximate sampling distribution of the two-sample Kolmogorov-Smirnov criterion D_{mn} $(m<n)$. *J.Am.Stat.Assoc.***64**,1625-1635.
Kim,P.J. & Jennrich,R.I.(1970).Tables of the exact sampling distribution of the two-sample Kolmogorov-Smirnov criterion D_{mn} $(m<n)$.In: Harter & Owen (1970),qv, 79-170.
Kimberley,M.M.(1986).Sketching a cross section of folded terrain with a microcomputer.In: Thornton & Merriam(1986),qv,165-188.
King,H.F.,McMahon,D.W. & Buttor,G.J.(1982).*A guide to the understanding of ore reserve estimation.* Aust.Inst. Ming.Metall., 21pp.
King,L.J.(1969). *Statistical analysis in geography.* Prentice-Hall, Englewood-Cliffs, NJ, 288 pp.
Kinnison,R.R. (1984). *Applied extreme value statistics.* Battelle Press, Columbus, Ohio.
Kiraly,L.(1969).Statistical analysis of fractures (orientation and density). *Geol.Rdsch.***59**,125-151.
Klastorin,T.D.(1983).Assessing cluster analysis methods.*J.Market.Res.***20**,92-98.
Klecka,W.R.(1980).*Discriminant analysis.* Sage,Berveley Hills, 71 pp.
Kleeman,A.W.(1967).Sampling error in the chemical analysis of rocks.*J.Geol. Soc.Aust.***14**,43-47.
Kleeman,W.R.(1980)*Discriminant analysis*.Sage,Beverly Hills,71pp.
Kleiner,B. & Hartigan,J.A.(1981).Representing points in many dimensions by trees and castles. *J.Am.Stat.Assoc.* **76**,260-269.
Klimley, S. (1986). Strengthening geological libraries through cooperative collection developments. *Proc.3rd Int.Conf. Geosci.Info. (Adelaide,S.Australia, June '86)*, Vol.1, 100-111.
Klovan,J.E.(1966).The use of factor analysis in determining depositional environments from grain-size distributions. *J.Sed.Petrol.***36**,115-125.
Klovan,J.E.(1975).R- and Q-mode factor analysis.In: McCammon(1975a),qv,21-69.
Klovan,J.E.(1981).A generalisation of extended Q-mode factor analysis to data matrices with variable row sums.*J.Math.Geol.***13**,217-224.
Klovan,J.E. & Imbrie,J.(1971).An algorithm and FORTRAN IV program for large-scale Q-mode factor analysis and calculation of factor scores.*J. Math.Geol.***3**,61-78.
Klovan,J.E. & Miesch,A.T.(1976).Extended CABFAC and QMODEL computer program for Q-mode factor analysis of compositional data.*Comput.Geosci.***1**,161-178.
Knapp,T.R.(1982).Canonical correlation analysis: a general parametric significance testing system.In: Fornell(1982a), qv,Vol.I,pp.22-35.
Knight,H.,Wright,D. & Burgess,D.(1985).*Volcanoes.* Macdonald,London,48pp.
Knowles, C.R. (1987). A BASIC program to recast garnet end-members. *Comput.Geosci.* **13**, 655-659.
Knudsen,H.P. & Kim,Y.C. (1977). A short course on geostatistical ore reserve estimation. Dept.Mining & Geological Engineering, University of Arizona, 259 pp.
Knuth,D.E.(1969).*The art of computer programming*.Addison-Wesley,Reading,Mass.
Koch,G.S.(1981).Computer applications in exploration and mining geology: ten years of progress.In: Merriam(1981b), qv,181-198.
Koch,G.S.(1987).*Exploration-geochemnical data analysis with the IBM-PC.* Van Nostrand Reinhold, New York,128pp.
Koch,G.S. & Link,R.F.(1970).*Statistical analysis of geological data*.Wiley,New York,Vol.I, 375 pp.
Koch,G.S. & Link,R.F.(1971).*Statistical analysis of geological data*.Wiley,New York,Vol.II, 438 pp.
Koch,G.S.,Link,R.F. & Schuenemeyer,J.H.(1972).*Computer programs for geology.* Artronic Information Systems,New York,142pp.
Koelling,M.E.V. & Whitten,E.H.T.(1973).Fortran IV program for spline surface interpolation and contour map production.*Geocom.Bull.* **6**.
Kohut,J.J.(1978).Information education in the geosciences: anatomy of a course. *Proc.Geosci.Inf. Soc.***8**:80-95.
Kolmogorov,A.N.(1933).Sulla determinazione empirica di una legge di distribuzione.*Gionale dell'Istituto Italiano degli Attuari* **4**,83-91.
Kork,J.O.(1977).Examination of the Chayes-Kruskal procedure for testing correlations between proportions. *J.Math.Geol.* **9**,543-562.
Korsch,R.J.(1977).MODES: a FORTRAN IV program to calculate modal analyses from raw point-count data. *Comput.Geosci.***3**,107-114.

Kosinowski,M.H.F.(1982).MSONRM: a FORTRAN program for the improved version of mesonorm calculations. *Comput.Geosci.***8**,11-20.

Kotz,S. & Johnson,N.L.(1982-8).*Encyclopedia of statistical sciences.*Wiley,New York,6 vols (at present).

Kowalski,C.J.(1972).On the effects of non-normality on the distribution of the sample product-moment correlation coefficient. *Applied Stat.***21**,1-12.

Krajewski,S.J.(1986).Microcomputers for explorationists.In: Hanley & Merriam(1986), qv,9-16.

Krauchenko,S.M.(1982).Quantitative prediction of niobium reserves of carbonatite plutons on correlation of their indicators. *Dok.Acad.Sci.USSR* **263**, 42-45.

Kremer,M.,Lenci,M. & Lesage,M.T.(1976).SIGMI: a user-oriented file-processing system. *Comput.Geosci.***1**,187-193.

Kretz,R.(1985).Calculation and illustration of uncertainty in geochemical analyses*J.Geol.Educ.* **33**,40-44.

Krige,D.G.(1951).A statistical approach to some basic mine valuation problems on the Witwatersrand. *J.S.Afr.Chem. Metall.Ming.Soc.*119-139.

Krige,D.G.(1960).On the departure of ore value distributions from the lognormal model in S.African gold mines. *J.S.Afr.Inst.Ming.Metall.***61**,231-244.

Krige,D.G.(1981).Lognormal-de Wijsian geostatistics for ore evaluation. *S.Afr.Inst.Ming.Metall.Geostats.Monogr.***1.**

Krige,D.G.(1984).Geostatistics and the definition of uncertainty. *Tr.Inst.Ming.Metall.*,A41-47.

Krige,D.G.(1985).The use of geostatistics in defining and reducing the uncertainty of grade estimates. *Tr.Geol.Soc.S.Afr.***88**,69-72.

Krige,D.G.(1986).Matheronian geostatistics — quo vadis? *J.Math.Geol.* **18**,501-502.

Krige,D.G. & Magri,E.J.(1982a).Geostatistical case studies of the advantages of lognormal-de Wijsian kriging with mean for a base metal mine and a gold mine.*J.Math.Geol.***14**,547-555.

Krige,D.G. & Magri,E.J.(1982b).Studies of the effects of outliers and data transformation on variogram estimates for a base metal and a gold ore body.*J.Math.Geol.***14**,557-565.

Krinsley,D.(1960).Magnesium,strontium and aragonite in the shells of certain littoral gastropods. *J.Palaeontol.* **34**,744-755.

Krishnaiah,P.R.(1980).*Analysis of variance.*North Holland,New York,1002p.

Kruhl,J.(1978).Current bedding in the Moinian quartzites at eastern Loch Leven,Scottish Highlands. *Neues Jb.Mineral. Abh.* **132**,52-66.

Krumbein,W.C.(1960).The geological population as a framework for analysing numerical data in geology.*Liverpool Manchester Geol.J.* **2**,341-368.

Krumbein,W.C.(1967).FORTRAN IV computer program for Markov chain experiments in geology. *Kansas Geol. Surv.Computer Contrib.***13,**138pp.

Krumbein,W.C.(1974).The pattern of quantification in geology.In: Merriam (1974),qv,51-66.

Krumbein,W.C.(1975).Markov models in the Earth Sciences.In: McCammon(1975a), qv,90-105.

Krumbein,W.C.(1978).Some recent development in the Mathematical Geology of stream-channel networks.In: Merriam(1978b),qv,19-22.

Krumbein,W.C. & Dacey,M.F.(1969).Markov chains and embedded Markov chains in geology. *J.Math.Geol.***1**,79-98.

Krumbein,W.C. & Graybill,F.A.(1965).*An introduction to statistical models in geology.* McGraw-Hill,New York, 475pp.

Krumbein,W.C. & Miller,R.L. (1953). Design of experiments for statistical analysis of geological data. *J. Geol.* **61** , 510-532.

Krumbein,W.C. & Tukey,J.W.(1963).Multivariate analysis of mineralogic, lithologic and chemical composition of rock bodies*J.Sed.Petrol.***26**,322-337.

Krumbein,W.C. & Watson,G.S.(1972).Effects on trend on correlation in open-closed three-component systems. *J.Math.Geol.***4**,317-330.

Kruskal,J.B.(1964).Nonmetric multidimensional scaling: a numerical method. *Psychometrika* **29**,115-129.

Kruskal,J.B. & Wish,M.(1978).*Multidimensional scaling.*Sage Publishers, Beverly Hills,93p.

Kruskal,W.H.(1958).Ordinal measures of association.*J.Am.Stat.Assoc.***53**,814-861.

Kruskal,W.H. & Wallis,W.A.(1952).Use of rank in one criterion variance analysis. *J.Am.Stat. Assoc.***47**,583-621 & erratum in **48**,910.

Kruskal,W.H. & Tanur,J.M.(1978).*International encyclopoedia of statistics.* Free Press,New York.

Kufs,C.T.(1979).Another view of the use of factor analysis in geology. *J.Math.Geol.***11**,717-720.

Kuiper,N.H.(1962).Tests concerning random points on a circle.*Proc.Koninkl. Nederl.Akad. Wetenschappen Ser.A.***63**,38-47.

Kulatilake,H.S.W.(1985).Fitting Fisher distributions to discontinuity orientation data*J.Geol.Educ.* **33(5)**:266-269.

Kulback,S.(1959)*Information theory and statistics.*Wiley,New York,395pp.

Kulback,S.,Kupperman,M. & Ku,H.H.(1962).Tests for contingency tables and Markov chains. *Technometrics* **4,**573-608.

Lachenbruch,P.A.(1975)*Discriminant analysis.*Hafner,New York,128pp.

Lafontain,L.J.(1970).Plotted and point-counted stereograms by computer, X-Y plotter or microfilm devices. *Bull.Geol.Soc.Am.***81,**1267-1271.

Lahiri,A. & Rao,S.V.L.N.(1974).Gradient analysis: a technique for the study of spatial variations. *Modern Geol.***5**,33-45.

Lahiri,A. & Rao,S.V.L.N.(1978).A choice between polynomial and Fourier trend surfaces.*Modern Geol.***6**,153-162.

Lance,G.N. & Williams,W.T.(1967).Mixed-data classificatory programs.I. Agglomerative systems.*Aust.Comput.J.* **1**,15-20.

Lang,H.M. & Rice,J.M.(1985).Regression modelling of metamorphic reactions in metapelites. *J.Petrol.***26**:857-887.

Langenberg,C.W.,Rondeel,H.E & Charlesworth,H.P.H.(1977).A structural study in the Belgian Ardennes with sections constructed using computer-based methods.*Geol.Mijnbouw* **56**,145-154.

Langley,R.(1970).*Practical statistics.*David & Charles,Newton Abbott,400pp.

Lapides,I.L. & Vladykin,N.V.(1975).Computerized calculation of crystallochemical formulae for complex rock-forming minerals.*Sov.Geol.Geophys.***16(10)**:116-122.

Laubscher,N.F.,Steffens,F.E. & Delange,E.M.(1968).Exact critical values for Mood's distribution-free test statistic for dispersion and its normal approximation.*Technometrics* **10**,497-508.

Laudon,R.C.(1986).Using spreadsheet software for gradebooks.*J.Geol.Educ.* **34**,106-107.

Launer,R.L. & Wilkinson,G.N.(1979).*Robustness in statistics.*Academic Press, New York,296pp.

Laurie,P. (1985). *Databases.* Chapman & Hall, 139 pp.

Law,D.(1987).Language standards and program presentation.*Comput.Geosci.***13**,93-4.

Lawley,D.N. & Maxwell,A.E.(1963).*Factor analysis as a statistical method.* Butterworth, London,117pp.

Lawley,D.N. & Maxwell,A.E.(1971).*Factor analysis as a geological method.* Butterworth, London,153pp.

Laycock,J.W. & Staines,H.R.E.(1979).Computer applications and information services at the Geological Survey, Queensland.*Ann.Rep.Under Sec.Mines Queensl.* 1978:101-102.

Lea,G.(1978).GEOARCHIVE: geosystems indexing policy.*Proc.Geosci.Inf.Soc.* **8**:42-56.

Lea,G.,Shearer,J. & Patterson,D.(1978).Computerized indexing of the Institute of Geological Sciences (UK) geological map collection. *Bull.Geogr.Map.Div. Spec.Libr.Assoc.***112**:27-46.

Lebart,L.,Morineau,A. & Warwick,K.M.(1984).*Multivariate descriptive statistical analysis.*Wiley,New York,231pp.

Le Bas,M.J.(1980).Petrological number-juggling.*Tr.Leics.Lit.Phil.Soc.***72**,70-96.

Le Bas,M.J.,Durham,J. & Plant,J.(1983).IGBA and the National Geochemical DataBank in the U.K.*Comput.Geosci.* **9**,513-521.

Le Bas,M.J.,Le Maitre,R.W.,Streckeisen,A. & Zanettin,B.(1986).A chemical classification of volcanic rocks based on the total alkali-silica diagram.*J.Petrol.***27**,745-750.

Lecarme,O. & Nebut,J-L.(1984).*Pascal for programmers.*McGraw-Hill,**272**pp.

LeComte,P.(1977).[Statistical treatment of data in geochemical prospecting.] *Mem.Inst.Geol. Univ.Louvain* **27**:349-356.

Lee,P.J.(1969).The theory and application of canonical trend surfaces.*J.Geol.* **77**,303-318.

Lee,W.H.K. & Habermann,R.E.(1986).Applications of personal computers in geophysics.*EOS* **67 (46)**,1321-3.

Lehmann,E.L.(1959).*Testing statistical inferences.*Wiley,New York,498pp.

Lehmann,E.L.(1975).*Nonparametrics:statistical methods based on ranks.* Holden-Day,San Fransisco.

Lehmann,R.S.(1986).Macintosh statistical packages.*Behavior Research Methods, Instruments & Computers* **18(2)**,177-187.

Lehmann,R.S.(1987).Statistics on the Macintosh.*Byte* **12**,201-214.

Leigh,B.(1981).Online searching of bibliographic geological databases and their use at the Science Reference Library.*J.Geol.Soc.Lond.*138,589-597.

Le Maitre,R.W.(1968).Chemical variation within and between volcanic rock series. — a statistical approach.*J.Petrol.* **9**,220-252.

Le Maitre,R.W.(1973).Experiences with CLAIR: a computerised library of analysed igneous rocks.*Chem.Geol.* **12**,301-308.

Le Maitre,R.W.(1976a).The chemical variability of some common igneous rocks. *J.Petrol.***17**,589-637.

Le Maitre,R.W.(1976b).A new approach to the classification of igneous rocks using the basalt-andesite-dacite-rhyolite suite as an example.*Contrib.Mineral. Petrol.***56**,191-203.

Le Maitre,R.W.(1979).A new generalised petrological mixing model.*Contrib. Mineral.Petrol.* **71**,133-137.

Le Maitre,R.W.(1981a).GENMIX — a generalised petrological mixing model.*Comput.Geosci.* **7**,229-247.

Le Maitre,R.W.(1981b).Some developments in computer applications in petrology. In: Merriam(1981b),qv,199-210.

Le Maitre,R.W.(1982).*Numerical Petrology.*Elsevier,Amsterdam,281pp.

Le Maitre,R.W. & Ferguson,A.K.(1978).The CLAIR data system.*Comput.Geosci.***4**,65-76.

Lemmer,I.C.(1985).A new way of estimating recoverable resources.*Tr.Geol.Soc.S.Afr.***88**,77-80.

Lenthall,D.H.(1972).The application of discriminatory and cluster analysis as as aid to the understanding of the acid phase of the Bushveld complex. *Inf.Circ.Econ.Geol.Res.Unit.Univ. Witwatersrand* **72**,33pp.

Lenthall,D.H., McCarthy,T.S. & McIver,J.R. (1974). A computer program for the construction of stereo pairs of the CMAS tetrahedron. *Tr.Geol.Soc.S.Afr.* **77**, 201-206.

Lepeltier,C.(1969).A simplified statistical treatment of geochemical data by graphical representation. *Econ.Geol.* **64**,538-550.

Lepeltier,C.(1977).The use of the computer in geochemical exploration. *Sci.Terre Inf.Geol.(Nancy),* **9**:15-39.

Le Roex,A.P.,Erland,A.J. & Needham,H.D.(1981).Geochemical and mineralogical evidence for the occurrence of at least 3 distinct magma-types in the 'Famous' region.*Contrib.Mineral.Petrol.* **77**,24-37.

Lessells,C.M. & Webster,R.(1984).A general text translation program for coded description. *Comput.Geosci.* **10**, 211-236.

Levine,P.A.,Merriam,D.F. & Sneath,P.H.A.(1981).Segmentation of geological data using the Kolmogorov-Smirnov test. *Comput.Geosci.***7**,415-426.

Lewis,B.R. & Ford,R.F.(1983).*Basic statistics using SAS*.West Publishing,St.Paul., Minnesota, 144pp.

Lewis,D.G.(1971).*The analysis of variance*.Manchester Univ.Press,32p.

Lewis,P.(1977).*Maps and statistics*.Methuen,London,318pp.

Lewis,T.G.(1986).*Macintosh hands-on Pascal*.Wadsworth,Belmont,CA.

Leymarie,P. & Isnard,P.(1977).[Cartography and statistical study of geochemical data: application to the northern Millevache granite].*Sci.Terre.Inf.Geol.(Nancy)* **21(2)**:151-186.

Li,T.M.,Handelsman,S.D. & Kovisaara,L.(1987).*Mineral resource management by personal computer*. Soc.Ming.Eng., Littleton,CO.,171pp.

Lilliefors,H.W.(1967).On the Kolmogorov-Smirnov test for normality with mean and variance unknown. *J.Am.Stat. Assoc.* **64**,399-402.

Lindquist,L.(1976).SELLO,a Fortran IV program for the transformation of skewed distributions to normality. *Comput. Geosci.* **1**,129-145.

Link,R.F. & Koch,G.S.(1975).Some consequences of applying lognormal theory to pseudolognormal distributions. *J.Math.Geol.***7**,117-128.

Lister,B.(1982).Evaluation of analytical data: a practical guide for geoanalysts.*Geostandards Newsletter* **6:2**,175-206.

Lister,B.(1984).A note on robust estimates.*Geostandards Newsletter* **8:2**,171-172.

Lister,B.(1985).Looking at analytical data.*Geostandards Newsletter* **9**,263-74.

Lister,B.(1986).Best estimates from interlaboratory data.*Anal.Chim.Acta* **186**,325-329.

Little,R.T.A.(1987). *Statistical analysis with missing data.* Wiley, New York.

Lohnes,P.R.(1961).Test space and discriminant space classification models and related significance tests. *Educ.Psychol.Measur.***21**,559-574.

Longe,P.V.(1976).Computers in mineral exploration:some uses,limitations and requirements. *Comput.Geosci.* **2**:325-329.

Longe,P.V.,Burk,C.F.,Dugas,J.,Erwing,K.A.,Ferguson,J.A.,Gunn,K.L.,Jackson,E.V.,Kelly,A.M.,Oliver,A.D., Sutterlin, P.G. & WIlliams,G.D.(1984). Computer-based files on mineral deposits: guidelines and recommended standards for data content. *Geol.Surv.Can.Pap.***78-26**, 72 pp.

Longley,J.W.(1967).An appraisal of least squares programs for the electronic computer from the point of view of the user. *J.Am.Stat.Soc.***62**,819-841.

Lonka,A.(1967).Trace elements in the Finnish Precambrian phyllites as indicators of salinity at the time of sedimentation.*Bull.Comm.Geol.Finlande* **209**.

Loudon,T.V.(1969).A small geological data library.*J.Math.Geol.* **1**,155-170.

Loudon,T.V.(1974).Analysis of geological data using ROK-DOC,a FORTRAN IV package for the IBM 360/65 computer.*Rep.Inst.Geol.Sci.***74/1**.

Loudon,T.V.(1979).*Computer methods in geology*.Academic Press,New York.

Ludwig,K.R. & Stuckless,J.S.(1979).Programs in Hewlett-Packard BASIC for storing retrieving and plotting rare-earth element data for geochemical studies, using HP-9831/9872 desk top computers and plotters.*U.S.G.S.Open File Rep.* **79/949**:24pp.

Lumsden,D.N.(1971).Markov chain analysis of carbonate rocks: application, limitations and implications as exemplified by the Pennsylvanian system in southern Nevada.*Bull.Geol. Soc.Am.***83**,447-462.

Lumsden,G.I. & Haworth,R.T.(1986).The BGS database. *J.Geol.Soc.Lond.* **143**, 379-380.

Lundstrom,I. & Bjork,L.(1979).Evaluation of map precision by means of the area of influence method. *J.Math.Geol.* **11**,701-716.

Lunneborg,C.E. & Abbott,R.D.(1983).*Elementary multivariate analysis for the behavioural sciences*.North-Holland,New York,522pp.

Lybanov,M.(1985).A simple generalised least-squares algorithm.*Comput.Geosci.* **11(4)**:501-508.

Ma,M.S. & Schmitt,R.A.(1976).Possible source materials for eucritic achondrites based on multi-linear regression analysis of trace elemental data. *Meteoritics* **11(4)**:324-325.

Maag,U.R.(1966).A *k*-sample analogue of Watson's U^2 statistic.*Biometrika* **53**, 579-584.

Maaløe,S.(1976).Quantitative aspects of fractional crystallisation of major elements*J.Geol.* **84**,81-96.

MacCaskie,D.R.(1984).Identification of petrogenetic processes using covariance plots of trace-element data. *Chem.Geol.* **42**,325-341.

Macdonald,R. & Bailey,P.K.(1973).The chemistry of the peralkaline oversaturated obsidians. *U.S.G.S.Prof.Pap.* **440-N-1.**

Magri,E.J. & Longstaff,W.S.(1979).The computerized system for the geostatistical determination of gold ore reserves

within the Anglo-Transvaal group.*J.Inst.Mine Surv.S.Afr.*,34-48.
Magri,E.J. & Longstaff,W.S.(1982).Answering the question: how much is in the ground? *Coal,Gold and base metals of southern Africa,March* 1982,61-74.
Magri,E.J. & Mostert,J.S.(1985).Some applications of geostatistics in mineral exploration. *Tr.Geol.Soc.S.Afr.* **88**,65-68.
Mahalanobis,P.C.(1936).On the generalized distance in statistics.*Proc.Nat.Inst. Sci.India* **12**,49-55.
Mahalanobis,P.C.(1949).A historical note on the D^2 statistic.*Sankhya* **9**,237-240.
Mahé,J.(1971). L'analyse factorielle des correspondences et son usage en paléontologie et dans l'étude de l'évolution. *Bull.Soc.Géol.Fr.Sér.7*, **16**, 336-340.
Malmgren,B.A. (1972). Morphometric studies of *Globorotalia pseudobulloides. Stokholm Contrib.Geol.* **24**, 33-49.
Malmgren,B.A.(1979).Multivariate normality tests of planktonic foraminiferal data.*J.Math.Geol.***11**, 285-297.
Malmqvist,L.(1978).An iterative regression analysis procedure for numerical interpretation of regional exploration geochemistry data.*J.Math.Geol.* **10**:23-41.
Mancey,S.J.(1982).Cluster analysis in geology.*Geol.Soc.Lond.Misc.Pap.***14**,89-102.
Mancey,S.J. & Howarth,R.J.(1980).Power-transform removal of skewness from large data-sets. *Tr.Inst.Ming.Metall.* **89**,B92-98.
Mancktelow,N.S.(1981).A least-squares method for determining the best-fit point maximum,great circle and small circle to multidirectional orientation data. *J.Math.Geol.***13**,507-522.
Manly,B.F.J. (1986). *Multivariate statistical methods: a primer.* Chapman & Hall, London, 159 pp.
Mann,C.J.(1976).The PLATO system, its language, assets and disadvantages in geological education. *Comput.Geosci.* **2**,41-50.
Mann,C.J.(1978).Randomness in nature.*Bull.Geol.Soc.Am.***81(1)**:95-104.
Mann,C.J.(1979).Obstacles to quantitative lithostratigraphic correlation. In: Gill & Merriam(1979),qv, 149-166.
Mann,C.J.(1981).Stratigraphic analysis: decades of revolution (1970-79) and refinement (1980-89). In: Merriam(1981b), qv,211-242.
Mann,C.J.(1987).Misuses of linear regression in Earth Sciences. In: Size(1987a),qv,74-107.
Manson,V.(1967).Geochemistry of basaltic rocks: major elements. In: Hess,H.H. & Poldevaart,A. (eds.) *Basalts,* 215-270. Wiley,New York.
Mann,H.B. & Whitney,D.R.(1947).On a test of whether one of two random variables is stochastically larger than the other. *Ann.Math.Stat.***18**,50-60.
Marascuilo,L.A. & Levin,J.R.(1983).*Multivariate statistics in the social sciences: a researcher's guide.* Brooks/Cole, Monterey, 530pp.
Marcotte,D. & David,M.(1981).Target definition of Kuroko-type deposits in Abitibi by discriminant analysis of geochemical data. *Can.Inst.Ming.Metall.Bull. (Montreal)* **74/828**,102-113.
Mardia,K.V.(1970).Measures of multivariate skewness and kurtosis with applications.*Biometrika* **57**,519-530.
Mardia,K.V.(1972).*Statistics of directional data.*Academic Press.
Mardia,K.V.(1981a).Directional statistics in geosciences.*Univ.Leeds Dep.Stat. Communications in Statistics,Theory & Methods* **A10 (15)**:1523-1543.
Mardia,K.V.(1981b).The evolution of directional models in geosciences since Fisher.In: Merriam(1981d),qv,39-46.
Mardia,K.V. & Zemroch,P.J.(1978).*Tables of the F and related distributions algorithms.*Academic Press,London.
Mareschal,J.C. & West,G.F.(1980).A model for Archaean tectonism,part 2: numerical models of vertical tectonism in greenstone belts.*Can.J.Earth Sci.* **17(1)**:60-71.
Mark,D.M.(1973).Analysis of axial orientation data including till fabrics. *Bull.Geol.Soc.Am.* **84**,1369-1374.
Mark,D.M. & Church,M.(1977).On the misuse of regression in Earth Science. *J.Math.Geol.***9**,63-76.
Mark,R.K.(1978).FORTRAN program for Shapiro-Wilk test for normality for Honeywell multics system. *U.S.G.S.Open File Rep.***78(1069)**:10pp.
Maron,M.J.(1982)*Numerical analysis-a practical approach.*Macmillan,New York,471pp.
Martin,E.L.(1975a).Computer programs for map grid reference conversion for Tasmania.Part 1.Australian national grid to Australian map grid. *Tech.Rep.Tasmania Dept.Mines* **18**:142-145.
Martin,E.L.(1975b).Computer programs for map grid reference conversion for Tasmania.Part 2.Australian map grid to Australian national grid. *Tech.Rep.Tasmania Dept.Mines* **19**:136-138.
Martin,G. & Gordon,T.(1977).Data-base management systems — data models and query languages. *Comput.Geosci.* **3**,387-394.
Martin,J.(1975).*Computer data-base organisation.*Prentice-Hall,558pp.
Martin,J.(1981).*An end-users guide to data-base.*Prentice-Hall,New Jersey,144pp.
Martin,J.K.(1978).Computer-based literature searching,USA.*Spec.Libr.(New York)* **69(1)**:1-6.
Maslyn,R.M.(1986).EXPLOR and PROSPECTOR: expert systems for oil and gas and mineral exploration.In: Hanley & Merriam(1986),qv,89-104.
Massart,D.L. & Kaufman,L.(1983).*The interpretation of analytical chemical data by the use of cluster analysis.* Wiley,New York.

Massey,W.J.(1951).The distribution of the maximum deviation between two sample cumulative step functions. *Ann.Math.Stat.***22**,125-128.

Massey,W.J.(1952).Distribution table for the deviation between two sample cumulatives.*Ann. Math.Stat.***23**,435-441.

Mather,P.M.(1976).*Computer methods for multivariate analysis in physical geography*.Wiley,New York.

Matheron,G.(1963).Principles of geostatistics.*Econ.Geol.***58**,1246-1266.

Matheron,G.(1967).Kriging or polynomial interpolation surfaces?A contribution to polemics in mathematical geology. *Bull.Can.Inst.Ming.Metall.***60**,1041-1045.

Matheron,G.(1971).The theory of regionalized variables.*Cah.Math.Fontainebleau,* **5**,211pp.

Matheron,G.(1981).Splines and kriging: their formal equivalence.In: Merriam(1981d),qv,77-98.

Matheron,G.(1986).Philipian/Watsonian high (flying) philosophy.*J.Math. Geol.***18**,503-504.

Matheron,G. & Armstrong,M.(1987).*Geostatistical case studies*.Kluwer Academic Press, Hingham,Mass.,264pp.

Matthews,J.A.(1981).*Quantitative and statistical approaches to geography: a practical manual*.Pergamon,204pp.

Maxwell,A.E.(1961).*Analysing quantitative data*.Methuen,London.

May,R.W. & Jones,B.(1982).Stochastic analysis of complex lithological successions. *J.Math.Geol.* **14**,405-417.

Mayo,W. & Long,K.A.(1976).Documentation of BMR geological branch computer programs. *Aust.Bur.Miner. Resour. Rec.* **1976/82,** Microfilm **MF4**.

Mazzullo,J. & Ehrlich,R.(1983).Grain shape variation in the St.Peter sandstone: a record of eolian and fluvial sedimentation of an early Palaeozoic cratonic sheet shand.*J.Sed.Petrol.* **53**, 105-119.

McCalla,T.R.(1967).*Introduction to numerical methods and FORTRAN programming*. Wiley,New York,359pp.

McCammon,R.B.(1966).Principal component analysis and its application in large-scale correlation studies. *J.Geol.***74**,721-733.

McCammon,R.B.(1968).The dendrograph: a new tool for correlation.*Bull.Geol.Soc. Am.***79**,1663-1670.

McCammon,R.B.(1973).Nonlinear regression for dependent variables.*J. Math.Geol.***5**,365-376.

McCammon,R.B.(1974).The statistical treatment of geochemical data.In: Levinson,A.A.(ed.), *Introduction to Exploration Geochemistry*, 469-508.Applied Publishing,Kalgary.

McCammon,R.B.(1975a).*Concepts in geostatistics*.Springer-Verlag,Berlin,168pp.

McCammon,R.B.(1975b).Statistics and probability.In: McCammon(1975a),qv,1-20.

McCammon,R.B.(1977).BINORM — a FORTRAN subroutine to calculate percentiles of a standardized binormal distribution. *Comput.Geosci.***3**,385-389.

McCann,C. & Till,R.(1976).The use of interactive computing in teaching geology and geophysics. *Comput.Geosci.* **2**,59-67.

McCarthy,T.S.(1981).A FORTRAN IV computer program for rapid graphic display of sedimentary borehole log data. *Tr.Geol.Soc.S.Afr.***84**,271-279.

McCarthy,T.S. & Hasty,R.A.(1976).Trace element distribution patterns and their relation to the crystallisation of granitic melts.*Bull.Geol.Soc.Am.***40**,1351-8.

McCray,A.W.(1975).*Petroleum evaluation and economic decisions*.Prentice-Hall,Englewood Cliffs,NJ, 448 pp.

McDonald,G.C. & Ayers,J.A.(1978). Some applications of the Chernoff face: a technique for graphically representing multivariate data. In: Wang(1978),qv,183-197.

McEachran,D.B. & Marshak,S.(1986).Teaching strain theory in structural geology using graphics programs for the Apple Macintosh.*J.Geol.Educ.***34**,191-194.

McGee,V.E. & Johnson,N.M.(1979).Statistical treatment of experimental errors in the fission track dating method. *J.Math.Geol.***11**:255-268.

McGill,R., Tukey,J.W. & Larsen,W.A.(1978).Variation of box plots.*Am.Statistician* **32**,12-16.

McHone,J.G.(1977).TRIPLOT: an APL program for plotting triangular diagrams. *Comput.Geosci.* **3**,633-666.

McIntyre,D.B.(1981).Developments at the man-machine interface.In: Merriam (1981b),qv,23-42.

McIntyre,G.M.,Brooks,C.,Compston,W. & Turek,A.(1966).The statistical assessment of Rb-Sr isochrons. *J.Geophys. Res.* **71**,5459-5468.

McKenzie,G.D.(1984).Using microcomputers to improve productivity in academia. *J.Geol.Educ.* **32**,18,171-174.

McLean,M.J.(1979).Graphics hardware in Australia.*Bull.Aust.Soc.Explor.Geophys.* **10(3)**:224-227.

McMillan,R.G.(1971).Tests for outliers in normal samples with unknown variance. *Technometrics* **13**,87-100.

McMillan,R.G. & David,H.A.(1971).Tests for one of two outliers in normal samples with known variance. *Technometrics* **13**,75-86.

McRae,D.J.(1971).MICKA, a FORTRAN IV iterative *K*-means cluster analysis program. *Behavl.Sci.* **16**,423-424.

McWilliams.P.C. & Tesarik,D.R.(1978).Multivariate analysis techniques with application in mining. *Inf.Circ.S.Afr.Bur. Mines* **8782**:40pp.

MDS (1980). *User's manual SV3.* Edinburgh University Program Library Unit, Edinburgh.

Meddis,R.(1984).*Statistics using ranks—a unified approach*.Blackwell,Oxford,449pp.

Melton,R.S.(1963).Some remarks on failure to meet assumptions in discriminant analysis. *Psychometrika* **28**,49-53.

Merriam,D.F.(1969).*Computer applications in the Earth Sciences*.Plenum,New York,281pp.

Merriam,D.F.(1970).*Geostatistics*.Plenum,New York,177pp.

Merriam,D.F.(1972).*Mathematical models of sedimentary processes*.Plenum,New York.

Merriam,D.F.(1974).The impact of quantification on geology.*Syracuse Univ. Geol.Contrib.***2,**104pp.

Merriam,D.F.(1975).*Computer fundamentals for geologists*.Earth Resour.Found., Univ.of Sydney, 295 pp.

Merriam,D.F.(1976a).*Random processes in geology*.Springer,Berlin,168pp.

Merriam,D.F.(1976b).CAI in geology.*Comput.Geosci.***2**,3-7.

Merriam,D.F.(1976c).*Quantitative techniques for the analysis of sediments*.Pergamon,New York, 174pp.

Merriam,D.F.(1978a).*Recent advances in geomathematics: an international symposium*.Pergamon, Oxford.

Merriam,D.F.(1978b).*Geomathematics: past,present and prospects*.Syracuse Univ.Geol.Contrib. 5,74pp.

Merriam,D.F.(1978c).The International Association for Mathematical Geology — a brief history and record of accomplishments.In: Merriam(1978b),qv,1-6.

Merriam,D.F.(1980).Computer applications in geology — two decades of progress.*Proc.Geol.Assoc.* **91**,53-58.

Merriam,D.F.(1981a).Use of computers by geologists in Australia.*Comput.Geosci.* **7**,323-326.

Merriam,D.F.(1981b).*Computer applications in the Earth Sciences: an update of the 70's*.Plenum,New York,385pp.

Merriam,D.F.(1981c).A forecast for use of computers by geologists in the coming decade of the 80s.In: Merriam(1981b),qv,369-380.

Merriam,D.F.(1981d).*Down-to-earth statistics: solutions looking for geological problems*.Syracuse Univ.Geol.Contrib. 8,97pp.

Merriam,D.F.(1981e).Roots of quantitative geology.In: Merriam(1981d),qv, 1-16.

Merriam,D.F.(1987).*Statistics for geoscientists*.(Translation from 2nd German Ed).Pergamon,New York,220pp.

Merriam,D.F. & Pena Daza,M.(1978).Influence on the chemical composition of Pennsylvanian limestones in Kansas.In: Merriam(1978b),qv,51-60.

Merriam,D.F. & Sneath,P.H.A.(1966).Quantitative comparison of contour maps. *J.Geophys.Res.* **71**,1105-1115.

Métais,D. & Chayes,F.(1964). Classification of lamprophyres — a possible petrographic application of multigroup discriminant function analysis. *Carn.Inst.Wash.Yrbk.* **63**, 182-185.

Metcalf,M.(1985a).*FORTRAN optimization*.(Revised Ed).Academic Press,253pp.

Metcalf,M.(1985b).*EffectiveFORTRAN 77 programming*.Clarendon Press, Oxford.

Meyer,S.L.(1975).*Data analysis for scientists and engineers*.Wiley,New York.

Miall,A.D.(1973).Markov chains applied to an ancient alluvial plain succession. *Sedimentol.***20**,347-64.

Michie,M.H.(1978).Data manipulation in cluster analysis and the concept of zero presence.*J.Math.Geol.***10(4)**:335-345.

Middleton,G.V.(1962).A multivariate statistical technique applied to a study of sandstone composition.*Tr.R.Soc.Can.* **111**,56,119-126.

Middleton,G.V.(1963).Statistical inference in geochemistry.*Spec.Pub.R.Soc.Can.* **6**,124-139.

Middleton,G.V.(1964).Statistical studies on scapolite.*Can.J.Earth-Sci.***1**,23-24.

Miesch,A.T.(1967).Theory of error in geochemical data.*U.S.G.S.Prof.Pap.***574-A**,1-17.

Miesch,A.T.(1969a).The constant sum problem in geochemistry. In: Merriam(1969),qv,161-176.

Miesch,A.T.(1969b).Critical review of some multivariate procedures in the analysis of geochemical data. *J.Math.Geol.* **1**,171-184.

Miesch,A.T.(1972).Sampling problems in trace element investigations of rocks. *Ann.N.Y.Acad. Sci.***199**:95-104.

Miesch,A.T.(1976a).Log transformations in geochemistry.*J.Math.Geol.* **9**:191-194.

Miesch,A.T.(1976b).Q-mode factor analysis of compositional data. *Comput.Geosci.***1**,147-159.

Miesch,A.T.(1976c).Interactive computer program for petrologic modelling with extended Q-mode factor analysis. *Comput.Geosci.***2**,439-492.

Miesch,A.T.(1976d).Q-mode factor analysis of geochemical and petrological matrices with constant row sums. *U.S.G.S. Prof.Pap.* **574-G,**1-47.

Miesch,A.T.(1979).Vector analysis of chemical variation in the lavas of Paracutin volcano,Mexico. *J.Math.Geol.* **11**,345-372.

Miesch,A.T.(1980).Scaling variables and interpretation of eigenvalues in principal component analysis of geologic data. *J.Math.Geol.***12**,525-538

Miesch,A.T.(1981a).Computer methods for geochemical and petrologic mixing processes.In: Merriam(1981b),qv,243-265.

Miesch,A.T.(1981b).Estimation of the geochemical threshold and its statistical significance. *J.Geochem.Explor.* **16**, 49-76.

Miesch,A.T.(1983).Correspondence analysis in geochemistry.*J.Math. Geol.***15**,501-504.

Miesch,A.T. & Reed,B.L.(1979).Compositional structures in two batholiths of circum-Pacific N.America. *U.S.G.S. Prof.Pap.* **574-H**.

Miesch,A.T.,Chao,E.C. & Cuttita,F.(1966).Multivariate analysis of geochemical data on tektites.*J.Geol.***74**,673-691.

Mihalasky,M.J., Mutschler,F.E., Etienne,J.E. & Gordon,T.L. (1987). *GOLDY — a geologic and economic database for giant gold lode camps of North America*. Eastern Washington University, Cheney, Washington (magnetic tape).

Millendorf,S.A.,Srivastava,G.S.,Dyman,T.A.,Brower,J.C.(1979). A FORTRAN program for calculating binary similarity coefficients. *Comput. Geosci.***4(3)**:307-311.

Miller,C.D.(1979).A statistical method for relative-age dating of moraines in the Sawatch range,Colorado.

Bull.Geol.Soc.Am.Part 2,**90**:1153-1164.

Miller,H.G.(1978).Three statistical tests for 'goodness-of-fit'.*Can.J.Earth Sci.***15(1)**:171-172.

Miller,M.K. & Myers,M.A.(1984).*Presenting the Macintosh.*Dilithium,Beaverton,OR.

Miller,R.G.(1974).The jackknife - a review. *Biometrika* **61**, 1-17.

Miller,R.G.(1986). *Beyond ANOVA: basics of applied statistics.* Wiley, New York, 317 pp.

Miller,R.L. & Goldberg,E.D.(1955).The normal distribution in geochemistry. *Geochim. Cosmochim.Acta.***8**,53-62.

Miller,R.L. & Kahn,J.S.(1962).*Statistical analysis in the geological sciences.* Wiley,New York,483pp.

Miller,S.L. & Morrison,D.L.(1985).What can be done with computerized gold data-values? *Tr.Geol.Soc.S.Afr.* **88**,99-108.

Minoux,L.(1986). *Mathematical programming: theory and algorithms.* Wiley, New York, 489 pp.

Minster,J.F.,Minster,J.B.,Treuil,M. & Allegre,J.(1977).Systematic use of trace elements in igneous processes.Part III: inverse problem: the fractional crystallisation process in volcanic suites.*Contrib.Mineral. Petrol.***61**,49-77.

Miranda,A.(1977).Os programas NORROCK e IUGS pela calculação normativa dalguns parámetros de rochas ígneas. *Comun.Geol.Serv.Port.***62**:325-334.

Miranda,A.(1979).Breva nota sobre a analise multivariate de dados químicos e rochas lamprofíricas portuguêses. *Comun.Geol.Serv.Port.***64**,61-62.

Miranda,A.M. & Roquette,M.J.(1977).A applicação dos gráficos de computador aos dados geoquímicos. *Comun. Geol. Serv. Port.* **62**:309-323.

Mitchell,R.H. (1986). *Kimberlites: mineralogy, geochemistry & petrology.* Plenum Press, New York, 442 pp.

Mock,C.M., Elliott,B.G.,Ewers,G.R. & Lorenz,R.P.(1987).Gold deposits of Western Australia: BMR datafile (MINDEP). *Resour.Rep.Aust.Bur.Miner.Resour.***3**, 34pp hard copy (+ 5.25" floppy disks or mainframe tapes).

Mogarovskii,V.V.(1962).Once again on the correlation among elements in natural objects.*Geochemistry* **9**,969-972.

Moiola,R.J. & Spencer,A.B.(1979).Differentiation of Eolian deposits by discriminant analysis. *U.S.G.S.Prof.Pap.* **1052**:53-58.

Molenaar,W.(1970).*Approximation to the Poisson, binomial and hypergeometric distribution functions.* Mathematisch Centrum, Amsterdam, 160 pp.

Moll,R.N.(1985).*Macintosh Pascal.* Houghton Mifflin,Boston.

Montgomery,D.C.(1982). *Introduction to linear regression analysis.* Wiley, New York, 504 pp.

Mood,A.M.(1954).On the asymptotic efficiency of certain nonparametric two-sample tests. *Ann.Math.Stat.***25**,514-522.

Mood,A.M.,Graybill,F.A. & Boes,D.C.(1974).*Introduction to the theory of statistics.*(3rd Ed). McGraw-Hill, 76pp.

Moore,C.B.,Pratt,D.D. & Parsons,M.L.(1977).Application of pattern recognition to the classification of metal-rich meteorites.*Meteoritics* **12(3)**:314-318.

Moore,C.E.,Grethen,B.,Rekoske,K.K. & Smith,B.L.(1986).Exploration software and data-sources for micros. *Spec.Publ. Geol.Soc.Houston (USA)*,167pp.

Moore,F.(1979).Some statistical calculations concerning the determination of trace constituents.*Geostandards Newsletter* **3**,105-108.

Moore,W.E.(1985).Key-word bibliography searches by personal computer. *J.Geol.Educ.***33**,38-39.

Moran,P.A.P.(1971).Estimating structural and functional relationships. *J.Multivar.Anal.***1**,232-255.

Morgan,C.L.(1985).*The hidden powers of the Macintosh.*New American Library,New York.

Morgan,C.O. & McNellis,J.M.(1971).Reduction of lithologic-log data to numbers for use in a digital computer. *J.Math.Geol.* **3**,79-86.

Morris,P.A.(1984).MAGFRAC: a BASIC program for least-squares approximation of fractional crystallisation. *Comput.Geosci.* **10**,437-444.

Morrison,D.F.(1967).*Multivariate statistical methods.*McGraw-Hill,New York,338pp.

Moses,L.E.(1952).A nonparametric test for scale.*Psychometrika* **17**,239-247.

Moshier,S.L.(1986).Computer approximations.*Byte* **11(4)**,161-176.

Mosteller,F. & Tukey,J.W.(1977).*Data analysis and regression.*Addison-Wesley, Reading,Mass.,588pp.

Mottana,A.,Sutterlin,P.G. & May,R.W.(1971).Factor analysis of garnets and omphacites: a contribution to the geochemical classification of eclogites. *Contrib.Mineral.Petrol.***31**,238-250.

Mulaik,S.A.(1972).*The foundations of factor analysis.*McGraw-Hill,New York.

Mulargia,F.,Tinti,S. & Boschi,E.(1985).A statistical analysis of flank eruptions on Etna volcano. *J.Volcanol. Geotherm.Res.* **23**,263-272.

Mullett,G.M. & Murray,T.W.(1971).A new method for examining rounding error in least-squares regression computer programs.*J.Am.Stat.Assoc.***66**,496-

Mundry,E.(1972).On the resolution of mixed frequency distributions into normal components. *J.Math.Geol.***4**,55-60.

Murdoch,J. & Barnes,J.A.(1974).*Statistical tables for science,engineering, management and business studies.* (2nd Ed.) Macmillan,London.

Mutschler,F.E.,Rougon,D.J. & Lavin,O.P.(1976).PETROS — a data-bank of major element chemical analyses of igneous rocks for research and teaching. *Comput.Geosci.***2**,51-57.

Mutschler, F.E., Griffin,M.E., Stevens,D.S. & Ludington,S. (1986). *A bibliography of alkaline roc nrltd ieradpsit,Noh American Cordillera.* Eastern Washington University, Cheney, Washington (magnetic tape).

Myers,D.E.(1986).Matheronian geostatistics: quo vadis?: *Comment.J.Math. Geol.***18**,699-700.

Myers,R.E.(1986).*Microcomputer graphics for the IBM-PC.*Addison-Wesley,Reading Mass.,268pp.

NAG (Numerical Algorithms Group, 1987). *NAG FORTRAN Library, Mini Manual Mark 12.* NAG, Oxford.

Naney,M.T.,Crowl,D.M. & Papike,J.J.(1976).The Apollo 16 drill core: statistical analysis of glass chemistry and the characterization of a high alumina-silica poor glass.*Geochim.Cosmochim.Acta Supp.*(Proc.7th.Lunar Sci.Conf.), **7(1)**:155-184.

Nanninga,P.M. & Davis,J.R.(1984).*GENERAL, a microcomputer-based expert system.* CSIRO Inst.Biol.Resour., Canberra.

Naylor,G.F.K. & Enticknap,E.(1981).*Statistics simplified.*Harcourt Brace Jovanovich.

NBS (National Bureau of Standards, 1953). *Probability tables for the analysis of extreme values.* US Govt.Printing Office, Washington DC.

Neave,H.R.(1978).*Statistics tables for mathematicians,engineers,economists and the behavioural and management sciences.*Allen & Unwin,London,89pp.

Neave,H.R.(1979).Quick and simple tests based on extreme observations. *J.Quality Technol.***11**, 66-79.

Neave,H.R.(1981).*Elementary statistics tables.*Allen & Unwin,London,48pp.

Neffendorf,H.(1983).Statistical packages for microcomputers: a listing. *The American Statistician* **37**,83-85.

Neilson,J.(1980).The European Documentation Centre as a source of information for geology. *Geol.Soc.Lond.Misc.Pap.* **12**,5-12.

Neilson,M.J. & Brockman,G.F.(1977).The error associated with point counting. *Am.Mineral.* **62**:1238-1244.

Neumann,H.,Mead,J. & Vitialano,C.J.(1984).Trace element variation during fractional crystallisation as calculated from the distribution laws. *Geochim.Cosmochim.Acta* **6**,90-100.

Nichol,I. & Webb,J.S.(1967).The application of computerized mathematical and statistical procedures to the interpretation of geochemical data.*Proc. Geol.Soc.Lond.***1642**,186-99.

Nichol,I.,Garrett,R.G. & Webb,J.S.(1966).Automatic data plotting and mathematical and statistical interpretation of geochemical data. *Geol.Surv.Can.Pap.***66-54**,195-210.

Nicolayson,L.O.(1971).Graphic interpretation of discordant age measurements of metamorphic rocks. *Ann.New York Acad.Sci.***91**,198-206.

Nidd,E. & Ambrose,J.W.(1971).Computerized solution for some problems of fold geometry.*Can.J.Earth Sci.***8**,688-693.

Nie,N.H.,Hull,G.H.,Jenkins,J.G.,Steinbrenner,K. & Bent,D.H.(1975).*SPSS: statistical package for the social sciences.* McGraw-Hill, 675pp.

Nielsen,R.L.(1985).EQUIL: a program for the modelling of low-pressure differentiation processes in natural mafic magma bodies. *Comput.Geosci.***11**:531-546.

Nigam,R.(1987).Distribution,factor analysis and ecology of benthic forams within inner shelf regime of Venguria-Bhatkal sector,West Coast,India. *J.Geol.Soc.India* **29**,327-335.

Ningsheng,W. (1986). Creation of the Chinese geological bibliographic database. *Proc.3rd Int.Conf.Geosci.Info. (Adelaide, S.Australia, June '86)*, Vol.1, 95-99.

Nishiwaki,N.(1979).Simulation of bed-thickness distribution based on waiting time in the Poisson process.In: Gill & Merriam(1979), qv,17-32.

Nishiwaki,N. (1986). Database on fossil specimens deposited in Japan. *Proc.3rd Int.Conf.Geosci.Info. (Adelaide, S.Australia, June '86)*, Vol.1, 62-70.

Noble,B.(1964).*Numerical methods.*Oliver & Boyd,Edinburgh.

Noether,G.(1956).Two sequential tests against trend.*J.Am.Stat.Assoc.***51**,440-450

Noether,G.(1971).*Introduction to statistics-a fresh approach.*Houghton Miffin,Boston,253pp.

NORSIGD (1980). *GPGS-F users' guide.* 4th Ed. Tapir Press.

North,G.W. (1986). Advances in geoscience information handling. *Proc.3rd Int.Conf.Geosci.Info. (Adelaide,S.Australia, June '86)*, Vol.1, 217-222.

Norusis,M.(1983).*SPSS^x introductory statistics guide.* McGraw-Hill,New York.

Nosal,M. & Vrbik,J.(1982).Stratigraphic analysis and the asymptotic distribution of the coefficient of cross-association. *J.Math.Geol.***14**,11-36.

Novak,T., Sanford,R.L. & Wang,Y.J. (19832). *Use of computers in the coal industry.* Soc.Ming.Eng.AIME.

Nystrom, D.(1987). Co-operative geographic information systems within the U.S.G.S. *Proc. 7th Ann. ESRI User Conf.* ESRI, California.

Obial,R.C.(1970).Cluster analysis as an aid in the interpretation of multi-element geochemical data. *Tr.Inst.Ming.Metall.* **79**,B175-180.

Odell,J.(1976).An introduction to the LSDO2 system for rock description. *Comput.Geosci.* **2**,501-505.

Odell,J.(1977a).Description in the geological sciences and the lithostratigraphical descriptive system LSDO2. *Geol.Mag.* **114**,81-163.

Odell,J.(1977b).LOGGER,a package which assists in the construction and rapid display of stratigraphical columns from

field data. *Comput.Geosci.* **3**,347-379.
Olea,R.A.(1975). Optimum mapping techniques using regionalized variable theory. *Kansas Geol. Surv.Series on Spatial Anal.* **2**, 137 pp.
Olea,R.A.(1977).Measuring spatial dependence with semivariograms.*Kansas Geol. Surv.Series on Spatial Anal.* **3**, 29 pp.
Olhoeft,G.R.(1978).Algorithm and BASIC program for ordinary least-squares regression in two and three dimensions. *U.S.G.S.Open File* **78/876**:8pp.
Olmstead,P.S. & Tukey,J.W.(1947).A corner test of association.*An..Math.Stat.***18**, 495-513.
Olson,A.C.(1986).Mapping applications for the microcomputer.In: Hanley & Merriam(1986),qv,215-224.
Oltz,D.F.(1971).Cluster analysis of late Cretaceous-Early Tertiary pollen and spore data. *Micropalaeontology* **17**,221-232.
Open University (UK) course team. (1973). *Continuous distributions, the EDCALC program, bivariate distributions.* Notes for Statistics course (MDT241), units 4,5 & 6. Open Univ. Press.
Open University (UK) course team. (1974a). *Estimation, regression, descriptive and economic statistics.* Notes for Statistics course (MDT 241), units 10,11 & 13. Open Univ. Press.
Open University (UK) course team (1974b). *Correlation, multivariate distributions, sampling.* Notes for Statistics course (MDT241), units 7,8 & 9. Open Univ. Press.
Open University (UK) course team. (1985). *Microcomputers in action in the classroom (kit).* Open Univ. Press.
Orris,J.B.(1982).The role of microcomputers and statistical computing. In: Rustagi & Wolfe(1982), qv,485-496.
Osborne,R.H.(1967).The American Upper Ordovician standard/R-mode factor analysis of Cincinnatian limestone. *J.Sed.Petrol.***37**,649-657.
Osborne,R.H.(1971).The American Upper Ordovician standard XIV.Markov and of typical Cincinattian sedimentation, Hamilton Cty,Ohio.*J.Sed.Petrol.***41**,444-9.
Otsu,H.,Matsuda,Y. & Vubota,R.(1985).[Determination of two-dimensional frequency distributions of geochemical data.] In Japanese,English summary.*Ming.Geol.(Tokyo)***35**,67-75.
Ottonello,G.(1983).Trace elements as monitors of magmatic processes: I) Limits imposed by Henry's Law problem; & II) Compositional effect of silicate liquids.In: Augustithis(1983),qv,39-82.
Ousey,J.R.(1984).Using a microcomputer-driven digitizer for laboratory courses and student research projects. *J.Geol.Educ.***32**,182-183.
Owen,D.B.(1962).*Handbook of statistical tables.*Addison-Wesley,Reading,Mass,580pp.
Page,B.G.N.,Bennett,J.D.,Cameron,N.R.,Bridge,D.McC.,Jeffrey,D.H.,Keats,W. & Thaib,J.(1978). Regional geochemistry, geological reconnaissance mapping and mineral exploration in northern Sumatra,Indonesia.*Proc.11th Commonwealth Ming.Metall.Congr.,Hong Kong*,455-462.
Palacios,T. & Faria,A.F.(1982).BASIC programs for the 'Apple2+' microcomputer for use in petrochemical studies and diagram projection.*Garcia Orta Ser.Geol.* **5(1-2)**.101-8.
Panofsky,H.A. & Brier,G.W.(1965). *Some applications of statistics to meteorology.* Penn.State Univ.PA, 224pp.
Pardee,M. (1984). *Pascal primer for the IBM PC.* Phone/Waite, New York, 292 pp.
Park,R.A.(1974).A multivariate analysis strategy for classifying palaeoenvironments. *J.Math.Geol.* **6**,333-352.
Parker,R.J.(1981).GEOIC: an interactive terminal-based geochemical data-processing system. *Comput.Geosci.* **7**,287-296.
Parkinson,D.(1954).Quantitative studies of brachiopods from the lower Carboniferous of England*J.Palaeont.***28**,367-81.
Parks,J.M.(1966).Cluster analysis applied to multivariate geological problems. *J.Geol.***74**,703-715.
Parsley,A.J.(1971).Application of autocorrelation criteria to the analysis of mapped geologic data from the Coal Measures of Central England.*J. Math.Geol.***3**,281-295.
Parsley,A.J. & Doveton,J.D.H.(1969).The role of some statistical and mathematical methods in the interpretation of regional geochemical data. *Econ.Geol.***64**,830.
Parzen,E.(1962).*Stochastic processes.*Holden-Day,San Fransisco.
Patel,J.K.(1982).*Handbook of the Normal distribution.* Dekker, New York, 337 pp.
Patil,G.P.,Kotz,S. & Ord,J.K.(1975).*Statistical distributions in scientific work.*Reidel,Dordrecht.
Pauncz,I.(1978).English language publications on the French school of geostatistics*J.Math. Geol.***10(2)**:253-260.
Pavlidis,T.(1982)*Algorithms for graphic and image processing.*Computer Sci. Press,Rockville (Maryland),416pp.
Pavlova,I.G. & Agukina,Z.V.(1967).Application of the discriminant function m method for separating metasomatic rocks of similar composition*Dokl.Acad.Sci. USSR* **177**:194-196.
Peach,P.A. & Renault,J.R.(1965).Statistical analysis of some characteristics of British Columbia Molybdenum occurrence. *Econ.Geol.***60**,1510-15.
Pearce,J.A.(1976).Statistical analysis of major element patterns in basalts. *J.Petrol.***17**,15-43.
Pearce,J.A.(1987).An expert system for the tectonic characterization of ancient volcanic rocks. *J.Volcanol.Geotherm.Res.* **32**,51-66.
Pearce,J.A. & Cann,J.R.(1971).Ophiolite origin investigated by discriminant analysis using Ti,Zr and Y.*Earth Planet. Sci.Lett.* **12**,339-349.
Pearce,J.A. & Cann,J.R.(1973).Tectonic setting of basic volcanic rocks determined using trace element analysis. *Earth Planet. Sci.Lett.***19**, 290-300.
Pearce,T.H.(1970).A contribution to the theory of variation diagrams. *Contrib.Mineral.Petrol.* **19**,142-157.

Pearce,T.H.(1983).An interactive computer program for simulating the effects of olivine fractionation from basaltic and ultrabasic liquids. *J.Geol.Educ.***31**,206-207.

Pearson,E.S.(1936). Note on probability levels for $\sqrt{b_1}$. *Biometrika* **28**, p.306.

Pearson,E.S. & Hartley,H.O.(1966).*Biometrika tables for statisticians, Vol.I.* (3rd Ed).Cambridge Univ. Press.

Pericott,R.H. & Allison,D.C.S.(1984).*Pascal for FORTRAN programmers.* Computer Science Press,335pp.

Peron,J. & Nemec,W.(1977).[Statistical computation of geological orientation data — a computer program for density diagrams]. In Polish, English summary p.647.*Kwart.Geol.* **21**,633-647.

Peschel,G. & Zezel,J.N.(1977).[Study of the statistical properties of measures of similarity for comparing geological objects]. In German. *Z.Geol.Wiss.(Berlin)*,**5(8)**:1021-1027.

Petitpierre,E. & Boivin,P.(1983).CRYSTALLISATION: a computer program for modelling the crystallisation of a magmatic liquid.*Comput.Geosci.* **9**,455-462.

Pettofrezzo,A.J.(1978).*Matrices and transformations.*Dover,New York,133pp.

Philip,G.M. & Watson,D.F.(1985). Theoretical aspects of grade-tonnage calculations. *Int.J.Ming.Eng.***3**,149-154.

Philip,G.M. & Watson,D.F.(1986a). A method for assessing local variations among scattered measurements*J.Math.Geol.* **18**, 759-764.

Philip,G.M. & Watson,D.F.(1986b).Matheronian geostatistics — quo vadis? *J.Math.Geol.***18**,93-117 & 505-509.

Philip,G.M. & Watson,D.F.(1987a). Some speculations on the randomness of Nature. *J.Math.Geol.* **19**, 571-573.

Philip,G.M. & Watson,D.F.(1987b).Probabilism in geological data analysis. *Geol.Mag.* **124**, 577-583.

Philip,G.M. & Watson,D.F.(1988). Determining the representative composition of a set of sandstone measurements. *Geol.Mag.* **125** (in press).

Phipps,T.E.(1976).The inversion of large matrices.*Byte* **11(4)**,181-188.

Piboule,M.(1977).[The use of discriminant analysis to determine the nature of the parent magmas of amphibolites,and its application to metabasic rocks from Rouergue and Limousin (Massif Central Francaise).]*Bull.Soc.Geol.Fr. Ser.*7, **19(5)**:1133-1143.

Piispanen,R.(1977).A FORTRAN IV program for the calculation of the petrological Niggli-Barth norms — a brief communication. *Bull.Geol.Soc.Finlande* **49(1)**:37-38.

Piispanen,R. & Alapieti,T.(1977).A factor analysis study of the major element geochemistry of granitic rocks. *Bull.Geol.Soc.Finlande* **49(2)**:143-150.

Pillai,K.C.S.(1960).*Statistical tables for tests of multivariate hypotheses.* Statistics Center,Univ.Phillipines, Manila.

Pincus,H.J.(1953).The analysis of aggregates of orientation data in the earth sciences.*J.Geol.* **61**,482-509.

Pincus,H.J.(1956).Some vector and arithmetic operations on two-dimensional orientation variates, with application to geologic data.*J.Geol.***64**,533-557.

Pirie,W.R.(1982).A graduate service level course in nonparametric methods: the technique of subject matter reports using a computer package. In: Rustagi & Wolfe(1982),qv,497-502.

Pirie,W.R. & Hamden,M.A. (1972).Somre revised continuity corrections for discrete data.*Biometrics* **28**,693-701.

Pirkle,H.,Howell,J.O.,Wecksung,G.W.,Duran,B.S. & Stablein,N.K.(1984).An example of cluster analysis applied to a large geological data-set: aerial radiometric data from Copper Mountain, Wyoming.*J.Math.Geol.***16**,479-498.

Placet,J.G.(1977).Description et usage d'un databank dans le géologie des mines. *Sci.Terre Inf.Geol.(Nancy)***10**:41-82.

Plackett,R.L.(1981).*The analysis of categorical data.*Griffin,London,207pp.

Plant,J.A. & Moore,P.J.(1979).Regional geochemical mapping and interpretation in Britain. *Phil.Tr.R.Soc.Lond.* **B288**, 95-112.

Plant,J.A., Watson,J.V. & Green,P.M.(1984). Moine-Dalradian relationships and their palaeotectonic significance. *Proc.Soc.Lond.***A395**,185-202.

Plant,J.A.,Forrest,M.D.,Hodgson,J.F.,Smith,R.T.S. & Stevenson,A.G.(1986).Regional geochemistry in the detection and modelling of mineral deposits.In: Thornton & Howarth(1986),qv,103-139.

Ploquin,A. & Royer,J.J.(1977).[The anorthosite-farsundite suite: an example of processing a geochemical file (major elements)].*Bull.Soc.Géol.Fr.Sér.*7,**19**: 959-964.

Podol'sky,Y.V.(1962).Linear parageneses of the main rock-forming minerals of the alkalic rocks in the central part of the Kola Peninsula.*Dokl.Acad.Sci. USSR* **146**:155-158.

Podol'sky,Y.V.(1975).Variation of chemical composition of alkaline granitic rocks in the central part of the Kola Peninsula. *J.Math.Geol.***7**, 215-236.

Poole,L.(1984).*MacWork,MacPlay: creative ideas for fun and profit on your Apple Macintosh.* Microsoft Publishing,Washington/Penguin,Harmondsworth.

Porter,T.M., Thorne,J.W., Radke, S.G. & Middleton,C.A. (1986). An integrated approach to data and information management for mineral exploration at CRA Exploration Pty Ltd., *Proc.3rd Int.Conf.Geosci.Info. (Adelaide, S.Australia, June '86)*, Vol.1, 192-201.

Potenza,R.(1973).A geomathematical investigation of syntexis in a gabbroic formation.*J.Math. Geol.***5**,321-340.

Potter,P.E. & Blakeley,R.F.(1968).Random processes and lithological transition. *J.Geol.***76**, 154-170.

Potter,P.E.,Shimp,N.F. & Witters,J.(1963).Trace elements in marine and freshwater argillaceous sediments. *Geochim.Cosmochim.Acta* **27**,669-694.

Potthoff,R.F.(1974).A non-parametric test of whether two simple regression lines are parallel. *Ann.Stat.***2**,295-310.

Pouclet,A.,Menot,R.P. & Piboule,M.(1981).Classement par l'analyse factorielle discriminante des laves du rift de l'Afrique Centrale (Zaire,Rwanda, Uganda).*C.R.Acad.Sci.Paris,Ser.*2, **292:8**,679-684.

Pouclet,A.,Menot,R.-P. & Piboule,M.(1984).Differénciation des laves de l'Afrique Centrale (Rift Ouest).Contribution de l'analyse statistique multivariée.*Neues Jb.Mineral.Abh.***149**,283-308.

Powell,F.C.(1982).*Statistical tables for social,biological and physical sciences.* Cambridge Univ.Press, 96 pp.

Powell,R.(1984).Inversion of the assimilation and fractional crystallisation (AFC) equations: characterization of contaminants from isotope and trace element relationships in volcanic suites.*J.Geol.Soc.Lond.***141**,447-452.

Powell,R.(1985).Regression diagnostics and robust regression in geothermometer/ geobarometer calibration — the garnet-clinopyroxene geothermometer revisited. *J.Metamorph.Geol.***3**,231-244.

Prelat,A.E.(1977).Discriminant analysis as a method of predicting mineral occurrence potentials in central Norway. *J.Math.Geol.***9(4)**:343-367.

Press,S.J. & Wilson,S.(1978).Choosing between logistic regression and discriminant analysis. *J.Am.Stat.Assoc.* **73**,699-705.

Press, W.H.(1986). *Numerical recipes example book, Pascal.* Cambridge Univ. Press.

Press,W.H.,Flannery,B.P.,Teukolsky,S.A. & Westerling,W.T.(1986).*Numerical recipes: the art of scientific computing.* Cambridge Univ.Press: 818pp.

Price,R.A. (1986). Geoscience information — a framework for formulating and implementing policies on resource development. *Proc.3rd Int.Conf.Geosci.Info. (Adelaide,S.Australia, June '86)*, Vol.2, 1-14.

Price,R.J. & Jorden,P.R.(1979).A FORTRAN IV program for foraminiferid stratigraphical correlation and palaeoenvironmental interpretation. *Comput.Geosci.***3**,601-615.

Prinz,M.(1967).Geochemistry of basaltic rocks: trace elements.In: Hess,H.H. & Poldevaart,A. (eds.) *Basalts,* 271-323. Wiley,New York.

Proschan,F.(1953).Rejection of outlying observations.*Am.J.Phys.***21**,520-525.

Provost,A. & Allègre,C.J.(1979).Process identification and search for optimal differentiation parameters from major element data: general presentation with emphasis on fractional crystallisation processes.*Geochim.Cosmochim.Acta* **43**,487-501.

Pruett,N.J. (1986). State-of-the-art of geoscience libraries and information systems. *Proc.3rd Int.Conf.Geosci.Info. (Adelaide, S.Australia, June '86)*, Vol.2, 15-30.

Puri,M.L. & Sen,P.K.(1971).*Non-parametric methods in multivariate analysis.* Wiley,New York.

Puri,M.L.,Sen,P.K. & Gokhale,D.V.(1970).On a class of rank order tests for independence in multivariate distributions. *Sankya*,**A.32**.

Quade,D.(1966).On the analysis of variance for the k-sample problem.*Ann.Math. Stat.***37**:1747-1758.

Radford,N.W.(1987).Assessment of error in sampling.*Bull.Aust.Inst.Geoscientists* **7**,123-144.

Rahmann,N.A.(1972).*Practical exercises in probability and statistics.* Hafner,New York.

Raikar,P.S.(1987).Univariate and multivariate stochastic modelling and forecasting of silica content in iron ores of northern Goa. *J.Geol.Soc.India* **29**,349-356.

Ramden,H.A.(1977).*JCL and advanced FORTRAN programming.*Elsevier Methods in Geomathematics,2. Elsevier, Amsterdam.

Ramsayer,G.R. & Bonham-Carter,G.F.(1974).Numerical classification of geological patterns characterised by binary variables. *J.Math.Geol.***6**,59-72.

Randles,R.H. & Hogg,R.V.(1973).Adaptive distribution-free tests.*Communications in Statistics* **2**,337-356.

Randles,R.H.,Ramberg,J.S. & Hogg,R.V.(1973).An adaptive procedure for selecting the population with largest location parameter.*Technometrics* **15**,769-778.

Randles,R.H.,Brofitt,J.D.,Ramberg,J.S. & Hogg,R.V.(1978).Discriminant analysis based on ranks.*J.Am.Stat.Assoc.* **73(362)**,379-384.

Rao,A.B. & Sial,A.N.(1972).Observations on alkaline plugs near Forteleza city,Brazil. *Proc.24th Int.Geol.Congr.***14**, 56-61.

Rao,C.P. & Naqvi,I.H.(1977).Petrography,geochemistry and factor analysis of a Lower Ordovician face sequence, Tasmania, Australia.*J.Sed.Petrol.***47**: 1036-1055.

Rao,C.R.(1952).*Advanced statistical methods in biometric research.* Wiley,New York.

Rao,C.R.(1964).The use and interpretation of principal components analysis in applied research. *Sankhya* **A26**,329-358.

Rao,S.V.L.N.(1971).Correlation between regression surfaces based on direct comparison of matrices.*Modern Geol.* **2**,173-177.

Rao,S.V.L.N. & Rao,M.S.(1970).A study of residual maps in the interpretation of geochemical anomalies. *J.Math.Geol.***2**,15-24.

Rassam,G.N.(1983).The Georef online bibliographic database as an educational tool.*J.Geol.Educ.* **31**,26-30.

Rassam, G.N. & Gravesteijn, J. (1986). Factual and bibliographical information in geoscience: the time for integration. *Proc.3rd Int.Conf.Geosci.Info. (Adelaide,S.Australia, June '86)*, Vol.2, 129-139.

Raup,D.M.(1981).Computers as a research tool in palaeontology.In: Merriam(1981b),qv,267-282.

Rayner, J.N.(1971). *An introduction to spectral analysis.* Pion,London, 174pp.
Read,W.A.(1969).Analysis of simulation of Namurian sediments in central Scotland using a Markov process model. *J.Math.Geol.* **1**, 199-219.
Read,W.A.(1976).An assessment of some quantitative methods of comparing lithological succession data.In: Merriam(1976c),qv,33-52.
Read,W.A. & Dean,J.M.(1972).Principal components analysis of lithological variables from some Namurian paralic sediments in central Scotland. *Bull.Geol.Surv.G.B.***40**,83-99.
Read, W.A.,Dean,J.M. & Cole,A.J.(1971).Some Namurian paralic sediments in central Scotland: an investigation of depositional environments and facies changes using iterative fitting trend surface analysis.*Q.J.Geol.Soc.Lond.* **127**,137-176.
Read,W.A. & Sackin,M.J.(1971).A quantitative comparison using cross-association of vertical sections of Namurian (E1) paralic sediments in the Kincardine basin.*Inst.Geol.Sci.(UK)* Rept.**71/14**,21pp.
Reeves,M.J. & Jucas,J.M.(1982).Microcomputer assisted learning applications in geological sciences. *Geol.Soc.Lond. Misc.Pap.* **14**,198-223.
Reidel,S.P. & Long,P.E.(1980).Discriminant analysis as a method of flow identification and correlation in layered basalt provinces (Columbia River Basalt Group).*Abstr.Progr.(Boulder)* **12:7**,p.507.
Reinhardt,J. & Van Driel,J.N.(1978).An abbreviation key for the DASCH system: a lithostratigraphic information storage and retrieval system. *U.S.G.S.Open File Rept.***78(961)**:28pp.
Rendu,J.M.(1978).An introduction to geostatistical methods of mineral evaluation.*S.Afr.Inst. Ming.Metall.Geostatistics Monograph* **2**,84pp.
Rendu,J.M.(1979).Normal and lognormal estimation.*J.Math.Geol.***11**: 407-422.
Rendu,J.M.(1984a).Geostatistical modelling and geological controls. *Tr.Inst.Ming.Metall.***166**,B172.
Rendu,J.M.(1984b).Interactive graphics for semivariogram modelling. *Mining Engineering,* **Sept.1984**,1223-1340.
Reyment,R.A.(1966).Preliminary observations on gastropod predations in the Western Niger delta. *Palaeogeogr. Palaeoclim. Palaeoecol.***2**,81-102.
Reyment,R.A. (1968). Multivariate statistical analysis in geology. *Publ.Palaeontol.Inst.Univ.Uppsala* **74**, 18 pp.
Reyment,R.A.(1971a).*Introduction to quantitative palaeoecology.* Elsevier,Amsterdam, 226 pp.
Reyment,R.A.(1971b).Vermuteter Dimorphismus bei der Ammonitengattung *Benuites. Bull.Geol.Inst.Univ.Uppsala N.Ser.***3**, 1-18.
Reyment,R.A.(1972). Models for studying the occurrence of lead and zinc in a deltaic environment. In: Merriam (1972), qv, 233-246.
Reyment,R.A.(1973). Factors in the distribution of fossil cephalopods. 3. Experiments with exact models of certain sheel types. *Bull.Geol.Inst.Univ.Uppsala N.Ser.***4**, 7-41.
Reyment,R.A.(1974).The age of zap.In: Merriam(1974),qv,19-26.
Reyment,R.A.(1975a). Analysis of a generic level transition in Cretaceous ammonites. *Evolution* **28**, 665-676.
Reyment,R.A.(1975b).Canonical correlation analysis of hemicytherid and trachyleberinid ostracods in the Niger delta. *Bull.Am.Assoc.Palaeontol.* **65**,141-145.
Reyment,R.A.(1976).Analysis of volcanic earthquakes of Asamayama (Japan). In: Merriam(1976a), 87-95.
Reyment,R.A.(1978).New trends in the multivariate analysis of palaeontologic data.In: Merriam (1978b),qv,57-64.
Reyment,R.A.(1979). Variation in ontogeny in *Bauchioceras* and *Gomboceras. Bull.Inst.Univ.Uppsala N.Ser.***8**, 89-111.
Reyment,R.A.(1981).The computer in palaeoecology.In: Merriam(1981b),qv, 283-304.
Reyment,R.A. & Banfield,C.F.(1976).Growth-free canonical variates applied to fossil foraminifera. *Bull.Geol.Inst. Univ. Uppsala* **7**,11-21.
Reyment,R.A. & Van Valen,L. (1969). *Buntonia olukundidin* sp.nov. (Ostracoda, Crustacea): a study of meristic variation in Palaeocene and recent ostracods. *Bull.Geol.Inst.Univ.Uppsala Ser.1*, 83-94.
Reyment,R.A.,Berthou,P.Y. & Moberg,B.A.(1976).Statistical recognition of terrestrial and marine sediments in the Lower Cretaceous of Portugal. In: Merriam(1976c),qv,53-60.
Reyment,R.A.,Blackith,R.E. & Campbell,N.A.(1984).*Multivariate morphometrics.* (2nd Ed).Academic Press,New York, 233pp.
Rhodes,J.M.(1969).The application of cluster and discriminant analysis to mapping granite intrusions.*Lithos* **2**,223-238.
Ribando,R.J.,Torrance,K.E. & Turcotte,D.L.(1978).Numerical calculations of the convective cooling of an infinite sill.*Tectonics* **50**:337-348.
Ribeiro,J.C. & Merriam,D.F.(1979).Quantitative analysis of depositional environments in the Reconcava Basin, Bahia, Brazil. In: Gill & Merriam (1979),qv,219-234.
Rich, C. (1987). Application of ARC/INFO to the national earthquake hazards reduction program. *Proc. 7th Ann. ESRI User Conf.* ESRI, California.
Richards,L.E.(1972).Refinement and extension of distribution-free discriminant analysis.*Appl.Stat.* **21**,174-176.
Rickwood,P.C.(1983).The use of cluster analysis in diverse geological problems: cluster analysis of geochemical data.In: Augustithis (1983),qv,115-124.
Rickwood,P.C.,Colon,E.,Dobos,V.J.,Guy,J.V.,Gwatkin,G.,Poole,R. & Wainwright,I.E. (1983).The use of cluster

analysis in diverse geological problems.The origin of basalt in the Blue Mountains near Sydney,New South Wales, Australia. In: Augustithis(1983),qv,124-135.

Ridgley,J. & Schnabel,R.W.(1978).A computer program designed to produce tables from alphanumeric data. *U.S.G.S. Open File Rept.***78/875**, 65pp.

Rieder,M.(1977).Micas: calculation of crystochemical formulas by a FORTRAN IV computer program. *Vestn.Ustred. Ustavu Geol.(Prague)* **52(6)**:333-342.

Ripley,B.P.(1981).*Spatial statistics.* Wiley,New York, 252 pp.

Robb,L.J.(1983).Trace element trends in granites and the distinction between partial melting and crystal fractionation processes: case studies from two granites in southern Africa.In: Augustithis (1983),qv,279-294.

Robinove,C.J.(1986).Principles of logic and the use of digital geographic information systems. *U.S.G.S.Circ.* **977**,19pp.

Robinson,E.A.(1981).*Time series analysis and applications.*Goose Pond Press,Houston.

Robinson,J.E.(1982).*Computer applications in petroleum geology.*Hutchinson Ross,164pp.

Robinson,J.E. & Carroll,S.(1977).Software for geologic processing of LANDSAT imagery. *Comput.Geosci.* **3(3)**,456-465.

Robinson,S.C.(1966).Interim Report of the Committee on storage and retrieval of geological data in Canada. *Pap.Geol.Surv.Can.* **66-43.**

Robinson,S.C.(1974).The role of a database in modern geology.In: Merriam (1974),qv,67-82.

Rock,N.M.S.(1976).Fenitisation around the Monchique alkaline complex,Portugal.*Lithos* **9**,263-279.

Rock,N.M.S.(1977).The nature and origin of lamprophyres: some definitions, distinctions and derivations. *Earth-Sci.Rev.***13,**123-169.

Rock,N.M.S.(1978).Petrology and petrogenesis of the Monchique alkaline complex, southern Portugal. *J.Petrol.***19**,171-214.

Rock,N.M.S.(1979).Petrology and origin of the type monchiquites and associated lamprophyres of Serra de Monchique, Portugal. *Tr.R.Soc.Edinburgh* **70**,149-70.

Rock,N.M.S.(1983).The Permo-Carboniferous camptonite-monchiquite dyke-suite of the Scottish Highlands and Islands: distribution,field and petrological aspects. *Rep.Inst.Geol.Sci.U.K.* **82/14.**

Rock,N.M.S.(1984a).Comparative geochemistry of nepheline syenites,tinguaites, phonolites and fenites from southern Portugal and East Africa.*Bol.Soc.Geol. Portugal* **21**,421-433.

Rock,N.M.S.(1984b).Nature and origin of calc-alkaline lamprophyres: minettes, vogesites,kersantites and spessartites. *Tr.R.Soc.Edinburgh: Earth-Sci.* **74,**193-227.

Rock,N.M.S.(1986a).The nature and origin of ultramafic lamprophyres: alnöites and related rocks.*J.Petrol.***27**,155-196.

Rock,N.M.S.(1986c).The geochemistry of Lewisian marbles.In: *The Lewisian and comparable high-grade gneiss terranes,* J.Tarney & R.G.Park (eds.) *Spec.Publ.Geol.Soc. London,* **27**,109-126.

Rock,N.M.S.(1986d).Value of chemostratigraphical correlation in metamorphic terrains: an illustration from the Colonsay Limestone,Inner Hebrides, Scotland.*Tr.R.Soc.Edinburgh: Earth Sci.***76**, 515-517.

Rock,N.M.S.(1986e).Chemistry of the Dalradian (Vendian-Cambrian) metalimestones, British Isles. *Chem.Geol.* **56**,289-311.

Rock,N.M.S.(1986f).NPSTAT: a FORTRAN-77 program to carry out non-parametric two- and multi-sample tests on geological data.*Comput.Geosci.***12**,757-777.

Rock,N.M.S.(1987a).A global database of analytical data for undersaturated syenitoids,trachytoids and phonolitoids. *Modern Geol.***11**,51-67.

Rock,N.M.S.(1987b).ANGLE: a FORTRAN-77 program to carry out one sample goodness-of-fit,two-sample and multi-sample tests on two-dimensional orientation data.*Comput.Geosci.* **13,**185-208.

Rock,N.M.S.(1987c).ROBUST: an interactive FORTRAN77 package for exploratory data analysis via robust estimates of location and scale,tests for normality and outlier assessment.*Comput.Geosci.* **13**, 463-494.

Rock,N.M.S.(1987d).CORANK: a FORTRAN-77 program to calculate matrices of Pearson,Spearman and Kendall correlation coefficients with pairwise treatment of missing data.*Comput.Geosci.* **13**, 659-662.

Rock,N.M.S.(1987e).The nature and origin of lamprophyres: an overview.In: *Alkaline igneous rocks.* J.G. Fitton & B.G.J.Upton (eds.) *Spec.Publ. Geol.Soc.London,* **30**,191-226.

Rock,N.M.S.(1987f). A general-purpose FORTRAN-77 program to generate normalized multi-element abundance diagrams in petrology and geochemistry. *J.Geol.Educ.* (in press).

Rock,N.M.S.(1987g).A FORTRAN program for tabulating and naming amphibole analyses according to the International Mineralogical Association Scheme. *Mineralogy & Petrology* **37**, 79-88.

Rock,N.M.S. (1988). Summary statistics in geochemistry: a study of the performance of robust estimates. *J.Math.Geol.* **20**, 243-275.

Rock,N.M.S. & Carroll,G.W. (1988). General-purpose and metamorphic ACF-AKF-AFM triangular diagram plotting programs for the Apple Macintosh microcomputer (in press).

Rock,N.M.S. & Duffy,T.R.(1986).REGRES: a FORTRAN-77 program to carry out non-parametric and 'structural' parametric solutions to bivariate regression equations.*Comput.Geosci.***12**,807-818.

Rock,N.M.S. & Hunter,R.H.(1987).Late Caledonian dyke-swarms of northern Britain: spatial intimacy between lamprophyric and granitic magmatism around the Ross of Mull pluton. *Geol.Rundschau* **76**,805-826.

Rock,N.M.S. & Macdonald,R.(1986).Petrology and origin of a peculiar lens of pelites,'limestones' and possible para-amphibolites from the Moines of the Ross of Mull,Scotland.*Proc.Geol.Assoc.* **97**,249-258.

Rock,N.M.S. & Waterhouse,K.H.(1986).The value of chemostratigraphical correlation in metamorphic terrains: an illustration from the Shinness and Armadale marbles,Sutherland,Scotland. *Proc.Geol.Assoc.***97**,347-356.

Rock,N.M.S. & Wheatley,M.R. (1988). Some experiences in integrating the use of micros and mainframes. *Comput.Geosci.* (in press).

Rock,N.M.S.,Jeffreys,L.A. & Macdonald,R.(1984).The problem of anomalous local limestone-pelite successions within the Moine outcrop; I: metamorphic limestones of the Great Glen area,from Ardgour to Nigg.*Scott.J.Geol.***20**, 383-406.

Rock,N.M.S.,Macdonald,R.,Walker,B.H.,May,F.,Peacock,J.D. & Scott,P.(1985). Intrusive metabasite belts within the Moine assemblage west of Loch Ness, Scotland: evidence for metabasite modification by country-rock interaction. *J.Geol.Soc.London* **142**,643-661.

Rock,N.M.S.,Macdonald,R. & Drewery,S.A.(1986a).Pelites of the Glen Urquhart serpentinite-metamorphic complex,W of Loch Ness.(Anomalous local limestone-pelite successions within the Moine outcrop; III). *Scott.J.Geol.***22**,179-202

Rock,N.M.S.,Macdonald,R.,Szucs,T. & Bower,J.(1986b).The comparative geochemistry of some Highland pelites. (Anomalous local limestone-pelite successions within the Moine outcrop,II). *Scott.J.Geol.***22**,107-126.

Rock,N.M.S.,Webb,J.A.,McNaughton,N.J. & Bell,G.(1987).Nonparametric estimation of averages and errors for small data-sets in isotope geoscience: a proposal.*Isot.Geosci.* **66**, 163-177.

Rock,N.M.S.,Hattie,J. & Brown,T.(1988). Macintosh statistical packages — a n end-users' resurveyview. *Wings for the Mind* (Australian Apple Consortium newsletter), **2** (in press).

Rocke,D.M.,Downs,G.W. & Rocke,A.T.(1982).Are robust estimators really necessary? *Technometrics* **24**,95-101.

Rogers,A.(1974).*Statistical analysis of spatial dispersions.*Pion,London,164pp.

Rogers,D.F. & Adams,J.A.(1976).*Mathematical elements for computer graphics.* McGraw-Hill,New York,239pp.

Rohlf,F.J. & Sokal,R.R.(1969).*Statistical tables.*W.H.Freeman,San Fransisco.

Rohrlich,V.,Lin,C. & Harbaugh,J.V.(1985).MARKTRAN II: an interactive FORTRAN program for calculating Markov transition probabilities of two-dimensional patterns.*Comput.Geosci.***11**:215-228.

Roloff,L. & Browder,G.T.(1985).*Geoscience software directory.*D.Reidel,Dordrecht.

Romaniuk,A.S. & Macdonald,R.J.C.(1979).A national information service in mining, mineral processing and extractive metallurgy.*Can.Inst.Ming.Metall.Bull.(Montreal)* **72**:138-142.

Romanova,M.A. & Sarmanov,O.V.(1970).*Topics in mathematical geology.*Consultants Bureau,New York,281pp.

Romesberg,H.B. & Marshall,K.(1985).CHITEST: a Monte-Carlo computer program for contingency table tests. *Comput.Geosci.* **11**,69-78.

Romesberg,H.B.,Marshall,K. & Mauk,T.P.(1981).FITEST: a computer program for "exact chi-square" goodness-of-fit significance tests.*Comput.Geosci.***7**,47-58.

Root,M.R.(1978).BULK: a computer program for calculating bulk chemical analyses from mineral phases observed in thin section.*Comput.Geosci.***4**,199-204.

Rose,A.W.,Hawkes,H.G. & Webb,J.S.(1979).*Geochemistry in mineral exploration.* Academic Press,657pp.

Rosenblatt,D.K.,Chayes,F. & Trochimczyk,J.(1978).An algebraic explanation of closure correlation among the coefficients of a principal component. *Carn.Inst.Wash.Yrbk.***77**:901-902.

Rosenfeld, A. & Huang, H.K. (1968).An application of cluster analysis to text and picture processing. *Techn.Rept.Computer Sci.Center Univ.Maryland, MD* **68-68**,64pp.

Royer,J.J. (1986). From statistics to expert systems: applications in geoscience. *Sci.de la Terre Sér.Informatique Géol.* **25**, 87 pp.

Royle,A.G.(1978).A probabilistic basis for ore reserve statements.*Mineral. Mag.***139(2)**:142-143.

Royle,A.G.(1980).Why geostatistics? *Eng.Ming.*,**May 1979**,92-120.

Royle,A.G.,Clark,I.,Brocker,P.I.,Parker,H.,Journel,A.,Rendu,J.M.,Sandefur,P.,Grant,D.C. & Mousset-Jones,P. (1980). *Geostatistics.* McGraw-Hill,New York.

Ruan,T.,Howarth,R.J. & Hale,M.(1985).Numerical modelling experiments in vapour geochemistry. I.Method and FORTRAN program.*Comput.Geosci.***11**:55-68.

Ruegg,N.R.(1976).Característicos da distribuição e direção dos elementos principais em rochas basálticos do basin Paraná. *Bol.Ist.Geol.(Sao Paulo)* **7**:81-106.

Russell,J.K.(1986).A FORTRAN77 computer program for the least squares analysis of chemical data in Pearce variation diagrams.*Comput.Geosci.*12,327-338.

Rustagi,J.S. & Wolfe,D.A.(1982).*Teaching of statistics and statistical computing.* Academic Press,548pp.

Ryan,T.A.,Joiner,B.L. & Ryan,B.F.(1976).*MINITAB student handbook.*Duxbury Press, North Scituate,Mass.,USA.

Ryan,T.A.,Joiner,B.L. & Ryan,B.F.(1982).*MINITAB reference manual.*Duxbury Press, North Scituate,Mass.,USA.

Saager,R. & Sinclair,A.J.(1974).Factor analysis of stream sediment data from the Mount Nansen area,Yukon Territory. *Mineral.Deposita* **9**,243-252.

Saager,R.,Meyer,M. & Muff,R.(1972).Gold distribution in supracrustal rocks from Archaean greenstone belts of southern Africa and from Palaeozoic ultramafic sequences of the European Alps: metallogenic and geochemical implications.

*Econ.Geol.***77**,1-24.

Sabine,P.A.(1953).The petrography and geological significance of the post-Cambrian minor intrusions of Assynt and the adjoining districts of North-West Scotland. *Q.J.Geol.Soc.Lond.***91**, 137-172.

Sackin,M.J. & Merriam,D.F.(1969).Autoassociation,a new geological tool. *J.Math.Geol.*1,7-16.

Saha,A.K. & Rao,S.V.L.N.(1971).Quantitative discrimination between magmatic units of the Singhbhum granite. *J.Math.Geol.* **3**,123-134.

Saha,A.K.,Bhattacharyaa,C. & Lakshmipathy,S.(1974).Some problems in interpreting the correlation between the modal variables in granitic rocks*J.Math. Geol.***6**,245-258.

Saha,A.K.,Sarkar,S.N.,Basu,S. & Ganguly,D.(1986).A multivariate statistical study of copper mineralisation in the central section of Masaboni mine, Eastern Singhbhum,India*J.Math.Geol.* **18**,215-239.

Sahu,B.K.(1974).Probabilistic solution to thin section size measurement. *Indian J.Earth Sci.***1**,22-36.

Salamon,M.D.G. & Lancaster,F.K.(1973).*Application of computer methods in the the mineral industry.* S.Afr.Inst.Ming.Metall.,Johannesburg,441pp.

Sampson,R.J.(1975).*SURFACE II graphics system.*In: Davis & McCullagh(1975),qv,244-266.

Sanford,R.L.,Myers,T.L. & Stiehr,J.F.(1978).A directory of computer programs applicable to US mining practices and problems. *U.S. Bur.Mines Open File Rep.* **131**:548pp.

Sanford,R.L.,Myers,T.L. & Stiehr,J.F.(1979).Technical note:directory of digital computer programs for the mining industry. *Ming.Eng.***31(9)**:1383-1386.

Sankar,D.M.(1979).Geometric means as probable values for compiled data on geochemical reference samples. *Geostandards Newsletter* **3**,199-205.

Santos Oliveira,J.M.(1978).Application of factor analysis to geochemical prospecting data from the Arouca-Castro region,northern Portugal. *Comun.Geol.Serv.Port.***63**:367-384.

Saprykin,E.P.,Tsitsiashvili,G.S.,Mezdrich,B.M. & Temkin,S.I.(1977).Method of linear discriminant functions in the problem of processing geological information given as an ordered set of objects. *Geophys.Surv.(Dordrecht)* **3**:69-73.

Sarazin,G.(1978).Multivariate correlation method for calculation of magmatic rocks weathering budget. *Geochem.J. (Tokyo)* **12(2)**:107-113.

SAS (Statistical Analysis Systems, 1980). *Applications guide.* SAS Inc., Cary, NC.

SAS (Statistical Analysis Systems, 1985a). *Introductory Guide.* 3rd Edn. SAS Inc., Cary, NC.

SAS (Statistical Analysis Systems, 1985b). *User's guide: basics.* Version 5 Edn. SAS Inc., Cary, NC.

SAS (Statistical Analysis Systems, 1985c). *ISAS User's guide: statistics.* Version 5 Edn. SAS Inc., Cary, NC.

SAS (Statistical Analysis Systems, 1986). *SAS companion guide for the VMS operating system.* SAS Inc., Cary, NC.

Savage,I.R.(1962).*A bibliography of nonparametric statistics.*Harvard Univ.Press.

Savazzi,E.(1985).SHELLGEN: a BASIC program for the modelling of molluscan ontogeny and morphogenesis. *Comput.Geosci.***11**:521-530.

Sawn,S. & Barry,A.(1972). *How vital are statistics?* Open University audio cassette & notes (course 201).

Saxena,S.K.(1969).Silicate solid solutions and geothermometry.4.Statistical study of chemical data on garnets and clinopyroxenes.*Contrib.Mineral.Petrol.* **23**,140-156.

Saxena,S.K.(1970).A statistical approach to the study of phase equilibria in multicomponent systems*Lithos* **3**,25-36.

Saxena,S.K. & Ekström,T.K.(1970).Statistical chemistry of calcic amphiboles. *Contrib.Mineral. Petrol.***26**,276-284.

Saxena,S.K. & Walter,L.S.(1974).A statistical-chemical and thermodynamic approach to the study of lunar mineralogy. *Geochim.Cosmochim.Acta* **38**,79-95.

Schach,S.(1969).Nonparametric symmetry tests of circular distributions. *Biometrics* **56**,571-577.

Schaeben,H.(1984).A new cluster algorithm for orientation data.*J.Math. Geol.***16**,139-154.

Scheffé,H.(1959).*The analysis of variance.*Wiley,New York.

Scheidegger,A.E.(1965).On the statistics of the orientation of bedding planes, grain axes and similar sedimentological data. *U.S.G.S.Prof.Pap.***525-C**:164-167.

Schervish, J. (1985). Review of Systat. *The American Statistician* **39**, 69-70.

Schiffelbein,P.(1987).Calculation of confidence limits for geologic measurements.In: Size(1987a),qv,21-32.

Schiffman,S.S.,Reynolds,M.L. & Young,F.W.(1981).*Introduction to multidimensional scaling.*Academic Press, NY.

Schlecht,W.G.(1951).Cooperative investigation of precision and accuracy in chemical analysis of silicate rocks. *Anal.Chem.***23**,1568-1571.

Schlecht,W.G. & Stevens,R.E.(1951).Results of chemical analysis of samples of granite and diabase.*Bull.U.S. Geol.Surv.* **980**,7-24.

Schmieman,S.(1985).*MacPaint: drawing,drafting,design.*Brady,Bowie,MD.

Schnapp,R.L.(1986).*Macintosh graphics in Modula-2.*Prentice-Hall,Englewood Cliffs,New Jersey.

Schonfeld,E.(1973).Lunar rock types and weighted least-squares mixing models. *Meteoritics* 8(4):432-435.

Schuenemeyer,J.H.,Koch,G.S. & Link,R.F.(1972).Computer program to analyse directional data,based on the methods of Fisher and Watson*J.Math. Geol.***4**,177-202.

Schwarzacher,W.(1969).The use of Markov chains in the study of sedimentary cycles*J.Math.Geol.* **1**,17-40.

Schwarzacher,W.(1975).*Sedimentation models and quantitative stratigraphy.* Elsevier, Amsterdam,382pp.

Schwarzacher,W.(1978).Mathematical geology and sedimentary stratigraphy. In: Merriam (1978b),qv,65-72.

Schweitzer,E.L.,Papike,J.J. & Bence,A.E.(1979).Statistical analysis of clino-pyroxenes from deep-sea basalts. *Am.Mineral.***64**,501-513.

Searight,T.K.(1985).A computer-assisted petroleum exploration and development exercise for undergraduate geology students. *J.Geol.Educ.***33**,45-52.

Selby,J. & Day,M.(1977).A database system for the storage and retrieval of geological data from boreholes (South Australia). *Miner.Resour.Rev.Dep. Mines S.Aust.*140:59-69.

Selley,R.C.(1969).Studies of sequence in sediments using a simple mathematical device.*Q.J.Geol. Soc.Lond.* **125**,557-581.

Sen,P.K.(1968).Estimates of the regression coefficient based on Kendall's tau. *J.Am.Stat.Assoc.* **63**,1379-1389.

Sen,P.K.(1985). *Theory and applications of sequential nonparametrics.* Society for Industrial & Applied Mathematics, Philadelphia, PA, 92 pp.

Sepkoski,J.J.(1974).Quantified coefficients of association and measurement of similarity. *J.Math.Geol.***6**,135-152.

Sepkoski,J.J.(1981).A factor analytic description of the Phanerozoic marine fossils record. *Palaeobiol.* **7**, 36-53.

Serra,J.(1987).Comments on "Geostatistics: models and tools for the Earth Sciences"*J.Math.Geol.* **19**,349-356.

Serra,S.(1973).A computer program for calculating and plotting of stress distributions and faulting. *J.Math.Geol.* **5**,397-407.

Shackleton,W.G.(1977).Computer assisted instruction in mineral exploration. *J.Geol.Educ.* **25(5)**:152-154.

Shafer,D.(1986).*Artificial intelligence programming for the Macintosh.* H.W.Sams,Indianapolis,IN.

Shanley,R.J. & Mahtab,M.A.(1973).A computer program for clustering data points on the sphere. *U.S. Bur.Mines. Inf.Circ.* **8624.**

Shanley,R.J. & Mahtab,M.A.(1975).FRACTAN: a computer code for analysis of clusters defined on the unit hemisphere. *Gov.Rep.Announce, Springfield,*VA,**PB-240-685**,54pp.

Shanley,R.J. & Mahtab,M.A.(1976).Delineation and analysis of clusters in orientation data. *J.Math.Geol.***8**,9-24.

Shapiro,S.S. & Wilk,M.B.(1965).An analysis of variance test for normality. *Biometrika* **52**,591-611.

Shapiro,S.S.,Wilk,M.B. & Chen,H.J.(1968).A comparative study of various tests for normality. *J.Am.Stat.Assoc.* **63**,1343-1372.

Sharma,K.N.M.,Laurin,A.F.,& Wynne-Edwards,H.R.(1972).Computer-based geological field data-system.*Quebec Geol. Explor. Service,*146pp.

Shaw,B.R. & Cubitt,J.M.(1979).Stratigraphy correlation of well logs: an automated approach. In: Gill & Merriam (1979), qv,127-148.

Shaw,D.M.(1961).Element distribution laws in geochemistry.*Geochim.Cosmochim. Acta* **23**,116-124.

Shaw,D.M.(1964).Test of the discriminant function in the amphibolite problem. *Bull.Am.Assoc. Pet.Geol.***48**,546-547.

Shaw,D.M.(1969).Evaluation of data. In: Wedepohl (1969–), qv, vol.I, 324-354.

Shaw,D.M.(1970).Trace element fractionation during anatexis.*Geochim.Cosmochim. Acta* **34**,237-243.

Shaw,D.M.(1974).R-mode factor analysis on enstatite chondrite analyses. *Geochim.Cosmochim. Acta* **38**,1607-1613.

Shaw,D.M. & Bankier,J.D.(1954).Statistical methods applied to geochemistry. *Geochim.Cosmochim.Acta.***5**,111-123.

Shaw,D.M. & Harmon,R.S.(1975).Factor analysis of elemental abundances in chondritic and achondritic meteorites. *Meteoritics* **10(3)**:253-282.

Shaw,D.M. & Harrison,W.P.(1955).Determination of the mode of a metamorphic rock. *Am.Mineral.***40**,614-623.

Shaw,D.M.,Dupuy,C.,Fratta,M. & Helsen,J.(1974).An application of factor analysis to basic volcanic rock geochemistry.*Bull.Volc.***38,**1070-1089.

Shaw,D.M.,Lipschutz,M.E.,Binz,C.M. & Kurimoto,R.K.(1973).Factor analysis applications to enstatite chondrite geochemistry. *Meteoritics* **8(4)**,p.438.

Sheeley,J.H.(1977).Tests for outlying observations.*J.Quality Technol.***9**,38-41.

Shelley,E.P.(1985).Directory of Government Geoscience databases in Australia. *Rept.Aust.Bur.Miner.Resour.* **269.**

Shepherd,J. & Gaskell,J.L.(1977).Trend analysis of fractures and fissure vein mineralization in the Drake volcanics of NSW,Australia. *Tr.Inst.Ming.Metall.* **86**,B9-16.

Shreider,Y.A.(1964).*Methods of statistical testing: Monte Carlo method.* Elsevier,Amsterdam,303pp.

Shreider,Y.A.(1966).*The Monte Carlo Method.*Pergamon,Oxford,381pp.

Shurtz,R.F.(1985).A critique of A.Journel's "The deterministic side of geostatistics". *J.Math.Geol.***17**,861-868.

Sichel,H.S.(1966).The estimation of means and associated confidence limits for small samples from lognormal populations.*J.S.Afr.Inst.Ming.Metall.***1-17.**

Siegel,J.B.(1985).*Statistical software for microcomputers.*Horth-Holland, New York.

Siegel,S.(1956)*Non-parametric statistics for the behavioral sciences.* McGraw-Hill,New York,312pp.

Siegel,S. & Tukey,J.W.(1960). A nonparametric sum of ranks procedure for relative spread in unpaired samples. *J.Am.Stat.Assoc.***55**, 429-444.

Siegenthaler,R.(1986).The use of database systems.*Terra Cognita* **6(1)**:83-88.

Silichev,M.K.(1976).Experience in using statistical methods to study the character of gold mineralisation. *Sov.Geol.Geophys.***17(9)**:41-45.

Simon,G.(1977).Multivariate generalization of Kendall's tau with application to data reduction. *J.Am.Stat.Assoc.* **72**,367-376.

Simonett,D.S.(1967).Landslide distribution and earthquakes in the Bewani and Torricelli Mountains,New Guinea: statistical analysis.In: Jennings,J.N. & Mablutt,J.A. (eds.) *Landform studies for Australia and New Guinea*, 64-84. Cambridge Univ.Press.

Sinclair,B.M.(1973). *Programming in BASIC, with applications.* McGraw-Hill, New York, 193 pp.

Sinclair,A.J.(1974).Selection threshold values in geochemical data using probability graphs. *J.Geochem.Explor.* **3**,129-149.

Sinclair,A.J.(1977).Application of probability graphs in mineral exploration. *J.Geochem.Explor.* **8(3)**:586-587.

Sinclair,I.R.(1986).*Collins Gem dictionary of computing*.Collins,London,380pp.

Sinding Larsen,R.(1977).Comment on the statistical treatment of geochemical exploration data. *Sci.Terre Inf.Geol. (Nancy)*, **9**:73-90.

Singer,B.(1979).Distribution-free methods for nonparametric problems: a classified and selective bibliography. *Br.Psychol. Soc., Leicester*,66pp.

Singh,T.R.P.(1978).Analytical studies on some statistical properties of mean values and their application to sampling and ore estimation. *J.Math.Geol.***10**,169-194.

Size,W.B.(1973).Interpretation of factor analysis on modal data from the Red Hill syenite complex. *J.Math.Geol.* **5**,191-198.

Size,W.B.(1987a).Use and abuse of statistical methods in the Earth Sciences. Int.Assoc.Math.Geol., Studies in Math.Geol.No.**1**. Oxford Univ.Pres, New York, 169 pp.

Size,W.B.(1987b).Use of representative samples and sampling plans in describing geological variability and trends. In: Size(1987a),qv, 3-20.

Skala,W.(1977).A mathematical model to investigate distortion of correlation coefficients in closed arrays. *J.Math.Geol.* **9**,519-528.

Skala,W.(1979a).Some effects of the constant-sum problem in geochemistry. *Chem.Geol.***27**,1-9.

Skala,W.(1979b).A Monte-Carlo model for reconstructing open arrays from closed arrays.In: Gill & Merriam (1979), qv,51-58.

Skelton,R. & Franklin,P.(1984).DMIPS: a computer based dynamic mine and quarry planning system. *S.Afr.Ming. World, April* **1984**,42-66.

Slakter,M.J.(1965).A comparison of the Pearson chi-square and Kolmogorov goodness-of-fit tests with respect to validity. *J.Am.Stat.Assoc.***60**,854-858.

Smirnov,N.V.(1939).[Estimation of deviation between empirical distribution functions in two independent samples].In Russian.*Bull.Muscow Univ.***22**.

Smirnov,N.V.(1948).Tables for estimating the goodness of fit of empirical distributions. *Ann.Math.Stat.***19**,280-281.

Smith,D.G.W. & Leibovitz,D.P.(1986).MinIdent: a database for minerals and a computer program for their identification. *Can.Mineral.***24**,695-708.

Smith,E.I. & Stupak,W.A.(1978).A FORTRAN IV program for the classification of volcanic rocks using the Irvine and Baragar classification scheme.*Comput. Geosci.***4(1)**:89-99.

Smith,E.L.(1978).Storage and retrieval systems for the reference minerals collection in the Georgina Basin project.*BMR J.Geol.Geophys.***3**,248-253.

Smith,F.G.(1966).*Geological data processing*.Harper & Row,New York,284pp.

Smith,F.G.(1969).Machine contouring in a three-variable closed array. *Can.J.Earth Sci.***6**,187-190.

Smith,F.G.(1972).A computer program to plot random or real proportions of 3 components of a mixture. *J.Math.Geol.***4**,263-268.

Smith,F.G. & Watson,D.F.(1972).A minimal test of significance of nonuniform density of data forms in a three variable closed array.*Can.J.Earth Sci.* **9**,1124-1128.

Smith,I.E.M.(1983).Some problems with numerical modelling of trace element abundances in igneous rocks.In: Augustithis(1983),qv,151-160.

Smith,J.V. & Waltho,A.(1988). *Austalian Geoscientific Software Catalogue.* Dept.Applied Geology,University of Technology, Sydney, 74 pp.

Smith,K. (1986). Geoscience information control using existing standards: a case study. *Proc.3rd Int.Conf.Geosci.Info. (Adelaide,S.Australia, June '86)*, Vol.1, 10-20.

Smith,M.J.,Hood,W.C. & Davis,R.W.(1978).MINEXP,a computer-simulated mineral exploration program. *J.Geol.Educ.* **26(3)**:93-95.

Smith,R.E.,Campbcll,N.A. & Perdrix,J.L. (1982). Identification of some Western Australian gossans by multi-element geochemistry. In: *Geochemical Exploration in deeply weather terrains*, Smith,R.E. (ed.) CSIRO, Floreat Park, W.Australia, 190 pp.

Smith,R.E., Campbell,N.A. & Litchfield,R. (1984). Multivariate statistical techniques applied to pisolitic laterite geochemistry at Golden Grove, Western Australia. *J.Geochem.Explor.* **22**, 193-216.

Smosna,R. & Warshauer,S.M.(1979).A scheme for multivariate analysis in carbonate petrology with an example from the

Silurian Tonoloway limestone. *J.Sed.Petrol.***49**,257-272.
Smyth,J.R.(1981).MANTLE: a program to calculate a 30kb norm assemblage. *Comput.Geosci.* **7**,27-34.
Sneath,P.H.A.(1977).A method for testing the distinctness of clusters and a test of the disjunction of 2 clusters in Euclidean space as measured by their overlap*J.Math.Geol.***9**,123-143.
Sneath,P.H.A.(1979a).The sampling statistic of the *W* statistic of disjunction for the arbitrary division of a random rectangular distribution*J. Math.Geol.***11**:423-429.
Sneath,P.H.A.(1979b).BASIC program for a significance test for clusters in UPGMA dendrograms obtained from squared Euclidean distances.*Comput.Geosci.* **5**:127-137.
Sneath,P.H.A.(1985).DENBRAN: a BASIC program for a significance test for multivariate normality of clusters from branching patterns in dendrograms. *Comput.Geosci.***11**:767-786.
Snedecor,G.W. & Cochran,W.G.(1967).*Statistical methods.*Iowa State Univ.Press, 593pp.
Snow,J.W.(1975).Association of proportions*J.Math.Geol.***7**,63-74.
Sokal,R.R. & Rohlf,F.J.(1962).The comparison of dendrograms by objective methods.*Taxonomy* **11**,33-40.
Sokal,R.R. & Rohlf,F.J.(1969)*Biometry.*Freeman,San Fransisco,776pp.
Sokal,R.R. & Rohlf,F.J.(1973)*Introduction to biostatistics.*Freeman, San Fransisco,368pp.
Sokal,R.R. & Sneath,P.H.A.(1963).*Principles of numerical taxonomy.*Freeman, San Fransisco,359pp.
Solety,P.,Beauchemin,Y. & Dumont,R.(1978).[Methods of using algorithms for the analysis of data and for classification in applied geochemistry]*.Bull.Bur.Miner.Resour.(France), Ser.2 Sect.3*,**3**:239-249.
Solomon,M.(1963).Counting and sampling errors in modal analysis by point counter. *J.Petrol.* **4**,367-382.
Somers,R.H.(1962).A new asymmetric measure of association for ordinal variables. *Am.Sociolog. Rev.***27**,799-811.
Southam,J.R. & Hay,W.W.(1978).Correlation of stratigraphic sections by continuous variables.*Comput.Geosci.* **4**, 257-260.
Sowerbutts,W.T.C.(1982).The use of interactive computer graphics for displaying spatial geological information. *Geol.Soc.Lond.Misc.Pap.***14**,139-154.
Sowerbutts,W.T.C.(1983).A surface-plotting program suitable for microcomputers. *Computer-aided design* **4485**, 324-327.
Sowerbutts,W.T.C.(1985)*.Register of geological microcomputer programs.* Geol.Information Group, Geol.Soc.London.
Späth,H.(1980).*Cluster analysis algorithms for data reduction and classification of objects.*Ellis Howard,Chichester.
Spearman,C.(1904).The proof and measurement of association between two things. *Am.J. Psychol.***15**,72-101.
Specht,D.F.(1967).Generation of polynomial discriminant functions for pattern recognition*IEEE Tr.Electr.Comput.* **16**,308-319.
Spencer,A.B. & Clabhugh,P.S.(1967).Computer programs for fabric diagrams. *Am.J.Sci.***265**,166-172.
Spicer,R.A. & Hill,C.R.(1979).Principal components and correspondence analysis of quantitative data from a Jurassic plant bed.*Rev.Palaebot.Palynol.***28**,273-289.
Sprent,R.(1977).*Statistics in action.*Penguin,London.
Sprent,R.(1981).*Quick statistics: an introduction to non-parametric methods.* Penguin,London.
SPSS (1986).*SPSS Users manual.*SPSS Inc.,Chicago.
Spurr,B.D.(1981).On estimating the parameters in mixtures of circular normal distributions.*J.Math. Geol.***13**,163-174.
Srivastava,G.S. & Carter,E.M.(1983).*An introduction to applied multivariate statistics.* North-Holland,New York, 394pp.
Srivastava,G.S. & Merriam,D.F.(1976).Computer constructed optical-rose diagrams. *Comput. Geosci.***1**,179-186.
Srivastava,G.S. & Sackin,M.J.(1971).AUTO,a FORTRAN IV program for autoassociation. *J.Math.Geol.***3**,193-202.
Srivastava,R.H.(1986).Philip & Watson — quo vadunt? *J.Math.Geol.***18**,141-146.
Starkey,J.(1974).The quantitative analysis of orientation data obtained by the Starkey method of X-ray fabric analysis.*Can.J.Earth Sci.***11**,1507-1516.
Starkey,J.(1976).The contouring of orientation data represented in spherical projection.*Can.J.Earth Sci.***14**,268-277.
Starks,T.H. & Fang,J.H.(1982).The effect of drift on the experimental semivariogram. *J.Math.Geol.* **14**,309-319.
Starks,T.H.,Behrens,N.A. & Fang,J.H.(1982).The combination of sampling and kriging in the regional estimation of coal resources.*J.Math.Geol.* **14**,87-106.
Stauffer,M.R.(1966).An empirical statistical study of three-dimensional fabric diagrams as used in structural analysis. *Can.J.Earth Sci.***3**,473-498.
Steck,G.P.(1969).The Smirnov two-sample tests as rank tests.*Ann.Math.Stat.***40**, 1449-1466.
Steel,R.G.D. & Torrie,J.H.(1980).*Principles and procedures of statistics: a biometrical approach.*(2nd Ed). International Student Edition,633pp.
Steele,T.W.(1978).A guide to the reporting of analytical results relating to the certification of geological reference materials.*Geostandards Newsletter* **2(1)**:31-33.
Steele,T.W.,Wilson,A.,Goudvis,R. & Ellis,P.J.(1978).Trace element data (1966-1977) for the six 'NIMROC' reference samples.*Geostandards Newsletter* **2(1)**: 71-106.
Steffens,F.E.(1985).Introduction to geostatistics.*Tr.Geol.Soc.S.Afr.***88**,61-64.
Steiger,R.H. & Jäger,E.(1977).Subcommission on geochronology: convention on the use of decay constants in geo- and

cosmochronology. *Earth Planet.Sci.Lett* 39, 359-362.

Steigerwald, C.H., Mutschler,F.E. & Ludington,S. (1983a).GRANNY, a data bank of chemical analyses of Laramide and younger high-silica rhyolites and granites from Colorado and north-centyal New Mexico. *USGS Open File Report OF83-0516*, 562 pp.

Steigerwald, C.H., Mutschler,F.E. & Ludington,S. (1983b). A bibliography of stockwork molybdenite deposits and related topics (wth an emphasis on the North American literature). *USGS Open File Report OF83-0382*, 112 pp.

Steinmetz,R.(1962).Analysis of vectorial data. *J.Sed.Petrol.* **32**,801-812.

Stephens,M.A.(1965).Significance points for the two-sample statistic $U^2_{M,N}$. *Biometrika* **52**,661-663.

Stephens,M.A.(1970).Use of Kolmogorov-Smirnov,Cramér-Von Mises and related statistics without extensive tables. *J.R.Stat.Soc.***B32**,115-122.

Stephens,M.A.(1974).EDF statistics for goodness of fit and some comparisons. *J.Am.Stat.Assoc.* **69**:730-737.

Stephens,W.E.(1976).The display of three-factor models.In: Merriam(1976c),qv, 169-172.

Stephens,W.E. & Dawson,J.B.(1977).Statistical comparison between pyroxenes from kimberlites and their associated xenoliths.*J.Geol.***85**,433-449.

Sterling,T.D. & Pollack,S.V.(1968).*Introduction to statistical data processing*. Prentice-Hall,New Jersey,663pp.

Stesky,R.M.(1985).Least-squares fitting of a noncircular cone.*Comput.Geosci.* **11(4)**:357-368.

Stevens,R.E. & Niles,W.W.(1960).Second report on a cooperative investigation of the composition of two silicate rocks.*U.S.Geol.Surv.Bull.***1113**,3-43.

Stewart,D. & Love,W.(1982).A general canonical correlation index.In: Fornell (1982a),qv,Vol.I.,55-63. Also *Psychol.Bull.*(1968),**70**,160-163.

Stewart,D.C.,Frizado,J. & Cummins,L.E.(1983).Error recognition in published chemical analyses of igneous rocks: consistency evaluation.*Comput.Geosci.* **9**,527-536.

Stiff,D.L.(1985).*Multiplan for the Macintosh with Microsoft Chart*.Brown,Dubuque,IO.

Stigler,S.M.(1977).Do robust estimates work with real data? *Ann.Stat.***5**, 1055-1098.

Stone,M. & Exley,C.S.(1978).A cluster analysis of chemical data from the granites of SW England.*Proc.Ussher Soc.* **4(2)**:172-181.

Stormer,J.C. & Nicholls,J.(1978).XLFRAC: a program for the interactive testing of magmatic differentiation models. *Comput.Geosci.***4**,143-159.

Stormer,J.C.,Dehn,M.H.,Leeman,W.P. & Matty,D.J.(1986).Programs ROCALC — norms and other geochemical calculations for igneous petrology.In: Hanley & Merriam(1986),qv,141-164.

Streckeisen,A.(1976).To each plutonic rock its proper name. *Earth-Sci.Rev.***12**,1-33.

Streckeisen,A.(1979).IUGS Subcommission on the systematics of igneous rocks: classification and nomenclature of volcanic rocks, lamprophyres, carbonatites and melilitic rocks.*Geology* **7**,331-335 and *Geol.Rundsch.***69**,194-207.

Stuckless,J.S.,Miesch,A.T.,Goldich,S.S. & Weiblen,D.W.(1979).A Q-mode factor model for the petrogenesis of the volcanic rocks from Ross Island and vicinity,Antarctica.*Am.Geophys.Union.Mem.Antarct.Res.Series.*

Sturgul,J.R. & Aiken,C.(1970).The best plane through data.*J. Math.Geol.***2**,325-333.

Sugranes,L.S.(1977).Alguns consideraciones sobre el tratamiento estadístico de dados direccionales en geologia. *Acta Geol.Hisp.***12(4-6)**:69-77.

Sutherland,D.G. & Dale,M.A.(1984).Method of establishing mineral sample size for sampling alluvial diamond deposits. *Tr.Inst.Ming.Metall.***93**,B55-58.

Sutterlin,P.G.(1981).The future of information systems in the Earth Sciences. In Merriam(1981b), qv,305-322.

Sutterlin,P.G. & Hastings,J.P.(1986).Trend-surface analysis revisited — a case history.*Comput. Geosci.***12**,537-562.

Sutterlin,P.G. & Sondergard,M.A.(1986).WSU-MAP: a microcomputer based reconnaissance mapping system for Kansas subsurface data.*Comput.Geosci.***12**,563-596.

Sutterlin,P.G.,Aaltonen,R.A. & Cooper,M.A.(1974).Some considerations in management of computer-processable files of geologic data.*J. Math.Geol.***6**,291-310.

Sutterlin,P.G.,Jeffery,K.G. & Gill,E.M.(1977).FILMATCH: a format for the interchange of computer-based files of structured data.*Comput.Geosci.* **3**,429-442.

Svane Petersen,T.(1977).FISK: a FORTRAN program to estimate the mode of hornblende/biotite gneisses and amphibolites from chemical analyses. *Comput.Geosci.***3**,19-24.

Swineford,A.,Frye,J.C. & Leonard,A.B.(1955).Petrography of the late Tertiary volcanic ash fall in the Central Great Plains. *J.Sed.Petrol.***25**,243-261.

Taranik,J.V.(1978).Principles of computer processing of LANDSAT data for geological applications. *U.S.G.S.Open File Rep.* **78(117)**:98pp.

Tatsuoka,M.M.(1971).*Multivariate analysis*.Wiley,New York,310pp.

Tatsuoka,M.M.(1975).The general linear model: a "new" trend in analysis of variance.Institute for personality and ability testing, Champaign, Illinois, 64pp.

Taylor,D. & Henderson,C.M.B.(1978).A computer model for the cubic sodalite structure.*Phys. Chem.Minerals* **2(4)**, 325-336.

Taylor,D.,Dempsey,M. & Henderson,C.(1985).The structural behaviour of the nepheline family:2:distance least-squares modelling of Sr and Ba aluminates MAl_2O_4.*Bull.Mineral.***108**: 643-652.

Taylor,G.R.,Gwatkin,C.C. & Chork,C.Y.(1984).Statistical interpretation of element distribution within distal volcanic exhalative mineralization near Mount Chalmers,Queensland,Australia. *Tr.Inst.Ming.Metall.***93**,B99-108.

Taylor,R.K.(1973).Compositional and geotechnical characterization of a 100-year old colliery spoil heap. *Tr.Inst.Ming.Metall.***82**,A1-14.

Teil,H.(1975).Correspondence factor analysis: an outline of its method. *J.Math.Geol.***7**,3-12.

Teil,H.(1976).The use of correspondence analysis in the metallogenic study of ultrabasic and basic complexes. *J.Math.Geol.***8**,669-682.

Teil,H. & Cheminée,J.L.(1975).Application of correspondence factor analysis to the study of major and trace elements in the Erta Ale Chain (Afar,Ethiopia). *J.Math.Geol.***7**,13-30.

Tellis,D.A. & Gerdes,L. (1986). Innovative developments in national and regional earth science reference databases in Australia. *Proc.3rd Int.Conf.Geosci.Info. (Adelaide,S.Australia, June '86)*, Vol.1, 71-94.

Temple,J.T.(1979).The use of factor analysis in geology.*J.Math.Geol.* **10**,379-381.

Temple,J.T.(1980).Ordination of palaeontological data: principal components analysis,correspondence analysis and non-linear mapping.*J.Geol.Soc.Lond.* **137**:p.106.

Tennant,C.B. & White,M.L.(1959).Study of the distribution of some geochemical data.*Econ.Geol.***54**,1281-1290.

Thom,R.(1978).Plate tectonics and catastrophe theory.*Catastrophist Geol. (Rio de Janeiro)* **3(1)**:30-48.

Thompson,M.(1978).DUPAN3,a subroutine for the interpretation of duplicated data in geochemical analysis. *Comput.Geosci.***4**,333-340.

Thompson,M. & Howarth,R.J.(1978).A new approach to the estimation of analytical precision. *J.Geochem.Explor.***9**, 23-30.

Thorndike,R.M.(1982).*Data collection and analysis.*Gardner Press,478pp.

Thornton,I. & Howarth,R.J.(1986).*Applied geochemistry in the 1980's.* Graham & Trotman, London,347pp.

Thrivikramaji,K.P. & Merriam,D.F.(1976).Trend analysis of sedimentary thickness data: the Pennsylvanian of Kansas.In: Merriam(1976c),qv,11-22.

Thurstone,L.L.(1947).*Multiple-factor analysis.*Univ.Chicago Press,535pp.

Tien,P.L.(1968).Differentiation of Pleistocene deposits in northeastern Kansas by clay minerals.*Clays & Clay Minerals* **16**,99-107.

Tietjen,G.L. & Moore,R.H.(1972).Some Grubbs-type statistics for the detection of outliers.*Technometrics* **14**,583-597.

Till,R.(1971).Are there geochemical criteria for differentiating reef from non-reef carbonates? *Bull.Am.Assoc. Pet.Geol.* **55**,523-528.

Till,R.(1973).The use of linear regression in geomorphology.*Area* **5**,303-308.

Till,R.(1974).*Statistical methods for the Earth scientist.*Macmillan,London,154pp.

Till,R.(1977).Programs in BASIC for simple correlation and regression (with worked examples). *Geol.Rep.Univ.Reading, UK,* **11**:63pp.

Till,R. & Colley,H.(1973).Thoughts on use of principal components analysis in petrogenetic problems. *J.Math.Geol.* **5**,341-350.

Till,R.,Hopkins,D.T. & McCann,C.(1972).A collection of computer programs in BASIC for geology and geophysics.*Geol.Rep.Univ.Reading,UK*,No.**5**.

Timcak,G. & Hroncova,E.(1978).[A FEL-FORTRAN program for the complete Niggli classification].In Slovak.*Geol.Pr.Spr.(Bratislava)* **70**:243-258.

Tinkler,K.J.(1971).Statistical analysis of tectonic patterns in areal volcanism in the Bunyarungu volcanic fields in W.Uganda. *J. Math.Geol.*3,335-356.

Tipper,J.C.(1976).The study of geological objects in 3 dimension by the computer reconstruction of serial sections. *J.Geol.***84**,476-484.

Tipper,J.C.(1977a).Three-dimensional analysis of geological forms.*J.Geol.* **85(5)**:591-611.

Tipper,J.C.(1977b).A method and FORTRAN program for the computerized reconstruction of three-dimensional objects from serial sections.*Comput. Geosci.***3(4)**:679-700.

Tipper,J.C.(1979a).An ALGOL program for dissimilarity analysis — a diversive omnithetic clustering technique.*Comput.Geosci.***5**,1-14.

Tipper,J.C.(1979b).Surface modelling techniques.*Kansas Geol.Surv.Ser.on Spatial Anal.No.***4**,108pp.

Titterington,D.M. & Halliday,A.N.(1979).On the fitting of parallel isochrons and the method of maximum likelihood.*Chem.Geol.***26**:183-195.

Tocher,F.E.(1978a).Some modification of a point-counting computer program for fabric analysis of axial orientations. *Comput.Geosci.***4(1)**:1-3.

Tocher,F.E.(1978b).Petrofabric point-counting program FABRIC (FORTRAN IV). *Comput.Geosci.* **4(1)**:5-21.

Tocher,F.E.(1979).The computer contouring of fabric diagrams.*Comput.Geosci.* **5**:73-126.

Tolstoy,M.I. & Ostafiychuk,I.M.(1965).The types of statistical distribution curves for chemical elements in rocks,and methods of computing their parameters.*Geochim.Cosmochim.Acta* **2**,993-1000.

Topley,C.G. & Burwell,A.D.M.(1984).TRIGPLOT: an interactive program in BASIC for plotting triangular diagrams.

*Comput.Geosci.***10**,277-310.

Torquato,J.R. & Frischkorn,H.(1983).The isotopes of oxygen,a geochemical indicator for Precambrian carbonated rocks.In: Augustithis(1983),qv,499-518.

Tough,J.G. & Leyshon,P.R.(1985).SPHINX: a program to fit the spherical and exponential models to experimental semivariograms.*Comput.Geosci.***11**:95-100.

Trapp,J.S.,Rockaway,J.D.(1977).Trend-surface analysis as an aid in exploration for Mississippi Valley-type ore deposits. *J.Math.Geol.***9(4)**:393-408.

Tripathi,V.S.(1979).Factor analysis in geochemical exploration.*J.Geochem. Explor.***11(3)**:263-275.

Trochimczyk,J. & Chayes,F.(1977a).Variation in the principal component coefficients of geochemical data. *Fortschr.Mineral.* **55(1)**:140-142.

Trochimczyk,J. & Chayes,F.(1977b).Sampling variation of principal components. *J.Math.Geol.* **9**,497-506.

Trochimczyk,J. & Chayes,F.(1978).Some properties of principle component scores. *J.Math.Geol.* **10**:1035-1036.

Troutman,B.M. & Williams,G.P.(1987).Fitting straight lines in the Earth Sciences. In: Size(1987a),qv,107-128.

Trustrum,K.(1971).*Linear programming.*Routledge & Kegal Paul,London,88pp.

Tryon,R.C. & Bailey,D.E.(1970).*Cluster analysis.*McGraw-Hill,New York,347pp.

Tsitsiashvili,G.S.,Mezorich,B.M. & Pushkar,V.S.(1977).Mathematical modelling of layered geological formations. *Sov.Geol.Geophys.***18(5)**:72-79.

Tukey,J.W.(1953).*The problem of multiple comparisons.*Princeton Univ.Press.

Tukey,J.W.(1977).*Exploratory data analysis.*Addison-Wesley,Reading,Mass.,688pp.

Türk,F.(1979).Transition analysis of structural sequences: discussion. *Bull.Geol.Soc.Am.Pt.I,* 90,989-999.

Twitty,W.B.(1986).*Programming the Macintosh: an advanced guide.*Scott,Glensview,IL.

Ullman,J.D.(1980).*Principles of database systems.*Computer Science Press.

Unan,C.(1983).The KAYDER information system for igneous rock data.*Comput. Geosci.***9**,503-512.

Unwin,D.(1981).*Introductory spatial analysis.*Methuen,212pp.

Unwin,D.J.(1986).DIGITS: a simple digitizer system for collecting and processing spatial data using an Apple II.In: Hanley & Merriam(1986),qv,225-236.

Unwin,D.J. & Dawson,J.A.(1985).*Computer programming for geographers.* Longmans,288pp.

Upadhyay,R., Srivastava,R.K. & Agarwal,V.(1986).Cluster analysis: its application to a volcanic suite.*Ind.J.Earth Sci.***13(4)**,311-316.

Upton,G.J.G.(1973).Single-sample tests for the Von Mises distribution. *Biometrika* **60**,87-99.

Usdansky,S.(1985).GRCHEM: a BASIC program to calculate granite chemistry from modal mineralogy. *Comput.Geosci.* **11**,229-234.

Usdansky,S.(1986).PERANORM: A BASIC program to calculate a modal norm for peraluminous granitoids. *Comput. Geosci.* **12**,13-20.

Valenchon,F.(1982).The use of correspondence analysis in geochemistry. *J.Math.Geol.***14**,331-342.

Valenchon,F.(1983).Correspondence analysis in geochemistry: reply.*J.Math.Geol.***15**,505-510.

Valentine,J.W. & Peddicord,R.G.(1967).Evaluation of fossil assemblages by cluster analysis. *J.Palaeontol.***41**,502-507.

Vallee,M.,Belisle,J.M. & David,M.(1982).Kriging as a tool to avoid overestimation of grade in sulfide ore bodies. *Tr.Inst.Ming.Metall.* **91**,A1-3.

Van der Waerden,B.L.(1952-3).Order tests for the two-sample problem and their power.*Proc. Kon.Ned.Akad. Wetenschappen (A),* **55**,453-458 & **56**,303-316.

Van der Wollenberg,A.L.(1977).Redundancy analysis — an alternative to canonical correlation analysis.*Psychometrika* **42**,207-219.

Van der Wollenberg,A.L.(1982).Redundancy analysis:an alternative for canonical correlation analysis.In: Fornell(1982a), qv, Vol.I.,64-81. Also *Psychometrika* (1977),**68**,207-219.

Van Duyn,J.(1982).*Developing a data dictionary system.*Prentice-Hall,204pp.

Vannier,M. & Woodtli,R.(1976).Teaching prospection for minerals by simulation techniques assisted by computer (abstr.) *25th Int.Geol.Congr.(Sydney)*:653-654.

Van Nouhuys,D.(1985).*Macintosh: a concise guide to applications software.* Wiley,New York.

Van Trump,G. & Miesch,A.T.(1977).The U.S.G.S. RASS-STATPAC system for management and statistical reduction of geochemical data.*Comput.Geosci.***3**,475-488.

Vasilenko,V.B. & Kholodova,L.D.(1977).Paired nonlinear regression of P_2O_5 with trace elements in rocks of the carbonate and chloride complexes of Seligdar. *Sov.Geol.Geophys.***18(7)**:49-56.

Vasilenko,V.B.,Khlestov,V.V. & Blinchik,T.M.(1978).Estimation of the errors of the mean values in geochemical samples with an arbitrary form of distribution. *Sov.Geol.Geophys.***19(5)**:18-23.

Velleman,P.F. & Hoaglin,D.C.(1981).*Application,basics and computing of exploratory data analysis.*Duxbury Press.

Venit,S.(1985).*Mac at work: Macintosh windows on business.*Wiley,New York.

Verly,G.,David,M.,Journel,A.G. & Mareschal,A.(1984).*Geostatistics for natural resources characterization.* Reidel, Dordrecht, Parts I & II,1092pp.

Verma,B.K.(1972).Efficiency of regression techniques for ore evaluation at Kolar gold fields,India.*J.Math.Geol.***4**,25-34.

Verma,M.P.,Aguilar-y-Vargas,V.H. & Verma,S.P.(1986).A program package for major element data-handling and CIPW norm calculations.*Comput.Geosci.*12,381-399.

Vinod,H.D. & Ullah,A.(1981)*Recent advances in regression methods.* Marcel Deliker,New York,361pp.

Vistelius,A.B.(1958).Paragenesis of sodium,potassium and uranium in volcanic rocks of Lassen Peak National Park, California. *Geochim.Cosmochim.Acta* 14,29-34.

Vistelius,A.B.(1960).The skew frequency distribution and fundamental law of the geochemical process.*J.Geol.*68,1-22.

Vistelius,A.B.(1961).Sedimentation time trend functions and their application for correlation of sedimentary deposition. *J.Geol.*69,705-728.

Vistelius,A.B.(1967).*Studies in mathematical geology*.Consultants Bureau,New York,294pp.

Vistelius,A.B.(1970).Statistical model of silicate analysis and results of investigation of G1-W1.*J.Math.Geol.*2,1-19.

Vistelius,A.B.(1971).Some lessons of the G1-W1 investigation.*J. Math.Geol.*3,323-326.

Vistelius,A.B. & Arango,G.B.(1973).Phosphorus in granitic rocks of Colombia. *J.Math. Geol.*5,127-148.

Vistelius,A.B. & Romanova,M.A.(1977).Model of degenerate case of crystallisation of ideal granite.*Dokl.Acad.Sci.USSR Earth Sci.Sect.*228:136-139.

Vistelius,A.B. & Ruiz Fuller,C.(1969).On the origin of variations in the composition of granitic rocks of Chile and Bolovia. *J.Math.Geol.* 1,113-114.

Vistelius,A.B. & Sarmanov,O.V.(1961).On the correlation between percentage values: major component correlation in ferromagnesian micas.*J.Geol.*69,145-153.

Vistelius,A.B.,Ivanov,D.N.,Kuroda,Y. & Ruiz Fuller,C.(1970).Variation of modal composition of granitic rocks in some regions around the Pacific.*J. Math.Geol.*2,63-80.

Von Mises,R.(1964).*Mathematical theory of probability and statistics.*Academic Press,New York.

Vrbik,J.(1985).Statistical properties of the number of runs of matches between two random stratigraphical sections.*J.Math.Geol.*17,29-40.

Vugrinovich,R.G.(1981).A distribution-free alternative to least-squares regression and its application to Rb/Sr isochron calculations. *J.Math.Geol.*13,443-454.

Waard,D.de,Mulhern,K. & Merriam,D.F.(1976).Mineral variation in anorthositic, troctolitic and adamellitic rocks of the Barth Island layered structure in the Nain anorthosite complex,Labrador.*J .Math.Geol.*8,561-574.

Wadatsumi,K.,Miyawaki,F.,Murayama,S. & Higashitani,M.(1976).GEODAS-DCRF: development of a relational database system and its application for storage and retrieval of complex data from research files. *Comput.Geosci.*2:357-364.

Wainwright,S.J. & Gilbert,R.I.(1981).A BASIC computer program which tests for normality and the presence of outliers in data.*Lab.Practice* 30,p.467.

Waite,M., Prata,S. & Martin,D.(1984). *C primer plus.* Howard Samd & Co., Indianapolis, 531 pp.

Wald,A. & Wolfowitz,J.(1939).Confidence limits for continuous distribution functions.*Ann.Math. Stat.*10,105-118.

Walker,H.M.(1940).Degrees of freedom.*J.Ed.Psych.*31,253-269.

Walker,R.D.(1977).Using GEOREF tapes as a current awareness in an academic setting.*Proc.Geosci.Inf.Soc.*7:64-71.

Walker,R.D. (1986). Patents as an important source of geoscience information with emphasis on developing countries. *Proc.3rd Int.Conf.Geosci.Info. (Adelaide,S.Australia, June '86)*, Vol.1, 128-142.

Wallbrecher,E.(1978).Ein Cluster-Verfahren zur richtungstatitischen Analyse Tektonischer Daten. [A cluster method for the analysis of directional tectonic data]. *Geol.Rdsch.* 67(3):840-857.

Wallbrecher,E.(1979).Methoden zum quantitativen Vergleich von Regelunsgraden und formenstrukturgeologischer Datenmengen mit Hilfe von Vektorstatistik und Eigenwert-Analyse. [Methods for quantitative comparison of preferred orientation and distribution forms from structural geologic data with the help of vector statistics and eigenvalues].*N.Jb.Geol.Palaeont.Abh.* 159:113-149.

Walpole,R.E. & Myers,R.H.(1978).*Probability and statistics for engineers and scientists.*MacMillan, New York,580pp.

Walters,R.F.(1969).Contouring by machine: a users' guide.*Bull.Am.Assoc. Pet.Geol.*53,2324-2340.

Waltz,E.C.(1986).Selecting statistical software for the Apple Macintosh: features of the major packages.*Bull.Ecol.Soc. Am.*68(2),175-183.

Wang,P.C.C.(1978).*Graphical representation of multivariate data.* Academic Press, New York.

Ward,D.C. & Walker,R.D. (1986). Twenty years of geoscience information — an analysis of coverage and content. *Proc.3rd Int.Conf.Geosci.Info. (Adelaide,S.Australia, June '86)*, Vol.2, 52-68.

Ward,R.F. & Werner,S.L.(1963).Analysis of variance of the composition of a migmatite.*Science* 140,978-979.

Ward,R.F. & Werner,S.L.(1964).Analysis of variance of a migmatite composition. II; comparison of two areas.*Science* 143,1032-1033.

Warshauer,S.M. & Smosna,R.(1981).On minimum spanning trees and the intergradation of clusters. *J.Math.Geol.* 13,225-236.

Watson,G.S.(1961).Goodness-of-fit tests on a circle.*Biometrika* 48,109-114.

Watson,G.S.(1966).The statistics of orientation data.*J.Geol.*74,786-797.

Watson,G.S.(1970).Orientation statistics in the earth sciences.*Bull.Inst.Geol. Univ.Uppsala* 2:73-89.

Watson,G.S.(1971).Trend-surface analysis*J.Math.Geol.***3**,215-226.

Watson,G.S.(1972).Trend-surface analysis and spatial correlation.*Geol.Soc.Am.Spec.Pap.***146**,39-46.

Watson,G.S.(1983).*Statistics on spheres.*Wiley,New York,238pp.

Watson,G.S.(1987).Confidence regions in ternary diagrams,2.*J.Math.Geol.* **19**,347-348.

Watson,G.S. & Nguyen,H.(1985).A confidence region in a ternary diagram from point counts. *J.Math.Geol.***17**,209-214.

Watson,G.S. & Williams,E.J.(1956).On the construction of significance tests on the circle and sphere.*Biometrika* **43**,344-352.

Watterson,K.(1985).Earth-science software in the public domain. *J.Geol.Educ.***33**,227-236.

Weaver,S.D. & Gibson,I.L.(1975).The origin of peralkaline obsidians: a discussion.*Mineral.Mag.***40**,415-416.

Weaver,S.D.,Sceal,J.S.c. & Gibson,I.L.(1972).Trace element data relevant to the origin of trachytic and pantelleritic lavas in the E.African Rift System.*Contrib.Mineral.Petrol.***36**,191-194.

Webb,J.S.,Thorton,I.,Thompson,M.,Howarth,R.J. & Lowenstein,P.L.(1978). *The Wolfson geochemical atlas of England & Wales.* Clarendon Press,Oxford,69pp.

Webb,W.M. & Briggs,L.I.(1966).The use of principal components analysis to screen mineralogical data. *J.Geol.* **74**,716-720.

Webster,R.(1980).DIVIDE: a FORTRAN IV program for segmenting multivariate one-dimensional spatial data. *Comput.Geosci.***6**,61-68.

Webster,T.(1983).*Australian microcomputer handbook.*Computer Reference Guides.

Webster,T. & Champion,R.(1986).*Microcomputer software buyers' guide.* Computer Reference/Coopers & Lybrand.

Webster,T.,Costelloe,L. & Mullins,M.(1983).*Australian terminals and printers handbook.* McGraw-Hill,New York, 255pp.

Wedepohl,K.H.(1969–). *Handbook of geochemistry.* Springer-Verlag, Berlin, 13 vols.

Weekes,A.J.(1985).*A GENSTAT primer.*University of York Press,UK,129pp.

Weil,D.F.,McKay,G.A.,Kridlebaugh,S. & Gruizeck,M.(1974).Modelling the evolution of Sm and Eu abundances during lunar igneous differentiation.*Geochim.Cosmochim.Acta.Suppl.* (*Proc.5th Lunar Sci.Conf.*) **5**,Vol.2,1337-1352.

Weisberg,S.(1985). *Applied linear regression.* 2nd Edn. Wiley, New York, 324 pp.

West,J.(1985).Towards an expert system for identification of minerals. *J.Math.Geol.***17(7)**:743-753.

Wetherill,G.B.(1966).*Sequential methods in statistics.*Methuen,London.

Wetherill,G.B.(1981).*Intermediate statistical methods.*Chapman & Hall,London,390pp.

Wheatley,M.R. & Rock,N.M.S. (1988). SPIDER: a Macintosh program to generate normalized multi-element diagrams in geochemistry. *Am.Mineral.* (in press).

Wheeler,S. & Watson,G.S.(1964).A distribution-free two-sample test on a circle. *Biometrika* **51**,256-257.

Wheildon,C.(1986).*Communicating — or just making pretty shapes* (revised Ed.) Newspaper Advertising Bureay of Australia Ltd.

Whitford,D.J. & Nicholls,I.A.(1976).Potassium variation in lavas across the Sunda Arc, Indonesia, and their petrogenetic implications. In: *Volcanism in Australasia,* Johnson, R.W. (ed.), 63-75. Elsevier, Amsterdam.

Whitten,E.H.T.(1961).Quantitative areal modal analysis of granitic complexes. *Bull.Geol.Soc. Am.***72**,1331-1360.

Whitten,E.H.T.(1962a).Areal variability of alkalis in the Malmsburg granite, Germany.*Neues Jb.Mineral.Monatsch.* **9**,193-200.

Whitten,E.H.T.(1962b).A new method for determining the average composition of a granite massif. *Geochim. Cosmochim.Acta* **26**,545-560.

Whitten,E.H.T.(1963).Application of quantitative methods in the geochemical study of granitic massifs. *Spec.Publ. R.Soc.Can.***6**,76-123.

Whitten,E.H.T.(1964).Models in the geochemical study of rock units.*Q.J.Colorado Sch.Mines* **59**,149-168.

Whitten,E.H.T.(1966).Sequential multivariate regression methods and scalars in the study of fold-geometry variability. *J.Geol.***74**,744-763.

Whitten,E.H.T.(1969). Trends in computer applications in structural geology. In: Merriam(1969),qv,233-249.

Whitten,E.H.T.(1970).Orthogonal polynomial trend surfaces for irregularly spaced data.*J.Math.Geol.***2**,141-152 & **3**,330.

Whitten,E.H.T.(1972).More on 'irregularly spaced data and orthogonal polynomial trend surfaces'.*J.Math.Geol.***4**,83.

Whitten,E.H.T.(1975).The practical use of trend surface analysis in the geological sciences.In: Davis & McCullagh (1975),qv,282-297.

Whitten,E.H.T.(1981).Trends in computer applications in structural geology: 1969-1979.In: Merriam(1981b),qv,323-368.

Whitten,E.H.T.(1983).Twenty-five years of mathematical geology: a new threshold.*J.Math.Geol.***15**,237-244.

Whitten,E.H.T.(1985).Suites within a granitoid batholith: a quantitative justification based on the Lachlan Fold Belt. *Geol.Zbornik* **36**,191-199.

Whitten,E.H.T.,Bornhorst,T.J.,Li,G.,Hicks,D.L. & Beckwith,J.P.(1987). Suites,subdivisions of batholiths and igneous rock classification: geological and mathematical conceptualization.*Am.J.Sci.* **287**,332-354.

Whitten,E.H.T. & Dacey,M.F.(1975).On the significance of certain Markovian features of granite textures. *J.Petrol.* **16**:429-453.

Whitten,E.H.T. & Koelling,M.E.V.(1972).Spline-surface interpolation,spatial filtering and trend surfaces for geologic

mapped variables.*J. Math.Geol.*5,111-126.
Whitten,E.H.T.,Li,G.,Bornhorst,T.J.,Christenson,P. & Hicks,D.(1987).Quantitative recognition of granitoid suites withint batholiths and other igneous assemblages. In: Size(1987a),qv,55-73.
Whittington,H.B. & Hughes,C.P.(1972).Ordovician geography and faunal provinces deduces from trilobite distribution. *Phil.Tr.R.Soc.Lond.***B263**,235-273.
Wickman,F.E.(1970).Determination of repose-period patterns of volcanoes from sequences of ash layers. *J.Math.Geol.***2**,277-284.
Wickman,F.E.(1976).Markov models of repose-period patterns of volcanoes. In: Merriam(1976a), 135-162.
Wigley,P.(1984).Commercially available geological databanks.*Geol.Soc. Lond.Spec.Publ.* **12**,329-341.
Wignall,T.K. & de Geoffroy,J. (1987). *Statistical models for optimizing mineral exploration.* Plenum,New York,444 pp.
Wilcoxon,F.(1945).Individual comparison by ranking methods.*Biom.Bull.***1**,80-83.
Wilcoxon,F.(1947).Probability tables for individual comparison by ranking methods. *Biom.Bull.***3**,119-122.
Wilkinson,J.H. (1965). *The algebraic eigenvalue problem.* Oxford Univ.Press, Oxford.
Wilkinson,L.(1986).*SYSTAT: the system for statistics.*Systat Inc.,Evanston,IL.
Wilks,S.S.(1938).The large-sample distribution of the likelihood ratio for testing composite hypotheses. *Ann.Math.Stat.***9**,60-62.
Wilks,S.S.(1962).*Mathematical statistics.*Wiley,New York,644pp.
Willen,D.C.(1985).*Macintosh programming techniques: Microsoft BASIC 2.0.* H.W.Sams, Indianapolis,IN.
Williams,E.A. & Gottman,J.M.(1982). *A users' guide to the Gottman-Williams time-series analysis computer programs for social scientists.* Cambridge Univ.Press, New Yor, 98pp.
Williams,E.J.(1959).*Regression analysis.*Wiley,New York,211pp.
Williams,E.J.(1961).Tests for discriminant functions.*J.Aust.Math.Soc.*2,243-252.
Williams,G.P.(1983).Improper use of regression equations in Earth Sciences. *Geology* **11**, 195-197 & **12**,125-127.
Williams,G.P. & Troutman,B.M.(1987).Algebraic manipulation of equations of best-fit straight lines. In: Size(1987a),qv,129-141.
Williams,R.B.G.(1984).*Introduction to statistics for geographers and earth scientists.*Macmillan, London,349pp.
Williams,S.(1986).*Programming the Macintosh in assembly language.*Sybex,Berkley,CA.
Williams,W.T. & Lambert,J.M.(1959).Multivariate methods in plant ecology. I.Association analysis in plant communities. *J.Ecol.***47**,83-101.
Williams,W.T. & Lambert,J.M.(1959).Multivariate methods in plant ecology. II.The use of an electronic digital computer for association analysis. *J.Ecol.***48**,689-710.
Williams,W.T. & Lambert,J.M.(1959).Multivariate methods in plant ecology. III.Inverse association analysis. *J.Ecol.***49**,717-729.
Wishart,D.(1969).FORTRAN II program for 8 methods of cluster analysis (Clustan I). *Kansas Geol.Surv.Comput. Contrib.* **38**,112pp.
Wishart,D.(1978).*CLUSTAN user manual.*(3rd Ed).Program Library, Edinburgh Univ.,UK, Rept.**47**.
Wolfenden,E.B.(1978).The information problems of the mining and petroleum geologist. *Geol.Soc.Lond.Misc.Publ.* **9**,1-14.
Wonnacott,T.H. & Wonnacott,R.J.(1977).*Introductory statistics.*Wiley,New York, 650pp.
Wood,I.D.(1985).An application of geostatistics in the Eastern Transvaal coalfields. *Tr.Geol.Soc.S.Afr.***88**,81-82.
Woronow,A.(1986).Power and robustness of Aitchison's test for complete subcompositional independence in closed arrays. *J.Math.Geol.***18**,563-76.
Woronow,A.(1987). Review of CODA (by J.Aitchison). *Comput.Geosci.* **13**, p.677.
Woronow,A. & Butler,J.C.(1985).Complete subcomposition independence testing of closed arrays. *Comput.Geosci.* **12**,267-279.
Woussen,G. & Côté, D.(1987). Hyperfunc: BASIC program to calculate hyperbolic magma-mixing curves for geochemical data. *Comput.Geosci.* **13**, 421-432.
Wray,W.B.(1973).A computer program to construct cross-sections of curved surfaces.*J.Math.Geol.***5**,149-161.
Wright,C.J.,McCarthy,T.S. & Cawthorn,R.G.(1983).Numerical modelling of trace element fractionation using diffusion controlled crystallisation. *Comput.Geosci.***9**,367-390.
Wright,R.M. & Switzer,P.(1971). Numerical classification applied to certain Jamaican Eocene nummulitoids. *J.Math.Geol.* **3**, 297-311.
Wright,T.L.(1974).Presentation and interpretation of chemical data for igneous rocks.*Contrib.Mineral.Petrol.***48**,233-248.
Wright,T.L. & Docherty,.P.C.(1970).A linear programming and least-squares computer method for solving petrologic mixing problems.*Bull.Geol.Soc.Am.* **81**,1995-2007.
Wright,T.L. & Hamilton,M.S.(1978a).A computer-assisted graphical method for identification and correlation of igneous rock chemistries.*Geology* **6**:16-20.
Wright,T.L. & Hamilton,M.S.(1978b).A computer-assisted graphical method for classifying igneous rocks and identifying unknown samples from their chemistry.*Fortschr.Mineral.***55(1)**:156-157.
Wrigley,N.(1985).*Categorial data analysis for geographers and environmental scientists.* Longman,London,392pp.

Yale,C. & Forsythe,A.B.(1976).Winsorized regression. ITechnometrics **18**,291-300.

Yamamoto,S.,Honjo,S. & Merriam,D.(1979).Quantitative chemical stratigraphy of the Niobrar Chalk in western Kansas. In: Gill & Merriam(1979),qv,235-244.

Yarka,P.J. & Cubitt,J.M.(1977).Data-base management software for computer applications on small computers. *Comput.Geosci.***3(3)**:443-448.

Yatabe,S.M. & Fabbri,A.G.(1986).The application of remote sensing to Canadian petroleum exploration: promising and as yet unexplored.*Comput.Geosci.***12**,597-610.

Yates,S.R.(1987).CONTUR: a FORTRAN algorithm for two-dimensional high quality contouring. *Comput.Geosci.* **13**,61-76.

Yates,S.R.,Warrick,A.W. & Myers,D.E.(1986).A disjunctive kriging program for two dimensions. *Comput.Geosci.* **12**,281-314.

Yeap,C.H. & Chan,S.H.(1981).Trend surface analysis of trace-element distribution in tin granites of Selangor and Negri Sembilan,Peninsular Malaysia.In: Chaston, I.R.M. (ed.), Asian Inst. Ming.Metall.(London),79-98.

Yevjevich,V.(1972).*Stochastic processes in hydrology.* Water Resources Publications,Fort Collins,Colorado, 276 pp.

Yoder,H.S. & Chayes,F.(1982).Alkali correlation in some alkali basalts.*Carn. Inst.Wash.Yearbk.* **81**,309-314.

Yoder,H.S. & Chayes,F.(1986).Linear alkali correlation in oceanic alkali basalts.*Bull.Geol.Soc. Finland* **58**,81-94.

York,D.(1966).Least-squares fitting of a straight line.*Can.J.Phys.***44,**1074-1086.

York,D.(1967).The best isochron.*Earth Planet.Sci.Lett.***2**,479-482.

York,D.(1969).Least-squares fitting of a straight line with correlated errors. *Earth Planet.Sci.Lett.* **5**,320-324.

Young,F.W.(1968).TORSCA-9,a FORTRAN IV program for nonmetric multidimensional scaling.*Behavioural Science* **13**,343-344.

Yufa,Y. & Gurvich,Y.M.(1964).The use of the median and quartiles in estimating normal and anomalous values of a geochemical field.*Geochemistry*,801-807.

Yule,G.U. & Kendall,M.G.(1950).*An introduction to the theory of statistics.* Griffin,London.

Zhao,P., Hu,W. & Li,Z. (1986). Variable selection in statistical prediction for mineral deposits. In: Royer (1986), qv.

Zhou,D.,Chang,T. & Davis,J.C.(1983).Dual extraction of R-mode and Q-mode factor solutions*J.Math.Geol.***15**,581-606.

Zi,S.Z. & Chayes,F.(1983).A prototype database for IGCP Project 163 — IGBA.*Comput.Geosci.***9,**523-526.

Zimmermann,S. & Brown,B.W.(1972).Performance of a rank distance procedure in discriminant analysis.*Biometrics* **28,**p.275 (abstr.)

Zizba,Z.V.(1953).Mean and standard deviation of certain geologic data: a discussion. *Am.J.Sci.*251,899-906.

Zodrow,E.L.(1976).Empirical behaviour of Chayes' null model*J.Math. Geol.*8,37-42.

Zupan,J.(1982).*Clustering of large data sets.*Wiley,New York,122pp.

Zwillenberg,H.J. (1986). Value and utility of information and information services in science and technology. *Proc.3rd Int.Conf.Geosci.Info. (Adelaide,S.Australia, June '86)*, Vol.1, 159-170.

GLOSSARY and INDEX

Mathematical, statistical and computing terms appearing in this glossary are **emboldened** at their first appearance in the main text. The glossary is intended to complement (1) The wide range of published *geological* dictionaries and glossaries (e.g. Bates & Jackson 1980), which cover all the traditional geosciences but not numerical geology. (2) The range of published dictionaries of *statistics* and *computing*, which are probably less than fully comprehensible to many geoscientists, because of their statistical jargon and mathematical notation, their lack of geological examples, and their minimal coverage of important geological specialities such as geostatistics.

This is the first glossary of its kind. Of previous geomathematical texts, only Cheeney (1983) and David et al.(1975) include even short glossaries: Cheeney's covers matrix algebra alone, while that of David et al. (1975) covers only geostatistics (and then only 'tentatively'). The present glossary particularly aims to explain terms in ways geologists of limited mathematical/statistical background can understand. Six groups of terms have received special attention, as likely to cause most confusion: (1) terms over whose meaning statisticians disagree (e.g. **nonparametric**); (2) terms whose common geological and statistical meanings differ (e.g. **sample**); (3) non-trivial synonyms (e.g. few geologists would automatically surmise that **reduced major axis** was synonymous with **unique line of organic correlation**; (4) acronyms and abbreviations, which are indexed under both abbreviation and full name (e.g. **S-N-K-test**); (5) statistical terms which resemble geological terms but refer to quite different things (e.g. **mode analysis** versus petrographic *modal analysis*; also **non-linear mapping**, which is nothing to do with the wobbly traverses of an inebriated field geologist!); and (6) statistical terms which are very similar to each other (e.g. **principle components analysis; principle coordinates analysis**).

This glossary resulted from a thorough search of (1) over 50 textbooks in the Bibliography; (2) large statistical dictionaries; (3) the main geomathematical journals (*Computers & Geosciences, Journal of Mathematical Geology, Journal of Geological Education,* together with publications of the *Computer-oriented geological society,* and the British and Australian geological society specialist geomathematical groups). This glossary is, however, one of computing, mathematical and statistical terms *as applied to geoscience*. It is not a dictionary of computing *per se*, so users should not expect to find herein the interminable jargonese, acronyms and abbreviations for software and hardware so beloved of computer scientists.

For reasons of space, entries have been kept as short as possible; the extensive referencing (explained overleaf) guides users through the maze of other literature. Further information on many terms is given in the main text.

EXPLANATION OF GLOSSARY ENTRIES

Indexed sources of further information: Each entry is accompanied by a set of references intended to enable interested readers to find further information on the particular topic, together with the page number of its main reference in the present text. The references listed are necessarily subjective choices of those the author found most helpful in compiling this glossary, but, as far as space permits, at least one geologically oriented reference is balanced by at least one reference to a treatise or monograph from the original discipline (computing, mathematics, statistics, etc.), so that both theoretical and applied information can be sought. The references here also complement rather than duplicate those in the main text, where possible: the original expositions are cited for statistic named after particular individuals, and entries are annotated according to whether they appear in two of the best-known statistical dictionaries. Although entries in Kendall & Buckland (1982) are in general no lengthier than those in the present glossary, but have a more mathematical/theoretical bias, entries in Kotz & Johnson (1982-8) are invaluable for more comprehensive information, including bibliographies, theoretical explanations and applications. However, this mammoth tome was available at the time of writing this book only as far as the letter 'R' (volume 8), so that citations are necessarily incomplete.

Note that the order of listing in these dictionaries is often different — for example, **coefficient of non-determination** may be listed as 'non-determination, coefficient of'.

Alphabetical order: ignores *all* non-alphanumeric symbols (hyphens, brackets, spaces, apostrophes, etc.) and is *case insensitive*. For example, to find **H-spread** look under hs and to find **t test** look under tt in normal ABC alphabetical order. Letters from the Greek alphabet are given in Greek in the body of entries but also in their conventional English forms (e.g. 'beta' for ß) in **headings**, to avoid possible confusion as to alphabetical placement. 'Mc' and 'Mac' are given in strict alphabetical order, not at the beginning of the Ms as in some telephone directories.

Brackets: indicate where entries are referred to in the literature by abbreviated names. For example, **box plot** is a common abbreviated term for **box-and-whisker plot**; this item is therefore entered as **box(-and-whisker) plots.** The abbreviate form **box plot** is also entered in its correct alphabetical place, for convenience.

Cross-references: between entries in the glossary are given in CAPITALS.

Mathematical formulae: are written in one-line format for brevity. In most cases, however, formulae are given in the main text.

The word sample: is used throughout in the *geological* sense of a SINGLE physical entity, (rock specimen, fossil, etc.), not the *statistical* sense of a collection of entities (see explanation under corresponding entry). **Group (of data)** is used in the conventional statistical sense of 'sample'.

INDEX

The Glossary also acts as an Index, in that page numbers to all those entries considered elsewhere in the book are given to the right of each emboldened entry.

Page

Absolute moment [Kendall & Buckland 1982]
The MOMENT of a frequency distribution in which the deviations about a fixed point (usually the mean or median) are taken without regard to sign. E.g. MEAN DEVIATION, MEDIAN DEVIATION.

Accuracy of data [Kendall & Buckland 1982; Le Maitre 1982,p.5] 102ff
The closeness of measured values to true values. Cf. PRECISION.

Adaptive pattern recognition [Mather 1976, p.453] 312
A form of MULTIGROUP DISCRIMINANT ANALYSIS, adapted to deal with DICHOTOMOUS data.

Admixed Normality [Dixon & Massey 1969, p.323)
Data mostly coming from a Normal distribution, but also containing individual values from other admixed distributions. Similar to CONTAMINATED.

Agglomerative cluster analysis [Everitt 1974, p.8] 281ff
Cluster analysis which proceeds by successively fusing INDIVIDUALS into clusters. Includes the SINGLE LINKAGE, COMPLETE LINKAGE, CENTROID CLUSTER, GROUP AVERAGE,WARD'S and LANCE & WILLIAM'S methods. Cf. DIVISIVE.

AI 59
See ARTIFICIAL INTELLIGENCE.

AID (Automatic interaction detector) method [Kendall & Buckland 1982;Everitt 1974, p.23] 278
A DIVISIVE method of CLUSTER ANALYSIS which proceeds by dividing the initial set of INDIVIDUALS through a series of binary splits into mutually exclusive MONOTHETIC classes.

Algorithm [Kendall & Buckland 1982]
A systematic sequence of instructions or rules for solving a problem. Usually implies a particular method of solving a problem (e.g. matrix inversion) in computer programming.

Aliasing [Davis 1986, p.257]
Refers to the incorporation of irresolvable high frequencies among lower frequencies, in a TIME-SERIES.

Allocation
Synonymous with ASSIGNMENT.

Alpha (α) coefficient of reliability [Mather 1976, p.247]
A coefficient used in ALPHA FACTOR ANALYSIS: the squared correlation between a common factor in the selection of variables and the corresponding universe common factor.

Alpha (α) error 73ff
In hypothesis testing, the probability of rejecting a hypothesis when it is actually true; corresponds to the SIGNIFICANCE LEVEL set by the user.

Alphanumeric data
A contraction of alphabetic + numeric. Data comprising numbers ± letters; may or may not include symbols (!*$ etc.)

Alternative hypothesis [Kendall & Buckland 1982] 73ff
See H_1.

Analysis of dispersion [Miller & Kahn 1962, p.249]
Multivariate analysis of variance.

Anderberg's *D* [SPSS 1986, p.740]
A PROBABALISTIC CONTINGENCY COEFFICIENT, defined as $\{\max(a,b)+\max(c,d)+\max(a,c)+\max(b,d) - \max(a+c,b+d) - \max(a+b,c+d)\}/2N$ — see Table 13.5. Ranges from 0 to 1.

Anderson's (metro)glyphs [Anderson 1960] 315
A pictorial method for displaying multivariate data in two dimension, with a similar purpose to KLEINER-HARTIGAN TREES and CHERNOFF FACES. A circle has 1–7 variable length 'whiskers' at the top, which code the variable values.

Andrews' estimate [Andrews et al. 1972] 88
A ROBUST ESTIMATE of LOCATION based on a sine function.

ANOVA (analysis of variance) [Miller & Kahn 1962, p.214; Le Maitre 1982, p.29] 151ff
The separation of total variation in data into components associated with defined sources of variation. See also ONE-WAY ANOVA, TWO-WAY ANOVA, DURBIN TEST, FRIEDMAN TEST, KRUSKAL-WALLIS TEST, BALANCED COMPLETE & INCOMPLETE BLOCK DESIGNS, LATIN SQUARE, REPEATED MEASURES.

Ansari–Bradley test [Kotz & Johnson 1982-8] 129
See FREUND-ANSARI-BRADLEY-DAVID-BARTON TEST.

ARE
See ASYMPTOTIC RELATIVE EFFICIENCY.

ARIMA (autoregressive integrated moving average) [NAG 1987,p.G13/2]
TIME-SERIES are often determined by a mixture of DETERMINISTIC components (e.g. seasons, lunar months) following fixed mathematical equations, and STOCHASTIC components (random statistical fluctuations). ARIMA models are appropriate where it is difficult to separate these two components. They use difference equations relating

Page

present and past values of the series.

Arithmetic mean 86

Sum of values/Number of values.

Artificial intelligence (AI) 59

General term referring to the use of computers to mimic human thoughts and actions. AI is used in EXPERT SYSTEMS.

Assignment [Everitt 1974, p.2] 275

The process of deciding into which of a number of 'known' groups an 'unknown' INDIVIDUAL should be allocated. Synonymous with IDENTIFICATION. One might, for example, wish to assign a fossil to one of a number of species.

Association 173

Term indicating the presence of a dependence (correlation, relationship) between two or more VARIABLES. Often confined to relationships between ATTRIBUTES, while CORRELATION covers those between quantitative variables.

Association analysis [Everitt 1974, p.21; Buchbinder & Gill 1976] 292

(1) A DIVISIVE method of CLUSTER ANALYSIS, applicable to DICHOTOMOUS data, and based on 2x2 CONTINGENCY TABLES. (2) Occasionally used synonymously with CORRELATION or REGRESSION.

Association coefficient [Everitt 1974, p.21]

A synonym for MATCHING COEFFICIENT.

Associatogram [Davis 1986, p.239] 213

A plot of AUTO-ASSOCIATION against LAG.

Asymptotic relative efficiency (ARE) [Gibbons 1971, p.273; Conover 1980, p.89] 118

If n_1 and n_2 are the sample sizes required for two tests T_1 and T_2 to have the same POWER under the same SIGNIFICANCE LEVEL; the limit of n_1/n_2, as n_1 approaches infinity, is the *ARE* of the first relative to the second test, provided that limit is independent of both power and significance level. For example, the MANN-WHITNEY test has an *ARE* of $3/\pi$ or 95.5% relative to the t test.

Attribute 68

A qualitative, NOMINAL or DICHOTOMOUS characteristic of an INDIVIDUAL (e.g. male/female; +ve/-ve optical sign; colour). Cf. VARIABLE.

Autoassociation [Davis 1986, p.255] 213

As AUTOCORRELATION but concerned with sequences of NOMINAL or DICHOTOMOUS ATTRIBUTES (STATES), such as lithologies in boreholes. Cf. CROSS-ASSOCIATION.

Autocorrelation [Davis 1986, p.232] 215

The 'self-comparison' (internal comparison) of a time or spatial sequence of RATIO DATA (e.g. electrical resistivities in boreholes), to determine whether there are cyclicities or periodicities in the sequence. Cf. CROSS-CORRELATION. Differently used by some authors to refer to the parameter of a parent population (see SERIAL CORRELATION).

Autocovariance [Davis 1986, p.218] 215

In AUTOCORRELATION, the covariance between a sequence and itself displaced by a LAG l, i.e. the covariance between all observations Y_i and Y_{i+l}.

Autocovariogram (autocovariance function) [Davis 1986, p.219] 215

A plot of AUTOCOVARIANCE against LAG, which indicates how cyclic a sequence of data is.

Auxiliary *F* functions

Functions used in GEOSTATISTICS to calculate variances of samples in a block; not to confuse with FISHER's F.

Average 86

A vague term often understood to refer to the ARITHMETIC MEAN, but better used as a general term for all ESTIMATES of LOCATION, including MEDIAN, MODE, etc.

Average entity stability [Everitt 1974, p.27]

A criterion used in PARTITIONING methods of CLUSTER ANALYSIS, which optimises the 'attraction' of an individual to a group, based on the average similarity between that individual and the members of the group.

Average variable complexity [Cuneo & Feldman 1986, p.124]

See VARIABLE COMPLEXITY.

Axial data [Cheeney 1983, p.10] 226

Orientation data lacking 'sense' (direction). E.g. azimuths of dykes, axes of folds. Cf. DIRECTIONAL DATA.

Bahadur efficiency [Bahadur 1960; Kendall & Buckland 1982]

An approximate measure of ASYMPTOTIC RELATIVE EFFICIENCY.

Balanced [Kotz & Johnson 1982–8] 154

Assignment of TREATMENT combinations to the experimental objects in such a way that a symmetric configuration is achieved. Thus in a 2-way ANOVA, 'balanced' indicates an equal number of responses for each *A-B* combination.

Balanced complete block design [Kotz & Johnson 1982-8] 155

A form of ANOVA in which there are subequal numbers of measurements available for *all* possible combinations of BLOCKS and TREATMENTS. Can be analyzed by a TWO-WAY ANOVA or by FRIEDMAN's TEST.

Page

Balanced factorial model [Cuneo & Feldman 1986]
Synonymous with BALANCED COMPLETED BLOCK DESIGN.

Balanced incomplete block design [Kotz & Johnson 1982–8; SPSS 1986 p.513] 159
A form of ANOVA in which, although there are no measurements at all available for some combinations of BLOCKS and TREATMENTS, the missing combinations are distributed through the table of measurements in a systematic manner, e.g. if combination BlockA/Treatment1 is undetermined, Block A has been measured with all other treatments, and Treatment1 with all other blocks. Can be analyzed by parametric ANOVA or by DURBIN'S (nonparametric) TEST.

Bartlett's *M*-test for homogeneity of variances [Bartlett 1937; Till 1974, p.105] 136
An approximate preparatory test to multigroup ANALYSIS OF VARIANCE, testing the equality of several population variances, assuming all are Normal. See also HARTLEY's & COCHRAN's tests. Multivariate versions exist [Kotz & Johnson, vol.6, p.29].

Bartlett's test for sphericity [SPSS 1986, p.501,533; Cuneo & Feldman 1986, p.118] 321
A test used to determine whether, in general, the coefficients in a matrix of PEARSON CORRELATION COEFFICIENTS differ significantly from zero. A multivariate analogue of the univariate *t-TEST* used for testing a single coefficient for significance, but resulting in a χ^2 rather than test statistic. Used in FACTOR ANALYSIS and other multivariate techniques.

Basic structure of a matrix
Synonymous with SINGULAR VALUE DECOMPOSITION.

Basis [Aitchison 1982,1984a,1986] 202
Used in the study of CLOSED DATA, to refer to the original open array from which a closed data-set (COMPOSITION) is generated by dividing by column or row totals. For many geological data (e.g. whole-rock analyses), the basis is purely hypothetical, i.e. does not exist. Synonymous with HYPOTHETICAL OPEN ARRAY.

Bessel functions [Kotz & Johnson 1982-8] 229
Cylindrical functions which appear in various statistical distribution formulae such as that of the VON MISES DISTRIBUTION and the CRAMÉR–VON MISES TEST.

Beta (ß) error 73
In hypothesis testing, the probability of accepting a hypothesis when it is actually false. Determined by the POWER of the test used.

Bias
Refers to the degree to which a statistical SAMPLE (group of data) is randomly chosen from a parent POPULATION. A random sample is unbiased, and as its size increases, it provides an increasingly good picture of its parent population. By contrast, a biased sample does not reflect its parent population, however large.

Binary data/variable/attribute
See DICHOTOMOUS attribute.

Binary explanatory variable [Mather 1976, p.197ff]
DICHOTOMOUS or MULTISTATE variables which are important in a REGRESSION study. For example, measurements might be taken of soil compositions over several different rock-types. The compositions can be divided into groups depending of the rock-type so that regression can be used to test hypotheses about inter-group differences, in a way similar to that used in ANALYSIS OF VARIANCE.

Bingham distribution [Kendall & Buckland 1982; Cheeney 1983, p.119] 240
A general distribution for 3-dimensional ORIENTATION DATA, with orthorhombic symmetry. Formally defined as the conditional distribution of a trivariate Normal vector of unit length with zero mean and arbitrary covariance matrix.

Binomial distribution [Kendall & Buckland 1982;Le Maitre 1982, p.12; Kotz & Johnson 1982-8] 80
The sampling distribution of frequencies expected in random samples drawn with REPLACEMENT from a two-class population. Mainly restricted in geology to pebble counts and the like, since most geological data are continuous. If an event has a probability p of appearing, the probability of r such events in n trials is ${}^{n}C_{r}(1-p)^{n-r}p^{r}$.

Binomial test [Siegel 1956, p.36; Conover 1980, p.96; Kendall & Buckland 1982; Kotz & Johnson 1982-8]
A ONE-GROUP test determining whether the frequencies of the group could have been drawn from a POPULATION having a specified probability. That is, a test for the value of p in the previous entry.

Birnbaum-Hall test [Birnbaum & Hall 1960; Conover 1980, p.377] 148
Effectively, an extension of the SMIRNOV test to comparing three or more INDEPENDENT groups of data. Of little use in geology at present as tables of critical values are only available for 3 groups of equal size.

Biserial correlation coefficient [Downie & Heath 1974,p.106; Kotz & Johnson 1982-8] 172
An inefficient coefficient associating one CONTINUOUS variable with one DICHOTOMOUS variable or one CONTINUOUS variable FORCED INTO A DICHOTOMY.

Bivariate
Involving two VARIABLES.

Biweight scale estimator [Hoaglin et al., 1983, p.417] 91
A ROBUST ESTIMATE of SCALE, based on M-ESTIMATES of LOCATION. Uses the MEDIAN DEVIATION

Page

(MAD) rather than standard deviation, and uses a factor c which determines that zero weight is assigned to outlying values more than $\approx 2c/3$ standard deviations from the median. Can also be used as a robust test for NORMALITY {Hoaglin et al., 1983, p.425].

Bizarre outlier [Powell 1985; Rock 1988] 113

An OUTLIER which is reproducible (i.e. accurate, 'correct' data) but still definitely spurious for some *geological* reason — i.e. forms no part of the data-set in question (e.g. an unsuspected among a set of limestones would yield bizarre outliers for Nb,P, REE and other elements). Such data are 'correct' but should still be eliminated when statistical analysis is performed. Their recognition, however, requires *geological* as well as purely statistical arguments — save only in the case of replicate data, where bizarre outliers are impossible. Cf. TRUE, FALSE OUTLIER.

Block [Kendall & Buckland 1982; SPSS 1986, p.573-4] 153

A group of RELATED items under observation. Used in experimental designs (especially ANOVA) to isolate sources of variability. Items within blocks are as far as possible homogenous, so that other uncontrolled variations can then be isolated by comparison between blocks. For example, in study of Pb isotope variations between different minerals (galena, sphalerite, etc.) across various geological terranes, the minerals would constitute the blocks (since each mineral is an immutable, constant species) and the terranes the TREATMENTS. See also BALANCED INCOMPLETE BLOCK DESIGN, RANDOMIZED COMPLETE BLOCK DESIGN.

Boolean data

Term sometimes used for DICHOTOMOUS data, e.g. in the computer language *Pascal*.

Bootstrapping [Efron 1982; Kotz & Johnson 1982-8]

Some statistical procedures are optimal only for specified values of parameters — e.g. Normal distributions with SKEW 0 and KURTOSIS 3. If these values are unknown, bootstrapped estimates are those calculated from the data *as if* they were the true estimates. Bootstrapping can be used, for example, with GOODNESS-OF-FIT tests, where the form of the population distribution is known (e.g. logNormal) but not its parameters (mean, variance, etc.)

Box-and-Cox transformation [Box & Cox 1964; Howarth & Earle 1979] 100

A generalized POWER TRANSFORMATION: $y = (x^{\lambda}-1)/\lambda$, which includes the reciprocal (λ=–1), square-root (λ=0.5), and logarithmic (λ=0) transformations. Has found widespread use in applied geochemistry.

Box(-and-whisker) plots [Tukey 1977, p.40] 94

Graphical plots using in EXPLORATORY DATA ANALYSIS to show the general spread of data. A vertical 'box' between 1st and 3rd QUARTILES (or LOWER and UPPER HINGES), barred at the MEDIAN, is joined to the values 1.5 times the quartiles by 'whiskers' (vertical lines).Values lying beyond the whiskers (outliers) are plotted individually. Variants are available, such as *notched box-plots*.

Box's M test [Davis 1986, p.497; Mather 1976, p.434; Kendall & Buckland 1982]

A multivariate analogue of BARTLETT's test, testing the homogeneity of several within-group dispersion matrices, preparatory to MANOVA and the comparison of several multivariate means. Tested for significance by conversion to an F ratio or a χ^2 test. Its values lies in its robustness to departures from Normality.

Bray–Curtis nonmetric coefficient [SPSS 1986, p.740]

A synonym for the LANCE & WILLIAMS COEFFICIENT.

Broadened median [Hoaglin et al., 1983, p.312]

A median based not merely on the one or two central values of an ordered data-set, but on between 3 and 6 of the central values, depending on the number of values, N. For example, with $5 \le N \le 12$, the broadened median is the average of the *three* central values.

Browne-Forsythe statistic [Browne & Forsythe 1974a,b]

A statistic used to test equality of several means in ONE-WAY ANOVA routines (for example in BMDP: Dixon 1975), which does not assume equality of variances, and reduces to the separate variance t test for 2 groups. Closely related to the WELCH STATISTIC. The statistic equals $\sum_i n_i(X_i - \bar{X})^2/\sum_i(1 - n_i/N)s_i^2$, critical values being obtained from the F distribution.

Brown–Mood median test [Kotz & Johnson 1982-8]

An uncommon name for the MEDIAN TEST.

Burnaby's procedure [Miller & Kahn 1962, p.361] 219

A test of whether two time-series are unrelated. Involves comparing matches and computing a χ^2 statistic.

CAD 52

Computer-assisted drafting (or drawing).

CAI 57

See COMPUTER-ASSISTED INSTRUCTION.

CAM 52

Computer-assisted mapping.

Canonical (correlation) analysis [Kendall & Buckland 1982; Kotz & Johnson 1982-8; Davis 1986,p.607] 261

A multivariate extension of REGRESSION, considering relationships between *two sets* of several variables (cf.regression, correlating 2 variables, and MULTIPLE REGRESSION, correlating one variable with *one* set of

Page

several). Involves transforming variates $X_1...X_p$ and $X_{p+1}...X_{p+q}$ linearly into variates $L_1....L_p$ and $L_{p+1}....L_{p+q}$ so that (a) members of each group are independent among themselves; (b) each member of one group is independent of all but one member of another group; (c) non-vanishing correlations between members of different groups are maximised. The correlations are called *canonical correlations*, and the sets of transformed variates, *canonical variates*. The largest or first canonical correlation is the maximum correlation attainable between the canonical variables, ie between the transformed original variables, the second is the maximum remaining correlation at right angles to the first, etc. The total number of canonical variables equals the number of variables in the smaller of the two original sets. Certain aspects of MULTIGROUP DISCRIMINANT ANALYSIS are special cases of canonical analysis; many discriminating plots in the geological literature label the axes 'canonical variables' rather than 'discriminant functions'. Could be used, for example, to determine the relationship between the chemical and optical properties of minerals — i.e. which chemical variables most closely relate to which optical properties.

Canonical variate analysis [Reyment et al. 1978] 309

Synonymous with MULTIGROUP DISCRIMINANT ANALYSIS.

Card-image

The analogue in a modern computer file (on a screen, tape or disc) of the old 80-column punched cards.

Cartet count method [Everitt 1974, p.32]

A DENSITY SEARCH technique of CLUSTER ANALYSIS, in which the hyperspace is partitioned and the number of individuals in each 'cartet' (hypercube) is counted. Fixing a significantly high density count, relative to the average total density, allows clusters to be delineated.

Case [SPSS 1986]

Generally synonymous with INDIVIDUAL.

Casewise omission [NAG 1987, p.G02/7] 206

The treatment of missing data by omission of entire INDIVIDUALS for which one or more variables is missing. E.g. if 10 rock analyses are to be compared on the 6 elements Si,Al,Fe,Mg,Ca,Ti, but one sample has no Ti value, then that sample is omitted altogether, and only the other 9 analyses are used. Allows multivariate techniques to be applied (though with considerable loss of information) if there are many missing data. Synonymous with LISTWISE omission [SPSS 1986]. Cf. PAIRWISE omission.

Catastrophe theory [Henley 1976a; Kotz & Johnson 1982-8]

A theory which allows for discontinuities in the behaviour of variables, in contradistinction to classical statistics which normally assumes continuous variation. Of obvious application to volcanic eruptions, earthquakes, etc.

Categorical variable [Kotz & Johnson 1982-8; SAS 1985, p.21]

Sometimes used synonymously with NOMINAL, sometimes with DISCRETE variable. Probably best avoided.

Censored data [David 1981, p.136; Kendall & Buckland 1982; Kotz & Johnson 1982-8]

A group of data in which certain values, though known to exist, are quantitatively unknown, or are quantitatively known but are not used. Klotz & Johnson (1982) restrict 'censored' to data which are reduced according to their *ranks*: e.g. the highest *N* values are censored (in which case TRIMMED MEANS, MEDIANS and other ROBUST ESTIMATES are all based on 'censored' data), from TRUNCATED data, which are reduced according to some external value: e.g. all values <10ppm or outside 5-10%. By contrast, Kendall & Buckland (1982) use 'Type I' and 'Type II' censoring in the same sense, using 'truncation' to refer to populations.

Central limit theorem [Miller & Kahn 1962,p.65; Till 1974,p.53; Kendall & Buckland 1982; Kotz & Johnson 1982-8] 77

The theorem giving the NORMAL DISTRIBUTION its central place in statistics. In its simplest form, states that the distribution of a statistic (such as an arithmetic mean) taken from *N* random samples of a population, tends to a Normal distribution as *N* increases, irrespective of the population distribution. For example, means taken from logNormal, bimodal or wholly irregular populations are distributed more Normally than the parent population.

Central tendency, centre [Kendall & Buckland 1982]

Synonymous with LOCATION.

Centring [Joreskog et al. 1976, p.123]

A data transformation in which the origin of observations is shifted, usually to the mean. Row, and row-and-column centring are difficult to justify in geology.

Centroid

The analogue of 'centre' in multi-dimensional space, usually referring to the middle of the multi-dimensional ellipsoid formed by multivariate data.

Centroid cluster analysis [Everitt 1974, p.12] 281ff

An AGGLOMERATIVE method of CLUSTER ANALYSIS, in which groups with the smallest distance between their CENTROIDS are fused first.

Chaining [Everitt 1974, p.30]

The tendency of CLUSTER ANALYSIS to allocate new INDIVIDUALS to existing rather than new clusters.

Character(istic)

Page

Can be synonymous with either ATTRIBUTE or VARIABLE.

Characteristic analysis [Botbol et al. 1977] 175

A method of analyzing relationships among many DICHOTOMOUS (0/1) variables. Can be regarded as a form of CLUSTER ANALYSIS, to compare similarities between objects measured on many dichotomous variables, or as a variety of CONTINGENCY TABLE analysis, assessing the association of one dichotomous variable of major interest (e.g. 1 = mineralization present, 0 = no mineralization) with a series of other dichotomous variables.

Characteristic (latent) root/vector [Kendall & Buckland 1982]

Synonyms respectively for EIGENVALUE/EIGENVECTOR.

Chebyshev's inequality [Steel & Torrie 1980, p.548; Kendall & Buckland 1982]

States that the area under any distribution curve which is farther away from the mean than X standard deviations is < $1/X^2$. E.g. 0.25 area lies farther than 2 standard deviations from the mean. An unusual example of a statistic which is DISTRIBUTION-FREE but PARAMETRIC.

Chernoff faces [Klotz & Johnson 1982; Barnett 1982, p.258; Green 1985, p.200] 315

Stylised ('cartoon') faces which are very effective in displaying certain types of multivariate data on a piece of paper, because of the human brain's ability to scan and read faces instantaneously. Up to 18 variables can be linked to facial features (angle of mouth and eyebrow, etc.), and relationships between variables can be displayed because of the connections between facial features. Has found widest application in social sciences and business (e.g. a series of increasingly gloomy faces tracing the bankruptcy of a company as reflected in its balance sheet). Faces have also found uses in palaeontology. Because Chernoff faces automatically impose an interpretation on the data ('sad', 'angry' etc.) they are best used to display data with some kind of 'quality' aspect (e.g. water purity, data accuracy). More neutral pictorial methods such as KLEINER-HARTIGAN TREES, ANDERSON'S METROGLYPHS and WEATHERVANE SYMBOLS are usually preferable for displaying entirely numerical data, such as rock or stream sediment compositions.

Chi-square (χ^2) distribution [Kendall & Buckland 1982; Kotz & Johnson 1982-8] 84

The distribution of the sum of squares of independent standardized Normal variates.

Chi-square (χ^2) statistic/test [Kendall & Buckland 1982; Kotz & Johnson 1982-8] 95

A statistic more-or-less following the χ^2 DISTRIBUTION, which relates observed to expected frequencies. Used in CONTINGENCY TABLES, GOODNESS-OF-FIT tests and several multivariate tests (e.g. BOX's *M*, WILKS' LAMBDA).

Circular histogram [Kendall & Buckland 1982; Mardia 1972, p.4; cf.Cheeney 1983, p.23] 227

A 'spoke diagram', derived by simply wrapping a linear histogram round a circle, so leaving gaps between the tops of the bars (Mardia 1972). Cheeney (1983), by contrast, uses the term as a synonym for ROSE DIAGRAM.

Circular mean [Mardia 1972, p.20] 228

Analogue of the linear MEAN for ORIENTATION DATA, but defined vectorially. Equals $\cos^{-1}(C/R)$ or $\sin^{-1}(S/R)$, where C = ($\sum$cosines of all angles in data-set)/number of angles; S = ($\sum$sines of all angles in data-set)/number of angles; $R = \sqrt{(C^2+S^2)}$, the RESULTANT LENGTH of the sample. As in the linear case, the sum of deviations about the circular mean is zero, and the CIRCULAR VARIANCE measured about it is a minimum.

Circular kurtosis [Mardia 1972, p.76] 228

Analogue of linear KURTOSIS for orientation data. The formula is usually taken so as to be zero for a VON MISES DISTRIBUTION.

Circular mean deviation [Mardia 1972, p.30] 228

Analogue of the linear MEAN DEVIATION for ORIENTATION DATA. Minimised when measured from the CIRCULAR MEDIAN DIRECTION.

Circular median direction [Mardia 1972, p.28; Cheeney 1983, p.22] 228

Analogue of the linear MEDIAN for ORIENTATION DATA. Given a set of points on a CIRCULAR PLOT, the median P is such that *(1)* half of the sample points are on each side of the diameter PQ through P, and (*2)* most points are nearer to P than Q. For a symmetrical distribution, the median will be along the axis of symmetry. For polymodal data, there is more than one possible median.

Circular mode [Mardia 1972, p.30] 228

On the diameter of a CIRCULAR PLOT which divides a sample into two equal parts, the direction which points to the greater density of points.

Circular plot [Cheeney 1983, p.22]

A circle on which 2-dimensional, directional data are plotted around the circumference. Corresponds in 2-D to the stereographic projection in 3-D.

Circular range [Mardia 1972, p.34; Kendall & Buckland 1982] 228

The length of the smallest angular arc which encompasses all of a set of ORIENTATION DATA.

Circular skew [Mardia 1972, p.74] 228

Analogue of linear SKEW for ORIENTATION DATA, equal to zero for symmetrical distributions such as the VON MISES.

Circular variance [Mardia 1972, p.21] 228

Page

Analogue of the linear VARIANCE for ORIENTATION DATA. Equal to [1-R], where R is the RESULTANT LENGTH. Reaches a minimum when measured about the CIRCULAR MEAN, as in the linear case. The resultant length is more commonly used for statistical inference than the variance itself. See also CONCENTRATION.

City block [SPSS 1986, p.736] 278

A DISTANCE COEFFICIENT defined as $\Sigma|X_i - Y_i|$. MINKOWSKI'S MEASURE with $p = 1$.

Classical statistics

Usually refers to the well-known PARAMETRIC methods such as the t-test and F-test based on the NORMAL DISTRIBUTION.

Classification [Everitt 1974, p.1; Kotz & Johnson 1982-8] 275

The placing of INDIVIDUALS or VARIABLES into GROUPS from first principles. The grouping of similar objects. Sometimes restricted to classification of individuals (Q-MODE; e.g. Everitt 1974, p.1). Techniques include CLUSTER ANALYSIS and ORDINATION. Cf.ASSIGNMENT.

Classificatory scale/variable [SAS 1985,p.21]

A synonym for NOMINAL SCALE/VARIABLE.

Closed data/array [Le Maitre 1982, p.45] 202ff

Data with constant totals (e.g. percentage and ppm data such as whole-rock analyses which total ≈ 100%). A very wide range of geological data are closed, yet their BASIS does not exist. For example, it is meaningful to say "this rock has 50.5% SiO_2", but not "there are 50.5 gms of SiO_2 in this rock". This causes very serious problems in statistical analysis (see COMPLETE SUBCOMPOSITIONAL INDEPENDENCE).

Closure

The property of CLOSED DATA.

Clumping techniques [Everitt 1974, p.35]

CLUSTER ANALYSIS techniques which allow overlapping clusters.

Cluster [Everitt 1974, p.43]

A group of contiguous elements of a statistical population.

Cluster analysis [Davis 1986, p.502; Le Maitre 1982, p.163; Everitt 1974] 281ff

CLASSIFICATION techniques dividing a group of INDIVIDUALS into either several discrete or overlapping clusters (non-hierarchical methods), or a DENDROGRAM (hierarchical methods). Cf. ORDINATION.

Cochran's *C* test for homogeneity of variances [Kendall & Buckland 1982; Kotz & Johnson 1982-8] 136

A preparatory, parametric test to multigroup ANALYSIS OF VARIANCE, testing whether one of several population variances may be significantly larger than the others. Based on the ratio of the largest estimate of variance to the total of all estimates. Sensitive to non-Normality. See also HARTLEY's & BARTLETT's tests.

Cochran's *Q* test [Siegel 1956, p.161; Conover 1980, p.96; Kotz & Johnson 1982-8] 179

A NONPARAMETRIC test for comparing more than two GROUPS of RELATED data measured on a DICHOTOMOUS or NOMINAL scale. Effectively, an extension of McNEMAR'S TEST to more than 2 groups of data. It could be used, for example, to test whether geology students have learned from a course, if their response to a particular question (marked as 'yes' or 'no', i.e. 0 or 1) is examined on days 1,2,3...N.

Coefficient of agreement [Kendall & Buckland 1982]

A coefficient correlating PAIRED X-Y values. Similar in aim to KENDALL's COEFFICIENT OF CONCORDANCE. Reduces to KENDALL'S τ RANK CORRELATION COEFFICIENT for 2 variables.

Coefficient of alienation (non-determination) [Kendall & Buckland 1982; Kotz & Johnson 1982-8]

Defined as $\sqrt{(1-r^2)}$ where r is the PEARSON CORRELATION COEFFICIENT. The proportion of the variance in one variable *not* explained by the other, in a linear REGRESSION model. The converse of COEFFICIENT OF DETERMINATION.

Coefficient of communality

A synonym for JACCARD'S COEFFICIENT.

Coefficient of concordance [Kendall & Buckland 1982; Kotz & Johnson 1982-8] 180

A measure of strength of ASSOCIATION between several NOMINAL or ORDINAL variables. Can be thought of either as an extension of the CORRELATION COEFFICIENT to more than two variables, or of the CONTINGENCY COEFFICIENT to ORDINAL data. Examples include KENDALL'S τ, GAMMA (γ) and SOMER'S d.

Coefficient of determination [Kendall & Buckland 1982; Kotz & Johnson 1982-8] 188

The square of the PEARSON CORRELATION COEFFICIENT between two variables, expressing the proportion of the variance in one variable explained by the other, in a linear REGRESSION model. Can be extended to the multivariate case in MULTIPLE REGRESSION. Varies from 0 to 1.

Coefficient of excess 92

Occasionally used for [b_2 - 3], b_2 being the KURTOSIS. A Normal Distribution then has the value 0 instead of 3.

Coefficient of proportional similarity [Le Maitre 1982, p.131]

Synonym for COSINE THETA.

Coefficient of similarity [Till 1974, p.134]

Page

See SIMILARITY COEFFICIENT.

Coefficient of variation [Kendall & Buckland 1982; Kotz & Johnson 1982-8] 91
The ratio: standard deviation/arithmetic mean. A useful relative measure of DISPERSION to compare variables whose absolute magnitudes differ widely (e.g. major and trace elements for a rock analysis).

Collinear [Kotz & Johnson 1982-8]
Indicates vectors which are highly correlated, i.e. form small angles with each other and are hence near-linear.

Communality [Kotz & Johnson 1982-8; Davis 1986, p.551] 321
In FACTOR ANALYSIS, expresses the proportion of total variance explained by the extracted factors or, equivalently, the extent to which the original variables are explained by the extracted factors. Will equal unity where the number of factors equals the number of variables. Termed *common factor variance* by Kendall & Buckland (1982).

Complete link(age) technique [Everitt 1974, p.11] 278
An AGGLOMERATIVE method of CLUSTER ANALYSIS, in which the distance between groups is defined as that between their most remote pair of INDIVIDUALS.

Complete subcompositional independence [Woronow & Butler 1985] 205
A condition in CLOSED DATA, in which there are no correlations present other than those induced by closure alone.

Completely randomized group (block) design [SPSS 1986, p.813] 138
A synonym for ONE-WAY ANOVA.

Components analysis [Kotz & Johnson 1982-8]
Sometimes used as an abbreviation for PRINCIPAL COMPONENTS ANALYSIS.

Computer-assisted instruction (CAI) 57
The use of computers in teaching. One of the most widely used CAI systems is called PLATO, developed originally at the University of Illinois, USA. It includes some 50,000 lessons in a myriad of subjects.

Concentration parameter [Kendall & Buckland 1982; Cheeney 1983, p.98 & 113] 228
The analogue of DISPERSION for 2- and 3-dimensional distributions such as VON MISES, FISHER and BINGHAM, analogous to an *inverse* standard deviation for linear data. The further concentration increases from zero, the further the distribution departs from circles/spheres, the greater the preferred orientation, and the SMALLER the dispersion.

Concordance
See KENDALL's COEFFICIENT OF CONCORDANCE.

Condition of a matrix
See ILL-CONDITIONED and CONDITION NUMBER.

Condition number [Kennedy & Gentle 1980, p.279]
The ratio of the largest to smallest diagonal element (singular values) of a matrix. Large condition numbers usually result in inaccurate solutions to techniques such as matrix INVERSION.

Confidence bands [Koch & Link 1971, p.14; Kotz & Johnson 1982-8] 192
Curved lines, surfaces or hypersurfaces that define upper and lower confidence limits for a regression line in 2,3 or more dimensions.

Confidence cone [Koch & Link 1971 p.139; Cheeney 1983, p.116] 239
The analogue of CONFIDENCE INTERVAL for 2- and 3-dimensional ORIENTATION DATA.

Confidence interval [Kendall & Buckland 1982; Kotz & Johnson 1982-8] 109ff
An interval determining the probability that a variable lies within it. E.g. if we state that the 95% confidence limits on a value of 5ppm are -3, +2ppm, we are 95% sure that the true value lies in the range 2–7ppm.

Confidence limits [Kendall & Buckland 1982] 109ff
The limits to a CONFIDENCE INTERVAL, i.e. 7 and 2 in the above example.

Conservative [Conover 1980, p.90]
A test (or confidence limit) is conservative if the actual SIGNIFICANCE LEVEL is smaller than that stated, i.e. the risk of making a TYPE I ERROR is not as great. Used to err on the side of caution, when computation of the exact significance levels is difficult, and approximations are made.

Contagious distribution [Miller & Kahn 1962, p.373; Kendall & Buckland 1982; Kotz & Johnson 1982-8]
A distribution in which there is a natural tendency towards clumping or clustering of points.

Contaminant
Synonymous with BIZARRE OUTLIER.

Contaminated
Usually refers to data containing CONTAMINANTS, i.e. in which most values come from a single population (with mean μ_1 and variance σ_1^2) but a small number some from another population (with mean μ_2 and variance σ_2^2).

Contingency coefficients [Siegel 1956, p.196; Lewis 1977, p.94; Kendall & Buckland 1982] 176
Coefficients measuring the strength of relationship between two DICHOTOMOUS or NOMINAL scale variables. Some (e.g. CRAMER's) are designed to lie between 0 (indicating no association) and 1 (indicating complete dependence), others (e.g. PEARSON's) are not so constrained, but all are designed to be less controlled by the number of data-values than is χ^2 itself. Contingency coefficients attempt to measure the *strength* of an association, whereas χ^2 itself merely

Page

indicates whether the association is statistically significant. Many contingency coefficients are nevertheless simply related to χ^2, whereas others (e.g. GOODMAN & KRUSKAL's τ) are *probabilistic*, that is, determine the predictability of one variable, given the other. Contingency coefficients are in effect R-mode equivalents of Q-mode MATCHING COEFFICIENTS, and some coefficients (e.g. YULES' Q) have been used in both ways. Both are based on the values *a,b,c,d* in a 2x2 contingency table (Table 13.5). See also FOURFOLD, LAMBDA, TETRACHORIC, TSCHUPROW'S *T*.

Contingency table [Kendall & Buckland 1982; Kotz & Johnson 1982-8] 173
A table showing the numbers of INDIVIDUALS showing 2 or more ATTRIBUTES. See previous entry. Conventionally, *a,c* refer to the number of cases of the attribute in the 2 groups of individuals, while *b,d* refer to the number of cases lacking the attributes. Can be regarded as ANOVA dealing with DICHOTOMOUS/NOMINAL data.

Continuity corrections [Kotz & Johnson 1982-8] 174
See YATES' CORRECTION.

Continuous spectrum [Davis 1986, p.258]
A spectrum calculated from a TIME-SERIES (which is normally sampled at discrete points), in which the VARIANCE is apportioned among set of frequency bands. A continuous plot of variance against frequency.

Continuous variable
A VARIABLE which can take any value. Implies measurement on an INTERVAL or RATIO scale. E.g. mass, volume.

Continuous variable forced into a dichotomy 67
A CONTINUOUS variable separated by cut-off values; e.g. 1-100% exam marks used to separate 'pass' and 'fail' at 50%, optic signs of minerals which are measurements of $2V\gamma$ separated into '+' ($2V\gamma < 90°$) and '–' ($2V\gamma > 90°$).

Cook's *D* [Cook 1977,1979; SAS 1985,p.10]
A statistic used to assess INFLUENTIAL OBSERVATIONS in linear REGRESSION. Measures the change of the slope and intercept estimates that results from deleting each influential observation.

Cophenetic correlation coefficient [Farris 1969; Davis 1986, p.507; Mather 1976, p.325]
A use of PEARSON'S CORRELATION COEFFICIENT to measure the efficacy of a DENDROGRAM produced by CLUSTER ANALYSIS. Measures the degree of distortion introduced by the clustering algorithm. Generally ranges from 0.6 to 0.95; values above 0.75 to 0.8 are felt to be acceptable.

Corner test for association [Dixon & Massey 1969, p.352; Steel & Torrie 1980, p.552] 181
A synonym for OLMSTEAD-TUKEY TEST. A type of MEDIAL TEST.

Correlation [Kendall & Buckland 1982; Kotz & Johnson 1982-8] 167ff
Analysis of the strength of association/dependence between two or more variables, usually measured by a *correlation coefficient*. See PEARSON'S, SPEARMAN'S, KENDALL'S COEFFICIENTS.

Correlation matrix [Le Maitre 1982, p.109; Kendall & Buckland 1982] 279
A matrix of correlation coefficients. The starting point for many multivariate techniques (e.g. PRINCIPAL COMPONENTS ANALYSIS, FACTOR ANALYSIS).

Correlation ratio, E or ζ (zeta) [Kendall & Buckland 1982; Kotz & Johnson 1982-8] 182
A measure of non-linear relationships: the ratio [between-groups variance (sum-of-squares)/total variance]. Analogous to a PRODUCT-MOMENT CORRELATION COEFFICIENT, *r*, except for its lack of sign. ($E^2 - r^2$) may be regarded as measuring the non-linear part of a relationship. Varies from 0 to 1.

Correlogram [Miller & Kahn 1962, p.349; Kendall & Buckland 1982; Davis 1986, p.217] 215
A plot of AUTOCORRELATION versus LAG, in the analysis of sequences of data.

Correspondence analysis [Joreskog et al. 1976, p.107; Kotz & Johnson 1982-8; Davis 1986,p.579] 327
Formally, a variety of PRINCIPAL COMPONENTS ANALYSIS allowing elasticity of scale, but also analogous to a combination of both Q- and R-MODE FACTOR ANALYSIS. It is also equivalent to a special case of CANONICAL CORRELATION. Klotz & Johnson(1982) describe it as principal components analysis on NOMINAL rather than RATIO data. Has been championed by French Canadian geologists in petrology and geochemistry.

Cosine theta [Davis 1986, p.563; Kotz & Johnson 1982-8]
A similarity coefficient used in Q-MODE FACTOR ANALYSIS, measuring the angle between the multivariate vectors of two INDIVIDUALS.

Covariance [Kendall & Buckland 1982]
The first PRODUCT-MOMENT of two variables about their mean values. Equals $\sum (Y-\bar{Y})(X-\bar{X})/N$.

Cox & Stuart's test for trend [Cox & Stuart 1955; Lewis 1977, p.195; Kendall & Buckland 1982] 210
An adaptation of the SIGN TEST examining trends on RATIO or ORDINAL measurements against position in a sequence in space or time. Particularly well suited to looking for trends in data which are known to be cyclical (e.g. rainfall, temperature, which vary annually), with which correlation coefficients cannot be used.

C_p statistic [Mather 1976, p.187; Kotz & Johnson 1982-8]
An alternative to the COEFFICIENT OF DETERMINATION used to compare the adequacy of MULTIPLE REGRESSION functions of different orders.

Cramer's *V* [Blalock 1972, p.297; Lewis 1977, p.93; Conover 1980, p.181; Coxon 1982, p.25; SAS 1985,p.414]

Page

A test used to determine whether a distribution is skewed (logarithmic), such that the GEOMETRIC MEAN would be a better estimate of LOCATION than the ARITHMETIC MEAN. Uses a coefficient of skewness (K) defined as: K = [log(LQ) + log(UQ) – 2log(MQ)]/[log(UQ)–log(LQ)] where UQ, LQ and MQ refer respectively to the UPPER, LOWER and MIDDLE (MEDIAN) QUARTILES. If K > +0.2, the data are symmetrical enough to use the arithmetic mean, but if K < 0.2, the distribution is near-logarithmic, and the geometric mean is preferable. The test works well provided the distribution is definitely and POSITIVELY skewed, with >50 values.

Degree of freedom (*df*) [Kendall & Buckland 1982]
Used on various senses, but most generally, the number of values which can be assigned arbitrarily within the specification of a system. The number of independent co-ordinate values necessary to determine a system. A group of N values for some variable has N degrees of freedom, but $(N-K)$ degrees of freedom if K functions of the values are held constant. Many test statistics, such as FISHER's F, STUDENT's t, χ^2 always have associated *df*.

Dendrogram/dendogram [Kendall & Buckland 1982; Davis 1986, p.505] 283ff
A diagram produced by HIERARCHICAL CLUSTER ANALYSIS, with a branching ('tree') structure, in which the degree of affinity between INDIVIDUALS is shown by their nearness in the structure, and each individual is part of another group of individuals at the next level of the hierarchy.

Density search clustering techniques [Everitt 1974, p.30] 278
CLUSTER ANALYSIS in which clusters are formed by searching for regions containing relatively dense concentrations of individuals.

Departure from Normality, test for [Kotz & Johnson 1982-8]
See NORMALITY TEST.

Dependent variable [Kendall & Buckland 1982] 187
A variable whose value is determined by another variable. E.g. in the variation of refractive index with plagioclase composition, refractive index is the dependent variable.

Determinant
A scalar number derived from operations on a SQUARE matrix. For a 2x2 matrix, the determinant is the value $[a_{11}a_{22} - a_{12}a_{21}]$ where a represents a matrix element. Equals the product of the EIGENVALUES of the matrix.

Deterministic model [Krumbein & Graybill 1965, p.15; Kendall & Buckland 1982] 75
A model in which there are no random elements, whose future course is determined by position, etc. at some fixed point in time.

Deterministic variable
A variable whose value is completely determined at any particular point by some mathematical function. A variable which has no random fluctuations. Cf. RANDOM VARIABLE.

De Wijsian model [Clark 1979, p.9] 272
A form of SEMI-VARIOGRAM without a SILL defined by the equation $\gamma(h) = 3\alpha\ln(h)$

Diagonal matrix [Joreskog et al. 1976, p.18]
A SYMMETRICAL matrix in which all off-diagonal elements are zero.

Dice coefficient [Mather 1976, p.314; SPSS 1986, p.739]
A MATCHING COEFFICIENT, defined as $2a/[2a+b+c]$ (see Table 13.5). Gives MATCHES twice the weight of MISMATCHES.

Dichotomous attribute [Steel & Torrie 1980, p.508] 63
A measurement which can only take two values (e.g. yes/no, presence/absence of a mineral, + or - ve optical sign).

Dichotomized variable
See CONTINUOUS VARIABLE FORCED INTO A DICHOTOMY.

Dimensionality [Le Maitre 1982,p.174]
(1) A qualitative term indicating the general size of a data-array. An array has higher dimensionality, the larger the number of INDIVIDUALS or VARIABLES. (2) More formally, in matrix algebra, the number of linearly independent vectors in a set.

Dimroth-Watson's distribution [Mardia 1972, p.233] 240
A symmetrical GIRDLE distribution for 3-dimensional AXIAL ORIENTATION DATA. A special case of the BINGHAM DISTRIBUTION.

Directional data [Kotz & Johnson 1982-8; Cheeney 1983, p.10] 226
ORIENTATION DATA with a 'sense' (usually symbolised by a line with arrowhead). E.g. palaeomagnetic directions, palaeocurrent directions, glacial striae. Cf. AXIAL DATA.

Discontinuous variable
Synonym for DISCRETE variable.

Discrete power spectrum [Davis 1986, p.257]
Synonymous with PERIODOGRAM.

Discrete variable
A variable which can only take certain (usually integer) values (e.g. subscripts in oxide formulae: SiO_2, Al_2O_3;

Page

A CONTINGENCY COEFFICIENT, designed to lie between 0 (for no association) and 1 (for complete association). Equals $\sqrt{\{\chi^2/N(\mathrm{MIN}-1)\}}$, where MIN is the lesser of the number of rows and columns in the contingency table. Equal to TSCHUPROW's *T* for a square table, but numerically larger with differing numbers of rows and columns, and may reach unity, so probably preferable to *T* for most geological applications.

Cramér-von Mises test [Conover 1980, p.373; Kendall & Buckland 1982; Kotz & Johnson 1982-8]

A test very similar to the KOLMOGOROV–SMIRNOV test in both principle and method of calculation, but only TWO-SIDED. Versions exist to test censored and multivariate data [Kotz & Johnson, vol.6, p.35ff]. Not so far widely used in geology.

Cressie–Hawkins semivariogram [Miller & Morrison 1985]

A ROBUST type of SEMIVARIOGRAM, less distorted than the more common MATHERONIAN or other types by OUTLIERS.

Critical region [Kendall & Buckland 1982]

Usually taken to mean the region in which a hypothesis is rejected (i.e. in which the calculated test statistic is < or > the critical value for the chosen significance level).

Critical value [Kendall & Buckland 1982]

The theoretical value of a test statistic corresponding to a given significance level as determined from its known sampling distribution, e.g. $\Pr(t > t_0) = 0.05$, t_0 is the critical value of *t* at the 5% level. Critical values are compared with actual calculated statistics from real data in HYPOTHESIS TESTING.

Cross-association [Davis 1986, p.234] 217

As CROSS-CORRELATION but comparing two separate sequences of NOMINAL or DICHOTOMOUS ATTRIBUTES (STATES), e.g. lithologies in boreholes. Cf. AUTO-ASSOCIATION.

Cross-correlation [Davis 1986, p.225] 218

The comparison of two separate time or spatial sequences of RATIO data (e.g. electrical resistivities in boreholes, density traverses of ore bodies). The sequences are moved past one another and the degree of correspondence between overlapping segments is calculated. Cf. AUTO-CORRELATION.

Cross-covariance [Kendall & Buckland 1982]

The numerator of the equation which determines a coefficient of CROSS-CORRELATION.

Cross median [Tukey 1977, p.668]

A 2 or multi-dimensional MEDIAN; i.e. that point (often not among the given points), whose coordinates are the medians of each coordinate of the given points.

Cross-product ratio measure of association [Lewis 1977, p.96]

A CONTINGENCY COEFFICIENT which is not dependent on the number of objects.

Cross-tabulation [SPSS 1986,p.337]

The preparation of CONTINGENCY TABLES.

Cumulative frequency distribution [Kendall & Buckland 1982]

A function F(X) measuring the total frequency of objects with values less than or equal to *X*. Varies from 0 to 100%.

Czekanowski coefficient [SPSS 1986, p.739]

A synonym for DICE COEFFICIENT.

Daniel's test for trend [Lewis 1977, p.201; Kendall & Buckland 1982] 209

Use of SPEARMAN's RANK CORRELATION COEFFICIENT with naturally increasing data-sequences (e.g. rainfall with time, resistivity with depth in a borehole).

Databank

Generally synonymous with DATABASE.

Database [Date 1981; Fidel 1987; Martin 1975, 1978, 1981] 37ff

A collection of data organized on a computer in a structured way, such that retrievals can be performed by linking various pieces of information together. Usually (though not always) based on more than one DATA-FILE.

Database management system (DBMS) 43ff

A piece of software for implementing and managing a DATABASE.

Data closure

See CLOSURE.

Data-file 18

Any collection of data held as a single named entity (usually implying storage on a computer).

David, Pearson & Hartley's test for outliers [David et al. 1954; Kotz & Johnson 1982-8]

A test to determine whether either or both of the two extreme values (maximum and minimum) or a given data-set are statistical outliers. Uses the sample STUDENTIZED RANGE, i.e. $[X_{max} - X_{min}]$/sample standard deviation.

David's test [NAG 1987, p.G08/5] 129

A NONPARAMETRIC test comparing the DISPERSION of two groups of data (samples) which makes allowance for possible differences of MEDIAN. Cf. MOOD'S TEST. Nonparametric analogue of an *F*-test.

Davie's test [Langley 1970, p.78]

Page

numbers of counts in, say, point-counting or XRF analysis). Applies to all NOMINAL, most ORDINAL, and some RATIO (counted) data.

Discriminant (function) analysis [Kendall & Buckland 1982; Kotz & Johnson 1982-8] 304ff

A MULTIVARIATE statistical technique used in CLASSIFICATION. In N-dimensional space, it maximises the separation between 2 or more groups of data, using discriminant function(s) which are (generally linear) functions of the original variables. For 2 groups, (1 & 2) a single function is calculated $Z = f(x)$, and a cut-off value is Z_0 chosen such that Group *A* is taken as all individuals with $Z > Z_0$, and Group *B* all those with $Z < Z_0$. Discriminant analysis maximises the correspondence between the *a priori* groups (1 & 2) and the *statistical* groups (*A* & *B*): i.e. if it is successful, the *a priori* groups will be reproduced to near 100% by the statistical groups. Other individuals ('unknowns') can then be classified as belonging to either Group *A* or Group *B* by their own *Z* value. When comparing *q* (>2) groups, (multigroup discriminant analysis), *q*–1 discriminant functions are compared. Widely used in petrology and mineral exploration.

Disjunction [Sneath 1977]

Usually used as the opposite of *overlap*, to refer to the extent to which clusters of data are separated from each other. See W STATISTIC OF DISJUNCTION.

Disjunctive kriging [Clark 1979, p.123; David 1977, p.323]

KRIGING which allows for the fact that grade/tonnage curves based on block estimates are biased towards lower tonnage and over-optimistic average grades, because estimates of block values do not equal actual block values and usually have larger VARIANCES.

Dispersion [Kendall & Buckland 1982] 91

The 'spread' or variability of a set of data. Usually measured by the STANDARD DEVIATION, MEAN DEVIATION, MEDIAN DEVIATION etc. Synonymous with SPREAD and essentially synonymous with SCALE.

Dispersion matrix

Synonymous with VARIANCE-COVARIANCE MATRIX.

Dispersion similarity measure [SPSS 1986, p.742]

A MATCHING COEFFICIENT defined as $(ad - bc)/N^2$ — see Table 13.5. Ranges from –1 to +1.

Distance coefficient [Davis 1986, p.503] 275

A measurement of the separation between INDIVIDUALS in CLUSTER ANALYSIS. Cf. SIMILARITY COEFFICIENT.

Distribution-free tests [Kendall & Buckland 1982; Kotz & Johnson 1982-8] 69

Tests which do not depend on the form of the underlying distribution. Often used synonymously with NONPARAMETRIC, but slightly distinct. Such tests are often much better suited to geological data than classical PARAMETRIC methods, because the latter assume (and are sensitive to departures from) Normality for the underlying POPULATION; such an assumption is generally either impossible to prove (for *small* geological data-sets, e.g. in isotope geoscience), or demonstrably erroneous (for larger data-sets, e.g. in geochemistry).

Divisive cluster analysis [Everitt 1974, p.18] 278

Techniques which split the initial group of individuals into successively smaller clusters. Includes ASSOCIATION ANALYSIS and AID methods. Cf. AGGLOMERATIVE.

Dixon's criteria [Dixon 1953; Kendall & Buckland 1982] 115

Criteria for assessing OUTLIERS of the form (for the largest value of an ordered set X_N) $r_{ij} = [X_N - X_{(N-i)}]/[X_N - X_{(j+1)}]$, where $1 \le i \le 2$, $0 \le j \le 2$ varies with *N* for up to 25 data-values.

Dominant cluster mode [Ellis et al. 1977; Lister 1982] 89

A ROBUST ESTIMATE of LOCATION in which the mean and standard deviation are calculated, outliers more than *K* deviations from the mean are rejected and the cycle repeated with smaller *K* until (i) a predetermined number of cycles has been completed; (ii) the estimate achieves stability; (iii) a predetermined number of results are left. *K* is not necessarily an integer. Mass spectrometer software used in isotopic geoscience often performs this kind of procedure to determine an average ratio.

Double absence

Synonym for NEGATIVE MATCH.

Double exponential filter [Davis 1986]

A FILTER in which weights decline exponentially aware from the point being estimated.

Double-tailed (sided) test

A synonym for TWO-TAILED TEST.

Drift

A term used variously to indicate *trend* in ore deposits (see UNIVERSAL KRIGING), or movement of XRF and other analytical machines away from their calibrations.

Duncan's multiple range test [Duncan 1955; Kendall & Buckland 1982; Kotz & Johnson 1982-8] 143

A PARAMETRIC test based on the STUDENTIZED RANGE, for comparing means of 3 or more independent groups once ONE-WAY ANOVA has determined that not all the means are equal.

Page

Dunnett's test [Steel & Torrie 1980, p.188] 144
A PARAMETRIC test for comparing means of 3 or more groups with a control group.

Durbin's test [Conover 1980, p.310; Kotz & Johnson 1982-8] 160
A NONPARAMETRIC test for differences in a BALANCED INCOMPLETE BLOCK DESIGN.

Durbin-Watson test [Durbin & Watson 1950; Kendall & Buckland 1982; Kotz & Johnson 1982-8] 189
A statistic in REGRESSION, testing for AUTOCORRELATION (trend, serial correlation) in LEAST-SQUARES residuals (errors) which might tend to invalidate the proposed regression model. Very useful in time-series analysis.

Eckart-Young theorem
See SINGULAR VALUE DECOMPOSITION.

EDF tests [Kotz & Johnson 1982-8]
See EMPIRICAL DISTRIBUTION FUNCTION TESTS.

Edgington's number of runs up and down test [Lewis 1977, p.189] 210
A NONPARAMETRIC test sensitive to trends in ORDINAL or RATIO data.

Eigenvalue [Le Maitre 1982, p.115; Kendall & Buckland 1982] 316
Geometrically, the √length of one of the *p* major axes of the ellipsoid formed in space by *p*-dimensional data.

Eigenvector [Le Maitre 1982, p.115] 316
Geometrically, the vector defining one of the *p* major axes of the ellipsoid formed in space by *p*-dimensional data.

Element
One row-column value in a matrix.

Empirical distribution function (EDF) tests [Kotz & Johnson 1982-8]
Tests based on distribution functions (especially cumulative frequency curves), which are used to test whether (i) a group of data comes from a specified population (ONE-GROUP test), or (ii) two or more groups of data come from the same POPULATION (two- and multi-group tests). Includes the KOLMOGOROV–SMIRNOV, CRAMÉR–VON MISES, KUIPER'S and WATSON's U^2 TESTS.

Ensemble [Davis 1986, p.259]
Applied to TIME-SERIES in the same way that POPULATION is applied to other types of data; i.e. a time-series is a sample from an ensemble.

Entity (Rock 1988] 102
Sometimes used synonymously with INDIVIDUAL, OBJECT, etc., but here reserved for objects on which there is expected to be no component of geological (ass opposed to sampling or measurement) error in the measurement of a particular variable. For example, an igneous intrusion emplaced (geologically speaking) instantaneously would be a geochronological entity with respect to its age: the geological error associated with its age would be zero.

Equal spacing procedure [Davis 1986]
The re-estimation of irregularly spaced values in a sequence of data at equal spacing (e.g. recasting of stratigraphical measurements every metre instead of bed-by-bed).

Equal spacing test [Mardia 1972, p.189; Kendall & Buckland 1982]
A NONPARAMETRIC test used to check UNIFORMITY in 2-dimensional ORIENTATION DATA. Uses SAMPLE ARC LENGTHS, T_i, which should be constant in a uniform distribution, and the test statistic is: $L = 0.5 \sum |(T_i - 2\pi/N)|$. Large values of L indicate clustering of observations.

Equamax solution (rotation) [Cuneo & Feldman 1986, p.114,121] 321
A type of ORTHOGONAL ROTATION in FACTOR ANALYSIS, in which the VARIANCE is allocated evenly across all the defined factors. Cf. QUARTIMAX, VARIMAX.

Ergodic [Davis 1986, p.259]
Refers to an ENSEMBLE (population) of TIME-SERIES within which all statistics (means, variances, etc.) are invariant, that is, all the sequences are (strongly) STATIONARY.

Error of the first kind [Kendall & Buckland 1982]
Synonymous with α ERROR.

Error of the second kind [Kendall & Buckland 1982]
Synonymous with ß ERROR.

Estimate of location/scale
See location, scale.

Euclidean distance 278
Analogue in N-dimensional space of the linear distance between two points in 2 or 3-dimensions. Equals the sum of squares of differences of the *N* variable values between the two points.

Event
Generally corresponds to INDIVIDUAL for time-related data.

Excess of kurtosis [NAG 1987, G01/1]
See COEFFICIENT OF EXCESS.

Expectation [Kendall & Buckland 1982]

Page

The mean value of a variable after repeated sampling.

Experimental design 102

Refers to the planning of experiments so that subsequent ANALYSIS OF VARIANCE will be able to separate out different sources of variation and home in on the results of interest. Relatively little use in geology relative to the experimental and social sciences, since most geological data are 'where you find them', and *not* obtained by experiment.

Expert system 59

A system in which quasi-human thought and decision-making processes are mimicked by a computer on a base of accumulated knowledge. Expert systems necessarily link DATABASES with ARTIFICIAL INTELLIGENCE. They are designed to achieve the accumulated expertise of a human specialist in the particular field, while avoiding human errors and inconsistencies. Expert systems have successfully in locating previously unknown mineral deposits (e.g. molybdenum in the USA), by applying all the geological reasoning exploration geologists would normally use to the usual spectrum of geological, geochemical and geophysical data.

Explanatory variable

Synonymous with INDEPENDENT VARIABLE.

Exploratory data analysis [Tukey 1977; Velleman & Hoaglin 1981; Kotz & Johnson 1982-8] 92

Techniques for the preliminary numerical and graphical examination of data, pioneered by the American J.W.Tukey, and based on RESISTANT or ROBUST ESTIMATES such as the MEDIAN and HINGES. Includes plots now widely used in geology, such as BOX-AND-WHISKER and STEM-AND-LEAF.

Extended Q-mode factor analysis [Le Maitre 1982, p.132]

A modification of FACTOR ANALYSIS used for compositional data, in which factor loadings and scores are transformed to sum to unity, so that they can be interpreted as end-member compositions. Uses the transformation $Z_{ij} = (X_{ij} - Xmin_j)/(Xmax_j - Xmin_j)$ where $Xmax_j$ and $Xmin_j$ are the maximum and minimum values of the jth variable. This allows equal weights to be given to all variables whatever their variance (e.g. to major, minor and trace elements, which have very different absolute magnitudes).

Extreme value

A rather vague term, probably best used synonymously with TRUE OUTLIER. Cf. FALSE, BIZARRE OUTLIER.

Extreme value statistics 116

Statistics concerned with rare (extreme) events such as volcanic eruptions, earthquakes, floods and other catastrophes.

Factor analysis [Davis 1986, p.546; Kendall & Buckland 1982; Kotz & Johnson 1982-8] 318ff

A technique studying the structure of MULTIVARIATE data, in which total variation is expressed in terms of a pre-determined number of factors. R-MODE factor analysis is mostly used to reduce the DIMENSIONALITY of data, and to explain observed variations in terms of a smaller number of unobserved factors, themselves interpretable in terms of geological processes (e.g. analyses of amphiboles can be used to identify element substitutions). Q-MODE factor analysis is mostly used to identify end-members from series of INDIVIDUALS (e.g. analyses of scapolites can be used to identify marialite and meionite as the end-members). Strictly, most geological applications are *principal factor analysis* in which residual error terms are ignored.

Factor loading

In FACTOR ANALYSIS, the loading of variable **X** onto Factor **I** is the coefficient in EIGENVECTOR **I** which corresponds to variable **X**. If we plot loadings, we view relationships between variables (cf. factor SCORE).

Factor score (weights)

The score of an OBJECT on a factor. Thus to compare the positions of *rocks,* we plot scores on factor 1 vs. 2, etc.

Fager's coefficient [Mather 1976, p.314]

A MATCHING COEFFICIENT used with DICHOTOMOUS data.

False outlier [Rock 1988] 113

A mistake — wrong data — showing up as an OUTLIER which is non-reproducible and definitely inaccurate in some way (due to bad sampling, or sample preparation poor quality control, machine failure, random statistical 'spikes', etc.) Such data are *always* to be eschewed. Cf. BIZARRE OUTLIER, TRUE OUTLIER.

Fat-tailed [Kotz & Johnson 1982-8]

Synonymous with LONG-TAILED.

Fence [Tukey 1977, p.44]

See INNER and OUTER FENCE.

Fiducial limits [Kendall & Buckland 1982]

Commonly used as a synonym for CONFIDENCE LIMITS, but sometimes (e.g. Cheeney 1983, p.91) refers to confidence BANDS on bivariate plots.

Field [Date 1981]

In DATABASE work, refers to a group of columns in an 80-column computer file card-image, which contain the data-values for one VARIABLE. E.g. columns 1-10 might be the field containing the sample number.

Filliben's (Normal probability plot correlation coefficient), *r* [Filliben 1975; Lister 1982] 97

A NORMALITY TEST (indicating Normality when $r = 1$). Formally, the PRODUCT-MOMENT CORRELATION

Page

between the ordered observations X_i and the order statistic medians from a STANDARD NORMAL distribution. It is basically a mathematical equivalent of the graphical NORMAL PROBABILITY PLOT, where a linear plot (i.e. high linear correlation) indicates Normality. Sensitive to non-Normal SKEW and KURTOSIS.

Filter [Davis 1986, p.261] 225
A mathematical device used to separate 'noise' (random fluctuations) from 'signal' (cyclicity, trend) in one-dimensional sequences of data, by smoothing the fluctuations whilst retaining the trends. See e.g. MOVING AVERAGE FILTER, DOUBLE EXPONENTIAL FILTER, SPENCER'S or SHEPPERD's FILTERS. Values are calculated at successive points which tend to smooth out the fluctuations.

Fisher (spherical Normal) distribution [Kendall & Buckland 1982; Cheeney 1983, p.112] 240
A 3-dimensional distribution for ORIENTATION DATA, with rotational symmetry. The analogue of the VON MISES DISTRIBUTION in 2-D and of the NORMAL DISTRIBUTION in 1-D.

Fisher's *F* (variance ratio) distribution [Kendall & Buckland 1982] 85
The distribution of the ratio of two independent quantities, each of which is distributed like a variance in Normal samples. Widely used in ANALYSIS OF VARIANCE, DISCRIMINANT ANALYSIS, etc.

Fisher's protected least significant difference (PLSD) test [Till 1974, p.111] 141
A PARAMETRIC test used to determine which of several means are unequal in ONE-WAY ANOVA.

Fisher (–Yates) exact (probability) test [Kendall & Buckland 1982; Kotz & Johnson 1982-8] 175
An alternative to the χ^2 test for 2x2 CONTINGENCY TABLES, giving the exact probability that a relationship exists between the VARIABLES, ATTRIBUTES or OBJECTS. Another *Fisher-Yates test* is almost identical in purpose (and method) to the VAN DER WAERDEN test [Kotz & Johnson 1982-8].

Fisher's Z transformation [Kendall & Buckland 1982; Kotz & Johnson 1982-8] 177
The transformation of PEARSON'S CORRELATION COEFFICIENT r by: $Z = \tanh^{-1}(r)$, used in comparing correlation coefficients with each other or with fixed values >0. Its distribution tends to Normality more rapidly than that of r, with increasing numbers of values.

Four-dimensional trend surface [Davis 1986, p.430]
A surface representing variation of a variable, such as the Au content of an ore body in 3-dimensional (*XYZ*) space.

Fourfold (φ or phi) coefficient [Blalock 1972, p.295; Downie & Heath 1974, p.108; Conover 1980, p.184] 176
A CONTINGENCY COEFFICIENT defined as $\phi^2 = \chi^2/N = (ad - bc)/\sqrt{(a+b)(a+c)(b+d)(c+d)}$ — see Table 13.5. A special case of the PRODUCT-MOMENT CORRELATION COEFFICIENT, applied to NOMINAL DATA. No association is indicated by a value of 0, perfect association by a value of 1 (in a 2x2 table), or by a higher value in a general rxc table.

Fourier series [Davis 1986, p.248] 224
The expression of an irregularly cyclical sequence of observations as the sum of a number of sine or cosine functions.

Frequency distribution curve [Cheeney 1983, p.14] 93
A curve joining the frequencies of a particular VARIABLE through its range of values. A 'smoothed histogram'.

Friedman's (two-way ANOVA by ranks) test [Friedman 1937; Siegel 1956, p.166; Conover 1980, p.299] 157
A NONPARAMETRIC test for comparing 3 or more RELATED groups of data. The test statistic is closely related both to KENDALL'S COEFFICIENT OF CONCORDANCE, and to SPEARMAN'S RANK CORRELATION COEFFICIENT. Termed *Friedman's* χ^2 *test* by Kotz & Johnson (1982-8).

F-spread (Fourth-spread) [Hoaglin et al. 1983] 91
Synonymous with H-SPREAD.

***F* test** [Kotz & Johnson 1982-8] 120
The standard PARAMETRIC test used for example to compare the dispersion (VARIANCE) of two or more groups of data, or the means of several groups of data, using FISHER'S *F* DISTRIBUTION.

Full Normal plot (FUNOP) [Koch & Link 1970, p.238; Kotz & Johnson 1982-8] 116
A plot used in the assessment of OUTLIERS, which considers the top and bottom third of a set of values (i.e. those beyond the upper and lower TERTILES). The JUDD of each of these values is calculated, ranked, and the Judd values plotted against their ranks. Any Judd which is more than 2 Judds above the median Judd value is regarded as suspect. The procedure is similar to making a plot on Normal probability paper.

Functional regression [Le Maitre 1982, p.54]
'Regression' involving 2 mathematical VARIABLES both subject to error, i.e. where there is no INDEPENDENT variable. E.g. analysis of the relationship between mass and volume of a rock-type, using replicate measurements on a single specimen. Cf. STRUCTURAL REGRESSION.

FUNOP [Kotz & Johnson 1982-8] 116
See FULL NORMAL PLOT.

Further neighbour technique [Everitt 1974, p.11]
Synonymous with COMPLETE LINKAGE TECHNIQUE.

Fuzzy clusters/sets [Everitt 1974, p.32; Gordon 1981, p.58] 292
Overlapping clusters produced by certain forms of CLUSTER ANALYSIS.

Page

Gamma (γ) [Blalock 1972, p.424; SAS 1985,p.414]
A COEFFICIENT of CONCORDANCE which attains its upper value of 1 when any of the cells is zero, and usually has a higher value than KENDALL'S *W*. Can also be used as a CONTINGENCY COEFFICIENT for ORDINAL data.

Gap statistic [Miesch 1981b] 117
A statistic designed specifically to estimate geochemical THRESHOLDS an determine their significance. Equals the maximum gap between adjacent, ordered data-values after adjustment for expected frequencies.

Gastwirth median [Gastwirth 1966; Andrews et al. 1972; Ellis 1981] 88
0.4 of the median added to 0.3 of both upper and lower TERTILES. A ROBUST ESTIMATE of LOCATION.

Gaussian (Normal) distribution [Kendall & Buckland 1982] 76
The fundamental 'bell-shaped' distribution in probability and statistics.

Gauss-Jordan elimination [Kotz & Johnson 1982-8]
A method for solving *N* simultaneous linear equations, used for MATRIX INVERSION in many statistical computer programs and packages covering techniques such as MULTIPLE REGRESSION.

Geary's *a* ratio [Geary 1935,1936; Lister 1982; Kendall & Buckland 1982] 98
A TEST OF NORMALITY: the ratio MEAN DEVIATION/STANDARD DEVIATION. Has the value $\sqrt{2/\pi} = 0.7979$ for a Normal distribution. Sensitive to departures from MESOKURTOSIS, but preferable to testing the actual kurtosis value because its sampling distribution tends to Normality for smaller samples.

Generalised inverse [Le Maitre 1982, p.175; Kotz & Johnson 1982-8]
A special INVERSE of a matrix, applicable to SINGULAR MATRICES, and defined as **X^**, where **X.X^.X** = **X**.

Generalised least-squares (GLS) [Joreskog et al. 1976, p.82; Mather 1976, p.79] 257
MULTIPLE REGRESSION in which the variances of the errors on the DEPENDENT VARIABLES due to each of the INDEPENDENT VARIABLES are not assumed to be equal (as in ordinary least-squares) but are given individual weights. Also, one of the methods of estimating FACTOR LOADINGS in FACTOR ANALYSIS.

Generalised linear model [Clark 1979, p.9] 272
A form of SEMI-VARIOGRAM without a SILL, defined by the equation $\gamma(h) = ph^{\alpha}$, where $0 < \alpha < 2$.

General linear model (GLIM) [Horton 1978; Krumbein & Graybill 1965] 151
Not to be confused with previous entry. A term used in statistics for an equation which actually covers many of the techniques in this book: ANOVA, CONTINGENCY TABLES, DISCRIMINANT ANALYSIS, REGRESSION, etc.

Geological occurrence model [Koch & Link 1971, p.201]
A method for screening large areas to select those most favourable for mineral exploration. The probability of a deposit is taken to be a function of a large set of geological variables, such as lithology, age of terrain, faulting, percentage of igneous intrusions, etc. Available information for known areas is translated by MULTIGROUP DISCRIMINANT ANALYSIS into a set of discriminant functions, which can then be applied to unknown areas.

Geometric mean [Kendall & Buckland 1982] 86
The *N*th root of the product of *N* values, or the exponent of the sum of their logarithms. Where it exists (i.e. all data-values are +ve), it lies between the harmonic and arithmetic means. Equals the median for a LOGNORMAL distribution.

Geostatistics [Clark 1979; Kotz & Johnson 1982-8] 271
NOT 'statistics in geology', but usually restricted to the application of MOVING AVERAGE techniques such as KRIGING and SEMIVARIOGRAMS to mining.

Girdle distribution [Mardia 1972, p.223; Kendall & Buckland 1982; Cheeney 1983, p.108] 239
A descriptive term for spherical distributions in which points tend to cluster around a great circle. Cf. UNIFORMITY.

Glyphs
See ANDERSON'S METROGLYPHS.

Goodman & Kruskal's gamma (γ) [Kotz & Johnson 1982-8; Coxon 1982, p.20; SPSS 1986, p.740] 184
A RANK CORRELATION COEFFICIENT, similar to τ (next entry), but measuring WEAK MONOTONICITY.

Goodman & Kruskal's tau (τ) [Goodman & Kruskal 1954; Blalock 1972, p.300; Kendall & Buckland 1982] 176
A series of probabilistic CONTINGENCY COEFFICIENTs, not based on χ^2. They involve the notion of predicting one value given the other, and are therefore asymmetric (i.e.$\tau_{xy} \neq \tau_{yx}$). τ_b is defined, for predicting *B* from *A*, as [{no.of errors not knowing *A* – no.of errors knowing A}/ no.of errors not knowing *A*]. For a 2x2 table, equals χ^2/N. Sometimes misleadingly called LAMBDA (λ; e.g. SPSS 1986,p.740] or KENDALL's τ_b [SAS 1985,p.414]. τ_b is similar to γ (GAMMA), but incorporates a correction for ties. Usually implies data measured on an ORDINAL scale.

Goodness-of-fit test [Kendall & Buckland 1982; Kotz & Johnson 1982-8] 94
Any test that a distribution of real data-values approximates a specified, theoretical distribution. Includes NORMALITY TESTS.

Gower's similarity coefficient [Everitt 1974, p.54] 300
A general coefficient commonly used in CLUSTER ANALYSIS and ORDINATION, which is uniquely suitable for DICHOTOMOUS, NOMINAL, ORDINAL or RATIO data alike.

Gramian matrix [Joreskog et al. 1976, p.51]

Page

A matrix with non-negative EIGENVALUES.

Gram-Schmidt method

A method of calculating ORTHOGONAL POLYNOMIALS in POLYNOMIAL REGRESSION.

Group

A collection of INDIVIDUALS. See also SAMPLE.

Group average method [Everitt 1974, p.15] 278ff

An AGGLOMERATIVE method of CLUSTER ANALYSIS, in which distance between groups is defined as the average of distance between all pairs of INDIVIDUALS in the groups.

Grouped regression [Mather 1976, p.293ff] 258

MULTIPLE REGRESSION in which one or more of the variables is DICHOTOMOUS or MULTISTATE (see BINARY EXPLANATORY VARIABLE). Dummy variables are introduced to divide the data into several groups depending on the value (0 or 1; 1,2,3, etc.) of this variable, so that its effect can be assessed.

Grubbs' maximum Studentized residuals [Lister 1982; Kendall & Buckland 1982; Hampel et al. 1986] 116

A parametric test for OUTLIERS, assessing the maximum value as $[X_{max}-\overline{X}]/s$, and the minimum as $[\overline{X}-X_{min}]/s$. Breaks down for substantial (> 10%) contamination of a Normal distribution.

Guttman's μ_2 coefficient [Kotz & Johnson 1982-8] 184

A CORRELATION COEFFICIENT which measures WEAK MONOTONICITY.

H_0 (null hypothesis) 73

The hypothesis under test, which determines the α ERROR. Generally one of equality (e.g. H_0: $\mu_1 = \mu_2$ meaning two means are hypothesised as being equal).

H_1 (alternative hypothesis) 73

Any admissible alternative hypothesis to H_0.

Hamann coefficient [Coxon 1982, p.27; SPSS 1986, p.740]

A MATCHING COEFFICIENT, defined as $\{(a+d)-(b+c)\}/N$ and interpretable in terms of conditional probabilities. Ranges from −1 to +1. Measures the probability that an ATTRIBUTE has the same state in both OBJECTS, minus the probability that it has different states.

Hampel's (3-part redescending) *M*-estimates [Andrews et al. 1972; Lister 1982; Hampel et al. 1986] 88

A family of extremely efficient ROBUST ESTIMATES of LOCATION, including 12*A*, 17*A*, 21*A* and 25*A*, based on a 3-part line-segment function.

Hanning [Tukey 1977, p.670; Kendall & Buckland 1982; Kotz & Johnson 1982-8]

A smoothing process alternatively describable as the result of (1) two repetitions of RUNNING MEANS of 2; (2) taking the mean of each value and its SKIP MEAN; (3) a MOVING AVERAGE with weights 0.25,0.5,0.25.

Harmonic mean [Kendall & Buckland 1982; Kotz & Johnson 1982-8] 86

The reciprocal of the mean reciprocal of a set of data-values, i.e. $N/\Sigma(1/X_i)$ for i =1 to *N*. Breaks down if any $X = 0$. For all-positive data-sets, it is less than either the ARITHMETIC or GEOMETRIC means.

Harmonic number [Davis 1986, p.253]

The number of cycles per basic interval, in a TIME-SERIES.

Harris image analysis [Cuneo & Feldman 1986, p.112] 319

A type of IMAGE ANALYSIS, requiring a NONSINGULAR CORRELATION MATRIX, and based on a theory of variable sampling as oppose to subject sampling. Uses a modification of the input correlation matrix, called the *image variance covariance matrix.* It has a tendency to extract more factors than non-image extraction algorithms, such as ITERATED PRINCIPAL AXIS.

Hartley's maximum *F* test [Till 1974, p.105; Kendall & Buckland 1982; Kotz & Johnson 1982-8] 136

A test for homogeneity of several VARIANCES, like BARTLETT'S or COCHRAN'S, but based on the ratio of the largest/smallest variances.

Harvey's test [Harvey 1974; Lister 1982] 116

A variation of GRUBBS' TEST for OUTLIERS, which assesses the outlier furthest from the mean relative to the mean and standard deviation of results *minus* this outlier. Critical values are only available for 3-7 results.

Heavy-tailed [Kotz & Johnson 1982-8]

Synonymous with LONG-TAILED.

Heteroscedasticity [Kendall & Buckland 1982; Kotz & Johnson 1982-8; Davis 1986, p.378] 191

The condition where, in the relationship between 2 variables, the VARIANCE of one varies for fixed values of the other. Cf.HOMOSCEDASTICITY. A heteroscedastic relationship invalidates certain statistical treatments such as classical REGRESSION.

Heywood case [SAS 1985,p.354; Cuneo & Feldman 1986, p.121] 321

An unusual result in FACTOR ANALYSIS, in which the final COMMUNALITY estimate for a particular variable equals ("Heywood case") or exceeds ("ultra-Heywood case") unity. Since communality is a squared correlation, it should be ≤ 1 .Specialists disagree as to whether a Heywood case factor analysis is meaningful, but an ultra-Heywood case is definitely not meaningful, and indicates bad prior communality estimates, too many or too few factors, not enough data

Page

to provide stable solutions, or a fundamentally inappropriate factor model.

Hierarchical (nested) analysis of variance 166

ANOVA in which the sources of variation being separated form some logical hierarchy: e.g. if variations due to specimens from a lithostratigraphical Member are being compared with those within the parent Formation and Group.

Hierarchical clustering techniques [Everitt 1974, 1974; Kotz & Johnson 1982-8] 281ff

CLUSTER ANALYSIS in which groups are successively subdivided to form a DENDROGRAM. No reallocation of early allocations of individuals is allowed. Divided into AGGLOMERATIVE and DIVISIVE methods.

Hierarchical database system [Date 1981] 43

A database in which data-items are arranged in a tree-structure, such that access between one item and another can only be attained by suitable movements up and down 'branches'.

Higher ordered metric scale [Coxon 1982, p.6] 64

A higher form of ORDINAL and ORDERED METRIC scales, in which not only the objects but also the differences between every possible pair of objects can be ranked. Rare in geology.

Hinge [Tukey 1977, p.33]

See UPPER and LOWER HINGE.

Histogram [Kendall & Buckland 1982; Kotz & Johnson 1982-8] 93

A univariate frequency diagram (bar chart) in which rectangles proportional in area to the class frequencies are erected on the horizontal axis, with widths corresponding to class intervals.

Hodges-Ajne test [Ajne 1968; Mardia 1972, p.182; Kendall & Buckland 1982; Cheeney 1983, p.94] 230

A rapid but not powerful NONPARAMETRIC test of UNIFORMITY for two-dimensional ORIENTATION DATA. Counts the number of points μ occurring on one side of the circle diameter, and finds the diameter for which μ is a minimum; μ then becomes the test statistic, with small values indicating preferred orientation.

Hodges-Lehmann estimates [Andrews et al.1972; Kendall & Buckland 1982; Kotz & Johnson 1982-8]

A family of ROBUST ESTIMATE of LOCATION. The best-known is the median of all possible pairwise medians of a data-set. Becomes cumbersome for large *N*. Cf. WALSH AVERAGE.

Homogeneous/homogeneity [Kendall & Buckland 1982]

Generally used as a multi-group analogue of 'equal'. I.e. 2 means are 'equal' or 'unequal', 10 means are 'homogeneous' (i.e. all 10 are equal), or 'inhomogeneous' (one or more are different from the others).

Homoscedasticity [Kendall & Buckland 1982; Davis 1986, p.195] 191

The condition where, in the relationship between 2 variables, the VARIANCE of one is fixed for all fixed values of the other. Cf.HETEROSCEDASTICITY. Certain classical statistical techniques (e.g. REGRESSION) assume homoscedasticity and are invalid if it does not hold.

Honestly significant difference test

See TUKEY's *w* PROCEDURE.

Horseshoe plot [Joreskog et al. 1976, p.115]

A plot sometimes generated by PRINCIPAL COMPONENTS or FACTOR ANALYSIS, showing a quadratic relationship between two principal components or factors. This tends to occur where correlations are high and subequal.

Hotelling–Lawley trace [Hotelling & Pabst 1936; SAS 1985,p.12]

A ratio of matrices statistic used to test the overall significance of a multivariate REGRESSION. Equivalent to an *F* test on a simple bivariate regression.

Hotelling–Pabst statistic [Lewis 1977, p.199; Kotz & Johnson 1982-8]

A rarely-used term for the numerator of SPEARMAN's ρ statistic, i.e. the sum of squared rank differences.

Hotelling's T^2 statistic [Miller & Kahn 1962, p.248; Kendall & Buckland 1982; Kotz & Johnson 1982-8] 305

A multivariate analogue of STUDENT's *t*, used for comparing multivariate means, Simply related to MAHALANOBIS' D^2 as follows: $T^2 = n_1 n_2 D^2/N$; and to FISHER'S F thus: $F = (N-\nu-1)T^2/[(N-2)\nu]$, where n_1, n_2 are the numbers of values in the two groups of data, $N = [n_1 + n_2]$, and ν is the number of variables.

HSD test

See TUKEY's w PROCEDURE.

H-spread (Hinge-Width) [Tukey 1977, p.44] 91

The range between the UPPER and LOWER HINGES. Almost identical to the INTERQUARTILE RANGE, but is defined slightly differently.

Huber's estimates [Andrews et al. 1972] 89

A family of ROBUST ESTIMATES of LOCATION.

Hypergeometric distribution [Blalock 1972, p.248; Kendall & Buckland 1982] 81

A distribution generally associated with sampling from a finite population without REPLACEMENT. Tends to the BINOMIAL DISTRIBUTION for an infinite population.

Hyperplane [Le Maitre 1982, p.76]

A plane in multidimensional space.

Hypersurface [Davis 1986, p.430]

Page

A synonym for a FOUR-DIMENSIONAL TREND SURFACE.

Hypothesis testing [Kotz & Johnson 1982-8] 73ff

The process of setting up a statistical hypothesis on some data and then subjecting it to some statistical tests usually with the expectation of rejecting the hypothesis. See NULL HYPOTHESIS, α and ß ERRORS.

Hypothetical open array 202

Any array of numbers, free of CLOSURE, from which a given CLOSED data-array may be generated. For many closed geological data (e.g. rock analyses) the open array is a purely imaginary, mathematical concept. Also termed a BASIS.

Identification

Synonymous with ASSIGNMENT.

Identity matrix

A SYMMETRICAL matrix in which all diagonal elements are 1 and all off-diagonal elements are 0. Serves the same function in linear algebra as unity in scalar arithmetic.

Ill-conditioned matrix [Kennedy & Gentle 1980, p.34; Kotz & Johnson 1982-8, vol.5, p.325]

A matrix, manipulations of which are sensitive to small changes in elements within that matrix. Refers especially to matrices which give very different INVERSES from different inversion algorithms, and hence unstable results in multivariate techniques such as MULTIPLE REGRESSION and FACTOR ANALYSIS. Can be measured by the P-CONDITION NUMBER, or by the DETERMINANT of the CORRELATION MATRIX. Very common in geology especially with compositional data.

Illusory correlation [Kendall & Buckland 1982] 185

Any correlation which is statistically significant but does *not* imply causal connection between the variables. E.g. specific gravity correlates closely with refractive index in plagioclases, not because of a direct causal connection but because both in turn correlate with anorthite content. Termed by some 'nonsense correlation'. Cf. SPURIOUS CORRELATION.

Image analysis [Mather 1976, p.272] 319

A variant of FACTOR ANALYSIS in which multiple correlations between the variables and factor are considered, rather than separate partial correlations as in ordinary factor analysis.

Incomplete block designs [Kotz & Johnson 1982-8] 159

A varied family of ANOVA designs which has attracted a large literature. The data relate to BLOCKS and TREATMENTS, but not all treatments are applied to all subjects in every block. BALANCED INCOMPLETE BLOCK DESIGNS are the easiest to analyze, but many other types (e.g. *partially balanced*) are also sometimes used.

Independent groups of data (samples) 72

GROUPS of data which are not directly related. Cf. RELATED SAMPLES.

Independent variable [Kendall & Buckland 1982] 187

A variable whose value determines that of another variable. E.g. in the variation of refractive index with plagioclase composition, composition is the independent variable.

Individual 68

Any discrete item (e.g. rock sample, fossil, mineral grain). Generally synonymous with ENTITY, OBJECT, ITEM.

Inefficient statistic [Kendall & Buckland 1982] 123

A statistic with less than the maximum possible precision. E.g. SUBSTITUTE *T*- AND *F*-TESTS based on the range rather than standard deviation.

Influential observation

Most commonly refers to individual observations which strongly influence the result of a REGRESSION: that is, removal of such values will have a disproportionately large effect on the slope, intercept and other estimates. Usually implies that the observation is a real part of the data-set being regressed, whereas OUTLIERS may not be.

Inner fence [Tukey 1977, p.44] 94

A value 1.5 H-SPREADS away from the MEDIAN, which forms the outer edge of the box on a BOX PLOT.

Interquartile coefficient of skewness [Weekes 1985, p.24] 92

A measure of SKEW defined as $[(Q_3+Q_1-2M)/(Q_3-Q_1)]$, where Q_3 and Q_1 are the third and first QUARTILES, and M is the MEDIAN.

Interquartile range/spread [Kendall & Buckland 1982] 92

The range between the 1st and 3rd QUARTILES, which includes half the total values of a set of data. A simple, ROBUST measure of DISPERSION, useful in EXPLORATORY DATA ANALYSIS. Cf. H-SPREAD

Interval scale/variable 65

CONTINUOUS variable in which the ratio of any two values is independent of the unit of measurement and of the arbitrary zero point. An interval variable can be manipulated with the operations =, ≠, <, >, +, -,*,/. E.g. Fahrenheit and Centigrade (but not Kelvin) temperatures. It is meaningful to say that 40°-35°C = 30°-25°C = 5 *degrees C* but *not* that 40°-35°C = 5°C; similarly 40°C is not in any sense 'twice' 20°C because 0°C is purely arbitrary.

Intraclass correlation coefficient [Blalock 1972, p.355]

A PRODUCT-MOMENT CORRELATION between all possible pairs of cases within categories for NOMINAL data.

Page

Mostly lies between 0 and 1, but may take negative values in special cases.

Inverse of a matrix [Joreskog et al. 1976, p.24]
The analogue in matrix algebra of a reciprocal in scalar arithmetic. That matrix which, multiplied by the original matrix, yields the IDENTITY MATRIX. The inverse exists and is unique if the original matrix is SQUARE and non-SINGULAR.

Isopleth envelope/surface [Davis 1986, p.430]
A synonym for a FOUR-DIMENSIONAL TREND SURFACE.

Item 68
Synonymous with INDIVIDUAL.

Iterated principal axis [Cuneo & Feldman 1986, p.112]
A method for extracting factors in FACTOR ANALYSIS, which requires initial estimates of the COMMUNALITIES, and modifies these with each iteration until the estimates stabilize.

Jaccard's coefficient [Jaccard 1908; Till 1974, p.135; SPSS 1986,p.739] 276
A MATCHING COEFFICIENT defined as $a/[a+b+c]$ (see MATCHES): the proportion of attributes in common among all the attributes recorded in the two objects being compared. Ignores NEGATIVE MATCHES (cf. SOKAL & SNEATH'S COEFFICIENT).

Jackknife [Miller 1974; Mosteller & Tukey 1977; Kendall & Buckland 1982; Kotz & Johnson 1982-8] 110
Methods of estimating the error and bias of an estimate when the underlying distribution is uncertain or unknown. It involves calculating the estimate with one data-value omitted in turn, and then averaging these 'trimmed' estimates. The resulting average is less biased than the overall average. Closely related to INFLUENCE FUNCTIONS that are widely used in ROBUST STATISTICS, and also to some NONPARAMETRIC methods.

John's adaptive estimate [Andrews et al. 1972; Johns 1974] 89
A nonparametric ROBUST ESTIMATE of LOCATION which takes the actual form of a distribution into account.

Judd [Koch & Link 1970, p.238] 116
A parameter used in FULL NORMAL PLOTS to assess outliers. The Judd is equal to [Value – median value]/Standard Normal deviate corresponding to the PLOTTING-POINT PERCENTAGE of the value.

Kaiser image analysis [Cuneo & Feldman 1986, p.112] 319
A type of IMAGE ANALYSIS, developed from HARRIS IMAGE ANALYSIS, in which the factor solution is rescaled so that it represents a FACTOR ANALYSIS of the original CORRELATION MATRIX. This allows the user to impose the traditional interpretive models applied to conventional factor analysis. Will result in the same number of factors as HARRIS IMAGE ANALYSIS, but the loadings may be quite different.

Kaiser's varimax rotation [Davis 1986, p.555]
See VARIMAX ROTATION.

Kendall's coefficient of concordance, *W* [Siegel 1956, p.229; Kendall & Buckland 1982] 180
A coefficient measuring the agreement among 3 of more sets of rankings. Complete agreement gives $W = 1$, complete disagreement $W = 0$. Simply related to the test statistic in FRIEDMAN'S test. One might, for example, use it to compare the extent of agreement or disagreement between 5 geologists asked to rank 10 possible areas for mineral exploration in order of their likely potential. A high value of W would give greater confidence in the prediction concerning the area ranked number 1.

Kendall's partial rank correlation coefficient [Siegel 1956, p.223; Conover, 1980, p.260] 252
A special case of KENDALL'S τ measuring strength of association between two variables with others held constant. Could be used, for example, to study the correlation between the refraction index of apatite and its F content, by isolating F from the effects of Cl, OH, CO_3 etc. which substitute for it in the apatite formula $Ca_5(PO_4)_3(F,Cl,OH,CO_3)$.

Kendall's phi (ϕ) coefficient [Lewis 1977, p.95]
A synonym for FOURFOLD COEFFICIENT.

Kendall's rank correlation coefficient, Tau (τ) [Siegel 1956, p.213; Conover 1980, p.257] 171
A 'coefficient of disarray', measuring the extent to which the rankings of 2 variables agree: the differences in the proportion of pairs of ranks that 'agree' A from those that 'disagree' D or, formally, $2(A - D)/N(N - 1)$ where N is the number of values. Uses the same information (and is hence as powerful) as SPEARMAN'S ρ, but measures a different property, usually has a lower absolute value, and can be used in a wider range of circumstances: e.g. (1) as KENDALL'S *PARTIAL* RANK CORRELATION COEFFICIENT, and (2) in estimating confidence intervals for slopes in NONPARAMETRIC REGRESSION. Varies from -1 to +1. A special case of a product-moment correlation coefficient.

Kendall's τ_b [SAS 1985,p.414]
Occasionally used (?erroneous) synonym for GOODMAN & KRUSKAL's τ_b.

Kleiner-Hartigan trees [Kleiner & Hartigan 1981; Barnett 1982] 315
Branching diagrams which can be used to display values for multivariate data pictorially. Trees have been employed in stream sediment geochemistry, one tree being placed at each confluence sampling point. Unlike CHERNOFF FACES, they do not impose any immediate interpretation on the data, and are thus more suited for 'neutral' or 'abstract'

Page

data-types.

Klotz test [Conover 1980, p.321; Kendall & Buckland 1982; Kotz & Johnson 1982-8] 131

A NONPARAMETRIC test for equal DISPERSION in two INDEPENDENT groups, which, like the VAN DER WAERDEN test, uses NORMAL SCORES instead of simple RANKS.

***K*-means cluster analysis** [Everitt 1974,p.25; Kotz & Johnson 1982-8] 293

A PARTITIONING technique in which K points are chosen in p-dimensional space which act as initial estimates of K cluster centres. The K points may be any K chosen systematically or at random (e.g. the first K in the data-set).

Kolmogorov one-group test [Kolmogorov 1933; Siegel 1956, p.47; Lewis 1977, p.220] 95

A powerful NONPARAMETRIC, EDF test for GOODNESS-OF-FIT, based on the maximum deviation between two CUMULATIVE DISTRIBUTIONS. See also SMIRNOV'S TEST. Often used as a NORMALITY TEST.

Kolmogorov-Smirnov tests [Siegel 1956,p.127; Kendall & Buckland 1982;Kotz & Johnson 1982-8] 95

Best used as a collective name for the one-group KOLMOGOROV and two-group SMIRNOV tests, but sometimes used vaguely to refer to one or other of these.

Kriging [Clark 1979; Davis 1986, p.383; Kotz & Johnson 1982-8] 273

A MOVING-AVERAGE method used in mineral exploration, named after D.G.Krige, a noted S.African mining geologist and statistician. Kriging attempts to interpolate the values of a spatially distributed variable between measured points, and to estimate the errors associated with interpolation. For example, the Au content of an ore might be estimated at an intermediate point between a series of boreholes. Regarded by many as superior to TREND SURFACE ANALYSIS and other competing techniques because: (1) it allows estimates for the errors of the estimated values to be determined; (2) it minimises these errors under a range of conditions. See also DISJUNCTIVE KRIGING, SIMPLE KRIGING, UNIVERSAL KRIGING.

Kruskal-Wallis test [Kruskal &Wallis 1952; Kendall & Buckland; Kotz & Johnson 1982-8] 145

A NONPARAMETRIC technique for comparing the LOCATION (median) of 3 of more groups of data. Performing this test is equivalent to performing a parametric ONE-WAY ANOVA (using the F TEST) on the ranks of the data rather than on the data themselves. The test is thus sometimes called *Kruskal-Wallis one-way ANOVA.*

Kuiper's test [Kuiper 1962; Mardia 1972, p.173; Lewis 1977, p.229; Cheeney 1983, p.95] 231

A NONPARAMETRIC test for UNIFORMITY in ORIENTATION DATA, analogous to the KOLMOGOROV TEST for LINEAR DATA. More powerful than the HODGES-AJNE TEST.

Kulczynski coefficients [Coxon 1982, p.27; SPSS 1986, p.739-740] 168

Two MATCHING COEFFICIENTS, equal to (1) $a/(b+c)$ and (2) $[a/(a+b) + a/(a+c)]/2$: see Table 13.5. (1) has a minimum of 0 but no upper limit, and is undefined where there are no MATCHES ($b = c = 0$). (2) ranges from 0 to 1 and yields the average conditional probability that an attribute is present in one object, given its presence in the other.

Kurtosis [Miller & Kahn 1962, p.43; Kendall & Buckland 1982; Kotz & Johnson 1982-8] 92

A measure of the 'peakedness' of a unimodal distribution, i.e. the steepness of ascent near the mode. When calculated as $NS_4/(S_2)^2$ where S_4 and S_2 are the 4th and 2nd MOMENTS (and then usually symbolised as 'b_2'), has a value of 3 for a NORMAL DISTRIBUTION. However, some statistical packages on computers (e.g. Statworks™ on the Macintosh) subtract the value 3 from the calculated result, so that 'kurtosis' is then zero for a Normal distribution. See MESOKURTIC, PLATYKURTIC, LEPTOKURTIC. The calculated kurtosis for a data-set can be used as a TEST FOR NORMALITY and also as a test for OUTLIERS. Kurtosis can be generalized to MULTIVARIATE distributions [Kotz & Johnson 1982-8, vol.6,p.122].

Lag [Kendall & Buckland 1982] 215

The amount of offset between two sequences of data being compared. When we compute the correlation between elements 1,2,3...(N-1) of one sequence with elements 2,3,4...N of another, we have computed the CROSS-CORRELATION of lag 1. Comparing one sequence with itself is similarly AUTOCORRELATION of lag 0.

Lambda, λ

Used in various senses in statistics. For example: (1) The ratio of variances of the random errors in X and Y, in 'STRUCTURAL' REGRESSION. (2) The exponent in POWER TRANSFORMATION. (3) A probabilistic CONTINGENCY COEFFICIENT related to, but easier to compute than, GOODMAN & KRUSKAL's τ [Blalock 1972, p.302]; this is inferior to τ where the marginal totals in a CONTINGENCY TABLE (Table 13.5) are not subequal, and may take on a numerical value of zero where one would not wish to refer to the variables as uncorrelated.

Lance & Williams coefficient [SPSS 1986, p.740] 168

A MATCHING COEFFICIENT defined as $(b+c)/4(a+b+c+d)$ — see Table 13.5 — and ranging from 0 to 1.

Lance & Williams technique [Everitt 1974, p.16] 278

A family of AGGLOMERATIVE methods of CLUSTER ANALYSIS.

Large-tailed [Kotz & Johnson 1982-8]

Synonymous with LONG-TAILED.

Latent root

A synonym for EIGENVECTOR.

Latent trait analysis/latent variable models [NAG 1987, p.G11/1]

Page

Terma sometimes used, notably in educational testing, to refer to FACTOR ANALYSIS of BINARY data.

Latin square [Kendall & Buckland 1982; SPSS 1986, p.514]
A special form of 2-way ANOVA design, aimed at removing from an experiment the variation from sources identified as the rows and columns of the square.

Least normal squares, LNS [Williams & Troutman 1987]
Synonymous with MAJOR AXIS REGRESION.

Least-squares method [Kendall & Buckland 1982; Kotz & Johnson 1982-8] 187
A technique of optimization in which the quantities under estimate are determined by minimising a certain quadratic form in the observations and those quantities. Generally, the squares of the deviation of the data from a model are minimized. Where the model involves Normally distributed errors, least-squares estimation becomes equivalent to the MAXIMUM LIKELIHOOD METHOD, but in general the method does not require the assumption of multivariate Normality.

Leptokurtic 92
Distribution curves whose KURTOSIS is >3 (i.e. more 'peaked' than a Normal distribution curve).

***L*-estimate** [Hoaglin et al., 1983, p.306] 88
A ROBUST ESTIMATE OF LOCATION based on a linear combination of the ordered (ranked) data-values. Includes the BROADENED MEDIAN, GASTWIRTH MEDIAN, TRIMEAN and TRIMMED MEANS.

Levelling [Davis 1986, p.259]
The process of subtracting a linear trend from a TIME-SERIES to make it STATIONARY.

Likelihood (function) [Kendall & Buckland 1982; Kotz & Johnson 1982-8]
If the distribution function of continuous variates $X_1...X_n$, dependent on parameters $Y_1...Y_k$ is expressed as $dF = f(X_1...X_n; Y_1...Y_k)dX1...dX_n$ the function $f(X_1...X_n; Y_1...Y_k)$ considered as a function of the *Y*'s for fixed *X*'s, is called the likelihood function, L(*X*).

Likelihood ratio [Kendall & Buckland 1982; Kotz & Johnson 1982-8]
If $X_1...X_n$ is a random sample from a population $f(X; Q_1...Q_k)$ the LIKELIHOOD for this particular sample is $L = \Pi f[X_i, Q_1...Q_k]$ This will have as maximum with respect to the Q's somewhere in the parameter space *S* which can be written L(*S*). For a subspace *s* of the parameter space (i.e. the set of populations corresponding to some restrictions on the parameters) there will also be a corresponding maximum value L(*s*). The null hypothesis that the particular population under test belongs to the subspaces of *S* may be tested by using the likelihood ratio $\lambda = L(s)/L(S)$, where $0 < \lambda < 1$, or some simple function of it. The most widely used likelihood ratio in geological applications is WILKS' LAMBDA.

Lilliefors' tests [Lilliefors 1967; Lewis 1977, p.220; Conover 1980, p.357; Kotz & Johnson 1982-8] 98
Special forms of the KOLMOGOROV test, used to test goodness-of-fit of a sample distribution to either a NORMAL or EXPONENTIAL DISTRIBUTION. Uses modified tables of critical values so that the test statistic *W* may be based on the mean and standard deviation calculated from the data, rather than the corresponding POPULATION parameters.

Linear data
Any data (such as densities, refractive indices, chemical analyses) which are scalar, i.e. not vectorial. Cf. ORIENTATION DATA.

Linear discriminant analysis [Le Maitre 1982, p.141] 304
The usual form of DISCRIMINANT ANALYSIS in which all the discriminant functions are linear combinations of the original variables.

Linear model [Clark 1979, p.8] 272
A form of SEMI-VARIOGRAM without a SILL, defined by the equation $\gamma(h) = kh$, where *k* is a constant.

Linear programing [Koch & Link 1971 p.276ff; Le Maitre 1982, p.102; Kotz & Johnson 1982-8]
An alternative to LEAST-SQUARES for solving mixing problems.

Line power spectrum [Davis 1986, p.257]
Synonymous with PERIODOGRAM.

Listwise omission [SPSS 1986]
Synonymous with CASEWISE.

Loading
See FACTOR LOADING.

Location 86
The 'typical', 'average' or 'most common' value of a VARIABLE, usually measured by the MEAN, MEDIAN etc (which are called *estimates of location* or *measures of centre*). If all values in a data-set are multiplied (or increased) by *c*, the location is also multiplied (or increased) by *c*. Synonymous with CENTRAL TENDENCY.

Logarithmic transformation 99
A general transformation of a VARIABLE, *X* to a new variable, *Y* using a relation such as $Y = a + b \log(X-c)$. Used to Normalise frequency functions, to stabilise variances, and to reduce curvilinear to linear relationships.

Logistic model/regression [Chung 1978; Chung & Agterberg 1980] 259

Page

A specialized form of MULTIPLE REGRESSION, in which the dependence of a DICHOTOMOUS (yes/no) event, such as mineralization, is determined from the interplay of several geological variables (lithology, etc) measured on higher (ORDINAL, NOMINAL, RATIO) scales.

Logit [Kotz & Johnson 1982-8] 259

A transformation of the form $p' = p/(1\{00\} - p)$, where p is a proportion {or percentage}. It has the effect of limitlessly extending the scale of variation from 0→1 {or 0→100%} to $-\infty \rightarrow +\infty$.

Loglinear model 259

A method of analyzing CONTINGENCY TABLES, based on the fact that contingency tables are ANOVA on DISCRETE data, and that, by expressing discrete data relationships in logarithmic form, a loglinear equation can be derived for a contingency table in the same way that a linear regression can be used to formulate an ANOVA. Can be regarded as an extension of ANOVA to non-linear relationships in the same way that POLYNOMIAL REGRESSION extends regression to non-linear relationships.

LogNormal distribution [Le Maitre 1982, p.11; Kotz & Johnson 1982-8] 77

A distribution in which the logarithms of the VALUES are Normally distributed. Tends to occur with variables having a natural lower or upper limit and is extremely common in geochemistry (e.g. Au values in ores). Can be identified by plotting data-values on logNormal probability paper, by determining that the sample GEOMETRIC MEAN equals the MEDIAN, or by performing TESTS FOR NORMALITY on the log-transformed data.

Long-tailed [Kotz & Johnson 1982-8]

Refers to distributions whose probabilities tend to zero relatively slowly away from the centre; that is, distributions which have more numerous extreme (high and/or low) data-values in the 'tails' of the distribution curve than is 'normal'. 'Normal' is usually taken to be measured by the NORMAL DISTRIBUTION itself. Because the tails have relatively many values, the centre of the distribution has concomitantly few, i.e. the central peak is flatter. Thus long-tailed distributions are also generally PLATYKURTIC. *Many* (perhaps even most) real data distributions are either long-tailed or SHORT-TAILED; i.e. the Normal distribution is a theoretical model which barely exists in nature. Terms such as *fat-tailed, long-tailed, thick-tailed,* and *large-tailed* are synonymous.

Lord's tests [Langley 1970, p.367; Kotz & Johnson 1982-8] 144

Forms of SUBSTITUTE *t*-TEST, applicable to one or more groups of data. The one-group example replaces the standard deviation with the sample range, the two-group example uses the 'range of the sample means' ($\overline{X}_{max} - \overline{X}_{min}$) divided by the 'sum of the sample ranges' ($\omega_1 + \omega_2 + ... \omega_n$).

Lower hinge [Tukey 1977, p.33] 94

The value midway between the minimum and MEDIAN values; crudely, the first quartile.

Mahalanobis D^2 distance [Mahalanobis 1936,1949; Kendall & Buckland 1982; Kotz & Johnson 1982-8] 306

A generalized multivariate measure of the 'distance' between two populations with differing means but identical DISPERSION matrices, used in DISCRIMINANT ANALYSIS. Simply related to HOTELLING's T^2 and hence a multivariate analogue of STUDENT's *t*. For multivariate Normally distributed data, Mahalanobis' D^2 is directly related to the probability of misclassification.

Major axis solution

See PEARSON's MAJOR AXIS SOLUTION.

Major product moment [Joreskog et al. 1976, p.21 & 124]

The post-multiplication of a matrix by its TRANSPOSE, i.e. $\mathbf{XX'}$. Can be used as a matrix of SIMILARITY COEFFICIENTS in FACTOR ANALYSIS.

Manhattan distance [SPSS 1986, p.737] 278

Synonymous with CITY BLOCK.

Mann-Whitney *U* test [Mann & Whitney 1947; Conover 1980, p.216; Kotz & Johnson 1982-8] 128

A NONPARAMETRIC test comparing the CENTRAL TENDENCY (essentially, median) of two independent GROUPS of univariate data. Performing the test is equivalent to performing a parametric *t* TEST on the ranks of the data rather than on the data themselves. A variant of the test is known as WILCOXON'S RANK SUM TEST. A MULTIVARIATE version of the test also exists [Kotz & Johnson 1982-8, vol.6, p.82].

MANOVA [Mather 1976]

An abbreviation for MULTIVARIATE ANALYSIS OF VARIANCE.

Mantel-Haenszel χ^2 [SAS 1985,p.413]

A statistic used in CONTINGENCY TABLES to test the ALTERNATIVE HYPOTHESIS that there is a *linear* (rather than just unspecified) association between the row and column variables. Equals $(N - 1)r^2$, where r is the PEARSON CORRELATION COEFFICIENT between the row and column variables.

Mardia's uniform scores test [Mardia 1972, p.197 & 206; Lewis 1977, p.231] 234

A NONPARAMETRIC test for comparing two or more groups of two-dimensional ORIENTATION data. The rank (R) of each angular measurement is converted into a 'uniform score', ($2\pi R/N$), where N is the number of measurements. The test statistic is based on the RESULTANT LENGTHS of the uniform scores.

Markov chain [Till 1974, p.14; Davis 1986, p.150] 222

Page

A MARKOV PROCESS in which the probability of transition from one discrete state to the next state (e.g. from sandstone to coal) depends on the previous state only.

Markov process [Till 1974, p.15; Kotz & Johnson 1982-8] 222

A natural process which has a random element, but also exhibits an effect in which one or more previous events influence, but do not rigidly control, subsequent events. Widely used in analysing stratigraphical sequences, where the deposition of a given lithology is often controlled to some extent by the nature of the lithology just deposited.

Matched-pairs [Kotz & Johnson 1982-8]

Pairs of data which are in some way RELATED. E.g. responses by the *same* set of geology students to two different questions, or to the *same* question by male/female students; values of some parameter 'before' and 'after' an experiment or treatment (e.g. fresh and weathered samples from the same outcrop).

Matched-pairs sign(ed-ranks) test

See WILCOXON MATCHED-PAIRS SIGNED-RANKS test.

Matches

See POSITIVE, NEGATIVE MATCHES and MISMATCHES.

Matching coefficient 168

Sometimes used synonymously with SIMILARITY COEFFICIENT, but better restricted to coefficients applicable to NOMINAL or DICHOTOMOUS data; based on numbers of MATCHES. See ANDERBERG, DICE, DISPERSION, HAMANN, JACCARD, KULCZYNSKI, LANCE & WILLIAMS, OCHIAI, OTSUKA, PATTERN DIFFERENCE, ROGERS & TANIMOTO, RUSSELL & RAO, SIMPSON, SIZE DIFFERENCE, SOKAL & SNEATH coefficients. These are all based on different combinations of a,b,c,d (see CONTINGENCY TABLES and Table 13.5). Applied e.g. to comparisons between faunas defined by the presence/absence of a number of species; thus if two faunas have 4 species in common, and 6 which occur in one but not the other, the JACCARD matching coefficient for the two faunas is 0.4. Effectively Q-mode equivalents of R-mode CONTINGENCY COEFFICIENTS, which are also based on a,b,c,d.

Matheronian semivariogram [Clark 1979, p.6] 272

Synonymous with SPHERICAL SEMIVARIOGRAM.

Matrix inversion [Joreskog et al. 1976, p.23]

The analogue in matrix algebra of division in scalar arithmetic, involving the formation of INVERSES of matrices.

Matrix sampling adequacy (MSA) [Cuneo & Feldman 1986, p.118] 320

A concept in FACTOR ANALYSIS, essentially the VARIABLE SAMPLING ADEQUACY for all variables in the matrix being factored. It measures the extent to which the variables form a homogeneous set (with minimal partial internal correlations), intrinsically suitable for factoring. Values >0.5 are generally assumed necessary for meaningful factor analysis, while values of 1.0 indicate zero partial correlations.

Maximum likelihood method [Joreskog et al. 1976,p.82; Kotz & Johnson 1982-8; Kendall & Buckland 1982]

An alternative to LEAST-SQUARES methods in optimization (minimization or maximization) of a function. Can be used for example in estimating FACTOR LOADINGS in FACTOR ANALYSIS. Involves maximizing the LIKELIHOOD FUNCTION of the given data.

Maximum Studentized residuals

See GRUBBS' MAXIMUM STUDENTIZED RESIDUALS.

McNemar test for the significance of changes [Siegel 1956, p.63; Kotz & Johnson 1982-8] 124

A NONPARAMETRIC test comparing two RELATED groups of at least NOMINAL data. Suited to 'before and after' experiments in which each individual is its own control, such as testing a group of geology students for their mineral identification abilities *before* and *after* some course of instruction.

Mean [Kendall & Buckland 1982]

See ARITHMETIC MEAN.

Mean (absolute) deviation [Kendall & Buckland 1982] 91

A fairly non-ROBUST measure of DISPERSION. The mean absolute value of the deviations from either the MEAN or MEDIAN. Formally, the first ABSOLUTE MOMENT. Not widely used in geology.

Mean square weighted deviate

See MSWD.

Mean-squares

A synonym for ANALYSIS OF VARIANCE.

Measures of centre

See LOCATION.

Medial correlation coefficient

In a MEDIAL TEST, the coefficient $2d/(N-1)$, where N is the total number of data-pairs, and d the total number in the positive and opposite quadrants.

Medial test [Kendall & Buckland 1982]

A simple graphical test for ASSOCIATION between two INTERVAL/RATIO VARIABLES. The X-Y scatter diagram is divided into four quadrants by lines parallel to the median X and Y values (Fig.13.2). Association is indicated if the

Page

number of points falling into the positive (top-right) quadrant differs significantly from 0.25, the number expected for no association. Medial tests do not indicate the *strength* of association, merely its presence or absence (cf. CORRELATION COEFFICIENTS). Examples include the OLMSTEAD-TUKEY test. Not so far widely used in geology, but they allow correlations to be determined from scatter plots, even where the original data are unavailable.

Median [Kendall & Buckland 1982] 86

That value of a variable which divides a total data-set of N values into two halves. The middle value in a set of ranked data (if N is odd), or the average of the two middle values (N even). The MEAN DEVIATION is a minimum when taken about the median. A highly RESISTANT measure of LOCATION.

Median (absolute) deviation 91

A measure of DISPERSION, highly robust to both SKEW and OUTLIERS. The median values of the absolute deviations about the median. The abbreviation MAD sometimes used for it is regrettably ambiguous, since it could refer to the MEAN DEVIATION.

Median test [Siegel 1956, p.111; Conover 1980, p.171; Kendall & Buckland 1982] 127

A NONPARAMETRIC test determining whether two or more independent groups of data differ in central tendency (median).

***M*-estimate** [Andrews et al. 1972] 88

Applies to estimates of CENTRAL TENDENCY (location) based on Maximum likelihood estimation, such as HUBER's ESTIMATES. The estimate T is an iterative solution of an equation of the form $\Sigma\varphi(X_i - T)/s$, where φ is some function X_i refers to the individual data-values, and s is a SCALE estimate such as the MEDIAN DEVIATION.

Method of mixtures [Everitt 1974, p.34]

A DENSITY SEARCH technique of CLUSTER ANALYSIS, which attempts to separate clusters by identifying their distinct probability distributions.

Metroglyphs

See ANDERSON'S METROGLYPHS.

Midmean [Andrews et al. 1972; Kendall & Buckland 1982; Kotz & Johnson 1982-8] 88

Synonym for 25% TRIMMED MEAN. A ROBUST ESTIMATE of LOCATION. If the number of values is not a multiple of 4, the two extreme remaining observations after trimming are appropriately weighted.

Midrange (midpoint) [Kendall & Buckland 1982; Kotz & Johnson 1982-8] 86

A measure of CENTRAL TENDENCY = 0.5 [maximum value + minimum value]. Completely non-ROBUST to outliers and of little value in geology. Sometimes misleadingly used as a synonym for MEDIAN.

Minimum spanning trees, MST [Kotz & Johnson 1982–8]

A concept derived from graph theory and used in CLASSIFICATION to determine distances between objects or clusters. For example, SINGLE LINKAGE CLUSTER ANALYSIS uses MST when connecting objects to clusters.

Minkowski's distance [SPSS 1986, p.736] 278

A generalized distance measure, $\sum\{|X_i - Y_i|^p\}^{1/p}$. Special cases include the EUCLIDEAN DISTANCE ($p = 2$) and MANHATTAN DISTANCE ($p = 1$).

Minor [Joreskog et al. 1976, p.23]

A kind of DETERMINANT formed from a SQUARE MATRIX in which the ith row and jth column have been deleted.

Minor product moment [Joreskog et al. 1976, p.21]

The pre-multiplication of a matrix by its TRANSPOSE, i.e. $\mathbf{X'X}$.

MINRES (minimum residuals) [Mather 1976, p.261]

An ALGORITHM permitting the use of LEAST-SQUARES in FACTOR ANALYSIS.

Mismatch, *m* [Till 1974, p.134]

In comparison of DICHOTOMOUS data for two INDIVIDUALS, the absence of a particular ATTRIBUTE on one individual coupled with its presence in the other (e.g. absence of olivine in one of two compared rock specimens and its presence in the other). Symbolisation is confusingly inconsistent in different texts: for example, Till (1974) uses m for $\{b + c\}$ as in SPSS (1986). Cf. POSITIVE MATCH, NEGATIVE MATCH.

Mode [Kendall & Buckland 1982] 86

That value of a VARIABLE which is possessed by the greatest number of INDIVIDUALS in a GROUP.

Mode analysis [Everitt 1974, p.33] 278

A DENSITY-SEARCH technique of CLUSTER ANALYSIS, which searches for natural sub-groupings by estimating disjoint density surfaces in the sample distribution. Not to be confused with petrographical mod*al* analysis.

Moments [Kendall & Buckland 1982] 91

The mean value of the power of a VARIABLE. The pth moment is $\sum_i (X_i - \bar{X})^p$ for all i. The 1st and 2nd moments are the MEAN DEVIATION and STANDARD DEVIATION. SKEW and KURTOSIS relate to the 3rd and 4th moments.

Monothetic [Everitt 1974, p.18] 292

DIVISIVE methods of CLUSTER ANALYSIS based on the possession or otherwise of a *single* specified ATTRIBUTE. Cf. POLYTHETIC, OMNITHETIC.

Monotonic(ity) [Kotz & Johnson 1982-8] 184

Page

Term indicating an *X-Y* relationship where, as *X* increases, *Y* always moves in the same direction (increases or decreases, but not both). Lines, reciprocal and exponential curves are all monotonic, whereas parabolae, cubic and higher order curves are not. Cf. POLYTONICITY. See also WEAK, STRONG, SEMIWEAK, SEMISTRONG.

Monte Carlo methods [Shreider 1964,1966; Kendall & Buckland 1982; Kotz & Johnson 1982-8]
The solution of mathematical problems arising in a STOCHASTIC context by sampling experiments.

Mood's *W*-test [Kendall & Buckland 1982; Kotz & Johnson 1982-8] 129
A NONPARAMETRIC test comparing the DISPERSION of two groups of univariate data about a common MEDIAN. Cf. DAVID'S TEST. Analogous to a classical '*F*-TEST'. A MULTIVARIATE analogue exists [Kotz & Johnson 1982-8, vol.6, p.85].

Moses test of extreme reactions [Siegel 1956, p.145]
A NONPARAMETRIC test comparing two INDEPENDENT GROUPS of data, specifically designed to detect extreme scores in either direction.

Moving average [Davis 1986, p.269; Kotz & Johnson 1982-8]
A type of FILTER, which calculates successive means over a specified number of data points.

MSWD (mean square of weighted deviates) [McIntyre et al. 1966; Brooks et al. 1972]
A statistic used in the assessment of Rb-Sr, Nd-Sm and other isochrons. Compares the geological spread of data-points about the STRUCTURAL REGRESSION LINE with the analytical error for the *X* and *Y* variates. MSWD = SUMS (sum of squares)/$(N-2)$, where *N* is the number of data-values. [Note that Brooks et al. incorrectly quote MSWD as $\sqrt{\{SUMS/(N-2)\}}$]. An MSWD < 3 is usually taken as a true 'isochron' (i.e. the geological scatter is small), MSWD > 3 as an 'errorchron' or 'scatterchron'. MSWD = 1 means the geological error equals the analytical error.

Multicollinearity [Kendall & Buckland 1982; Kotz & Johnson 1982-8] 255
A situation, e.g. in MULTIPLE REGRESSION, where the INDEPENDENT VARIABLES are themselves correlated: one variable is nearly a linear combination of other variables. The coefficients of the regression on these variables are then indeterminate. Some authors have used it as a multivariate extension of COLLINEARITY, but the two are in reality slightly distinct. Common in geology, and needs careful attention.

Multidimensional scaling (MDS) [Mather 1976, p.332; Coxon 1982; Kotz & Johnson 1982-8] 296
A powerful ORDINATION technique. NON-METRIC MDS is based on ranks and is hence applicable to ORDINAL data; *metric* MDS is applicable to RATIO data. A multivariate group of data is expressed iteratively as a two-dimensional scatter plot; the extent to which this fails to express the original multidimensional relationships is expressed by STRESS.

Multigroup discriminant analysis [Le Maitre 1982, p.141] 309
DISCRIMINANT ANALYSIS applied to more than 2 groups of data.

Multinomial distribution [Blalock 1972, p.170; Kendall & Buckland 1982; Kotz & Johnson 1982-8] 81
The discrete distribution associated with events which can have more than two outcomes. A generalisation of the BINOMIAL DISTRIBUTION. Of very restricted applicability in geology.

Multiple comparison (tests) 140
Usually refers to tests used after ONE-WAY ANOVA has proved that several means are INHOMOGENEOUS, to check which particular means differ from one another. Includes SCHEFFÉ's TEST, FISHER'S PLSD, TUKEY'S *w*, etc.

Multiple correlation coefficient [Dixon & Massey 1969, p.215; Kendall & Buckland 1982] 254
An extension of the CORRELATION COEFFICIENT to associations between more than two variables.

Multiple discriminant analysis [Le Maitre 1982, p.141] 304
A rather ambiguous term, sometimes used synonymously with MULTIGROUP DISCRIMINANT ANALYSIS, sometimes to include both TWO-GROUP and MULTIGROUP analysis. Best avoided.

Multiple factor analysis [Kendall & Buckland 1982]
A synonym for FACTOR ANALYSIS.

Multiple regression [Davis 1986, p.469] 253ff
REGRESSION in which the DEPENDENT VARIABLE is controlled by more than one INDEPENDENT VARIABLE. For example, the refractive index of pyroxene (the dependent variable) varies with the contents of En, Fs, Wo, Ac etc, etc. (the independent variables).

Multistate data 64
A synonym for NOMINAL data such as lithologies (*state 1* = limestone, *state 2* = sandstone etc.)

Multivariate
Involving several (generally, more than 2) VARIABLES. Covers a great many geological data-types (rock, mineral and stream sediment analyses, fossil size measurements, etc.).

Multivariate analysis-of-variance (MANOVA) [Kotz & Johnson 1982-8] 305
The method for comparing means from several MULTIVARIATE Normal populations having common variance.

Nearest neighbour analysis [Davis 1986, p.308] 266
An alternative to QUADRAT ANALYSIS in the study of distributions of points on maps, which considers the distances between adjacent points using the NEAREST NEIGHBOUR STATISTIC. Superior to quadrat analysis in that

Page

it provides an indication of the nature of the point distribution beyond a simple test of random distribution.

Nearest neighbour statistic [Davis 1986, p.308; Mather 1976, p.122] 266

The ratio of the mean distance *M* between every point and its nearest neighbour, in a 2-dimensional areal data distribution, to the expected mean distance. Equals $2M\sqrt{p}$, where *p* is the density of points per unit area under consideration. Varies from 0 where all points coincide, to 1 for a random distribution, to 2.15 where mean distances are maximised in a hexagonal pattern.

Nearest neighbour technique [Everitt 1974, p.9; Kotz & Johnson 1982-8]

Synonymous with SINGLE LINKAGE TECHNIQUE.

Negative match [Till 1974, p.134]

In comparison of DICHOTOMOUS data for two INDIVIDUALS, the absence of a particular ATTRIBUTE in both individuals (e.g. absence of olivine in two rock specimens). The value *d* in a conventionally notated CONTINGENCY TABLE (Table 13.5), but commonly notated as *n*. Cf. MISMATCH, POSITIVE MATCH.

Nested analysis of variance 166

See HIERARCHICAL ANALYSIS OF VARIANCE.

Network database system [Date 1981] 43

An enhanced HIERARCHICAL DATABASE SYSTEM, in which closer links are incorporated between 'branches' of the tree-structure than are possible in a strict hierarchical system.

Neuman-Keuls procedure [Kotz & Johnson 1982-8]

See S-N-K.

Neymann-Pearson-Bartlett test

A synonym for BARTLETT's *M*-TEST.

Noether's test for cyclic change/trend [Lewis 1977, p.193; Kendall & Buckland 1982] 210

A NONPARAMETRIC test sensitive to both monotonic and cyclic trend (i.e. a trend that periodically reverses direction) in data, based on grouping the data-values into triplets.

Noise [Kotz & Johnson 1982-8]

Something that interferes with a desired signal: e.g. in geochronology — sampling/analytical errors, age discordances due to resetting, weathering, varying blocking temperatures, etc.

Nominal scale/variable [Kotz & Johnson 1982-8] 64

Measurement at its weakest level, where numbers, letters or other symbols are used simply to classify an ATTRIBUTE (e.g. labels on rock specimens; 'P-type' and 'N-type' MORB; 'Group 1' and 'Group 2' kimberlite). The only permissible mathematical operation is '=' (or '≠'). For example specimen number 4 is in no sense 'twice' specimen 2, but merely different from it.

Non-hierarchical cluster analysis [Mather 1976, p.327 & 343] 292ff

CLUSTER ANALYSIS whose aim is not to produce a hierarchical DENDROGRAM, but merely a list of *N* non-overlapping clusters, with the membership of each cluster among the original group of OBJECTS.

Non-linear mapping (NLM) [Mather 1976, p.339] 302

An ORDINATION technique similar to MULTIDIMENSIONAL SCALING.

Non-metric data

Qualitative (NOMINAL or ORDINAL) data, as opposed to metric (*quantitative*, INTERVAL or RATIO) data.

Non-metric multidimensional scaling [Kruskal 1964; Mather 1976, p.331]

See MULTIDIMENSIONAL SCALING.

Nonparametric statistics [Conover 1980, p.91-3; Singer 1979; Kendall & Buckland 1982] 69

Various conflicting definitions are available (see summary in Conover). Statistics which may be used on NOMINAL or ORDINAL data, and on RATIO data without explicitly making any assertions about either the form of the distribution governing the data, or about the parameters of that distribution (means, standard deviation, etc.), even if its form is known. Some writers distinguish nonparametric from DISTRIBUTION-FREE, citing, for example CHEBYSHEV'S INEQUALITY, which is parametric but makes no assumptions about the parent distribution form. All writers however agree that 'parametric' tests are mostly very sensitive to departures from the assumed form of the population distribution function, whereas 'nonparametric' tests are not at all sensitive; 'robust' tests occupy an intermediate position.

Non-statistical factor analysis [Mather 1976, p.247]

See α FACTOR ANALYSIS.

Normal (Gaussian) distribution [Kendall & Buckland 1982] 76

The fundamental 'bell-shaped' distribution in statistics and probability theory, with the equation:

$$y = \frac{1}{\sqrt{2\pi}} e^{-(x-\mu)^2/2\sigma^2} \quad \text{for } -\infty < x < +\infty$$

A limiting form of the BINOMIAL DISTRIBUTION.

Normalisation

(1) Conversion of data such that they approximate a Normal distribution. (2) In matrix algebra, conversion of vectors to

Page

unit length.

Normality test [Kotz & Johnson 1982] 96

Tests that a given set of data approximate a NORMAL DISTRIBUTION. Includes GEARY'S RATIO, FILLIBEN's *r*, and tests based on SKEW and KURTOSIS (all PARAMETRIC tests), together with the SHAPIRO–WILK TEST, LILLIEFOR's TEST and a test based on the BIWEIGHT ESTIMATOR (all ROBUST or NONPARAMETRIC). Extremely important in geology, since many geological data-sets are *not* Normally distributed, so that classic (parametric) methods are of dubious application.

Normal probability plot [Koch & Link 1970, p.236] 96

A plot of a set of observations on special, Normal probability, graph paper. Cumulative frequency against value forms a straight line for Normally distributed observations.

Normal scores [Conover 1980,p.317] 126

Used in various senses in the literature (see also RANKITS). Used here to denote scores which convert RANKS into numbers more nearly resembling observations from a Normal Distribution. The Normal score for the *k*th of *N* ranks is the *k/(N+1)*th quantile of the Standard Normal Distribution; e.g. for the 1st of 5 values, the 1/6th quantile or –0.9661.

Normal scores tests [Kendall & Buckland 1982; Kotz & Johnson 1982-8]

NONPARAMETRIC tests based on converting raw data first into ranks and then into NORMAL SCORES (e.g. KLOTZ, VAN DER WAERDEN , VAN EEDEN tests). Instead of using mere ranks, these tests use NORMAL SCORES, which more nearly resemble observations from a Normal Distribution. For example, the ranks 1,2,3,4,5,6,7,8,9 are replaced by Normal scores of -1.2816, -0.8416,-0.5244,-0.2533, 0.0,+0.2533, +0.5244,+0.8416, +1.2816, which are perfectly symmetrically distributed around zero (provided there are no ties), and spread out in the same way as a truly Normal distribution. Normal scores tests have the same ARE as classical parametric tests, where the latter are performing optimally with Normally distributed data; where data are not Normal, Normal scores tests have higher ARE, and may be infinitely better. As Conover (1980, p.316) puts its, Normal scores tests "are always *at least as good as* the usual parametric tests, such as the *t* test and the *F* test" (italics added). Overall, they combine useful properties of both parametric and nonparametric approaches.

Notched box plot [McGill et al. 1978] 94

A BOX PLOT in which the positions of the CONFIDENCE LIMITS (usually 95%) about the MEDIAN are indicated by triangular 'notches', extending from the median line to the sides of the box. Allows medians from different boxplots to be visually compared pairwise, for equality.

Null correlation 168

The correlation expected when there is no relationship between two VARIABLES, against which the value of a CORRELATION COEFFICIENT is tested for significance. Not necessarily zero, if the data are subject to CLOSURE.

Null hypothesis [Kendall & Buckland 1982; Kotz & Johnson 1982-8] 73

See H_o.

Numerically stable [Kennedy & Gentle, 1980, p.34]

Data analysis methods which tend to be little affected by rounding errors and cancellations in computations.

Numerically unstable [Kennedy & Gentle, 1980, p.34]

Data analysis methods in which serious accumulations of rounding errors and/or severe cancellations may take place, so that final results may appear correct but are actually unsatisfactory.

Nyquist frequency [Davis 1986, p.257]

The highest frequency that can be estimated in a TIME-SERIES; equals twice the separation between observations.

Object 68

Synonymous with ENTITY, INDIVIDUAL, SPECIMEN, ITEM. A rock, fossil, sediment horizon, river, volcano etc.

Oblique-projection method [Joreskog et al. 1976, p.139]

A variety of OBLIQUE ROTATION in FACTOR ANALYSIS, in which the factors are forced to be collinear with the most divergent of the original VARIABLES. Cf. PROMAX ROTATION.

Oblique rotation [Joreskog et al. 1976, p.133]

Rotation of axes in FACTOR ANALYSIS without the constraint that the rotated factors be orthogonal. Allows for correlations between factors, which thus plot at angles <90° to one another.

Observation

Generally synonymous with (data-)VALUE.

Observational vector

Formal statistical term for INDIVIDUAL.

Ochiai coefficient [Coxon 1982, p.27; SPSS 1986, p.741] 168

A MATCHING COEFFICIENT defined as $\sqrt{\{a/(a+b).a/(a+c)\}}$ — see Table 13.5. The binary form of the cosine, ranging from 0 to 1.

Olmstead-Tukey test for association [Olmstead & Tukey 1947; Kotz & Johnson 1982-8] 181

A MEDIAL test which gives special weight to extreme (outlying) values. Also called the CORNER TEST.

Omnithetic [Tipper 1979a] 292

Page

DIVISIVE methods of CLUSTER ANALYSIS based on the possession or otherwise of *all* specified attributes. Cf. MONOTHETIC, POLYTHETIC.

One-sample (group) test 101

Often used synonymously with GOODNESS-OF-FIT test, but probably better used to distinguish tests which concern a *specified parameter* (e.g. mean, variance) of a distribution, rather than the class of distribution itself (Normal, Poisson, etc.) The ONE-GROUP *t*TEST does *not* test that a distribution is Normal, but rather assumes this and goes on to test that the population has a specified mean. 'One-sample' does *not* imply that data for only one rock 'sample' are available (see SAMPLE). It arises because one *statistical* sample (group of data) is compared with an assumed population.

One-sided (tailed) test [Kendall & Buckland 1982] 73

A hypothesis test in which the region of rejection is wholly located at one end of the distribution of the test statistic. A test which assumes the direction in which the compared values differ. E.g. H_0: $\mu_1 = \mu_2$ as against H_1: $\mu_1 > \mu_2$ is a one-sided test (*see* H_0, H_1), whereas against H_1: $\mu = \mu_2$ it is TWO-SIDED, since in the latter case the direction is not stated. The probability associated with a given value of a test statistic in a one-sided test is *half* that of the *same* value for a two-sided test. For example, a one-sided probability of 0.025 (equivalent to a SIGNIFICANCE LEVEL of 97.5%) corresponds to a two-sided probability of 0.05 (95% significance level).

One-way analysis-of-variance (ANOVA) [Kotz & Johnson 1982-8] 138

ANOVA in which only one determining factor is involved. The commonest application is to test the equality of means for 3 or more independent groups of data (samples). For example, if the initial Sr isotopic ratios for several sets of randomly collected specimens of serpentinite, basalt and gabbro were to be compared, one-way ANOVA would suffice, but if each set of rock-types included specimens from each of several distinct geological terranes (e.g. greenstone belt, ophiolite, island-arc), then *TWO-WAY* ANOVA would be required.

Optimization-partitioning clustering techniques [Everitt 1974, p.24] 278

CLUSTER ANALYSIS in which mutually exclusive clusters are formed by optimization of a 'clustering criterion'.

Optimized matrix [Hennebert & Lees 1985, p.124]

A SQUARE SYMMETRIC matrix in which all diagonal elements are unity, and other elements decrease in value monotonically away from the diagonal. Used for example in interpreting matrices of MATCHING COEFFICIENTS.

Order of a matrix [Joreskog et al. 1976, p.9]

The size of a matrix in terms of its number of rows and columns.

Order statistics [Kendall & Buckland 1982; Kotz & Johnson 1982-8]

Any statistic based on an ordered (i.e. ascending or descending) set of data, and using the *actual data-values* themselves: e.g. range, median, minimum, maximum, interquartile range. Cf. Rank-order statistics.

Ordinal (ranking) scale/variable [Kotz & Johnson 1982-8] 64

Measurements which can be ranked qualitatively but not quantitatively. E.g. Moh's scale of hardness, where diamond (scale 10) is harder, but not exactly 10 times harder, than talc (scale 1). Permissible mathematical operations on ordinal data are $=$, $\neq$, $>$ and $<$. Most ordinal variables are DISCRETE, i.e. take only whole-number values, but fractional or decimal values are sometimes used in a qualititative way (e.g. fayalite has a Moh hardness of "6.5").

Ordinal scaling analysis

Synonymous with *non-metric multidimensional scaling*. See MULTIDIMENSIONAL SCALING.

Ordinary least-squares, OLS [Mather 1976; Williams & Troutman 1987]

(1) Used in contrast to GENERALIZED LEAST SQUARES: MULTIPLE REGRESSION in which the variances of the DEPENDENT VARIABLE due to each of the INDEPENDENT VARIABLES are assumed to be equal. (2) Alternatively used to refer to classical, linear, *Y* on *X* REGRESSION, where *X* is assumed to be without error, in contradistinction to LEAST NORMAL SQUARES or REDUCED MAJOR AXIS regression.

Ordination [Mather 1976, p.329] 296

CLASSIFICATION techniques which do not seek to formulate either CLUSTERS or DENDROGRAMS, but merely seek to display relationships between a set of INDIVIDUALS graphically, in as few dimensions as possible. Similar objects are plotted near each other, dissimilar objects far apart. Groupings can then be recognised by visual examination of the resulting scatter plots. Includes NONLINEAR MAPPING, MULTIDIMENSIONAL SCALING, QUADRATIC LOSS FUNCTIONS, and certain uses of PRINCIPAL COMPONENTS ANALYSIS. Potentially extremely useful in geology, in showing the 'best' way to display relationships among multivariate data on a piece of paper. Variation diagrams in petrology and geochemistry are an extremely crude type of ordination, which merely use visual inspection to check which is the 'best' plot.

Orientation data 226ff

Angular, vectorial, 2- or 3-dimensional data such as strikes. Includes DIRECTIONAL and AXIAL (non-directional) data.

Orthogonal [Kendall & Buckland 1982]

In statistical terms, *uncorrelated* (at right angles).

Orthogonal matrix [Joreskog et al. 1976, p.25]

A matrix whose MINOR PRODUCT MOMENT is a DIAGONAL matrix.

Orthogonal polynomials [Mather 1976, p.96]

Page

A device used in POLYNOMIAL REGRESSION. The power terms of X in the equation $Y = \beta_o + \beta_1 X + \beta_2 X^2 + \beta_3 X^3 ..... + \beta_n X^n$ are correlated, i.e. non-orthogonal, which leads to ILL-CONDITIONED matrices in solving the equation. Orthogonal polynomials, Z, are projections of the originals X's onto arbitrary orthogonal coordinate systems; they carry the same information but are not correlated with each other. The new equation becomes $Y = A + B^*Z_1 + C^*Z_2 + D^*Z_3 + ...$

Orthogonal solution (rotation) [Cuneo & Feldman 1986]
Rotation of the initially calculated axes in FACTOR ANALYSIS, constraining the rotated axes to be at 90° to one another (i.e. making them uncorrelated).

Orthogonal vector [Joreskog et al. 1976, p.25]
A vector whose MINOR PRODUCT is zero.

Orthonormal [Joreskog et al. 1976, p.25]
A matrix whose MINOR PRODUCT MOMENT is an IDENTITY matrix.

Orthotran solution (rotation) [Cuneo & Feldman 1986, p.114] 319
An OBLIQUE ROTATION of axes in FACTOR ANALYSIS.

Otsuka's coefficient [Mather 1976, p.314] 168
A MATCHING COEFFICIENT used with DICHOTOMOUS data, the binary equivalent of COSINE THETA for ratio data. Defined as $c/\sqrt{(n_1 n_2)}$— see Table 13.5.

Outer fence [Tukey 1977, p.44] 94
A value 3 H-SPREADS away from the median, used in EDA. Values beyond the outer fence are taken as OUTLIERS.

Outliers [Kendall & Buckland 1982; Kotz & Johnson 1982-8] 113ff
Observations far removed from the remainder of a data-set (i.e. extremely high and/or low values). Outliers are readily identified by purely statistical means; e.g. HARVEY's and DIXON's tests, but for most data, geological reasoning is then required to determine whether the identified outliers are TRUE, FALSE or BIZARRE, and hence whether they should be rejected or not.

Paired data
See MATCHED PAIRS.

Pairwise omission [NAG 1987, p.G02/7] 206
The treatment of missing data for an INDIVIDUAL by omitting that individual only from calculations actually involving its missing values. E.g. if we had values for Rb for 3 rocks, and of Sr,Ba for 4, pairwise correlation coefficients would be based on 3 Rb-Sr and Rb-Ba pairs, but 4 Sr-Ba pairs. Pairwise omission allows a limited number of techniques to be executed, but results may be difficult to interpret (due to differing numbers of values). Resulting matrices of correlation coefficients, for example, need not be POSITIVE DEFINITE. True multivariate techniques (e.g. discriminant analysis) can only be performed using CASEWISE omission. On the other hand, pairwise omission uses more of the available information in the data than does casewise.

Parameter [Kendall & Buckland 1982] 68
Generally used for values defining POPULATIONS, as opposed to statistics defining sample distributions. For the mean and standard deviation, for example, μ and σ are population *parameters* whereas $\overline{X}$ and s are sample *statistics.*

Parametric statistics 69
Denotes statistics based on a given distribution (almost always the NORMAL DISTRIBUTION), such as STUDENT'S t, FISHER's F, etc. They are valid for testing hypotheses only insofar as the given data follow that distribution closely.

Partial correlation [Kendall & Buckland 1982] 252
Correlation between two variables when one or more additional variables is held constant. See KENDALL'S AND PEARSON'S PARTIAL CORRELATION COEFFICIENTS.

Partially balanced incomplete block design (PBIB) [SPSS 1986, p.742]
A form of ANOVA similar to a BALANCED INCOMPLETE BLOCK DESIGN, except that the missing measurements are not as systematically spread through the design table.

Partitioning techniques [Everitt 1974, p.24ff] 278
CLUSTER ANALYSIS which allows reallocation of individuals, to allow poor initial partitions to be corrected at a later stage in the analysis. Cf. HIERARCHICAL TECHNIQUES.

Pattern recognition
A general term for techniques which seek patterns (groupings, trends) in data without any *a priori* assumptions. Includes most forms of CLASSIFICATION, ORDINATION and, in some cases, FACTOR ANALYSIS.

***P*-condition number** [Mather 1976, p.49]
The ratio of the largest to smallest EIGENVALUES of a matrix. Varies from 1 (ideally conditioned or orthogonal matrix) to ∞ (singular matrix). See ILL-CONDITIONED.

Peak test [Kotz & Johnson 1982-8]
A NONPARAMETRIC test for HETEROSCEDASTICITY, useful in REGRESSION if the usual NORMALITY assumptions cannot be made.

Page

Pearson's contingency coefficient [Kotz & Johnson 1982-8] 176
A CONTINGENCY COEFFICIENT defined as $\sqrt{\{\chi^2/(N+\chi^2)\}}$. Lack of association is indicated by a value of 0, but the upper limit depends on the numbers of rows and columns in the CONTINGENCY TABLE, so the strength of association is difficult to gauge.

Pearson's major axis solution [Le Maitre 1982, p.55] 192
A 'STRUCTURAL' REGRESSION linking two variables, both with associated errors, in which the *perpendicular distances* from data-points to the regression line are minimised. Corresponds to a LAMBDA value of 1.

Pearson's *r* product-moment linear correlation coefficient 168
See PRODUCT-MOMENT CORRELATION COEFFICIENT.

Percentiles [Kendall & Buckland 1982]
Values which divide a total frequency distribution into 100 equal parts. The MEDIAN is the 50th percentile, and the QUARTILES are the 25th and 75th.

Periodogram [Davis 1986, p.256]
A plot of POWER versus HARMONIC NUMBER, for a TIME-SERIES. Also called *discrete* or *line power spectrum.*

Phi (φ)coefficient [Downie & Heath 1974, p.108; Kotz & Johnson 1982-8]
A synonym for FOURFOLD COEFFICIENT.

Pitman efficiency [Conover 1980, p.89; Kotz & Johnson 1982-8]
Synonym for ASYMPTOTIC RELATIVE EFFICIENCY.

Platykurtic [Kendall & Buckland 1982] 92
Distributions whose KURTOSIS is < 3 (i.e. less 'peaked' than a NORMAL DISTRIBUTION).

Plotting point percentage [Koch & Link 1970, p.236]
In the production of NORMAL PROBABILITY plots, cumulative frequency is with advantage replaced (for fewer than 100 observations) by a plotting percentage obtained by the formula 100(3*cumulative frequency-1)/(3*N+1), where N is the number of observations. Also used in the calculation of JUDDS.

Point-biserial correlation coefficient [Downie & Heath 1974, p.101] 172
A coefficient measuring association between one CONTINUOUS and one truly DICHOTOMOUS VARIABLE.

Point density [Cheeney 1983 p.108]
A descriptive term for UNIMODAL spherical distributions in which points tend to cluster around one single, preferred orientation: e.g. measurements in a rock with a pronounced lineation.

Poisson distribution [Le Maitre 1982, p.11; Kendall & Buckland 1982] 82
A discrete distribution applicable to numbers of events (e.g. X-ray counts or volcanic eruptions) occurring over a specified region or time. Usually applies where the events are rare but the region/time interval large. Its mean and variance are equal. A limiting case of the BINOMIAL DISTRIBUTION. Mainly applies in geology to errors in X-ray counting techniques, but also applicable to numbers of meteorite falls over given areas, etc. Has the equation:

$$y = \frac{e^{-m} m^x}{x!}, \text{ for } x = 0,1,2\ldots,$$ where m is the rate or average number of times the event occurs.

Poisson test [Langley 1970, p.230]
A straightforward application of the POISSON DISTRIBUTION, to compare (a) the number of isolated occurrences (e.g. meteorite falls) in a random sample of a certain size or duration, with the expected numbers for such a sample from a much larger set of observations; (b) the proportions of a random binomial sample and an average derived from a much larger set of observations. In both cases, the *zI* TEST is used instead where there are >40 observations in the smaller sample.

Polar wedge diagram [Mardia 1972, p.6; Kendall & Buckland 1982]
A joint synonym for CIRCULAR HISTOGRAM and ROSE DIAGRAM.

Polynomial (curvilinear, non-linear) regression/trend [Kendall & Buckland 1982] 196
The extension of bivariate REGRESSION to the non-linear case, via an equation of the form $Y = ß_o + ß_1X + ß_2X^2 + ß_3X^3 \ldots + ß_nX^n$.

Polythetic [Everitt 1974, p.18] 292
DIVISIVE methods of CLUSTER ANALYSIS based on the possession or otherwise of *several* specified ATTRIBUTES. Cf. MONOTHETIC, OMNITHETIC.

Polytomous [SPSS 1986,p.576]
An unusual synonym for NOMINAL (used in contradistinction to DICHOTOMOUS).

Polytonic(ity) [Kotz & Johnson 1982-8]
An X-Y relationship in which Y may increase or decrease while X increases. Cf. MONOTONIC.

Pooled variance-covariance matrix
The sum of VARIANCE-COVARIANCE matrices for 2 or more groups of data.

Population [Kendall & Buckland 1982; Cheeney 1983, p.7] 69
The total number of all possible SPECIMENS in a study. E.g. all the pebbles on a beach, all the rock comprising a

Page

stratigraphical unit, all the fossils that ever occurred in a bed.

Positive definite matrix [Joreskog et al. 1976, p.23; Mather 1976, p.31]
A matrix in which all the MINORS associated with the elements of the principal diagonal are > 0. Alternatively, a matrix with no zero or negative EIGENVALUES.

Positive match [Till 1974, p. 134]
In comparison of DICHOTOMOUS data for two INDIVIDUALS, the presence of a particular ATTRIBUTE in both individuals (e.g. presence of olivine in *both* of two rock specimens). Cf. MISMATCH, NEGATIVE MATCH.

Positive semidefinite matrix
A matrix with no negative (but some zero) EIGENVALUES.

Power-efficiency (of a statistical test) [Kendall & Buckland 1982] 74
The POWER of the test relative to its most powerful alternative. For example, the MANN-WHITNEY test has a power-efficiency of about 95% relative to the *t*-TEST; this means it rejects a false H_1 on a sample of 100 as many times as the *t*-test does on a sample of 95. Usually equals the RELATIVE EFFICIENCY of a test against the *t* or *F* test (whichever is the nearest equivalent).

Power (of a statistical test) 74
The probability of rejecting H_1 when it is false: [1 – probability of ß ERROR]. Thus when the NULL HYPOTHESIS H_0 is accepted, the probability of being correct equals the power of the particular test being used.

Power (of a time-series) [Davis 1986, p.257]
The VARIANCE, which for sinucoidal functions is simply half the squared amplitude.

Power transformation
TRANSFORMATION of the form $Y = X^n$. See also GENERALIZED POWER TRANSFORMATION.

Precision [Le Maitre 1982, p.5; Kendall & Buckland 1982] 102ff
The conformity of REPLICATE data to themselves. Precision is least when variability (DISPERSION) is greatest. Cf. ACCURACY. In general varies with $\sqrt{N}$ where *N* is the number of measurements/observations.

Presence/absence data
DICHOTOMOUS data such as the presence or absence of a mineral in a rock.

Primary key [Date 1981] 46
One or more FIELDS in a RELATIONAL DATABASE SYSTEM which uniquely identify every RECORD in the RELATION being examined. Typically comprises a data-item such as the sample number (which is presumably unique to each sample), but may comprise several such items (= *composite primary key*).

Primary pattern solution [Cuneo & Feldman 1986]
An OBLIQUE ROTATION of axes in FACTOR ANALYSIS, defining loadings that are regression coefficients predicting the STANDARD SCORE of a variable in terms of the calculated factors. Closely related to the REFERENCE STRUCTURE SOLUTION.

Principal axes [Cheeney 1983, p.119]
The 3 mutually perpendicular planes of orthorhombic symmetry for a BINGHAM DISTRIBUTION.

Principal axis solution [Le Maitre 1982, p.55]
See PEARSON's MAJOR AXIS SOLUTION.

Principal components analysis [Le Maitre 1982, p.106; Kendall & Buckland 1982] 315
A MULTIVARIATE technique for examining the structure of large data-arrays, and for reducing their DIMENSIONALITY. Expresses the total variance of the array in terms of a small number of *principal components* (EIGENVECTORS). Can be based on a VARIANCE-COVARIANCE or CORRELATION matrices, the latter eliminating the influence absolute size of different variables may have on the result. Strictly, principal components analysis is a mathematical manipulation and not a statistical procedure, for nothing is said in it about probability, hypothesis testing, etc., and no assumptions are made about the data distributions treated. Cf. FACTOR ANALYSIS. Use of principal components analysis to reveal groups of INDIVIDUALS is a form of ORDINATION. Now widely used in geology.

Principal coordinates analysis [Joreskog et al. 1976, p.100; Gordon 1981; Mather 1976, p.330] 300
Effectively, Q-MODE PRINCIPAL COMPONENTS ANALYSIS on a matrix of MATCHING or SIMILARITY COEFFICIENTS, allowing INDIVIDUALS to be classified into groups by the visual examination of the resulting scatter of points. A form of ORDINATION. The alternative to Q-MODE FACTOR ANALYSIS for DICHOTOMOUS, NOMINAL and ORDINAL data.

Principal diagonal
That diagonal of a matrix running from top-left to bottom-right.

Principal factor method [Joreskog et al. 1976, p.78]
A largely obsolete method of estimating FACTOR LOADINGS in FACTOR ANALYSIS.

Principle of analytical linear dependence [Joreskog et al. 1976, p.66]
The idea that factors in FACTOR ANALYSIS should account for all linear relationships among VARIABLES.

Principle of simple structure [Joreskog et al.1976, p.66]

Page

The idea in FACTOR ANALYSIS that only those factors for which the VARIABLES have simple representations are meaningful — i.e. the matrix of FACTOR LOADINGS should have as many zero elements as possible, and one factor should only be connected with a small proportion of the variables.

Probability density function (pdf) [Cheeney 1983, p.14]
Synonym for FREQUENCY DISTRIBUTION CURVE.

Product-moment correlation coefficient (Pearson's *r*) [Kendall & Buckland 1982] 168
A PARAMETRIC, BIVARIATE correlation coefficient defined as COVARIANCE(X,Y)/√ [VARIANCE(X) * VARIANCE(Y)]. Lies within the range -1 to +1, unity indicating perfect correlation, 0 indicating no correlation.

Promax rotation [Joreskog et al. 1976, p.137]
A variety of OBLIQUE ROTATION in FACTOR ANALYSIS, in which the (abstract) factor axes are located so as to minimize some function. Cf. OBLIQUE-PROJECTION.

Q-analysis
A synonym for CLUSTER ANALYSIS.

***q*-group Smirnov test** [Conover 1980, p.379] 149
An extension of the SMIRNOV TEST to >2 groups of data. Of limited use in geology, as critical values are only available for small numbers of groups, with equal numbers of data-values in each.

Q-mode analysis 72
That version of any technique which concentrates on relationships between INDIVIDUALS, rather than VARIABLES. Cf. R-MODE.

Quade test [Quade 1966; Conover 1980, p.295] 158
A NONPARAMETRIC test to determine whether one or several TREATMENTS in a RANDOMIZED BLOCK DESIGN tends to yield larger values. Tests the same hypothesis as the FRIEDMAN TEST, but requires that comparisons to at least the ORDINAL level can be made between BLOCKS.

Quadrant sum test
A synonym for OLMSTEAD-TUKEY TEST.

Quadrat [Kendall & Buckland 1982] 264
Equal-sized subareas on a map.

Quadrat analysis [Davis 1986] 264
The testing of a distribution of points on a map for randomness by dividing the area into equal-sized subareas ('quadrats'), and testing the observed numbers of points in each quadrat against the expected numbers (which are equal for a random distribution) using a χ^2 test. Cf. NEAREST NEIGHBOUR ANALYSIS.

Quadratic loss function [Mather 1976, p.340] 302
An ORDINATION technique.

Qualitative variable [SAS 1985,p.21]
A synonym for a NOMINAL variable.

Quantile test [Conover 1980, p.106]
A form of the BINOMIAL TEST used to test the hypothesis that the *p*th population QUANTILE of a random set of data has a given value.

Quartile [Kendall & Buckland 1982]
There are 3 values which separate a total frequency distribution into 4 equal parts. Other than the middle value (MEDIAN), these are termed the *Lower* and *Upper quartiles.*

Quartile deviation (semi-interquartile range/spread) [Kendall & Buckland 1982] 91
Half the range between the first and third QUARTILES.

Quartimax solution (rotation) [Cuneo & Feldman 1986] 319
A type of ORTHOGONAL ROTATION in FACTOR ANALYSIS, in which most of the VARIANCE is allocated to a few factors (especially the first), resulting in less simple structures for the 2nd and subsequent factors. Cf. EQUAMAX, VARIMAX solutions.

Quasi-midrange [David 1981, p.181]
A quick but non-ROBUST estimate of LOCATION, $0.5[X_i + X_{N+1-i}]$, where i is chosen for maximum efficiency.

Quasi-range [David 1981, p.190]
A quick but non-ROBUST estimate of SCALE, $W_i = [X_{N+1-i} - X_i]$, where $2 \leq i \leq N/2$ is chosen for maximum efficiency. W_1 (the ordinary range) is most efficient for $N \leq 18$, W_2 for $18 < N < 32$, W_3 for $N \geq 32$, and so on.

Random errors [Kendall & Buckland 1982]
Errors due to machine fluctuations and other stochastic processes, causing data to be less PRECISE. Cf. SYSTEMATIC ERRORS.

Randomization test for matched-pairs [Siegel 1956, p.88; Conover 1980, p.330; Kendall & Buckland 1982]
A NONPARAMETRIC test revealing the exact probability that a set of MATCHED-PAIRS of data are stochastically equal.

Randomization test for two independent groups [Siegel 1956, p.152; Conover 1980, p.328]

Page

A NONPARAMETRIC test revealing the exact probability that two independent groups of data are equal.

Randomized complete block design [Conover 1980, p.294; Kendall & Buckland 1982] 155

An experimental design in which each and every TREATMENT is allocated randomly between each BLOCK, allowing unbiased estimates of error to be made. Not often possible in geology because of the lack of laboratory control over data availability. Tested by 2-WAY ANOVA, including the NONPARAMETRIC, FRIEDMAN and QUADE tests.

Random variable

A key concept in statistics. A variable with random fluctuations, such that its value at any particular point cann6ot be described or predetermined by any mathematical function.

Random walk [Davis 1986, p.314; Raup 1981, p.271; Schwarzacher 1978,p.66,128 etc.]

A STOCHASTIC PROCESS in which a zigzag pattern is traced on some diagram or map by the movement of an object or changing of the values of a variable. For example, if a particular object can either move upwards or downwards from any given point with equal probability, the zigzag trace of its actual position on an X-t (position-time) graph over a period of time will form a *simple random walk*. Widely used in sedimentology and palaeontology. Random walks can also be used to set up random (unbiased) sampling traverses. Related to MARKOV PROCESSES.

Range [Kendall & Buckland 1982]

The difference between the maximum and minimum values of a data-set.

Range of influence [Clark 1979, p.6] 272

A distance in GEOSTATISTICS which indicates how far spatial dependence extends between data-points. That is, the ore-value at a point A only affects the ore-value at point B if point B lies within its range of influence. Beyond separation a, grades are independent of one another. It is determined by the distance at which the SEMI-VARIOGRAM reaches a SILL.

Range test [Mardia 1972, p.188]

A test for UNIFORMITY in 2-dimensional ORIENTATION DATA, based on the fact that the CIRCULAR RANGE for non-uniform data will be small.

Rank [Kendall & Buckland 1982]

The ordinal number of a datum in a data-set ordered according to some criterion. E.g. the smallest content of Sr in a group of 10 limestones has rank 1, the highest has rank 10.

Rank correlation [Kendall & Buckland 1982] 170

CORRELATION based on the degree of correspondence between two sets of ranks. See SPEARMAN's ρ and KENDALL's τ.

Ranking scale

A synonym for ORDINAL SCALE.

Rankits [Powell 1982,p.32]

If N random observations from a STANDARD NORMAL DISTRIBUTION are ranked, the expected values $E(N,k)$ for each rank k are called rankits. If $N = 20$, for example, the expected value of the smallest observation ($k = 1$) is -1.8675 and of the third largest ($k = 18$)1.1309. The values are symmetrical about the median, i.e.$E(N,k) = E(N,[N+1-k])$. A plot of observed values against rankits will yield a straight line for Normally-distributed data, and can thus be used as a simple, graphical Normality test. Sometimes termed NORMAL SCORES, but this is used in a related but different sense here. Used in the TERRY-HOEFFDING TEST.

Rank of matrix [Joreskog et al. 1976, p.36; Mather 1976, p.32]

The lowest common ORDER among all pairs of smaller matrices whose product is the matrix in question. E.g. if a 12 x 8 matrix can be expressed as the product of one 4 x 8 and one 12 x 4 matrix, but by no matrices of order <4, then its rank is 4. Alternatively, the number of non-zero EIGENVALUES of the matrix, or the number of linearly independent rows and columns it possesses. A fundamental concept in FACTOR ANALYSIS, as it relates to the possibility of expressing a large data-array in terms of smaller arrays.

Rank-order statistics [Kendall & Buckland 1982]

Statistics based only on the ranks of data, not their actual values: e.g. rank correlation coefficients. Cf. ORDER STATISTICS.

Rank sum test (Wilcoxon's) [Dixon & Massey 1969, p.344]

A variant of the MANN-WHITNEY TEST which uses a slightly different formula but produces the same results.

Rao's *R* statistic [Le Maitre 1982, p.152]

A statistic distributed approximately as FISHER's F, used to test the equality of MULTIVARIATE means. A function of WILKS' LAMBDA.

Rao's test for homogeneity of correlation coefficients [Miller & Kahn 1962, p.224] 178

A PARAMETRIC test suitable for comparing 2 or more PRODUCT-MOMENT CORRELATION COEFFICIENTS for groups of data of different sizes.

Ratio scale/variable 66

The highest scale of measurement, with all the characteristics of an INTERVAL scale, plus a true zero point. Mass, weight, chemical composition, volume, etc. are all measured on ratio scales. The ratio of any two ratio scale points is

Page

independent of the unit of measurement. All mathematical operations (=,<,>, *,/, etc.) can be performed on ratio measurements.

Rayleigh's test [Mardia 1972, p.133; Kendall & Buckland 1982; Cheeney 1983, p.100] 232

A PARAMETRIC test of UNIFORMITY for ORIENTATION data. The uniformly most powerful test where the mean direction is unknown.

Record

Sometimes used synonymously INDIVIDUAL, OBJECT, etc., and sometimes with CARD-IMAGE: i.e. may refer to one *or more* 80-column CARD-IMAGES containing all or parts of the data for one individual in a computer data-file (especially in DATABASE work).

Reduced major axis, RMA [Till 1974, p.99; Le Maitre 1982, p.54; Williams & Troutman 1987] 193

A 'STRUCTURAL' REGRESSION linking two variables, both with associated errors, in which the triangular areas between the data-points and the regression line are minimised in size; corresponding LAMBDA (λ) value = [Variance (Y)/Variance(X)].

Redundancy analysis [Van der Wollenberg 1977]

An alternative to CANONICAL CORRELATION in measuring relationships between two *sets* of variables. Measures the average proportion of variance in one set which is predictable by the other set. Redundancy measures, unlike canonical correlations, are *not* symmetric (measures calculated for each direction will not be equal).

Reference structure solution [Cuneo & Feldman 1986, p.122]

An OBLIQUE ROTATION of axes in FACTOR ANALYSIS, defining loadings that are correlations. Closely related to the PRIMARY PATTERN SOLUTION.

Regionalized variable [Matheron 1971; Davis 1986, p.239] 272

A variable which has properties intermediate between a RANDOM and a DETERMINISTIC variable. Typically refers to geological phenomena that have geographic distributions, such as topographic height, grade of an ore body, or K_2O content of a granite batholith. Unlike random variables, regionalized variables have continuity from place to place, but their variations cannot be described by any simple deterministic mathematical function. The fundamental concept behind GEOSTATISTICS.

Region of rejection [Kendall & Buckland 1982]

In hypothesis testing, a region of sample space such that if a sample point falls within it the hypothesis under test is rejected.

Regression [Kendall & Buckland 1982] 187ff

The detailed study of the way in which one DEPENDENT VARIABLE is controlled by one or more INDEPENDENT VARIABLES. CORRELATION merely measures the strength of association between variables, whereas regression estimates the actual equation governing the relationship, with its CONFIDENCE LIMITS, etc. Includes BIVARIATE and MULTIPLE, linear and POLYNOMIAL, GROUPED REGRESSION and various other types. Many geological relationships, in which neither variable is dependent, are not strictly assessed by classical regression methods but by STRUCTURAL solutions.

Related data (samples, groups) 72

Broadly, groups of data consisting of MATCHED PAIRS/triplets, ie in which something is held constant between the measurements. E.g. Compositions of weathered and unweathered rock samples from the same outcrop, responses of the same set of students to a given set of questions before and after a training course.

Relation 45

Equivalent in a RELATIONAL DATABASE SYSTEM to a 2-dimensional table of data on a sheet of paper.

Relational database system [Date 1981] 45

A database in which RECORDS are grouped into one or more tables ('relations'), each record being uniquely identified by a PRIMARY KEY. Data to be incorporated into such a system have to comply with the concept of 'Third normal form', which effectively requires that all data-items be independent.

Relative efficiency of an estimator [Kendall & Buckland 1982]

If estimator E_1 has the same precision (i.e. sampling variance) for sample size n_1 as estimator E_2 has for sample size n_2, then the relative efficiency of E_1/E_2 is n_2/n_1.

Relative efficiency of a test, RE [Conover 1980, p.88; Kendall & Buckland 1982]

If T_1 and T_2 represent two tests of the same H_0 against the same H_1, with the same SIGNIFICANCE LEVELS and POWERS (i.e. the same α and β ERRORS), then the *RE* of T_1 to T_2 is n_2/n_1, the ratio of the two sample sizes involved. See also POWER-EFFICIENCY and ASYMPTOTIC RELATIVE EFFICIENCY.

Relative error

A synonym for COEFFICIENT OF VARIATION.

Remaining space variables [Chayes 1981,1983c] 206

Transformations of closed data which consider variations only within the 'remaining space' not occupied by the predominant and most variable constituent. For example, in rock analyses, SiO_2 has the highest value and variance, and the 'remaining space' amounts to [100% - SiO_2]. Niggli numbers are effectively remaining space variables, being the

Page

molar proportion of a particular oxide in that part of the rock which is *not* SiO_2. They help to reduce the CLOSURE effect in percentage data. Further, component A of a closed array is said to be 'neutral' with respect to component B if A and B/(1-A) are independent. Their mutual correlation can then be tested against a NULL CORRELATION of zero.

Repeated measures design [Cuneo & Feldman 1986] 164

A type of ANOVA in which the same TREATMENT is repeated several times on one or more BLOCKS. For example, one might make repeated measurements of the Ba content of one set of rocks by several analytical methods (XRF, AAS, etc.)

Replacement (sampling with) [Kendall & Buckland 1982] 81

Sampling from a finite population in which the individual is returned to that population after its characteristics have been recorded, and before the next individual is withdrawn. Its converse is '*Sampling without replacement*'.

Replicate data

Repeat measurements of a particular variable on the *same* specimen (e.g. Sr determinations on an international standard rock at different laboratories). Such values should be stochastically identical and Normally distributed in practice.

Resistant [Kotz & Johnson 1982, Vol.2, p.579]

Refers to the insensitivity of a test or statistic to localized misbehaviour in data — changing a small part of the data produces only minimal changes in the test or statistic. Often confused with ROBUST, which is similar but slightly different in meaning. For example, the MEDIAN is more *resistant* than *robust* as an 'average', since other estimates of LOCATION are more efficient over a wider range of distributions, but also more susceptible to bad behaviour. The ARITHMETIC MEAN, by contrast, is neither resistant nor robust.`

Response [Davis 1986]

A plot of the weights placed on each observation adjacent to the one of interest, in a FILTER. Illustrates the 'shape' of the filter.

Response variable

Synonymous with DEPENDENT VARIABLE.

Resultant length, *R* 228

See under CIRCULAR MEAN.

Ridge parameter

See next entry.

Ridge regression [Mather 1976, p.74] 257

A form of MULTIPLE REGRESSION which attempts to avoid problems caused by MULTICOLLINEARITY, that is, highly correlated independent variables. The resulting estimates of the slope and intercept of the regression line may be slightly biased, but often more precise than those obtained via ordinary LEAST-SQUARES, while the correlation coefficient may be closer to the true parameter. The procedure introduces a *ridge parameter*, θ, which controls the extent of the bias introduced. For $\theta = 0$, the estimates are the same as for least-squares. The method looks for a value, as θ increases (usually remaining < 1), where the resulting estimates change as slowly as possible.

R-mode analysis 72

That version of any technique which concentrates on relationships between VARIABLES, rather than INDIVIDUALS. Cf. Q-MODE.

Robust estimates [Andrews et al. 1972] 87

Estimates of LOCATION, SCALE etc. which are not as sensitive as conventional estimates to OUTLIERS, or to deviations from Normality, but use more information in the data than NONPARAMETRIC estimates based on ranks.

Robust(ness) [Kendall & Buckland 1982] 69

Refers to the extent to which the POWER of a test or statistic is resistant to (and unaffected in efficiency after) deviations from its inherent assumptions. E.g. PARAMETRIC tests carry several assumptions (Normal distributions, equal variances, etc.) and are far less robust to deviations from Normality than NONPARAMETRIC tests (which mostly carry few or no such assumptions). The following are placed in order of robustness as estimates of DISPERSION: STANDARD DEVIATION < MEAN DEVIATION < MEDIAN DEVIATION. Cf. RESISTANT.

Robust regression [Huber 1981] 194

A form of PARAMETRIC REGRESSION in which OUTLIERS are eliminated by TRIMMING, WINSORIZING, etc., to reduce the instability of the solution.

Rogers & Tanimoto coefficient [Coxon 1982, p.27; SPSS 1986, p.739] 168

A MATCHING COEFFICIENT, defined as $(a+d)/[a+d+2(b+c)]$ -see Table 13.5.

Rose diagram [Mardia 1972, p.5; Kendall & Buckland 1982] 227

A circular plot displaying 2-dimensional DIRECTIONAL data, in which sectors are constructed with apices at the centre of the plot, whose radii (normal rose diagram) or areas (equi-areal rose diagram) are made proportional to the class frequency. Some authors (e.g. Cheeney 1983) advocate the equi-areal rose diagram, others (Mardia 1972) the normal one. Cheeney also uses the term CIRCULAR HISTOGRAM synonymously with rose diagram, whereas Mardia makes a distinction between them.

Roy's maximum root [SAS 1985, p.12]

Page

A statistic used to test the overall significance of a multivariate REGRESSION. Equivalent to an F test on a simple bivariate regression.

Runs test [Siegel 1956, p.136] 132

A NONPARAMETRIC ONE-SAMPLE or TWO-SAMPLE TEST based on grouping all data-values together and counting the number of 'runs' (contiguous subsets of values from one group). Usually less powerful than the MANN-WHITNEY or SMIRNOV tests, but one of few tests with is valid equally for LINEAR and ORIENTATION data.

Russell & Rao coefficient [SPSS 1986, p.730] 168

A MATCHING COEFFICIENT, defined as $a/(a + b + c + d)$ — see Table 13.5.

Sample [Kendall & Buckland 1982] 69

Statistically, a finite part of a POPULATION provided by some selection process to study the properties of the parent population. Distinct from the common geological use of the word to refer e.g. to a single *physical* rock sample (better termed a SPECIMEN) from an outcrop. However, the distinction in some geological textbooks between a SPECIMEN (e.g. a single fossil) and a SAMPLE (a collection of fossils from a bed) is rarely adhered to in the geological literature, especially as, for example, stream sediment 'samples' are never described as 'specimens' but are as much single entities as rock 'specimens'. Generally, 'sample' as used in statistics texts corresponds to 'group of data' in geology, and 'sample' in geology to 'individual' 'object', 'case' or 'item' in statistics.

Sample arc length [Mardia 1972, p.172 & 187]

In an ordered set of 2-dimensional ORIENTATION DATA, the spacings between successive measurements, i.e. $X_n - X_{n-1}$. Used in tests for uniformity such as the RANGE and EQUAL SPACINGS tests.

Sampled population [Koch & Link 1970, p.45]

The population actually sampled, about which we can make statistical inferences from study of the sample itself. Cf. TARGET POPULATION.

Sampling adequacy [Cuneo & Feldman 1986, p.117]

See MATXIR SAMPLING ADEQUENCY, VARIABLE SAMPLING ADEQUACY.

Scalar data

A synonym for LINEAR DATA.

Scale 91

The spread or variability of data, most commonly measured by the STANDARD DEVIATION. Essentially synonymous with DISPERSION in common usage, but actually refers to the *absolute* rather than *relative* spread of data. For example, re-expressing a given set of data in *gms* rather than *kg* increases the scale (the data have been 'scaled up', in common parlance), but not the dispersion, i.e. the inherent *internal* variability. In general, if all values in a data-set are increased by adding a constant c, the scale and dispersion both remain unchanged, but if they are multiplied by a constant k, the scale is also multiplied by k, whereas the dispersion remains the same.

Scheffé's *F* test [Miller & Kahn 1962, p.173; Koch & Link 1971, p.196; Steel & Torrie 1980, p.183] 142

A conservative PARAMETRIC test simultaneously comparing any number of means in ONE-WAY ANOVA. Similar to a two-group t test but for >2 means. Normally employed to identify *which* of >2 means is unequal after a significant F-TEST has indicated inequality overall.

S-distribution curves [Lister 1982]

A simple graphical representation of data, in which N data-values are plotted in order along the X-axis, against a Y axis running from 1 to N. A NORMAL DISTRIBUTION will then yield an S-shaped curve.

Semi-interquartile range 91

Half the INTERQUARTILE RANGE, often preferred to the latter as a measure of DISPERSION as its absolute magnitude is nearer to that of other common measures such as the STANDARD DEVIATION.

Semistrong monotonicity [Kotz & Johnson 1982-8] 184

A MONOTONIC X-Y relationship in which, if $X_i > X_j$ then $Y_i \geq Y_j$ and if $X_i = X_j$ then $Y_i = Y_j$.

Semivariogram [Clark 1979, p.5ff; Davis 1986, p.239] 272

A plot of distance (X) against a form of the VARIANCE (γ) of an economic material over a deposit (e.g. Au content), used in GEOSTATISTICS. Indicates the distance over which the ore value at one particular point in the deposit influences values at adjacent points. Various types exist, including DE WIJSIAN, SPHERICAL, LINEAR, MATHERONIAN and CRESSIE–HAWKINS. Semivariograms are usually calculated for each major directions (E–W, N–S) in a deposit; only if the deposit is isotropic will these different semivariograms be equivalent.

Semiweak monotonicity [Kotz & Johnson 1982-8] 184

A MONOTONIC X-Y relationship in which, if $X_i > X_j$ then $Y_i > Y_j$ and if $Y_i = Y_j$ then $X_i = X_j$.

Serial correlation [Davis 1986, p.220]

An ambiguous term, used in at least 3 different senses by different authors: (1) synonymously with AUTO-CORRELATION; (2) to refer to the corresponding POPULATION PARAMETER of which autocorrelation is the SAMPLE STATISTIC; (3) synonymously with CROSS-CORRELATION. Best avoided.

Seriation [Davis 1986, p.214]

Page

Refers to the placing of observations in some logical (usually chronological) order by comparing their similarities. SLOTTING two stratigraphical columns together, by pairing each point in one sequence with the most similar point in the other, is an example, since the final arrangement is both lithologically and chronologically meaningful.

Shapiro–Wilk *W* test [Shapiro & Wilk 1965; Conover 1980, p.363] 97

One of the most powerful NONPARAMETRIC tests that a sample distribution is NORMAL. Has the advantage that several samples, too small to be tested individually, can be combined into one overall test.

Shepperd's filter [Davis 1986]

A FILTER which uses a 5-term equation (i.e. the 2 values either side of the value in question are taken into account).

Shorth (shortest half) mean [Kendall & Buckland 1982] 89

A ROBUST ESTIMATE of LOCATION, defined as the arithmetic mean of the 'shortest half' of the data, that is, that consecutive 50% of the data-values whose RANGE is least.

Short-tailed [Kotz & Johnson 1982-8]

Opposite of LONG-TAILED (qv). Refers to distributions with relatively few extreme values, i.e. few values in the 'tails' of the distribution curve. Short-tailed distributions tend to be LEPTOKURTIC.

Shotgun approach [Brower & Veinus 1964] 328

Exploratory data-analysis in which as many techniques as possible are applied to the data, in a 'suck it and see' manner. Cf. ZAP APPROACH.

Sichel's *t* estimator [David 1977 p.14,18]

An ESTIMATE of LOCATION specifically designed for the LOGNORMAL distributions commonly encountered in mining and exploration, to overcome the tendency of the ARITHMETIC MEAN to overestimate the 'average'. It is a function of the MEDIAN (equivalent for a logNormal distribution to the GEOMETRIC MEAN), *N*, the numbers of values, and $ß^2$, the logarithmic VARIANCE.

Significance level [Kendall & Buckland 1982] 73

The probability that a given hypothesis is true. E.g. if we state that X and Y are statistically equal at the 95% significance level, we believe ourselves to be correct in 19 out of 20 cases. The size of the REJECTION REGION. A 95% level is sometimes expressed as 0.95, as 5% or as 0.05; it is least confusing is we speak of *95% significance* and *5% probability* as synonymous alternatives.

Sign test [Siegel 1956, p.68; Conover 1980, p.122; Kendall & Buckland 1982] 124

A simple NONPARAMETRIC test for two RELATED SAMPLES. Not powerful, as it does not use the actual data-values, but only the sign of their differences. Also used in the determination of NONPARAMETRIC CONFIDENCE INTERVALS for the MEDIAN.

Sill [Clark 1979, p.6] 272

The VARIANCE, γ(h), value at which an ideal SEMIVARIOGRAM flattens off. Equals the ordinary variance of the ore grades.

Similarity coefficient [Everitt 1974, p.50; Till 1974, p.134; Davis 1986]

A coefficient comparing INDIVIDUALS (Q-mode) or VARIABLES (R-mode), on quantitative, RATIO or INTERVAL data. E.g. COSINE THETA, PEARSON's *r*. Values near 1 indicate close similarity. Cf. MATCHING COEFFICIENT.

Simple kriging 273

KRIGING which assumes no spatial trend in data. Cf. UNIVERSAL KRIGING.

Simple matching similarity measure [SPSS 1986, p.739]

A MATCHING COEFFICIENT, defined as $(a+d)/N$ —the ratio of the number of MATCHES to the total number of attributes (see Table 13.5).

Simplex space [Le Maitre 1982, p.46]

A restricted region of multidimensional space.

Simpson coefficient [Mather 1976, p.314] 168

A MATCHING COEFFICIENT used with DICHOTOMOUS data. Defined as p/n_1, where p is the number of POSITIVE MATCHES and n_1 is the smaller of the total numbers of attributes in the two samples. Conservative in the sense that it measures similarity rather than dissimilarity.

Single link(age) technique [Davis 1986, p.510] 292

An AGGLOMERATIVE method of CLUSTER ANALYSIS which fuses individuals to clusters on the basis of highest similarity between the individual and individuals already in the cluster.

Single-sided test [Kendall & Buckland 1982] 73

See ONE-SIDED TEST.

Singular matrix [Joreskog et al. 1976, p.23]

A matrix whose DETERMINANT is zero, i.e. which has at least one zero EIGENVALUE. Often occurs in matrices of CLOSED data. Its only INVERSE is a special form called a GENERALISED INVERSE.

Singular value decomposition [Joreskog et al. 1976; Kennedy & Gentle 1980, p.278, p.47]

The concept that a matrix of RANK *r* can be expressed as a linear combination of matrices of lesser rank (e.g. 1). Sometimes referred to as the ECKART-YOUNG THEOREM. Given the real n x p matrix **X**, there exists an n x n

ORTHOGONAL matrix **U**, a p x p orthogonal matrix **V**, and an n x p DIAGONAL matrix **S** having non-negative elements such that **U** x **V** = **S** or, equivalently, **X** = **USV**. Singular value decomposition determines the matrices **U**, **V** and **S**. Fundamental in FACTOR ANALYSIS.

Singular values [Kennedy & Gentle 1980, p.278]
The diagonal elements of a matrix, with equal row and column positions.

Size difference [SPSS 1986, p.741]
A MATCHING COEFFICIENT, defined as $(b-c)^2/N^2$ — see Table 13.5. It has a minimum of 0 but no upper limit.

Skew(ness) [Kendall & Buckland 1982] 92
Asymmetry in a frequency distribution. Positive skew indicates a long tail towards high values, negative skew a long tail towards low values. The most common measure, often notated as '$\sqrt{b_1}$', is defined as $\sqrt{N}S_3/S_2^{1.5}$, where S_3 and S_2 are the 3rd and 2nd MOMENTS. Another is defined as 3(Mean − Median)/Standard deviation. A third is the INTERQUARTILE COEFFICIENT OF SKEWNESS. Yet another is used in DAVIE'S TEST. Skew has zero value for a NORMAL distribution. It can be used as a simple TEST OF NORMALITY, preferably in combination with other tests, since it is only sensitive to asymmetry. Can be generalized to the MULTIVARIATE case [Kotz & Johnson 1982-8, vol.6,p.122ff].

Skip mean [Tukey 1977, p.674]
In smoothing, the mean of X_{n-1} and X_{n+1}, placed at n.

Slice [Tukey 1977, p.674]
Those (X,Y) pairs whose X-values lie between some chosen cut-off limits.

Slotting [Gordon & Reyment 1979; Davis 1986, p.214]
Usually refers to the matching of two borehole sequences, via pairing each point in one with the most similar lithology in the other, and shuffling the sections together like a deck of cards whilst preserving stratigraphical order.

Smirnov (two-group Kolmogorov-Smirnov) test [Smirnov 1939; Siegel 1956; Conover 1980, p.369] 133
A powerful NONPARAMETRIC test based on the maximum deviation between two cumulative distribution functions, for comparing two INDEPENDENT GROUPS of data.

Snedecor's *F* 85
A synonym for FISHER'S *F*.

S-N-K test
See STUDENT-NEUMAN-KEULS test.

Sokal & Sneath's coefficients [Till 1974, p.134; SPSS 1986, p.739] 168
A series of 5 MATCHING COEFFICIENTS (see Table 13.5), defined as: (1) $2(a+d)/([2(a+d)+b+c]$; (2) $a/[a+2(b+c)]$; (3) $(a+d)/(b+c)$,which is undefined with $b=c=0$; (4) $[a/(a+b)+a/(a+c)+d/(b+d)+d/(c+d)]/4$, yielding the conditional probability that an attribute is in the same state (presence or absence) in both objects; (5) $ad/\sqrt{(a+b)(a+c)(b+d)(c+d)}$. All except (3) range from 0 to 1; (3) has a minimum of 0 but no upper limit.

Somers' *d* coefficients [Somers 1962; Blalock 1972, p.424; SAS 1985,p.415] 168
Two asymmetric COEFFICIENTS OF CONCORDANCE, which can be though of as slope analogues in MULTIPLE REGRESSION. A modification of GOODMAN & KRUSKAL's τ_b, correcting for ties on the independent variable.

Spearman's rho (ρ) rank correlation coefficient [Spearman 1904; Kendall & Buckland 1982] 170
A NONPARAMETRIC measure of STRONG MONOTONICITY using only the ranks of data-values, not their absolute values. Equivalent to a PRODUCT-MOMENT CORRELATION COEFFICIENT calculated with ranks rather than raw data. Ranges from -1 to +1, but is particularly useful because values near unity can indicate trends other than merely linear (e.g. linear, exponential and reciprocal trends will all yield high values of ρ because they are monotonic).

Specimen 68
Generally synonymous with INDIVIDUAL, OBJECT ITEM etc. Better than SAMPLE in Numerical Geology as it avoids confusion with *statistical* usage of the term sample.

Spectral density function [Davis 1986, p.258]
See CONTINUOUS SPECTRUM.

Spencer's filter [Davis 1986]
A FILTER which uses a 21-term equation (i.e. the 10 values on either side of the value in question are taken into account).

Spherical semivariogram [Clark 1979, p.6] 272
A SEMIVARIOGRAM which rises in a smooth curve to a flat SILL.

Splines [Rogers & Adams 1976; Tipper 1979b; Davis 1986, p.204] 225
Continuous piecewise fitted curves, numerically reproducing the drafting function of 'french curves'. For example, a cubic polynomial $Y = aX + bX^2 + cX^3$ can be made to pass exactly through 4 points, but to fit a continuous curve through, say, 6 points, the curve can be divided into two segments, and *two* cubic polynomials made to fit through 3 points each, such that the slopes of the curves at the point of meeting are the same. The mathematical description of the segmented curve is called a *spline function*. Named from draftsmens' flexible plastic splines which can be bent to confirm to an irregular shape. Very important in computer graphics.

Page

Spread
Synonymous with DISPERSION.

Spurious correlation 185
A statistically significant correlation induced by the method of handling of the data, and not present in the data themselves. E.g. negative correlations between SiO_2 and most major oxides in igneous rock analyses are largely due to CLOSURE (100% totals), not to any intrinsic relationship.

Squared ranks test [Conover 1980, p.239] 130
A NONPARAMETRIC test assessing whether the DISPERSIONS of two or more INDEPENDENT groups of data are equal or HOMOGENEOUS. The nonparametric equivalent of the *F* TEST.

Square matrix
A matrix with an equal number of rows and columns.

Standard deviation [Kendall & Buckland 1982] 91
The most widely used measure of DISPERSION of a frequency distribution, equal to the positive square root of the VARIANCE.

Standard error (of the mean) (SE)
The value $s/\sqrt{N}$, where s is the standard deviation and N the number of data-values.

Standardization 77
Conversion to a STANDARD (NORMAL) DEVIATE.

Standard(ized) (Normal) deviate {snd} [Kendall & Buckland 1982] 77
The value $(X_i-\overline{X})/s$ where X_i is a data-value, $\overline{X}$ = mean data-value, and s = standard deviation. Also termed *Z-score.*

Standardized range [Powell 1982, p.34]
RANGE/σ, i.e. same as STUDENTIZED RANGE, but using sample rather than population standard deviation.

Standard Normal distribution 76
A NORMAL DISTRIBUTION with mean = 0 and standard deviation = 1. Synonymous with UNIT NORMAL DISTRIBUTION. Widely used in hypothesis testing.

Star symbol plot 315
A useful way of representing and comparing multivariate data graphically. Each star represents one OBJECT, and consists of a series of rays drawn from a central point, with each ray representing one variable and having length proportional to the value of that variable. Somewhat similar to a SUNRAY PLOT.

State
Generally synonymous with ATTRIBUTE, but usually applied to sequences. E.g. the 'state' of a stratigraphic succession is the lithology at a particular point.

Stationary/stationarity [Davis 1986, p.221,259]
Refers to a sequence of data showing no significant trend (in space or, for a TIME-SERIES, time). If a sequence is divided into small segments and their means are subequal, the sequence is *first-order stationary.* If in addition the AUTOCOVARIANCE is invariant along the sequence, it is *second-order stationary* ('weakly stationary'). If higher-order moments are invariant along the sequence, it is 'strongly stationary'. *Self-stationarity* refers to a time-series showing constant statistics from segment to segment.

Statistic [Kendall & Buckland 1982] 69
A summary value calculated from a GROUP of data (sample), often (but not necessarily) as an estimator of some POPULATION PARAMETER.

Statistical outlier 113
An outlier identified by purely statistical means, i.e. a test such as DIXON's, HARVEY's or GRUBBS'. Geological reasoning is then required to classify the outlier as TRUE, FALSE or BIZARRE, so as to decide what to do with it.

Stem-and-leaf plot [Tukey 1977; Hoaglin et al., 1983] 93
An alternative form of HISTOGRAM used in graphical EXPLORATORY DATA ANALYSIS.

Stepwise regression [Kennedy & Gentle 1980, p.335; Kendall & Buckland 1982] 257
Part of MULTIPLE REGRESSION analysis, which seeks to identify the most important INDEPENDENT VARIABLES in determining the variation of a DEPENDENT variable, by successive elimination of the former.

Stochastic [Kendall & Buckland 1982] 75
Implies the presence of a random factor; e.g. STOCHASTIC PROCESS. A stochastic variable/sequence/process cannot be exactly predicted or modelled at a particular point in time or space, owing to this random element.

Stochastic model [Krumbein & Graybill 1965, p.19; Kendall & Buckland 1982] 75
A model which incorporates some STOCHASTIC elements.

Stochastic process 75
A process controlled by random factors — e.g. Brownian motion.

Stochastic variable
A random variable.

Stress [Mather 1976, p.334] 298

Page

A measure used in MULTIDIMENSIONAL SCALING to assess the goodness-of-fit of the graphical representation in reduced space, to the original data-matrix. The technique aims to minimise this measure.

Strong monotonicity [Kotz & Johnson 1982-8] 184

A MONOTONIC *X-Y* relationship in which, if $X_i > X_j$ then $Y_i \geq Y_j$ and if $X_i = X_j$ then $Y_i = Y_j$. Commonly measured by SPEARMAN's ρ or KENDALLS' τ.

Structural regression [Le Maitre 1982, p.54] 192

'Regression' involving two random variables, both subject to error, i.e. where there is no INDEPENDENT variable. E.g. Rb-Sr isochron calculation, or the analysis of the relationship between mass and volume of a rock-type, using measurements on a set of different specimens. Cf. FUNCTIONAL REGRESSION.

Stuart's τ_c[SAS 1985,p.414]

A COEFFICIENT OF CONCORDANCE, related to GAMMA and to GOODMAN & KRUSKAL's τ_n, but making a correction for table size as well as ties.

Studentization

The process of removing complications due to the existence of an unknown parent scale-parameter, by constructing a statistic whose sampling distribution is independent of scale. Generally involves dividing by the standard deviation. See next entries.

Studentized range [Kendall & Buckland 1982]

The +ve range divided by the standard deviation, used in several PARAMETRIC tests such as TUKEY's *w* PROCEDURE and in tests for OUTLIERS.

Studentized residual [Hampel et al. 1986]

See GRUBBS' TEST.

Student-Newman-Keuls' (S-N-K) test [Steel & Torrie 1980, p.186; Kotz & Johnson 1982-8] 143

An iterative PARAMETRIC test for deciding which of several means is unequal, after a significant F ratio has been obtained (indicating overall inhomogeneity of the means) in a ONE-WAY ANOVA. The maximum and minimum mean values are first compared; testing stops if these are declared homogeneous, but continues with the next less-different means if they are declared unequal, until a homogeneous pair is found. The test uses the STUDENTIZED RANGE.

Student's *t* distribution [Kendall & Buckland 1982] 84

The distribution of the ratio of a sample mean to a sample variance, multiplied by a constant, in samples from a NORMAL DISTRIBUTION. See also *t*-TEST.

Sturge's rule [Kendall & Buckland 1982]

An empirical rule for assessing the desirable number of frequency classes into which a data-set should be cast when plotting a HISTOGRAM. If *N* is the number of values, and *K* the number of classes, then $K = 1 + 3.3 \log\{N\}$. I.e. 100 data-values should be divided into 8 or more classes.

Subarea analysis [Davis 1986]

Analysis of the distribution of points on a map by dividing the map into a number of equal subareas (quadrats).

Substitutability analysis [Davis 1986, p.276] 223

A technique determining whether two or more STATES are somehow related, in the sense that they can proxy or substitute for one another in a sequence. E.g. if the lithological sequences Sandstone-Limestone-Shale and Sandstone-Marl-Shale are repeatedly found in a stratigraphical succession, Limestone and Marl are showing substitutability. It is usually measured by a a coefficient ranging between 0 and 1 (for complete substitutability).

Substitute *F*-test [Dixon & Massey 1969, p.139] 123

An INEFFICIENT test in which the usual mean-square estimators of VARIANCE are replaced by estimates based on the RANGE.

Substitute *t*-test [Dixon & Massey 1969, p.139; Miller & Kahn 1962, p.103] 123

A modified form of the *t*-TEST in which the numerator and denominator of the *t*-statistic are replaced by more easily calculated statistics such as mean ranges. For example, LORD'S TEST.

Sunray plot 315

A graphical means of representing and comparing MULTIVARIATE data, somewhat similar to a STAR SYMBOL PLOT, except that polygon sides rather than rays represent the value of each variable. Each ray is scaled so that a side will intersect it in the middle if a variable is exactly equal to the sample mean. The extreme points on each ray represent the standard deviations at a specified multiple.

Symmetrical matrix [Joreskog et al. 1976, p.18]

A matrix in which elements in the ith row and jth column equal those in the jth row and ith column, for all i and j.

Systematic error [Kendall & Buckland 1982]

Errors due e.g. to incorrect machine calibration etc., causing loss of ACCURACY. Cf. RANDOM ERRORS.

Target population [Koch & Link 1970, p.45]

The population of interest, about which we wish to make geological inferences, as opposed to the SAMPLED POPULATION, which we have actually sampled. For example, in studies of the Skaergaard intrusion, the sampled population would presumably be the present outcrop of the intrusion. The target population could include the Hidden

Page

Border Series (beneath the present exposure level), that part of the intrusion already removed by erosion, etc., depending on the object of study. The difference is important, for we can only make statistical extrapolations from sample to sampled population, whereas we really want to extrapolate to the target population.

Tax map [Everitt 1974, p.31]

A DENSITY SEARCH technique of CLUSTER ANALYSIS, imitating human perception of clusters in 2- or 3-dimension, which compares relative distances between point points and searches for continuous, relative densely populated regions surrounded by continuous relatively empty regions.

Terry-Hoeffding Normal scores test [Powell 1982, p.73] 129

A NONPARAMETRIC test for comparing the LOCATION of two groups of data, using the same procedures as the MANN-WHITNEY TEST, but based on NORMAL SCORES instead of raw RANKS, and hence asymptotically more efficient. A TWO-GROUP equivalent of the VAN DER WAERDEN TEST.

Tertiles (upper and lower)

Extreme values left when the outer third of values from both extremes of a data-set have been removed.

Test of Normality

See NORMALITY TEST.

Tetrachoric correlation coefficient [Downie & Heath 1974, p.109; Dixon & Massey 1969, p.216]

A CONTINGENCY COEFFICIENT for continuous data forced into a dichotomy. In a 2x2 CONTINGENCY TABLE, the value *ad/bc*.

Theil's complete method [Sprent 1981, p.231] 194

A form of NONPARAMETRIC REGRESSION. See next entry.

Theil's incomplete method [Sprent 1981, p.227] 195

A form of NONPARAMETRIC REGRESSION, alternative to conventional PARAMETRIC, LEAST-SQUARES METHODS. It is based on a ranked set of slopes b_{ij}, formed between a subset of paired data-values, $\{Y_i,X_i\}$ and $\{Y_j,X_j\}$, where $b_{ij} = [Y_i - Y_j] / [X_i - X_j]$, and i,j extend over the values around the MEDIAN X and Y.

Theory of regionalized variables [Matheron 1971]

The statistical theory from which KRIGING derives.

Thick-tailed [Kotz & Johnson 1982-8]

Synonymous with LONG-TAILED.

Thickened range [David 1981, p.191]

A linear combination of QUASI-RANGES, $W = [W_1 + W_2 + ... W_i]$, used to increase the ROBUSTNESS of the quasi-range as a quick measure of SCALE.

Threshold [Miesch 1981] 117

A geochemical cutoff designed to separate 'background' values in a data-distribution from 'anomalous' values (OUTLIERS) which represent known or potential mineralization. The upper limit of normal background variation. Recognised via a GAP STATISTIC.

Time-series 209

Any sequence of data in which time is one of the variables (e.g. numbers of earthquakes over a number of years; a seismometer trace; monthly figures for discharge from a water-course).

Time-trend analysis [Davis 1986, p.268]

Synonymous with 'filtering', i.e the application of FILTERS to one-dimensional sequences of data.

Total matrix sampling adequacy [Cuneo & Feldman 1986, p.118]

See MATRIX SAMPLING ADEQUACY.

Trace of a matrix [Joreskog et al. 1976, p.17]

The sum of the elements of the principal diagonal of a SQUARE matrix.

Transformation [Le Maitre 1982, p.14] 99

Recalculation of variables by e.g. a logarithmic or power function, usually so as to make the distribution more Normal. Widely used in applied geochemistry, geostatistics, etc., so that statistical analysis which assumes Normal distributions can be applied to non-Normal (especially logNormal) geological data.

Transition (probability) matrix [Davis 1986, p.276; Till 1974, p.11] 222

A matrix expressing the probabilities of transitions from one state to another in a time or spatial sequence of ATTRIBUTES. E.g. the probabilities of transition from limestone to sandstone or coal in a sedimentary succession.

Transpose of a matrix

Interchange of the rows and columns of a matrix; i.e. 3 x 6 becomes 6 x 3.

Treatment [Conover 1980, p.294; Kendall & Buckland 1982; SPSS 1986, p.513] 153

A term used for experimental designs, to cover stimuli applied to observe their effect or to compare with other treatments. In geology, generally applies to things over which the geologist has some sampling control, as opposed to BLOCKS which are irrevocably determined by Nature. See BLOCK.

Trees and castles 315

An effective graphical method of displaying multivariate data. Has been used for example in applied geochemistry to

Page

summarise stream sediment compositions, a tree being placed on a locality map at each sampling point (Fig.23.1).

Trend surface 269

The two-dimensional regression surface arising from TREND SURFACE ANALYSIS.

Trend surface analysis [Davis 1986, p.405] 269

A technique used in map analysis to separate 'regional' trends/variations from local fluctuations. MULTIPLE REGRESSION in which the INDEPENDENT VARIABLES are locations (depths, latitudes, etc.)

Trimean [Andrews et al. 1972; Tukey 1977, p.675] 88

A ROBUST M-ESTIMATE OF LOCATION: 0.5 of the median added to 0.25 of the two QUARTILES (or HINGES).

Trimmed means [Andrews et al. 1972; Kendall & Buckland 1982] 88

Estimates which ignore the extremes of a data distribution. E.g. the 25% trimmed mean (MIDMEAN) ignores the upper and lower 25% of values. The 50% trimmed mean is the MEDIAN. Trimmed means are more ROBUST than arithmetic means. All are ROBUST L-ESTIMATES of LOCATION.

Trimming 88

The process of ignoring/eliminating EXTREME VALUES (OUTLIERS) altogether. Cf. WINSORIZING.

True outlier [Rock 1988] 113

A data-value which, though revealed as a STATISTICAL OUTLIER, is nevertheless both accurate ('good' data) and *geologically* a true part of the data in question. Regional exploration geochemical programmes essentially aim to detect true outliers, which are pointers to mineral deposits. True outliers should *not* in general be eliminated from data-sets when performing statistical analyses. Their recognition, however, requires subjective geological reasoning as well as purely statistical treatment — save only in the case of replicate data-sets, where true outliers *cannot* occur Cf. BIZARRE and FALSE OUTLIERS.

Truncation [Kendall & Buckland 1982]

The elimination or ignoring of all values above or below a fixed numerical value, producing a *truncated distribution.* Cf. CENSORING, WINSORIZING. Implicit in many geochemical data-sets, where values below machine detection limits occur. If values quoted as, say, '<5 ppm' are ignored altogether, the data become truncated.

Tschuprow's *T* coefficient [Blalock 1972, p.296; Lewis 1977, p.94] 176

A general CONTINGENCY COEFFICIENT designed to lie between 0 (for no association) and 1 (for complete association in a square table). The upper limit decreases as the numbers of rows and columns in the table diverge.

***t*-test** 121

A test based on the STUDENT's *t*-DISTRIBUTION. A ONE SAMPLE *t*-test compares the mean of a group of data against that of some known standard, whereas a TWO-SAMPLE test compares two measured groups. A paired *t*-test is also available for comparing MATCHED-PAIRS.

Tukey's *q* test [Kendall & Buckland 1982]

A synonym for TUKEY's *w* PROCEDURE.

Tukey's quick *T* test [Sprent 1981, p.130; Kendall & Buckland 1982] 127

A simple test for comparing the LOCATION of two groups of data, based on the overlap of data-values (i.e. the number in one group which exceed all those in the other, and vice-versa). One of few statistical tests where the CRITICAL VALUES can be readily memorised.

Tukey's *w* procedure [Miller & Kahn 1962, p.165; Steel & Torrie 1980, p.185; Koch & Link 1971, p.194] 142

A PARAMETRIC procedure for pairwise comparison of 3 or more means, once inhomogeneity of the means has been indicated by ONE-WAY ANOVA. Makes use of the STUDENTIZED RANGE. The procedure then determines which means differ and which do not. Sometimes called *Tukey's honestly significant difference (HSD) test.* Similar to the STUDENT-NEUMAN-KEULS test, but more conservative since it does not take step sizes into consideration.

Two-group discriminant analysis [Le Maitre 1982, p.141] 307

More precise description of the technique usually described simply as DISCRIMINANT ANALYSIS. Cf. MULTIGROUP DISCRIMINANT ANALYSIS.

Two-group (sample) test 72

Any test comparing two groups (samples) of real data (see SAMPLE).

Two-sided (tailed) test 73

A test for which the REGION OF REJECTION comprises areas at both extremes of the sampling distribution of the test statistic. It is usual (but not essential) to allot half the probability of rejection to each extreme, giving a symmetrical test. Cf. ONE-TAILED TEST.

Two-way analysis of variance (ANOVA) 155ff

Analysis of variance in which there are two sources of variability. E.g. Pb isotope ratios might be measured for a range of minerals (the BLOCKS) from various geological terranes (the TREATMENTS). ANOVA will then determine how much variation can be attributed to the different minerals and how much to the terranes.

Type I error 74

Synonymous with α ERROR.

Type II error 74

Page

Synonymous with ß ERROR.

Unbalanced factorial (block) design (model) [Cuneo & Feldman 1986, p.132] 162
A type of ANOVA in which the TREATMENTS are applied to BLOCKS consists of different numbers of items: i.e. the table giving the number of subjects on which each experiment was performed has unequal cell frequencies. There is no essential difference in the interpretation of these designs relative to BALANCED DESIGNS, but their computation usually takes longer.

Unbiased estimate (statistic)
An estimate of some population parameter, derived from a statistical SAMPLE (group of measurements), is unbiased if its average value, taken over a large set of samples, equals that parameter. For example, the sample mean is an unbiased estimate of the population mean.

Uniform distribution
(1) A 2- or 3-dimensional distribution of ORIENTATION DATA, in which there is no preferred orientation. (2) A spatial distribution with an equal number of points in any subarea of fixed size.

Uniformity [Cheeney, 1983, p.93] 230
Term applied to 2- and 3-dimensional ORIENTATION DATA, and implying that the density of directions is much the same over the total circumference of the circle (2-D) or sphere (3-D), i.e. that there is no preferred orientation. Several tests for uniformity exist — e.g. HODGES-AJNE and KUIPER'S.

Uniform scores test [Kendall & Buckland 1982]
See MARDIA'S TEST.

Unimodal [Kendall & Buckland 1982]
A distribution having a single MODE (peak).

Unimodal (spherical) distribution [Mardia 1972, p.222]
A distribution of 3-dimensional ORIENTATION DATA in which there is one major directionofclstring ofmasuremns. ouormeauemnts on a rock with a pronounced lineation.

Unipolar (spherical) distribution [Mardia 1972, p.222]
A UNIMODAL (SPHERICAL) DISTRIBUTION in which there is rotational symmetry about the modal direction.

Unique line of organic correlation [Miller & Kahn 1962, p.204]
Synonymous with REDUCED MAJOR AXIS SOLUTION.

Unit Normal distribution
Synonymous with STANDARD NORMAL DISTRIBUTION.

Univariate 72
Involving only one VARIABLE.

Universal kriging [David 1977, p.266; Davis 1986, p.393; Royle et al. 1980, p.121] 273
KRIGING which takes trend (drift, non-stationarity) of the regionalized variable into account. Cf. SIMPLE KRIGING.

Unweighted average method [Le Maitre 1982, p.167]
An AGGLOMERATIVE method of CLUSTER analysis, in which an object is linked to a cluster if it has the highest similarity with the average similarity measure of that cluster. A new average is the comuted fo te luser in hchall objets ae treated equally. Cf. WEIGHTED-PAIR GROUP AVERAGE METHOD.

Upper hinge [Tukey 1977, p.33] 94
The value midway between the MEDIAN and maximum value of a data-set; almost identical to the 3rd (upper) QUARTILE.

Value 68
The qualitative or quantitative figure attached to a measurement. An OBJECT may be described by its associated values of several VARIABLES — e.g. a rock specimen by its actual contents of major and trace elements.

Van der Waerden test [Van der Waerden 1952-3; Conover 1980, p.318; Kendall & Buckland 1982] 147
A NONPARAMETRIC test comparing the LOCATION of 3 or more INDEPENDENT groups of data. Similar to the KRUSKAL-WALLIS test, but more powerful because it uses NORMAL SCORES instead of simple ranks.

Variable [Kendall & Buckland 1982] 68
Any quantitative measurement on an OBJECT which varies — e.g. birefringence, specific gravity, resistivity, dip and strike. Cf. ATTRIBUTE. Some statisticians distinguish variable from *variate* , meaning a RANDOM VARIABLE.

Variable complexity [Cuneo & Feldman 1986, p.123] 324
In an ideal, simple solution from FACTOR ANALYSIS, each variable is accounted for (on average) by only one factor, alternatively expressed by saying: *average variable complexity* = 1. The more factors required to account for each variable (i.e. the less the ideal structure is achieved), the more the individual variable complexities increase above unity. The average variable complexity of an OBLIQUE SOLUTION will always be less than that of an ORTHOGONAL SOLUTION. If the variable complexities are high, this suggests that insufficient factors have been extracted.

Variable sampling adequacy (VSA) [Cuneo & Feldman 1986, p.117] 320
To obtain a meaningful FACTOR ANALYSIS, each single variable should ideally show minimal PARTIAL CORRELATIONS with other single variables, but maximal MULTIPLE CORRELATION with all other variables

Page

combined. I.e. the variation shown by every variable can be very largely explained in terms of *all* other variables combined, but by no means by any other single variable. Variable sampling adequacy measures the extent to which this situation is achieved for each individual variable, while MATRIX SAMPLING ADEQUACY (MSA) measures it for the whole variable set. Any individual variable which has a VSA less than about 0.5 will usually result in a similarly suppressed MSA; this variable will have an unpredictable effect on the FACTOR ANALYSIS, and may intrinsically be of a different type from the other variables, i.e. inappropriate for inclusion with them. For example, if factoring a dataset of water analyses to account for the sources of the water, the variable *temperature* was included along with, say, pH, total dissolved solids, salinity, etc., temperature would probably show a low VSA because it is a physical variable, unrelated to (and hence inappropriately factored along with) the other, chemical variables.

Variance [Kendall & Buckland 1982] 91

The second moment of a frequency distribution about the ARITHMETIC MEAN; the mean of the squared deviations about the mean. Square of the STANDARD DEVIATION.

Variance-covariance matrix [Kendall & Buckland 1982] 279

A SQUARE SYMMETRICAL matrix in which the diagonal elements contain the COVARIANCES, and the off-diagonal elements the VARIANCES.

Variance dissimilarity measure [SPSS 1986, p.742]

A MATCHING COEFFICIENT defined as $(b+c)/4N$ — see Table 13.5. Has a minimum of 0 but no upper limit.

Variate [Kendall & Buckland 1982]

Occasionally used to denote a random VARIABLE.

Varimax solution (rotation) [Joreskog et al.1976,p.130;Cuneo & Feldman 1986,p.114,121] 319

A type of ORTHOGONAL rotation of factor axes in FACTOR ANALYSIS. It compromises between the alternative, EQUAMAX and QUARTIMAX solutions. The new positions are such that projections from each VARIABLE onto the factor axes are either near the extremities or near the origin. This makes interpretation of the factors easier, in terms of the original variables.

Variogram

Inaccurately used as a synonym of SEMIVARIOGRAM.

Vector analysis [Imbrie & Van Andel 1964]

A variant of FACTOR ANALYSIS. Whereas factor analysis resolves multivariate data into several theoretical factors,vector analysis resolves into selected data vectors that represent actually observed, compositionally extreme end-member samples (Q-mode vector analysis), or into variables characterized by least inter-correlation (R-mode vector analysis).

Volume-variance relationship 106

The relationship that the variance of measurements on geological samples tends to decrease as the physical bulk (volume) of the samples themselves increases. Crucial in GEOSTATISTICS.

Von Mises distribution [Mardia 1972, p.57; Kendall & Buckland 1982; Cheeney 1983 p.98] 229

A close but not exact analogue of the NORMAL DISTRIBUTION, for ORIENTATION DATA. Exactly equivalent to the WRAPPED NORMAL DISTRIBUTION only for infinite or zero CONCENTRATIONS. Possesses some of the properties of the linear Normal distribution (e.g. maximum likelihood), but the wrapped Normal distribution possesses the others.

Wald-Wolfowitz Runs test [Wald & Wolfowitz 1939; Siegel 1956, p.136; Kendall & Buckland 1982] 132

See RUNS TEST.

Walsh averages [Ryan et al. 1976, p.269]

The average of every possible pair of values in a data-set (including each value paired with itself).

Walsh test [Siegel 1956, p.83; Conover 1980, p.292]

A powerful NONPARAMETRIC test for comparing two RELATED GROUPS of data, which are symmetrically distributed (i.e. where the mean is a reliable measure of CENTRAL TENDENCY, and equals the MEDIAN).

Ward's method [Everitt 1974, p.15] 278

An AGGLOMERATIVE method of CLUSTER ANALYSIS, in which, at each step in the analysis, union of every possible pair of clusters is considered and the two clusters are fused which result in the least increase in the error sum of squares (the sum of squared deviations of every individual from the cluster mean, representing the loss of information clustering).

Watson's U^2 test [Watson 1961; Mardia 1972, p.180 & 201; Cheeney 1983, p.97] 231

A NONPARAMETRIC test of UNIFORMITY for ORIENTATION DATA, generally more powerful than the HODGES–AJNE or KUIPER tests. Analogous to the CRAMER–VON MISES test for linear data. Can be extended into a 2-group test for equality of mean directions.

Watson–Williams test [Watson & Williams 1956; Kendall & Buckland 1982; Cheeney 1983, p.101] 236

A PARAMETRIC test for comparing mean directions of either 2- or 3-dimensional ORIENTATION DATA. Exists in both one-group and two-group forms (respectively testing that a group of data has some predetermined mean, and comparing two group means). The nearest (but by no means exact) equivalent of the *t*-test for linear data.

Page

Weak monotonicity [Kotz & Johnson 1982-8] 184

A MONOTONIC *X-Y* relationship in which, if $X_i > X_j$ then $Y_i \geq Y_j$ and if $X_i = X_j$ then $Y_i < Y_j$. Commonly measured by GUTTMAN's μ_2 or GOODMAN & KRUSKAL's γ.

Weathervane symbols [Barnett 1982, p.257] 315

A pictorial means of displaying multivariate data in two dimensions, similar in purpose to CHERNOFF FACES, KLEINER-HARTIGAN TREES and ANDERSON'S METROGLYPHS.

Weighted average [Kendall & Buckland 1982] 86

An average in which individual values are weighted according to their determined importance. Widely used in REGRESSION.

Weighted pair-group average method [Le Maitre 1982, p.167] 278

An AGGLOMERATIVE method of CLUSTER ANALYSIS, similar to the UNWEIGHTED AVERAGE METHOD, except that, each time an object is added to a cluster, the new average gives equal weight to the similarity measure of the new object as to the previous group average, so weighting the new group average in favour of the single added object.

Welch statistic, *W* [Browne & Forsythe 1974; Dixon 1975]

A statistic for comparing the equality of several means in ONE-WAY ANOVA, which does not assume equality of variances, and reduces to a separate variance *t* TEST for 2 groups of data. Closely related to the BROWN-FORSYTHE STATISTIC. When all population means are equal (even if variances are unequal),*W* is approximately distributed as FISHER's *F*.

Westenberg–Mood median test [Lewis 1977, p.202]

A rarely used name for the MEDIAN test, using FISHER'S EXACT PROBABILITY.

Wilcoxon–Mann–Whitney test [Siegel 1956, p.116; Conover 1980, p.216] 128

A synonym for the MANN-WHITNEY TEST.

Wilcoxon's matched pairs-sign(ed ranks) test [Wilcoxon 1945,1947; Conover 1980, p.280] 125

A NONPARAMETRIC test for comparing two RELATED groups of data, more powerful than the RUNS TEST but less powerful than the MANN-WHITNEY or SMIRNOV TESTS. Uses the magnitude of differences between paired observations, and so requires data to be measured on at least an ORDINAL scale. Also used in the determination of symmetric CONFIDENCE LIMITS for the MEDIAN.

Wilcoxon's rank sum test [Gibbons 1971, p.164; Lewis 1977, p.147]

A synonym for RANK SUM TEST and, effectively, for the MANN-WHITNEY TEST.

Wilcoxon's signed ranks test

A synonym for WILCOXON'S MATCHED PAIRS-SIGNED RANKS TEST.

Wilcoxon's stratified test [Langley 1970, p.190]

An adaptation of WILCOXON'S RANK SUM TEST to STRATIFIED data. It might be used to compare, say, data on the radioactivity of two nuclear waste burial sites (using different construction methods) after 1,2,3 etc. years.

Wilks' Λ (lambda) criterion [Wilks 1938,1962; Le Maitre 1982, p.152; Kotz & Johnson 1982-8] 310

A MAXIMUM LIKELIHOOD criterion used in multivariate analysis for testing equality of several multivariate means or DISPERSIONS: the ratio of determinants of two matrices of sums of squares and products; the numerator corresponds to a sum-within-groups and the denominator to total sum. Tested for significance by conversion to χ^2 or to FISHER'S *F* statistic Can be regarded as an extension of HOTELLING'S T^2 to several means, or of SCHEFFE'S TEST to multivariate data.

Wilson's *e* [Coxon 1982, p.20]

An ORDINAL measure of ASSOCIATION, related to KENDALL's τ and GOODMAN AND KRUSKAL'S γ.

Winsorized estimate [Dixon & Massey 1969, p.330]

A procedure for making estimates more ROBUST. Extreme values are replaced with the values of their nearest neighbours: in *first-level Winsorization*, the maximum and minimum values are replaced with the second highest and lowest values respectively; in *second-level Winsorization*, the two highest and lowest values are replaced, etc. E.g. original data-set: 3 8 12 13 14 16 18 34; 1st-level Winsorized: 8 8 12 13 14 16 18 18. An alternative to TRIMMED ESTIMATES, in which extreme values are ignored altogether, rather than replaced. Also useful in treating missing data, e.g the 3rd-level Winsorized mean of the following data-values: -- -- 108 111 119 121 125 -- -- -- (in which -- indicates missing data), is obtained as [(4 * 111) + 119 + 121 + (4 * 125)]/10 = 118.4.

Winsorizing

The process of deriving WINSORIZED ESTIMATES. Many geochemical data-sets are implicitly winsorized, where values below machine detection limits occur. If such values (quoted as, say, '< 5 ppm') are treated as real values either of zero or of the known detection limit (say, 3 ppm), the data then become winsorized. Cf. TRUNCATION, CENSORING.

Wrapped Normal distribution [Mardia 1972, p.55] 228

The distribution obtained by simply wrapping a linear NORMAL DISTRIBUTION around a circle. Possesses several of the properties (e.g. additive property) for ORIENTATION DATA that the Normal distribution possesses for linear data; most of the others are possessed by the VON MISES DISTRIBUTION.

Page

W statistic of disjunction [Sneath 1977]
A method of testing the significance of clusters resulting from CLUSTER ANALYSIS. It determines whether the overlap between clusters is less than some predetermine amount (the less the overlap, the better the clustering).

***W* test**
Generally synonymous with SHAPIRO–WILK TEST.

Yates' correction for continuity [Till 1974, p.123; Kendall & Buckland 1982; Davis 1986, p.237] 175
An adjustment used in the calculation of χ^2 for a 2x2 CONTINGENCY TABLE, in which 0.5 is subtracted from one of the 4 cells, and the others are adjusted so that row and column totals remain constant. The effect is to bring the distribution based on discontinuous frequencies nearer to the continuous χ^2 distribution used in testing for association.

Yules' *Q* [Blalock 1972, p.298; Coxon 1982, p.27; SPSS 1986, p.741] 168
A CONTINGENCY COEFFICIENT applicable only to 2x2 CONTINGENCY TABLES, and varying from 0 (indicating no association) to +1 or -1 (when one of the 4 cells is zero). Defined as $(ad - bc)/(ad + bc)$ — see Table 13.5. Most suited to measuring associations between CONTINUOUS VARIABLES FORCED INTO A DICHOTOMY. A special case of γ (see GAMMA). Can also be used as a MATCHING COEFFICIENT.

Yules'*Y* coefficient of colligation [SPSS 1986, p.741] 168
A CONTINGENCY COEFFICIENT, defined as $[\sqrt{(ad)} - \sqrt{(bc)}]/[\sqrt{(ad)} + \sqrt{(bc)}]$ — Table 13.5. Ranges from –1 to +1.

Zap approach [Brouwer & Veinus 1964] 328
Exploratory data-analysis in which one powerful technique is applied to extract maximum information. Cf. SHOTGUN APPROACH.

***zI* test** [Langley 1970, p.245]
An adaptation of the *zM* TEST to ORDINAL instead of RATIO data. Used instead of the POISSON or BINOMIAL TESTS for data outwith the range of tables published for these tests.

***zM* test** [Langley 1970, p.152] 101
The converse of STUDENT's *t*-TEST, used to compare a random sample of one or more measurements with a large parent group whose mean and standard deviation are known. The formula is the same as the *t*-TEST, except that the standard deviation of the parent group is used instead of that of the sample.

Zonation [Davis 1986, p.211]
Refers to dividing a sequence of observations into relatively uniform segments (e.g. lithological units in a layered igneous intrusion, well-log, borehole; stripes on a magnetic traverse).

Z score
(1) A synonym for STANDARD NORMAL DEVIATE. (2) A score on a DISCRIMINANT FUNCTION

Z transformation 177
See FISHER'S Z TRANSFORMATION.